14th International Symposium for the Advancement of Boundary Layer Remote Sensing 2008

(ISARS 2008)

IOP Conference Series: Earth and Environment Science Volume 1

Copenhagen, Denmark
23-25 June 2008

ISBN: 978-1-61738-295-6
ISSN: 1755-1307

Printed from e-media with permission by:

Curran Associates, Inc.
57 Morehouse Lane
Red Hook, NY 12571

CURRAN ASSOCIATES INC.
proceedings.com

Some format issues inherent in the e-media version may also appear in this print version.

Printed by Curran Associates, Inc. (2010)

For permission requests, please contact the Institute of Physics
at the address below.

Institute of Physics
Dirac House, Temple Back
Bristol BS1 6BE UK

Phone: 44 1 17 929 7481
Fax: 44 1 17 920 0979

techtracking@iop.org

Additional copies of this publication are available from:

Curran Associates, Inc.
57 Morehouse Lane
Red Hook, NY 12571 USA
Phone: 845-758-0400
Fax: 845-758-2634
Email: curran@proceedings.com
Web: www.proceedings.com

TABLE OF CONTENTS

The 14th International Symposium for the Advancement of Boundary Layer Remote Sensing (ISARS 2008) addresses acoustical, optical and microwave techniques to probe the lower part of the atmosphere. The symposium focuses on the physical basis of remote sensing techniques and new instruments. A theme for the conference is also various applications of remote sensing, this year with special emphasis on wind energy.

ISARS is an informal association of scientists from all over the world which organizes a symposium every second year. While the abbreviation ISARS has remained unchanged since the start in Calgary 1981, the words have changed from International Symposium on Acoustic Remote Sensing and Associated Techniques of the Atmosphere and Oceans because other techniques than the acoustic have become important for boundary layer remote sensing. Specifically lasers for remote wind sensing are developing rapidly.

By the end of each symposium the chairman of the next has been elected. So far the symposia have taken place in different countries each time with different chairs. The scientific organizing committee, which consists mainly of chair persons of previous symposia, maintains the continuity of themes and of the organization in general.

After the last symposium held in Garmisch-Partenkirchen, many of the papers appeared in revised and improved form in a special issue of Meteorologische Zeitschrift. A similar special issue is also planned to follow ISARS 2008.

I wish to express my gratitude to the scientific organizing committee for valuable advice and to the local organizing committee for all their effort with the conference papers and the conference itself.

Jakob Mann, Conference Chair, http://www.isars2008.dk/
Risø-DTU, Denmark

IOP Conf. Series: Earth and Environmental Science **1** (2008) 012001 doi:10.1088/1755-1307/1/1/012001

Backscattering and reflection of acoustic waves in the stable atmospheric boundary layer

M A Kallistratova

Obukhov Institute of Atmospheric Physics RAS, 3 Pyzhevskii, Moscow 119017, Russia

E-mail: margo@ifaran.ru

Abstract. Among different applications of sodars in studies of the atmospheric boundary layer (ABL), the remote sensing of the temperature structure parameter C_T^2 and the closely associated structure parameter of refraction index C_n^2 occupies an important place. These parameters are calculated from sodar return signal with the help of the theory of Bragg volumetric scattering of sound by small-scale turbulent irregularities of temperature. According to the Kolmogorov-Oboukhov conception, the irregularities are assumed to be locally homogeneous and isotropic. The theory was confirmed by experiments and universally recognized as a classical one. However, detailed comparisons are still being made between C_T^2 values derived from sodar data and from aircraft and meteorological-tower measurements. The results of such comparisons performed by different researchers under different conditions contradict to each other. In this paper, the results of comparisons between remote and *in situ* measurements of C_T^2 in the ABL are briefly reviewed. Some examples of a systematic overestimation of C_T^2 calculated from sodar data obtained under stable stratification of the ABL are given. Technical and physical causes of these discrepancies are discussed. The following factors are considered as possible physical factors: contribution of a partial (Fresnel) reflection of sound from layers with a sharp temperature jump, anisotropy of small-scale turbulence, intermittency, and contribution of quasi-wave air-density inhomogeneities. It is shown that these factors can explain some short-lived increase of the sodar return signal, but not a systematic overestimation of C_T^2. The role of sound-beam refraction (which regularly occurs in the stable ABL due to strong air-temperature and wind-speed gradients) is discussed. Due to a refraction turn of sodar beam, the angle of scattering changes, and scattering by turbulent wind-speed irregularities could permanently contribute to return signal of a monostatic sodar.

1. Introduction

The present-day uses of sodars in studying the atmospheric boundary layer (ABL) are listed in [1]. Among these uses, measurements of the temperature structure parameter C_T^2 with a vertical monostatic sodar are very important. The point is that C_T^2 is directly related to the structure parameter of the refractive index for visible light, C_n^2 [2]: C_n^2 [$m^{-2/3}$] $\cong$ C_T^2 10^{-12} [$K^2m^{-2/3}$]. Data on the vertical profiles of C_n^2 are necessary to estimate fluctuations in the parameters of optical waves for astroclimatic problems and for the development of optical wireless communication systems. Below, two examples illustrating the use of sodars with this end in view are given.

In the 1970s, sodar was first successfully used to search for astronomical sites (i.e. sites, where degradation of optical «seeing» due to turbulence in the ABL is minimal) in Hawaii. Figure 1a gives some examples of sodar measurements of C_n^2 over a mountain terrain, which show significant differences in night optical turbulence for different regions. Note that a tenfold increase in the integral (throughout the atmospheric thickness) value of C_n^2 results in a fourfold decrease in the effective resolution of a telescope. In recent years, interest in sodar astroclimatic studies has quickened due to the PILOT program in Antarctica [4, 5]. To measure the C_n^2 profiles over the Antarctic plateau, a minisodar with a high resolution, SNODAR, was especially developed [6].

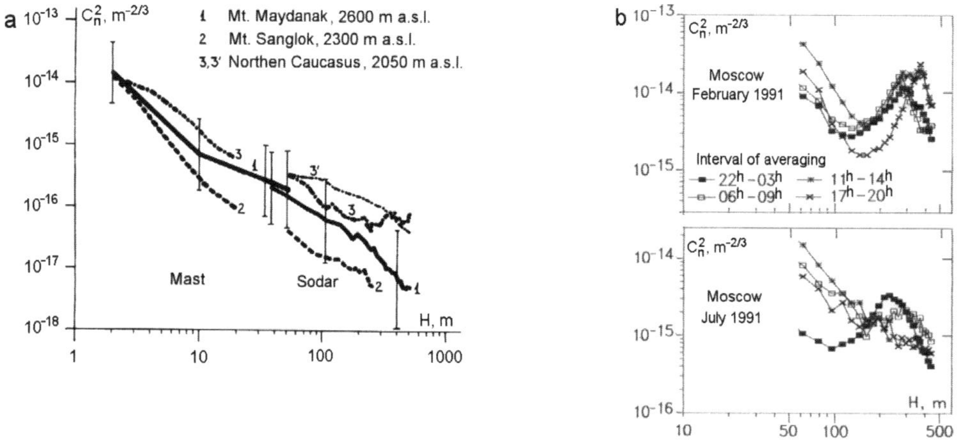

Figure 1. Sodar profiles of the optic refractive index structure parameter C_n^2. (a) – the nocturnal profiles at mountain astronomical sites in Uzbekistan and Tajikistan, Central Asia (1, 2), Northern Caucasus (3, 3'), and island Maui, Hawaii (4) [3]. (b) The monthly average C_n^2 profiles over the center of Moscow for winter and summer, and for different time of day [9]. Optimal heights for optical links is about 150 m AGL, where average annual value of Cn2 has a local minimum.

Atmospheric turbulence causes fluctuations in the intensity and phase of received signal, resulting in impaired free-space optical link performance. In the last decade, many studies have been devoted to this problem (see, for example, [7, 8]), but only few experimental data are available on the profiles of the structure parameter of refractive index for an urban air basin. Figure 1b shows the C_n^2 profiles measured with sodar over the center of Moscow. Due to a large amount of elevated inversions over an urban heat island, the vertical profiles of C_n^2 are non-monotonic, and it is possible to choose an optimal height for optical link, at which the yearly mean of C_n^2 is minimum.

To determine C_T2 from sodar data, the theory of the Bragg volume sound scattering from small-scale turbulent air-temperature inhomogeneities [2,10,11] is used. This theory laid by A.M. Oboukhov and developed by V.I. Tatarskii and A.S. Monin in the 1960s is based on the Kolmogorov model of locally homogeneous and isotropic turbulence. The relationships obtained theoretically were experimentally supported and universally recognized as classical. However, a detailed comparison between the C_T^2 values obtained from sodar data and from aircraft and meteorological-tower measurements is made to the present day. In some experiments, a good agreement was obtained. And, in others, it was found that, under a stable stratification of the ABL, a systematic overestimation of sodar data, as compared to in-situ measurements, is observed. Therefore, until the present time, the questions of applicability of the theory [2, 11] to the stable ABL and of necessary consideration for alternative mechanisms generating sodar echo-signals are discussed. Note that the same problem also arises in sounding the troposphere and lower stratosphere with the aid of radars [12, 13]. Thus, the problem of attainable accuracy in determining C_T^2 remotely still has no final solution. In this paper, the results of comparisons between remote and in-situ measurements of the temperature structure parameter C_T^2 in the ABL are briefly reviewed. Some examples of a systematic overestimation of C_T^2 calculated from the intensity of sodar return signal with the aid of the classical theory are given. Possible instrumental and physical causes of discrepancies are discussed, including, contribution of a partial (Fresnel) reflection of sound from the layers with a sudden temperature change, anisotropy of small-scale turbulence, turbulence intermittency, and contribution of quasiregular wave inhomogeneities of air density. It is shown that the above factors can explain a short-term increase in the intensity of sodar echo-signal, but not a systematic overestimation of C_T^2. The role of sound-beam refraction caused by strong air-temperature and wind-velocity gradients regularly observed in the stable ABL is discussed. A refractive turn of directional sodar diagram changes the angle of scattering, therefore, the scattering from turbulent wind-velocity inhomogeneities can systematically contribute to echo-signal of a monostatic sodar.

2. Classical formulae for sound backscattering

The classical theory of wave scattering in a turbulent medium [2] yields the following formula for the effective cross-section of sound scattering in the atmosphere, σ_θ:

$$\sigma_\theta = 2\pi k^4 \cos^2\theta \cdot \{\Phi_T(K) \cdot (2T)^{-2} + \cos^2(\theta/2)\Phi_V(K) \cdot (c)^{-2}\}, \quad (1)$$

where

$$\Phi_T(K) = 0.033 C_T^2 K^{-11/3} \quad (2)$$

$$\Phi_V(K) = 0.061 C_V^2 K^{-11/3} \quad (2')$$

are the Kolmogorov spectra of temperature and wind speed fluctuations for locally homogeneous and isotropic turbulence,

$$C_T^2 = \langle (T_1 - T_2)^2 \rangle r_{1,2}^{-2/3}; \quad (3)$$

$$C_V^2 = \langle (v_1 - v_2)^2 \rangle r_{1,2}^{-2/3}; \quad (3')$$

(the Kolmogorov-Obukhov "2/3 law"); T is the mean air temperature; c is mean speed of sound; $r_{1,2}$ is the distance between the points of measuring T_1 and T_2 are measured (averaging is denoted by the brackets $< >$); $k=2\pi/\lambda$; $K = 2\pi/l_t$ is the wave number of turbulence for the length scale l_t.

For backscattering ($\theta = 180°$) only the first term in the figure brackets (1) is retained:

$$\sigma_{180} = 2\pi k^4 \cdot \Phi_T(2k) \cdot (2T)^{-2}, \quad (1')$$

$K = 2k$ is the K value responsible for the backscattering (i.e. scattering at the angle $\theta = 180°$). The scheme of the Kolmogorov spectrum is shown in Figure 2.

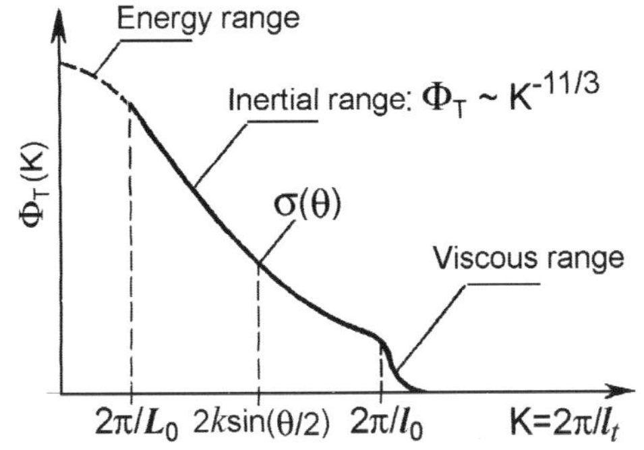

Figure 2. Scheme of the 3-D spectral density of temperature fluctuations in the atmosphere according to Kolmogorov. The bounds of the inertial range of turbulence are indicated:

$$2\pi/L_0 < K < 2\pi/l_0,$$

where L_0 is the outer scale, and l_0 is the inner scale of turbulence. Spectral density at $K = 2k \sin(\theta/2)$ governs scattering at the angle θ.

It follows from (1') and (2) that

$$\sigma_{180} = 3.9 \cdot 10^{-3} k^{1/3} C_T^2 T^{-2} = 7.15 \cdot 10^{-3} \lambda^{-1/3} C_T^2 T^{-2} \quad (4)$$

$$C_T^2 = 140\,\sigma_{180}\,T^2\lambda^{-1/3}. \quad (5)$$

Formulas (4) and (5) are valid only if the characteristic scale of scattering irregularities, $l_t \cong \lambda/2$, is within the "inertial range of turbulence", where the concept of locally homogeneous and isotropic turbulence is valid. This condition is always met for sodars with the carrier frequencies 1.5-6.0 kHz, which corresponds to $\lambda/2 \sim 2$ - 10 cm.

The value of σ_{180} is determined from the transmitted and received (by monostatic sodar) acoustical powers P_t and P_r, respectively:

$$\sigma_{180} = P_r R^2 \exp(2\alpha R)\,[P_t\,(c\tau/2)\,S_{ant}]^{-1}, \quad (6)$$

where R is the distance between acoustical antenna and scattering volume, α is the coefficient of sound absorption in the air, $c\tau/2$ is the vertical length of scattering volume (c is the speed of sound, τ is the duration of acoustic pulse), and S_{ant} is an effective area of transmitting-receiving antenna.

To calculate σ_{180}, an absolute acoustical calibration of sodar is needed. Calibration in an anechoic camera is most reliable (see, for example, [14]), but this procedure is very laborious. Therefore, a relative calibration is more often performed against the reflection of sodar radiation from an object with a known reflection coefficient. However, the accuracy of such calibration is low.

The theory of sound scattering, whose conclusions are given above, was experimentally supported by the results of measurements of scattering indicatrix within a wide range of scattering angles in the atmospheric surface layer [15, 16] and under laboratory conditions [17]. For the ABL, only the C_T^2 values obtained from the measurements with calibrated monostatic sodars and from in-situ measurements were compared.

3. Comparison C_T^2 by sodar with in situ measurements

By now, a few tens of papers devoted to the studies of the accuracy and reliability of sodar C_T^2 data have been published. The results obtained under different atmospheric conditions are contradictory. In some papers (see, for example, [18-23]), there is a good agreement between in-situ data on C_T^2 and its values obtained from sodar return signal interpreted on the basis of the Tatarskii scattering theory. Figure 3 gives some examples of such agreement. The deviations from the medial line are allowable, when the structure parameter generally varies over more than two orders of magnitude.

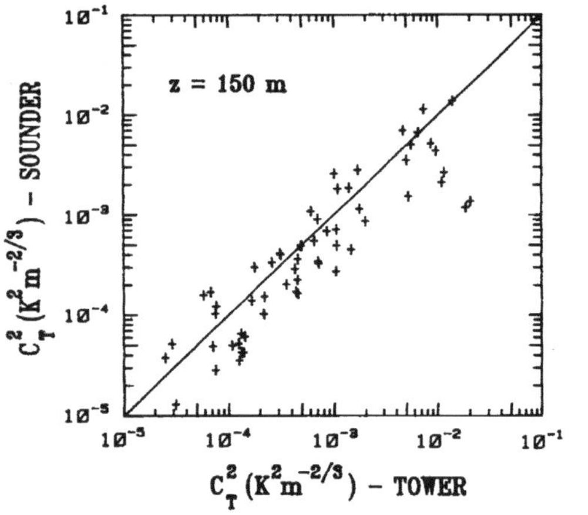

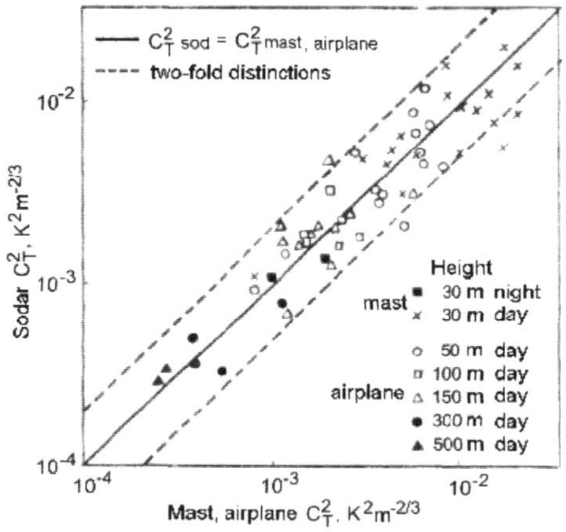

Figure 3. Examples of good agreement between sodar and in situ measurements. Left – over mountain terrain [21], right – over a flat plain [22]. The solid lines correspond to equality.

In other studies (see, for example, [14, 24-28], a systematic overestimation of sodar data is observed under the stable ABL. Such examples based on the $\langle C_T^2\rangle_{sodar} \cong (1.4\text{-}2.0)\langle C_T^2\rangle_{in\ situ}$ data are given in Figure 4.

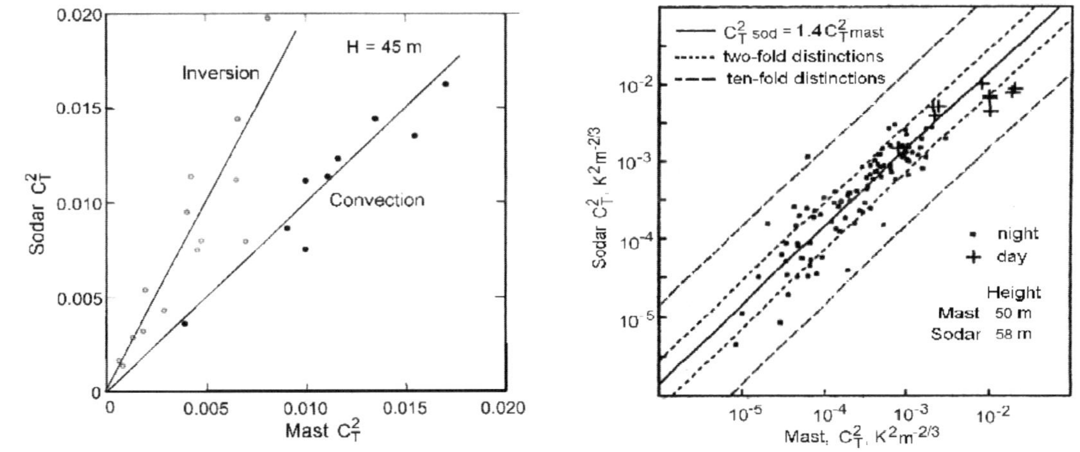

Figure 4. Examples of overestimation of C_T^2 by sodar. Left – over a slightly irregular terrain; right - over a mountain summit [14].

In some cases, significant discrepancies were observed. Figure 5 gives two examples of time variations in C_T^2, when the discrepancy exceeded one order of magnitude. Note that, in both the examples, the pattern of time variations in sodar data properly follows that for in-situ data. Unfortunately, such data are usually poorly documented, and, therefore, it is difficult to identify the conditions of temperature and wind stratifications, which result in such discrepancies.

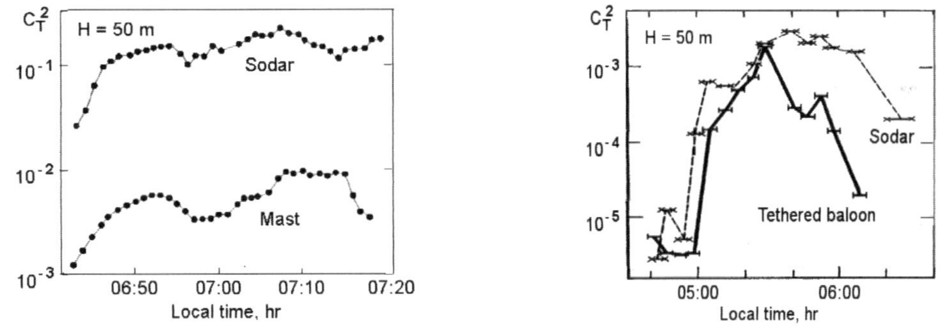

Figure 5. Time series of C_T^2, in which a 10-fold excess of sodar data was observed during more than 1 hour. Left – over a slightly irregular terrain [24]; right – over a mountain summit [27].

The diurnal recording of C_T^2 values and the corresponding sodar echogram presented in Figure 6 demonstrate two interesting features of the comparison results. First, during the period of existence of surface and raised inversions (identified from the character of return signal), different regions of the behavior of the low-frequency (with a period of the order of one hour) component of C_T^2 variations are observed. Thus, from 19:00 to 22:00 and from 10:00 to 11:00, almost an ideal agreement is seen between the values under comparison; from 22:00 to 05:00 and from 09:00 to 10:00, a moderate overestimation of sodar values is observed; and from 05:00 to 09:00, there is almost a 100-fold increase. Second, within all these regions (as well as in Figure 5), there is a good correlation of the high-frequency component (with a period of several minutes).

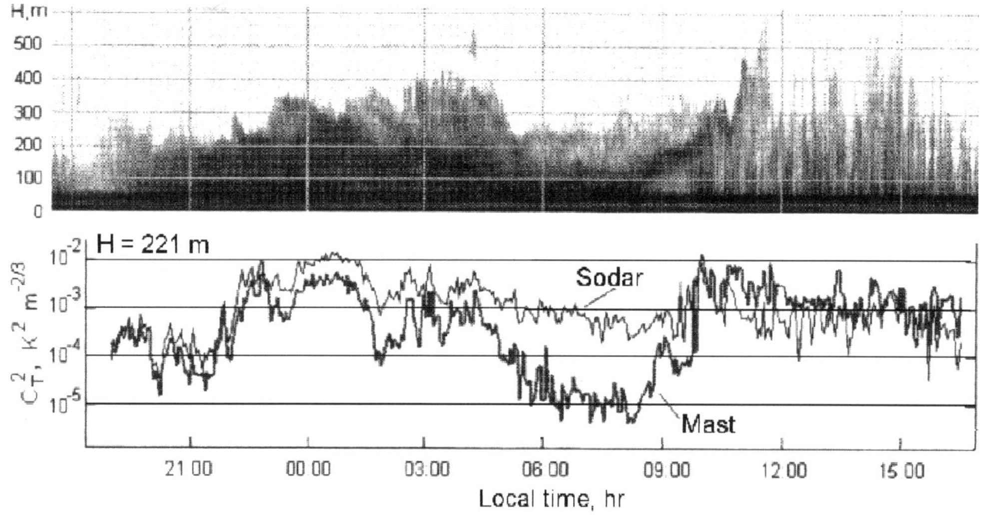

Figure 6. Diurnal time series of sodar and in-situ C_T^2 (bottom) and corresponding registration of sodar return intensity (top) [28].

4. Possible causes of the discrepancies

Among possible causes of the discrepancies, methodical and technical errors should be distinguished from physical factors that affect sodar return signal. A very credible explanation of the discrepancy observed by some authors was made in [23]: "In nights the C_T^2 value is small, and SNR is poor, therefore the corresponding smallest values are not used in the averaged echo, and it is shifted to higher than the actual average value".

The validity of such an explanation is supported by Figure 7, which gives time variations in the sodar and in-situ values of C_T^2 and the dependence of their ratio on C_T^2 magnitude. Some physical factors, which can result in overestimated sodar data, will be considered below.

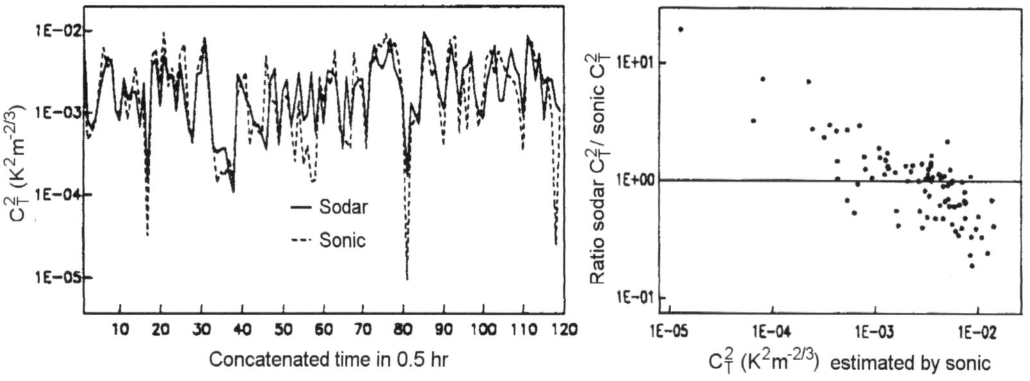

Figure 7. (left) Time series of C_T^2 values from sodar and sonic anemometer at the height 100 m. (right) $(C_T^2)_{sodar}$ / $(C_T^2)_{sonic}$ versus $(C_T^2)_{sonic}$ [23].

4.1. Partial (Fresnel) reflection

The role of reflection from layers with a sudden change of the refractive index (the so-called "sheets" or "pancakes") in the formation of radar return has already been discussed in literature for almost half a century. Figure 8a illustrates the heart of the problem, and Figure 8b shows the ABL temperature profile obtained with a high resolution. The form of this profile is very similar to picture C) in Figure 8a, which justifies the reality of the mechanisms presented in this picture.

However, the sound waves within the frequency range used in sodars cannot be reflected from such sheets, because, for an effective reflection, it is necessary that the transition-layer thickness should be less than wavelength, and the thickness of the layers presented in Figure 8b is about 30 cm.

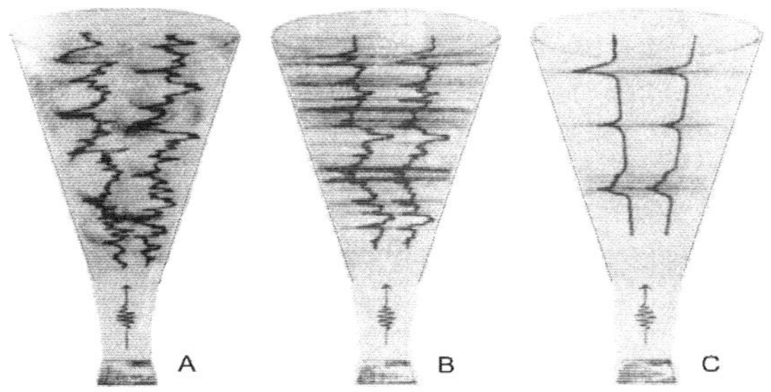

Figure 8a. Artist's conception of atmospheric refractivity structure pertinent to A) isotropic turbulence scatter, B) Fresnel volumetric scatter, C) Fresnel reflection from isolated layers [29].

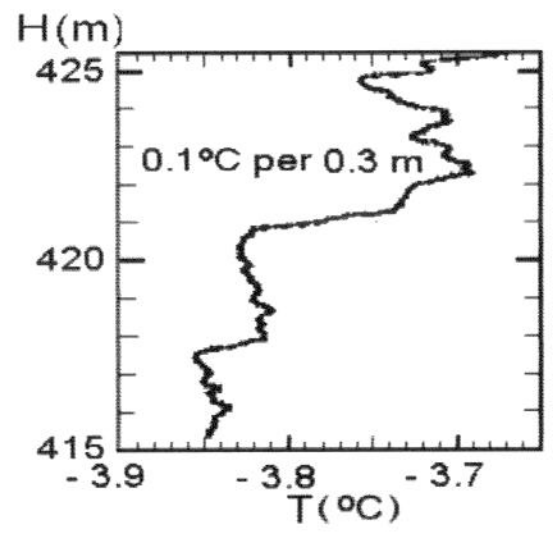

Figure 8b. A group of submeter temperature sheets observed from helicopter [30].

In literature, there is an example of more sudden changes (see Figure 9). In [31], the intensity of reflection of sound with a frequency of 2000 Hz at a sudden change of air temperature (1 degree per 3 cm) was estimated. This reflection proved to be equivalent to the scattering from isotropic turbulence at the value of C_T^2, which exceeds its maximum observed in the ABL by two orders of magnitude. Thus, the Fresnel reflection can, in principle, explain the overestimation of sodar C_T^2 value. However, in [31], such a high temperature gradient was observed only once during several nights and lasted only one minute (at a wind velocity of 12 m/s). The whole time period, within which sufficiently high temperature gradients were recorded, was equal to 20 min.

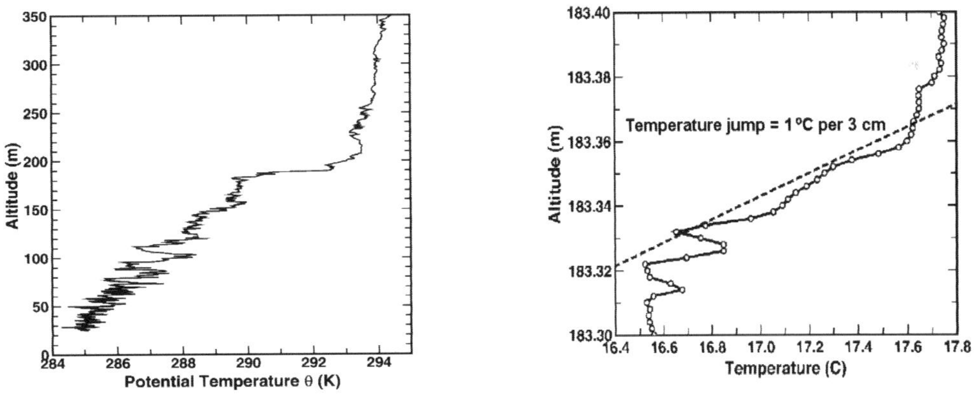

Figure 9. (left) High-resolution temperature profile obtained from ascending kite. (right) Detailed view of the steep temperature gradient shown in the left panel [31].

Longer periods of significant temperature gradients at inversion boundaries were observed [33]. Figure 10 shows the temperature profiles and C_T, which were obtained from measurements with the aid of a carriage moving along a meteorological tower. In these profiles, the temperature jumps are smaller than those in Figure 9 (only 5^0 per 5 m), but, inside the regions with strong gradients, multilayer structures are seen. Such a stratification structure is very similar to picture B) in Figure 8a and may cause the occurrence of an additional volume reflection from multiple weak temperature jumps. Moreover, Figure 10 shows narrow but sufficiently strong peaks of C_T within the regions with high gradients, which also must increase a sodar return signal. When comparing sodar signals with conventional measurements from meteorological towers, where the distance between temperature sensors is several tens of meters, such narrow regions may occur between the sensors, and this will distort comparison results.

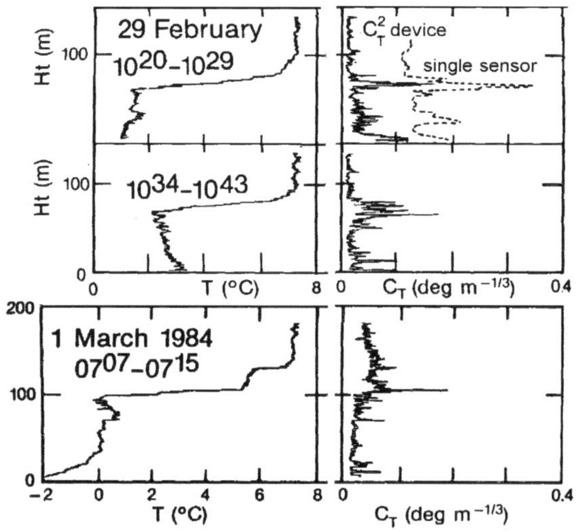

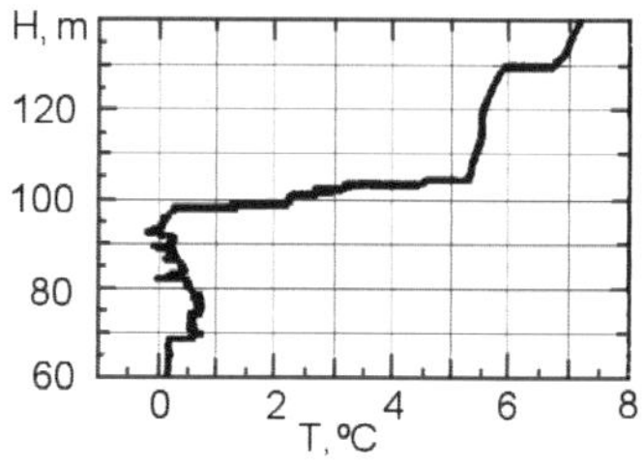

Figure 10. (left) Profiles of temperature and C_T from a carriage moving along a meteorological tower [33]. Detailed view of the temperature inversion on March 1 is shown atop. Multilayer structure is seen within the sector of strong temperature gradient.

Very few profiles of a high resolution (as in Figures 8b-10) have been obtained up to now. The frequency of occurrence of temperature jumps in the ABL is unknown; therefore, it is difficult to draw a definite inference about their influence on the regular overestimation of sodar C_T^2 data.

4.2. Effect of turbulence anisotropy

Many researchers assume that overestimation of sodar C_T^2 values is related to the anisotropy of small-scale turbulent inhomogeneities under inversion conditions, which is caused by buoyancy forces. The pioneering studies [24, 18] of anisotropy in the stable ABL were conducted through measurements of the zenithal aspect sensitivity of sodar return signal. Scatterer anisotropy, as well as reflection from a horizontal layer, must result in a decrease of the return signal intensity, P_r, at a deviation of probing beam from the vertical, i.e., $K_a \equiv P_{rv}/P_{rt} > 1$ (the subscripts rv and rt correspond to the vertical and tilted sodars, respectively). With the aid of similar method, turbulence anisotropy was observed during radar measurements in the stratosphere [29, 34].

To take into account anisotropy during wave scattering, it was proposed in [35, 36] to replace the Kolmogorov spectrum (2) by the phenomenological model of anisotropic spectrum Φ_{Ta} of the form:

$$\Phi_{Ta}(K_x, K_y, K_z) = \Phi_{Ta}(\eta K_{hor}, K_z) = AC_T^2\eta^2(\eta^2 K_{hor}^2 + K_z^2)^{-11/6}, \quad (7)$$

where η is the coefficient of turbulence anisotropy, and A is a numerical coefficient.

The calculated in [36] dependence of K_a on η is graphically given in Figure 11a, and Figures 11bcd show some results of experimental studies of sodar aspect sensitivity in the night ABL [37].

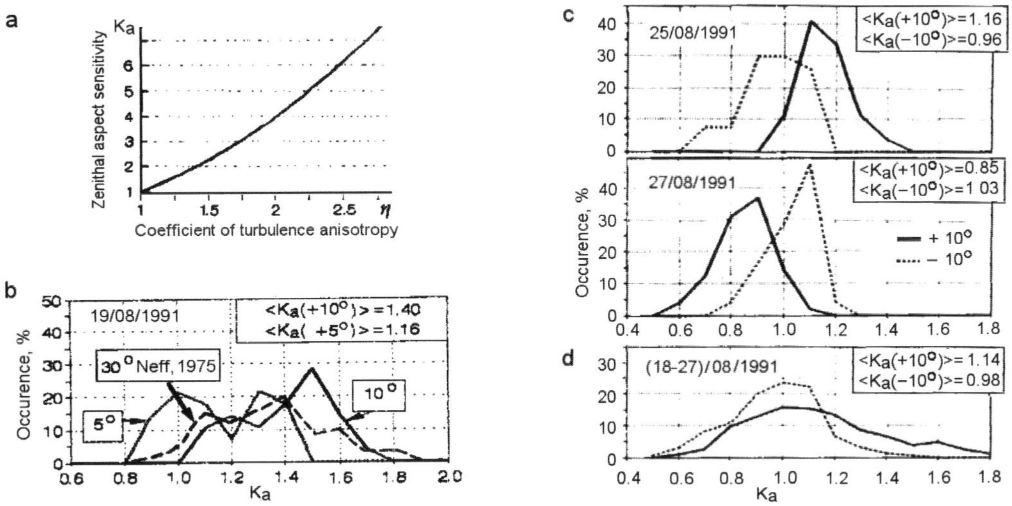

Figure 11. (a) Dependence of sodar aspect sensitivity, *Ka* , on turbulence anisotropy, η, for the parameters of sodar used in the experiment [37]. (b) Histograms of *Ka* for two sodar tilt antenna angles (5° and 10°), which were obtained during one of the nights; histogram for the angle 30°, which was calculated from data [18], is shown as well. (c) As (b) but for the antenna tilts in two opposite directions, +10° and -10° , for two nights. (d) As (c) but averaged over 5 nights. The mean values, <*Ka* >, for each histogram are indicated at the plots (b), (c) and (d) in the rectangular frames.

It follows from Figures 11b and 11d (with consideration for Figure 11a) that the maximum anisotropy coefficient for the night ABL is $\eta \cong 1.4$. Sometimes the backscatter anisotropy can double the sodar value of CT2, but, on average, this effect is small (much smaller than that for radiowaves in the stratosphere). Besides, it follows from Figures 11c and 11d that the backscatter intensity also depends on the azimuth angle of sodar beam. The cause of the azimuthal dependence can lie either in the asymmetry of the inclination of anisotropic inhomogeneities due to wind, or in the wind refraction of sodar beam.

4.3. Intermittency

The influence of turbulence intermittency on time variations in the intensity of sodar return signal is evident. This problem was considered in several papers [38, 39]. Some authors (see, for example, [26]) assume that intermittency can also affect the mean intensity of signal. However, there is no need for a special consideration of intermittency in the theory of scattering, because intermittency can affect the intensity of signal only to the extent to which it affects the spectrum of turbulence within the range of scales comparable to the length of sound wave.

In his work (1962), A. M. Obukhov [40] considered this problem and came to the conclusion that the consideration for intermittency "gives a very weak deviation from the (2/3)-power law". This is supported by the results of numerous measurements of the spectrum of temperature in the ABL, which, within the inertial interval, always follow the (-5/3)-power law.

Apparently, the work [41] finally clarified this problem. In this work, experimental and model studies of the spectrum of temperature fluctuations in the night ABL were conducted. The authors made a sharp distinction between small-scale intermittency and large-scale intermittency and showed that C_T^2 "changes only as a result of the latter". The energy spectrum of temperature fluctuations obtained by these authors is shown in Figure 12.

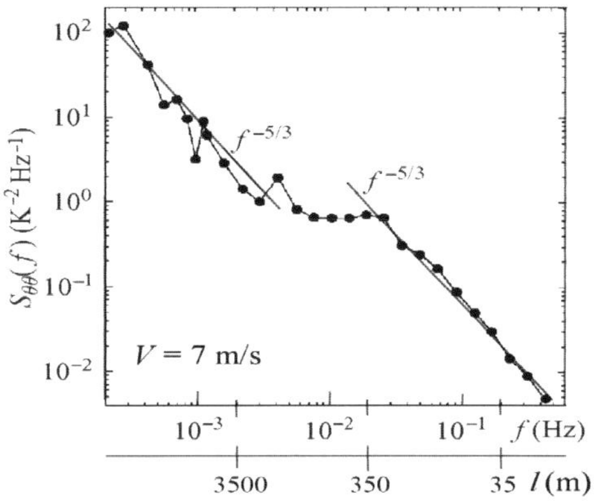

Figure 12. Frequency spectra of temperature fluctuations in double-logarithmic representation. The data were collected from 03:00 LST to 05:00 LST at H = 74 m. Two regimes showing a (-5/3)-power law are separated by a plateau between 0.005 Hz and 0.02 Hz, corresponding to time scales ranging from 50 s to 3.3 min and (at a mean wind speed of U = 7 m s-1) to length scales between 350 m and 1400 m [41].

4.4. Influence of wave activity

In sodar echograms, internal gravity waves (IGWs) and Kelvin-Helmholz waves (KHWs) in the nocturnal ABL are often recorded. In the echograms, IGWs are seen well, because they cause wave-like motions of layered turbulized structures. The KHWs, which themselves form turbulized structures resembling (in form) cat's eyes, are also seen well. The question arises as to what extent the cross-section of sound scattering in these regions obeys the Tatarskii theory.

An analytical calculation of the spectrum of temperature fluctuations under the influence of IGWs was performed in [42]. With the use of this spectrum, an expression was obtained for the effective cross-section of sound-wave scattering from IGWs [43]. The value of the σ_{180} estimated by us from formula (22) in [43] (at vertical sounding at the carrier frequency f = 2000 Hz) proved to be 4 orders of magnitude lower than that obtained from the above formula (4) for scattering from small-scale turbulence. Therefore, an additional contribution of scattering from IGWs into sodar return signal is negligibly small.

We are not aware of similar calculations or measurements of the energy spectrum of temperature fluctuations in the KHW region. Significant characteristic scales of KHWs observed in the ABL (wavelength 0.5-3.0 km) suggest that a direct contribution of scattering from these waves into sodar signal is also small. However, we cannot exclude the possibility of formation of thin layers with significant temperature gradients inside KHWs. Figure 13 gives an example of comparison between the sodar and microtemperature profiles of CT2 in the presence of KHWs, and the corresponding echogram. At heights between 100 and 150 m, the profiles are in sufficiently good agreement. However, below 70 m, where "cat eyes" are more clearly seen in the echogram, the disagreement between the profiles reaches one order of magnitude. It follows from this example that KHWs in some way intensify return signal.

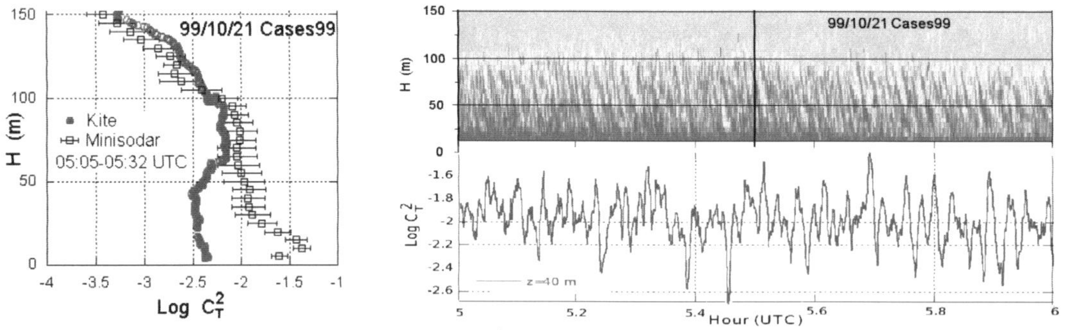

Figure 13. (left) Comparison of a profile of C_T^2 measured with a minisodar with that obtained from a kite; the two devices were roughly 1 km apart. (right) KHWs in the facsimile record during the measurements of C_T^2 profiles and fluctuations at a height of 40 m (Courtesy of R.L. Coulter)

4.5. Contribution of wind speed fluctuations

Measurements of C_T^2 with a monostatic sodar are possible, because turbulent wind-velocity inhomogeneities do not scatter sound at an angle of 180°. The gap in the scattering indicatrix is very narrow, but not to an extent that it can strongly distort C_T^2 measurements. If we assume that the wind and temperature terms in formula (2) are equal at a scattering angle of 45° and that the width of the sodar directivity diagram at a level of 3 dB is $2\gamma \sim 6°$, the contribution of wind fluctuations into backscattering will be about 2 10^{-3} of the contribution of temperature fluctuations. Even with consideration for the fact that the value of the wind term in (1) is usually one-two orders of magnitude larger than that of the temperature term, such a contribution can be neglected within an obtainable accuracy in determining C_T^2.

In the real stratified atmosphere, sound-wave refraction occurs under the influence of vertical temperature and wind-velocity gradients. Wind refraction is usually much stronger than temperature refraction. Sodar beam affected by wind velocity is curved. In this case, the real scattering angle becomes different from the angle determined by the geometry of sodar in a stationary homogeneous medium, and the contribution wind fluctuations to backscattering increases.

Numerical estimates of the wind refraction influence on excess of a backscattered signal show that this effect is small [44]. However, it should be noted that only mean climatic gradients of wind speed were used for such estimates.

The results of current studies of low-level jets (LLJs) with lidars [45] and sodars [46] can force a significant revision of these estimates. At jets, which are regularly observed in the stable ABL, the wind-velocity gradients in the lower part of LLJs several times exceed their means. This must result in stronger (than expected earlier) deviations of scattering angle from 180° . Simultaneously, in the lower part of LLJs, turbulent kinetic energy increases, i.e., the wind structure parameter C_V^2 increases.

Therefore, it is quite possible that just the nocturnal jets are responsible for overestimation of sodar C_T^2 values. It is essential that, under such an assumption, the two features of the results of comparisons between sodar and micro-meteorological measurements (which were noted at the end of Section 3) are naturally explained. First, jets, as well as the discrepancy, usually intensify during the second half of night. Second, under inversions, the values of C_T^2 and C_V^2 are usually well correlated; therefore, in parallel with an increase in the mean intensity of sodar return signal due to the contribution of C_V^2, the correlation of signal intensity with C_T^2 values measured with micro-thermometers from a meteorological tower is conserved.

5. Discussion and conclusion

Most of the results of comparisons between sodar and in-situ data on C_T^2 indicate a surprisingly good agreement with the theory of Bragg scattering from the Kolmogorov turbulence. We suggest that deviations from the agreement, which is observed sometimes, are caused by the influence of complementary factors on the return signal rather than by a violation of the classical regularities of sound scattering by small-scale turbulence.

Note that, due to methodical difficulties associated with such comparisons and possible technical errors, all the published data concerning this problem cannot be considered reliable. Nevertheless, a systematic overestimation of sodar C_T^2 values, on average, by factor 1.5 is undoubtedly observed in the stable ABL. In addition, many cases are noted, where, during a few hours (mainly after midnight and before dawn), sodar C_T^2 values exceed the results of microtemperature measurements by one or even two orders of magnitude. A special feature of such episodes is the conservation of visible correlation between sodar and micro-temperature time variations over periods of ~ 5-20 min. The intensity of small-scale temperature turbulence is apparently closely related to the factors that lead to an increase in the mean level of return signal.

The following two causes of an increase in the intensity of return signal seem to be most probable: 1) the partial reflection of sound from sharp temperature jumps at the boundaries of inversion layers and "cat-eyes"; 2) contribution of sound scattering from wind-velocity fluctuations due to the wind refraction of sodar beam, which rapidly increases at low-level jet streams characteristic of the night ABL. For a rough estimate of the ABL state, when an overestimation of sodar C_T^2 values is possible, a visual analysis of echograms is helpful. Note also that the Bragg scattering by isotropic turbulence can reliably be distinguished from other sources of return signal only on the basis of the results of measurements of the full indicatrix of sodar return signal. The performance of such measurements in the ABL is difficult.

Acknowledgments
Author is grateful to R.L Coulter, R.D. Kouznetsov, and V.F. Kramar for fruitful discussions.
This work was supported by the Russian Foundation for Basic Research through grants 06-05-08086, 07-05-00521, and 08-05-00671.

References

[1] Kallistratova M A Coulter RL 2004 Application of sodars in the study and monitoring of the environment. *Meteor. Atm. Phys.* **85** 21-38
[2] Tatarskii V I 1971 *The Effects of the Turbulent Atmosphere on Wave Propagation* (Springfield: US Dept. Commerce).
[3] Gur'janov A E, Kallistratova M A, Kutyrev A S, Petenko I V, Tokvinin A B and Scheglov P V 1992 The contribution of the lower atmospheric layers to the seeing at some mountain observatories *Astron. and Astrophys.* **262** 373-81
[4] Lawrence J S, Ashley M C B, Tokovinin A and Travouillon T 2004 Exceptional astronomical seeing conditions above Dome C in Antarctica *Nature* **431** 278-81
[5] Lawrence J S, Ashley M C B, Bonner C S, Bradley S, Luong-van D and Storey J W V 2007 Characterisation of the Dome C atmospheric boundary layer turbulence with a non-Doppler acoustic radar EAS Public. Series **25** 31-34
[6] Bonner C, Ashley M C B, Lawrence J, Luong-van D, Storey J W V, Xu Z, Zhu Z, Zhang Z, Bradley S and Wang L 2008 SNODAR: measuring the height of the Antarctic atmospheric boundary layer Geoph. Res. Abstr. **10** EGU2008-A-12429
[7] Ricklin J C and Davidson F M 2002 Atmospheric turbulence effects on a partially coherent gaussian beam: implications for free-space laser communication *JOSA* **19** 1794-802
[8] Polynkin P, Peleg A, Klein L, Rhoadarmer T and Moloney J 2007 Optimized multiemitter beams for free-space optical communications through turbulent atmosphere *Opt. Lett.* **32** 885-87
[9] Kallistratova M A and Pekour M S 1994 Structure of optical turbulence over large city *SPIE* **2222** *Atmospheric Propagation and Remote Sensing III* Eds. W.A. Flood and W.B.P.Miller 894-905
[10] Isimaru A 1978 *Wave Propagation and Scattering in Random Media* (London, NY: Acad. Press)
[11] Ostashev V E 1997 *Acoustics in Moving Inhomogeneous Media* (London: E&FN SPON)
[12] Luce H, Crochet M and Dalaudier F 2001 Temperature sheets and aspect sensitive radar echoes Annales Geoph. **19** 899 920
[13] Chau J L, Doviak R J, Muschinski A and Holloway C L 2000 Tropospheric measurements of turbulence and characteristics of Bragg scatterers using the Jicamarca VHF radar *Radio Sci.* **35** 179-193
[14] Danilov S D, Gur'yanov A E, Kallistratova M A, Petenko I V, Singal S P, Pahwa D R and Gera B S 1994 Simple method of calibration of conventional sodar antenna system *Int. J. Remote Sensing* **153** 07-12
[15] Kallistratova M A 1962 Experimental investigation of the sound wave propagation in the atmosphere *Trudy Inst. Fiz. Amos., Atmos. Turbullentnost* No 4, Moscow pp 203-256 (English translation, U.S. AirForce FTD TT-63-441)
[16] Kallistratova M A 1994 Acoustic and radioacoustic remote sensing studies in Former USSR *Int. J. Remote Sensing* 15 251-66
[17] Baerg W and Schwarz W H 1965 Measurements of the scattering of sound from turbulence *J.*

Acoust. Soc. Amer. **39** 1125-32

[18] Neff W D 1975 Quantitative evaluations of acoustic echoes from the planetary boundary layer *NOAA Technical Report ERL 322-WPL 38,* Boulder, CO pp. 1-34

[19] Thomson D W, Coulter R L and Warhaft Z 1978 Simultaneous measurements of turbulence in the lower atmosphere using sodar and aircraft *J. Appl. Met.* **17** 725-34

[20] Asimakopoulos D N, Cole R S, Caughey S J and Crease B A 1976 A quantitative comparison between acoustic sounder returns and the direct measurement of atmospheric temperature fluctuations Bound.-Layer Met. **10** 137-47

[21] Kunkel K E 1981 Two-dimensional behavior of the temperature structure parameter near an escarpment *Proc. 1st Int. Symp. Acoustic Remote Sensing (Calgary, Canada, 21-24 May 1981)* pp 4.77-4.83

[22] Gur'yanov A E, Zubkovskii S L, Kallistratova M A, Karyukin G A, Kukharets V P and Petenko I V 1981 On the reliability of temperature structure characteristic vertical profile determination in the atmosphere by acoustic sounding method *Izvestia Acad. Sci. USSR, Atmos. Ocean Phys.* **17** 146-52

[23] Vogt S, Kalthoff N and Thomas P 1992 Estimations of the sensible heat flux and the temperature structure parameter using sodar and a sonic anemometer – an intercomparison *Proc. 6th Int. Symp. Acoustic Remote Sensing (Athens, Greece, 26-29 May 1992)* ed D N Asimakopoulos pp. 75-89

[24] Beran D W, Hooke W H and Clifford S F 1973 Acoustic echo-sounding techniques and their application to gravity wave, turbulence and stability studies Bound.-Layer Met. **4** 133-53

[25] Haugen D A and Kaimal J C 1978 Measuring temperature structure parameter profiles with an acoustic sounder. *J. Appl. Meteor.* **17** 895-99

[26] Zhou M, Chen Y, Lu N and Li S 1982 A comparative study between the temperature structure coefficient detected by sodar and direct methods *Acta Geophys. Sinica* **25** 492-99

[27] Gur'yanov A E, Kallistratova M A, Martvel F E, Pekour M S, Petenko I V and Time N S 1987 Comparison of sodar and microfluctuation measurements of the temperature structure parameter in mountainous terrain. *Izv. Acad. Sci. USSR, Atmos. Ocean Phys.* **23** 685-91

[28] Rusakov Yu 2006 On the consistency between "plume-like" echograms and temperature pulsation field in a convection atmosphere *Extend Abstr. 13th ISARS, (Garmisch-Partenkirchen, Germany, 18-20 July, 2006*) ed S Emeis pp 21-22

[29] Gage K S and Balsley B B 1981 Recent advances in Doppler radar probing of the clear atmosphere *Atmos. Technology* **13** 321-32

[30] Muschinski A and Wode C 1998 First in situ evidence for co-existing submeter temperature and humidity sheets in the lower free troposphere *J. Atmos. Sci.* **55** 2893-906.

[31] Balsley B B, Jensen M L, Frehlich R G, Meillier Y and Muschinski A.2003 Extreme gradients in the nocturnal boundary layer: structure, evolution, and potential causes *J. Atmos. Sci.* **60**, 2496-508.

[32] Kouznetsov R 2008 Acoustic backscattering at strong temperature gradients in the ABL *this collection*

[33] Gossard E E, Gaynor J E, Zamora R J and Neff W D 1985 Finestructure of elevated stable layers observed by sounder and in situ tower sensors *J. Atm. Sci.* **42** 2156-69

[34] Tsuda T, Sato T, Hirose K, Fukao S and Kato S 1986 MU radar observations of the aspect sensitivity of back-scattered VHF echo power in the troposphere and lower atmosphere. *Radio Sci.* **21**, 971-80

[35] Doviak R J and Zrnic D S 1984 Reflections and scatter formula for anisotropic air *Radio Sci.* **19** 325-36

[36] Gurvich A S and Kon A I 1992 The back-scattering from anisotropic turbulent irregularities J. Electromagn. Waves and Applic. **6**, 107-18

[37] Kallistratova M A and Petenko I V 1993 Aspect sensitivity of sound back-scattering in the atmospheric boundary layer *Appl. Phys. B* **57** 41-48

[38] Gurvich A S and Kukharets V P 1985 Turbulence intermittency influence in the dispersion of

radiowave in the atmosphere *Electrodyn. Radiowave Prop.* **8** 1531-37

[39] Tieme N S, Shurygin E A and Nesterova T N 1987 The intermittency of the turbulence and the fluctuations of echo return intensity in acoustic sounding in the convective atmosphere *Izv. Acad. Sci. USSR Atmos. Ocean. Phys.* **23** 21-30

[40] Obukhov A M 1962 Some specific features of atmospheric turbulence *J. Fluid Mech.* **13** 77-81

[41] Muschinski A, Frehlich R G and Balsley B B 2004 Small-scale and large-scale intermittency in the nocturnal boundary layer and the residual layer *J. Fluid Mech.* **515** 319-51

[42] Chunchuzov I P 2002 On the high-wave number form of the Eulerian internal wave spectrum in the atmosphere *J. Atmos. Sci.* **59** 1753-74

[43] Ostashev V E, Chunchuzov I P and Wilson D K 2005 Sound propagation through and scattering by internal gravity waves in a stably stratified atmosphere *J. Acoust. Soc. Am.* **118** 3420-29

[44] Brown E H and Hall F F Jr. 1978 Advances in atmospheric acoustics *Rev. Geophys. Space Phys.,* **16** 47-110

[45] Banta R M, Newsom R K, Lundquist J K, Pichugina Y L, Coulter R L and Mahrt L 2002: Nocturnal low-level jet characteristics over Kansas during CASES-99. Bound.-Layer Met. **105** 221-52

[46] Kallistratova M A 2008 Investigation of Low-Level-Jets over rural and urban areas using two sodars *this collection*

IOP Conf. Series: Earth and Environmental Science **1** (2008) 012002 doi:10.1088/1755-1307/1/1/012002

Application of sodar to interpret CO_2 and CO profiles and their dependence on boundary layer structure

W. Neff, A. Andrews, D. Wolfe

NOAA Earth System Research Laboratory, 325 Broadway, Boulder Colorado 80305

william.neff@noaa.gov

Abstract. The Boulder Atmospheric Observatory (BAO) tower was constructed and became operational in 1977. This 300-m tower, although originally supporting the development and improvement of ground-based remote sensing devices, has been used extensively in the study of the atmospheric boundary layer as well as plume dispersion and air quality. It was used in studies of the Denver Brown Cloud during the winters of 1987-1988 and 1996-1997. Located about 20 km east from the foothills of the Rocky Mountains, it is subject to a wide range of weather conditions ranging from night-time drainage winds with a low-level jet structure, to down-slope wind storms and upslope snow storms. During the summer of 2007, three levels of CO_2 and CO gas sampling (at 22, 100, and 300 m) were added as the tower became part of the NOAA ESRL/Global Monitoring Division CO_2 tall-tower network. The tower's location in complex terrain and its proximity to urban areas will provide a number of challenges in the interpretation of the data it provides. In this paper, we will describe some of the history of the tower in past air quality studies, examples of its complex meteorological setting and initial examples comparing diurnal variation in CO_2 and CO with boundary layer depths and structure observed with an acoustic sounder.

1. The use of sodar to characterize near-surface chemical processes at the BAO tower

Sodars have long been used to characterize the behavior of the planetary boundary layer. In many cases the characterizations have been qualitative in nature or, in some analyses, sodar facsimile records and Doppler-derived wind fields have been used to describe changes in near-surface chemistry with time. Although automatic detection of the mixing layer depth using sodars has been problematic [1, 2], some progress has been made in idealized situations such as those at the South Pole where there is no diurnal cycle and the boundary layer is largely stable and slowly changing [3]. In addition, sodars have been used in combination with balloon profiling of chemical species to better elucidate boundary layer mixing processes that affect surface concentrations [3-5].

ESRL's Global Monitoring Division (GMD) began making measurements from tall towers across the US in the 1990s (http://www.esrl.noaa.gov/gmd/ccgg/towers/index.html). Recently, the 300-m tower at the Boulder Atmospheric Observatory (BAO) was added to the network (Figure 1). The BAO tall tower site is the first to be located near complex terrain and a large urban area. However, extensive past studies of the meteorology and air quality along the Colorado Front Range will aid in the analyses of the complex data expected at this site. Completed in 1977, the BAO has been a unique research facility for studying the planetary boundary layer and for testing and calibrating atmospheric sensors. The centerpiece of the facility is a 300-m tower instrumented at five levels with slow-response temperature and wind sensors, a variety of remote sensing systems, and a real-time processing and display capability that greatly reduces analysis time for scientists (http://www.esrl.noaa.gov/psd/technology/bao/site/). The BAO has been the host of several large national and international experiments and numerous smaller ones. The field programs at the BAO have included 1) instrument evaluations, 2) plume dispersion studies, and regional air quality studies including the 1987-1988 Denver Brown Cloud Study and the 1996-1997 Northern Front Range Air Quality Study (NFRAQS) [6].

Figure 1. The 300-m BAO tower located east of the Rocky Mountains. Although located initially in 1979 in a predominately rural area, urbanization in the area is increasing with housing developments, gas wells, and a high-traffic interstate highway a few km to the east (evident near the top of the figure.) Located in an area dominated by nighttime drainage flows, small urban areas to the west of the tower (to the right of the figure) can contribute to the emissions observed at the tower. Southerly winds can bring emissions from the major Denver metropolitan area to the tower. Easterly winds can bring either aged pollution or relatively clean air from the Great Plains to the site.

2. CO_2 and CO in the surface layer

Carbon dioxide (CO_2) concentrations in the atmosphere depend on a number of sources and sinks, anthropogenic and natural. CO_2, in particular, is subject to a strong diurnal cycle with uptake by vegetation in the daytime and release from the surface through ecosystem respiration at nighttime. Because of the surface character of the sources and sinks of CO_2, boundary-layer mixing plays a key role governing its concentration. Carbon monoxide (CO), on the other hand, has its source in combustion sources, both at the surface and elevated. Examining the covariance of CO and CO_2 at various heights can suggest the origin of the observed CO_2. For example, a distinct diurnal cycle in CO_2 near the surface in the absence of a CO signal would indicate a dominance of biogenic sources and sinks of CO_2. The complications occur when surface sources of CO_2 and CO, such as vehicle and other combustion sources, comingle with the biogenic sources and sinks. This paper addresses the potential value of sodar profiling of the boundary layer in the interpretation of these various sources in combination with fixed level measurements of tall towers such as that at the BAO.

3. CO_2 and CO Measurements

CO_2 is measured using a non-dispersive infrared gas analyzer [Li-Cor 7000 CO_2/H_2O], and CO is measured using gas filter correlation [48C Trace Level, Thermo Electron Corporation]. These detectors are embedded in a customized trace gas analyzer system such that both gas analyzers are housed in temperature-controlled enclosures and are equipped with active pressure and flow control ["A High-Precision CO_2 and CO Analysis System for Automated Semi-Continuous Profile Sampling from Tall Towers", Andrews et al., manuscript in preparation]. The analyzers are calibrated two to four times per day through an external sample selection manifold using a set of 8 on-site reference gases (5 for CO_2, 3 for CO). The CO baseline drift is monitored every 30 minutes using ambient air that is scrubbed of CO using a catalytic reagent [Sofnocat]. CO_2 baseline drift is monitored every one to two hours by measuring a standard with near ambient concentration. The CO_2/CO analyzer is housed in a building at the base of the tower. Air is drawn down from the tower through three sampling lines positioned at 22, 100 and 300 meters above ground level. The tubes are flushed continuously at a flow rate of five to ten liters per minute. The analysis system operates on a five minute cycle. The average value for the final 30 seconds of each five minute interval is used to compute CO_2 and CO mixing ratios, so a complete profile is obtained every 15 minutes. This sampling sequence allows for full equilibration of the CO_2 mixing ratio even for large amplitude transitions between inlet heights.

4. The meteorology of the BAO Tower site

Four regimes dominate air quality along the eastern slopes of the Rocky Mountains:

1) Nocturnal drainage flows that follow the South Platte River from the southwest to the northeast through Denver. The nocturnal drainage jet structure [7], because of a nearly laminar layer that forms between 100 m and 200 m, may result in the trapping of urban surface emissions in a thin layer below the jet and may isolate elevated emissions (from point sources) in the air flow above the wind maximum: This drainage system extends well into northeastern Colorado during summer [8]. *Implications for CO_2 observations:* At the BAO these drainage winds are generally from the southwest or west and do not originate from major urban areas. In these cases, comparison of CO and CO_2, particularly at nighttime, should allow distinguishing combustion versus biogeochemical sources.
2) Thermally and/or dynamically driven northeasterly winds (upslope, toward the foothills), often associated with a shallow front-like or surge structure only a few hundred meters deep, that can transport cool air from the lowlands of the South Platte, northeast of Denver, southwestward into the foothills. During the Brown Cloud Study[6], these winds were most likely to occur during the afternoon but were also observed at many other times of the day and night [7, 9, 10], sometimes as a result of mesoscale eddies that form along the Front Range (e.g.[11, 12]). These upslope and recirculating flows enable aged aerosol and/or precursor gases such as ammonia to return to Denver and may contribute to a rapid degradation of visibility [13]. The stability of the shallow air mass limits vertical mixing and allows further buildup of pollution. When alternating with a nocturnal drainage wind, they may lead to a day-to-day recycling of the same airmass. *Implications for CO_2 observations:* Under these conditions, urban sources are likely to dominate boundary layer CO_2 and CO behavior. When below the 300 m level, only the 22-m and 100-m levels may show these urban influences.
3) Moist, cool northeasterly upslope winds, usually in response to lee cyclogenesis southeast of Denver and/or cold, surface high pressure developing over the Great Plains to the northeast of Denver, sometimes result in snowfall along the base of the mountains, but also in fogs and low clouds. Such conditions can support rapid chemical transformations, such as SO_2 to sulfate, that depend on the presence of clouds [14]. A related area that merits further investigation is melting and evaporation into the shallow boundary layers that often follow snowstorms. *Implications for CO_2 observations:* During the initial phase of these upslope conditions, urban residue may dominate; as the upslope continues, cleaner air from the plains may change the CO/CO_2 behavior significantly.
4) Downslope westerly winds that usually are strongest near the foothills west and north of Denver and which are associated with falls of pressure along the foothills in combination with higher surface pressure to the east over the plains, contributing to shallow upslope flows along the Platte

River. Warm westerly winds several hundred meters aloft and light, cool easterly winds near the surface enhance the low-level temperature inversion creating strong trapping conditions unless there is a strong differential acceleration of the wind across the inversion layer. During the Brown Cloud field study, the inversion often proved remarkably resistant to erosion by the strong westerly winds above it. *Implications for CO_2 observations:* Because of the position of the BAO, closer to the foothills, it usually is dominated by high winds from the mountains which should show minimal urban influence. However, sometimes the boundary between the clean mountain air and the polluted air masses over the plains may oscillate back and forth through the BAO site as the mountain wave phase shifts. These will provide interesting cases for analysis.

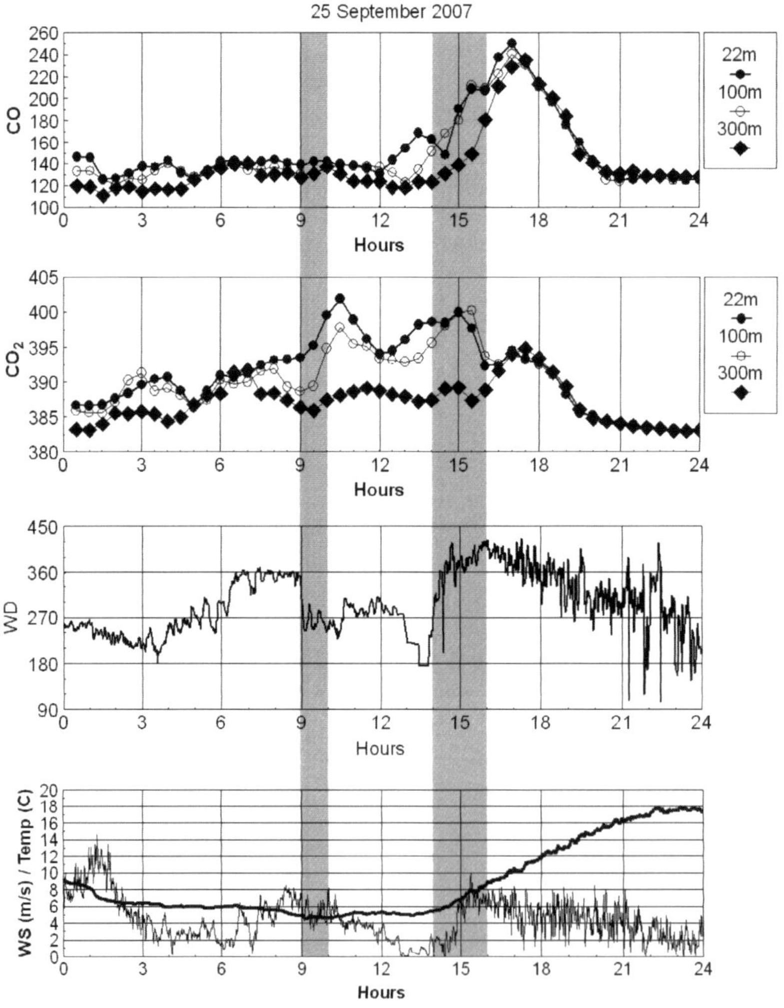

Figure 2: Time series of a) CO (ppb), b) CO_2 (ppm), c) Wind Direction, and d) Wind Speed (thin line) and Temperature (thick line) for September 25, 2007. CO and CO_2 data are smoothed with a 3-point filter. Gray shading indicates specific periods called out for discussion.

5. Case Study

Figure 2 shows an example over a full 24 hour period on 25 September 2007 (UTC) of CO, CO_2, temperature, and surface wind speed and wind direction. Wind directions on this day varied from southwest to northeast with daytime convection beginning around 1500 UTC with an increase in wind speed and temperature together with a shift of wind direction to the northeast (upslope toward the mountains). During the night, winds were predominately from the west and decreasing in speed from 0800 UTC to 1400 UTC, characteristic of typical nocturnal drainage flows.

During the night period, the sodar facsimile record (Figure 3) shows a mixing layer increasing in depth between 0900 and 1000 UTC from about 50 m to about 200 m. The 300m-level CO_2 is nearly constant or decreases from its initial value at 0800 UTC (local midnight) to 1600 UTC in the early morning whereas lower levels increase in relative value. This increase in concentrations of CO_2 at lower levels coincides with a wind shift from northerly to westerly at 0900 UTC. Because there is no sudden collapse of the boundary layer after 0930 UTC, one might assume that the lowest tower levels are simply seeing a different source of CO_2, perhaps associated with downslope flow off the mountains to the west.

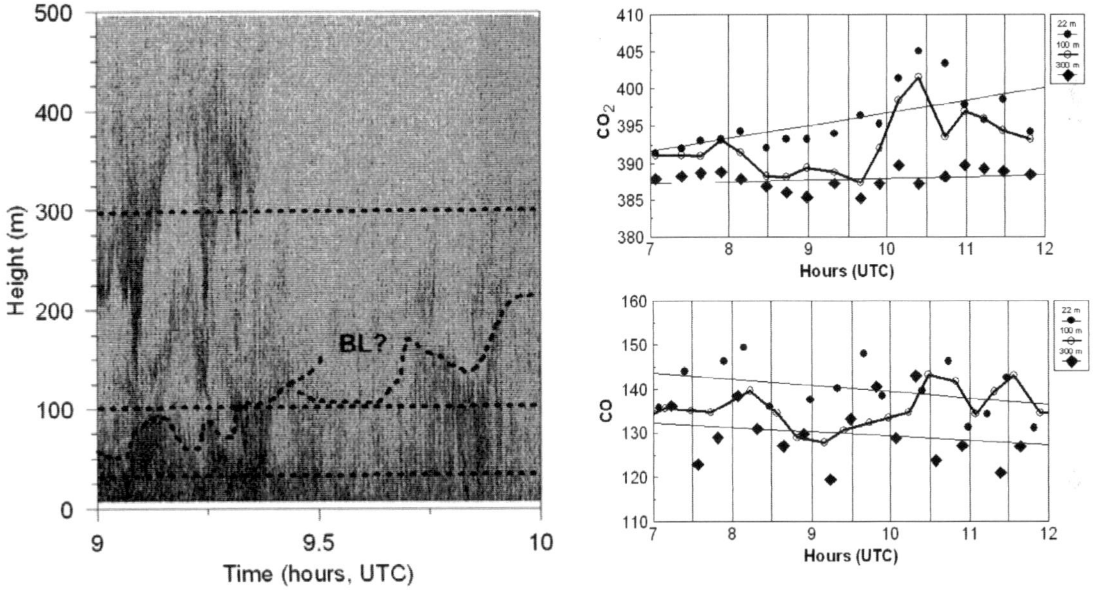

Figure 3. Sodar facsimile record from 25 September 2007 between 0900 and 1000 UTC (left) with expanded time series of CO and CO_2 (right) bracketing the period of interest.

Figure 2 showed the effect of the transition from nocturnal inversion conditions to a convective regime as the traces for CO and CO_2 at all levels converge by 1700 UTC. Figure 4 shows the sodar record for this period with a residual of the nocturnal inversion still present at 1400 UTC and then transitioning to fully developed convection by 1600 UTC. The increase in CO during this period coincides with the morning rush hour and with a wind direction that would bring automobile emissions toward the tower from the interstate highway just to the east of the tower site.

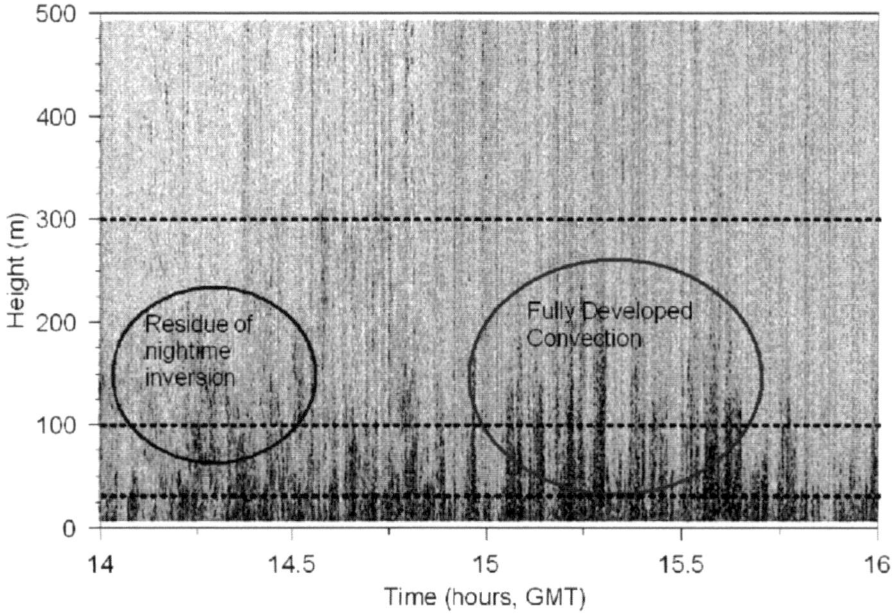

Figure 4. Sodar record during the morning transition from 1400 to 1600 UTC on 25 September 2007.

6. Future plans and challenges

CO and CO_2 data in combination with routine sodar facsimile recordings have been archived beginning in late summer of 2007. During 2008, we expect to be able to examine the full seasonal cycle from winter through the spring and summer growing period, into the next fall and winter. The challenge in this analysis will to develop, if possible, an automatic detection method for the nocturnal surface mixing-layer depth. Our earlier results from the South Pole suggest that this is quite feasible for slowly evolving stable layers. Examination of data from this last winter suggest that under very stable, light wind conditions, short-term wave modulation of the inversion layer depth leads to rapid variation in CO and CO_2 concentrations that will be difficult to link in a one-to-one fashion with observe depths.

References

[1] Beyrich F 1997 Mixing height estimation from sodar data - A critical discussion. *Atmos. Env.* 31(23):3941-3953.
[2] Seibert P, Beyrich F, Gryning SE, Joffre S, Rasmussen A, Tercier P 2000 Review and intercomparison of operational methods for the determination of the mixing height. *Atmos. Env.* 34(7):1001-1027.
[3] Neff W, Helmig D, Grachev A, Davis D 1998 A study of boundary layer behavior associated with high NO concentrations at the South Pole using a minisodar, tethered balloon, and sonic anemometer *Atmos. Env.* 42(12):2817-2830.
[4] Beyrich F, Weisensee U, Sprung D, Gusten H 1996 Comparative analysis of sodar and ozone profile measurements in a complex structured boundary layer and implications for mixing height estimation. Bound.-Layer Meteorol. 81(1):1-9.
[5] Helmig D, Johnson BJ, Warshawsky M, Morse T, Neff WD, Eisele F, et al. 2008 Nitric oxide in the boundary-layer at South Pole during the Antarctic Tropospheric Chemistry Investigation (ANTCI). *Atmos. Env.* 42(12):2817-2830.

[6] Neff WD 1997 The Denver Brown Cloud studies from the perspective of model assessment needs and the role of meteorology. *J. Air & Waste Manag. Assoc.* 47(3):269-285.
[7] Neff WD, Eberhard WL, King CW, 1990 Remote sensor observations of thermally and dynamically driven circulations in the South Platte River basin. *Preprints: Fifth Conf. on Mountain Meteorology*, Boston: American Meteorological Society 42-49.
[8] Toth JJ, Johnson RH 1985 Summer Surface Flow Characteristics Over Northeast Colorado. *Mon. Wea. Rev.* 113(9):1458-1469.
[9] Neff WD, 1990 Remote Sensing of Atmospheric Processes over Complex Terrain. In: Blumen W, editor. *Atmospheric Processes over Complex Terrain*. Boston: American Metoorological Society pp. 173-228.
[10] Crow LW 1973 Airflow study related to EPA field monitoring program Denver metropolitan area, November 1973. In: *Denver Air Pollution Study: Proceedings of a Symposium; Vol. I* Environmental Sciences Research Laboratory, Office of Research and Development, U.S. EPA
[11] Levinson DH, Banta RM. 1995 Observations Of A Terrain-Forced Mesoscale Vortex And Canyon Drainage Flows Along The Front Range Of Colorado. *Mon. Wea. Rev.* 123(7):2029-2050.
[12] Wilczak JM, Christian TW 1990 Case-Study Of An Orographically Induced Mesoscale Vortex (Denver Cyclone) *Mon. Wea. Rev.* 118(5):1082-1102.
[13] Sloane CS, Watson JG, Chow JC, Pritchett LC, Richards LW 1990 Size distribution and optical properties of the Denver Brown Cloud. In: Mathai CV editor *Transactions: Visibility and Fine Particles* Pittsburgh: Air & Waste Management Association 384-393
[14] McHenry JN, Dennis RL 1994 The Relative Importance Of Oxidation Pathways And Clouds To Atmospheric Ambient Sulfate Production As Predicted By The Regional Acid Deposition Model. J. Appl. Meteor. 33(7):890-905.

IOP Conf. Series: Earth and Environmental Science 1 (2008) 012003 doi:10.1088/1755-1307/1/1/012003

Inter-comparison of retrieved wind fields from large-eddy simulations and radar measurements in the convective boundary layer

Danny Scipión[1,2,3], Aaron Botnick[2], Robert Palmer[2], Phillip Chilson[2], and Evgeni Fedorovich[2]

[1] School of Electrical and Computer Engineering, University of Oklahoma, Norman, OK, USA
[2] School of Meteorology, University of Oklahoma, Norman, OK, USA

E-mail: dscipion@ou.edu

Abstract. Estimates of the wind retrieved from boundary layer radars are inherently averaged in space and time. Using large eddy simulation (LES) results, we are able to examine to what extent these measurements are representative of the underlying parameters of the convective boundary layer (CBL) and how they are impacted by various averaging methodologies. Comparisons between an LES with real forcing and Boundary Layer Radar (BLR) wind retrievals have been analyzed. Good agreement between the LES winds and those estimated using the BLR is observed. After removing an instrumental bias, an estimate of the probability density function (PDF) and skewness of the radar and LES vertical velocity estimates was calculated as a function of height. The PDFs of vertical velocity and the height profiles of the skewness of vertical velocity exhibit good agreement for both the LES and BLR calculations and have the expected canonical structure.

1. Introduction

Three-dimensional virtual radar wind retrievals based on numerical large-eddy simulations (LES) have been previously obtained in a Multi-Radar Experiment (MRE) described in [1]. Five virtual boundary layer radars (BLRs) were equidistantly spaced within the LES domain (see Section 3). These radars were directed at sampling volumes located at different heights. The retrievals at each height were then estimated every minute and were further used to evaluate turbulence parameters. Although the obtained wind estimates provided satisfactory agreement with those from the LES, they were obtained for an idealized five-radar configuration. A more realistic setup discussed in this study is realized by using a single radar pointing in three non-coplanar directions. The three wind components are obtained employing the Doppler beam swinging (DBS) technique [2]. Two convective boundary layer (CBL) cases are considered. The first is an idealized case of sheared CBL [1; 3–5], for which spaced averaged radial velocity estimates from a virtual BLR are obtained every second. These estimates are compared with radial velocities from the LES. Because estimates of radial velocity every second are not common, a study of different time-averaging intervals is conducted to determine the minimum average time required

[3] Present address: University of Oklahoma, School of Meteorology, 120 David L. Boren Blvd., Rm 5900, Norman, OK 73072-7307

for spaced and time averages estimates from radar and time averages from LES to converge. The second case is a CBL with realistic environmental forcing [6], simulated over 12-hr interval. Realistic forcing is simulated through the incorporation into LES of environmental soundings, observed surface flux data, and geostrophic wind forcing. Numerical data for this case are compared with actual radar estimates of the CBL wind fields.

2. Large Eddy Simulation (LES)

The LES code employed in our study is described in [3] and [4]. This code was extensively tested in comparison with several other representative LES codes and with experimental data for clear CBLs with and without wind shear. The code was found to confidently reproduce turbulence structure for a broad variety of flow regimes observed in the clear CBL. The LES setup used for the idealized CBL case is described in [1; 5]. In addition, a more realistic CBL case observed at the Southern Great Plains Atmospheric Radiation Measurement Climate Research Facility (SGP ACRF) in Lamont, OK, beginning 11:29 UTC on 8 June 2007 is considered. The LES setup for this case, described in [6], uses SGP ACRF soundings along with external data retrieved from a large scale atmospheric model. Examples of instantaneous LES fields for realistic CBL case are presented in Figure 1.

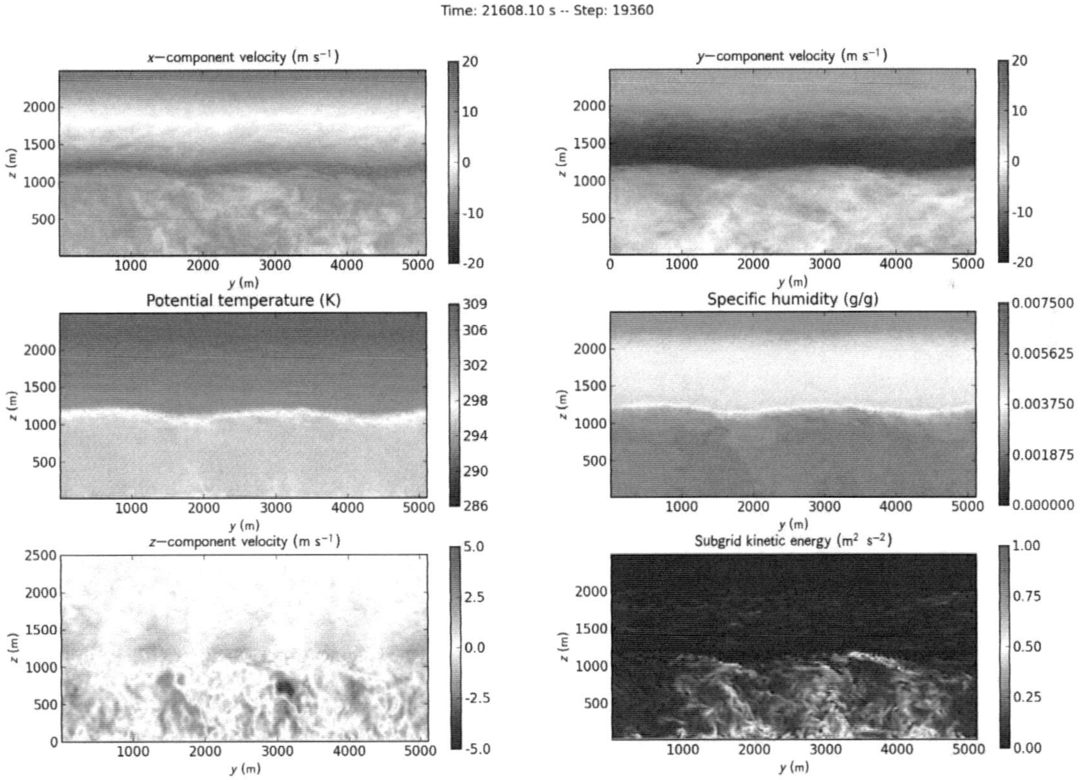

Figure 1. Examples of LES fields in the sub-domain of the radar simulator. Top-left: zonal wind. Top-right: meridional wind. Middle-left: potential temperature. Middle-right: specific humidity. Bottom-left: vertical wind. Bottom-right: sub-grid kinetic energy. All data presented refer to the same single realization in time (one LES time step).

3. LES-Based Radar Simulation

The method to create the simulated radar signals within the LES fields is described in [5] and is based on the work of [7]. The time-series data for the virtual BLR are compiled by summing the contribution from each LES point within the radar resolution volume. This is determined by the radar pulse width and beam width. The virtual radar is patterned after a Vaisala UHF BLR-LAP3000 and operates at a central frequency of 915 MHz, with a half-power beam width of 9°. It is possible to direct the radar beam vertically or electronically steered 15.5° off-vertical along 4 different azimuth angles: 0°, 90°, 180°, 270°. In the present study, a range resolution of 50 m is used. Additive white Gaussian noise was added to the time-series data with an SNR of 10 dB to produce realistic signals. The three moments are estimated after conventional spectral analysis of the time-series data. The maximum time processed for the idealized case is ~4.3 hr. The radial velocity retrieved from the first moment of the radar was compared with the radial velocity calculated along a similar radial component within the LES domain (an LES "pencil thin" beam). A comparison of both quantities is presented in Figure 2 where good agreement was found between the two estimates. The time interval of each estimate is 1 s.

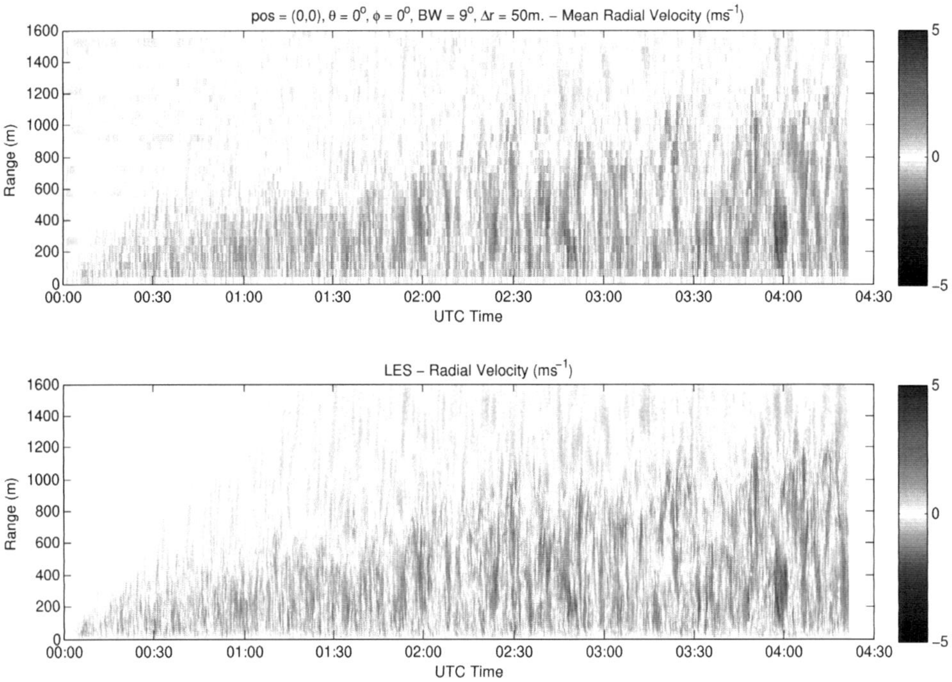

Figure 2. Comparison of the radial velocity retrieved from the virtual BLR (top) and the LES (bottom). The time spacing between subsequent velocity estimates is 1 s.

3.1. *Time and Spatial Averaging*

Several analyses were made in order to assess the agreement between radial velocities from the radar and LES. As we know, radar estimates are obtained from a time and spatial average within the resolution volume defined by the antenna beamwidth and radar pulse width. The focus of this section is to study the *spatial ergodicity* of the signal and to determine the minimum time average required for both estimates to converge. This averaging time is dependent on the mean wind velocity during the analysis period. The LES radial velocities are obtained at every time step (1 s). However, it is not realistic to obtain radar estimates every second; in the present study, wind velocity estimates are obtained every 12 min or more. Averaged radial velocity estimates from the radar are obtained by first averaging a number of spectra and then calculating the moments. LES profiles are obtained by temporal averaging of the velocity fields directly over a corresponding interval of time. Different time averaging periods have been employed during this study (see Figure 3). As can be observed the correlation coefficient (ρ) increases as the averaging period increases, reaching its maxima at 30 s ($\rho = 0.93$). After 120 s ($\rho = 0.92$) the correlation coefficient begins to decrease indicating a lack of statistical stationary. The peak in the correlation curve corresponds to the time required for atmospheric parcels to be advected through the radar sampling volume at a height of 1000 m and for the prescribed wind speed of 5 m s^{-1} [1; 5]. As in all averaging schemes of real data, a trade-off exists between temporal averaging and stationarity.

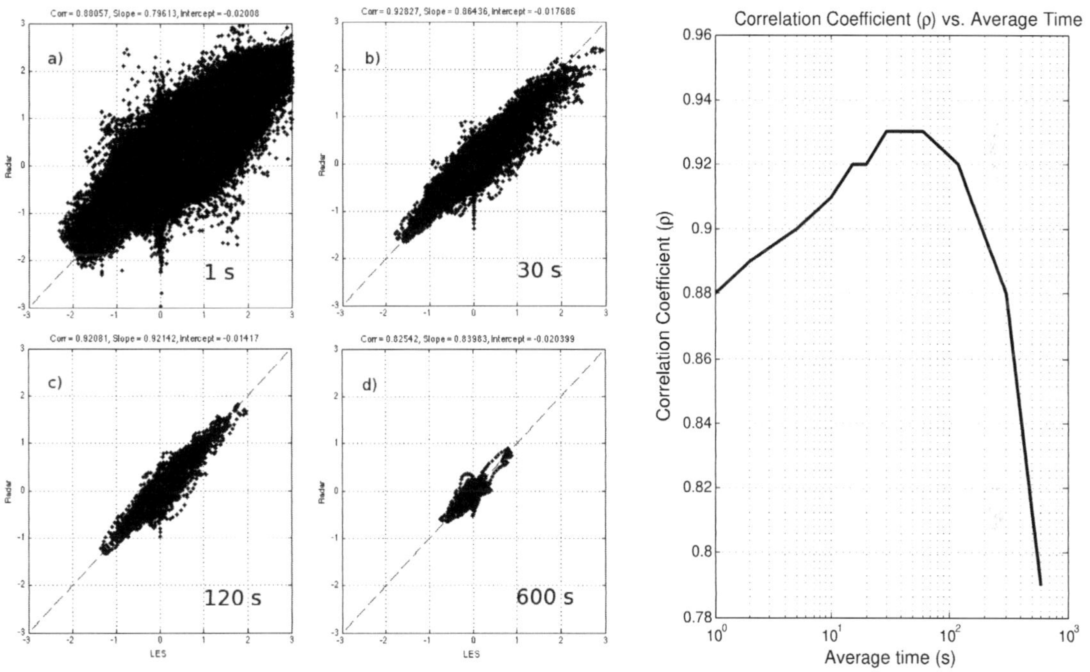

Figure 3. Left: scatter diagrams comparing radial velocity estimates from the virtual BLR with radial velocity profiles obtained from the LES at different averaging times: a) 1 s, b) 30 s, c) 120 s, and d) 600 s. Right: logarithmic plot of correlation coefficient (ρ) vs. averaged time. After 120 s, the correlation coefficient ρ has a value of 0.92, which indicates that the time required for the signal to be *spatially ergodic* has been reached

4. Data Analysis

Data in the following sections are from the realistic LES case, simulating CBL growth on 8 June 2007 at the SGP ACRF [6]. Approximately 11 hr of this 12-hr simulation are used in the analysis. In this case, data were output in slices through the center of the LES domain in the zonal and meridional directions. Radial velocities were then calculated from the LES slices pointing vertically and at 15.5° zenith angle in the four orthogonal directions. These data were used later to estimate the wind components using DBS.

In order to obtain fair comparisons with the 915 MHz radar located at the SGP ACRF site, the averaging time for the LES data was chosen to be 12 min. The mean wind components were estimated from these 12-min averages. For each of the wind components, four quantities were analyzed: DBS values of the radial velocities from the LES (LES_{DBS}), output velocities along a vertical profile at the center the LES domain (LES_{point}), averaged LES value over beam directions (LES_{avg}), DBS estimates from the SGP ACRF radar ($Radar_{ARM}$). Results for the three wind components are presented in Figures 4 (zonal), 5 (meridional), and 6 (vertical).

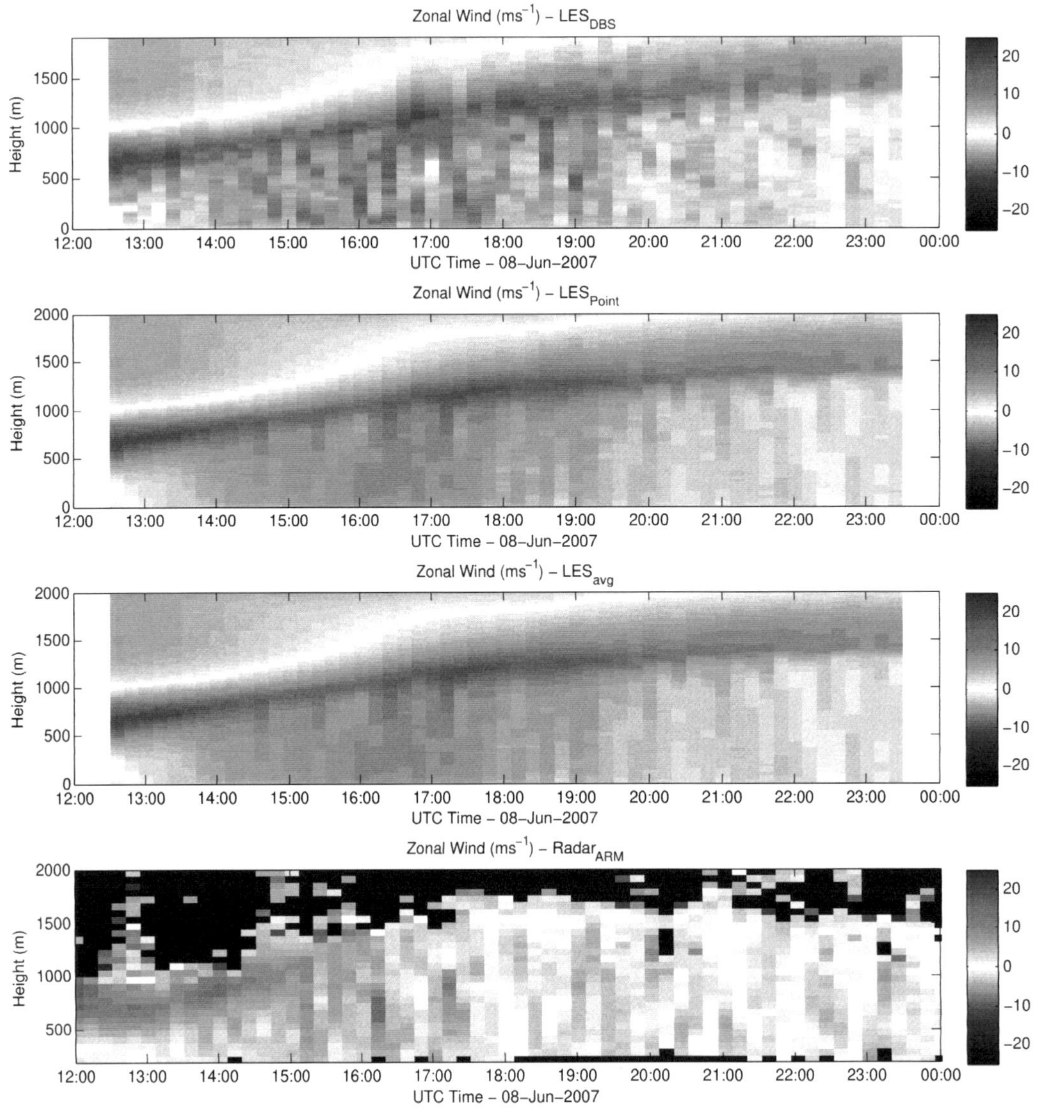

Figure 4. Zonal Wind retrievals averaged every 12 min. Top: LES-DBS. Middle-top: LES-Profile at center of domain. Middle-bottom: spatially averaged profile including the five LES-DBS beams. Bottom: DBS estimates from the SGP ACRF radar.

To obtain the LES_{DBS} values, the following procedure was used. First, the radial velocities from the LES were averaged every 20 s to emulate the dwell time from the radar. Second, velocity samples were taken every 6 min, corresponding to the time required for the radar to return to the same beam position within one DBS cycle. Third, the velocities were averaged

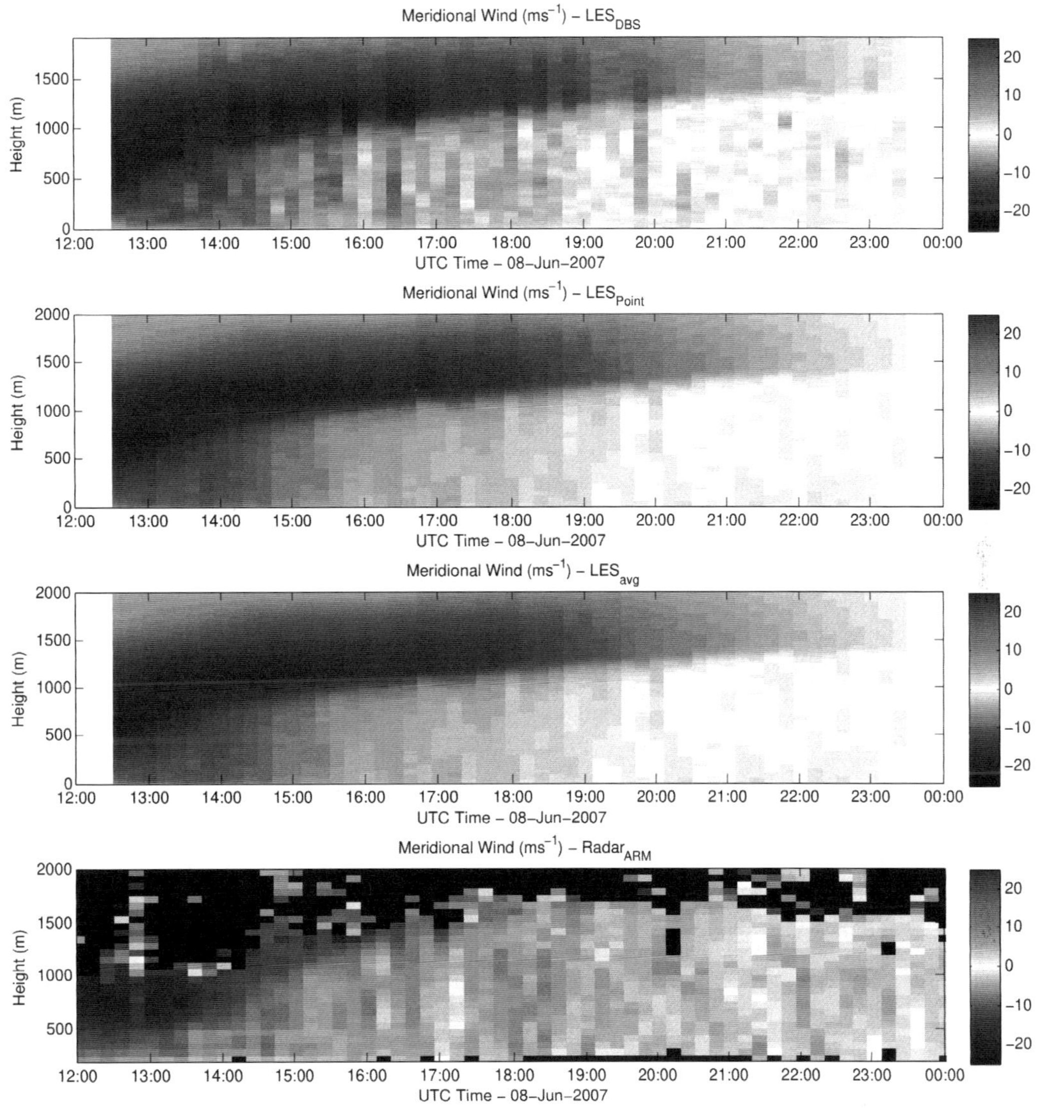

Figure 5. Meridional Wind retrievals averaged every 12 min. Top: LES-DBS. Middle-top: LES-Profile at center of domain. Middle-bottom: spatially averaged profile including the five LES-DBS beams. Bottom: DBS estimates from the SGP ACRF radar.

over 12-min periods representing the DBS sampling time. Finally, DBS was used to retrieve all three wind components. For LES_{point} and LES_{avg}, LES data with 10 s increments were averaged in time to obtain 12 min estimates used in the comparisons. Radial velocities from the radar were averaged over the same time period.

Time-averaged wind profiles obtained by LES_{point} and LES_{avg} methods, qualitatively appear similar. This suggests that our earlier assumption of ergodicity is appropriate for this case. Therefore, in further comparisons, only the winds from LES_{point} are used. The estimates from LES_{DBS} in this realistic case do not appear as smooth as those from the idealized case. In general, the simulated wind fields from the LES realistic case [6] agree rather well with the

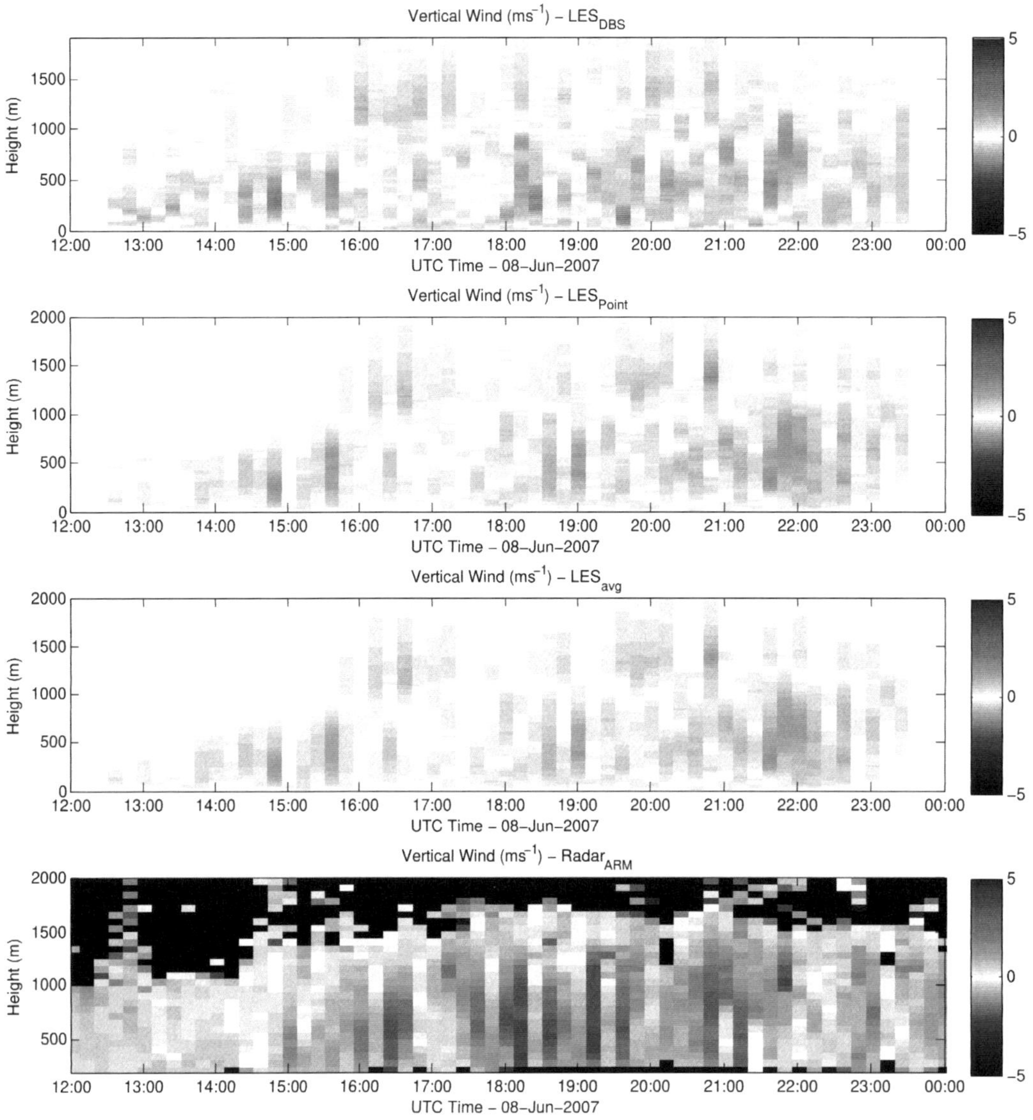

Figure 6. Vertical Wind retrievals averaged every 12 min. Top: LES-DBS. Middle-top: LES-Profile at center of domain. Middle-bottom: spatially averaged profile including the five LES-DBS beams. Bottom: DBS estimates from the SGP ACRF radar.

radar estimates. The same agreement can also be observed in the zonal and meridional winds at ~530 m (see Figure 7). In addition to the LES comparisons, sounding data for 12z, 18z, and 0z are also presented in the same figure. These results complement the comparison of the structure parameter of refractivity (C_n^2) presented in [6].

As can be observed in Figures 6 and 7, radar estimates of vertical velocity do not compare well with the estimates obtained from the LES. A negative bias in vertical velocity can be observed in the lower panel of Figure 7. The bias has a maximum of approximately -1 m s^{-1} and varies over the observation time; suggesting it is not the result of a constant offset error. Presently, the cause of this bias has not been determined; however, the bias is likely due to a slight error in

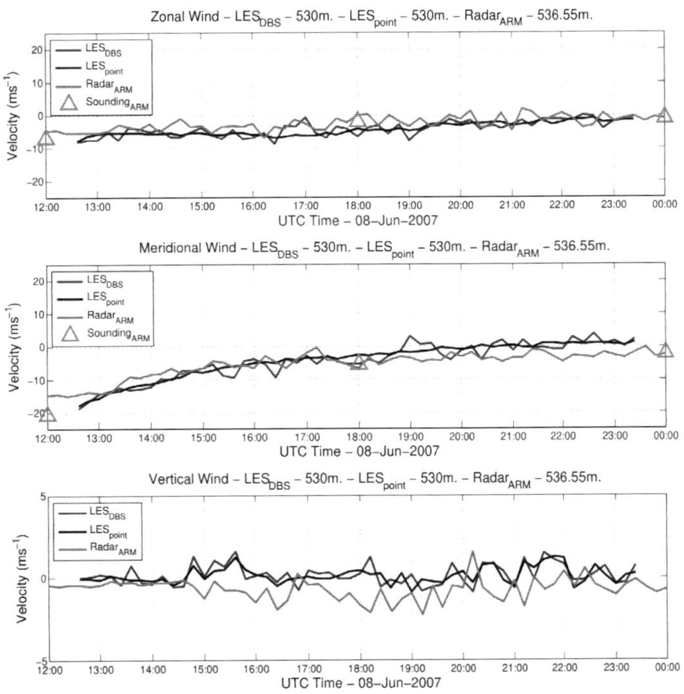

Figure 7. Wind Comparison from the LES_{DBS}, LES_{point}, $Radar_{ARM}$, and a sounding (launched at three different times: 12z, 18z, and 0z) at ~530 m. Top: zonal wind. Middle: meridional wind. Bottom: vertical velocity. Vertical velocity from the radar show a negative bias which is not constant over the observation time.

the zenith pointing angle. The zonal and meridional components of the winds are on the order of tens of m s^{-1}, so they would not be significantly affected by the proposed zenith angle error. Furthermore, the radar retrieved horizontal wind estimates are consistent with the sounding data. However, a pointing error could cause a component of the horizontal wind to erroneously be reflected in the vertical velocity estimates and the effects would be expected to vary in time and altitude. This is being further investigated.

The probability density function (PDF) of all the vertical velocities of the LES_{DBS}, LES_{point}, and $Radar_{ARM}$ are compared in Figure 8. The mean vertical velocity sampled over a sufficiently long period of time is expected to approach zero. However, the mean vertical velocity from the radar has a bias of ~-0.65 m s^{-1}. As a first order approximation at correcting for the bias, we have at this point simply removed the bias before performing the PDF calculations. It should be noted that the LES_{point} PDF was calculated without any time averaging scheme, which means the time resolution of ~10 s was used. This explains the smoothness of the curve as well as the wider velocity range. The floor around 5×10^{-3} in the radar data can be attributed to the measurement noise inherently present within the estimates. The PDFs from the three signals agree well. Furthermore, each exhibits a positive skewness. The skewness is indicative of overall structure of the observed velocity patterns [8]: when the skewness is positive, updrafts are narrower and more intense than the surrounding downdrafts.

The height dependence of the vertical velocity skewness in the CBL has been previously analyzed using LES and observations [8] and LES and wind tunnel experiments [9; 10]. Here, we calculate vertical profiles of skewness using vertical velocity data from LES and radar for the actual CBL case of 8 June 2007 applying averaging time of three hours (from 20:00 to 23:00). The

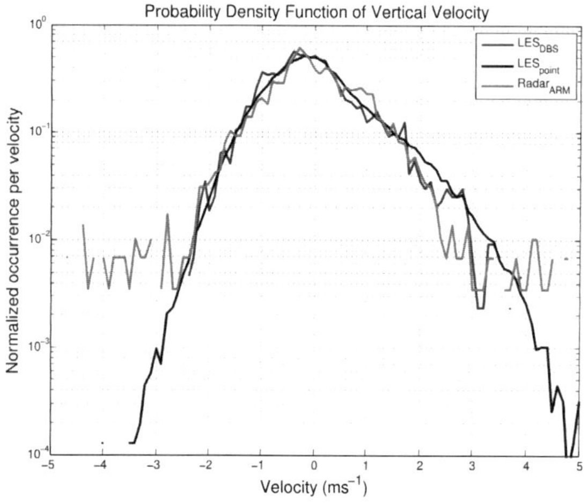

Figure 8. Probability density function of vertical velocity from LES_{DBS}, LES_{point}, and $Radar_{ARM}$. The bias in the vertical velocity estimates from the radar was removed, which simply shifted the PDF and did not affect the shape. A good agreement is observed among the three quantities. The floor around 5×10^{-3} in the radar data can be attributed to the noise inherently present within the estimates.

comparison results are presented in Figure 9. In the main portion of CBL, large scale turbulence structures are represented by narrow, fast updrafts and broader, slower downdrafts, so skewness of the vertical velocity in the mixed-layer is positive. Its grows with height indicates the increase in magnitude and range of vertical positive values towards the CBL top. The Skewness reaches the maximum around the base of the inversion, and then drops to zero in the interior of the inversion layer where the upward component of turbulent motion is in approximate equilibrium with the downward component [10]. The radar estimates of skewness agree reasonable well with the LES estimates and are similar to those presented in [9; 10].

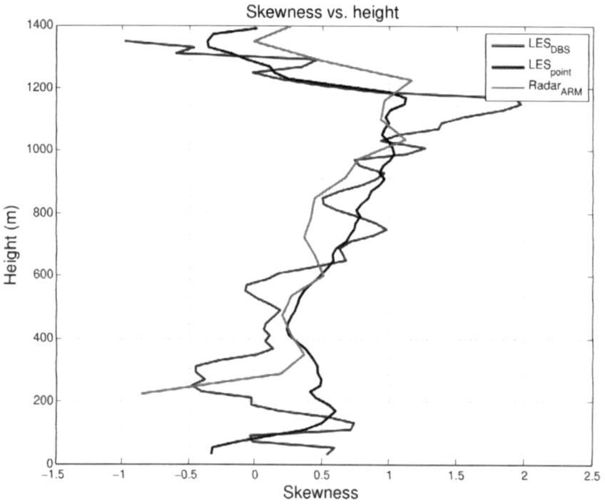

Figure 9. Skewness as a function of height for the vertical velocity between the 20:00 and 23:00 for the case of 8 June 2007. The skewness of the vertical velocity in the mixed-layer is mainly positive and grow with height.

5. Conclusions

Comparisons of the virtual BLR's radial velocities with LES velocity estimates were presented. The minimum averaging time required for the signal to approximate *spatial ergodicity* was estimated to be approximately between 30 s and 120 s for the idealized CBL case. Longer averaging produced decorrelation from the breakdown of statistical stationarity in the signal. Realistic comparisons between LES simulated velocities and actual radar data were also shown for the case of 8 June 2007. Good qualitative agreement was found in the zonal and meridional wind fields retrieved from the radar and estimated with LES. The vertical wind comparisons were studied in more detail after noticing an initial disagreement between LES and radar data. This disagreement between estimates may be attributed to an instrument error; although, further study is necessary to confirm this. After removing the bias, the PDFs from the radar and realistic LES cases were in agreement. The skewness of the vertical velocity has been analyzed as a function of height. In the well-mixed portion of the CBL is mainly positive and grows with height. The promising agreement between wind retrievals from LES and the radar encourages us to continue our research to further characterize the CBL wind and turbulent fields through a combination of both techniques.

Acknowledgments

The authors of this work wish to acknowledge the National Science Foundation (NSF) for their support via grant ATM-0553345. Observational data for this study were obtained from the Atmospheric Radiation Measurement (ARM) Program sponsored by the U.S. Department of Energy, Office of Science, Office of Biological and Environmental Research, Environmental Science Division.

References

[1] Scipión D E, Palmer R D, Fedorovich E, Chilson P B and Botnick A M 2007 *33rd International Conference on Radar Meteorology* (American Meteorological Society, Cains, Australia)

[2] Balsley B B and Gage K 1982 *Bull. Amer. Meteor. Soc.* **63** 1009–1018

[3] Fedorovich E, Conzemius R and Mironov D 2004 *J. Atmos. Sci.* **61** 281–295

[4] Conzemius R J and Fedorovich E 2006 *J. Atmos. Sci.* **63** 1151–1178

[5] Scipión D E, Chilson P B, Fedorovich E and Palmer R D 2008 *J. Atmos. Oceanic Technol.* **In Press**

[6] Botnick A M and Fedorovich E 2007 *Workshop on Quality and Reliability of Large-Eddy Simulations* (Leuven, Belgium)

[7] Muschinski A P, Sullivan P P, Hill R J, Cohn S A, Lenschow D H and Doviak R J 1999 *Radio Sci.* **34** 1437–1459

[8] Moeng C H and Rotunno R 1990 *J. Atmos. Sci.* **19** 1149–1162

[9] Fedorovich E, Kaiser R, Rau M and Plate E 1996 *J. Atmos. Sci.* **53** 1273–1289

[10] Fedorovich E, Nieuwstadt F T M and Kaiser R 2001 *J. Atmos. Sci.* **58** 70–86

IOP Conf. Series: Earth and Environmental Science **1** (2008) 012004 doi:10.1088/1755-1307/1/1/012004

Mean wind field in the urban atmospheric boundary layer by sodar data

V P Yushkov

Faculty of Physics, Lomonosov Moscow State University, Lenin Hills, Moscow, 119992, Russia

E-mail: yushkov@phys.msu.ru

Abstract. Two-year uninterrupted sodar measurements of vertical profiles of wind speed and direction in the height range 40 to 250-500 m were carried out in two points of Moscow megalopolis. Seasonal and diurnal courses of mean speed profiles, wind speed distributions, wind roses and occurrence of strong and weak wind speed events are shown. Influence of urban heat island and orography on wind field is discussed.

1. Introduction

Long-term continuous measurements of wind velocity profiles over urban area by acoustic sounding are important to research of diurnal, seasonal and synoptic variability in UBL. Experimental studies of wind velocity and turbulence in urban canopy layer and above use both contact (with cap and ultrasonic anemometers at meteorological masts) and ground-based remote sensing(with sodars, radars, RASS, lidars) methods. Sodars are convenient and low-cost remote tools for continuous observations and they have high resolution in lower part of ABL where the maximum of wind shear is placed [1]. Sodar measurements in urban areas were carried out, for example, in Turin, Milan, Vienna, Hanover [2, 3].

This paper presents results of continuous measurements of wind speed profiles and direction distributions over Moscow megalopolis (more then 10 millions inhabitants) [4]. Using of two identical sodars in two points of Moscow allows also measuring of spatial heterogeneity of wind and turbulence in large city.

2. Instrumentation and measurement sites

Measurements in Moscow are carried out by three-component mono-static Doppler sodars LATAN-3 developed at the Obukhov Institute of Atmospheric Physics (IAPh). Detailed description of this tool and its accuracy can be found in [5]. Two sodars are synchronously operated at two sites: on the roof of IAPh building in Moscow downtown and in Moscow State University (MSU) on the roof of Physics Faculty building in the south-west district of Moscow. Uninterrupted measurements are carried out since April 2005.

Distance between two sites is about 7 km. Difference of sodars placement heights is about 30 m (roof level of Physics Faculty building is 40 m agl and sodar on IAPh building roof are placed at the height of 13 m agl). Mean height of Lenin Hills above Moscow River is about 70 m. This common difference is the same as difference in wind profiles (will be presented below) and difference of surface inversion heights which was determined from echograms in our and earlier experiments [6].

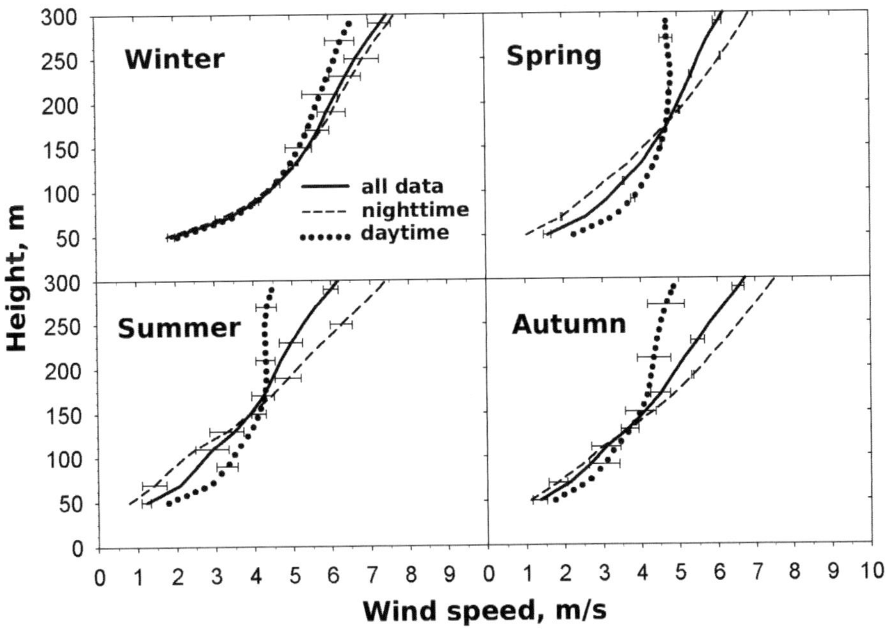

Figure 1. Seasonal means wind profiles for two-year measurements (June 2005 - May 2007) in Moscow downtown. Horizontal lines show a difference between the first and the second year data.

There are two more sites of sodars measurements in Moscow and comparison with their data can be useful for more complete research.

3. Mean wind velocity. Diurnal and annual variability

Mean vertical wind speed profiles and wind roses in atmospheric boundary layer (ABL) and distribution of wind speed at different levels, in diurnal and annual courses present a common picture of regional climatology of ABL. It may be useful for air pollution control and for understanding and investigating of heat and momentum exchange in urban areas.

Fig.1 demonstrates mean seasonal wind profiles in centre of Moscow observed for two year period (from June 2005 to May 2007 for each seasons (3 months). To show diurnal course the night-time (from 0 to 6 hour LST) and daytime (from 10 to 16 hour LST) profiles are also displayed. Four seasons show the annual cycle of means and of mean diurnal course. Difference between night-time and daytime profiles is larger in summer. It indicates the turbulence intensity difference between night- and daytime.

Mean wind profiles over Moscow are very similar to those found in Hanover [3]. From 50 to 300 meters mean wind speed increase is about 4-6 times. Above 300-500 meters wind shear lesser because of the top border of stable layer is displaced at these heights and wind speed increase is limited by larger momentum flux in daytime. Wind speed measuring by sodar at heights above 300 m is limited by backscattering conditions and by urban noise influence, so monthly or seasonally averaged data must be calculated carefully. In the lower 50 m the wind speed measurements may be supplemented by in-situ ones.

As stratification for a long period is varying considerably monthly or seasonal means really demonstrate a synoptic state distribution and distribution of wind speed. Figures 2 and 3 show these distributions for different heights, seasons and time intervals. Fig.2 displays distributions at 100, 200 and 300 m heights from all data for each season (3 months) as well as Fig. 3 shows

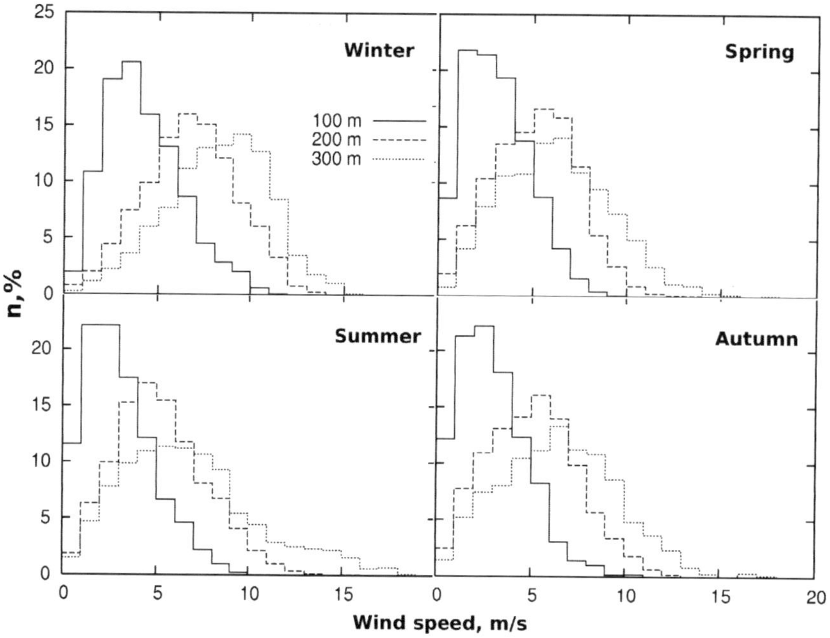

Figure 2. Wind speed distributions for each season at 3 heights: 100,200 and 300 m in Moscow downtown in 2007 (December 2006 - November 2007).

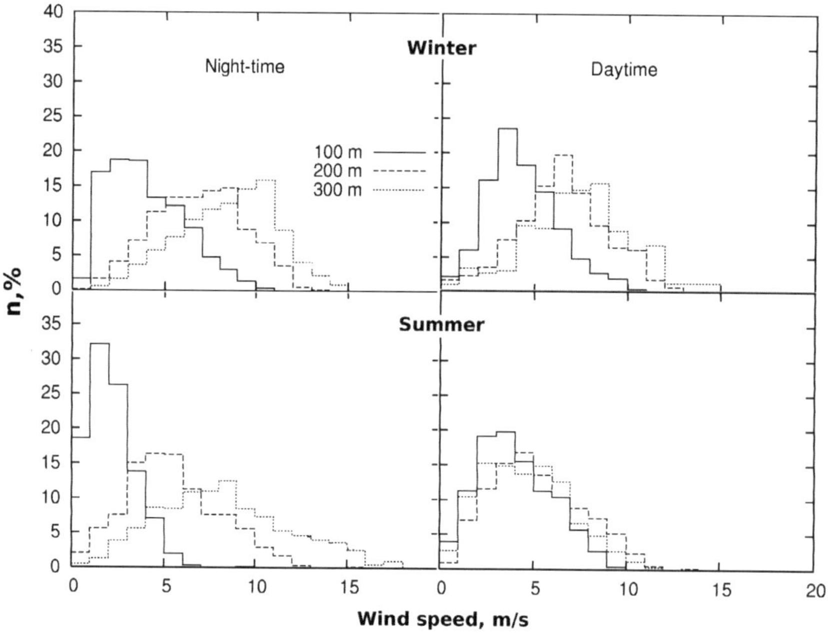

Figure 3. Wind speed distribution for night-time (left) and daytime (right) in summer (lower) and winter (upper). Measurements in Moscow downtown, 2007.

distributions for night- (from 0 to 6 LST) and daytime (from 10 to 16 LST) intervals in cold and warm seasons.

Figures show that speed distribution for a long time can be analytically described, mean

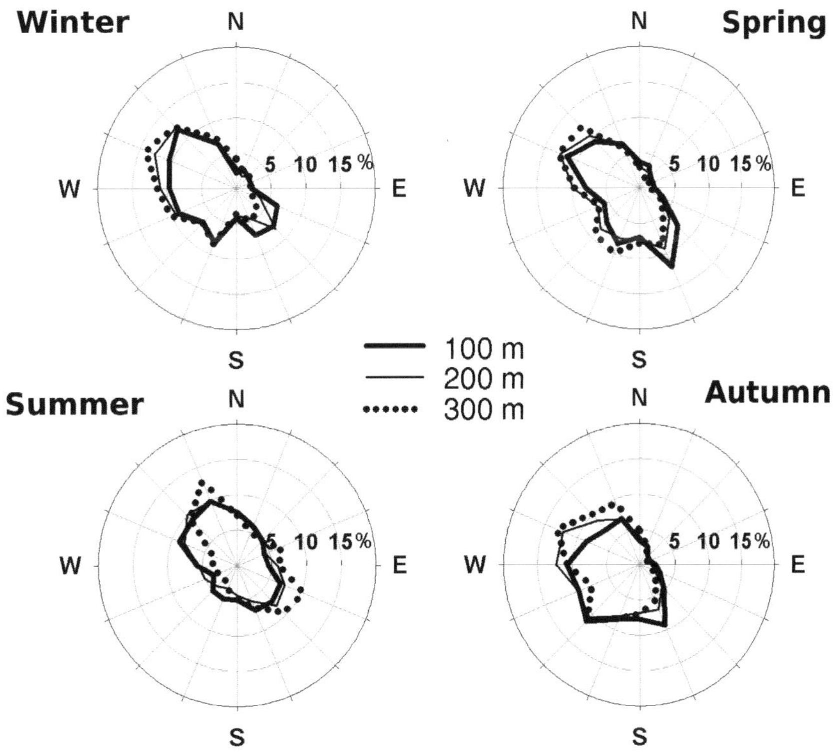

Figure 4. Wind rose in Moscow downtown at different heights calculated for two-year (June 2005-May 2007) measurements.

speed and variations are increased with height and diurnal means have no significant annual course. Mean wind speed, for example at 100 m height, is only about 4 m/s and mean speed for half-hour averaging doesn't exceed 10 m/s. Evidently, wind flaws which have duration in a few minutes are observed more frequently. They demonstrate meso-scale wind variability.

Comparison of day- and night-time distributions demonstrates annual cycle of diurnal course. Fig.3 shows also the variability of wind profiles at different stratification.

Wind rose for every season at the same heights is presented in Fig.4. Variability of wind direction distribution, both seasonal and annual, linked with synoptic variations amounts up to 5-10% that is sometimes 50% of mean value. Generally western directions are prevailed at Moscow latitude. Wind turn with height in Ekman layer isn't noticeable and visible differences may be connected with synoptic variability and wind rose discreteness. Though noticeable regular Ekman turning isn't observed higher then 50 meters, strong wind direction turn is observed for fronts passage (about 20-40 degrees and more).

4. Vertical gradients of wind speed in urban ABL

Whereas wind speed shear is main feature of ABL and it is connected with thermal stratification we calculate averaged diurnal course of vertical wind speed gradient (Fig.5). Fig.6 shows its distribution. To emphasize difference of diurnal course at lower and higher levels in Fig.5 the lowest level is displayed (50 m) and other levels are also shown via 100 m.

There is rapid increase of gradient at lower levels and decrease at higher ones when ABL passes from stable to unstable state. Distribution of gradients demonstrates wide range because of synoptic variability. Thus both average profiles and average diurnal course only generalize

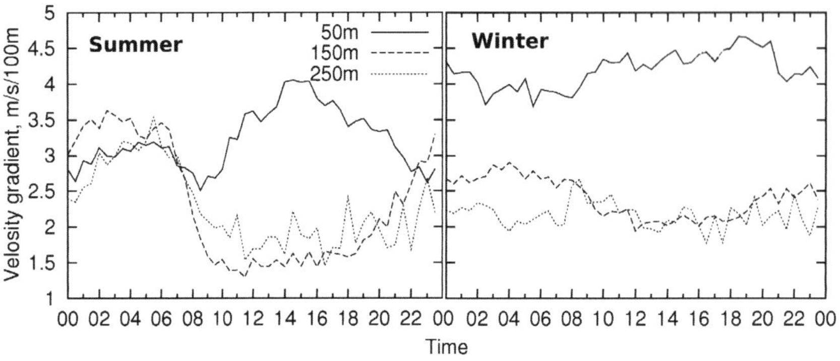

Figure 5. Averaged diurnal course of vertical wind speed gradients at 3 heights: 50, 150 and 250 m in Moscow downtown in summer and winter 2007.

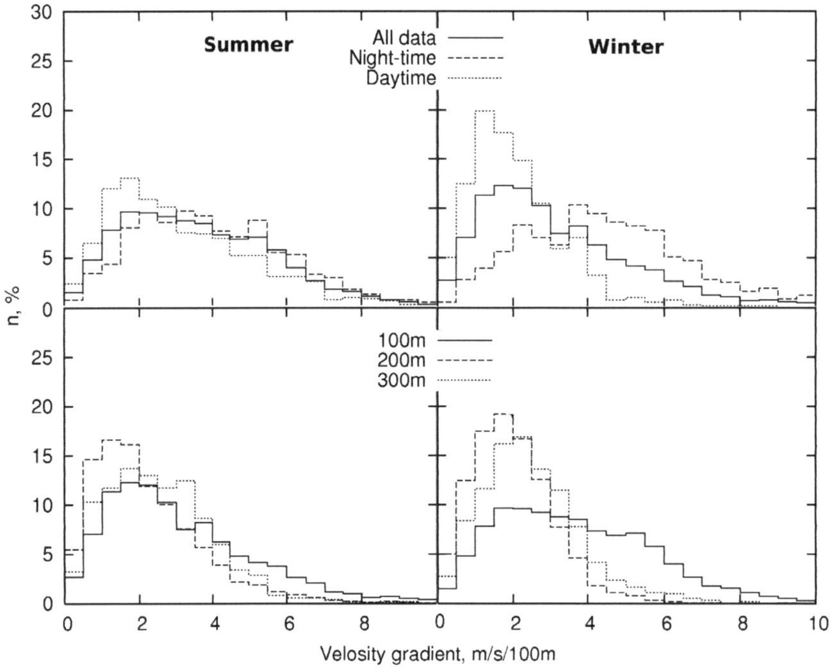

Figure 6. Distribution of vertical wind speed gradients for summer and winter (measurements in 2007) for all-, day- and night-time observations (above) and for different heights: 100,200 and 300 m (all-time data, below).

observing statistics.

5. Occurrence and duration of rare events

Common practice in meteorology is calculating of average occurrence and duration of meteorological phenomena so these estimations for strong and weak wind speed are often calculated. In the table our results of such analysis are presented with dividing on three-hour observing intervals from two-year measurements for different seasons and for year period.

However, careful analysis of results and comparison with distributions show that for rare events this approach cannot be satisfactory. Confidence interval for estimations of rare events is

Table. Diurnal course of occurrence (in %) and duration of strong(more then 8 m/s at 150 m level) and week (<2 m/s) wind speed for each season.

Season	Time interval, hours								Day	Duration, hours	
	0-3	3-6	6-9	9-12	12-15	15-18	18-21	21-24		mean	max
	Strong wind frequency (>8 m/s at 150 m level), %										
Winter	15.0	13.5	11.4	9.3	12.1	13.5	17.2	15.7	13.5	2.4	44.5
Spring	4.1	3.7	3.9	4.7	7.5	9.1	7.2	3.9	5.5	1.3	8.5
Summer	2.1	3.3	1.7	3.8	5.9	5.9	2.8	2.2	3.5	1.4	10.0
Autumn	4.1	3.8	3.3	3.5	3.6	2.5	3.5	4.1	3.6	1.9	19.0
Year	6.3	6.0	5.1	5.3	7.3	7.8	7.7	6.5	6.5	1.8	44.5
	Weak wind frequency (<2 m/s at 150 m level), %										
Winter	6.8	7.6	6.9	8.4	7.2	5.5	4.4	4.6	6.4	1.7	10.5
Spring	14.6	16.2	18.9	17.7	7.2	4.7	9.9	15.3	13.0	1.7	12.0
Summer	18.0	18.0	26.2	21.4	9.9	10.2	14.7	22.9	17.6	1.8	19.0
Autumn	13.3	17.3	17.7	17.2	13.9	13.0	18.8	14.1	15.6	2.4	31.0
Year	13.2	14.8	17.4	16.2	9.5	8.3	11.8	14.1	13.2	1.9	31.0

relatively wide and estimations for different sites or observing intervals cannot be compared and used for further conclusions. Calculated numbers are sensitive to criterion of weak and strong wind that is arbitrary. More accurate and stable value is parameter (or a few parameters) of distribution, if distribution is stable to variations. For example, approximation of wind speed distribution by circular normal distribution allows using only one parameter, giving good accuracy, annual and diurnal course, height dependence.

6. Wind speed comparison at two urban sites

Moscow is example of a large city in the middle of Russian plain for which influence of urban heating and roughness doesn't mix with influence of mountain topography or seabreeze so comparison of measurements at nearby sites (distance is about 7 km) is greatly important.

Fig. 7 shows seasonal mean wind profiles in Moscow downtown and at MSU site for summer 2007. To see differences in detail one week of data on heights: 100,200 and 300 meters are

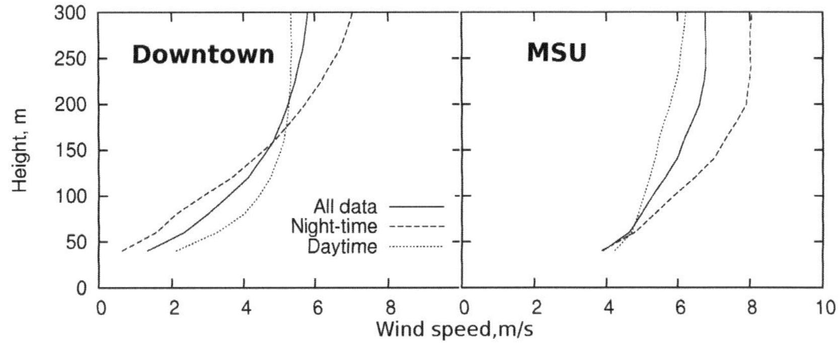

Figure 7. Mean seasonally wind speed profiles in summer in Moscow downtown and in MSU. July-August 2007.

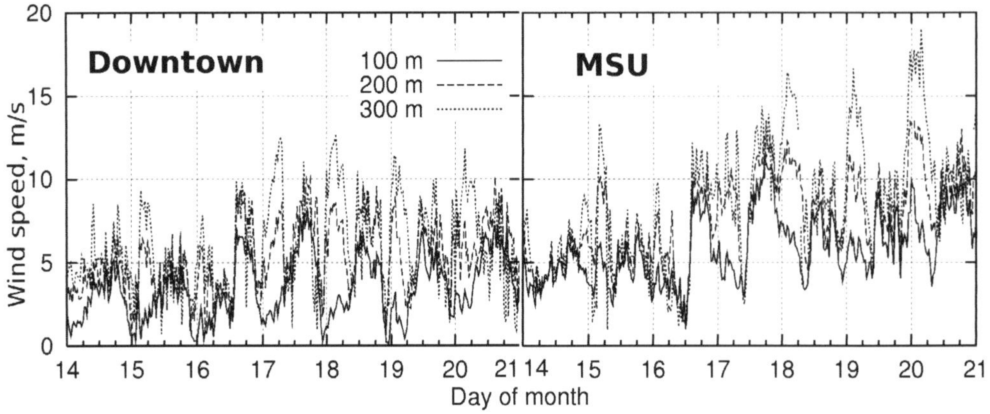

Figure 8. Example of synchronous wind speed course in IAPh and MSU. July 2007. Larger wind speed is demonstrated at elevated site at Lenin hills.

presented on Fig.8. Evidently there is difference not only between average velocity profiles, but between nocturnal and diurnal time intervals too.

Detailed analysis shows that the most agreement of wind speed at two sites is obtained for height difference in 80 m, that more then twice larger then difference between sodar placements above ground level and roughly corresponds to common height difference between observing sites. This difference not only confirms measurements in the 1990s [6], but taking this difference into account closeness of diurnal courses at two sites fits in details. Fig.9 demonstrates this for two accordant heights (50 meters in MSU and 130 meters in IAPh) in warm and cold period. Comparison with Fig.8 shows that some differences connected with meso-scale circulation, lesser then difference at one height.

Distributions of wind speed at the one (100 m) and at the accordant heights (50 m in MSU and 130 m in IAPh) in summer for day and night (Fig.10) show same tendency. Possibly, this difference is the effect of urban heating and roughness, however its stability in annual and diurnal course permit to suspect urban circulation peculiarity.

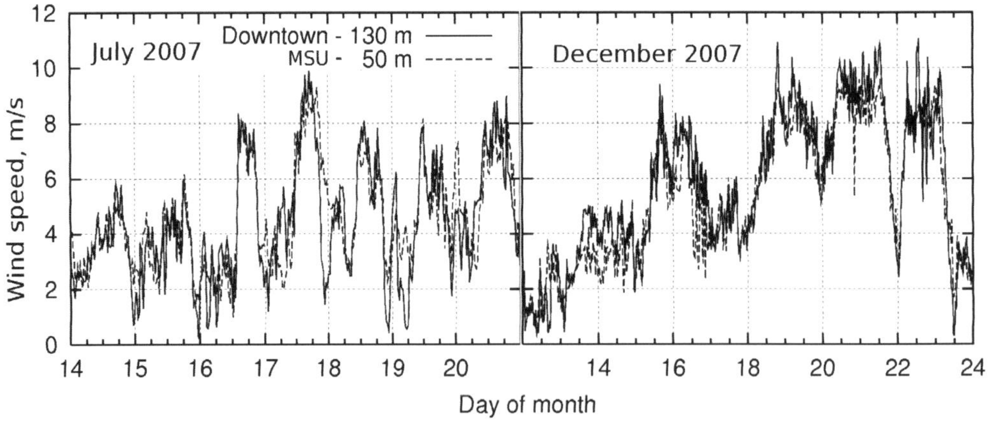

Figure 9. Comparison of wind speed at accordant heights in summer and winter.

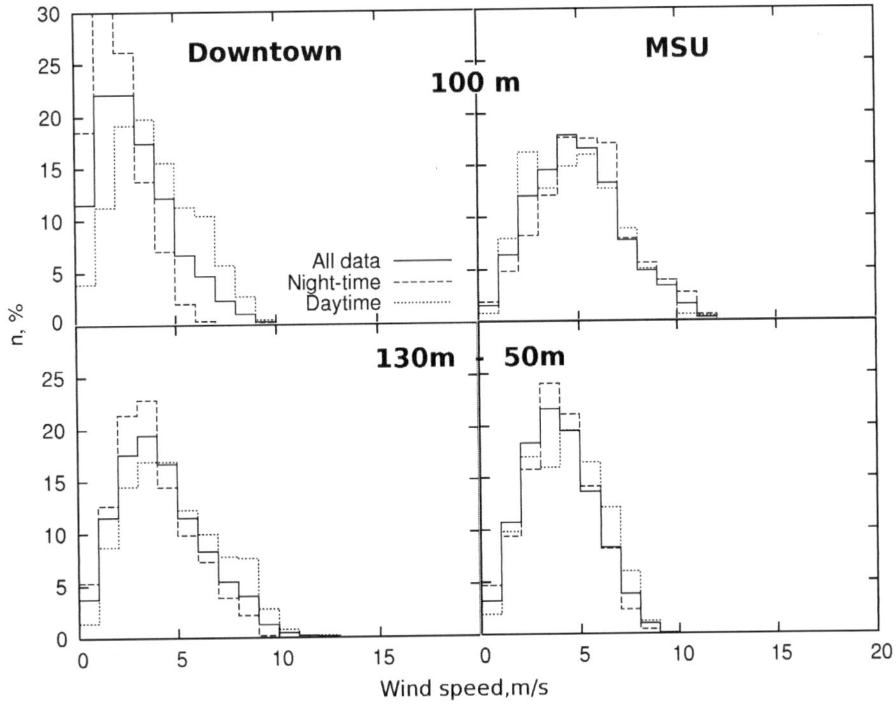

Figure 10. Distribution of wind speed in Moscow downtown an at Lenin Hills at the same (100 m, above)and at the accordant heights (50 m at MSU site, 130 m in centre of Moscow, below)

7. Limitations of acoustic sounding by data availability

Decrease of data availability at higher levels and increase of statistical heterogeneity should be taken into account for acoustic remote sensing, because of conditions for acoustic backscattering depend on measured wind speed and stability. But criterion of data validity is often subjective, and decision making can vary for different devices. At heights where wind velocity measurements by acoustic sounding are difficult, dynamic processes are very interesting: below there is roughness sub-layer (canopy and wake layers), above there is transitional to geostrophic wind layer and the top of Low-Level-Jets [7].

Open architecture of sodar LATAN-3 permits to test different algorithms for calculation of average wind profiles and for control of data validity. For inaccuracy estimation produced by data missing two scenarios were tested: "rigorous", when only reliable data (signal is much more noise) are included in means and "liberal", when "possible" data (signal is about noise) are also included with extrapolation of data above top limit. Tests show that at higher levels mean wind speed can vary on 2-3 m/s for different scenarios that consists 20-30% of mean value.

8. Conclusion

Mean monthly and seasonally wind speed profiles depend from wind speed distribution that is determined by synoptic variability. Parameters of distribution depend from height, time and observing place. Thus, different sites in urban area have own distributions and their "effective" height may be calculated as well as influence of urban heating and roughness change on these distributions may be estimated.

Mean wind speed profiles and distributions above towns in one climatic zone are alike: in Moscow and Hanover [3] in layer from 50 to 300 meters mean wind speed increases from 2 to 8 m/s, height of the smaller diurnal variability is about 100 m. In winter period slightly-stable stratification is often observed with minimum of diurnal course.

Wind speed distributions show that occurrence and duration of rare events are sensible both to criterion and to synoptic variability.

At last, measuring of mean wind characteristics by sodar at heights, where data availability decrease below 50%, is sensible to method of missing data "filling".

Acknowledgments

We are grateful to M.A. Kallistratova for research initialization and to R.D. Kouznetsov for continuous support. Measurements in IAPh were carried out by Rostislav Kouznetsov. This study was supported by Russian Foundation for Basic Research, projects nos. 05-05-64786, 07-05-13610 and 08-05-00984.

References

[1] Coulter R L, Kallistratova M A 2004 Two-decade progress in sodar techniques: a review of 11 ISARS proceedings *Meteorol. Atmos. Physics* **85** 3–19.

[2] Kallistratova M A, Coulter R L 2004 Application of sodars in the study nad monitoring of the environment *Meteorol. Atmos. Physics* **85** 21–37.

[3] Emeis S, Baumann-Stanzer K, Piringer M, Kallistratova M, Kouznetsov R, Yushkov V 2007 Wind and turbulence in the urban boundary layer - analysis from acoustic remote sensing data and fit to analytical relations *Meteorologische Zeitschrift* **16** No 4 393–06

[4] Yushkov V P, Kallistratova M A, Kuznetsov R D, Kurbatov G A, Kramar V F 2007 Experience in measuring the wind-velocity profile in an urban environment with a Doppler sodar *Izvestiya Atmospheric and Oceanic Physics* **43** No. 2 168–80

[5] Kouznetsov R D 2006 The new PC-based sodar LATAN-3. *Extended Abstr. 13th Int. Symp. On Advancement of Boundary Layer Remote Sensing(ISARS-2006), 18-20 July 2006, Garmisch-Partenkirchen, Germany* 97–98.

[6] Pekour M S, Kallistratova M A 1993 Study of the boundary layer over Moscow for air-pollution application *Appl. Phys. B* **57** 49-55

[7] Kallistratova M A 2008 Investigation of low level jets over rural and urban areas using two sodars *Extended Abstr. 14th Int. Symp. On Advancement of Boundary Layer Remote Sensing(ISARS-2008), 23-25 June 2008, Riso, Denmark*

RASS sound speed profile (SSP) measurements for use in outdoor sound propagation models

S G Bradley[1,2], S v Hünerbein[2,3], D Waddington[2],
[1]Physics Department, University of Auckland, NZ
[2]Research Institute for the Built and Human Environment, University of Salford, UK

E-mail: s.vonhunerbein@salford.ac.uk

Abstract. The performance of outdoor sound propagation models depends to a great extent on meteorological input parameters. In an effort to improve speed and accuracy, model output synthetic sound speed profiles (SSP) are commonly used depending on meteorological classification schemes. In order to use SSP measured by RASS in outdoor sound propagation models, the complex profiles need to be simplified. In this paper we extend an investigation on the spatial and temporal characteristics of the meteorological data set required to yield adequate comparisons between models and field measurements, so that the models can be fairly judged. Vertical SSP from RASS, SODAR wind profiles as well as mast wind and temperature data from a flat terrain site and measured over a period of several months are used to evaluate applicability of the logarithmic approximation for a stability classification scheme proposed by the HARMONOISE working group.

1. Introduction

Simple sound speed profiles (SSP) are widely used as data input into sound propagation models for applications such as the forecast of environmental noise. Within the HARMONOISE Project [3] the authors looked in detail at the properties of sound speed profiles for different wind and stability classes. They also investigated how to derive simple SSP from rudimentary weather information such as data from a nearby AWS and observations of cloud cover. From these observations they proposed the use of fifteen different SSP according to 15 different stability classes ranging from very convective in low winds to very stable in low winds. With the increasing wind speed the mixing in the boundary layer leads to more and more neutral stratification.

Given that RASS instruments measure SSP directly the authors [2] have also looked at how much information is necessary for the input of SSP into sound propagation models and which fitting algorithms should be used. Using the three stability classes convective, neutral and stable they found that simple logarithmic fitting gives sufficiently accurate profiles. Using this methodology the current study explores whether the use of a larger number of stability classes will give increased information given that the variability in measured profiles is high.

2. Method

The data set was collected within the framework of the Profiler Inter-comparison Experiment (PIE) during the WISE project. The coastal site is characterised by flat land. The RASS was situated about

[3] To whom any correspondence should be addressed.

70m from the meteorological mast which was instrumented with sonic anemometers, cup anemometers and other instruments at heights ranging from 1 – 116 m. Further details about the site and instrumentation can be found in [1]. For the categorisation of the stability classes we used 10 m mast wind speeds and inverse Monin-Obukhov lengths from the sonic anemometer at 20m height. The stability classification used is listed in Table 1.

Table 1: Stability classes adapted from Heimann 2007: W1-W5 are wind values given in ms^{-1}, S1-S5 are given in inverse Monin-Obukhov lengths.

		S1	S2	S3	S4	S5
W1	0-1	<-0.065	<-0.025		<0.02	<0.065
W2	1-3	<-0.035	<-0.01		<0.01	<0.035
W3	3-6	<-0.015	<-0.005		<0.005	<0.015
W4	6-10	<-0.005				<0.005
W5	>10					

W1-5 refers to wind speed in ms^{-1}, S1-5 to stability as defined by the inverse Monin-Obukhov Length L^{-1} in m^{-1}. Dark shaded areas refer to convective situations on the left hand side and to stably stratified situations on the right hand side. Data were then compared to SODAR reflectivity (not shown). For all stability classes in the wind category W2 and for neutral cases RASS SSP, measured sound speed data were used to fit average logarithmic profiles. S1-5W2 was chosen as a compromise when both strongly stable and strongly convective situations occur and the classification can be assumed to be correct as inertia problems from the cup anemometer are starting to decrease. SSP were measured every 30 seconds. Table 2 gives an overview over the data used.

Table 2. Data overview: Stability classes are denoted according to Table 1, *denotes intervals when stability fits the respective class intermittently.

Stability class	Day	Time	No of profiles
S1W2	2/5/2004	8:00-9:00	143
	10/5/2004	9:00-10:20	194
S2W2	6/5/2004	5:30-9:50*	117
	11/5/2004	1:40-13:50*	241
S4W2	4/5/2004	0:40-4:20*	211
	13/5/2004	2:17-20:33*	166
S5W2	1/5/2004	20:30-23:59	495
	2/5/2004	19:30-20:40	157
	13/5/2004	3:00-4:10	155
Neutral	1/5/2004	4:00-7:00	414
	7/5/2004	5:00-9:00	518
	12/5/2004	4:30-7:00	306

The fitted profiles take the form of

$$c(z) = \eta + a \ln z + \varepsilon(z) \qquad (1)$$

where c is the sound speed (SS) at height z, and a and η are constants throughout a particular profile. The height-dependent function $\varepsilon(z)$ is the residual between the actual profile and the log-linear approximation.

3. Results

Figures 1-5 show the average observed profiles for different days and the respective logarithmic fits in a separate plot for each stability class. The error bars denote the standard deviation of the observed data points. In general, data availability decreases with height. Fig. 1 contains the profiles for strongly convective situations of stability class S1W2. Both profiles show a very similar SSP gradient and, in agreement with expectation, higher SS near the ground.

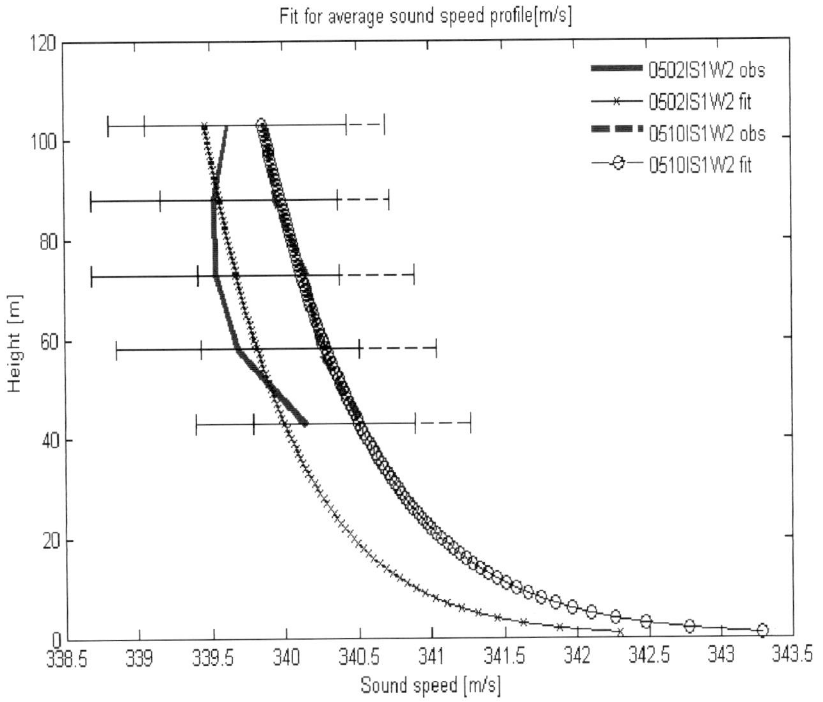

Figure 1. Log-fits for convective stability class S1W2 compared to measured average profile and standard deviation.

In Fig. 2 when convection is weaker the SS gradient is smaller than in Fig.1. The two profiles show a greater spread in SS gradient. Figures 3-5 show the classes moderately stable S4W2, strongly stable S5W2 and neutral, respectively. In Fig. 3 the profiles look more like convective SSP and much like the profiles in S2W2. For the strongly stable case the profile from 13/5 in Fig 4 looks stable whereas the other two look slightly convective. In Fig 5 the supposedly neutral profiles all have a stable SSP gradient. Note the large standard deviations which are common to all data. They suggest that a variety of profile shapes are possible.

IOP Conf. Series: Earth and Environmental Science 1 (2008) 012005 doi:10.1088/1755-1307/1/1/012005

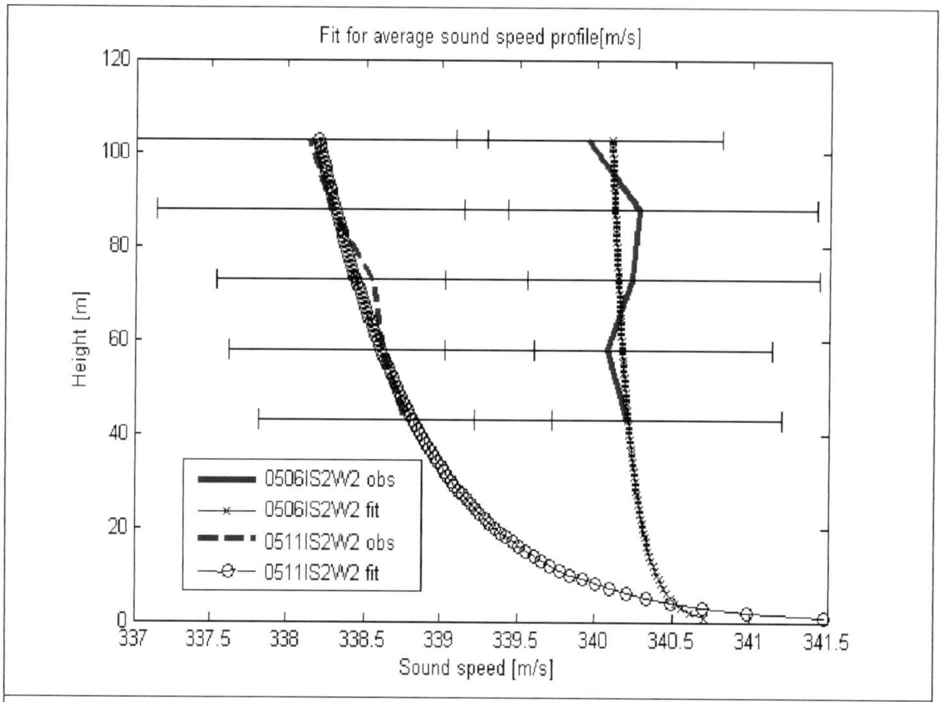

Figure 2. Log-fits for moderately convective stability class S2W2 compared to measured average profile and standard deviation.

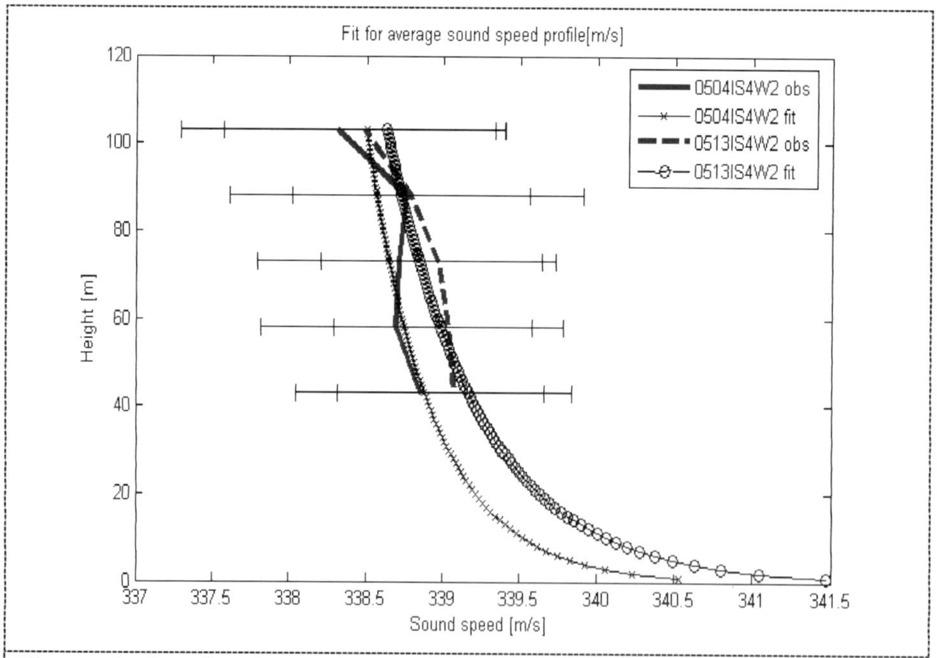

Figure 3. Log-fits for moderately stable stability class S4W2 compared to measured average profile and standard deviation.

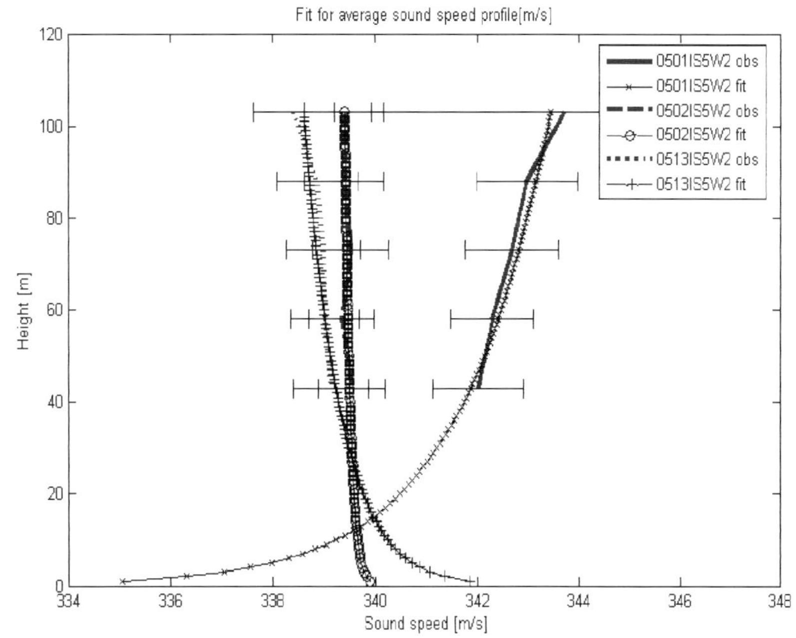

Figure 4. Log-fits for stable stability class S5W2 compared to measured average profile and standard deviation.

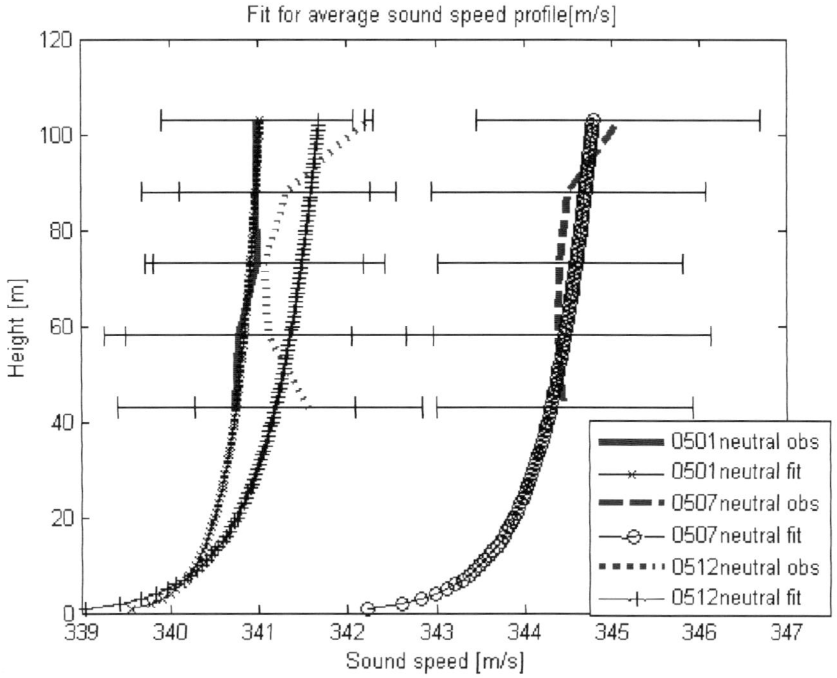

Figure 5. Log-fits for neutral stability class compared to measured average profile and standard deviation.

4. Conclusions

In a first step, logarithmic fits of RASS measured sound speed profiles (SSP) have been created as meteorological input parameters into sound propagation models. This was done for a number of boundary layer stability classes to answer the question whether there would be a distinct SSP for each stability class that is statistically significantly different from the profiles in another stability class. A first glance the results for stable and neutral situations seem to suggest that the answer is no, as the neutral fitted profiles look more stable than some of the supposedly moderately stable profiles. This is due to the lack of data below 43m AGL and also to the large standard deviation of the observed data The large standard deviation is probably due to the use of 30 second data as compared to the commonly used 10 minute averages. Therefore the current analysis should be compared to the 10minute averages to establish in how far data quality and therefore standard deviation and goodness of fit can be improved. However, the use of high temporal resolution data is desirable to establish the spatial representativeness of the SSP measurements. To supplement the profile with data points from lower heights the sound speeds measured by the sonic anemometer at 20m above ground level should be taken into account for the analysis.

In a next step, a rigorous statistical analysis of the SSP gradients and they residuals needs to be done as well as a χ^2 test to decide whether two logarithmic profiles belong statistically to the same population. Both of these analyses will be presented at the conference.

5. References

[1] Bradley S G, Antoniou I, Hünerbein S v, Kindler D, Jorgensen H E and Noord M d 2005 SODAR Calibration for Wind Energy Applications, Final Reporting on WP3, EU WISE project NNE5-2001-297. (Salford: Research Institute of the Built and Human Environment) p 68

[2] Bradley S G, Mursch-Radlgruber E and Hünerbein S v 2007 SODAR Measurements of Wing Vortex Strength and Position *J. Atmos. Ocean. Tech.* **24** 141-55

[3] Heimann D, Bakermans M, Defrance J and Kühner D 2007 Vertical Sound Speed Profiles Determined from Meteorological Measurements Near the Ground *Acta Acustica united with Acustica* **93** 228 - 40

IOP Conf. Series: Earth and Environmental Science 1 (2008) 012006 doi:10.1088/1755-1307/1/1/012006

Investigating a valley atmosphere with a Sodar/RASS and comparison to a flatland site

Martin Piringer, August Kaiser

Central Institute for Meteorology and Geodynamics, A-1190 Vienna, Austria

martin.piringer@zamg.ac.at

Abstract. Two 1290 MHz Sodar/RASS instruments have been operated between November 2006 and June 2007 in an industrially used narrow valley with an average crest height of approx. 300 m in Central Austria and in the Austrian flatlands east of Vienna. Vertical profiles of temperature, wind direction and speed, vertical velocity and sigma-w up to a height of 600 m above ground, i. e. well above average crest height at the valley site, have been obtained. Specific features of the valley atmosphere, like the change of wind direction and wind speed with height and the seasonal changes of wind speed and temperature are presented. A comparison of average vertical profiles of wind speed, temperature, and σ_w shows characteristic differences between the two sites.

1. Introduction

A Sodar/RASS is a combination of an acoustic radar (Sodar for "Sound detection and ranging") and two electromagnetic grid antennas (RASS for "Radio Acoustic Sounding System"). The Sodar antenna emits signals (2 kHz) every few seconds. It records the intensity of acoustic backscatter S of the air. Only inhomogeneities in the temperature field in the order of half of the wavelength of the sound waves from the Sodar contribute to S. Normally this is within the "inertial sub-range" of the turbulence spectrum. Therefore S contains information on the structure of the turbulence in the boundary layer, but not on the sign of the vertical temperature gradient [1]. As the elements of thermal turbulence move at the speed of the wind, a Doppler effect results, which is used to calculate the three-dimensional wind. More information on the measurement principles of Sodars is given e. g. in [2].

The RASS is an extension to a wind profiler or Sodar that allows for the measurement of the virtual temperature profile above the radar. It uses an acoustic source to generate a vertically propagating sound wave in the radar's scattering volume. RASS temperature retrievals are sensitive to the wind component parallel to the sound beam and the humidity [1]. In strong winds, the vertical range of the RASS can be reduced compared to the Sodar.

A 1290 MHz Sodar/RASS manufactured by Metek company, Germany (http://www.metek.de/), was operated between November 2006 and June 2007 in an industrially used narrow valley with an average crest height of approx. 300 m in Central Austria. Its wind and temperature measurements up to a height of 600 m are analyzed here to discuss the characteristics of the wind and temperature structure within and above the valley. The results are compared to a site with another Sodar/RASS of the same type which was operated in the flatlands of Eastern Austria during the same period.

2. Description of the sites

One Sodar/RASS was positioned in a NW-SE oriented Alpine side valley at 15° 02' 57" E and 47° 23' 14" N at a height of 583 m above sea level (Figure 1a). Data were retrieved at 25 m height intervals, starting 50 m above valley floor. At the site, the valley is only 500 m wide, and the average crest height is approx. 300 m. The other Sodar/RASS was positioned in a rather flat area at the Eastern border of Austria near Bratislava (Fig. 1b) at 17° 04' 29" E and 48° 06' 24" N at a height of 135 m. Data were retrieved at the same levels as at the valley site. Data availability is near 100 % near the ground and over 90 % up to 400 m. Higher up, it decreases rapidly to less than 50 % in 600 m.

a) b)

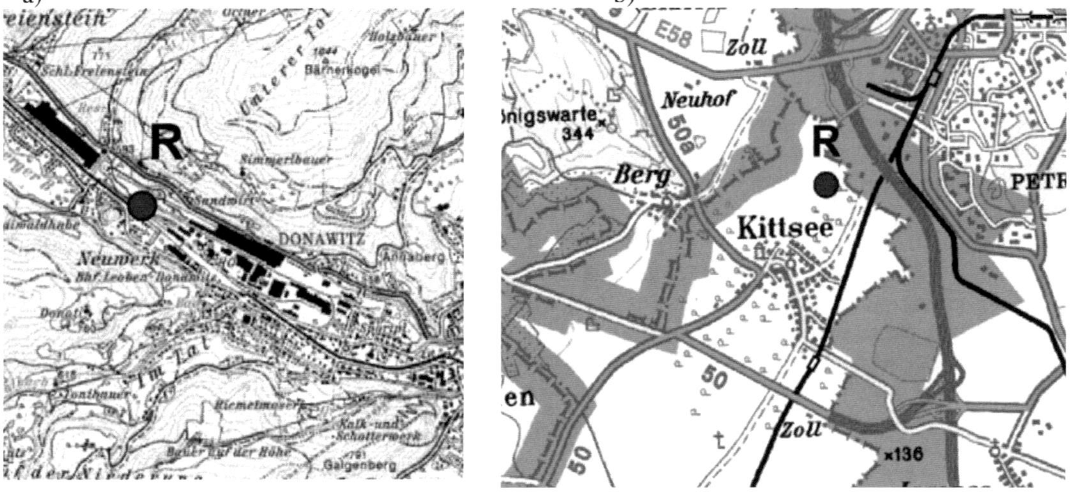

Figure 1: Areas of the Sodar/RASS sites (blue circles) a) in an Alpine side valley in Central Austria (Source: Austrian map, scale 1:50.000); b) in Eastern Austria at the Slovak border (Source: Austrian map, scale 1:200.000)

3. Results and discussion

3.1. Wind and temperature structure at the valley site

Measurements at the valley site 100 m above the ground (Fig.1a) show that the flow is strongly channeled along the valley axis (Fig. 2a). As the measurement period comprises winter and spring months, down-valley flow dominates, because of the shading effects from the valley sidewalls; at night, 50 % of all winds are from NW. Even during daytime, NW winds are more common than the up-valley flow because of the shading effect and because this is also the dominant wind direction in cyclonic weather conditions. Thus, daytime up-valley winds (from E to SE) amount to only about 25 % of all daytime wind directions. Daytime up-valley winds show an average wind speed of about 3 m/s, night-time down-valley flow of 2 m/s (Fig. 2b).

At crest height 300 m above the valley floor, the strong channeling of the flow disappears (Fig. 3a). The flow regime is now more W-E-oriented. The influence of the valley on the flow is weakened in favor of that of the large-scale pressure gradients which in Austria means a dominance of westerly winds. Wind speeds are increased compared to 100 m height (Fig. 3b), but remain lowest for NE winds that are shaded by the highest hills in the area. At a height of 500 m above the ground winds are predominantly from W to SW (Fig. 4a) with an average speed of 5 m/s (Fig. 4b); day-night-differences are further weakened, revealing that there is no valley influence at that height.

IOP Conf. Series: Earth and Environmental Science **1** (2008) 012006 doi:10.1088/1755-1307/1/1/012006

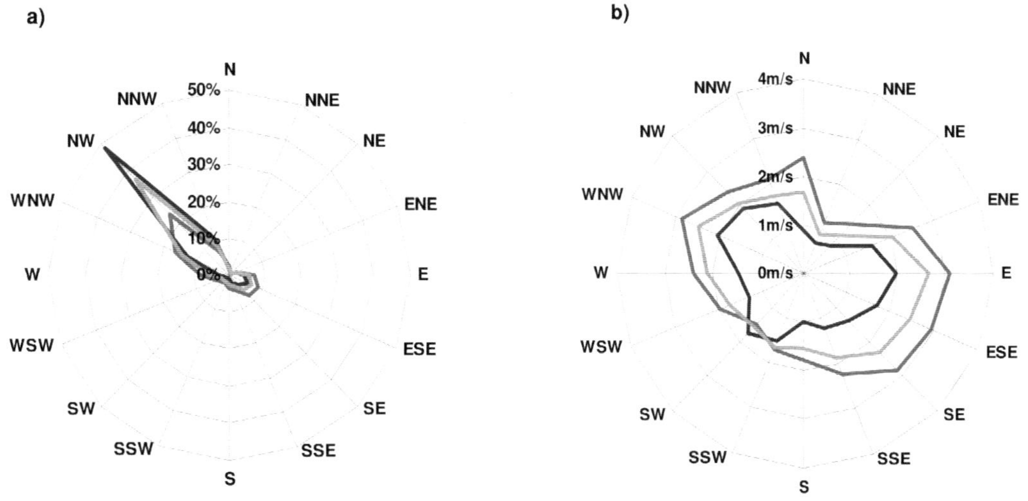

Figure 2: Wind rose (a) and average wind speeds (b) at 100 m above valley floor. Red lines: daytime; blue lines: night time, green lines: all data.

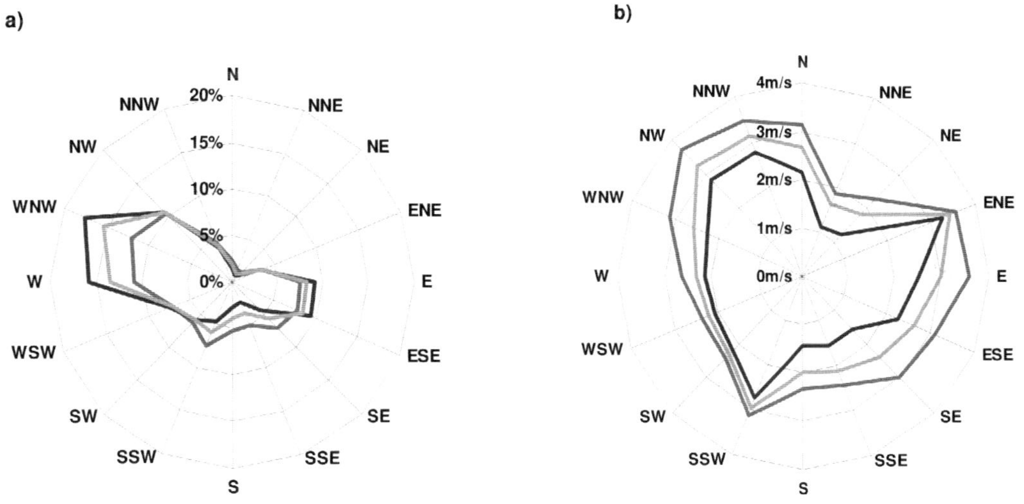

Figure 3: Wind rose (a) and average wind speeds (b) at 300 m above valley floor. Red lines: daytime; blue lines: night time, green lines: all data.

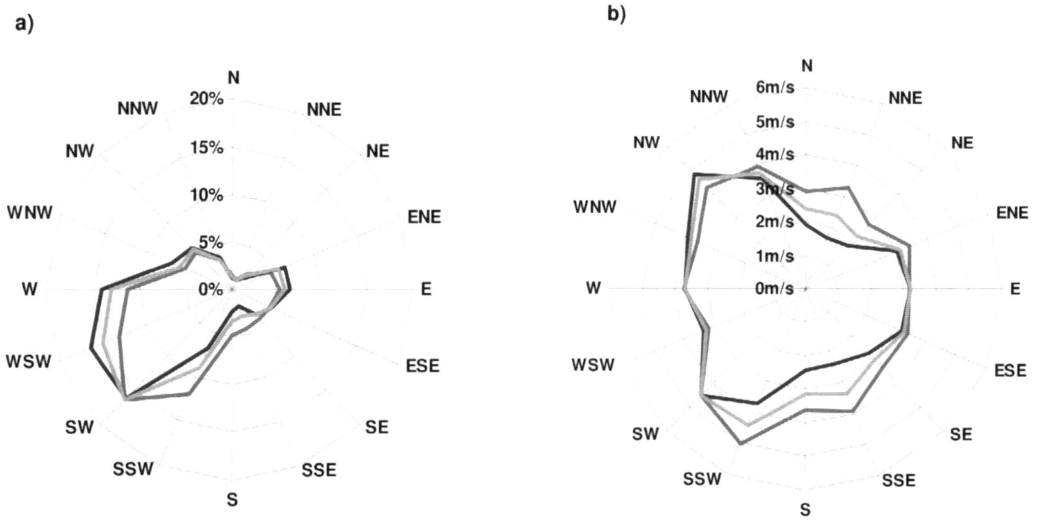

Figure 4: Wind rose (a) and average wind speeds (b) at 500 m above valley floor. Red lines: daytime; blue lines: night time, green lines: all data.

As just explained and also seen in Fig. 5a, day-night-differences in average wind speed are larger within the valley (up to 300 m above the ground) than above it. During night-time, the wind speed increase in the valley with height is very moderate, due to the predominantly stable stratification near the ground. Starting at crest height, night-time wind speeds increase from about 2 m/s to about 4 m/s at 600 m. The night-time flow in the valley is thus strongly decoupled form the air aloft. During the well-mixed conditions around noon, wind speeds increase more steadily with height throughout the whole layer investigated. Wind speed differences with height depend strongly on the season, as revealed in Fig. 5b: especially in November and December 2006, wind speed changes in the valley were almost non-existent but slightly increase in the following months. Above the valley, wintertime wind speed increase is very large, decreasing towards spring. The decoupling of the valley atmosphere from the air aloft is thus most pronounced in wintertime, as expected, decreases towards spring and has – on average – disappeared in May and June 2007.

a) b)

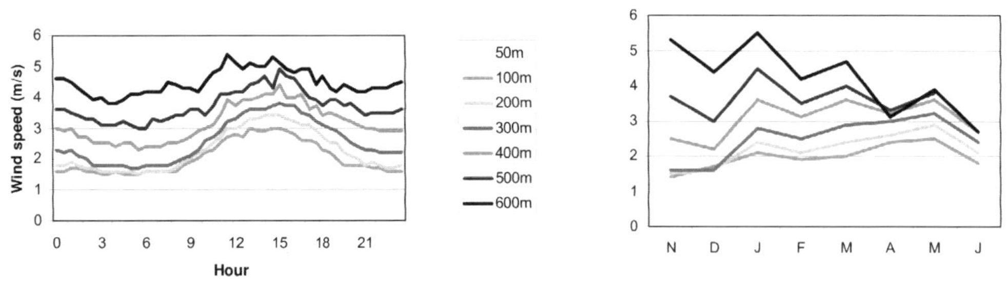

Figure 5: Daily course (a) and monthly means (b) of Sodar/RASS wind speed at selected heights at the valley site

The findings from the wind speed are confirmed by the Sodar/RASS temperature measurements. The daily course of temperature (Fig. 6a) shows the strong changes between the night-time stable situation with the highest temperatures 600 m above ground and the reversal during daytime. However, even at daytime, temperature decrease with height within the valley is only moderate, indicating an on average stable layer within the valley, well in accordance with the findings from the wind direction regime (Fig. 2a). Well-mixed conditions during daytime, on average, are found only above crest height. The monthly means of air temperature at the different levels confirm again the difference between winter and spring months: from November to January, the air is on average stably stratified up to 600 m, whereas from February on, the air within the valley becomes warmer than aloft. The average stability in and above the valley thus depends not only on daytime, but also on the season.

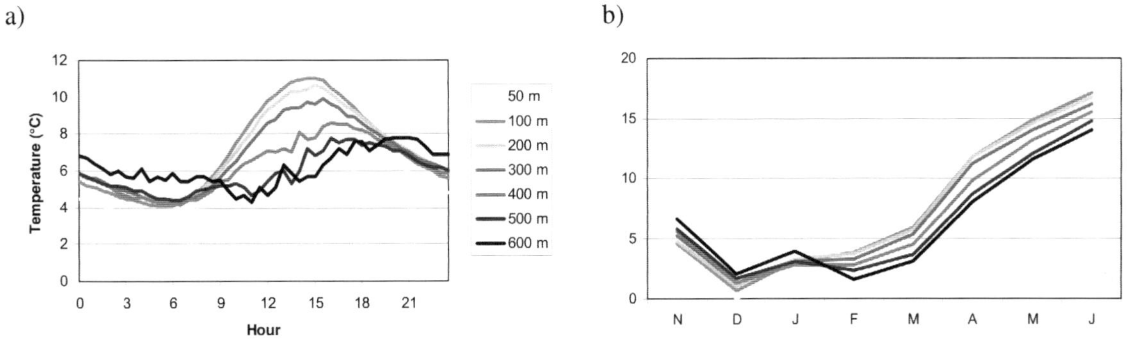

Figure 6: Daily course (a) and monthly means (b) of Sodar/RASS temperature at selected heights at the valley site

3.2. Comparison between the valley and the flatland site

The comparison between the valley site and the site in the Austrian flatlands near the border to Slovakia (Fig. 1b) reveals some interesting features characteristic for the types of landscape studied. At the flatland site, average wind speeds are a factor 2 to 3 higher than at the valley site at all levels (Fig. 7a). The normalized profiles show that the wind speed in the valley (red line in Fig. 7b) increases less than above crest height as already discussed in Section 3.1. At the flatland site, in contrast, the increase is more rapid up to 400 m and then stops. This means that, as expected, the flat area exerts its influence on the wind speed profile to a lower height than is seen in the valley and its adjacent hills.

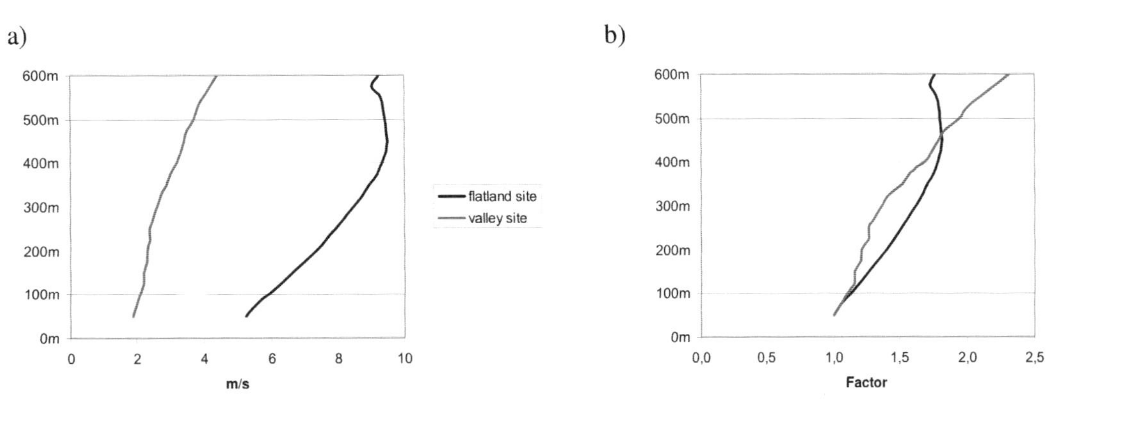

Figure 7: Vertical profiles of wind speed at the flatland site (blue) and the valley site (red). a) averaged profiles; b) normalized profiles

The flatland site is on average about 2°C warmer than the valley site (Fig. 8a), except above 500 m. Both sites show on average a ground-based inversion of similar height and magnitude. Above the ground inversion, the average temperature profile is slightly more stable at the flatland site than at the valley site; this stable layer at the flatland site coincides with the layer of the wind speed increase (Fig. 7). Above 400 m, there is a rather sharp temperature decrease. At the valley site, the temperature continues to drop until 500 m and then rises slightly again. This may be due to frequent fog or a frequently occurring low-level cloud layer caused by night-time cooling the base of which might often be found at this height. These features are even better seen from Fig. 8b which shows the temperature change with height for each site, taking the average temperature at 50 m as a reference value. At the valley site, the air within the valley up to crest height is on average warmer than near ground; above crest height, it is cooler. There is less change with height up to about 400 m at the flatland site, before a rapid temperature decrease sets in.

a)

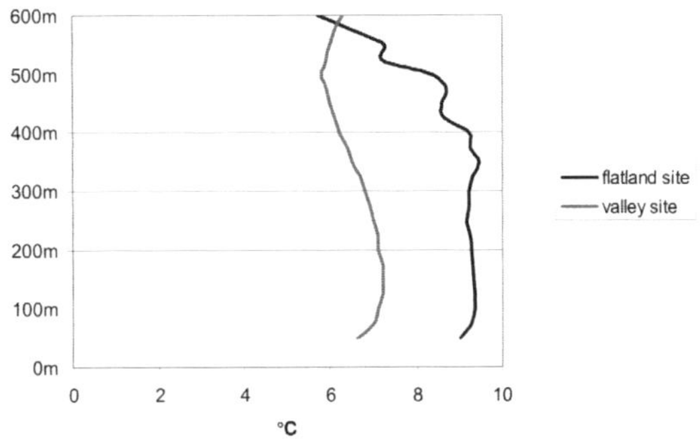

b)

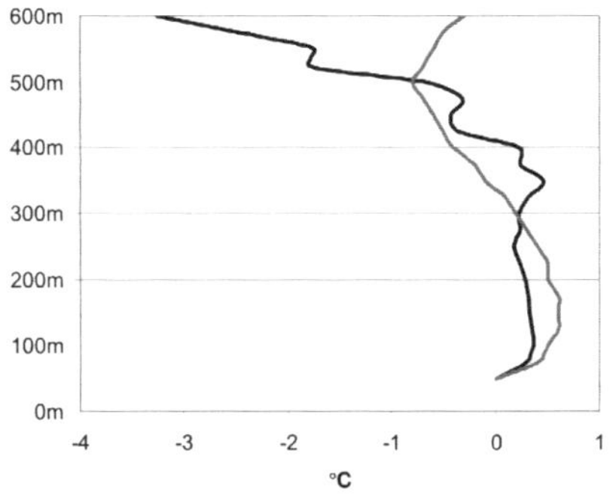

Figure 8: Vertical profiles of temperature at the flatland site (blue) and the valley site (red). a) averaged profiles; b) temperature change with height

The valley site is in all layers less turbulent on average than the flatland site, revealed by the lower σ_w values (Fig. 9a). The lower σ_w values at the valley site are expected as also the wind speeds are on average much lower (Fig. 7a). Due to the lower wind speeds, less mechanical turbulence is to be expected; and due to the shadowing by the valley sidewalls, also thermal turbulence is restricted compared to the flatland site. The normalized profiles (Fig. 9b) show a very similar behaviour up to about 300 m, i. e. the increase in σ_w is comparable at the two sites. Above that height, σ_w continues to increase at the flatland site, whereas it remains at the same level at the valley site. However, the increase throughout the whole layer is about 10 % only at the valley site and 20 % at the flatland site; this change is not considered very strong, and these differences should therefore not be over-interpreted.

a) b)

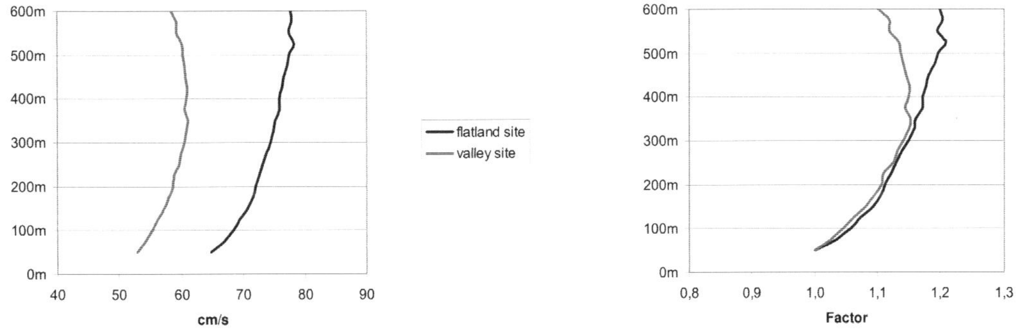

Figure 9: Vertical profiles of σ_w at the flatland site (blue) and the valley site (red).
a) averaged profiles; b) normalized profiles

4. Concluding remarks

Between November 2006 and June 2007, two Sodar/RASS instruments have been operated at two locations in Austria under quite different ambient conditions (Section 2, Figure 1). Due to the winter and spring seasons predominantly covered, the atmosphere in the valley (Section 3.1) tends to be stable (Figure 6), with markedly channelled flow below crest height (Fig. 2a) and a pronounced seasonal change in wind speed and temperature with height (Figs. 5b and 6b), indicating the expected improved vertical mixing conditions towards spring. The comparison with the flatland site (Section 3.2) reveals on average larger wind speeds, higher temperatures, and more turbulence at the flatland site compared to the valley site. The investigation has shown that the Sodar/RASS is well suited to detect characteristic features of the boundary layer up to several 100 m.

Metek company provides the users with a graphical software package and with quality-flagged data files. The graphics package allows for a stepwise retrieval of outliers and other data of poor quality. Furthermore, data can be smoothed with height and time to support the visual interpretation. Some care and exercise as well as a documentation of the steps undertaken is necessary to enable for the reproduction of the same figures. The quality-flagged data files are used for statistical analysis. Having deleted all data of poor quality by the automatic routine, time series plots will still reveal some outliers which have to be removed manually. Therefore, a fully automated data control and evaluation routine is not possible, and some time has to be invested to create a reliable Sodar/RASS data base.

References

[1] Piringer, M., S. Joffre (Eds.), 2005: The urban surface energy budget and mixing height in European cities: data, models and challenges for urban meteorology and air quality – Final report of Working Group 2 of COST Action 715, ISBN 954-9526-29-1. Demetra Ltd. Publisheers. Printed in Bulgaria, 239 pp.

[2] Beyrich, F., 1997: Mixing height estimation from sodar data – a critical discussion. Atmos. Environ. **31**, 3941 – 3953.

IOP Conf. Series: Earth and Environmental Science 1 (2008) 012007 doi:10.1088/1755-1307/1/1/012007

Sodar-derived structural functions of the wind velocity field

L G Shamanaeva
Institute of Atmospheric Optics of the SB RAS 1, Akademicheskii Ave., 634021 Tomsk, Russia,

E-mail: sima@iao.ru

Abstract. The structural functions of the wind velocity field reconstructed from the vertical profiles of the wind velocity vector measured with the Zvuk-2 three-channel monostatic Doppler sodar (1700 Hz) and a commercial minisodar of the Atmospheric Systems Corporation (4900 Hz) are presented in the report. The Doppler sodars allow long-term series of instantaneous values of the three wind velocity components in the atmospheric boundary layer to be measured. The temporal longitudinal and transverse structural functions of the wind velocity field, calculated from these time series in the altitude ranges of 75-535 m with a step of 20 m (Zvuk-2 sodar) and 15-200 m with a step of 5 m (minisodar) are presented. With the use of the hypothesis of frozen turbulence, the spatial longitudinal and transverse structural functions are calculated.

1. INTRODUCTION

At present, acoustic radars (sodars) are widely used to measure vertical profiles of the wind velocity vector and characteristics of the dynamic turbulence in the atmospheric boundary layer. To analyze random wind velocity fields, it is expedient to use the velocity structural functions [1]. Application of sodars allows long-term data series to be obtained and the velocity structure function for separation distances up to several hundred meters to be calculated. The structural functions of the wind velocity field reconstructed for different altitudes from series of vertical profiles of the wind velocity vector measured with the Zvuk-2 three-channel monostatic Doppler sodar (1700 Hz) [2] and a commercial minisodar of the Atmospheric Systems Corporation (4900 Hz) [3] are presented in the report.

2. INSTRUMENTATION AND DATA COLLECTION

One of the most important characteristics of the random wind velocity field in the atmosphere is the structural tensor of the second rank [4]

$$D_{ik}(\mathbf{r},\Delta\mathbf{r}) = \langle [V_i(\mathbf{r}+\Delta\mathbf{r}) - V_i(\mathbf{r})] \rangle \times \langle V_k(\mathbf{r}+\Delta\mathbf{r}) - V_k(\mathbf{r}) \rangle, \quad (1)$$

where $\mathbf{r} = (x, y, z)$ specifies the observation point, $\Delta\mathbf{r}$ is the increment of $\mathbf{r}$, and $V_i(\mathbf{r})$ is the projection of the wind velocity vector $\mathbf{V}(\mathbf{r})$ onto the ith coordinate axis. In general, the tensor $D_{ik}(\mathbf{r},\Delta\mathbf{r})$ can have 9 independent components. In the theory of atmospheric turbulence [4], the wind velocity field is assumed locally isotropic, which reduces the number of independent components to two referred to as the longitudinal ($D_{rr}(\mathbf{r},\Delta\mathbf{r})$) and transverse $(D_{tt}(\mathbf{r},\Delta\mathbf{r}))$ structural functions.

The Zvuk-2 sodar design and operation were described in detail in [2]. The sodar operated at a frequency of 1700 Hz, its pulse repetition period was 11.5 s, and its pulse length was 150 ms. One transceiving antenna was pointed vertically, and two others were tilted at angles of 20° from the vertical in the orthogonal planes. The instantaneous wind velocity components $V_{xj}(z_k)$, $V_{yj}(z_k)$, and $V_{zj}(z_k)$ were stored every 11.5 s for 24 successive range gates z_k of 20 m each starting from $z_{0V} = 55$ m, where j denotes the serial number of the wind velocity profile in a measurement series, $j = 1, ..., N$, $N = 53$.

The vertical profiles of the backscattered signal power proportional to $C_T^2(z_l)$ were stored every 11.5 s in successive range gates z_l 2 m each starting from $z_0 = 58.9$ m. The proportionality coefficient was determined by sodar calibration against $C_T^2(z_t)$ measured in developed convection with the ultrasonic thermometer/anemometer.

The temporal longitudinal and transverse velocity structural functions $D_{rr}(z_k, n\Delta t)$ and $D_{tt}(z_k, n\Delta t)$ were then calculated in the altitude ranges of 75-535 m with a step of 20 m for each range gate z_k, with $\Delta t = 11.5$ s, $n = 1, 2, ..., N/5$. Accepting the hypothesis of frozen turbulence, the corresponding spatial velocity structural functions $D_{rr}(z_k, n\Delta r)$ and $D_{tt}(z_k, n\Delta r)$ were then calculated, where $\Delta r = \Delta t<\mathbf{V}(z_k)>$ and $<\mathbf{V}(z_k)>$ is the wind velocity vector for the range gate z_k averaged over the entire measurement period $T = N\Delta t$.

Long-term series of instantaneous values of the three wind velocity components measured with an autonomous AV4000 Doppler minisodar equipped with a solar battery and a 50-element phased antenna array [3] kindly provided by Dr. Ken Underwood, President of the Atmospheric Systems Corporation, were also processed. The sodar operated at a frequency of 4900 Hz, and measurements were performed in the altitude range of 15-200 m with a step of 5 m.

3. RESULTS AND THEIR DISCUSSION

Figure 1a shows 10-min vertical profiles of the three components of the wind velocity vector reconstructed from sodar measurements on June 25, 1997 from 10 till 10:10, Tomsk local time. Here crosses indicate $V_x(z)$, squares $V_y(z)$, triangles $V_z(z)$, and the dashed curve shows the average wind velocity profile $V_{av}(z)$. The wind shear region is clearly seen between 215 and 315 m, in which $V_{av}(z)$ increased from 5 to 15 m/s at $z = 295$ m and then subsequently decreased to 4 m/s.

The corresponding 10-min vertical profiles of the structural parameters are illustrated by figures 1b and 1c. The reconstructed values of $C_V^2(z)$ are shown by black squares in figure 1b, and their approximation by the least-squares method is shown by the red solid curve. The values of $C_T^2(z)$ shown in figure 1c were normalized by $C_T^2(z_0)$ at the first sensing altitude, indicated in the figure.

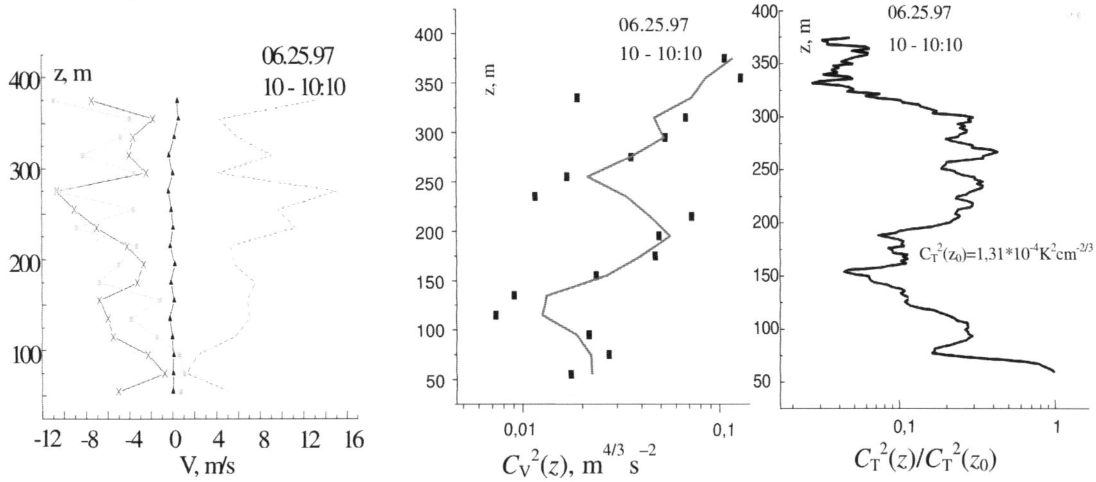

a *b* *c*

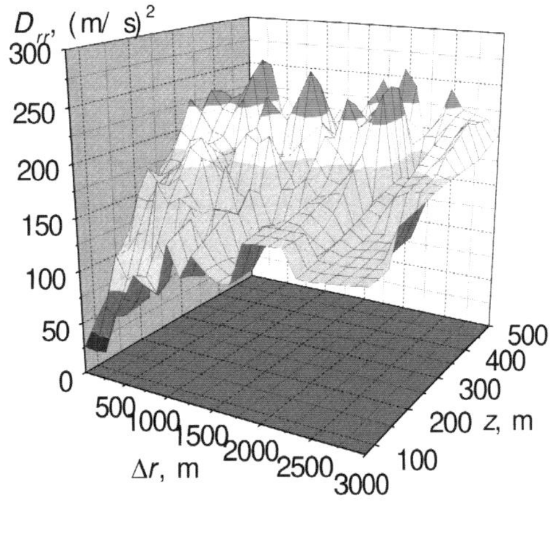

d

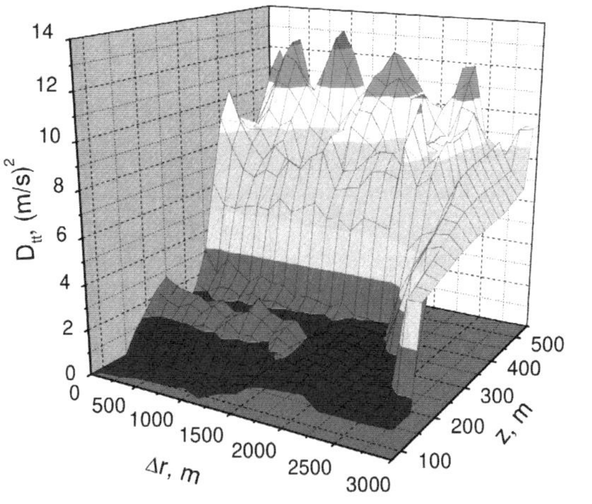

e

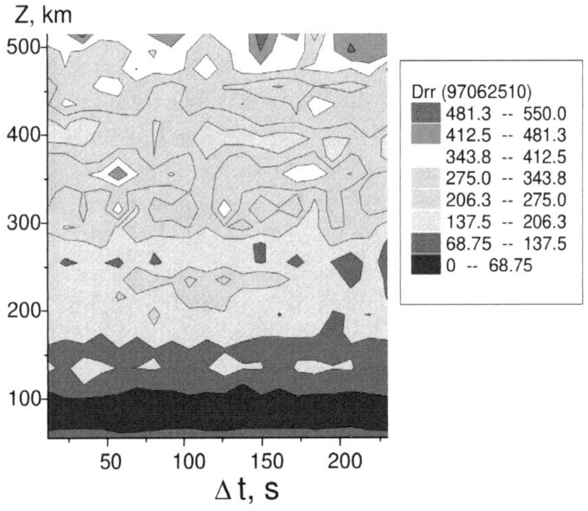

f

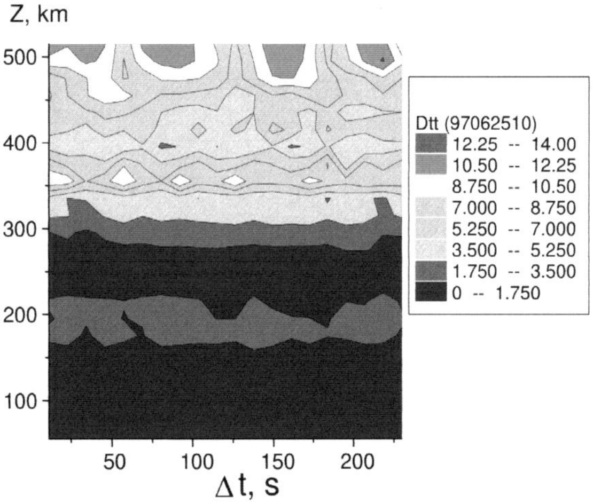

g

Figure 1. Vertical profiles of the three components of the wind velocity vector (a), velocity (b) and thermal (c) structural parameters, and spatial longitudinal (d and f) and transverse (e and g) velocity structural functions reconstructed from measurements with the Zvuk-2 three-channel monostatic Doppler sodar on June 25, 1997 from 10 till 10:10, Tomsk local time.

A wide maximum can be seen in the $C_T^2(z)$ profile, which indicates the presence of an elevated temperature inversion. Below the inversion, C_T^2 follows a $z^{-4/3}$ dependence typical of convection. Above the inversion, C_T^2 becomes small. From Fig. 1b it can be seen that C_V^2 increases with altitude from ~0.02 $m^{4/3}s^{-2}$ at $z = 55$ m to ~0.1 $m^{4/3}s^{-2}$ at $z = 375$ m, that is, by a factor of 5. Its altitude dependence is much weaker in comparison with C_T^2 which changes almost by two orders of magnitude. This is in agreement with the data presented in [5, 6]. In addition, sodar measurements indicated that the main contribution to the refractive index structure parameter $C_n^2(z) = C_T^2(z)/(4T)^2 + C_V^2(z)/C_0^2$ comes from the dynamic turbulence. This is in agreement with the data presented in [5].

The most typical behavior of the spatial longitudinal and transverse structural functions in the morning is illustrated by Figs. 1d and e. Here, the structural functions are shown for separation distances up to 3000 m. The longitudinal structure function first increases with the separation distance Δr and then saturates; the greater the sensing altitude z, the smaller Δr at which D_{rr} saturates. This can be explained by larger values of the average wind velocity at these altitudes [6].

The transverse structure function monotonically increases with Δr. It also increases with the sensing altitude. As expected, it is much smaller in comparison with D_{rr}. Below the wind shear region, D_{rr} and D_{tt} are strongly suppressed for all Δr. Above the wind shear layer, the structural functions noticeably increase. The similarity and synchronism in variations of D_{rr} and D_{tt} can also bee seen.

Figures 2a and 2b illustrate the temporal and spatial longitudinal structural functions $D_{rr}(z, \Delta t)$ and $D_{rr}(z, \Delta r)$ reconstructed from sodar measurements on October 5, 1996 at night hours from 01 till 01:10, Tomsk local time. It can be seen that the velocity structural functions are spatially inhomogeneous not only in the vertical, but also in the horizontal plane, with intermittence regions in which they differ significantly.

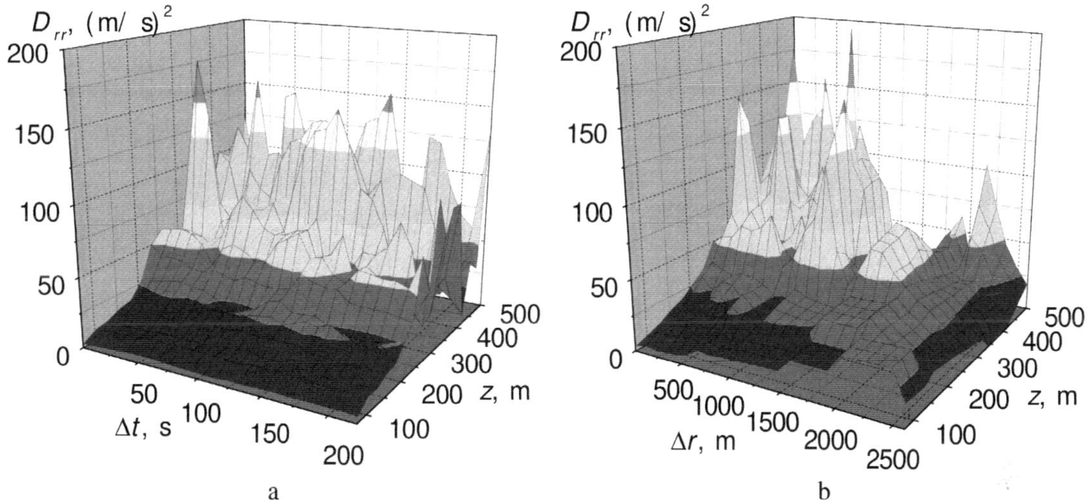

Figure 2. Longitudinal temporal, $D_{rr}(z, \Delta t)$ (a), and spatial, $D_{rr}(z, \Delta r)$ (b), velocity structural functions reconstructed from measurements with the Zvuk-2 sodar on October 5, 1996 from 01 till 01:10, Tomsk local time.

Figure 3a-b shows the longitudinal temporal and transverse spatial structural functions calculated from the vertical profiles of the wind velocity vector measured with the minisodar on May 25, 2006 from 16 till 16:10, LT. Two elevated layers of enhanced turbulence at altitudes of 160-170 m and 140 m are clearly pronounced. The horizontal homogeneity of the wind velocity field can be seen from figure 3a. Figure 4 shows the vertical profiles of the structural characteristic $C_V^2(z)$ of the wind velocity field retrieved from the calculated values of $D_{rr}(z, \Delta t)$ and $D_{tt}(z, \Delta r)$ shown in figure 3. The model profile of the structural characteristic is also shown in figure 4. It can be seen that the model profile describes well the altitude behavior of the structural characteristics retrieved from the sodar data. From figure 4 it can be seen that the C_V^2 values retrieved from the longitudinal velocity structural functions are smaller than those calculated from the transverse structure function.

4. CONCLUSIONS

In conclusion it should be noted that the suggested algorithm of sodar data processing can be used to visualize large-scale inhomogeneities of the wind velocity field in the atmospheric boundary layer, thereby providing information about spatial structure and intermittence of atmospheric turbulence in real time. It also allows the structural functions to be calculated for long separations between the observation points. An analysis of the structural functions demonstrates a highly inhomogeneous

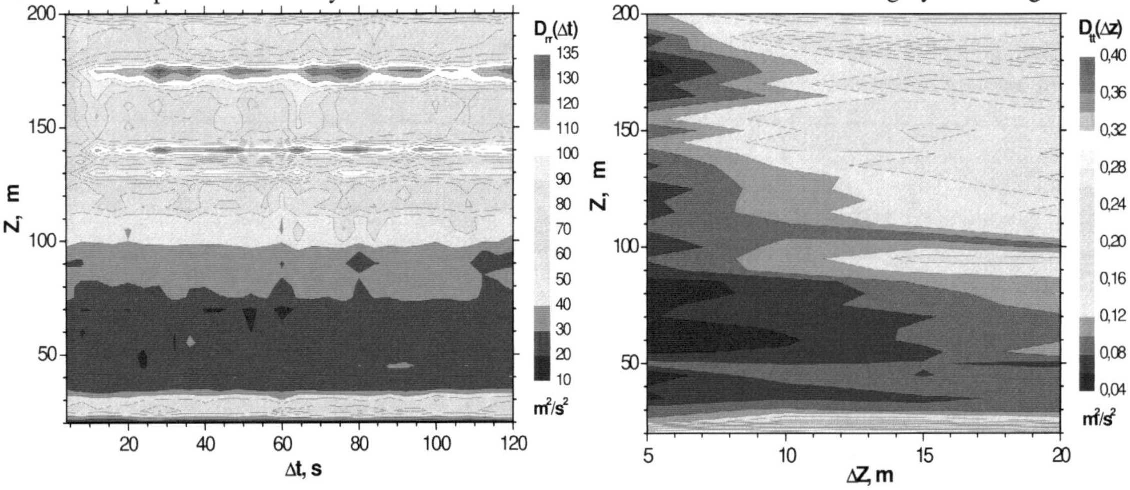

a *b*

Figure 3. Longitudinal temporal, $D_{rr}(z, \Delta t)$ (a), and transverse spatial, $D_{tt}(z, \Delta r)$ (b), velocity structural functions reconstructed from minisodar measurements on May 25, 2006 from 16 till 16:10, LT.

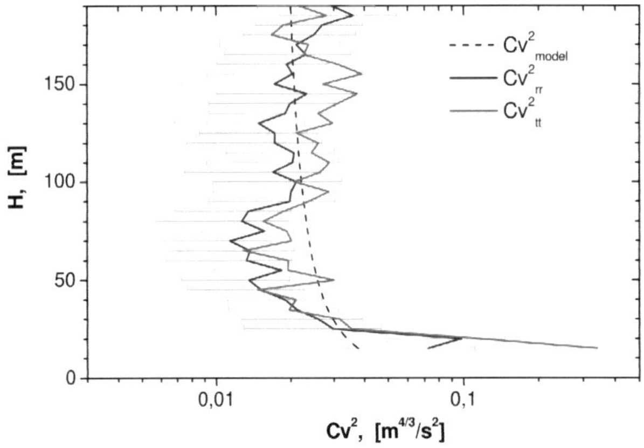

Figure 4. Vertical profiles of the structural characteristics of the wind velocity field retrieved from the velocity structural functions $D_{rr}(z, \Delta t)$ and $D_{tt}(z, \Delta r)$ shown in figures 3a and 3b.

spatial field structure. At night, in the presence of a wind shear, a decrease in the longitudinal structural functions was observed under the wind shear region together with its significant increase above the shear region. Under conditions of daytime convection, the structural functions first increased with the separation of the observation points and then were saturated.

ACKNOWLEDGMENTS

The author would like to acknowledge Professor N. P. Krasnenko and Ph. D. Ken Underwood for kindly providing sodar data.

REFERENCES

[1] Rytov S M, Kravtsov Yu A, and Tatarskii V A 1978 *Introduction to Statistical Radiophysics*, Part 2 (Moscow: Nauka)

IOP Conf. Series: Earth and Environmental Science **1** (2008) 012007 doi:10.1088/1755-1307/1/1/012007

[2] Gladkikh V A, Krasnenko N P, and Fedorov V A 1997 *Extended Abstracts of the COST-76 Profiler Workshop* (Engelberg, Switzerland) p 174-177
[3] http://minisodar.org
[4] Tatarskii V I 1961 *Wave Propagation in a Turbulent Medium* (New York: Dover)
[5] Petenko I V, Kallistratova M A, and Bedulin A N 1996 *Proc. 8th Int. Symp. Acoustic Remote Sensing and Associated Techniques of the Atmosphere and Oceans* (Moscow, Russia) p 3.55-3.60
[6] Bogushevich A Ya and Shamanaeva L G 1999 *Atmos. Oceanic Opt.* **12** 54-57

IOP Conf. Series: Earth and Environmental Science **1** (2008) 012008 doi:10.1088/1755-1307/1/1/012008

Recent progress in acoustic tomography of the atmosphere

V E Ostashev,[1, 2] **S N Vecherin,**[2] **D K Wilson,**[3] **A Ziemann**[4] **and G H Goedecke**[2]

[1] NOAA/Earth System Research Laboratory, Boulder, CO 80305
[2] Physics Department, New Mexico State University, Las Cruces, NM 88003
[3] U.S. Army Engineer Research and Development Center, Hanover, NH 03755
[4] University of Leipzig, Institute for Meteorology, Leipzig, Germany, 04103

E-mail: vladimir.ostashev@noaa.gov

Abstract. Acoustic tomography of the atmospheric surface layer is based on measurements of travel times of sound propagation among different pairs of sources and receivers usually located several meters above the ground on a horizontal scale of about 100 m. The measured travel times are used as input data in an inverse algorithm for reconstruction of temperature and wind velocity fields. Improved knowledge of these fields is important in boundary layer meteorology, theories of turbulence, and studies of electromagnetic and acoustic wave propagation in the atmosphere. In this paper, a short overview and current status of acoustic travel-time tomography of the atmosphere are presented. A brief description of a 3D array for acoustic tomography of the atmosphere which is being built at the Boulder Atmospheric Observatory is given. Furthermore, different inverse algorithms for reconstruction of temperature and velocity fields are discussed, including stochastic inversion and a recently developed time-dependent stochastic inversion. The latter inverse algorithm was used to reconstruct temperature and wind velocity fields in acoustic tomography experiments. Examples of the reconstructed fields are presented and discussed.

1. Introduction

For many decades, tomography has been widely used in physics, medicine, and technology for remote sensing of different media. Acoustic tomography of the atmosphere enables one to reconstruct the temperature T and wind velocity $\vec{v}$ fields and to monitor their evolution in time and space [1, 2]. These fields are important in many practical applications such as boundary layer meteorology, theories of turbulence, wave propagation through a turbulent atmosphere, etc. Using various schemes of acoustic tomography, the temperature and wind velocity fields have been reconstructed in different regions of the atmosphere: in the atmospheric surface layer (ASL), e.g. see [3-5], which extends to just a few meters above the ground; in the atmospheric boundary layer (ABL) up to the heights of a few hundreds meters (e.g. [6]), and in the upper atmosphere up to the heights of about 100 km [7-9].

The main goal of the paper is to present the status of acoustic travel-time tomography of the ASL and recent progress in the field. To this end, the forward and inverse problems in acoustic tomography of the ASL will be formulated. Different approaches to solution of the inverse problem will be considered, including a recently developed time-dependent stochastic inversion (TDSI) algorithm [10, 11]. Results obtained in numerical simulations of acoustic tomography of the ASL will be outlined. Acoustic tomography experiments will be discussed, including some already done and others planned for the future. The temperature and wind velocity fields reconstructed with the use of TDSI in acoustic tomography experiments will be presented.

The paper is organized as follows. In section 2, a brief historical overview of acoustic tomography of the atmosphere is presented. Forward and inverse problems of acoustic tomography of the ASL are formulated in section 3. In section 4, the TDSI algorithm is presented, and its applications to numerical simulations of acoustic tomography and reconstruction of T and $\vec{v}$ fields in tomography experiments are discussed. Lastly, conclusions are summarized in section 5.

2. Brief historical overview

Probably the first experiments in acoustic tomography of the atmosphere were carried out in the early 1900's; see [7] and references therein. Note that the term *tomography* was not yet used at that time. Travel times and angles of arrival of sound and infrasound waves generated by large explosions on the ground were recorded by ground-based acoustic sensors (or observers) located at different distances from the explosions. Using these travel times and angles of arrival, scientists tried to reconstruct the vertical profiles of temperature $T(z)$ and wind velocity $\vec{v}(z)$ up to the height z of turning points of the sound waves that could be as high as 100 km. Until World War II, many such acoustic experiments were carried out in Europe in the attempt to reconstruct the vertical profiles $T(z)$ and $\vec{v}(z)$ in the upper atmosphere.

This effort was significantly reduced after World War II because of the use of rockets in studies of the upper atmosphere. Nevertheless, some research in this area continued after World War II (e.g., [8, 9]), mainly for detection and localization of nuclear explosions on the ground [12]. Currently, there are about 30 infrasound stations (with the same number to be deployed) which are a part of the International Monitoring System for verifying compliance with the Comprehensive Nuclear-Test-Ban Treaty, see www.ctbto.org. Although deployed for a different purpose, these infrasound stations could, in principle, be used for acoustic tomography of the atmosphere since they routinely measure sound and infrasound signals from mining, meteors, weather, and other natural or human-caused phenomena [13, 14]. Note that seismic stations, which have a broader network, often measure sound or infrasound waves generated by some of these phenomena [15]. Therefore, a combined network of infrasound and seismic stations could be used for acoustic tomography of the atmosphere on a global scale.

Schemes of acoustic tomography of the ABL, somewhat similar to those used for remote sensing of the upper atmosphere before World War II, were implemented or proposed in [6, 16, 17]. The schemes are intended to work in situations when there is a temperature inversion in the ABL or the effective sound speed $c_{eff}(z)$ reaches its maximum at some height above the ground. ($c_{eff}(z)$ is a sum of the adiabatic sound speed and the component of wind velocity in the direction of sound propagation.) In these schemes, sources of sound signals can be different: loud-speakers, propane cannons, etc.

Another scheme of acoustic tomography of a stratified atmosphere was proposed in [1, 18]. In the scheme, the time delays of sound signals from commercial airplanes received by microphones located on the ground are measured. Then, the vertical profiles $T(z)$ and $\vec{v}(z)$ can be reconstructed up to the heights of the airplanes (about 10 km).

In the rest of the paper, acoustic tomography of the ASL will be considered. It has advantages in comparison with point measurements of the temperature and wind velocity fields by conventional meteorological devices, e.g. sonic anemometers. First, these devices have a distinct inertia, can be affected by radiation, and can disturb the temperature and wind velocity fields while acoustic tomography does not have these shortcomings. Second, acoustic tomography requires fewer sensors per unit of data than do conventional meteorological devices. Also note that acoustic tomography has an advantage in comparison with volume-imaging lidars, which cannot be used directly for remote sensing of temperature fluctuations. Possible applications of acoustic tomography of the ASL include: validation and input data for atmospheric models, studies of turbulence over complex topography and heterogeneous surfaces, input data for wave propagation models, etc.

The first experimental implementation of acoustic tomography of the ASL was reported in [3]. The tomography array consisted of three speakers and five microphones located a few meters above the ground along a perimeter of a rectangle with dimensions 200 x 240 m. The array enabled measurement of the travel times of sound propagation among different pairs of speakers and microphones. The temperature $T(\vec{r})$ and wind velocity $\vec{v}(\vec{r})$ fields within the tomographic area were then reconstructed using a stochastic inversion (SI) algorithm. Here, $\vec{r}$ =(x, y) are horizontal coordinates.

Similar experimental implementations of acoustic tomography of the ASL have been done by scientists from the University of Leipzig, Germany. In the experiments, the size of a tomographic array was variable, of the order of several hundred meters. Speakers and microphones were located on tripods 2 m above the ground. The number of speakers and microphones varied and could be as large as 8 and 12, respectively. Since 1996, this tomography array has been used in several experiments [5, 19-26]. The German scientists have also built a small-scale tomography array for indoor experiments [27, 28]. A size of the small-scale array varied from 1 m to 20 m. The array was built to test acoustic tomography in controlled environments and for studies of turbulence in wind channels.

An innovative scheme of acoustic tomography of the ASL was proposed in [29] and experimentally demonstrated in [4]. The sound sources were calling animals (e.g. birds) and their sounds were recorded by microphones situated close to the ground. The time differences of call arrivals at different microphones were measured. Using these time differences, it was possible to locate calling animals and to reconstruct the temperature and wind velocity fields inside a tomographic volume.

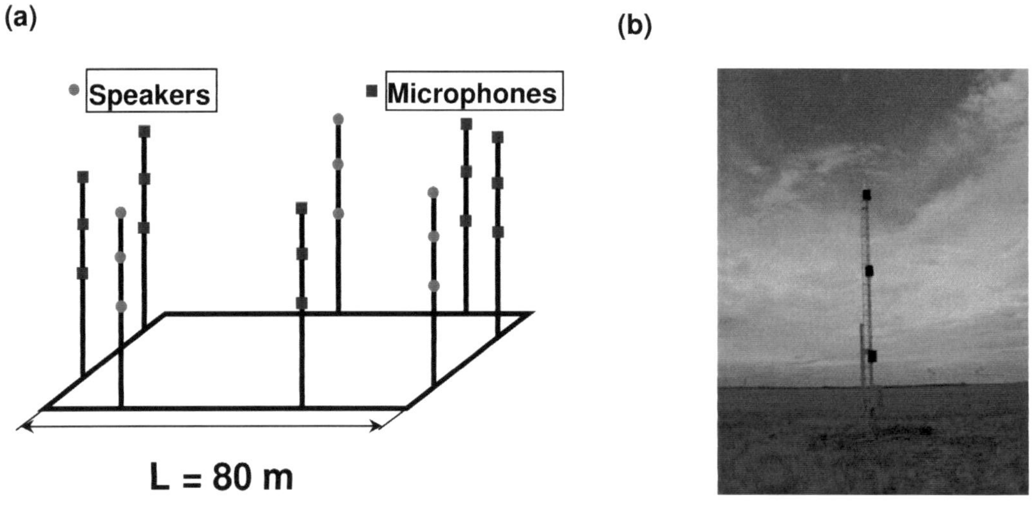

Figure 1. (a) Schematic of the 3D array for acoustic tomography which has been built at the Boulder Atmospheric Observatory. (b) A speaker tower of the array.

Finally note that a first 3D array for acoustic tomography of the ASL has been built [30,31] at the Boulder Atmospheric Observatory located near Boulder, CO and operated by the Earth System Research Laboratory of NOAA. The array consists of three speaker and five microphone towers located along the perimeter of a square with the side length $L = 80$ m, as shown schematically in figure 1(a). Speakers and microphones are located on the towers at three adjustable levels. The towers are 9.1 m height and can be bent over to install equipment. One of the speaker towers is shown in figure 1(b). In comparison with previous tomography arrays, the array shown in figure 1 has a smaller horizontal size which would enable more detailed reconstruction of the temperature and velocity fields.

3. Forward and inverse problems in acoustic tomography of the atmospheric surface layer

The arrays for acoustic tomography of the ASL enable measurements of travel times τ_i of sound propagation from speakers to microphones. Here, the index $i = 1,2,...,I$ indicates a particular speaker-microphone pair, and I is a number of such pairs. For example, if only three speakers and five microphones located at the upper level of the tomography array depicted in figure 1(a) are used (resulting in a 2D, horizontal-slice tomography), the number I is given by $I = 3\times 5 = 15$. Note that, unless it is stated otherwise, horizontal-slice tomography is considered in the rest of the paper.

The forward problem of acoustic travel-time tomography of the atmosphere is to express τ_i in terms of temperature and wind velocity fields. When doing this, it is worthwhile to represent the fields as follows: $T(\vec{r}) = T_0 + T'(\vec{r})$ and $\vec{v}(\vec{r}) = \vec{v}_0 + \vec{v}'(\vec{r})$. Here, T_0 and $\vec{v}_0 = (v_{0,x}, v_{0,y})$ are the mean values (within the tomographic area) of the temperature and wind velocity vector, and $T'(\vec{r})$ and $\vec{v}'(\vec{r}) = (v'_{0,x}(\vec{r}), v'_{0,y}(\vec{r}))$ are their fluctuating parts. The subscripts x and y indicate the components of the velocity vector along the x- and y- axis, respectively. Then, a solution of the forward problem is given by [10]:

$$\tau_i = \frac{L_i}{c_0}\left(1 - \frac{v_{0,x}\cos\varphi_i + v_{0,y}\sin\varphi_i}{c_0}\right) - \frac{1}{c_0}\int_{L_i} dl \left[\frac{T'(\vec{r})}{2T_0} + \frac{v'_x(\vec{r})\cos\varphi_i}{c_0} + \frac{v'_y(\vec{r})\sin\varphi_i}{c_0}\right] + \varepsilon_i \,. \quad (1)$$

Here, L_i is the length of the ith sound propagation path, the integration is along this path, φ_i is the angle between the path and the x-axis, ε_i correspond to errors in measurements of travel times and transducer coordinates, and $c_0 = \sqrt{\gamma R_a T_0}$ is the mean value of the adiabatic sound speed, where $\gamma = 1.4$ and R_a is the gas constant for air. Strictly speaking, acoustic tomography enables reconstruction of the acoustic virtual temperature T which is related to the thermodynamic temperature T_{th} by [32]: $T \cong T_{th}(1 + 0.51q)$, where q is the specific humidity. Therefore, throughout the text, temperature actually means acoustic virtual temperature.

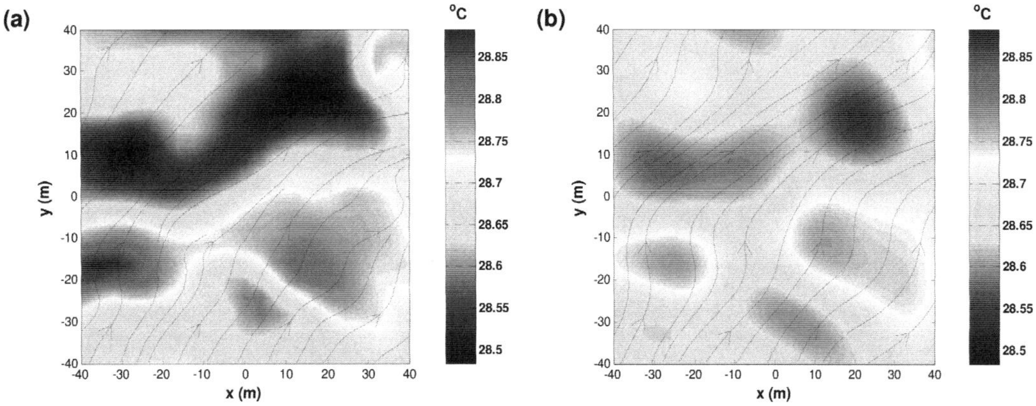

Figure 2. Temperature fields (a) obtained with LES, (b) reconstructed with TDSI. Solid lines with arrows indicate the direction of wind velocity.

The inverse problem in acoustic travel-time tomography of the atmosphere is to reconstruct (estimate) the temperature $T(\vec{r}) = T_0 + T'(\vec{r})$ and velocity $\vec{v}(\vec{r}) = \vec{v}_0 + \vec{v}'(\vec{r})$ fields given the travel times τ_i and the values of L_i and φ_i, which can be calculated if transducer coordinates are known. This is a challenging problem since, in the tomography experiments carried out so far and those that are planned for the future, the number I of sound propagation paths is relatively small, less than 100 or so. On the other hand, a detailed visualization of small-scale turbulence in the ASL requires the values of T and $\vec{v}$ at different spatial points whose number should be significantly greater than 100. For example, figure 2(a) depicts a horizontal slice of a temperature field from a large eddy simulation (LES). To plot this figure and similar ones for two components of wind velocity, 1200 values of T and $\vec{v}$ at spatial points separated by 4 m along the x - and y - axis are required.

In the literature, different approaches to solution of the inverse problem have been used. In one group of the approaches [19-28, 33, 34], which is called *algebraic reconstruction*, a tomographic area (e.g. that shown in figure 2(a)) is divided into cells with constant values of T_j and $\vec{v}_j$, where $j = 1,2,...,J$ and J is the number of tomographic cells. In this case the integrals in equations (1), where $i = 1,2,...,I$, can be evaluated and the equations take an algebraic form. Then, the resulting equations are solved for T_j and $\vec{v}_j$ using the least squares solution or similar techniques. Algebraic reconstruction enables accurate estimation of the temperature T and velocity $\vec{v}$ fields if the inverse problem is overdetermined, i.e. $3J < I$. However, since the number I of sound propagation paths is less than 100, the T and $\vec{v}$ fields can estimated only at a limited number of spatial points so that detailed reconstruction of turbulence is not feasible.

To overcome this issue, it was suggested in [3] to use a SI algorithm in acoustic tomography of the atmosphere. The SI algorithm is particularly well suited for solutions of underdetermined inverse problems when $I < 3J$. In SI, it is worthwhile to separate the reconstruction of the mean values of temperature and wind velocity, T_0 and $\vec{v}_0$, from the reconstruction of their fluctuating parts, $T'(\vec{r})$ and $\vec{v}'(\vec{r})$. The reconstruction of T_0 and $\vec{v}_0$ is done by the following approach. First, T' and $\vec{v}'$ are set to zero in equations (1). Then assuming that the travel times τ_i are known, the resulting equations are solved for T_0 and $\vec{v}_0$ using the least squares solution. Numerical simulations of 2D and 3D acoustic tomography of the ASL [10, 35] showed that this approach results in a very accurate reconstruction of the mean temperature and velocity fields.

After the mean fields T_0 and $\vec{v}_0$ are estimated, SI is used to reconstruct the fluctuations $T'(\vec{r})$ and $\vec{v}'(\vec{r})$ using equations (1). The main idea of SI is to assume that $T'(\vec{r})$ and $\vec{v}'(\vec{r})$ are random fields with given spatial correlation functions of temperature and wind velocity fluctuations, $B_T(r)$ and $B_{n,m}(\vec{r})$, where $n,m = 1,2$. Then using these known correlation functions and the measured travel times τ_i, the $T'(\vec{r})$ and $\vec{v}'(\vec{r})$ fields are obtained with formulas well-known in the literature, e.g. [3, 10]. Furthermore, SI enables estimation of expected errors of the reconstruction.

At a first sight, SI has a disadvantage in comparison with algebraic reconstruction since it requires *a priori* information about the correlation functions of the fields to be reconstructed. In reality, however, only the variances of the temperature and velocity fluctuations need to be known, which can be easily measured with a sonic anemometer situated inside or close to a tomography array. Indeed in [36], different forms of $B_T(r)$ and $B_{n,m}(\vec{r})$ were used to reconstruct the $T'(\vec{r})$ and $\vec{v}'(\vec{r})$ fields in numerical simulations of acoustic tomography of the atmosphere: the exponential, Gaussian, von Kármán, and a new correlation function having dependence on the distance r between two points of observation given by $\exp(-r^{2/3}/a^{2/3})$, where a is the outer length-scale of turbulence. In these numerical simulations, $T(\vec{r})$ and $\vec{v}(\vec{r})$ fields were modeled with LES and the travel times τ_i were calculated with equations (1) for locations of transducers as those at the upper level of the tomography array shown in figure 1(a). Then, the travel times were distorted by noise ε_i. Finally, the distorted travel times $\tau_i + \varepsilon_i$ were used for reconstruction of the temperature and velocity fields with SI algorithm. The results obtained in such numerical simulations showed that the reconstructed temperature and velocity fields are practically independent of the forms of the correlation functions $B_T(r)$ and $B_{n,m}(\vec{r})$ if the outer length-scales a in these functions are chosen properly. Thus, SI is not sensitive to exact forms of the correlation functions. In our studies, $B_T(r)$ and $B_{n,m}(\vec{r})$ are chosen as Gaussian correlation functions since, in this case, some calculations can be done analytically.

The numerical simulations also showed that the smallest turbulence eddies that can be reconstructed with SI are of order a. Therefore, in SI the outer length-scale a actually plays a role of an adjustable parameter: The smaller a, the better the spatial resolution. However, decreasing a results in increasing errors in reconstruction. Thus, the value of a should be chosen as a tradeoff between these opposite tendencies. Finally, the numerical simulations showed that SI provides a better reconstruction of the $T'(\vec{r})$ and $\vec{v}'(\vec{r})$ fields than the algebraic reconstructions which use the least squares or damped least squares solutions [34]. This result is due to the fact that, in the atmosphere, the $T'(\vec{r})$ and $\vec{v}'(\vec{r})$ fields do correlate in space, which is accounted for in SI but not in the algebraic reconstruction. A disadvantage of SI is that it does not employ the fact that the temperature and velocity fields also correlate in time. These temporal correlations are taken into account in the TDSI algorithm [10, 11], which can be considered as a generalization of SI and is discussed in section 4.

Finally, we point out that some velocity fields can be "invisible", i.e. cannot be reconstructed in acoustic travel-time tomography. Examples of such fields are given in [37, 38] (also see [33]). However, the invisible fields presented in these papers are relatively simple and do not resemble atmospheric turbulence. Therefore, it would be interesting to study to what extent realistic turbulence fields (e.g. LES fields) can be reconstructed with travel-time tomography.

4. Time-dependent stochastic inversion

The main idea of the TDSI is to assume that the temperature and velocity fluctuations, $T'(\vec{r},t)$ and $\vec{v}'(\vec{r},t)$, are random functions in both space and time with known spatial-temporal correlation functions $B_T(\vec{r},t)$ and $B_{n,m}(\vec{r},t)$, where t is time [10]. To take the full advantage of the TDSI, the travel times τ_i need to be measured repeatedly at the time moments $t = t_n$, where $n = 1,2,...,N$ and N is the number of travel time measurements (tomography scans). Then, the temperature and velocity fields are reconstructed given N sets of τ_i and assuming that $B_T(\vec{r},t)$ and $B_{n,m}(\vec{r},t)$ are known. This allows one to increase the amount of input data (in comparison with SI) without increasing the number of transducers and, hence, to improve tomographic reconstruction.

To obtain the spatial-temporal correlation functions $B_T(\vec{r},t)$ and $B_{n,m}(\vec{r},t)$, a theory of locally frozen turbulence was employed [11]. This theory generalizes the widely used hypothesis of (perfectly) frozen turbulence to the case when turbulence is advected with the wind velocity, which can change in space and time. Using the theory of locally frozen turbulence, the spatial-temporal correlation functions $B_T(\vec{r},t)$ and $B_{n,m}(\vec{r},t)$ were expressed in terms of the spatial correlation functions $B_T(\vec{r})$ and $B_{n,m}(\vec{r})$. The specific forms of $B_T(\vec{r},t)$ and $B_{n,m}(\vec{r},t)$ were obtained assuming that $B_T(\vec{r})$ and $B_{n,m}(\vec{r})$ are given by Gaussian correlation functions.

In TDSI, the forward problem is still given by equations (1), where the travel times and the temperature and velocity fields now depend on time t: $\tau_i = \tau_i(t)$, $T(\vec{r},t) = T_0(t) + T'(\vec{r},t)$, $\vec{v}(\vec{r},t) = \vec{v}_0(t) + \vec{v}'(\vec{r},t)$, and $c_0 = c_0(t)$. In the inverse problem, the mean fields $T_0(t)$ and $\vec{v}_0(t)$

are reconstructed at the time moments t_n of the tomography experiments by the same approach as in SI. Then, the fluctuations $T'(\vec{r},t)$ and $\vec{v}'(\vec{r},t)$ are reconstructed using formulas derived in [10, 11, 35, 39].

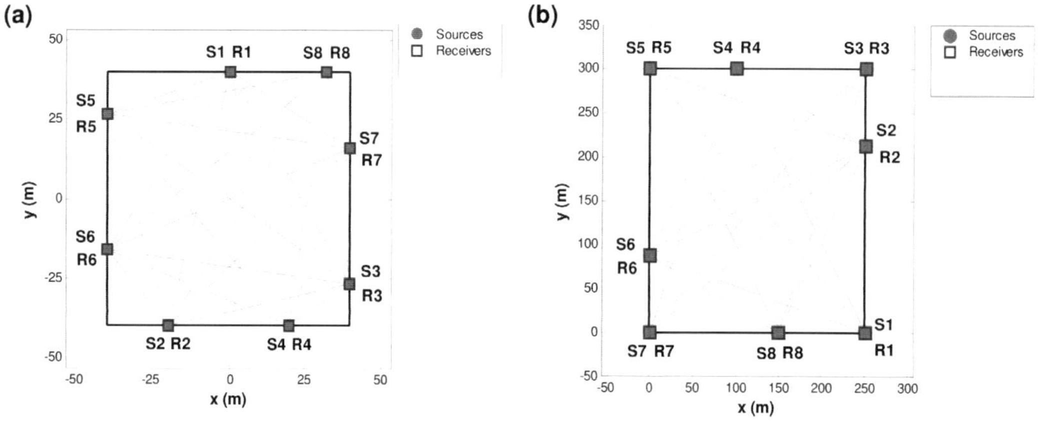

Figure 3. Tomography arrays used in (a) numerical simulations, (b) outdoor experiment STINHO. S1,S2,…,S8 are sound sources (speakers) and R1,R2,…,R8 are receivers (microphones). Dashed lines are sound propagation paths.

The developed TDSI was applied in 2D numerical simulations of acoustic tomography of the atmosphere [10, 39] and in 3D simulations [35]. Figure 3(a) depicts a tomography array used in one of the 2D simulations [39]. The transducers are located at the same positions as those at the upper level of the array in figure 1(a). Note that each transducer in figure 3(a) can work as a source and a receiver. (This can be done, e.g., by placing a speaker and a microphone close to each other). This tomography scheme enables one to achieve reciprocal sound transmission and significantly increase the number of travel times τ_i. The scheme will be used in further developments of the 3D tomography array shown in figure 1(a). In numerical simulations, the $T(\vec{r},t)$ and $\vec{v}(\vec{r},t)$ fields were modeled with LES fields. The travel times $\tau_i(t_n)$ were calculated for five time moments t_1, t_2, …, t_5. The time interval between these time moments was 4.6 s. Then, the travel times $\tau_i(t_n)$ were distorted to simulate errors in travel time measurements and uncertainties in transducer positions. Finally, the TDSI algorithm was used to reconstruct the temperature and velocity fields. Figures 2(a) and (b) depict the LES temperature field and the reconstructed field at the time moment t_3. Solid lines with arrows indicate directions of wind velocity. It follows from the figure that "cold" and "warm" temperature eddies of the LES field are reliably reconstructed with the TDSI; the actual spatially averaged root mean squared error of the reconstruction is just 0.05 ^{o}C. Though not presented here for brevity, similar results were obtained in reconstruction of the velocity field.

Note that the TDSI reconstruction depicted in figure 2(b) is based on equations (1). There is a different approach to reconstruction of T and $\vec{v}$ fields in acoustic tomography with reciprocal transmission. The travel times of sound propagation in opposite directions are added and subtracted. The resulting sums contain only temperature fluctuations T', while the differences contain only velocity fluctuations $\vec{v}'$. (This is a principle of operation of sonic anemometers.) Then, the inverse problem can be split into two parts: to reconstruct the temperature fluctuations and to reconstruct the velocity fluctuations. The TDSI algorithm can be used to solve both inverse problems. This approach was applied in the numerical experiment described above. Interestingly, the reconstructed temperature field was almost identical to that shown in figure 2(b). The same was true for the velocity field and for many other numerical simulations performed.

The developed TDSI algorithm was also applied to reconstruction of temperature and wind velocity fields in outdoor and indoor tomography experiments carried out by scientists from the University of Leipzig. The results obtained are presented in [11, 39, 40, 41]. As an example, figure 3(b) depicts a schematic of the outdoor tomography array with reciprocal sound transmission employed in the STINHO experiment [22]. In the experiment, the travel times $\tau_i(t_n)$ were measured every minute on June 17, 2002. Three sets of $\tau_i(t_n)$, measured at 8:54, 8:55, and 8:56 a.m., were used in the TDSI. The temperature and velocity fields reconstructed at 8:55 a.m. are shown in figure 4(a) and (b). Cold and warm temperature eddies and slow and fast velocity eddies are clearly seen in the figure. The estimated averaged root mean squared errors in reconstruction are 0.17 °C and 0.27 m/s.

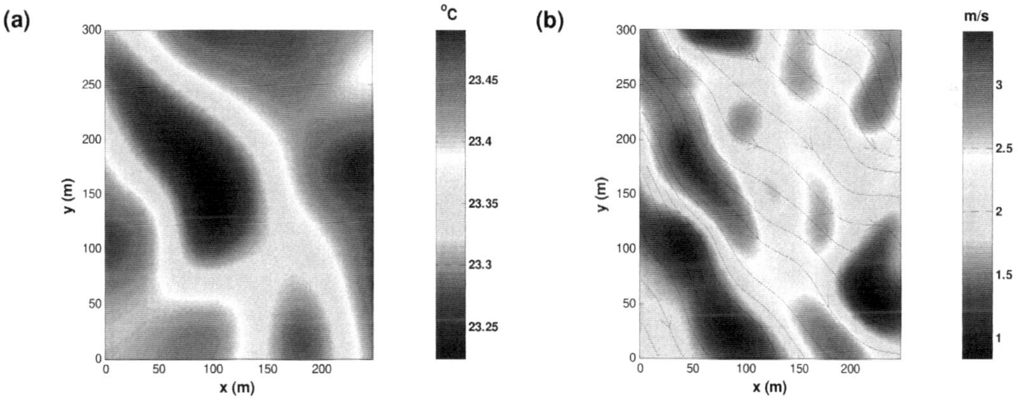

Figure 4. (a) The temperature and (b) velocity fields reconstructed with TDSI in acoustic tomography experiment STINHO on June 17, 2002.

5. Conclusions

A brief historical overview of acoustic tomography of the atmosphere was first presented. The rest of the paper was devoted to a presentation of acoustic travel-time tomography of the ASL. Acoustic tomography experiments in the ASL were discussed, including some already performed and some planned for the future. Travel times of sound propagation among sources and receivers in a tomography array were expressed in terms of temperature and wind velocity fields and coordinates of transducers, see equation (1). The inverse problem of tomography was formulated: to reconstruct the temperature and velocity fields from the travel times and transducer coordinates. Different approaches to solution of the inverse problem were outlined: algebraic reconstruction, SI, and TDSI.

It was indicated that, among these approaches, TDSI gives the best reconstruction of temperature and velocity fields. Numerical simulations of acoustic travel-time tomography clearly show that TDSI gives reliable reconstruction of turbulent eddies modeled with LES fields. TDSI was also applied to reconstruction of temperature and wind velocity fields in outdoor and indoor tomography experiments carried out by scientists from University of Leipzig, Germany. Some of the results obtained in reconstruction of these fields were presented.

Acknowledgments. This material is partly based upon work that was supported by the U.S. Army Research Office under contracts W911NF-06-1-0007 and DAAD19-03-1-0341.

References

[1] Wilson D K, Ziemann A, Ostashev V E and Voronovich A G 2001 An overview of acoustic travel-time tomography in the atmosphere and its potential applications *Acust. Acta Acust.* **87** 721-730

[2] Wilson D K, Ostashev V E, Vecherin S N, Voronovich A G, Collier S L and Noble J M 2004 Assessment of acoustic travel-time tomography of the atmospheric surface layer *Proc. AMS Symposium on Boundary Layers and Turbulence* (Portland, ME)

[3] Wilson D K and Thomson D W 1994 Acoustic tomographic monitoring of the atmospheric surface layer *J. Atmos. Ocean. Tech.* **11** 751-769

[4] Spiesberger J L 1999 Locating animals from their sounds and tomography of the atmosphere: Experimental demonstration *J. Acoust. Soc. Am.* **106** 837-846

[5] Arnold K, Ziemann A, Raabe A and Spindler G 2004 Acoustic tomography and conventional meteorological measurements over heterogeneous surfaces *Meteor. Atmos. Phys.* **85** 175-186

[6] Chunchuzov I, Kulichkov S N, Perepelkin V, Ziemann A, Arnold K and Kniffka A 2007 Acoustic tomographic study of the mesoscale coherent structures in the lower atmosphere *Proc. Meetings Acoust.* http://scitation.aip.org/confst/ASA/pub/8/4pPA5.pdf

[7] Dukkert P 1931 *Über die Ausbreitung von Explosionswellen in der Erdatmosphäre* (Leipzig: Akademische Verlagsgesellschaft)

[8] Crary A P 1950 Stratosphere winds and temperatures from acoustical propagation studies *J. Meteorol.* **7** 233-242

[9] Kulichkov S N and Bush G A 2001 Rapid variations in infrasonic signals at long distances from one-type explosions *Izvestiya, Atm. Ocean. Phys.* **37** 306-313

[10] Vecherin S N, Ostashev V E, Goedecke G H, Wilson D K and Voronovich A G 2006 Time-dependent stochastic inversion in acoustic travel-time tomography of the atmosphere *J. Acoust. Soc. Am.* **119** 2579-2588

[11] Vecherin S N, Ostashev V E, Ziemann A, Wilson D K, Arnold K and Barth M 2007 Tomographic reconstruction of atmospheric turbulence with the use of time-dependent stochastic inversion *J. Acoust. Soc. Am.* **122** 1416-1425

[12] Pierce A D, Posey J W and Iliff E F 1971 Variation of nuclear explosion generated acoustic-gravity wave forms with burst height and with energy yield *J. Geophys. Res.* **76** 5025-5041

[13] Bedard A J and Georges T M 2000 Atmospheric infrasound *Phys. Today* **53** 32-37

[14] McKenna M H, Stump B W, Hayek S, McKenna J R and Stanton T R 2007 Tele-infrasonic studies of hard-rock mining explosions *J. Acoust. Soc. Am.* **122** 97-106

[15] Gibbons S, Ringdal F and Kværna T 2007 Joint seismic-infrasonic processing of recordings from repeating source of atmospheric explosions *J. Acoust. Soc. Am.* **122** EL158-164

[16] Greenfield R J, Teufel M, Thomson D W and Coulter R 1974 A method for measurement of temperature profiels from refractive transmission of sound *J. Geophys. Res.* **79** 5551-5554

[17] Ostashev V E 1982 Returned-inclined acoustic sounding of the atmosphere *Izv. Acad. Scienc. USSR. Atmos. Ocean. Phys.* **18** 695-698

[18] Ostashev V E, Voronovich A G and Wilson D K 2000 Acoustic tomography of the atmosphere *IEEE Intern. Geoscience and Remote Sensing Symposium* pp 1186-1188

[19] Arnold K, Ziemann A and Raabe A 1999 Acoustic tomography inside the atmospheric boundary layer *Phys. Chem. Earth (B)* **24** 133-137

[20] Arnold A, Ziemann A and Raabe A 2001 Tomographic monitoring of wind and temperature at different heights above the ground *Acust. Acta Acust.* **87** 703-708

[21] Raabe A, Raasch S, Arnold K, Letzel M, Weinbrecht S and Ziemann A 2003 Akustische Tomography der atmosphärischen Grenzschicht zur Validierung eines LES Modells *Abschlussbericht zu den durch die DFG geförderten Projekten RA 569/4-1,2 und RA 617/4-1,2* (Leipzig und Hannover)

[22] Raabe A, Arnold K, Ziemann A, Schröter M, Raasch S, Bange J, Zittel P, Spiess Th, Foken Th, Göckede M, Beyrich F and Leps J-P 2005 STINHO - STructure of turbulent transport under

INHOmogeneous conditions - the micro-α scale field experiment and LES modeling *Meteorol. Z.* **14** 315-327

[23] Tetzlaff G, Arnold K, Raabe A and Ziemann A 2002 Observations of area averaged near-surface wind- and temperature fields in real terrain using acoustic travel time tomography *Meteorol. Z.* **11** 273-283

[24] Ziemann A, Arnold K and Raabe A 2001 Acoustic tomography as a method to identify small-scale land surface characteristics *Acust. Acta Acust.* **87** 731-737

[25] Ziemann A, Arnold K and Raabe A 1999 Acoustic travel-time tomography - a method for remote sensing of the atmospheric surface layer *Meteorol. Atmos. Phys.* **71** 43-51

[26] Ziemann A, Arnold K and Raabe A 2002 Acoustic tomography as a remote sensing method to investigate the near surface atmospheric boundary layer in comparison with in situ measurememts *J. Atmos. Oceanic Technol.* **19** 1208-1215

[27] Barth M, Raabe A, Holstein P, Müller R, Ziemann A, Arnold K, Mackenzie D, Starke E and Seliger M 2004 Acoustic travel time tomography as a tool to investigate temperature distributions on different spatial scales *Proc. 12th Intern. Symposium on Acoustic Remote Sensing* (Cambridge, UK)

[28] Holstein P, Raabe A, Müller R, Barth M, Mackenzie D and Starke E 2004 Acoustic tomography on the basis of travel-time measurements *Meas. Sci. Technol.* **15** 1420-1428

[29] Spiesberger J L and Fristrup K M 1990 Passive localization of calling animals and sensing of their acoustic environment using acoustic tomography *Am. Nat.* **135** 107-153

[30] Ostashev V E and Goedecke G H 2007 *Acoustic Tomography of the Atmosphere* (*ARO project* DAAD19-03-1-0341)

[31] Ostashev V E, Bedard A and Voronovich A G 2002 Array for acoustic tomography of the atmosphere *Intern. Geoscience and Remote Sensing Symp.* (Toronto, Canada) pp 862-864

[32] Ostashev V E 1997 *Acoustics in Moving Inhomogeneous Media* (London: E&FN SPON)

[33] Jovanovic I, Sbaiz L and Vetterli M 2006 Acoustic tomography method for measuring temperature and wind velocity *IEEE Intern. Conf. on Acoust., Speech and Signal Processing* (Toulouse, France)

[34] Vecherin S N, Ostashev V E, Wilson D K, Voronovich A G, Goedecke G H, Collier S L, Noble J M and Ligon D 2004 Forward and inverse problems of acoustic tomography of the atmosphere *Proc. 11th Intern. Symp. on Long Range Sound Propagation* (Fairlee, VT) pp 293-312

[35] Vecherin S N, Ostashev V E and Wilson D K 2008 Three-dimensional acoustic travel-time tomography of the atmosphere *Acust. Acta Acust.* (in print)

[36] Ostashev V E, Vecherin S N, Wilson D K and Collier S L 2006 Correlation functions of temperature and velocity fluctuations in a turbulent atmosphere *J. Acoust. Soc. Am.* **119** 3264

[37] Johnson S A, Greenleaf J F, Hansen C R, Samayoa W F, Tanaka M, Lent A, Christensen D A and Woolley R L 1977 *Acoustic Holography* ed L W Kessler (New York: Plenum **7**) pp 307-326

[38] Braun H and Hauck A 1991 Tomographic reconstruction of vector fields IEEE *Transactions on Signal Processing* **39** 464-471

[39] Vecherin S N, Ostashev V E, Wilson D K and Ziemann A 2008 Time-dependent stochastic inversion in acoustic tomography of the atmosphere with reciprocal transmission *Meas. Sci. Technol.* (submitted)

[40] Vecherin S N, Ostashev V E, Goedecke G H, Wilson D K, Ziemann A, Barth M and Arnold K 2006 Reconstruction of turbulent fields in acoustic tomography experiments *Proc. 12th Intern. Symp. on Long Range Sound Propagation* (New Orleans, LI) pp 364-384

[41] Ostashev V E, Vecherin S N, Wilson D K, Ziemann A, Barth M and Arnold K 2006Acoustic travel-time tomography of temperature and wind velocity fields in the atmosphere *J. Acoust. Soc. Am.* **120** 3336

Robust, "blind", *in-situ* calibration of SODARs

Stuart Bradley

Physics Department, The University of Auckland, Private Bag 92019, Auckland, New Zealand,

and

School of Computing, Science and Engineering, University of Salford, M5 4WT, UK

s.bradley@auckland.ac.nz

Abstract. Doppler spectra for SODARs are an average weighted over a volume according to the generally unknown beam pattern. By tilting the SODAR at a number of known angles, the beam pattern "kernel" can be obtained and hence an absolute (field) calibration. No assumptions are required and the method does not require any comparison against mast or other installations

1. Introduction

Commercially available SODARs transmit a short pulse in at least three upward directions. Scattering from atmospheric turbulent refractive index fluctuations results in a time series signal from each direction. Spectral analysis of time-gated segments of these time series gives a spectral peak which is a measure of the turbulent strength and whose frequency is a measure of the Doppler shift experienced by the moving scatterers. Using at least three independent acoustic beams assures a system of at least three equations in the vector wind Cartesian components $\boldsymbol{V} = (u, v, w)$. Solving this set of equations then gives a wind profile with estimates at the centre of each height represented by the centre of each time gate [1].

There is very little that can 'go wrong' with such a simple remote-sensing arrangement since the the only real decision required is the position of each spectral peak. Peak position estimation in the presence of some background noise is a well-established procedure.

Nevertheless, large efforts have been expended on comparisons between mast-mounted anemometers and SODARs in such experiments as PIE [2]. There are a number of reasons for this work, including the desire to ultimately have remote-sensing recognized as a viable replacement for well-established mast instrumentation, and the problems with variable treatment of background noise problems. The latter include random acoustic noise and unwanted reflections from rain or fixed objects near to the SODAR. Unwanted reflections can, however, be largely removed by judicious use of software filters, leaving mostly the random noise component. Some of this random component is actually due to atmospheric fluctuations (i.e the assumption that there should be a very narrow spread in wind velocity over an averaging period may be false).

The most important findings of the WISE/PIE work were

- that a SODAR gives similar variability in wind speeds to a cup anemometer (Fig. 1a), and

- there remain (small) *systematic* errors in wind speeds estimated by a SODAR (Fig 1b).

The major challenge for SODAR calibration is therefore to remove these systematic biases. Such biases can obviously be detected through SODAR-mast comparisons, but these are in general rather inconvenient. Therefore, in this paper, we consider a new method for doing *in-situ* field calibrations of a SODAR. This method has the huge advantage of not requiring comparison against some other 'standard', nor requiring any assumptions regarding SODAR geometry and operation.

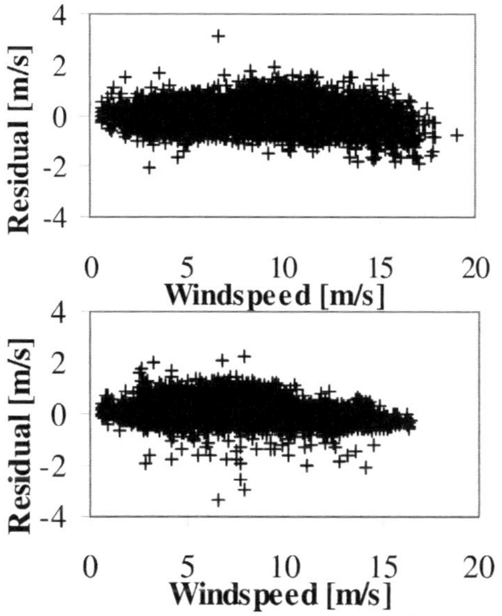

Figure 1a. Residual plots for a SODAR compared with a mast-mounted cup anemometer (upper plot) and for two cup anemometers (lower plot) at 60m. Superimposed line: running average of 100 points.

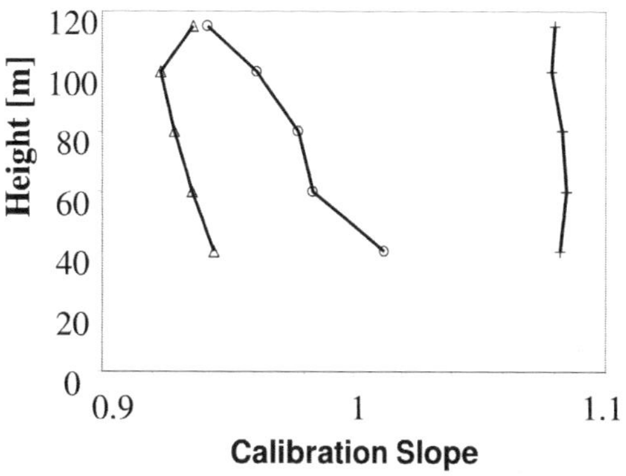

Figure 1b. Variation in calibration slope with height for three different SODAR models.

2. SODAR calibration error

It is known that the Doppler shift from a horizontal wind in the plane of a beam tilted at angle θ is proportional to sinθ. Unfortunately, θ is in practice a weighted average over the scattering volume, and may not be accurately known, given the presence of acoustic baffles. The acoustic radar equation covers this in principle [1], but we will first briefly go through the basics because the frequency-dependence is not usually evident.

The acoustic power transmitted has a frequency-dependent spectral density which can be approximated by

$$\frac{dP_T}{df} = E \exp\left[-\frac{1}{2\sigma_f^2}(f - f_T)^2 \right] \tag{1}$$

where E is the spectral amplitude, f is frequency, σ_f is the spectral width, and f_T the frequency of the transmitted pulse. The power per unit frequency interval transmitted into solid angle $d\Omega$ is

$$\frac{dP_T}{df} G d\Omega,$$

the sound intensity at range r is

$$\frac{dP_T}{df} G \frac{e^{-\alpha r}}{r^2}$$

(where α is the acoustic absorption), the power scattered into solid angle $\Delta\Omega_s$ is

$$\frac{dP_T}{df} G \frac{e^{-\alpha r}}{r^2} \sigma_s \Delta\Omega_s \Delta V$$

(where σ_s is the scattering cross-section area per unit volume and per unit solid angle, and ΔV is the scattering volume), and the spectral density of received power at the same (mono-static) antenna

$$\frac{dP_R}{df} = \int_{\Omega} \frac{dP_T}{df} G \frac{e^{-\alpha r}}{r^2} \sigma_s \left(\frac{A_e}{r^2}\right)\left(r^2 d\Omega c\tau\right) e^{-\alpha r} \tag{2}$$

(where A_e is the effective receiving antenna area, allowing for efficiency and orientation, c is the speed of sound, and τ is the pulse duration). Allowing for Doppler shifting of the spectrum to be centered on f_D rather than f_T,

$$\frac{dP_R}{df} = Ec\tau \frac{e^{-2\alpha r}}{r^2} \sigma_s \int_{\Omega} \exp\left[-\frac{1}{2\sigma_f^2}(f - f_D)^2\right] K d\Omega \tag{3}$$

where $K = GA_e$ is an angle-dependent sensitivity kernel. The atmospheric absorption and scattering parts have been taken outside of the scattering volume integral since they are only weakly frequency-dependent and it is assumed that they do not vary much within a typical scattering volume. However, the Doppler-shifted echo frequency f_D, and the antenna angular sensitivity K will vary with beam pointing direction.

For example, if the acoustic beam has sensitivity K at a zenith angle θ and azimuth angle ϕ, then the integral is

$$\int_0^{2\pi} \left\{ \int_{-\frac{\pi}{2}}^{\frac{\pi}{2}} \exp\left[-\frac{1}{2\sigma_f^2}\left[f - f_T\left(1 - 2\frac{u}{c}\sin\theta\cos\phi - 2\frac{v}{c}\sin\theta\sin\phi - 2\frac{w}{c}\cos\theta\right)\right]^2\right] K(\theta,\phi)\sin\theta d\theta \right\} d\phi$$

where the vector wind is $\boldsymbol{V} = (u, v, w)$. The usual assumption is that the beam in the $+x$ direction is effectively an angular delta-function

$$K(\theta,\phi) = A_s \cos\theta_0 \delta(\theta - \theta_0)\delta(\phi) \tag{4}$$

where A_s is the physical SODAR antenna area, adjusted for microphone efficiency. Then the above integral becomes

$$A_s \exp\left[-\frac{1}{2\sigma_f^2}\left[f - f_T\left(1 - 2\frac{u}{c}\sin\theta_0 - 2\frac{w}{c}\cos\theta_0\right)\right]^2\right] \sin\theta_0 \cos\theta_0$$

so that the spectrum peaks at

$$f_x = f_T\left(1 - 2\frac{u}{c}\sin\theta_0 - 2\frac{w}{c}\cos\theta_0\right). \tag{5}$$

Similarly, it is usually assumed that the beam in the +z direction has the form

$$K(\theta,\phi) = A_s\delta(\theta)\delta(\phi)$$

so that that spectrum peaks at

$$f_z = f_T\left(1 - 2\frac{w}{c}\right). \tag{6}$$

Solution of (5) and (6) gives estimates of u and w.

The uncertainty in spectral peak position arising from the integral in (3) has the potential to be the largest source of systematic error, particularly for 3-beam SODARs. If the effect of angular weighting is to shift θ_0 to $\theta_0+d\theta_0$ in (5) then, to first order, the estimated wind component $u+du$, when the actual wind component is u, can be estimated from

$$\frac{du}{u} = -\left(\frac{1}{\tan\theta_0} - \frac{w}{u}\right)d\theta_0. \tag{7}$$

Typically $\theta_0 = 20°$, giving a 5% error in estimation of u for each 1° error in estimation of θ_0. Careful analysis shows that there are unlikely to be other comparable calibration errors.

3. The effect of a finite-width beam on spectral position and shape

The geometry for a finite-width tilted beam is shown in Figure 2.

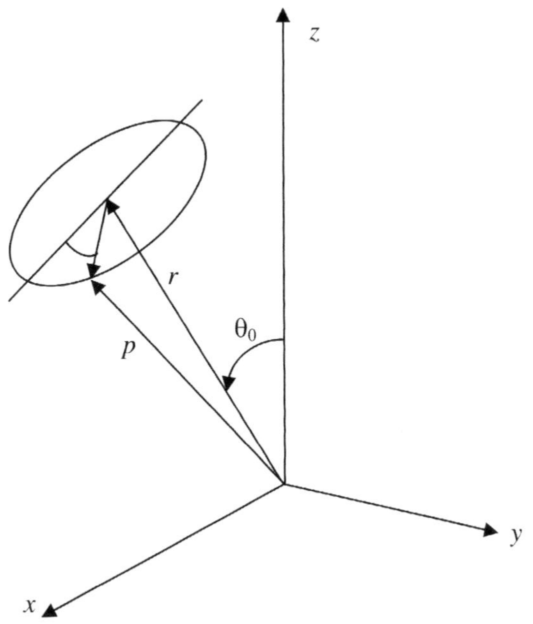

Figure 2. The geometry of a SODAR beam.

The influence of the term in v is to give a broader spectral peak but not to change the peak position, so will be ignored in the following. Transverse sensitivity in K also simply changes the volume averaging to give a broader spectrum. Finally, the effect of the $\sin\theta$ weighting on u is to bias the spectral peak to the equivalent of a larger effective θ_0. There is therefore a small change in the effective tilt angle, as expected. However, this does not change the methodology when the effective tilt angle is unknown anyway.

4. The principle of the new calibration method

The concept of the new calibration method can be illustrated using the simple example of a delta-function beam shape of unknown tilt angle θ_0 in the $+x$ direction. The entire SODAR is deliberately tilted a series of extra angle increments φ_n, which provides new measured

$$\alpha_n = \frac{c}{2}\left(1 - \frac{f_{x,n}}{f_T}\right) = u\sin(\theta_0 + \varphi_n) + w\cos(\theta_0 + \varphi_n) \tag{8}$$

and

$$\beta_n = \frac{c}{2}\left(1 - \frac{f_{z,n}}{f_T}\right) = u\sin(\theta_n) + w\cos(\theta_n). \tag{9}$$

From (9), we obtain from Least-Squares regression

$$\begin{aligned}\hat{u} &= \frac{\sum\beta_n\sin\varphi_n\sum\cos^2\varphi_n - \sum\beta_n\cos\varphi_n\sum\sin\varphi_n\cos\varphi_n}{\sum\sin^2\varphi_n\sum\cos^2\varphi_n - \left(\sum\sin\varphi_n\cos\varphi_n\right)^2} \\ \hat{w} &= \frac{\sum\beta_n\cos\varphi_n\sum\sin^2\varphi_n - \sum\beta_n\sin\varphi_n\sum\sin\varphi_n\cos\varphi_n}{\sum\sin^2\varphi_n\sum\cos^2\varphi_n - \left(\sum\sin\varphi_n\cos\varphi_n\right)^2}\end{aligned} \tag{10}$$

From these estimates we form

$$V = \sqrt{\hat{u}^2 + \hat{w}^2}$$

$$\begin{aligned}\gamma &= \tan^{-1}\frac{\hat{w}}{\hat{u}} \\ &= \tan^{-1}\frac{\sum\beta_n\cos\varphi_n\sum\sin^2\varphi_n - \sum\beta_n\sin\varphi_n\sum\sin\varphi_n\cos\varphi_n}{\sum\beta_n\sin\varphi_n\sum\cos^2\varphi_n - \sum\beta_n\cos\varphi_n\sum\sin\varphi_n\cos\varphi_n}\end{aligned} \tag{11}$$

These values are then applied in (8) in a second linear least-squares regression using

$$\begin{aligned}\alpha_n &= V\cos\gamma\sin(\theta_0 + \varphi_n) + V\sin\gamma\cos(\theta_0 + \varphi_n) \\ &= V\sin(\theta_0 + \gamma + \varphi_n) \\ &= V\cos(\varphi_n)\sin(\theta_0 + \gamma) + V\sin(\varphi_n)\cos(\theta_0 + \gamma)\end{aligned}$$

to obtain

$$\hat{\theta}_0 = \gamma + \eta \tag{12}$$

where

$$\eta = \tan^{-1}\frac{\sum\alpha_n\cos\varphi_n\sum\sin^2\varphi_n - \sum\alpha_n\sin\varphi_n\sum\sin\varphi_n\cos\varphi_n}{\sum\alpha_n\sin\varphi_n\sum\cos^2\varphi_n - \sum\alpha_n\cos\varphi_n\sum\sin\varphi_n\cos\varphi_n}.$$

5. Error analysis

In order to achieve a calibration accuracy of 1%, we need $\sigma_\theta \approx 0.2° \approx 4\text{x}10^{-3}$ radian. The above methodology is simulated with random errors in the φ_n values for $\theta_0 = 20°$, $N = 9$, and $\Delta\varphi_n = 2.5°$. It is found that tilt angles need to be determined to about 0.2° (i.e. comparable to the required precision in estimated θ_0), which is achievable with a linear actuator and a digital inclinometer (most digital inclinometers give σ_α=0.1°). The accuracy of the SODAR spectral peak estimation is not a problem.

6. Field measurements on a SODAR

Field measurements have been attempted on a trailer-mounted AeroVironment 4000 SODAR by tilting the entire trailer, as shown in Figure 3. This was not very successful because of difficulty with stability and also in achieving large enough angle changes.

Figure 3. Tilting a SODAR using an hydraulic jack.

The conference presentation will show results from both this AeroVironment 4000 and a Metek DSP, where the SODAR itself is precision-tilted using a linear actuator.

7. Conclusions

Since Doppler measurement is inherently calculable, the main source of systematic calibration errors for SODARs (and the reason so many mast comparisons have been performed) is uncertainty regarding the effective beam pointing angle.

A new method for *absolute* calibration of SODARs is described. The method makes no assumptions whatsoever about the SODAR operation and its hardware and software, other than the assumption that only one beam is transmitted at a time. Regardless of the complexity of the actual beam shape, the *effective* beam tilt angle is accurately estimated: this is the angle which must be used in estimations of velocity components.

No additional electronics is required except to operate the linear actuator and record the angle using a digital inclinometer. The linear actuator now being used is inexpensive and compact (can easily be carried in one hand). We are also currently designing a very simple mounting which can be used on any SODAR.

The only limitation evident at this stage is the requirement for horizontally stratified flow, since the regression methods use both a tilted beam and a vertical beam.

We are also considering the application for SODARs such as the AQS which do not have a vertical beam: this is a simple modification of the regression equations.

References

[1] Bradley S. G. *Atmospheric Acoustic Remote Sensing.* Florida, CRC Press/Taylor and Francis Group, 271pp., 2007. ISBN 978-0-8493-3588-4.

[2] Bradley S.G. (Ed), I. Antoniou, S. von Hünerbein, D. Kindler, M. de Noord, and H. E. Jørgensen, 2005: SODAR calibration for wind energy applications. Manchester, UK, *University of Salford*. 69pp.

IOP Conf. Series: Earth and Environmental Science 1 (2008) 012010 doi:10.1088/1755-1307/1/1/012010

A scanning bi-static SODAR

Paul Behrens [1], Stuart Bradley [1,2], Sabine Von Hunerbein [2]

[1] Physics Department, Auckland University, 38 Princes Street, Auckland, New Zealand.

[2] Acoustics Department, Newton Building, University of Salford, Greater Manchester M5 4WT UK

Abstract. Field results are given from a bi-static SODAR which uses a single central vertical transmission and three distributed microphone array receivers. Fourier transform delay methods are applied to data sampled from each microphone to retrospectively scan in angle and follow the transmitted pulse. Advantages of sampling a narrow atmospheric column, rather than distributed volumes are discussed.

1. Introduction

Traditional SODARs are mono-static i.e. speakers and receivers are co-located. A bi-static SODAR arrangement separates transmit and receive transducers into dedicated modules. The bi-static SODAR holds many advantages over traditional mono-static SODARs. Increased signal to noise, a single scattering volume, less side-lobe interference and reduced fixed echo problems are just some examples. Previous bi-static designs include Mickkelson (2007) [1] and Shamanaev (2003) [2] however both used fixed angle antennas, here a phased array antenna has been developed to concurrently sample a column of air up to 200m. Many phased array techniques utilise hardware time delay electronics, this design utilizes a software based, Fourier-domain shifting technique to locate a transmitted pulse.

2. Bi-static advantages

The amount of reflected acoustic energy is dependant upon the structure functions C_v and C_T which describe the velocity and temperature inhomogeneities in a moving field. Scattered acoustic power P_R is given by:

$$P_R \propto \frac{\sin^2 \beta}{r^2}\left[C_T^2 + 3.66(1-\sin\beta)\frac{T^2}{c^2}C_v^2 \right] \qquad (1)$$

where r is the total sound path distance, c is the speed of sound in air, T the absolute air temperature and β the angle made between transmitter, pulse height and receiver. Importantly, in the mono-static case β is equal to 90° which results in the cancellation of the C_V term and therefore leads to a reduction of received power and signal to noise. In the bi-static case the term contributes to the overall P_R according to the separation of transmitter and receiver. As C_v^2 is much greater than C_T^2 the improvements in signal to noise are significant. Additionally, since C_T^2 vanishes in neutral atmospheres mono-static SODARs can have reduced data availability.

Other advantages intrinsic to the bi-static geometry include:

- A larger detected Doppler shift, making shifted atmospheric echoes easier to detect over fixed echoes,
- The bi-static samples a single column of air, eliminating possible complex terrain inaccuracies of mono-static SODARs.

Disadvantages to the bi-static geometry are:

- The logistics of separating receiver and transmitter modules over 40-50m.
- Ensuring that the transmitting sound and steered receiving beam intersect.
- Three receivers are required, one for each wind velocity component.

3. Phased array theory

A linear phased array consists of a series of equally spaced microphones set along one axis. A phased array can be made to listen at any required angle by applying a progressive time shift along the microphones; this is equivalent to altering the phase and is commonly called beam steering (figure 1).

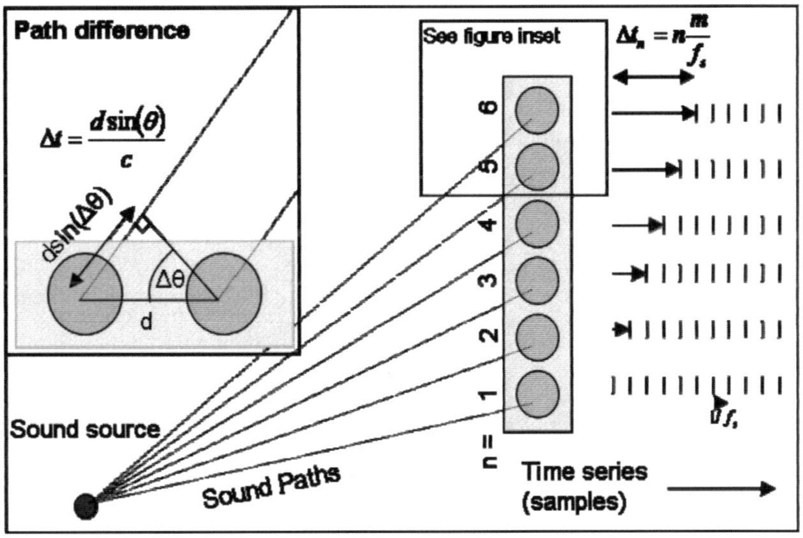

Figure 1. The phased array principle: a sound source is sampled at a frequency f_s across N microphones, Δt is the time delay, m relates to the angle investigated and c is the speed of sound. The extra path length from adjacent speakers is given in the figure inset.

In the far field the directivity for a phased array calculated as the summed intensities of N microphones can be expressed as:

$$A(\theta) = G(\theta)\frac{\sin\left[N\frac{1}{2}kd\left(\sin\left(\theta - \theta_{offset}\right)\right)\right]}{\sin\left[\frac{1}{2}kd\left(\sin\left(\theta - \theta_{offset}\right)\right)\right]} \qquad (2)$$

where k is the wavenumber ($2\pi/\lambda$), d the spacing between microphones, θ is the angle, N the total number of microphones and $G(\theta)$ the gain or shading function across the array. This directivity results in a large central peak of sensitivity and several associated side-lobes. These side-lobes may be removed through the use of hanning shading at the expense of the central peak strength and resolution (figure 2).

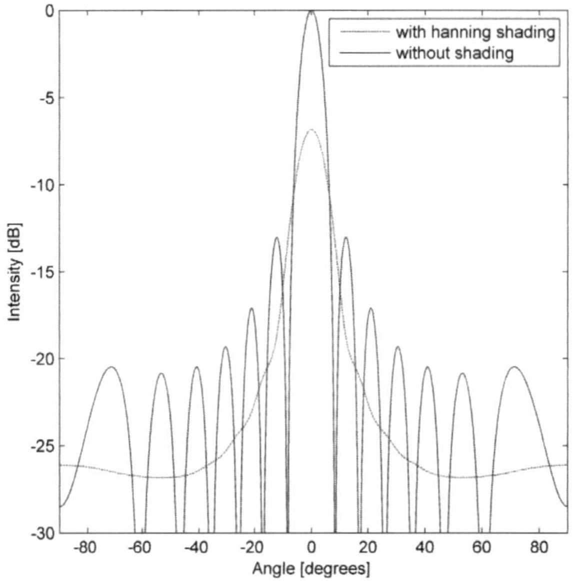

Figure 2. The beam pattern for 11 microphones spaced at 0.085m sampling a signal of 3.3kHz with and without beam shading.

The beam steering phase shift can be applied in hardware, as is traditionally done in the arrays of mono-static SODARs, or retrospectively in software. In the time domain the shift can be applied as a series of loops, delaying the signal progressively for each channel.

However, a more efficient method than time delay steering can be utilized by phase shifting in the frequency domain. If $V_n(\omega)$ is the Fourier transform of time series $V_n(t)$ we may apply a phase shift whereby:

$$\int V_n(\omega) e^{j\omega(t+\tau)} d\omega = V_n(t+\tau) \qquad (3)$$

where τ is the time delay applied and ω the angular frequency. Over N microphones this yields a shifted signal:

$$p(t) = \int \left[\sum_{n=0}^{N-1} w_n V_n(\omega) e^{j\omega(n-1)\tau} \right] e^{j\omega t} d\omega \qquad (4)$$

where w gives the required window shading (hann, hamming, blackmann etc.). This method holds a clear advantage over time domain delay shifting of speed. Fourier transform delay methods also have the added functionality of being able to apply non-integer steps of $1/f_s$ across a steered beam. These non-integer steps allow for signal interpolation in the frequency domain.

4. Bi – static theory

At the intersection of transmitter and receiver cones the scanning volume may become quite large (figure 4). The number of received Doppler shifts in this volume will increase and the resolution would therefore be reduced. To mitigate this loss in height resolution we have three options:

- Synchronized scanning – follow a short pulse of sound by phase shifting at the same speed as the propagating pulse (figure 3).

- Searching scanning – transmit a short pulse and searching the column of air for the region of largest signal (figure 3).
- Continuous scanning with modulated transmission (figure 4).

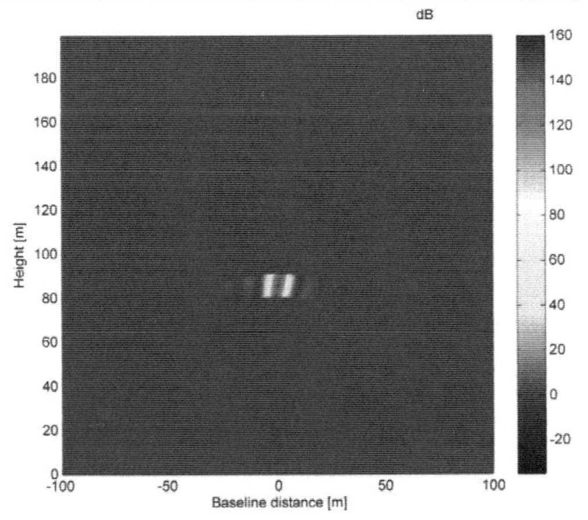

Figure 3. The unshaded bi-static sensitivity of a 30ms pulse received at a phased array located at 40m.

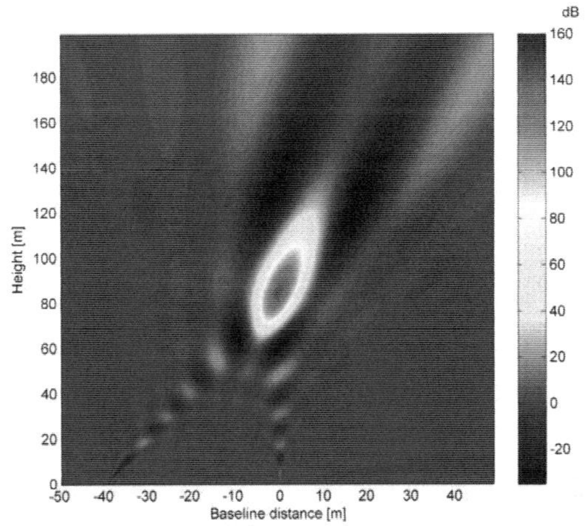

Figure 4. The unshaded bi-static sensitivity of a continuous column of sound at 3.3kHz.

4.1. Continuous scanning with continuous transmission

Continuous transmission allows for an increased signal energy which again increases signal to noise. However, the direct signal from transmitter to receiver can quickly overwhelm the atmospheric echo. For this reason the continuous transmission would have to be modulated in someway in order to isolate the interrogated heights. The signal could be modulated around a central frequency or a step-chirp could be used Bradley 1999 [3]. However, this modulation would alter the sound-turbulence interaction, thus changing the scattering properties; more work would have to be done to quantify this.

5. Instrument arrangement

The phased array used in this investigation consists of 11 KSN1005 Motorola tweeters in a linear geometry (figure 5). This feeds a data translation DT9836 12-channel analogue to digital converter. Data is synchronously recorded from the 11 channels at 225kHz with a Matlab routine. Currently data is processed slower than real time but efforts are underway to improve this including C++ routines, multi-core processing and code optimization. An 11 channel pre-amplifier is used to amplify the atmospheric echoes.

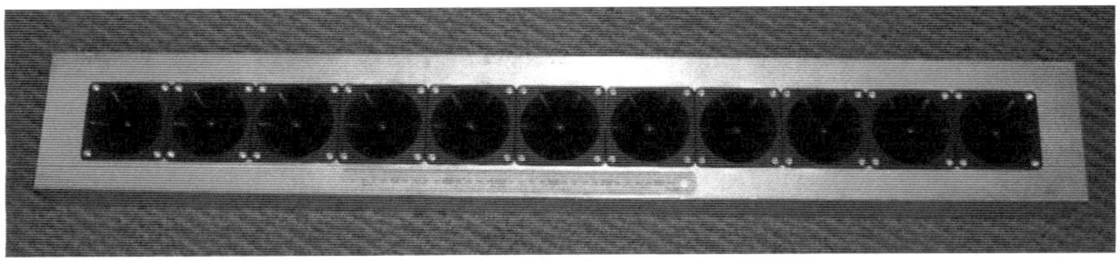

Figure 5. The experimental array consisting of 11 KSN1005 Motorola tweeters acting as microphones.

6. Field testing

Testing has been conducted at both the University of Auckland's anechoic chamber and Ardmore field site. This testing has verified the scanning ability of the array to locate a sound source at varying distances in the field. Testing indicates an ability to locate a sound source to within 1.16°.

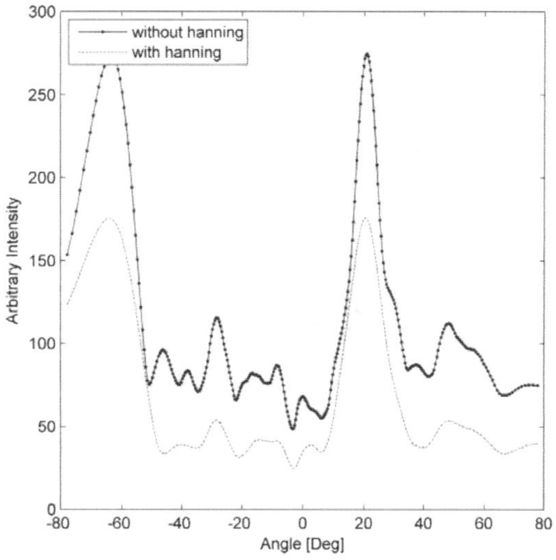

Figure 6a. Field testing: a sound source is located at 22°.

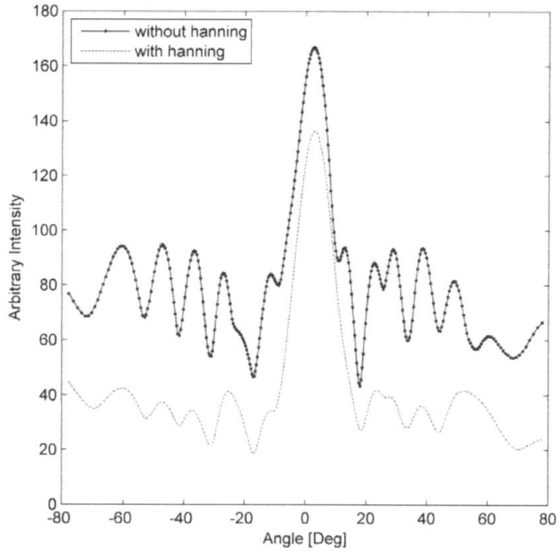

Figure 6b. Anechoic chamber testing: a sound source is located at 2.7°.

Field testing has been conducted next to an airport and the system has shown robustness to noisy environments.

Figure 7a. Field testing next to the University of Auckland's 30m meteorological mast.

Figure 7b. Anechoic chamber testing, work was performed in the near field.

7. Future work

Development of the scanning bi-static SODAR will now focus on developing the electronics to sample the wind velocity 'on sky'. Another experimental array is being constructed consisting of 3x12 Motorola tweeters; this will help aid alignment with the transmitted pulse. At present the routines performing the phase shifting are too slow for real time use. The speed of these routines will be increased to close to or at real time.

8. Acknowledgements

The authors would like to thank Mighty River Power for providing funding for this work.

References

[1] Mikkelsen, T. et al. (2007). "The Bistatic Sodar "Heimdall" – You blow, I listen,", Riso National Laboratory DTU, ISBN 87-550-3250-8,

[2] Shamanaev, S. V. (2003). "Acoustic Sounding of Raindrop Size Distribution." Journal of Atmospheric and Oceanic Technology 20(1): 152-158.

[3] Bradley (1999) "Use of coded waveforms for SODAR systems", Meterology and Atmospheric Physics 71(1-2): 15-22

IOP Conf. Series: Earth and Environmental Science **1** (2008) 012011 doi:10.1088/1755-1307/1/1/012011

All-fiber Coherent Doppler LIDAR technologies at Mitsubishi Electric Corporation

T Ando, S Kameyama, Y Hirano
Mitsubishi Electric Corporation, Information Technology R&D Center,
5-1-1 Ofuna Kamakura, Kanagawa 247-8501, Japan

E-mail: Ando.Toshiyuki@ap.MitsubishiElectric.co.jp

Abstract. A 1.5 micron pulsed Coherent Doppler LIDAR system using all fiber optical components has been attracted attention for remote wind sensing because of its eye-safety, reliability and easy deployment. We report on our key technologies such as fiber based MOPA (Master Oscillator Power Amplification) transmitter and real-time signal processing. Some performance results are also provided.

1. Introduction

A Coherent Doppler LIDAR (CDL) is an attractive sensor for wind sensing because it offers a method of remote wind speed measurement in the clear atmospheric condition. The CDL involves irradiations of coherent light and detection of weak backscattered from natural aerosols in the distant atmosphere, and provides the line-of-sight component of wind speed via Doppler frequency shifts from backscattered light. This technique was first demonstrated in 1970s, and has been many researches for various applications. All-fiber CDL using wavelength of 1.5 micron has such many advantages as its eye-safety, its reliability for various environmental conditions and its flexibility for components' layouts [1,2]. Mitsubishi Electric has been involved to develop a commercial prototype of all-fiber CDL system since 2004, which successfully made confirmations of its highly measuring accuracy based on simultaneous measurements with a conventional ultrasonic anemometer [3]. In 2005 commercial product of all-fiber CDL system (LR-05FC series) has been manufactured, which successfully achieved reductions not only of its dimensions but also for manufacturing cost less than 50% of the prototype system [4]. Our all-fiber CDL has been mainly used fiber-optic components which have been largely improved its cost, compactness and reliability in telecommunications industry. However some specific points needed to be considered for pulsed CDL system, generating optical seed pulses with highly on/off extinction ratio, obtaining high peak power using fiber amplification, and an optical heterodyning without any fluctuation in state of polarization. Non-linear optical effects in optical fiber have to be specially considered to improve measuring performances of all-fiber CDL. Stimulated Brillouin Scattering (SBS) is one of the most dominant non-linear effects to limit measuring distance because output peak power has to be kept a SBS threshold to avoid parasitic oscillations imposing damages on fiber components [5]. We have developed short-length fiber amplifier to increase SBS threshold in 2006[6]. Self-Phase Modulation (SPM) is also known as a non-linear effect which induces the frequency chirp within a transmitting optical pulse, leading to offset errors on the measured wind velocities. As to this SPM influences, it has been clearly confirmed that these errors can be compensated by subtracting offset values which analytically estimated from temporal profiles of output optical pulses [7]. High speed signal processing is also important aspects for improvement of real-time processing system because of wide bandwidth of ~100MHz for detection of Doppler shifts at wavelength of 1.5micron.

We describe in this paper main characteristics in our standard all-fiber CDL product, some key technologies we have developed, field experiments for verification of measuring accuracy in the CDL system.

2. System configuration

2.1 Packaging

Figure 1 shows that the outer look image of the standard all-fiber CDL system. The system consists of a main container (53(W) x65(H) x56(D) cm, 46kg), and an optical antenna (15(W) x15(H) x30(D) cm, 7kg) connected with an optical fiber cables, and a tripod (2kg). A main container includes a fiber-based optical transmitter/ receiver (T/R) unit, a PC based signal processing unit, and its power supply in a 19 inch 10U rack case with casters which enables us to be easily carried it as well as quickly setup(~5minutes) on observation sites. The total power consumption is designed less than 400VA, which makes it possible to be operated by an automobile's power outlet with a commercially available DC-AC converter. By changing the scanning scheme the system has four measuring modes as follows: Line Of Sight (LOS) wind velocity with a fixed beam direction, sectored (±20deg) Plan Position Indicator (PPI) mode with a horizontally linear beam scanning, sectored Range Height Indicator (RHI) mode with a vertically linear beam scanning, and Velocity Azimuth Display (VAD) mode with a conical beam scanning (±10deg). In this standard CDL it has been considerably improved its usability with simultaneously refreshed quick-looking screens (LOS velocity, PPI, RHI, Doppler spectra, and wind vector) as well as touch-panel user interfaces. Figure1 (b) shows that an example of screen image (anemometer mode), which indicates horizontal wind speed direction, and vertical wind speed with respect to the measuring range. The refresh rate of these screens is up to a few Hz depending on the scanning rate.

(a) (b)

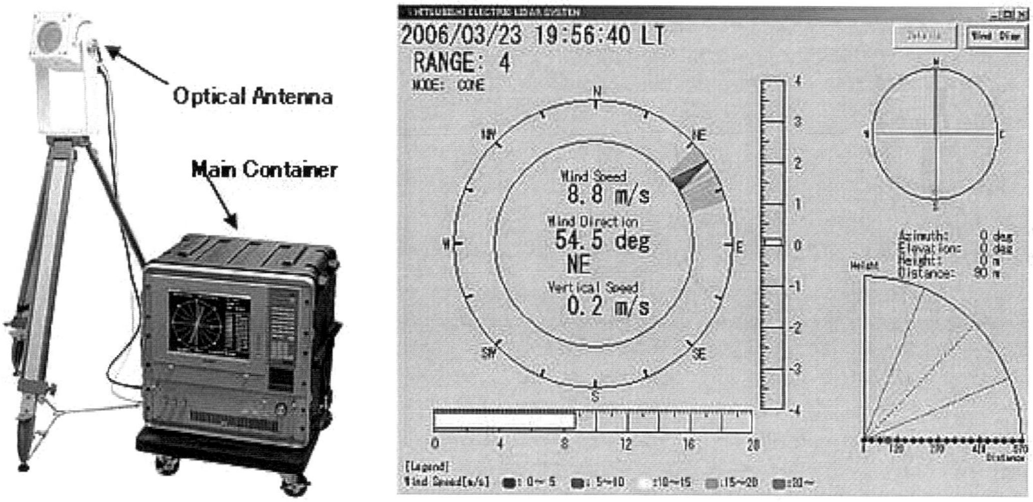

Figure 1. Outer look of the standard all-fiber CDL system

(a) Outer view (b) An example of quick look screen (Anemometer mode)

2.2 All-fiber optical transmitter / receiver

Figure 2 shows that the block diagrams of the standard CDL system. The optical TRX unit is combined a MOPA (Master Oscillator Power Amplifier) transmitter with an optical heterodyne receiver in which all optical components are connected by polarization maintaining fiber. The optical parameters of the system are follows: The wavelength of 1.54 μ m, pulse duration of 200, 500 and 1000ns, Pulse Repetition Frequency (PRF) of 4 kHz (16 kHz optional), transmitting pulse energy of 1.8~4.6 μ J. For this, range resolution can be selectable from 30, 75, and 150m by changing the pulse width.

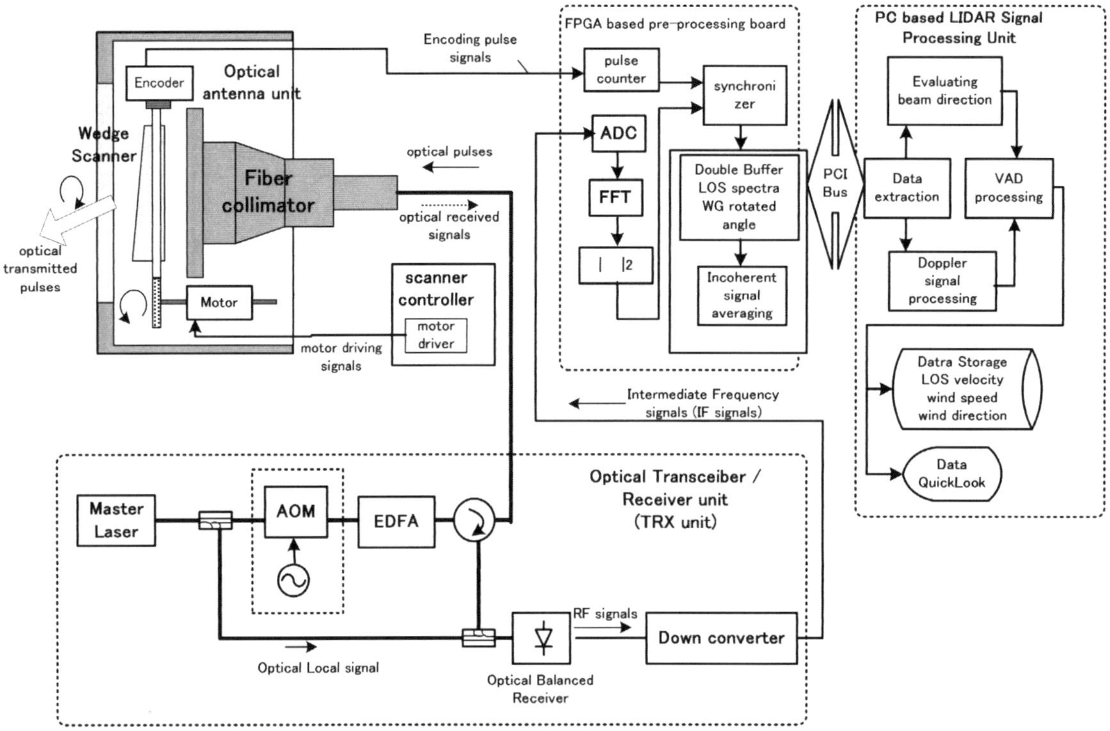

Figure 2. The block diagram of the standard all-fiber CDL system

2.3 Optical antenna unit

The optical antenna unit is consisting of a fiber collimator and a compact scanner which are covered by a waterproofing case. The fiber collimator has been designed as the refractive-type fiber collimator with an effective aperture diameter of 60mm, which can set focusing at from 100m to 1km without any truncation. It is noted that the passive-athermal design enables us to be kept the Wave Front Error from less than 1/14 of wavelength (Marechal criterion) over the temperature range of 20±30 deg, leading to the operation at outside laboratory without any special care.

A compact scanner enabled output beams to be conically scanned by simply rotating a silicon wedge prism and directly monitored its rotating angle. The angular information of output beams enabled to be obtained by combining an attainable wedge-rotating angle with a constant wedge-deflecting angle. Furthermore the optional mode has been readily prepared for linear scanning by inserting an extra wedge prism reversely rotated to a primary one.

2.4 FPGA based signal acquisition and pre-processor

Figure 3 (a) shows that the schematic block diagram of a PC based signal processing unit. In this signal processing unit, a Field Programmable Gate Array (FPGA) pre-processing board has been newly developed for achieving high speed Doppler signal processing after A/D acquisition as shown in fig.3 (b). This pre-processing board contain Fast Fourier Transform (FFT), data accumulation, and coupling with encoder data as beam direction from the scanner unit. In order to accommodate a bandwidth of 100MHz corresponding to Doppler velocity of +/- 38m/sec, input IF signals are sampled at rate of 216MS/s at the A/D converter with resolution of 8bit in the front-end of the pre-processing board. Then these sampled data are executed 256-points FFT with appropriate zero-padding to the range resolution simultaneously for each range as Doppler spectra in a single Pulse Repetition Interval (PRI), and also made incoherent accumulation up to 16000. This design enables to be used in range gate up to 1.19μs corresponding to range resolution of ~150m, leading to a flexible operation for range resolution by only changing parameter of zero-padding length.

As for the number of segmentation for processing range, 80-ranges Doppler spectra can be attainable in real-time at PRF of 4 kHz. The number of processing ranges and PRFs has a trade-off relationship; operation at PRF of 16 kHz can be also sustained by reducing number of ranges to 20 in a PRI. This enables to meet different PRF requirements for optical pulse sources in standard CDL system as well as middle range CDL system.

(a)

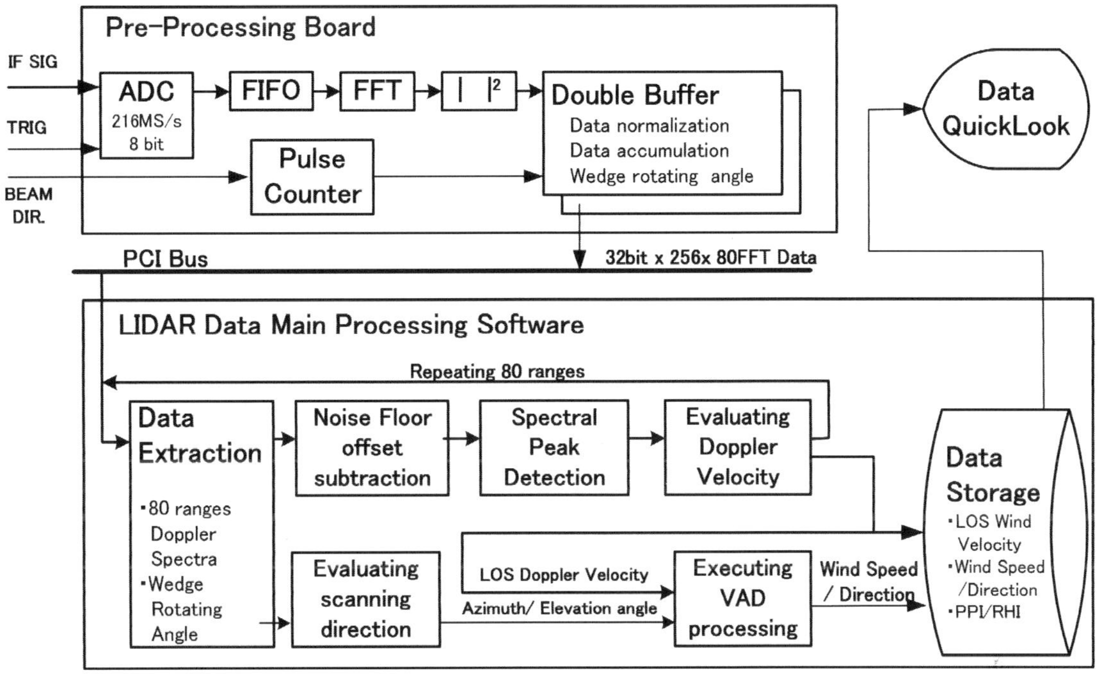

(b)

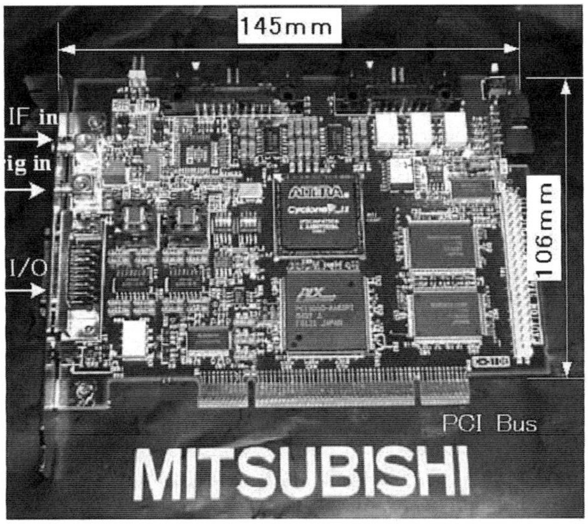

Figure 3. Schematic block diagram of a PC based signal processing unit
(a) Schematic block diagram, (b) outer view of FPGA based pre-processing board

The cumulative power spectra data with each beam direction are transferred to the double buffer as 32bit x 256 x 20 data and extracted by the LIDAR data main processing program in the Personal Computer (PC) via PCI bus. In the main program spectral peak detection, evaluating Doppler velocity has been performed after subtracting the noise floor offset at every range gate. From this processing Line Of Sight (LOS) velocity, velocity variance and detectability are evaluated with quick look screen images shown in figure 4 (a). It is noted that Doppler spectra can also be monitored with screen images shown in figure 4 (b), which enable us to be quickly comprehended whether the observations are correctly performed on site. The spatial distribution of LOS wind velocities can also be displayed under different beam scanning conditions, which enables us to be easily found wind shear. Figure 5(a) and (b) show the screen images for the sectored (±20deg) Plan Position Indicator (PPI) mode with a horizontally linear beam scanning, sectored Range Height Indicator (RHI) mode with a vertically linear beam scanning respectively.

(a) (b)

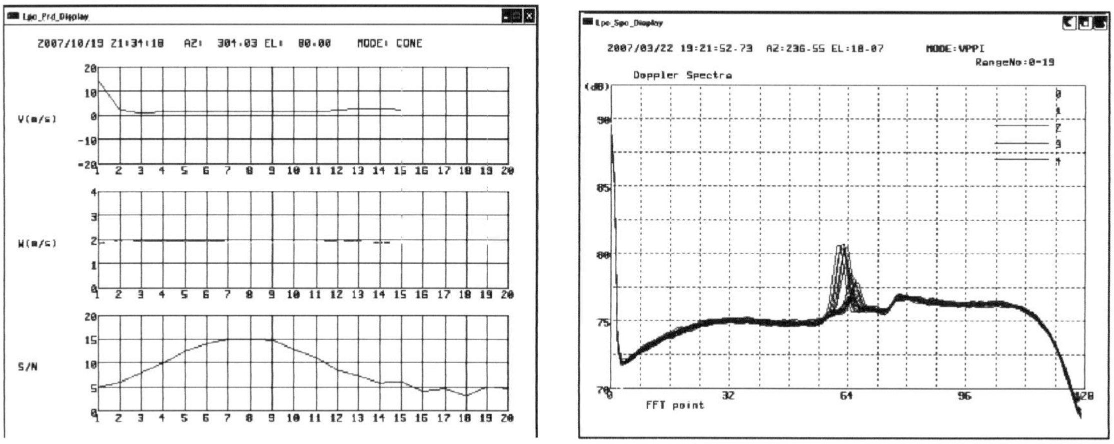

Figure 4. Screen images of quick look display for measured LOS data
(a) LOS wind velocity, velocity variance and detectability, (b) LOS Doppler spectra

(a) (b)

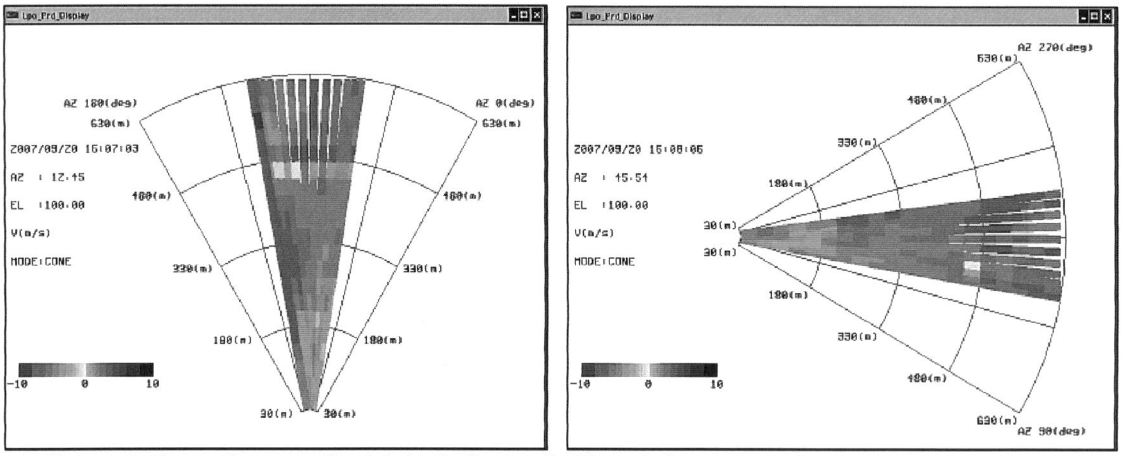

Figure 5. Screen images of quick-look display for linearly beam scanning
(a) Sectored Plan Position Indicator (PPI) mode , (b) sectored Range Height Indicator (RHI) mode

2.5 Wind vector calculations in the software

In order to obtain wind vector information it is necessary to acquire more than three LOS velocities with different beam direction to determine the three dimensional wind vector, leading to the true wind speed and direction. Provided spatially uniform wind distribution, LOS wind velocities periodically varies with azimuthal angle of conical beam scanning. Velocity Azimuth Display (VAD) uses the characteristics that Horizontal wind speed, direction and vertical wind speed can be evaluated from amplitude, phase and bias components of the periodical variation of LOS wind velocity.

Figure 6 shows the screen images for wind speed and direction on quick look display, (a) anemometer mode and (b) height distribution of wind vector components.

In this example the height distributions of horizontal wind speed, wind direction and vertical wind speed are obtained at a vertical resolution of 30m, with the height range between 30 – 600 m, in the data refresh rate of 1Hz. The height range can be expanded by changing the resolution wider or by increasing the number of range-segmentation. It is noted that our VAD based calculations for wind vector measurement are generalized to azimuth / elevation angles in order to be applied to tilted-center-angles to the zenith in conical beam scanning. This leads to flexible deployments on site to avoid the beam truncations with fixed obstacles, e.g. trees, buildings and electric power lines.

(a) (b)

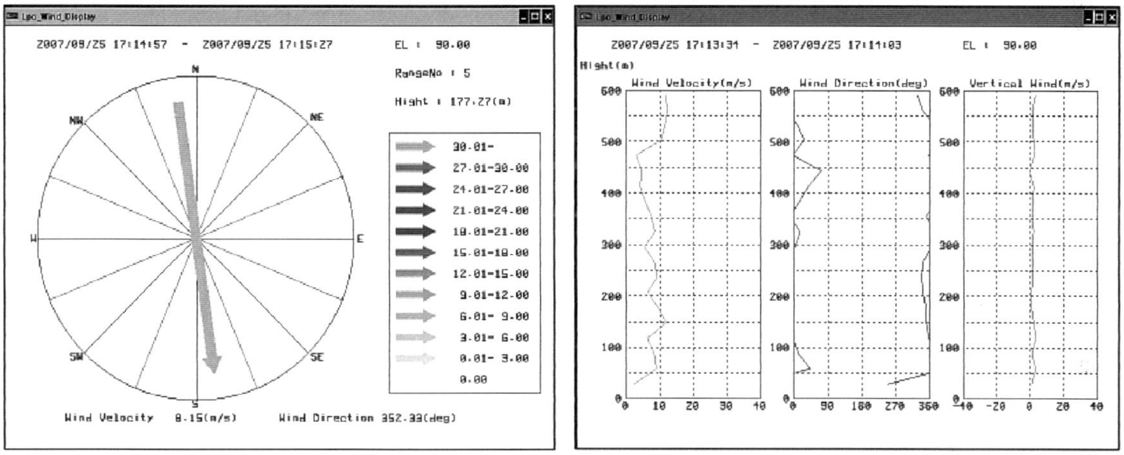

Figure 6. Screen images of quick-look display for wind speed, direction and vertical wind speed (a) Anemometry mode, (b) Height distribution of wind vector components

The refresh rate of calculated wind vector results is up to a few Hz depending on the scanning rate. Measuring results are simultaneously stored in the disk drive as time series data files with a spread-sheet format, and are easily displayed back again by loading these files.

Additionally horizontal wind vector calculations are also prepared for horizontally linear beam scanning. This observation mode enables to be attained the horizontal distribution of wind speed and wind direction close on the surface of the ground.

Figure 7 shows an example of screen image for horizontal wind speed and wind direction at elevation angle of nearly zero degree. In this example the horizontal wind speeds and wind vectors are indicated as colour feathers segmented by 20 ranges with a radial range between 130-1650m, scanning range of +/- 20 degrees. This measuring mode may be promising for wind flow analysis at urban region because of measurement of the horizontal wind field at specific areas surrounding many buildings with no ground clutter.

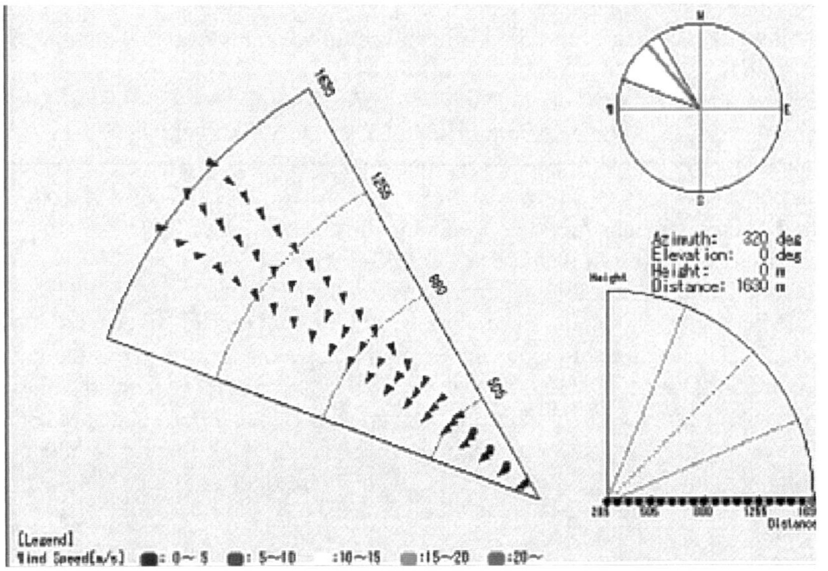

Figure 7. Screen images of quick-look display for horizontal wind speed and wind direction under horizontally linear beam scanning

3. Evaluation of measuring accuracy

Simultaneous measurement of the CDL prototype system and the ultrasonic anemometer has been performed to confirm measuring accuracy of wind velocity/ wind direction over one week from 17 September 2004 and one month from 24 November 2004. The measurement has been performed at the Tokyo Electric Power Corporation in Fukushima about 0.5 km away from the Pacific coast. This measurement site was an open space surrounded with woods, and had a 103-m height tower for metrological observation. An ultrasonic anemometer was mounted on the top of the tower which continuously observed horizontal wind speed and wind direction at a rate of 6 seconds and stored their data after averaging over 6 minutes. The measuring layout is shown in figure 8. The CDL was positioned about 50m south east of the tower. The CDL was operated in conical beam scanning with an angle of 15 degrees from zenith angle to obtain the horizontal wind speed and direction. 20 segmented LOS-wind velocities were simultaneously measured from 50 m to 650 m with a range resolution of 30 m. Every 1 second LOS wind velocities were obtained after 4000 incoherent integration of each Doppler spectrum. Scanning rate was 5 degrees / sec to the azimuthal angle, so the wind vector calculations with the VAD algorithm was performed with 72-direction of LOS data. In order to compare with the data from the ultrasonic anemometer, the specific segments at 110 m height were extracted from the calculated wind vector data by the CDL, and were averaged over 6 minutes.

Figure 9 shows the results of simultaneous observation of the wind speed and wind direction by an ultrasonic anemometer and the CDL system during one week on 10-17 September 2004.

In figure 9 (a) the regression plot of horizontal wind speed data displays good correlation with a correlation coefficient of 0.972. A polar plot of horizontal wind direction also indicates good agreement with residual error of less than 22.5 degrees shown in figure 9 (b).

Note that, there is no ambiguity of 180 degrees in wind direction of CLD data because of the bi-directional (upwind and downwind) measurement on LOS wind velocity.

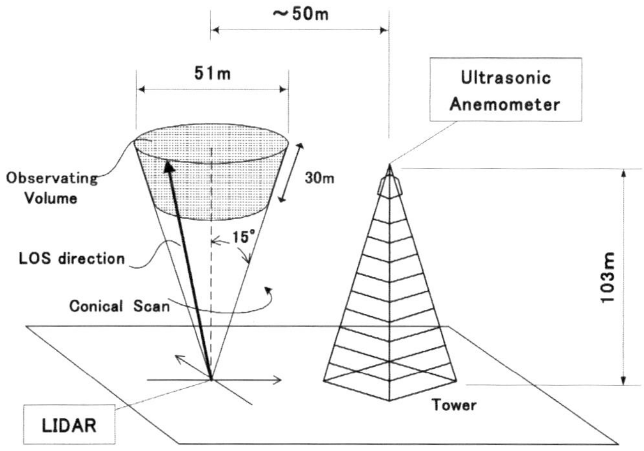

Figure 8. Experimental layout for simultaneous measurement of the CDL prototype system and the ultrasonic anemometer

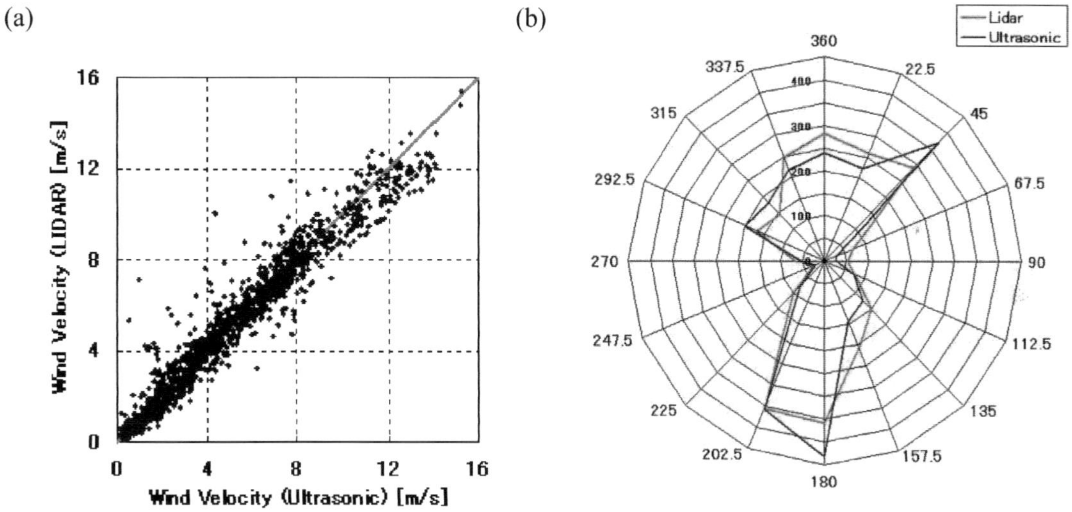

Figure 9. Experimental results on simultaneous measurement of the CDL and the ultrasonic anemometer 103 m above ground during one week on 10-17 September 2004, in Fukushima Japan.
(a) Regression plot of horizontal wind speed, (b) Polar plot of horizontal wind direction
The both data were averaged over 6 minutes.
These results display good agreement and correlation. Correlation coefficient was 0.972.

As to long term evaluation, figure 10 shows that the comparison of time series horizontal wind as measured with the CDL and the ultrasonic anemometer. The measured data were also averaged over 6 minutes both wind speed (shown in figure 10 (a)) and wind direction (in figure 10 (b)) display fairly good agreement. The CDL system has been continuously worked well without any trouble, but there were some periods the difference between data from both instruments has been increased. Figure 10 (c) shows the precipitation in Tomioka town near the observation site. It has been found that the differences have been increased for about one day after the rainfall. We speculate these differences may be caused by decreasing the amount of aerosols after the rainfall or by shading laser beam with small puddles on a window shield of the optical antenna. Figure 11 shows the regression plot of horizontal wind speed during one-month simultaneous measurement of the CDL and the ultrasonic anemometer. The correlation coefficient was 0.93.

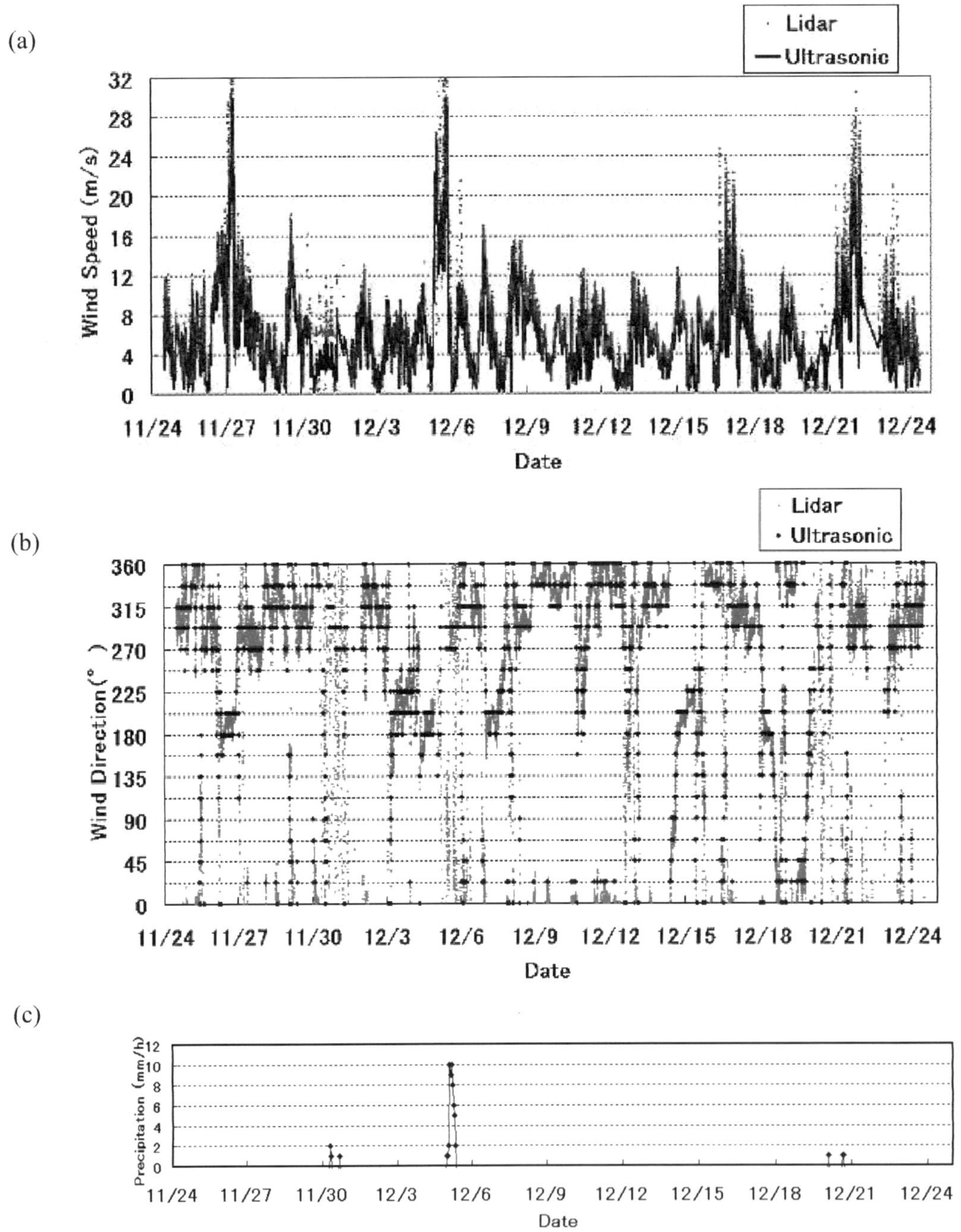

Figure 10. Experimental results on simultaneous measurement of the CDL and the ultrasonic anemometer 103 m above ground during one month from 24 November 2004, in Fukushima Japan. (a) Comparison of time series of horizontal wind speed, (b) comparison of time series of horizontal wind direction, and (c) time series of precipitation in Tomioka town near the observation site.

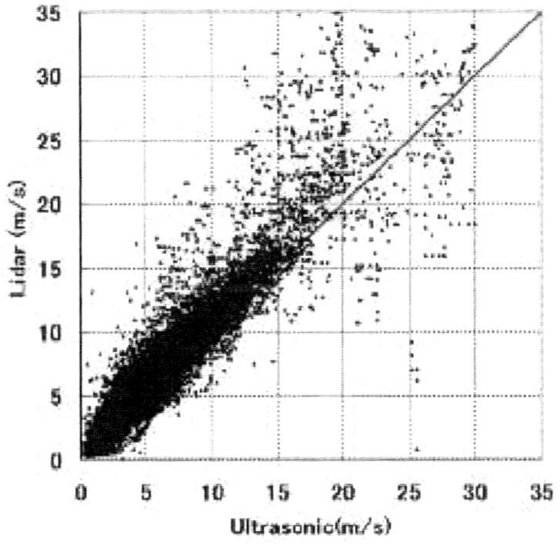

Figure 11. Regression plot of horizontal wind speed with simultaneous measurement of the CDL and the ultrasonic anemometer during one-month from 24 November 2004, in Fukushima, Japan. The correlation coefficient was 0.93.

4. Summery

We have described in this paper the works included in all-fiber Coherent Doppler LIDAR technologies for wind field sensing at Mitsubishi Electric. Combining an all-fiber optical transmitter / receiver with a FPGA based signal pre-processor made a number of improvements, reducing cost, down sizing, and flexible data processing. Through the long term field experiment the CDL has been successfully proved its robustness and its efficient measuring accuracy. Recently we have developed two systems based on the standard all-fiber CDL system described in this paper. The first system is the long-range wind sensing CDL by attaching a high peak power fiber amplifier to the standard system. The measurable distance has successfully enhanced up to 8 km at a ground experiment [6]. The airborne experiment has also carried out by using this CDL system installed in JAXA's aircraft (BC65) in the spring of 2007 [8]. The developing a long-range air borne CDL system is an ongoing project collaborated with JAXA to detect Clear Air Turbulence at high altitude. The second system is the very small CDL which allow to be operated by Lithium polymer battery without any external power supply [9]. Full assessment of these new CDL systems will be presented soon.

5. References

[1] Kameyama S, Ando T, Asaka K, Hirano Y and Wadaka S 2007 *Appl. Opt.* **46** 1953

[2] Kameyama S, Ando T, Asaka K, Hirano Y and Inokuchi H 2003 *Proc. 12th Coherent Laser Radar Conf. (Bar Harbor).* 15

[3] Furuta M, Matsuda T, Ando T, Nakayama T, Otsu J and Komiyama A 2005 *Proc. 13th Coherent Laser Radar Conf. (Kamakura)* 251

[4] Ando T, Furuta M, Tanaka H, Nagashima M, Kameyama S, Suzuki J and Hirano Y 2005 *Proc .13th Coherent Laser Radar Conf. (Kamakura)* 170.

[5] Agrawal G P 2001 *Nonlinear Fiber Optics* (New York: Academic Press)

[6] Ando T, Kameyama S, Sakimura T, Asaka K and Hirano Y 2007 *Proc. 14th Coherent Laser Radar Conf. (Colorado)*

[7] Ando T, Kameyama S, Sakamaki H, Wakayama T and Hirano Y 2005 *Proc. 13th Coherent Laser Radar Conf. (Kamakura)* 66

[8] Inokuchi H, Tanaka H and Ando T 2008 *Proc. Int. Council. Aeronatural Science (Alaska)*

[9] Ando T Yamaguchi K and Hirano Y 2008 *Proc. 55th Spring Meet. Japan. Soc. Appl. Phys. (Funabashi)* (in Japanese)

Acknowledgements

We greatly appreciates, the Tokyo Electric Power corporation Fukushima second nuclear power plant, had to offer the measurement place and the ultrasonic anemometer data.

IOP Conf. Series: Earth and Environmental Science **1** (2008) 012012 doi:10.1088/1755-1307/1/1/012012

Comparison of 3D turbulence measurements using three staring wind lidars and a sonic anemometer

J Mann[2], J-P Cariou[1], M S Courtney[2], R Parmentier[1], T Mikkelsen[2], R Wagner[2], P Lindelöw[2], M Sjöholm[2] and K Enevoldsen[2]

[1] Leosphere, Paris, France

[2] Risø National Laboratory, DTU, Roskilde, Denmark

E-mail: `jakob.mann@risoe.dk`

Abstract. Three pulsed lidars were used in staring, non-scanning mode, placed so that their beams crossed close to a 3D sonic anemometer. The goal is to compare lidar volume averaged wind measurement with point measurement reference sensors and to demonstrate the feasibility of performing 3D turbulence measurements with lidars. The results show a very good correlation between the lidar and the sonic times series. The variance of the velocity measured by the lidar is attenuated due to spatial filtering, and the amount of attenuation can be predicted theoretically.

1. Introduction

For wind energy research in particular and for micro-meteorology in general we want to be able to measure turbulence, not just the statistics, but time series of the fluctuating three-dimensional velocity. We do that with three pulsed lidars pointing towards the same position in space and investigate the how well the turbulence measurements compare with a co-located sonic anemometer.

2. Geometry of the experiment

Four Leosphere WindCubes were used in the experiment. Three of them were staring towards a Metek sonic anemometer 78 m above the ground. The first, WC1, pointed vertically, the two others, WC2 and WC3, pointed $\alpha = 56°$ away from vertical. The angle WC3 – WC1 – WC2 was 90°. In the horizontal direction WC2 pointed 26° North of West, while WC3 pointed 26° East of North, see figure 1.

The intention with the fourth WindCube, which was scanning conically, is to compare the computed time series of wind vectors assuming horizontal homogeneity with the real wind vector measured by the three staring lidars or the sonic anemometer. This analysis will be reported elsewhere.

The experiment took place at the test station for large wind turbines in Western Denmark at Høvsøre, operated by Risø/DTU. The experiment ran for almost a week in December 2007 and recorded wind speeds up to 21 m/s as seen from figure 2. Only a fraction of the data have been analyzed in this paper.

In figure 3 a ten minutes time series of wind speed from the three WindCubes are shown together with sonic measurements. If $\boldsymbol{n}$ is a unit vector in the direction of a lidar beam and $\boldsymbol{u}$ is the wind vector measured by the sonic, then shown is $\boldsymbol{n} \cdot \boldsymbol{u}$. The sonic speeds have been

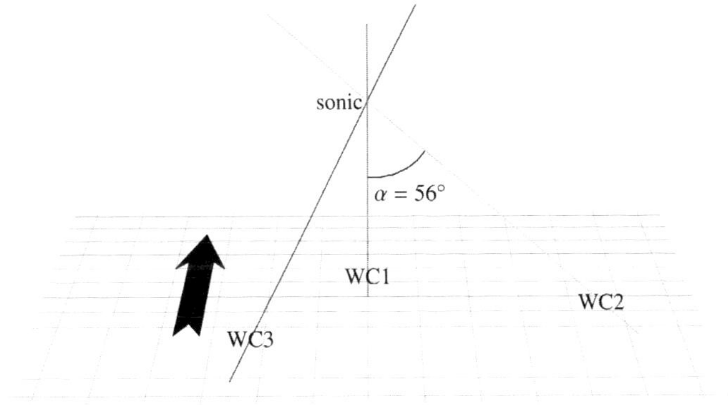

Figure 1. Sketch of the experiment as seen from the South. The arrow points towards the North.

block averaged over 1.5 s corresponding to the sampling rate of the lidar. Each wind speed determination by the lidar is based on Doppler spectra averaged over 0.5 s.

It is clear from figure 3 that the lidar follows the wind fluctuations very faithfully down to fluctuations with time scales of the order of some seconds. The fastest details are not completely reproduced. Furthermore, there seems to be an offset of around 1 m/s for vertically pointed lidar, WindCube 1. The reason for this discrepancy may be due to a misalignment of WindCube 1 or the sonic anemometer. Neither seems to have $\langle w \rangle = 0$ which should be expected from the rather flat site.

3. Detailed comparison of 3D turbulence measurements

We would like to understand the limitations in the lidar turbulence measurements compared to the sonic, which is very close to being a point measurement and is the preferred instrument for atmospheric turbulence. We assume that the wind speed measured by the lidar is

$$v(\boldsymbol{x}) = \int_{-\infty}^{\infty} \varphi(s)\boldsymbol{n} \cdot \boldsymbol{u}(s\boldsymbol{n} + \boldsymbol{x})ds \ , \tag{1}$$

where $\boldsymbol{n}$ is a unit vector in the direction of the laser beam, $\boldsymbol{u}(\boldsymbol{x})$ is the velocity field. The center of the lidar measuring volume is $\boldsymbol{x}$. We suppress the time dependence since we assume Taylor's frozen turbulence hypothesis to be valid [1]. The weighting function φ is normalized to unit integral and several suggestions of its shape are investigated. The most commonly accepted is

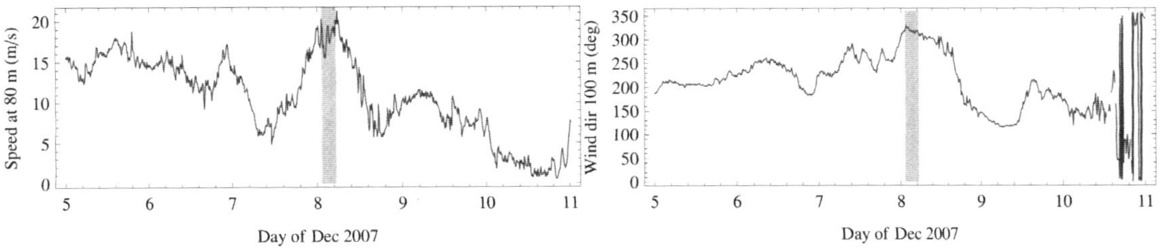

Figure 2. The wind speed (left) and wind direction (right) during the experimental campaign. The period Dec 8. from 01:10 to 05:30, which is used for detailed comparison is indicated.

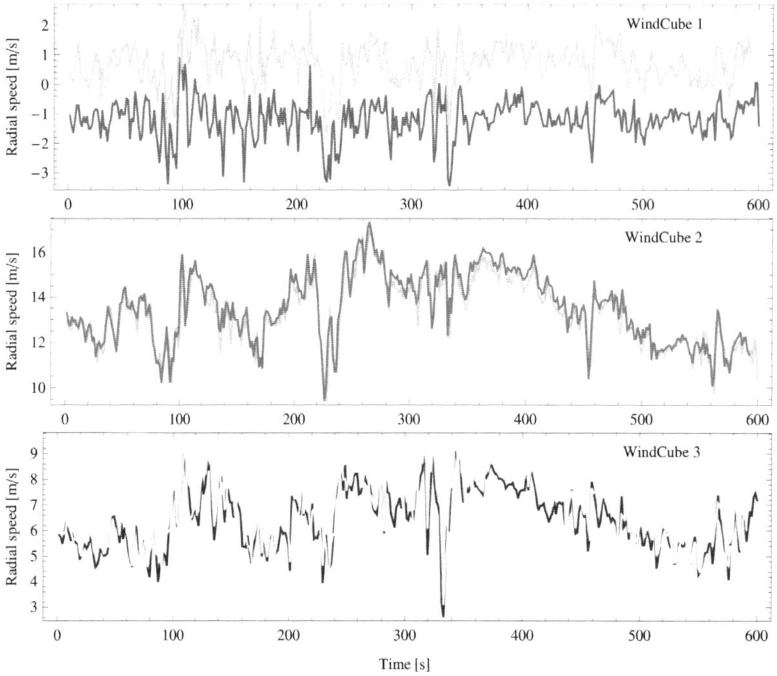

Figure 3. A ten minute period of wind data (Dec 8. 02:50 – 03:00). The red curve is the three-dimensional sonic data projected onto the direction of the beam of WindCube 1. The data from this lidar are shown in cyan. Likewise, the projects sonic speed and the corresponding WindCube 2 speed is shown in green and magenta. For WindCube 3 it is blue and .

(see figure 4)

$$\varphi(s) = \begin{cases} \frac{l-|s|}{l^2} & \text{for } |s| < l \\ 0 & \text{elsewhere} \end{cases} \quad (2)$$

which is valid if the Doppler frequency is determined as the first moment of the signal spectrum (with the background subtracted appropriately) [2]. However, not always is the Doppler shift determined in this way and there has been other suggestions. One of these, which is thought to have relevance for the way the radial velocity is determined from the WindCube lidar is [3]

$$\varphi(s) = \begin{cases} \frac{3(l-|s|)^2}{2l^3} & \text{for } |s| < l \\ 0 & \text{elsewhere} \end{cases} \quad (3)$$

which is displayed in figure 4. We ignore any effect of focusing of the pulsed lidar, which in a more detailed analysis should be taken into account [3]. In the case of the WindCube $l = 30$ m. The Fourier transforms of these weighting functions, which we are going to use later, are

$$\varphi(k) = \begin{cases} \operatorname{sinc}^2(kl/2) & \text{weighting function (2)} \\ \frac{6}{k^2l^2}\left(1 - \operatorname{sinc}(kl)\right) & \text{weighting function (3)} \end{cases}, \quad (4)$$

where $\operatorname{sinc}(x) \equiv (\sin x)/x$.

We are now interested in the spectrum of v and how it is influenced by the averaging in (1). It is clear that if the filter length l is sufficiently small the velocity measured by lidar would

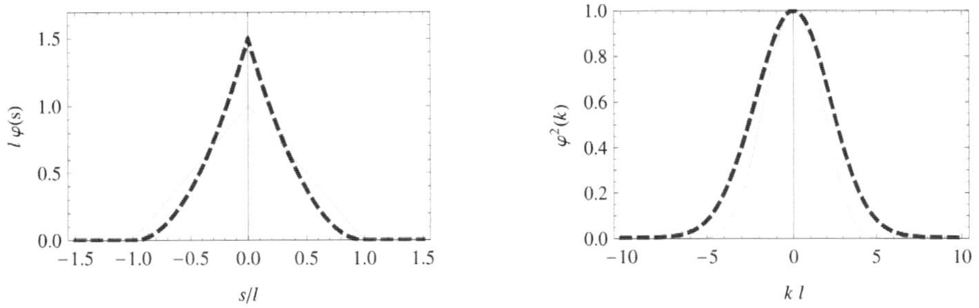

Figure 4. The weighting function $\varphi(s)$ (left) and the square of its Fourier transform $\varphi^2(k)$ (right). The solid, green curve is (2) and the thick, dashed blue curve is (3).

simply be $\boldsymbol{n} \cdot \boldsymbol{u}$ and the spectrum could be expressed as a combination of the unfiltered one-point spectra of the velocity.

The correlation function of v is

$$\begin{aligned} R_v(x_1) &= \langle v(\boldsymbol{x})v(\boldsymbol{x}+x_1\boldsymbol{e}_1)\rangle \\ &= \left\langle \int\int \varphi(s)\varphi(s')n_iu_i(s\boldsymbol{n}+\boldsymbol{x})n_ju_j(s'\boldsymbol{n}+\boldsymbol{x}+x_1\boldsymbol{e}_1)dsds'\right\rangle \\ &= n_in_j\int\int \varphi(s)\varphi(s')R_{ij}\left((s'-s)\boldsymbol{n}+x_1\boldsymbol{e}_1\right)dsds' \end{aligned} \quad (5)$$

where R_{ij} is the correlation tensor of the velocity field [4] and where summation over repeated indices is assumed. The spectrum of v can therefore be expressed as

$$\begin{aligned} F_v(k_1) &= \frac{1}{2\pi}\int \langle v(\boldsymbol{x})v(\boldsymbol{x}+x_1\boldsymbol{e}_1)\rangle\, \mathrm{e}^{-\mathrm{i}k_1x_1}dx_1 \\ &= n_in_j\int\int \varphi(s)\varphi(s')\int\int \Phi_{ij}(\boldsymbol{k})\exp\left(\mathrm{i}\boldsymbol{k}\cdot\boldsymbol{n}(s'-s)\right)dk_2dk_3dsds' \\ &= n_in_j\int\int |\varphi(\boldsymbol{k}\cdot\boldsymbol{n})|^2\,\Phi_{ij}(\boldsymbol{k})dk_2dk_3 \ . \end{aligned} \quad (6)$$

It is important to point out that it is not enough to know the second order turbulence statistics at the height of the sonic anemometer, in order to calculate the lidar spectra. We need three-dimensional, spatial statistics, or, in other words, the spectral tensor $\Phi_{ij}(\boldsymbol{k})$ [4], as seen from (6). It is difficult to measure directly, but fortunately, the spectral tensor model by [5] allows us to estimate $\Phi(\boldsymbol{k})$ through measurements from a sonic. This is done by fitting three parameters $\alpha\varepsilon^{2/3}$, L and Γ to the three one-dimensional spectra measured by the sonic. Here α is the spectral Kolomogorov constant [4], ε is the turbulent energy dissipation, L is a length scale proportional to the energy containing eddies in the turbulence, and Γ is a parameter describing the degree of anisotropy of the turbulence [5]. For the time period indicated by gray in figure 2 we get $\alpha\varepsilon^{2/3} = 0.072\ \mathrm{m}^{2/3}\mathrm{s}^{-1}$, $L = 42.5$ m and $\Gamma = 3.9$, and the fitted and measured spectra are shown in figure 5.

With these values we are now able to evaluate (6) and estimate theoretically the ratio between the velocity spectrum measured by the lidar and the sonic. To the left in figure 6 spectra from both instruments are shown and their ratio and theoretical expectations are shown to the right. In addition to (6) the theoretical ratios have been multiplied with $\mathrm{sinc}^2(k_1l/2)$ where $l = t_a\mathrm{s}U$ with $t_a = 0.5$ s, because the WindCube spectra are obtained by averaging over half a second.

What seems a bit unexpected is the ratio of WindCube 1 and the vertical component spectra at low frequencies, where it is significantly larger than one. At low frequencies the spectral

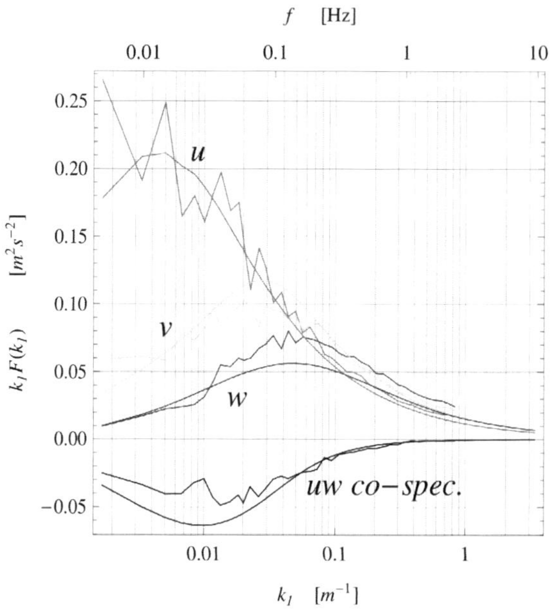

Figure 5. Measured and modeled spectra for the time period indicated by gray in figure 2. The u-spectrum is red, v green and w blue. The co-spectrum of u and w is shown in black.

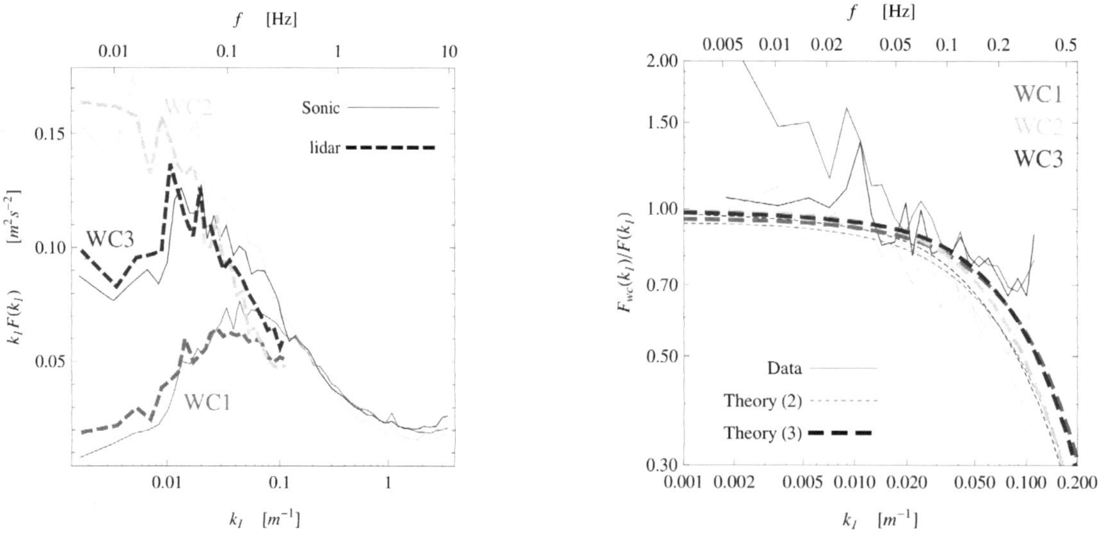

Figure 6. Left: Spectra measured by the three lidars and the spectra of the corresponding component of the sonic velocity vector. Right:The ratio of lidar spectra and sonic spectra and the corresponding theoretical predictions.

energy content of the vertical component is much lower than that of the two horizontal, as seen from figure 5. Therefore, if the lidar WindCube 1 is not pointing completely vertical, the spectrum may be influenced by the horizontal component at these low frequencies. Apart from that the spectrum ratios seem to follow quite well the theoretical expectations. It is predicted and observed that the spectrum from WindCube 2 is attenuated more than the other lidars for the particular wind direction used in the analysis. The statistics is, however, not good enough to distinguish between the filter functions (2) and (3). Close to the Nyquist frequency of the lidar measurements the spectra seem to rise probably due to noise.

4. Conclusion

The times series of the speed measured by the lidar and the corresponding component of the sonic agree very well down to time scales of the order of two seconds. The variance of the velocity measured by the lidar is attenuated due to spatial filtering, and the amount of attenuation can be predicted theoretically, however, not well enough to distinguish two different proposals for the spatial filter function.

Only a fraction of the data have been analyzed so far, and in the future we would like to do a more careful analysis where the focusing of the laser beam is also taken into account.

We would like to thank technicians at the test station at Høvsøre, Bjarne Sønderskov and Anders Ramsing Vestergaard, for their practical help during the experiment.

References

[1] Panofsky H A and Dutton J A 1984 *Atmospheric Turbulence* (New York: John Wiley & Sons)
[2] Frehlich R G and Yadlowsky M J 1994 **11** 1217–1230
[3] Lindelöw P 2008 *Fiber Based Coherent Lidars for Remote Wind Sensing* Ph.D. thesis Danish Technical University
[4] Pope S B 2000 *Turbulent Flows* (Cambridge University Press)
[5] Mann J 1994 *J. Fluid Mech.* **273** 141–168

IOP Conf. Series: Earth and Environmental Science **1** (2008) 012013 doi:10.1088/1755-1307/1/1/012013

Propagation of laser-induced shock waves in the atmosphere

N P Krasnenko,[1] S V Shamanaev[2] and L G Shamanaeva[2]

[1]Institute of Monitoring of Climatic and Ecological Systems of the SB RAS, 10/3, Akademicheskii Pr., 634055 Tomsk, Russia
[2]Institute of Atmospheric Optics of the SB RAS, 1, Akademicheskii Pr., 634055 Tomsk, Russia,

E-mail: krasnenko@imces.ru, shaman@iao.ru, sima@iao.ru

Abstract. Laser-induced shock waves in the atmosphere are unique sources of high-power wideband acoustic radiation that can be recommended for investigation of wideband high-power acoustic signal propagation in the atmosphere. The present report is aimed at a comparison of waveforms and spectra of laser-induced acoustic shock waves generated by single liquid and solid aerosol particles, monodisperse liquid and solid aerosol particle ensembles, and outdoor atmospheric aerosol and elucidation of their special features. Results of several measurement runs performed under laboratory conditions and in the outdoor atmosphere in winter and summer are discussed.

1. Introduction

Results of theoretical end experimental investigations of propagation of acoustic shock waves induced by focusing high-power pulsed laser beams into the atmosphere are presented in the report. Several measurement runs were performed under laboratory conditions with a pulsed GOS-1001 neodymium glass laser (λ = 1.06 μm, E_{max1} = 1 kJ, τ_{L1} = 1 ms) [1–5] and a pulsed CO_2 laser (λ = 10.6 μm, E_{max2} = 5 J, τ_{L2} = 3 μs) [6] and in the outdoor atmosphere with pulsed CO_2 lasers (λ = 10.6 μm, E_{max3} = 500 J, τ_{L3} = 0.3 μs, E_{max4} = 2kJ, and τ_{L4} = 1 μs) [7–11]. Amplitudes, waveforms and spectra of acoustic pulses from laser breakdowns initiated by a single liquid or solid aerosol particle or an ensemble of mono- or polydisperse particles were investigated. The present report considers the special features of waveforms and spectra of acoustic pulses induced by laser sparks compared with those of acoustic signals induced by laser radiation propagating in the atmosphere in the underbreakdown (thermooptical) regime [12] and with the corresponding parameters of shock waves induced by laser sparks generated by focusing Nd: YAG laser radiation into air in [13] under laboratory conditions.

2. Acoustic signals from laser breakdown initiated on liquid-droplet aerosols

Propagation of high-power pulsed laser radiation in the atmosphere is accompanied by sound generation. Mechanisms of laser generation of sound depend on the energetic parameters of acting laser radiation (its wavelength, energy density, pulse duration) and meteorological conditions. They include thermal self-action caused by laser radiation absorption on atmospheric aerosol and gaseous components, evaporation, and low-threshold optical breakdown on aerosol particles. The low-threshold optical breakdown in the atmosphere is caused by the presence of liquid-droplet and solid aerosol particles that intensively absorb laser radiation and initiate local plasma formations at laser radiation intensities one-two orders of magnitude smaller than in pure air [14].

In this section, we consider evaporation and breakdown mechanisms of sound generation by liquid-droplet aerosols of different microphysical compositions upon exposure to CO_2-laser radiation. With increase in the laser energy density, surface evaporation of liquid-droplet aerosol particles is followed by explosive evaporation of the entire particle volume and subsequent optical breakdown. We performed laboratory measurements with artificial monodisperse water-droplet aerosols and field measurements in the presence of water-droplet atmospheric aerosols.

A laboratory setup [15] comprised a transversely excited atmospheric pressure CO_2 pulsed laser with energy per pulse $E_L = 5$ J and pulse duration $\tau_L = 3$ μs. Laser radiation was focused by a spherical mirror with a 60-cm focal length. The beam cross-sectional area in the caustic region (by definition, its longitudinal coordinates are ends of the straight line segment into which a point source is imaged by the optical system) was 4 mm^2. Water-droplet aerosol particles with sizes of 2.7 μm generated by an ultrasonic aerosol particle generator were injected into the caustic region in the direction perpendicular to the laser beam axis. The concentration of droplets was determined from the aerosol extinction coefficient measured with the use of a cw ruby laser with a wavelength of 0.69 μm. As is well known, the water aerosol extinction coefficient depends on the water particle sizes and concentration. From the measured ratio of the transmitted to received ruby laser radiation power, the aerosol extinction coefficient was calculated with the use of the Bouguer law, and the particle concentration was then calculated for the given particle sizes. Acoustic signals were recorded by microphones 7 mm in diameter in the frequency range 20 Hz–0.1 MHz and precision impulse sound level meters.

Field measurements were performed with the same laser source. Laser radiation was focused by a mirror Cassegrain telescope into the atmosphere at distances of 50–150 m.

Figure 2.1 shows the dependence of the acoustic pressure level on the laser energy density measured in the laboratory experiment for three indicated water-droplet particle concentrations. Bends corresponding to the threshold of explosive vaporization of water droplets at $Q_{e.v} \cong 2$ J/cm^2 can clearly be seen in the curves.

The results obtained were interpreted based on a description of the acoustic signal waveform generated by a single water droplet by an N-wave whose energy can be expressed as follows:

$$W_N = V_0 P^2 / \rho_0 c_0^2 , \tag{2.1}$$

where V_0 is the interaction volume, P is the excess pressure inside the energy liberation zone, ρ_0 is the air density, and c_0 is the velocity of sound. The acoustic energy was then calculated from the formula

$$W_{ac} = 4\pi r^2 P_p^2 \tau / \rho_0 c_0 , \tag{2.2}$$

where r is the distance from the receiving microphone to the center of the interaction volume, τ is the acoustic pulse duration including the compression (τ_+) and rarefaction (τ_-) periods, and the peak acoustic pressure level P_p is proportional to the P value determined by the total mass of the liquid evaporated before the laser pulse termination and by the air temperature increase due to heating by hot droplets and water vapor. The excess pressure P can be found from the state equation for an air-vapor mixture:

$$P = R_{air}\rho_0 (T - T_0) + R_{vap} X q_0 T , \tag{2.3}$$

where T_0 and T are the initial and final temperatures of the medium; R_{air} and R_{vap} are the gas constants for air and water vapor, respectively; and $X = 1 - q / q_0$ is the degree of water droplet vaporization specified by the ratio of the initial (q_0) and final water content (q) of the aerosol by the time of laser pulse termination.

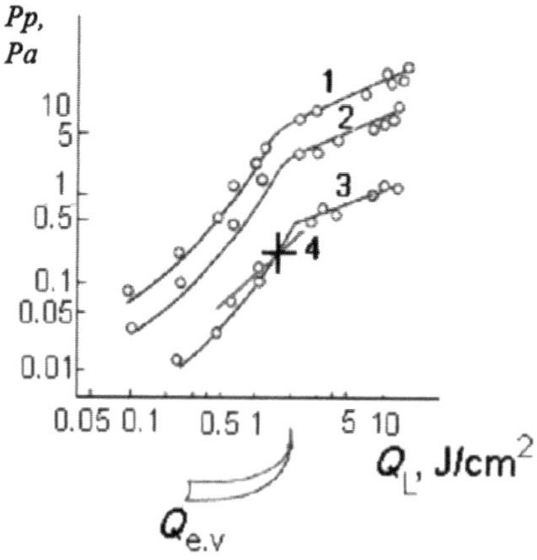

Figure 2.1. Dependences of the acoustic pressure level P on the laser energy density Q_L measured in the laboratory experiment with monodisperse water-droplet aerosol (curves 1–3) and in the outdoor atmosphere (red curve 4, summer haze conditions). Particle concentration $N = 1.2 \cdot 10^6$ (curve 1), $6 \cdot 10^5$ (curve 2), and $1.2 \cdot 10^5$ (curve 3).

For the degree of water droplet vaporization, we have

$$X = \begin{cases} 1 - \exp(-\beta Q_L), & Q_L < Q_{e.v}, \\ 1 - \exp(-\beta Q_{e.v})(1 - X_{e.v})\exp[-\beta_1 (Q_L - Q_{e.v})], & Q_L \geq Q_{e.v}, \end{cases} \quad (2.4)$$

where β specifies the droplet vaporization efficiency, β_1 is the energy parameter describing the vaporization efficiency of initial droplet fragmentation products, and $X_{e.v}$ is the degree of explosive vaporization. Based on our experimental data, we can write $W_{ac} = \beta_{al} W_{las}$, where the coefficient of transformation of laser energy into acoustic one is $\beta_{al} = (1.6–6.8) \cdot 10^{-7}$.

3. Acoustic signals from laser breakdown initiated on solid aerosol particles

Experiments on the study of propagation of high-power pulsed laser radiation with wavelengths in the range 1–10 μm in the atmosphere comprising solid aerosol particles systematized in [14] demonstrate that when the laser radiation density exceeded 6 J/cm^2, solid aerosol particles with sizes exceeding the critical size a_{cr}, initiate local spherical plasma formations. According to [14], $a_{cr} \approx (A/I)^{1/g}$, where I is the laser radiation intensity, and $A \approx 2.9 \cdot 10^9$ W/(cm·μm$^{1.5}$) and $g = 1-2$ are empirical constants. In the laser wavelength range 1–11 μm, a_{cr} depends only weakly on the material and shape of aerosol particles. Each plasma formation generates a spherical acoustic pulse whose waveform reproduces a blast N-wave. Acoustic signal from laser sparks is a linear superposition of pulses from individual luminous plasma formations. In section 3.1, characteristics of acoustic signals from laser breakdown initiated on a single solid aerosol particle are studied theoretically and experimentally, and in section 3.2, acoustic signals from laser breakdown initiated on an ensemble of monodisperse solid aerosol particles are considered.

3.1. Characteristics of acoustic signals from laser breakdown initiated on a single solid aerosol particle

The problem of sound generation by a sphere with a high pressure surrounded by an infinite region of homogeneous pressure was solved in [16] based on the equation of sound wave propagation which with allowance for the spherical symmetry of the problem assumes the form

$$\frac{\partial^2}{c^2\partial t^2}(r\varphi)=\frac{\partial^2}{\partial r^2}(r\varphi). \tag{3.1}$$

Here φ is the velocity potential depending on the distance to the observation point r and time t and related to the sound wave velocity components $u_i\ (i=1,2,3)$ and pressure P by the well-known expressions

$$u_i=\frac{\partial\varphi}{\partial x_i}, \tag{3.2}$$

$$P=-\rho\frac{\partial\varphi}{\partial t}, \tag{3.3}$$

and c is the velocity of sound. At the initial moment $t=0$, the system is at rest. The interface between the regions is fixed. At $t>0$, the interface is removed. Then the excess acoustic pressure p will be

$$p=-\frac{\rho\partial\varphi}{\partial t}. \tag{3.4}$$

At the initial moment, it is equal to p_0 for $r<r_p$, where r_p is the radius of the high-pressure region (in our case, plasma formation). With these initial conditions, a solution of (3.1) for the sound pressure at the observation point has the form [16]

$$p=-\frac{p_0}{2r}\left[(ct-r)H\left(t-\frac{r-r_p}{c}\right)-(ct-r)H\left(t-\frac{r+r_p}{c}\right)\right], \tag{3.5}$$

where $H(t)$ is a step function. That is, for the given $r>r_p$, the sound pressure is equal to zero before $t_1=(r-r_p)/c$, when the initial wave arrives from the compression region, and after $t_2=(r+r_p)/c$, when the wave from the most distant point of the compression region arrives. For $(r-r_p)/c<t<(r+r_p)/c$, the sound pressure decreases linearly with time:

$$p(t)=\begin{cases}0, & t<t_1,\\ \dfrac{p_0(r-ct)}{2r}, & t_1\le t\le t_2,\\ 0, & t>t_2.\end{cases} \tag{3.6}$$

Thus, the optical breakdown initiated on a solid aerosol particle induces an acoustic pulse shaped as an N-wave. The energy spectrum of this pulse can be written as

$$S(\omega)=\left|\int_{-\infty}^{\infty}p(t)e^{-i\omega t}dt\right|^2=\left|\frac{p_0}{2}\int_{t_1}^{t_2}e^{-i\omega t}dt-\frac{p_0}{2r}\int_{t_1}^{t_2}te^{-i\omega t}dt\right|^2=I_1+I_2, \tag{3.7}$$

where $\omega=2\pi f$, and f is the sound frequency.

The first integral is reduced to the tabular one [17]

$$\int e^{ax}dx=e^{ax}, \tag{3.8}$$

where $a=-i\omega$ and $x=t$, that is,

$$I_1=\frac{r}{2(-i\omega)}e^{-i\omega t}\Bigg|_{\frac{r-r_p}{c}}^{\frac{r+r_p}{c}}=\frac{r}{\omega}e^{-i\omega r/c}\sin\frac{\omega r_p}{c}. \tag{3.9}$$

The second integral is also reduced to the tabular one of the form [17]

$$\int xe^{ax}dx=e^{ax}\left(\frac{x}{a}-\frac{1}{a^2}\right). \tag{3.10}$$

Then

$$I_2 = \frac{c}{2} e^{-i\omega t} \left(\frac{1}{\omega^2} - \frac{x}{i\omega} \right) \Bigg|_{t_2}^{t_1} = ce^{-i\omega r/c} \left[\left(\frac{r}{\omega c} - \frac{1}{\omega^2} \right) \sin \frac{\omega r_p}{c} + \frac{ir_p}{\omega c} \cos \frac{\omega r_p}{c} \right]. \tag{3.11}$$

From (3.9) and (3.11), we have

$$S(\omega) = \frac{p_0^2}{r^2 \omega^2} \left| \frac{c^2}{\omega^2} \sin^2 \frac{\omega r_p}{c} + \frac{ir_p^2}{c^2} \cos^2 \frac{\omega r_p}{c} \right|. \tag{3.12}$$

The second term in (3.12) can be neglected, because in our case, the inequality $c/\omega >> r_p/c$ is always satisfied. We finally obtain

$$S(\omega) = \frac{p_0^2 c^2}{r^2 \omega^4} \sin^2 \frac{\omega r_p}{c}. \tag{3.13}$$

Hence, the spectrum of acoustic signal of a single spherical plasma formation has a maximum at the frequency ω_{max} for which

$$\sin \frac{\omega_{max} r_p}{c} = 1, \tag{3.14}$$

which corresponds to

$$f_{max} = \frac{\omega_{max}}{2\pi} = \frac{c}{4r_p} = \frac{1}{2\tau_p} = \frac{1}{4\tau_+}. \tag{3.15}$$

Here $\tau_p = 2r_p/c$ is the duration of the acoustic pulse generated by the local plasma formation and $\tau_+ = \tau_p/2$ is the duration of its positive half-period. That is, the spectrum of such pulse has a maximum at the frequency inversely proportional to the radius of the spherical plasma formation.

Experimental investigations were performed using the experimental setup described in detail in [5]. A laser breakdown was initiated by focusing radiation of a GOS-1001 laser ($\lambda = 1.06$ μm, $E_{Lmax} = 1$ kJ, and $\tau_L = 1$ms) with a short-focus lens at a distance of 25–50 cm. The focal beam diameter was 7 mm. A single aerosol particle initiating a plasma formation was suspended on the point of a thin metal needle in the beam focus. Acoustic signal was received by a microphone placed at a distance of 15 cm under the laser beam caustic. Figure 3.1 shows the waveform of the acoustic signal induced by optical breakdown on a 400-μm SiO_2 particle, and figure 3.2 shows its spectrum measured with a Robotron parallel spectrum analyzer. It can be seen that the acoustic signal waveform is well described by *N*-wave (3.6). This is also confirmed by laboratory experiments [13] in which radiation of a Q-

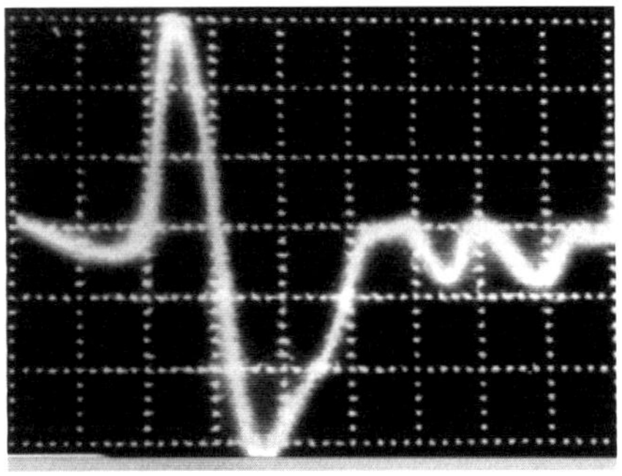

Figure 3.1. Waveform of the acoustic signal induced by an optical breakdown on a 400-μm SiO_2 particle. The abscissa scale division is 50 μs, and the ordinate scale division is 1 V.

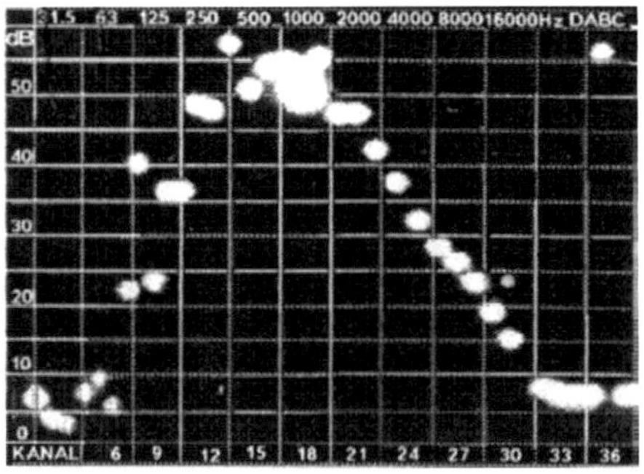

Figure 3.2. Spectrum of the acoustic signal shown in figure 3.1 measured with a Robotron spectrum analyzer.

switch Surelite III-10 Nd:YAG laser ($\lambda = 1.064$ μm) was focused at a distance of 10 cm into air, a breakdown was initiated, and a sound pulse from the laser breakdown was registered.

The acoustic signal spectrum shown in figure 3.2 is relatively smooth and has a wide maximum. Our experiments demonstrated that the peak acoustic pressure level changed from 45 to 58 dB and the width of these spectra at half maximum changed from 2 to 6 kHz. With increasing particle sizes, the maximum of the spectrum was displaced toward lower frequencies, as predicted by (3.15).

3.2. Characteristics of acoustic signals from optical breakdown initiated on an ensemble of monodisperse solid particles

In this case, a laser spark was initiated by an ensemble of solid monodisperse particles injected into the laser beam caustic region. The time history of an acoustic signal from laser breakdown initiated on an ensemble of monodisperse SiC particles with sizes of 7 μm is shown in figure 3.3. Our measurements confirm that each plasma formation generates a spherical sound pulse, and the total acoustic signal from the laser spark represents their linear superposition.

Since the distribution of monodisperse absorbing aerosol particles initiating plasma formations has a random nature, this causes time realizations of acoustic signals also to be random. The time histories of acoustic signals were digitized with a step of $\Delta t = 0.04$ ms, and their spectra were calculated with the use of the discrete Fourier transform from the formula

$$S\left(\frac{k}{T}\right) = \frac{1}{T}\left|\sum_{i=0}^{N-1} p(i\Delta t)e^{-j(2\pi ki/N)}\right|^2, \tag{3.16}$$

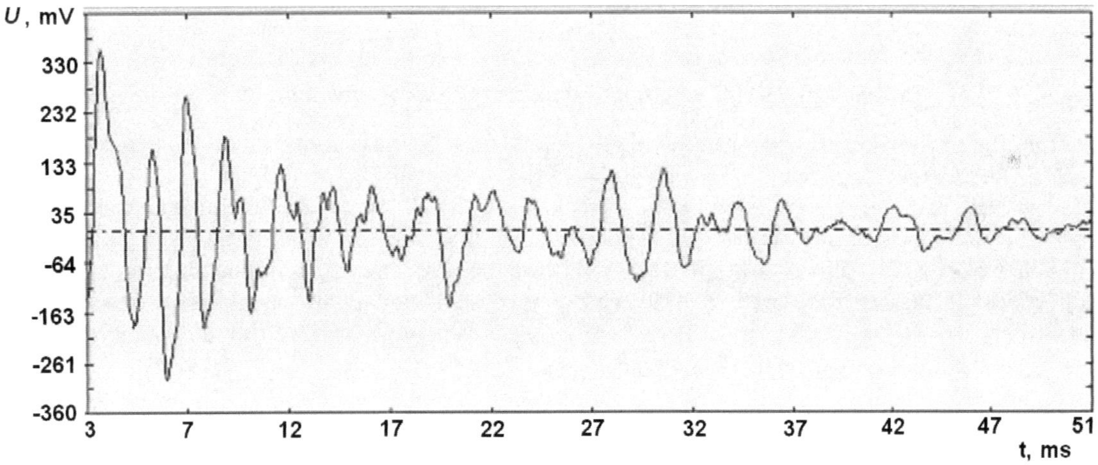

Figure 3.3. Time history of the acoustic signal induced by a laser breakdown initiated on an ensemble of monodisperse SiC particles with sizes of 7 μm.

where $i = 0,1,\ldots,N-1$, $N = \tau_{ac}/\Delta t$, and $k = 0,1,\ldots,N/2$.

Figure 3.4 shows the spectrum of the acoustic signal shown in figure 3.3. To increase the stability of the spectral estimate, it was smoothed with a spectral window. Here logarithm of *S*(*f*), in (ADC codes)^2/Hz, is plotted on the ordinate, and logarithm of *f*, in Hz, is plotted on the abscissa. Our measurements demonstrate that each individual plasma formation induces a single spherical sound pulse. The total acoustic signal from the breakdown initiated on an ensemble of monodisperse solid aerosol particles is their linear superposition.

Because monodisperse absorbing particles initiating plasma formations are randomly distributed in the laser beam caustic region, this causes acoustic signal time realizations to be also random.

The energy spectrum of a large number of pulses with random amplitudes p_{ac} and preset shapes $p(t)$ can be represented as [18, 19]

$$S_{sp}(\omega) = 2\pi N \left\langle |p_{ac}|^2 \right\rangle S(\omega), \tag{3.17}$$

where N is the number of pulses in the examined realization, $S(\omega)$ is the Fourier transform of the pulse shape (in our case, the N-wave, see (3.6)), and angular brackets denote ensemble averaging. From (3.17) it follows that the total spectrum has a maximum whose amplitude and width depend on the number of pulses and their average amplitude. As follows from (3.13) and (3.15), the frequency of the maximum is inversely proportional to the mean plasma formation radius.

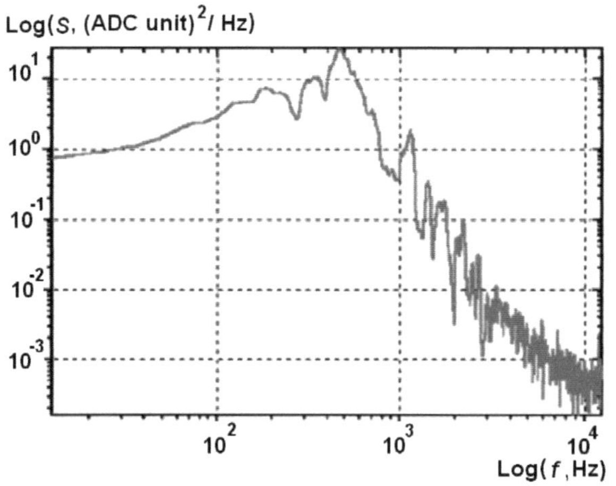

Figure 3.4. Calculated spectrum of the acoustic signal shown in figure 3.3.

From figure 3.4 it can be seen that the calculated spectrum has a wide maximum with an amplitude of 27.1 (ADC units)^2/Hz at a frequency of 469 Hz. Our analysis performed in [3, 4] also has demonstrated that the spectra of acoustic signals from breakdown initiated on an ensemble of monodiaperse solid aerosol particle are determined by the waveform of the acoustic pulse from individual plasma formation. In [4] it was found that the frequency of this maximum is inversely proportional to the average duration of acoustic pulses from individual plasma formations directly proportional to the mean size of particles initiating plasma formations with the proportionality coefficient determined by the particle material.

4. Acoustic signals from laser sparks in the atmosphere

This section presents the characteristics of acoustic signals from laser sparks initiated by laser radiation focusing in the outdoor atmosphere. In the first measurement run, the spark was initiated by focusing of radiation of an electroionization pulsed CO_2 laser (λ = 10.6 μm, E_{max} = 500 J, τ_L = 0.3 μs) into the atmosphere at distances of 50–150 m from the source. The laser energy density in the caustic region was varied from 9 to 15 J/cm^2. The spark length varied from 3 to 11 m. The receiving microphone was placed under the laser beam caustic at distances of 20cm –4 m from the laser beam axis. In the second measurement run, the spark was initiated by focusing of radiation of a pulsed CO_2 laser (λ = 10.6 μm, E_{max} = 2 kJ, τ_L = 1 μs) at distances of 100, 250, and 500 m into the atmosphere. The receiving microphone was placed under the laser beam caustic at distances of 20–50 cm from the laser beam axis. The acting laser energy density was varied from 8 to 20 J/cm^2.

Figure 4.1 shows a photograph of the laser spark initiated by laser radiation focusing at a distance of 120 m into the spot with a cross-sectional area of 10 cm^2. It can be seen that the laser spark is an aggregate of individual spherical luminous plasma formations initiated by aerosol particles that intensively absorb laser radiation. The laser spark length during measurements was varied from 0.5 to

20 m. The time history of the acoustic signal from this laser spark is shown in Fig. 4.2. An electric analog of the acoustic pressure, in mV, is plotted on the ordinate. N-waves with equal amplitudes of positive and negative half-periods generated by individual plasma formations are clearly seen. Experimental dependences of the peak sound pressure level on the acting laser energy density are shown in Fig. 4.3. Unlike the corresponding dependences for water-droplet aerosols shown in figure 2.1, they are well approximated by straight lines whose slope depends on the wind velocity *V*, namely, it decreases twice as *V* increases from 0 to 2 m/s.

Figure 4.1. Photograph of a laser spark initiated by focusing of CO_2 laser radiation into the atmosphere.

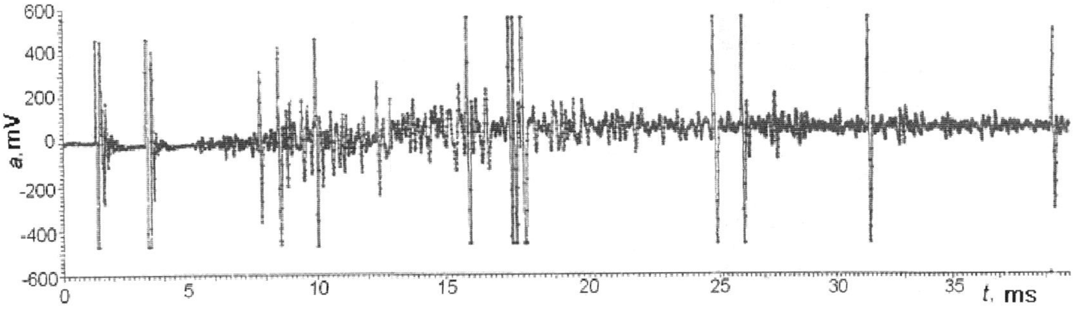

Figure 4.2. Time history of the acoustic signal from the laser spark shown in figure 4.1.

The spectra of acoustic signals for the second measurement run are shown by curves 1-3 in figure 4.4. The dashed curve here shows a spectrum of explosives borrowed from [20]. The vertical bars indicate a 80% confidence level for the calculated spectra. A minimum and a high-frequency maximum are clearly pronounced in all spectra. The presence of the minimum we explain by the interference of acoustic radiation from individual plasma formations. Its frequency correlated with the average distance between plasma formations, that is, with their concentration per unit length.

As demonstrated in [2], the frequency of the maximum is inversely proportional to the average radius of plasma formations, and its amplitude and width depend on the number of acoustic pulses and their average amplitude. The low-frequency part of the spectra is analogous to that of the explosive. The analogy between acoustic pulses from explosions and laser-induced acoustic sparks was also pointed out in [13]. During our measurements, the frequency of the maximum was in the range 2–5 kHz, its amplitude varied 20 times, and its half-width changed by a factor of 2.5. It can be seen that with increase in the distance to the receiving microphone, the amplitude of the high-frequency maximum decreases. This is most likely due to stronger absorption of acoustic radiation at higher frequencies.

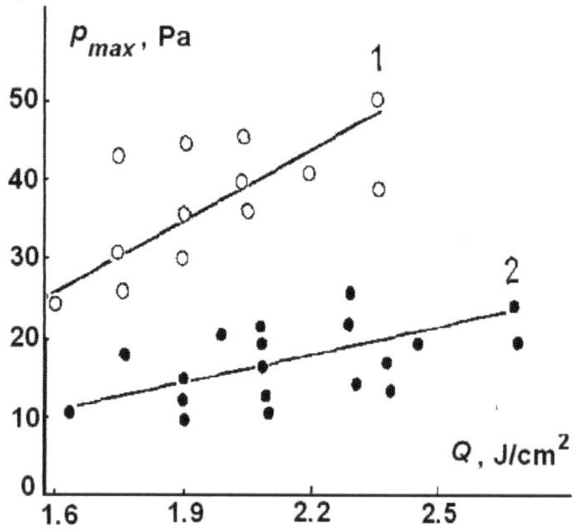

Figure 4.3. Peak sound pressure vs. the acting laser energy density recorded in summer at the air temperature $T = 19°C$, relative air humidity $u = 88\%$, and wind velocity $V = 0$ (1) and 2 m/s (2).

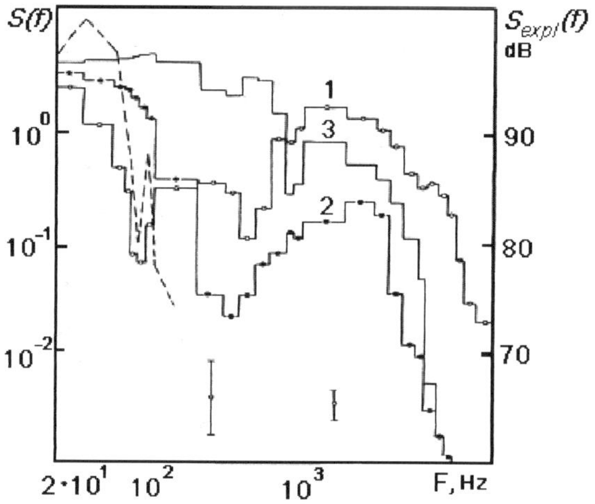

Figure 4.4. Histograms of spectra of acoustic signals induced by laser sparks in the atmosphere for $r = 20$ cm (1), 2 m (2), and 4 m (3), $W_{ac} = 1.18\cdot10^{-5}$ (1) and $2.9\cdot10^{-8}$ (2 and 3), and $T = -27.5$ (1) and $-11°C$ (2 and 3). Curve 4 shows the spectrum of an explosive borrowed from [20].

5. Conclusions

Results of theoretical end experimental investigations of propagation of acoustic shock waves from laser breakdowns initiated by focusing of high-power pulsed laser beams into the atmosphere have been presented in the report. Several measurement runs were performed under laboratory conditions and in the outdoor atmosphere in winter and summer. They have demonstrated that the optical breakdown is a unique source of high-power wideband acoustic pulses that can be recommended for investigations of wideband acoustic signal propagation in the atmospheric boundary layer. Our estimates showed that the laser-induced acoustic shock waves can be reliably recorded with a microphone placed at distances up to 1 km from the laser spark. The spectral range of laser-induced shock waves cannot be obtained by other sources of acoustic radiation. It was established that the dependence of the peak pressure level from laser breakdowns initiated by solid aerosol particles on the acting laser energy density is well described by straight lines whose slope is determined by the meteorological conditions. The corresponding dependence for the liquid-water aerosols bends at the laser energy density corresponding to the threshold of explosive vaporization of water-droplet aerosols. The laser breakdown initiated on a single aerosol particle generates an acoustic pulse whose waveform reproduces a blast N-wave. This allows us to estimate the aerosol particle concentration from the number of acoustic pulses in acoustic signals received from laser sparks. The spectra of acoustic pulses have wide asymmetric maxima at frequencies proportional to the size of particle initiating laser breakdown. This allows the spectral analysis of acoustic signals to be used for real-time diagnostics of laser spark parameters. The energy of laser-induced shock pulses was in the range 10^{-8}–10^{-3} J depending on the acting laser energy density. Acoustic signals from laser sparks are a linear superposition of acoustic pulses from individual spherical plasma formations. Their spectra have a strong maximum (whose frequency is determined by the average plasma formation size) and a minimum (whose frequency correlates with the concentration of local plasma formations). The amplitude of the high-frequency maximum decreases with increasing distance to the receiving microphone.

References

[1] Shamanaev S V 1998 *Opt. Atmos. Okeana* **11** 1352–1355
[2] Shamanaev S V and Shamanaeva L G 2005 *Russ. Phys. J.* 11 74–80
[3] Shamanaev S V 2000 *Proc. SPIE* **4341** 225–230
[4] Shamanaev S V 2000 *Proc. SPIE* **4341** 231–236
[5] Shamanaev S V 2001 *Russ. Phys. J.* **44** 1256–1262
[6] Bochkarev N N, Zuev V E, and Krasnenko N P 1988 *Lasers – Physics and applications: Proc. 5th Int. Schools on Quantum Electronics* ed A Y Spasov (Bulgaria: Institute of Electronics Bulgarian Academy of Sciences) 818–821
[7] Bochkarev N N, Krasnenko N P, and Sorokin Yu M 1990 *Opt. Atmos.* **3** 563–678
[8] Belyaev E B, Godlevskii A P, Kopytin Yu D, Krasnenko N P, Muravskii V P and Shamanaeva L G 1982 *Pis'ma Zh. Tekh. Fiz.* **8** 333–337
[9] Shamanaeva L 1996 *Proc. 8th Int. Symp. Acoustic Remote Sensing and Associated Techniques of the Atmosphere and Oceans* (Moscow: Institute of Atmospheric Physics of the Russian Academy of Sciences) pp1.35–1.40
[10] Shamanaev S V and Shamanaeva L G 2004 *Proc. 12th Int. Symp. Acoustic Remote Sensing of the Atmosphere and Oceans and Associated Techniques* (Cambridge: Clare College) pp 145–148
[11] Shamanaev S V and Shamanaeva L G 2007 *Russ. Phys. J.* 12 26–31
[12] Shamanaeva L G 2001 *Russ. Phys. J.* 11 49–55
[13] Attenborough K and Qin Q 2007 *Proc. 19th Int. Congr. Acoust.* (Madrid) 5 pp
[14] Kopytin Yu D, Protasevich E T, Chistyakova L K and Shishkovskii V I 1992 *Air Medium upon Exposure to Laser and High-Frequency Radiation* (Novosibirsk: Nauka)
[15] Bochkarev N N, Geints Yu E, Zemlyanov A A, Kabanov A M, Krasnenko N P 1988 *Opt. Atmos.* **1**, 111–112
[16] Jeffris G and Swirls B 1970 *Methods of Mathematical Physics* (Moscow: Mir)
[17] Gradshtein I S and Ryzhik I M 1971 *Tables of Integrals, Sums, Series and Products* (Moscow: Science)
[18] Rytov S M 1976 *Random Processes* (Moscow: Nauka)
[19] Rytov S M,Kravtsov Yu A and Tatarskii V I 1978 *Random Fields* (Moscow: Nauka)
[20] Schomer P D, Goff R J and Little L M 1978 *J. Acoust. Soc. Am.* **63** 1431–1443

IOP Conf. Series: Earth and Environmental Science **1** (2008) 012014 doi:10.1088/1755-1307/1/1/012014

SUMO: A Small Unmanned Meteorological Observer for atmospheric boundary layer research

J. Reuder[1], P. Brisset[2], M. Jonassen[1], M. Müller[3], and S. Mayer[1]

[1] Geophysical Institute, University of Bergen, Allegaten 70, 5009 Bergen, Norway
[2] Ecole Nationale de l'Aviation Civile (ENAC), 7 avenue Edouard Belin, 31055 Toulouse, France
[3] Orleansstrasse 26a, 31135 Hildesheim, Germany

E-mail:
joachim.reuder@gfi.uib.no
pascal.brisset@enac.fr
marius.jonassen@gfi.uib.no
martin@pfump.org
stephanie.mayer@gfi.uib.no

Abstract. A new system for atmospheric measurements in the lower troposphere has been developed and successfully tested. The presented Small Unmanned Meteorological Observer (SUMO) is based on a light-weighted commercially available model airplane, equipped with an autopilot and meteorological sensors for temperature, humidity and pressure.
During the 5 week field campaign FLOHOF (Flow over and around Hofsjökull) in Central Iceland the system has been successfully tested in July/August 2007. Atmospheric profiles of temperature, humidity, wind speed and wind direction have been determined up to 3500 m above ground. In addition the applicability of SUMO for horizontal surveys up to 4 km away from the launch site has been approved. During a 3 week campaign on and around Spitsbergen in February/March 2008 the SUMO system also proved its functionality under harsh polar conditions, reaching altitudes above 1500 m at ground temperatures of -20 ^{o}C and wind speeds up to 15 m s^{-1}.
With its wingspan of 80 cm, its length of 75 cm and its weight of below 600 g, SUMO is easy to transport and operate even in remote areas. The direct material costs for one SUMO unit, including airplane, autopilot and sensors are below 1200 Euro. Assuming at least several tenths of flights for each airframe, SUMO provides a cost-efficient measurement system with a large potential to close the existing observational gap of reasonable atmospheric measurement systems in between meteorological masts/towers and radiosondes.

1. Introduction

Detailed information on the vertical structure and the horizontal variability of the lower atmosphere, especially with respect to temperature, humidity and wind, is the key for the understanding of exchange processes in the Atmospheric Boundary Layer (ABL). Further improvement of this understanding requires a combination of fine-scale modeling efforts and innovative measurement strategies. Corresponding work is crucial for the validation and improvement of boundary layer parameterization schemes and thus for future progress both in numerical weather prediction and in climate modeling.
Up to now there is a lack of cost efficient measurement systems, applicable for ABL phenomena

covering horizontal scales between 100 m up to 10 km. First attempts on the basis of remotely controlled small unmanned aerial vehicles (SUAVs) have already been started some decades ago [1], and the systems have gradually been improved until today. A corresponding profiling system (KALI) has for example been developed at the University of Munich, Germany and has been successfully used during several field campaigns in Nepal, Bolivia and Germany to measure atmospheric profiles of pressure, temperature and humidity up to 3 km above ground (e.g. [2], [3]).
The two main shortcomings of that system are the need of experienced pilots for remote control of the SUAV and the lack of an appropriate wind measurement system. During the last years substantial progress in the field of miniaturization of electronic components has been achieved. As a consequence, miscellaneous autonomous SUAV systems, mainly used as drones for military reconnaissance purposes, have been developed. In addition, smaller, faster and cheaper sensors for meteorological parameters and positioning purposes (e.g. GPS) are available now. Sporadic attempts to use and combine this new technical potential for atmospheric measurements have shown some promising results [4], [5], [6].

2. The SUMO system

The Small Unmanned Meteorological Observer (SUMO) has been developed as a mobile and cost-efficient platform for the determination of the vertical distribution of temperature, humidity, wind speed and wind direction in the ABL and can be operated as "recoverable radiosonde" for boundary layer research. In its recent version, it is based on a commercially available model plane construction kit equipped with an autopilot system and meteorological sensors for the measurement of temperature, humidity and pressure. Figure 1 shows the aircraft and the laptop used as ground control station during SUMO operation.

Figure 1. The SUMO airframe based on the model airplane FunJet by Multiplex and the laptop used as ground control unit.

2.1. Airframe

The construction kit FunJet manufactured by Multiplex is used as airframe of the presented prototype of SUMO. It is a delta-wing pusher prop jet composed of lightweight foam material

EPP (expanded polypropylen), electrically powered by a brushless motor (AXI 2212/26) driving a 9"x 6" propeller. A lithium polymer (LiPo) battery pack with a capacity of 2100 mAh enables a flight endurance of around 20-30 minutes. Due to its size and weight, the SUMO system is highly mobile and easy to operate, even in remote areas and under harsh environmental conditions. The technical details are summarized in table 1.

Table 1. Technical details of the FunJet airframe used as SUMO platform.

length	75 cm
wingspan	80 cm
weight	580 g
average air speed	12-18 m s^{-1}
maximum air speed	35 m s^{-1}
average ascent rate	7-10 m s^{-1}
maximum ascent rate	15 m s^{-1}
maximum altitude above ground	3.5 km
endurance	up to 30 min

The FunJet airframe is quite robust and rather inexpensive (below 100 Euro). Its light weight and the corresponding low speed during landing minimize the risk of structural damage. Additionally EPP can be repaired by instant glue and activator spray within seconds if fractures of the fuselage or wings should occur.

2.2. Paparazzi autopilot system

Paparazzi is an open source autopilot system oriented toward inexpensive autonomous aircraft operation. The system has been designed to be easily adapted to any type of airframe and is currently used in both fixed and rotary wing systems.
It basically consists of:

- the airborne processor board with its required sensors for the determination of the aircrafts attitude (mainly GPS for position and speed and a set of infrared sensors for the horizontal alignment)
- the airborne autopilot software
- the ground control station
- the online communication hardware and communication protocols
- a standard remote control (RC) transmitter as safety link option

The main components of the system are schematically depicted in figure 2. A comprehensive overview of hard- and software setup and configuration as well as the basic principles of operation of the Paparazzi solution can be found in [7] and [8] or on the corresponding home page of the Paparazzi project (http://paparazzi.enac.fr/wiki/index.php/Main_Page).

A powerful flight plan language allows the operator to define complex autonomous missions and create a logical tree of autonomous decisions for the system to make while in flight to perform any task or adapt to any scenario. The ground station operator can also manually navigate the aircraft at any time using video (not installed on SUMO), real-time GPS data, and/or visual contact while relying on the autopilot to perform only the flight stabilization. The ground control station utilizes a powerful and flexible client/server architecture which allows the operator to

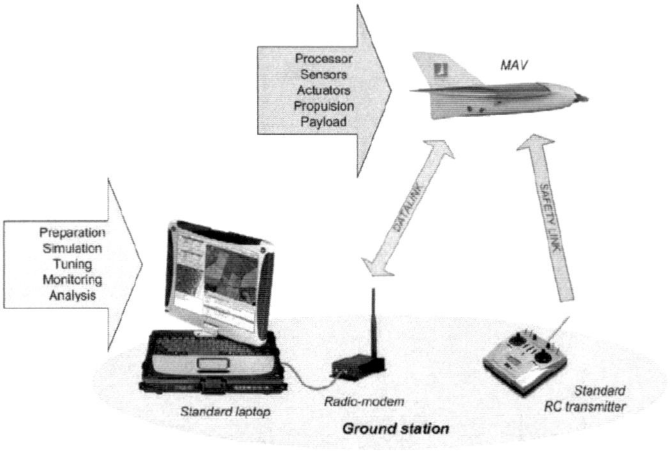

Figure 2. The main components of the SUMO system based on the Paparazzi autopilot solution.

control one or more aircraft from a single location or from multiple locations.
As mentioned above, Paparazzi is an open source project. Source code, hardware schematics and thorough documentation are released under the GNU Public License and are freely available for anyone to download from the project website mentioned above.

2.3. Meteorological sensors

SUMO is equipped with sensors for the measurement of pressure, temperature and relative humidity. The pressure sensor is mounted inside the fuselage, while the sensors for temperature and humidity are attached to the fuselage under the wings to provide good ventilation and to minimize heating by solar radiation. Pressure is measured by the miniaturized (diameter 6.1 mm, height 1.7 mm) SCP1000 Absolute Pressure Sensor from VTI Technologies. Its measurement range covers 300 to 1200 hPa with a resolution of 0.015 hPa and an absolute pressure accuracy of 1.5 hPa in the range 600-1200 hPa. The relative pressure accuracy relevant for atmospheric profiling is 0.5 hPa. The pressure sensor is equipped with an onboard temperature sensor that provides in-flight temperature inside the aircraft, an important information for the estimation of battery capacity especially under cold environmental conditions. Temperature and humidity is measured by the sensor DigiPicco I2C manufactured by IST. It combines a PT1000 temperature element with the capacitative humidity sensor P14 SMD. During operation it turned out that the temperature sensor shows a distinctly slower response than the humidity sensor, most likely due to its location on the circuit board behind the humidity sensor. Therefore a second, independent and faster sensor combination (SHT75 by Sensirion) has been additionally mounted in autumn 2007 for the latest measurements. The main specifications of the temperature and humidity sensors are summarized in table 2.

2.4. Operation of SUMO for atmospheric profiling

In general the SUMO system requires two persons for operation. One is preparing and controlling the autonomous mission on the ground control station (GCS), while the other is operating the standard remote control (RC) transmitter and acting as safety pilot during the flight. Take-off and landing are usually performed in manual mode, even though autonomous start is generally unproblematic and autonomous landing is possible at least in flat terrain and under weak wind conditions.

Table 2. Specifications of the temperature and humidity sensors used on.

sensor	DigiPicco		SHT 75	
parameter	RH	T	RH	T
resolution	0.003 %	0.005 K	0.03 %	0.01 K
range	0-100 %	-40 - 50 ^{o}C	0-100 %	-40 - 124 ^{o}C
abs. Accuracy	< 3 %	$\pm$0.5 K	$\pm$1.8 %	$\pm$0.5 K
rel. Accuracy	N/A	N/A	$\pm$0.1 %	$\pm$0.1 K
response time	5 s	N/A	6 s	5 s

The following favorable mode of operation has been developed during numerous meteorological ascents. SUMO is started manually and flown to a safe altitude of around 100 m above ground. Then the pilot switches control from the RC transmitter to the GCS for autonomous flight. The aircraft is first sent in a standby mode, i.e. flying circles around a defined home point at a given altitude of 150 m. This enables a check of proper aircraft functionality and allows for the adjustment of the temperature and humidity sensors to the environmental conditions. The next step of the flight plan is an upward profile to a user-defined altitude that can be changed in-flight. For that purpose, the airplane is operated with constant throttle and constant pitch angle, the autopilot uses variations of the roll angle to keep the aircraft on its circular upward spiraling track. After reaching the defined maximum altitude, the engine is switched off and the pitch angle set to a constant negative value. From that point the aircraft is gliding down on a spiral track until reaching the safety altitude of 150 m. From there, the pilot is taking control for manual landing. Fig. 3 shows an example of a typical helical flight pattern for boundary layer profiling, where different colors are used for ascent and descent. Typical ascent and descent rates of the SUMO aircraft are in the order of 7-10 m s^{-1}. The x and y values show the relative distance of SUMO to the launch site with respect to height above ground. The maximum height of this flight was 2567 m at an overall flight duration of 16 minutes. Especially at higher altitudes there are some data gaps in the track due to loss of telemetry connection for a few seconds.

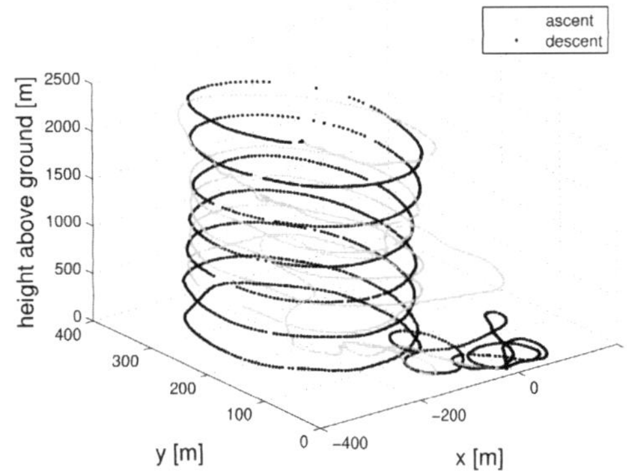

Figure 3. An example of a typical flight pattern during the FLOHOF campaign on Iceland. The ascent starts at 17:56 UTC on 18.08.2007.

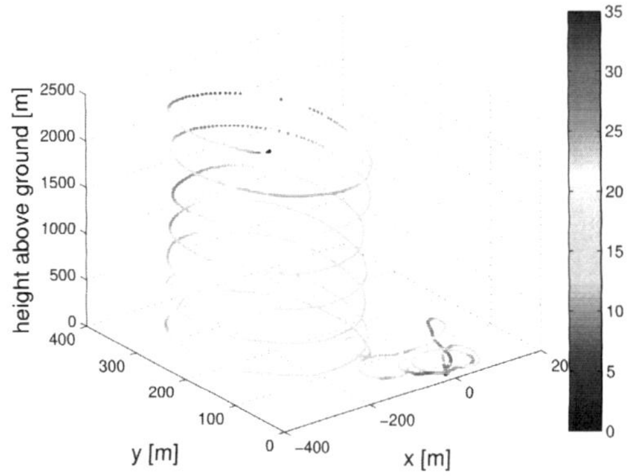

Figure 4. As in fig. 3, but only descent. The colors indicate the ground speed of the aircraft given in m s^{-1}.

Figure 4 shows the instant ground speed from on-board GPS at each position of the track by color code. In one section, the ground speed is decelerated by headwind, indicated by blueish and greenish colors, while it is accelerated by tailwind in the opposite section of the spiral, indicated by yellowish and reddish colors. Assuming constant true air speed, these ground speed differences can be used to determine the horizontal wind speed and direction. Operating the aircraft with constant throttle (zero during descent) and constant pitch angle, this assumption is fulfilled in good approximation. Wind speed can be determined from the difference between minimum and maximum ground speed over a full circle, the corresponding wind direction from the position of minimum and maximum along the track.

3. Measurements

SUMO has been extensively tested during two measurement campaigns. The first one, FLOHOF (FLOw over and around HOFsjokull) took place in July/August 2007 in Central Iceland and was mainly dedicated to the investigation of non-stationary gravity waves (for details see www.flohof.uib.no). During a 3 week campaign on and around Spitsbergen in February/March 2008 the system has been operated for the first time in polar environment to investigate the arctic ABL. Measurement flights have been performed from land near Longyearbyen, from sea ice, and from deck of the ice-breaking Norwegian coast guard vessel KV Svalbard. During both campaigns a total number of more than 70 meteorological profiling ascents have been performed with one airframe. Several ascents of the FLOHOF campaign on Iceland reached altitudes of above 3000 m. Under arctic conditions on and around Spitsbergen, with surface temperatures around -20 °C and boundary layer wind speeds of 10-15 m s^{-1} maximum altitudes of above 1500 m could be reached. The following section will exemplary present temperature, humidity and wind profiles, measured by the SUMO system.

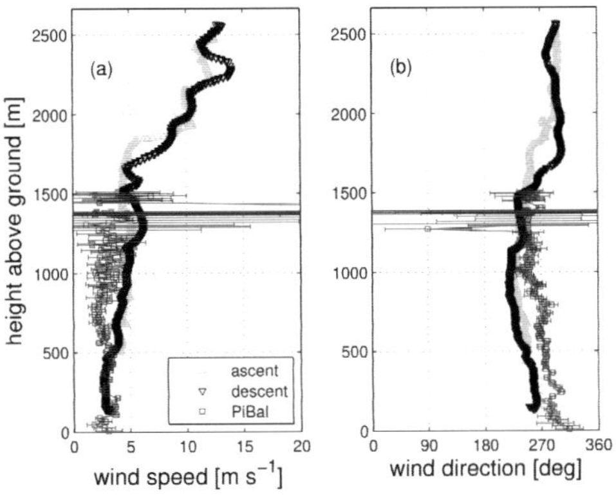

Figure 5. Profiles of wind speed (a) and wind direction (b) measured during the FLOHOF campaign on 18.08.2007. SUMO started at 17:56 UTC, the PiBal ascent at 18:54 UTC.

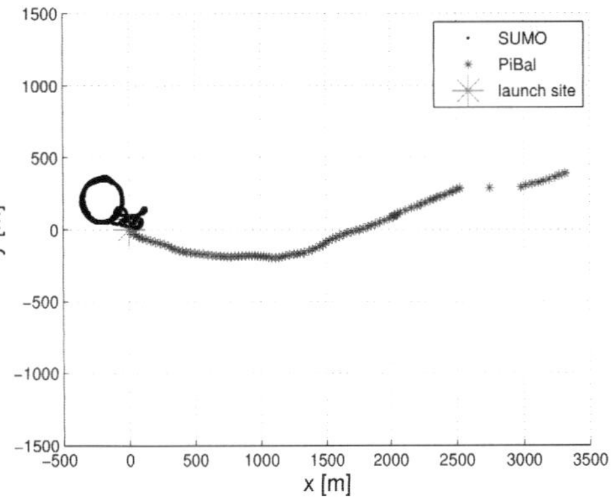

Figure 6. Projection of the SUMO and PiBal tracks to the horizontal x-y plane for the intercomparison presented in figure 5.

3.1. *Wind profiles*

Figure 5 presents an example of the profiles of wind speed and wind direction derived by the method described above. In accordance with the corresponding temperature and humidity profiles, showing a distinct capping inversion (figure 7), the ascent represents a well developed

boundary layer with rather week and constant wind speed of 3-5 m s^{-1} up to 1600 m above ground. The wind direction near the ground is W, slightly turning to the left with height. At the inversion the wind speed starts to increase to values above 10 m s^{-1}, while the wind direction changes to WNW.
The SUMO wind profile starts at 500 m above ground for the ascent (green) and ends at 150 m above ground in the descent (black). This is due to the fact that SUMO did not fly in autonomous mode below approximately 250 m. In order to obtain wind information via the optimization algorithm the first (last) circle in the auto mode flight pattern has to be completed in the ascent (descent). Therefore, the wind information has to be corrected with respect to height depending on the flight pattern's helical stretching (not shown).

In order to validate the wind estimation method, SUMO ascents have been compared to Pilot Balloon (PiBal) ascents. This method uses small helium filled ballons that are released and optically tracked with two theodolites (e.g. [9]). The corresponding measurement data and their error estimates are plotted in figure 5 in blue. Compared to SUMO, the PiBal method shows very similar wind profiles - both in speed and direction - between 150 up to 1250 m above ground. The large error bars above about 1300 m in the PiBal ascent show the limitation of this system due to a strong increase in uncertainty with increasing distance of the balloon from the observer [10]. Taking into account the delay of about one hour between SUMO and PiBal ascent and the spatial offset due to the drift of the ballon with the wind (see figure 6), the correspondence between the two measurement methods is quite well and promising.

3.2. Temperature and humidity profiles

Figure 7 shows the temperature and humidity profiles taken at the same time as the wind measurements from FLOHOF discussed in the previous section. It can be seen that the vertical structure of the associated parameters is well resolved. The most prominent feature is a marked inversion ranging from appr. 1800 to 2000 m, indicated by an increase in temperature of about 3 K and a corresponding decrease in humidity from above 80 % down to 25 %.

The ascent and descent profiles show systematic deviations at a given altitude that are in the

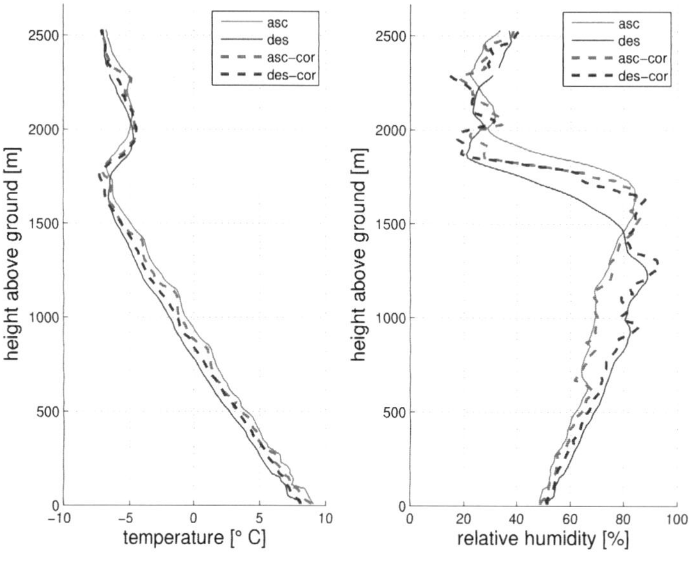

Figure 7. Profiles of temperature and relative humidity, 18.08.2007, start at 17:56 UTC.

order of 1.5 K for temperature and 5-10 % for relative humidity. A closer look reveals that it can be mainly seen as a systematic height displacement, especially evident in humidity around the inversion. This feature has to be attributed to the sensor response time, leading to a general warm bias in the ascent and cold bias in descent. Correspondingly, the humidity has a dry bias during ascent and a wet bias during descent. Due to its deterministic nature, this error can be compensated for to a large extent by the application of a numerical correction scheme. The profiles after application of such a scheme are plotted as dashed lines in figure 7. A main indication of the scheme's effectiveness can be seen by both its ability to bring the profiles closer together and by positioning the inversion heights for relative humidity and temperature to the same level. The remaining deviation in relative humidity has not necessarily to be addressed to sensor problems (e.g. hysteresis). Humidity profiles during the whole FLOHOF campaign, taken by the remotely controlled aircraft KALI and by a tethered balloon, have documented the large spatial and temporal variability of relative humidity in the region.

3.3. Measurements in the Arctic ABL

Figure 8 shows an example of temperature and humidity profiles obtained during the field campaign on and around Spitsbergen in late winter 2008. Here only the results from the descending part of profile are included, because it turned out that the ascent profiles were subject to a distinct warm bias in the lowest levels. This is caused by residual heat due to the large temperature differences between the laboratory for flight preparation (+15 ^{o}C) and the measurement environment itself (-20 ^{o}C). A new feature of the SUMO system during this campaign was the inclusion of an additional, faster sensor for temperature and relative humidity. The results from both sensor sets are presented and compared in figure 8.

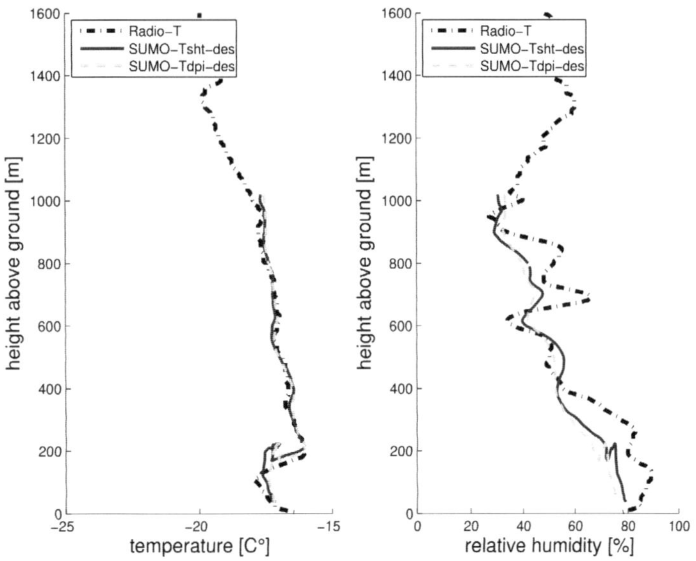

Figure 8. Profiles of temperature and relative humidity taken onboard the Norwegian Coast Guard vessel KV Svalbard at 76.74 ^{o}N and 18.25 ^{o}E in the marginal ice zone on 28.02.2008. (Take-off SUMO: 15:11; launch of radiosonde: 15:15 UTC)

In comparison to most profiles obtained at Iceland, the profiles from the Spitsbergen area, have generally signs of a more complex lower level structure often characterized by multiple

inversions. A central question is how well SUMO is able to resemble the true vertical structure of these fine scale features. The answer was sought through an intercomparison with a radiosonde (Vaisala RS92), which represents a well established and acknowledged measurement platform. SUMO (15:11 UTC) and the radiosonde (15:15) have been launched nearly simultaneously and the ascent rates of both systems are rather similar. It turns out that the temperature profiles of both SUMO sensors and the radiosonde are in excellent agreement of better than 0.5 K from 50 m to the maximum SUMO height of 1000 m during this ascent. The irregularity around 200 m altitude can be explained by a short term ascending motion of the aircraft after switching back from autopilot into manual mode. The relative humidity values have somewhat larger deviations, especially in comparison to the gradients, possibly indicating a generally longer response time for the SUMO instruments than for those of the radiosonde. Furthermore, the measurements from the newest sensor seem to reveal more details than for the old one, indicating a shorter time constant.
Wind profiles of the Spitsbergen campaign are not yet evaluated, but will also be compared to the radiosoundings soon.

4. Summary and outlook

The most important advantage of the SUMO system is its easy-to-hand and cost-efficient performance. During the Spitsbergen campaign, the ground control station was for the first time operated exclusively by scientists rather than aircraft specialists. This proves that the Paparazzi system has now reached a level of user-friendliness that enables operation after rather short training.
The new sensors for temperature and humidity provide the corresponding profiles in an accuracy comparable to that of the latest and well-established radiosonde systems. First validations of the SUMO wind profiles indicate that this information can be obtained at least satisfactory. By keeping strictly to the circling/helical flight pattern, no additional equipment is needed to require information about the horizontal wind. The method uses only modules which are part of the autopilot system anyway. It has to be stressed that the wind estimation method works only for a circling/helical flight pattern which can only be precisely achieved in auto mode. For safety reasons, the auto mode could not be used for heights closer than 150 m above ground during the field campaigns in 2007 and 2008. For the future it is desirable that additional circles in autonomous mode will be flown close to the ground. This will require additional SUMO sensors for continuous monitoring of the distance to the ground. The integration of such sensors will also be one important step toward fully autonomous landing.
The wind estimation method in use has clearly potential for improvements, thus one main focus in the future will be set on corresponding validation and intercomparison with the available parallel radiosonde ascents during the Spitsbergen campaign and by the performance of tailored validation campaigns, e.g. flying SUMO around high meteorological towers. In this context the ongoing Gufuskalar project, equipping a 412 m high mast in Northwest-Iceland with meteorological instrumentation, will be of large importance.

Acknowledgments

The authors are grateful to Christian Lindenberg for his work on the assembly and maintenance of the airplanes and for his assistance as safety pilot during the flights on Iceland and to Michael Kislat acting as safety pilot during the campaign on and around Spitsbergen.
The main financial support for development of SUMO was provided by the Meltzer University Foundation, Bergen, Norway. The comprehensive test of the system on Iceland in August 2007 would not have been possible without the FLOHOF field campaign funded by the German Research Foundation under contract EG 40/25-1. The arctic flights on and around Spitsbergen have been funded by the Norwegian IPY-THORPEX campaign and a personal stipend of

Marius Jonassen (2008-356) provided by the Norwegian Polar Institute. Many thanks to the crew of the Norwegian Coast Guard vessel KV Svalbard for assistance during the research cruise.

References

[1] Konrad T G, Hill M L, Rowland J R and Meyer J H 1970 A small, radio-controlled aircraft as a platform for meteorological sensors *Appl. Phys. Lab. Tech. Digest* 11

[2] Egger J and et al 2002 Diurnal winds in the Himalayan Kali Gandaki valley. Part III: Remotely piloted aircraft soundings *Mon. Wea. Rev.* **130** 2042

[3] Egger J and et al 2005 Diurnal circulation of the Bolivian Altiplano. Part I: Observations *Mon. Wea. Rev.* **133** 911

[4] Holland G J, Webster P J, Curry J A, Tyrell G, Gauntlett D, Brett G, Becker J, Hoag R and Vaglienti W 2001 The Aerosonde robotic aircraft: A new paradigm for environmental observations *BAMS* **82** 889

[5] Ma S, Chen H, Wang G, Pan Y and Li Q 2004 A miniature robotic plane meteorological sounding system *Adv. Atmos. Sci.* **21** 890

[6] Spiess T, Bange J, Buschmann M and Voersmann P 2007 First application of the meteorological Mini-UAV 'M2AV' *Meteorol. Z.* **16** 159

[7] ENAC 2008 Paparazzi user's manual http://paparazzi.enac.fr/wiki_images/Users_manual.pdf

[8] Brisset P 2006 The Paparazzi solution http://www.recherche.enac.fr/paparazzi/papers_2006/mav06_paparazzi.pdf

[9] Thyer N 1961 Double theodolite pibal evaluating by computer *J. Appl. Met.* **1** 66–68

[10] Schaefer J T and III C A D 1978 The inherent position errors in double-theodolite pibal measurements *J. Appl. Met.* **17** 911–915

IOP Conf. Series: Earth and Environmental Science 1 (2008) 012015 doi:10.1088/1755-1307/1/1/012015

CURIE: a new clear air Doppler radar dedicated to the lower part of the Atmospheric Boundary layer (20 m- 750m)

H Al-Sakka[1], A Weill[1], C Le Gac[1], R. Ney[1], E Dupont[2]
Presentation to the *14 th international Symposium of Advancement of Boundary layer Remote sensing,,* June 23-25, 2008, Risoe National Laboratory, Roskilde, Denmark

E-mail: Hassan.al-sakka@cetp.ipsl.fr

Abstract. A new X-band miniradar, the CURIE radar (*Canopy Urban Research on Interactions and Exchanges*), mainly adapted to low Atmospheric Boundary Layer sounding has been developed at CETP. After a brief description of the opportunity and working conditions in a turbulent atmosphere, main characteristics are presented. Though this radar works in presence of precipitation as all X-band radar can do, this paper is more dedicated to clear air used in the turbulent atmospheric boundary layer. We are presented comparisons with UHF observations and boundary layer information which can be inferred from CURIE as entrainment across the inversion layer.

1. Introduction

Air dynamics knowledge in the lower atmosphere is necessary for several kinds of investigations especially with the need for a fine spatial vertical resolution close to 20 m in the first 500 m of the *ABL* (*Atmospheric Boundary Layer*) and with a first level of observation as low as possible. For example pollution studies have to be connected to the UBL (*Urban Boundary Layer*) and exchange at the top of the city itself which concerns particularly the city canopy. This zone is usually covered by Sodar and Lidar measurement but:

- Sodar is not very efficient in very noisy environments as (edges of highway, some urban areas...). More, Sodar noise itself does not seem to be socially accepted by city inhabitants. However as shown by Little [1], acoustic reflectivity (for Sodar) is larger than electromagnetic reflectivity (for radar) in the ABL but this superiority vanishes in presence of large acoustic noise;
- Lidar may be unable to work correctly in very low atmospheric layers in case of too many particles and intense low level fogs...Above typically 500 m height, ST Radar (UHF, VHF) and Lidar are indisputably privileged instruments.

These considerations indicate an interest to develop an instrument equivalent to Sodar but not sensitive to ambient acoustic noise and with no acoustic noise generation. For that purpose, we have developed

[1] Centre d'étude des Environnements Terrestre et Planétaire, CETP/IPSL, 10-12 Avenue de L'Europe, 78140 Vélizy France.
[2] EDF-R&D - CEREA - 6 quai Watier - 78401 Chatou cedex - France.

a new X-band Radar CURIE (*Canopée Urbaine Radar pour l'étude des Interactions et Echanges*) for Urban Canopy Radar study of Interactions and Exchanges.

X-band was chosen because small antennas can be used, which is easy in an urban configuration. The possibility to develop low power and less expensive radar using solid state transmitter system was also another reason. However, an important question was the feasibility of clear air Bragg scattering for centimeter wave lengths in the inertial subrange i.e. close to the dissipation range. However several observations (for acoustic waves) using for example 6 kHz up to 20 kHz have shown convincing turbulent Bragg responses, (see Kallistratova [2], Monin [3]) corresponding to λ_{Turb} between 0.8 cm and 2.5 cm. For centimeter electromagnetic waves it must be the same, Tatarskii [4]. CURIE results will show that this assumption was relevant. Notice that the Kolmogorov dissipation scale in the low boundary layer is close to one mm.
Another important point is the minimum of turbulence to be detected. To be very sensitive to turbulence in the ABL, CURIE reflectivity, in term of Cn^2, should be at least of 10^{-16} m^2 $s^{-2/3}$, see Neff and Coulter [5].
In this paper after CURIE radar and data processing description in section 2, we present several CURIE observations in the low clear air ABL in section 3. Some conclusions and perspective of CURIE use are then presented.

2. Radar description and data processing

CURIE is a pulse coded X-band Radar. Two Prototypes were developed: the first prototype or demonstration prototype was implemented to demonstrate that CURIE can give relevant atmospheric results in clear-air ABL. Radar characteristics were: wavelength: $\lambda = 3.2$ cm, 10 W: peak power (solid state amplifier), pulse length: 200 ns, ambiguity distance: 150 m, antenna: parabolic , diameter 1.2 m, gain 40 dBi and beam angle 2°. More information about this prototype and results can be shown in Weill et al. [6].
The second prototype is implemented for scientific purpose: study the atmospheric properties (turbulence, precipitation ...). Description of this prototype is listed in table 1. An offset antenna is used to minimize the effects of secondary beams and ground clutter (fixed echoes). Orientation of this antenna is piloted thanks to an azimuth elevation positioning where we can change automatically the orientation using a positioning system. The received signal is amplified, and then sampled with the intermediary frequency.
60 MHz signal generation (BPRK and pulse modulation), digital receiver, coherent integration, decoding and emitting are programmed in a FPGA. Radial Doppler velocities averages are then processed which provides reliable measurements from an altitude of about 40 meters (to be improved) above ground level and its maximum range, depending on the chosen coded mode is about 720 m (if 16 momentum codes are used).
The Radar processes 4096 FFT points, 300 coherent integrations that correspond to a spectrum every 3 second. CURIE data are then analyzed using Matlab where Doppler spectrums are averaged, smoothed to discriminate and to eliminate noisy information.
Due to different kinds of scatter mechanisms associated with turbulence or with rain drops two types of reflectivity are used:

- In presence of turbulence:

$$\eta = 0.38 C_n^2 \lambda^{-1/3} \tag{1}$$

  Where η (m^{-1}) is the scattering cross section, Cn^2 ($m^{-2/3}$) is the refractive index structure function.
- In presence of hydrometeors:

$$\eta = \frac{\pi^5 |K|^2}{\lambda^4} Z \tag{2}$$

Where $|K|^2$ is a function of the complex index of refraction (depending on the electric properties of water) and Z is the radar reflectivity factor.

Notice that mixed cases can also occur.

Table 1. Description of scientific Prototype of CURIE

Characteristic	Value
Frequency	9.42Ghz
Pulse width : -Biphasic mode	160 ns
-Coded mode (Spano code)	4, 8 , 16 moments
Repetition period	0.6 μs to 4.8 μs
Spatial resolution	24 m
Antenna Gain	40 dBi
Aperture	1.9°

3. Results

3.1. CURIE Results

CURIE Radar is located inside SIRTA (an experimental site, for more information see *http://sirta.ipsl.polytechnique.fr/)* at Palaiseau. Measurements were performed mainly during 3 periods in 2007. The first period is between March, 21 and April, 15 2007, the second period in May (~10 days) and the last period is during September (2 days). Raw primary data were processed and stored, using Labview, in 1 hour files. Using the positioning system, CURIE can point automatically along different directions (elevation and azimuth). In the 2 first periods of data, radar was positioned in vertical direction, so velocity measured was the vertical velocity and the last period is to test the positioning system, for horizontal wind profiling, Every 2 minutes antenna was pointing in a different direction, then radar was fixed with an angle equal to 15° with the vertical axe.
Different parameters (power spectrum, Doppler velocity, standard deviation of the Doppler velocity, refractive index structure function ...) are calculated.

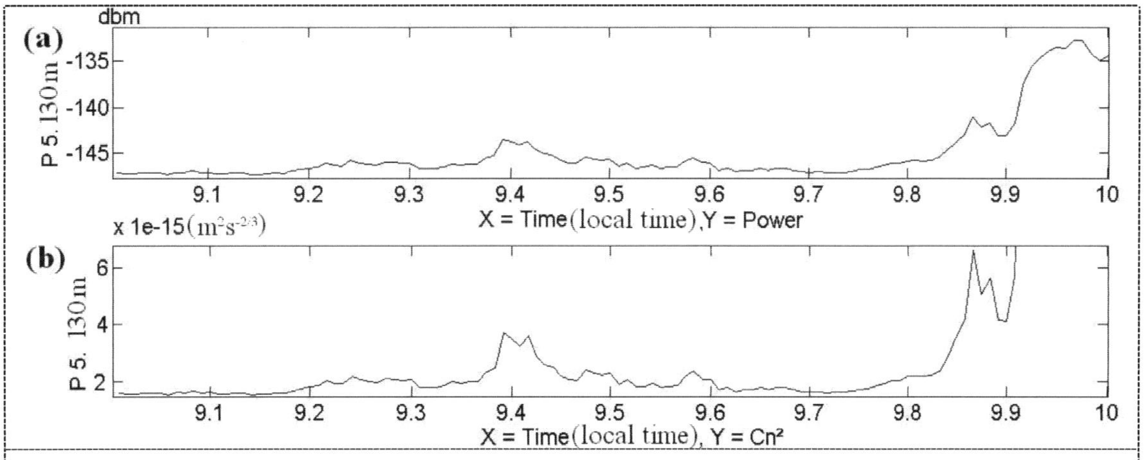

Figure 1. Response of CURIE in Clear-Air, (a) Received Power in dbm, (b) Cn^2 (x 10^{-15}), local time

Figure 1 show the response of CURIE in Clear-Air, this data was collected during March, 31, 2007 at 9h AM (France Time), at an altitude of 130 m (P 5: 5th port). In (a) we plot the power spectrum as a function of time. Signal remains relatively stationary up to the last 10 minutes, where the power increases rapidly due to initiation of a precipitation. In (b), refractive index structure function was calculated using equation (1). In this equation, the scattering cross section η is computed using the equation of radar in clear air. During the last 10 minutes, we cannot talk about Bragg scattering since Rayleigh scattering due to rain drops occurs. The order of magnitude of Cn^2 is larger than 10^{-16} m² s$^{-2/3}$.

3.2. A tentative to determine entrainment flux across inversion layer

It is well known that boundary layer evolution is determined by dynamics of exchange across inversion layers and is related to the entrainment fluxes. Ball [7] quoted by Tennekes [8] suggested that entrainment flux was related to vertical kinetic energy across the inversion layer and this concept was used with Sodar; see Weill and Lehmann [9]. A rough estimate of entrainment fluxes Q_i was proposed:

$$Q_i = -A\frac{\sigma_w^3}{h} \tag{3}$$

Where A is close to 1.4 and h is the height of the inversion.

This entrainment flux is reduced to the virtual sensible heat flux if wind shear vanishes (clear air convective case) but is generally associated with momentum production in the wind shear and buoyancy production at the inversion level h:

$$Q_i = -\left(A_1 U_*^2 \frac{dU}{dz} + B_1 \frac{g}{\theta_v} < w'\theta_v > \right) \tag{4}$$

Where A_1 and B_1 are constant, U_* is the friction velocity, U the mean wind, z the height, θ_v the mean virtual potential temperature and $<w'\theta'_v>$ the mean local virtual temperature flux.

Notice that virtual temperature is used due to wet air entrainment across the boundary layer. This mechanism is responsible for cloud formation; see Mathieu et al. [10].

Figure 2 shows estimates of entrainment fluxes from (3), just before precipitation, with data collected on March, 29, 2007, local time. Of course, these raw estimations of entrainment fluxes must be validated and compared to cloud parameterization and modeling with different mechanisms as suggested in [10] as bulk Richardson number. Comparisons with surface fluxes have also to be undertaken. More, dry and wet buoyancy components due respectively to sensible and latent heat fluxes must be taken into account.

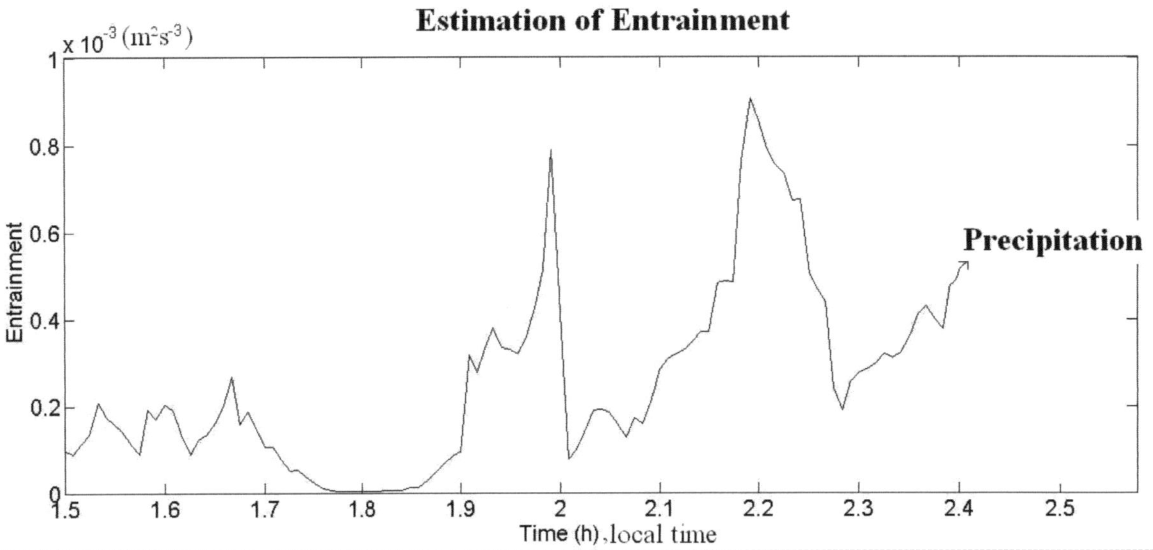

Figure 2. Estimation of the entrainment (m^2s^{-3}), local time.

3.3. Comparison with UHF Radar

The UHF Radar Deg rewind PCL1300 is used by EDF (Electricité De France) to study the wind profiling; it is located in SIRTA near CURIE Radar. Its average power is 3.5 watts (using an amplifier as power equal to 500w and pulse width equal to 350ns), broadcasts at a frequency close to 1.238 Gigahertz, (λ = 24 cm). The Radar antenna is made up of five panels necessary for the formation of the five beams (one vertical and four oblique beams). The other four oblique beams are made from four panels tilted 17° from the horizontal. Each panel consists of a network composed of eight antennas. The radar antenna is usually surrounded by a fence designed to reduce ground clutter from around fixed objects (trees, buildings ...). It detects the radial Doppler velocity within a volume resolution that has aperture of 8.5°. It provides reliable measurements from an altitude of about 85 meters above ground level, witch has been proved in campaign measurement in Lannemezan (South West of France), and its maximum range, which depends on the intensity of the echoes from the atmosphere, is about 1500 to 2000 m. Note that signal received has been reduced using an HS switch in the receiver.

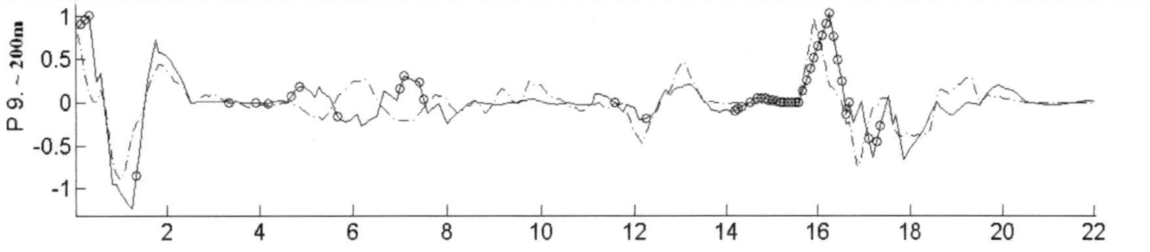

Figure 3. Comparison of vertical velocity fluctuations between CURIE and UHF radar, GMT time

Data coming from this Radar are stored in 5 minutes files, but each file from this radar corresponds to moving averages (averages on 30 minutes delayed every 5 mn).
Figure 2 shows a comparison of vertical velocity fluctuations between CURIE and the UHF Radar, after elimination of a negative mean average bias associated with each radar and of same order of magnitude. In this Figure, we plot CURIE vertical fluctuations (in dashed line) and UHF radar fluctuations as function of time (GMT time in hour); at an altitude of 200 m (P 9: gate 9th), this data was collected on March, 30, 2007. The black circles correspond to cases where data were not validated. We can observe a good correspondence between these two graphs. This relatively good correlation seems to be systematic at least for concomitant observations but differences must be understood due to the different wave lengths and the different radar characteristics.

4. Conclusion

CURIE is course in development and validation and seems to be adapted to «urban micrometeorology».
This radar has been indeed designed for low atmospheric boundary layer profiling. In this paper, we only present first interesting observations in clear air:

- Turbulence observation in the boundary layer and Cn^2 evaluation.

This shows that CURIE sensibility is what was awaited, but systematic comparisons during different conditions, with independent estimations remains fundamental to understand if turbulence response corresponds to inertial turbulence.

- Raw analysis of entrainment across a stable layer just before cloud evolution toward precipitation.

This type of analysis is necessary if we have to take into account that CURIE can be applied to clear air physics as well as precipitation physics in the ABL.

- A comparison of vertical velocity fluctuations with UHF Radar.

It shows convincing results about measured variables but suggests to go further into scatter mechanisms at different wave lengths to explain discrepancies.

Different works are in progress:

- Systematic validation of CURIE during different ABL conditions;
- Precipitation analysis;
- Discrimination of turbulence in presence of precipitation.

We have only described a few examples of results as obtained with a vertically pointing antenna, but in the future the orientation capabilities of CURIE will be used for horizontal wind profiling. For that purpose, real time processing will be improved using fast response relevant operational algorithms. If CURIE is intended to work for urban micrometeorology, CURIE has also some capabilities to work on MABL (Marine Boundary Layer).

References

[1] Little C G 1972, On the detectability of fog, cloud, and snow by acoustic echo sounding methods. *J. Atmos. Sci.* **28,** 748-755.

[2] Kallistratova M A 1959, Procedure for investigating sound scattering in the atmosphere, *Akad. Zh.,* **5, 5,** 496-498.

[3] Monin, A S 1962, Characteristics of the scattering of sound in a turbulent atmosphere, *Sov. Phys. Acoustics (English Traslation)* **7,** 370.

[4] Tatarski V I 1961, Wave propagation in a turbulent medium. *McGRAW-HILL book company*, INC.

[5] Neff W D and Coulter R L 1986, Acoustic Remote Sensing, in : Probing the Atmospheric Boundary Layer, D. H. Lenschow editor, p. 201-266, *Amer. Meteor. Soc.*, Boston, MA.

[6] Weill A, Legac C, Ney R and Chardenal L, 2003, A new mini Radar to investigate urban canopy: CURIE canopy urban Radar for investigation of exchanges, *MST10 Workshop*, Peru.

[7] Ball F K 1960, Control of inversion height by surface heating. *Quart .J. Roy. Meteor. Soc.*, **86,** 483-494.

[8] Tenekes H A 1973, Model for the dynamics of the inversion above a convective boundary layer. *J. Atm. Sci.*, **30,** 558-565.

[9] Weill A, and Lehmann H 1990, Already twenty years of acoustic sounding: some applications *Zeitschrist für Meteorologie* **40 (4)** 241-250.

[10] Mathieu A, Lahellec A and Weill A 2004, Evaluation of a numerical weather forecast model using using boundary layer cloud-top temperature retrieved from AVHRR, *Monthly weather review,* **132,** 915-928.

Acknowledgments

CURIE has received supports from INSU (CNRS) (Moyens Mi-Lourds), from SESAME project (funded by Ile de France region), from CETP and from IPSL.
It is a pleasure to acknowledge all our colleagues who have helped in the development of CURIE and particularly Martial Haeffelin, Scientific and Technical Head of SIRTA to have given us the opportunity to install CURIE inside SIRTA.

IOP Conf. Series: Earth and Environmental Science 1 (2008) 012016 doi:10.1088/1755-1307/1/1/012016

Sodar and RASS with the use of wideband acoustic impulses

Y Ulyanov[1], N Maksimova [2], V Misaylov[3]
[1] National Technical University "KhPI", Kharkov, Ukraine
E-mail: ulyanov@kpi.kharkov.ua

[2] Kharkov National University of Radio Electronics, Kharkov, Ukraine
E-mail: shifrin@kture.kharkov.ua

[3] Kharkov University of Air Force, Ukraine
E-mail: misaylov@ukr.net

Abstract. In the paper, possibilities of high-power wideband acoustic pulses for sodar and RASS are analyzed. It is taken into account that at propagation of unfilled acoustic impulses, effects of their nonlinear interaction with the air do not reveal themselves. An engineering mathematical model is built for propagation of the acoustic impulse from the gun shot in the neutral atmosphere with regard of the divergence, classic and relaxation absorption of sound waves in the air, and values of the impulse front cross-section for radio waves are obtained. Cross-sections of impulse fronts are compared with those of the atmospheric turbulent inhomogeneities. The results of comparison have shown that the echo-signal from the impulse front exceeds in level the echo-signal from the turbulent atmospheric inhomogeneities . This confirms the practical realizability of such a RASS method, that allows simplifying the RASS equipment, lowering its cost and decreasing its dimensions.

1. Introduction

As is well known, the sodar efficiency is connected with the negative influence of ambient noise, which shows itself in deterioration of signal-to-noise ratio at the receiver input. In the RASS hardware, such a lowering of the signal-to-noise ratio in the receivable signal is often stipulated by violation of the coherent reflection of electromagnetic waves from acoustic parcels. Such a decrease in snr can not be directly compensated when increasing the power of sinusoidal waves with the acoustic source, as the nonlinear properties of the air play a role of a limiting factor [1].

A goal of the paper is a study of possibilities to use powerful short unfilled acoustic impulses for acoustic and radioacoustic sounding of the atmosphere. As is known, at propagation of such acoustic impulses, their nonlinear interaction with the air appears to be of much lesser extension than for parcels of sinusoidal waves [2]. Therefore, for sounding, the unfilled acoustic impulses can be used, whose intensity level significantly exceeds those traditionally inherent in acoustic transmitters of modern sodar and RASS. Besides, a small spatial extension of such acoustic impulses will provide the essential increase in the height resolution of the sounding equipment. Simultaneously, it eliminates the limitations due to the selectivity of the RASS method.

In order to obtain practically meaningful estimates of such a method realizability, by possible sources of acoustic impulses for sounding, shots of the concrete rifle are used. Its technical characteristics are included in the consideration [3, 4]. A circle of possible sources of powerful acoustic impulses is not limited by devices of blasting type. In this circle, e.g., high-power acoustic arc discharge generators can be included [5]. A good compatibility of such acoustic generators with RASS equipment is confirmed in practice [6].

As applied to sodars, we see that when using in them sources of powerful unfilled acoustic impulses their noise immunity might be improved. In addition, the extension of the sodar dead zone will be significantly decreased, the height resolution will be made better, and a scope of the sodar output information will be widened. As applied to RASS, in this case, the condition of strict relation between operating frequencies of sound source and Doppler radar (the Bragg condition) might be removed. The real possibility appears to use any frequency radar, including that of incoherent type [7].

2. Main part

2.1. Energy indices of RASS methods

As the paper scope restricted let us dwell on consideration of RASS only, when proving the possibility of practical realization of sounding of the atmosphere with the use of wideband single acoustic impulses.

All the known methods of radioacoustic atmospheric sounding are based on the Bragg diffraction of electromagnetic waves on a spatial grating of the air density periodic inhomogeneities created by the moving acoustic wave packet with the finite amplitude [8]. A resonant character of the diffraction requires the precise implementation of the Bragg condition relating wavelengths of sound (λ_S) and electromagnetic (λ_E) waves with regard of a value of the angle of incidence of the latter on such a grating (φ)

$$2\lambda_S \cos\varphi = \lambda_E . \tag{1}$$

The increase in a level of electromagnetic radiation reflected by the grating is provided in known RASS methods, when satisfying condition (1), at the expense of using the extended acoustic sounding parcel. Its vertical size stipulates the spatial resolution of RASS as

$$\Delta H = \lambda_S \frac{m}{2} \tag{2}$$

where m is the number of wavelengths in the acoustic sounding parcel.

A value of disturbance of the refractivity index for electromagnetic waves (ΔN), created in the air by the directed sound source can be estimated by the Debay equation relating the value of N with the air temperature (T_a), and the atmosphere pressure (P_A). According to this equation, at T_a = 15°C and P_A= 1013 hPa, ΔN will constitute

$$\Delta N \approx 0.27\Delta p_A - 1.27\Delta T_a . \tag{3}$$

In view of the adiabaticity of processes at sound propagation in the air, using the known equation of the dry adiabat, it is possible to obtain the following expression for values ΔN and Δp_A being created by sound in the air

$$\Delta N \approx 0.17\Delta p_A . \tag{4}$$

The power reflection coefficient (ρ_e^2) of electromagnetic waves reflected by the acoustic parcel in the air, under the Bragg condition satisfied, according to [9], can be expressed as

$$\rho_e^2 = (\Delta N)^2 \frac{m^2\pi^2}{4} = 7.1\cdot 10^{-14} (\Delta p_A \cdot m)^2 . \tag{5}$$

In view of (2), expression (5) takes the form

$$\rho_e^2 = 7.1 \cdot 10^{-14} \left(\frac{\Delta p_A \cdot \Delta H}{\lambda_s} \right)^2 . \tag{6}$$

The increase in the acoustic power has a certain physical limit. Propagation of sinusoidal sound waves with a large amplitude leads to the nonlinear interaction of sound with the atmospheric air. This stipulates additional acoustic losses exceeding those due to the absorption of sound with a small amplitude. As a result, a major energy of the wave is consumed to creation of harmonics that intensively attenuate in the atmosphere. The distance to the acoustic radiator at which the sound wave may be considered as plane and beyond which nonlinear effects appear, according to [8], is

$$R_{nonl} = 16 C_S / \lambda_S . \tag{7}$$

For the engineering estimate they usually assume that the acoustic power of sinusoidal waves must not exceed the value $P_{s_{\max}}$ [10]

$$P_{S\max} \approx 1.88 \cdot 10^4 \lambda_a / R_{nonl} . \tag{8}$$

The existence of some threshold level of the sound pressure $P_{s_{\max}}$ for the radiator of periodic sound waves, whose excess does not cause the growth of sound pressure, is confirmed by the experimental investigations [11]. For sound radiators operating, e.g., at the frequency of the order of 2 kHz, By such a threshold level will be the sound pressure of about 80 Pa nearby the source. When increasing the operating frequency the value of $P_{s_{\max}}$ decreases.

So, one has to increase the energy potential of sodar and RASS at the expense of building up the number of wavelengths (*m*) in acoustic sounding parcels. This, however, leads to the decrease in the range resolution of the sounding equipment. Besides, for RASS, the additional increase in m sharpens the resonant properties of the acoustic sounding parcels, making difficult the work under conditions of the real temperature stratified atmosphere. Because of that they have either to use frequency-manipulated acoustic parcels or to introduce the automatic-frequency control for the electromagnetic radiation [8]. By cost of such technical supplements are, as is known, the loss of urgency of obtaining the information about temperature profiles and the decrease in the accuracy of measurements [12, 13]. In addition, note that the increase in the energy potential of the Doppler radar being part of RASS cannot be considered as acceptable. This will result in the lack of one of main advantages of RASS compared with meteorological radars – the lack of its good electromagnetic compatibility owing to the fact that Doppler radar has the energy potential lower by 40 dB than that of meteorological radar wind profilers [13].

As an alternative here can be used wideband unfilled acoustic sounding impulses instead of filled (sinusoidal) ones inherent in modern radioacoustic equipment. By an example of such an unfilled wideband acoustic impulse is a sonic clap accompanying the rifle shot.

First of all, the physical possibility of realization of the radioacoustic sounding method with the use of the unfilled acoustic impulse should be estimated. For this, compare values of power reflection coefficients for electromagnetic waves (ρ_e^2) when using filled and unfilled sound impulses. Let us take, e.g., values of technical characteristics of the known RASS described in [14]. Here, taking into account a value of the sound pressure of – 25 Pa produced by the acoustic antenna at the distance of 10 m and setting the minimal value of m of 30, as aresult of calculation by formula (6) we obtain the value $\rho_e^2 \approx 5 \cdot 10^{-12}$ for the sounding distance of 100 m. The same equipment for the same sounding distance, when using acoustic impulse accompanying a shot from the rifle [15] with the sound pressure of 1600 Pa, in two meters from the muzzle, will operate with the value $\rho_e^2 \approx 2 \cdot 10^{-12}$.

In calculations of ρ_e^2 for the unfilled acoustic impulse, we used a model of the refractivity index jump with the transitive zone smaller in extension than $\lambda_s/4$ [16] and both the expression obtained in from the Fresnel zone and relation (4) given above. In this case, we have obtained that

$$\rho_e^2 = \frac{(\Delta N)^2}{4} \cdot 10^{-12} = 7.2 \cdot 10^{-15} (\Delta P_A)^2 . \qquad (9)$$

As follows from the given numerical estimate, the reflection coefficients do not noticeably differ. This indicates the possibility of practical realizability of the RASS method with a source of sound as a single unfilled acoustic impulse analogous to the rifle shot one. More successfully there can be used the arc discharge sources of sound, capable to create, according to experimental data [5], the sound pressure of up to 3550 Pa at the distance of 1 m.

2.2. Modeling of propagation of unfilled acoustic impulses in air

When designing RASS operating in accordance with the proposed method , it is necessary to take into account the fact that nearby sources of unfilled acoustic impulses, there takes place degeneration of the acoustic impulse into the weak shock wave whose speed differs from the current sound speed. From a family of plots presented in figure 1, one can judge about how necessary is such an account under real conditions and at what part of the distance of sounding the acoustic impulse from the gun shot moves with a speed exceeding the adiabatic one. These numerical estimations are performed by Dr. Odintsov S.L. from IAO SB RAS, who kindly presented them to us.

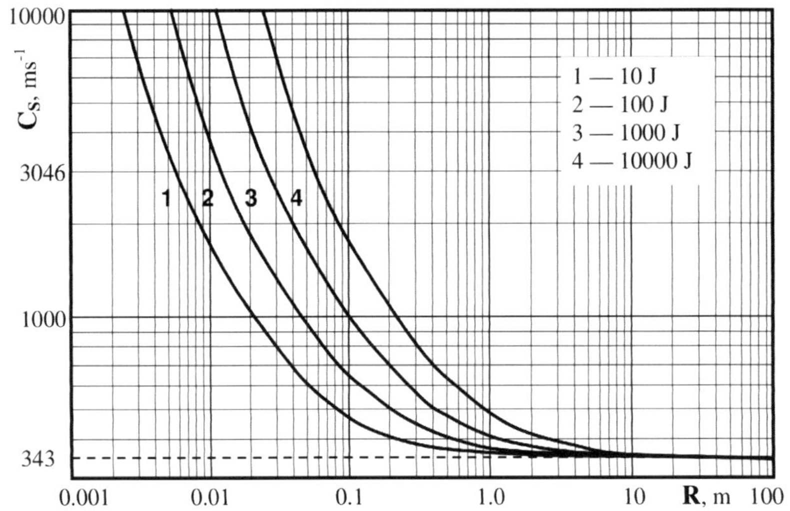

Figure 1. Dependence of the wave front speed on the distance from the source for different values of shot energy.

It is seen from figure 1 that a size of the area, wherein the nonlinear effects reveal themselves, depends on the energy characteristic of the source, and for powers of impulse sources of up to 10000 J does not exceed 20 m. An extension of a zone, for which one can tell about the evident shock wave, is restricted by units of meters.

The wideband acoustic signal is the superposition of harmonics of different frequencies. Therefore, when considering a model of its propagation, it is expedient to represent its shape as the Fresnel integral. When modeling propagation of such a signal, the energy losses should be taken into account. Since the absorption factor depends on its frequency, then the amplitude of each component of wideband signal harmonics will vary with the distance by its own law.

Thus, when such a signal propagates in the atmosphere, beside of the general decrease in its amplitude, its shape will continuously vary. In order to build the mathematical model, we have used a shape of the real sound impulse at the gun shot, as in [5]. The wideband acoustic impulse can be considered as an aperiodic function as the sound pressure versus time, which becomes null at infinity, and it can be presented as the Fourier integral. For simplificity, assume that sound propagates in the homogeneous air volume, wherein normal conditions are established, such as T_a=293.15 K, P_a=1013 mbar, H_a=80 %. In this case, the vertical gradient of the absorption factor will degenerate into the linear dependence on the distance.

In view of this assumption, the relations for modeling take the form

$$F_s(\omega) = \frac{1}{2\pi}\int_{-\infty}^{\infty} p(t, r=1)\, e^{i\omega t}\, 10^{-\frac{1}{20}\int_1^r \alpha_\omega(r)dr}\, dt; \tag{10}$$

$$p(t,r) = \frac{2}{r}\int_{-\infty}^{\infty} F_s(\omega)\, e^{-i\omega t}\, d\omega. \tag{11}$$

As a result of the modeling, the distance dependences of the spectrum and the shape of the acoustic impulse are obtained , which are presented in figures 2 ad 3, respectively.

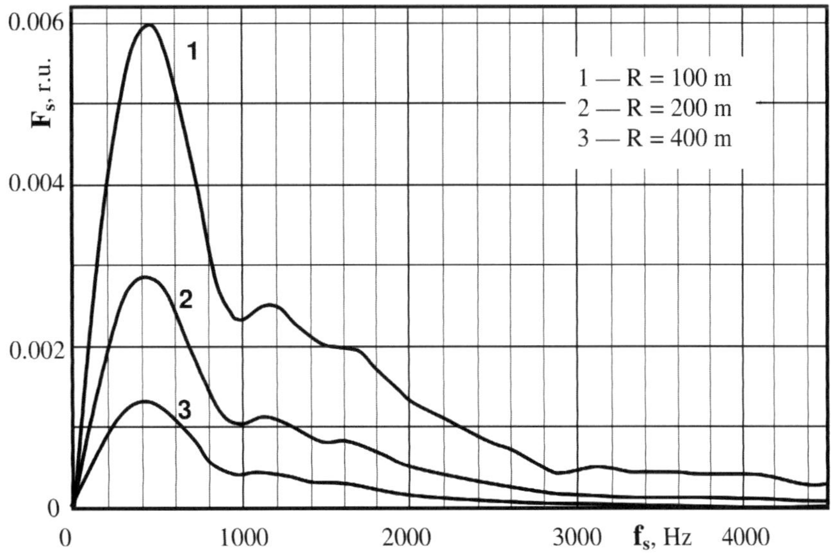

Figure 2. Variations of the rifle shot acoustic impulse spectrum versus the distance from the source.

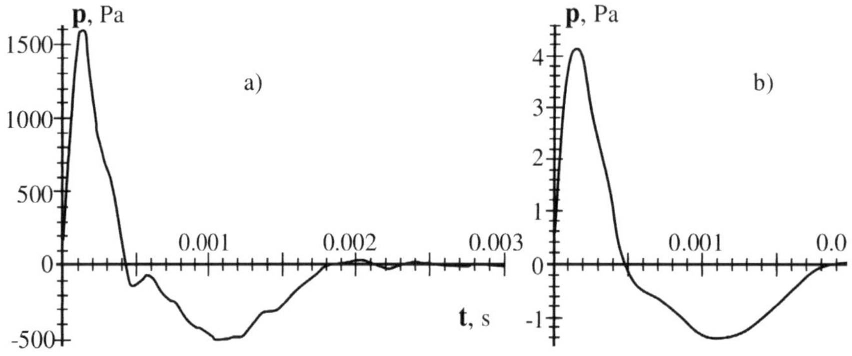

Figure 3. Variations of the acoustic impulse shape with the increase in the distance to the source: a) R = 2 m; b) R = 400 m.

The analysis of the dependence of the wideband signal spectrum on the distance to the source shows the strong frequency dependence of the sound absorption factor we calculated (α_ω) on the basis of data of the ISO 9613-2:1996 standard. Nevertheless one can easily note that at the set interval of heights for the most significant part of the spectrum, there is no its significant shift. Maximum of the pressure spectrum is at the frequency interval of 150 to 900 Hz, what is well agreed with the results of other known investigations in the area of the sound radiation from rifle shots [17, 18].
Plots of variations of the acoustic impulse shape in figure 3 show that only small fall of its reflectivity for electromagnetic waves can be expected, which is connected to the decrease in the pulse tilt at the distance under consideration.

2.3. *Estimation of cross-section of the gun shot*
Let us calculate the effective cross-section (ECS) of the front of the modulated acoustic impulse. Under assumption of the atmosphere inhomogeneity and of the absence of the cross wind, the front of the acoustic impulse will be a segment of the sphere, whose ECS, according to [19], looks like

$$\sigma_i = \rho_e^{\ 2} \frac{\pi a^2 R^2}{\left(R^2 - a^2\right)} \tag{12}$$

where a is the front curvature radius (the distance to the source); ρ_e^2 is the coefficient of power reflection of radio waves from the pressure differential created by the impulse front.
When suggesting that the source of the acoustic impulse has the radiation pattern f(φ), then the front impulse ECS takes the form

$$\sigma_i = \rho_e^{\ 2} \frac{\pi \left(a \sin\varphi/2\right)^2 R^2}{\left(R^2 - a^2\right)}, \tag{13}$$

When calculating the front impulse ECS, we assumed the following:
1) the source of acoustic impulses is shifted by the distance of 1 m with respect to the phase center of the radar antenna along its radiation pattern axis;
2) axes of radiation patterns of the acoustic source and radiolocation antenna coincide;
3) the value of the radiation pattern of the acoustic source is 20°.

Calculated ECS values of the impulse (σ_i) and those of turbulent inhomogeneities in the air volume closed in the radar radiation pattern (σ_t) are presented in table 1.

Table 1. Values of ECS for different heights.

Distance to source *R*, m	ECS of impulse, σ, m^2	ECS of turbulent inhomogeneities, σ, m^2
200	$4.1560 \cdot 10^{-5}$	$2.5839 \cdot 10^{-7}$
400	$7.5396 \cdot 10^{-5}$	$2.0517 \cdot 10^{-6}$
600	$1.0367 \cdot 10^{-4}$	$6.9072 \cdot 10^{-5}$
800	$1.2694 \cdot 10^{-4}$	$1.6352 \cdot 10^{-5}$
1000	$1.4593 \cdot 10^{-4}$	$3.1914 \cdot 10^{-5}$

Comparison of data in table 1 shows the excess of σ_i relative to σ_t by two orders at the height of up to about 200 m with the decrease down to one order at the height of up to 1000 m.

3. Conclusion

In the course of the work, the possibilities of using for radioacoustic sounding of the atmosphere the wideband unfilled acoustic signals similar to sound impulses at shots and to those created by powerful electro spark generators of sound are analyzed. The mathematical model is created for propagation of the acoustic impulse in the neutral atmosphere in view of divergence, classical and relaxation absorption of sound waves in the air. Values of impulse front ECS for radio waves are defined.
Comparison of ECS values of the impulse front with those of the atmospheric turbulent inhomgeneities shows that the echo-signal level from the impulse front exceeds that of the echo-signal from the atmospheric inhomogeneities. This confims the possibility of using the wideband unfilled acoustic signals in sodar and RASS for sounding the atmosphere.

4. References

[1] Danilov S D, Kallistratova M A 1987 Estimations of a nonlinear sound depletion at radioacoustic sounding of the atmosphere *Proceedings of the Soviet Academy of Sciences, FAO* **23 9** pp 997-999 (in Russian)
[2] Galiamina I P 1979 *Ultrasound. Small encyclopedia.* (Moscow Soviet encyclopedia) 400 p (in Russian)
[3] Markula T 2006 Propagation, measurement and assessment of shooting noise. *Master's thesis submitted in partial fulfillment of the requirements for the degree of Master of Science in Technology* Espoo 73 p At http://lib.tkk.fi/Dipl/2006/urn006257.pdf.
[4] Altmann J 2002 Acoustic Weapons – A Prospective Assessment *Science and Global Security* **2 9** pp 165-234
[5] Boesch H Edwin, Benwell Jr Bruce T, and Ellis Vincent J 2000 A High Power Electrically Driven Impulse Acoustic Source for Target Effects Experiments and Area Denial Applications *Army Research Laboratory* Adelphi MD At <http://www.dtic.mil/ndia/NLD3/boe.pdf>
[6] Ulyanov Y N, Podolyan A M 1982 On the use of unmonochromatic sonic radiation for radioacoustic sounding of the atmosphere *Proc. 7th All Union Simp. on Laser and Acoustic. Atm. Sounding* Tomsk **2** pp 141- 144 (in Russian)
[7] Ulyanov Y N, Prokopenko Y V, Maksimova N G 1996 Radioacoustic sounding with the use of single sonic impulses *Proc. of Second. Int. conf. on Theory and techniques of transmitting, receiving and processing of information* Tuapse – Kharkov **2** pp 53-54 (in Russian)
[8] Kallistratova M A, Kon A I 1985 *Radioacoustic sounding of atmosphere* (Moscow Nauka edition) 197p (in Russian)

[9] Cage K S, Balsley B B 1980 On the scattering and reflection mechanisms contributing to clear air radar echoes from troposphere, stratosphere and mesosphere *Radio Science* **15** **2** pp 243-257

[10] Rudenko OV et al 1982 *Foundations of nonlinear acoustics* (Moscow Nauka) 176 p (in Russian)

[11] Bochkarev N N, Klochkov V A, Krasnenko N P et all 1988 Powerful acoustic array for atmospheric researches *In the book "Sonic and optical waves propagation in the atmosphere"* Tomsk IOA pp 101-104 (in Russian)

[12] Gorelik A G, Zhezherin V R, Kallistratova M A et all 1987 Radioacoustic ABL sounding of with automatic frequency control *Meteorological investigations* **28** pp 82-92 (in Russian)

[13] Misajlov VL, Ulyanov YN 2005 Operational measuring of the refraction index for tropospheric ducts diagnostic *Collected. sci. papers on Information processing systems* Kharkiv Univ. of Air Force **6** **46** pp 82-109 (in Russian)

[14] Ulyanov Y N 1984 Double frequency RASS *Proc. of VIII All Union Symp. on Lidar and Acoustic. Atm. Sounding* Tomsk **2** pp 196-199(in Russian)

[15] Don C G and Cramond A J 1987 Impulse propagation in a neutral atmosphere *The Journal of the Acoustical Society of America* **81** **5** May pp 1341-1349

[16] Atlas D. 1964 Advances in radar meteorology *Advances in Geophysics* **10** pp 317-478

[17] Ylikoski M, Pekkarinen J O, Starck J P, Pääkkönen R J, Ylikoski J S. 1995 Physical characteristics of gunfire impulse noise and its attenuation by hearing protectors *Scandinavian Audiology* **24** **1** pp 3-11

[18] Bronkhorst A W 1996 Sound Exposure Caused by the Automatic Rifle Diemaco C7A1 and the Mortar 60 mm Hotchkiss-Brandt (Geluidbelasting veroorzaakt door het geweer diemaco C7A1 en het mortier 60 mm Hotchkiss-Brandt*) Final rept., ADA320703, METAALINSTITUUT TNO APELDOORN (NETHERLANDS)* **14** At <http://handle.dtic.mil/100.2/ADA320703

[19] Chernikov A A 1979 *Radar reflection from the clear atmosphere* (Leningrad Gidrometeoizdat) 46 P (in Russian)

Doppler Lidar measurements of winds and turbulence in the boundary layer

Rod Frehlich

CIRES, University of Colorado, Boulder, CO USA 80309

E-mail: rgf@cires.colorado.edu

Abstract. Coherent Doppler lidar is ideally suited for accurate measurements of the profiles of wind and turbulence using various beam scanning and processing techniques. The most attractive scanning pattern is a three dimensional pattern which samples the largest number of independent samples of the turbulent field. Careful processing of the radial velocity estimates is required to correct for the spatial filtering by the lidar pulse and to remove the bias introduced by the estimation error. The various processing techniques and fundamental assumptions will be presented. Lidar derived profiles of wind and turbulence will be compared with in situ data, when available. Results will also be presented from a field campaign focused on wind energy applications.

1. Introduction

Accurate profiles of wind speed, wind direction, and turbulence statistics are critical for fundamental studies of boundary layer processes. The most important practical applications are monitoring wind shear and turbulence at airports, homeland security, and in the wind energy industry for wind resource assessment and efficient operation of wind farms. Coherent Doppler lidar has become a mature technology that provides high spatial and temporal resolution measurements of wind speed and wind statistics [1]-[5] and are ideally suited to study the Atmospheric Boundary Layer (ABL). The performance of Doppler lidar depends on the lidar parameters, the atmospheric conditions, and the velocity estimation algorithm [6]-[24]. The most promising Doppler lidar design for providing accurate profiles of the mean and turbulent wind field statistics in the boundary layer is an eye-safe solid-state scanning coherent Doppler lidar [3]-[5]. A short laser pulse with a Gaussian temporal profile is transmitted as a collimated beam and the photons scattered by aerosol particles are converted into an oscillating signal with a Doppler frequency shift f that is proportional to the radial velocity v_r of those aerosol particles illuminated by the lidar pulse over the processing range-gate. Various processing algorithms have been developed to produce accurate radial velocity estimates [1]-[21]. The statistical accuracy of the radial velocity estimates is well documented by computer simulations [12]-[23] and from statistical analysis of Doppler lidar data [24].

Profiles of turbulent statistics have also been produced from Doppler lidar radial velocity statistics by removing the contribution from estimation error [12] and by careful correction for the spatial filtering of the velocity field by the lidar pulse [7], [10]-[13] as well as using the lidar signal spectral width [13]. Various lidar processing techniques have been proposed to extract useful information about the mean and turbulent statistics of the velocity field based on the spatial statistics of the radial velocity along the laser beam axis, i.e., the longitudinal velocity statistics. The energy dissipation rate ϵ_u, the radial velocity variance σ_u^2, and the outer scale L_{0u} of the turbulence have been estimated by fitting measured velocity structure functions along the lidar beam to simple theoretical models [7]-[14]. Since many adjacent range-gates are required to resolve the spatial scales of interest, this technique does not produce high resolution profiles of the turbulence statistics which are critical for understanding the structure and evolution of the stable boundary layer. However, this technique can be used to provide reasonable vertical resolution if the elevation angle is small. Another difficulty associated with boundary layer measurements of turbulence statistics is that, because of the inherent spatial variability or intermittency, accurate measurements of turbulence profiles require a sufficiently large number of independent samples N_I [25]. New lidar scanning configurations and processing algorithms were developed to increase N_I and to improve the vertical resolution of the profiles of turbulent statistics [14] by fitting the structure function of the radial velocity as a function of the arc distance in the azimuth direction for a collection of fixed elevation angles (Plan Position Indicator PPI) which provides estimates of the transverse velocity statistics ϵ_v, σ_v^2, and L_{0v}. The radial and azimuth structure function techniques will be compared with high resolution in situ data to evaluate profiles of the longitudinal and transverse velocity turbulence statistics.

In many cases, the spatial velocity statistics do not have a well defined length scale (outer scale of turbulence), especially above the boundary layer [14] and a rigorous definition of the turbulent velocity fluctuations and velocity variance is not possible. Typically, a spatial or temporal filter of data is performed to separate the local mean and turbulent fluctuations. High resolution scanning Doppler lidar is ideally suited for investigating the impact of this filtering because it provides a three dimensional field.

2. Doppler Lidar Measurements

The most accurate boundary layer profiles of wind statistics are produced by a three-dimensional scanning pattern which is chosen as a collection of Plan Position Indicator (PPI) scans, i.e., a variable azimuth angle ϕ for a set of fixed elevation angles θ (see [14] Fig. 2). For each radial velocity measurement at range R, the mean radial velocity is well approximated by [6, 7]

$$m_{wgt}(R,\phi,\theta) = \frac{1}{\Delta\phi}\int_{\phi-\Delta\phi/2}^{\phi+\Delta\phi/2}\int_{-\infty}^{\infty} v_r(s,\phi',\theta)W(R-s)\,ds\,d\phi' \tag{1}$$

where $v_r(r,\phi,\theta)$ is the random radial velocity along the lidar beam, $W(r)$ is a normalized weighting function that depends on the lidar pulse and process range-gate, and $\Delta\phi$ is the change in azimuth angle during the total measurement time.

The lidar transmits a nearly Gaussian pulse with an intensity temporal profile $\propto \exp(-t^2/\sigma_P^2)$ where t denotes time and σ_P defines the pulse length [9, 24]. The lidar pulse range-weighting function for each instant of time is given by

$$I_n(r) = \frac{1}{\sqrt{\pi} r_P}\exp\left[-\frac{r^2}{r_P^2}\right] \tag{2}$$

where the 1/e radius is

$$r_P = c\,\sigma_P/2\,, \tag{3}$$

c is the speed of light, and

$$\Delta r = \sqrt{\ln 2}\,c\,\sigma_P \tag{4}$$

is the Full Width at Half Maximum (FWHM) of the lidar pulse sensing volume in range which defines the extent of the illuminated aerosol targets. The lidar range-weighting function is then given by

$$W(r) = \frac{1}{\Delta p}\int_{-\frac{\Delta p}{2}}^{\frac{\Delta p}{2}} I_n(r+s)ds = \frac{1}{2\Delta p}\left(\mathrm{Erf}[(r+\Delta p/2)/r_P] - \mathrm{Erf}[(r-\Delta p/2)/r_P]\right) \tag{5}$$

where

$$\mathrm{Erf}(z) = \frac{2}{\sqrt{\pi}}\int_0^z \exp(-t^2)dt \tag{6}$$

is the error function.

The data presented here are from the 2-μm eye-safe WindTracer lidar manufactured by Coherent Technologies Inc. (CTI) [3, 4]. The lidar parameters were set to a pulse width $\Delta r=$ 66.0 m, a range-gate length Δp = 72.0 m, an azimuth scan rate of 2.5^o per second, a total measurement time per radial velocity profile T = 0.2 second, and therefore the azimuth spacing $\Delta\phi$ = 0.5^o. The range-weighting function for the given lidar parameters is shown in Fig. 1. Note that the FWHM of $W(r)$ is ΔR = 85.8 m, which defines the effective resolution of the lidar radial velocity measurements.

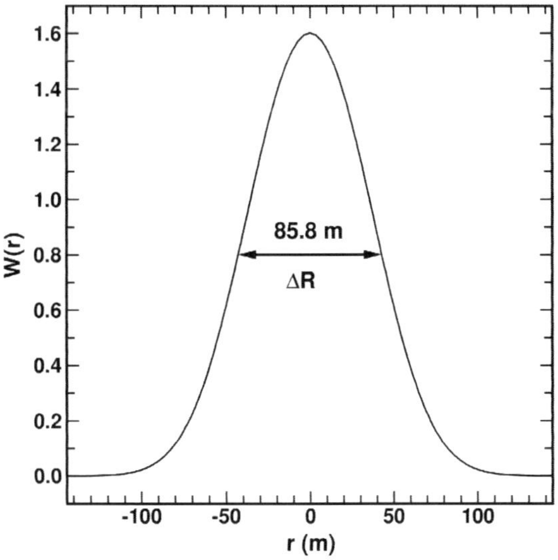

Figure 1. Lidar range weighting function $W(r)$ for Δp=72 m, Δr=66 m and ΔR=85.8 m.

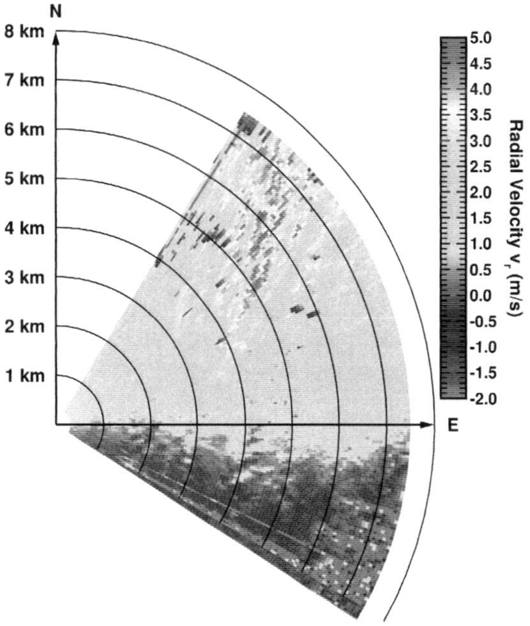

Figure 2. Radial velocity map for zero elevation angle.

An example of the radial velocities for a single PPI scan for early morning convection is shown in Fig. 2. The lidar is located 57 m above a flat region and the velocity statistics are approximately homogeneous for ranges up to 4 km. There are some obstructions at azimuth angles greater than 110°. Profiles of the mean wind vector are produced with a standard mean-square error algorithm based on sub sections of a PPI scan assuming the horizontal wind field is uniform and the average vertical velocity is zero. An example of the radial velocity as a function of azimuth angle for a fixed range-gate in Fig. 2 is shown in Fig. 3 as well as the best-fit mean velocity assuming a constant wind speed and direction [14, 26]. The turbulent fluctuation of Doppler lidar radial velocity is defined as the deviations from the best-fit model and is therefore based on the sptial variability of the velocity instead of the deviations from a time-average. The random scatter is mostly produced by the turbulence because the estimation error of each measurement is typically less than 0.1 m/s [24].

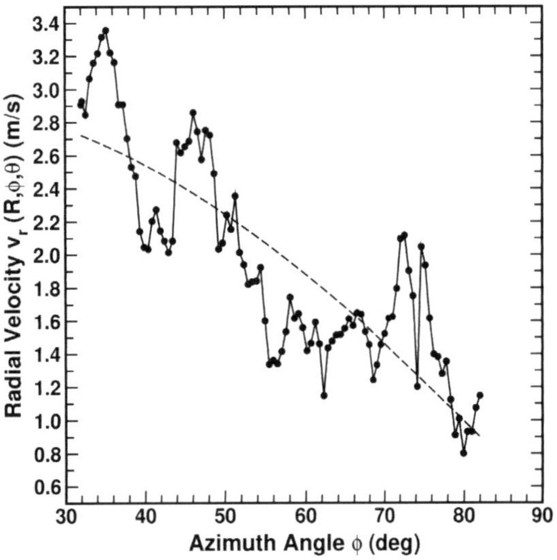

Figure 3. Radial velocity estimates $v_r(R, \phi, \theta)$ vs azimuth angle ϕ (bullets) for a range gate distance $R = 1200$ m from the data shown in Fig. 2. The radial velocity for the best-fit wind speed (2.94 m/s) and direction (189.8^o) is shown as a dashed line.

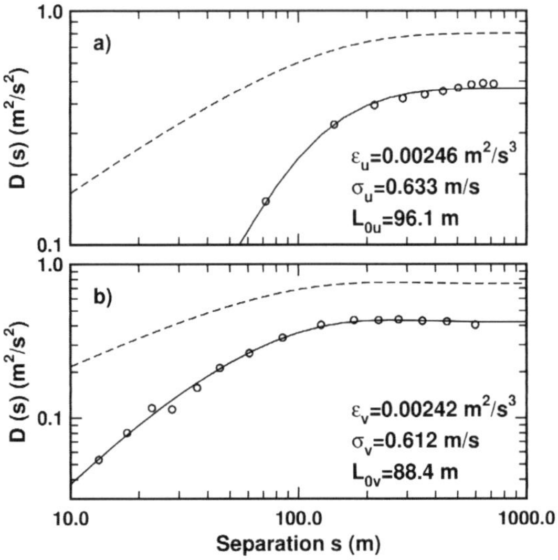

Figure 4. Structure function estimates of turbulence for 21 minutes of data (5 consecutive scans starting with the data shown in Fig. 2) with a) the longitudinal structure function technique and b) the azimuth structure function technique. The average Doppler lidar structure functions are shown as circles, the best-fit theoretical models are solid lines, and the corresponding best-fit von Kármán model are dashed lines.

Unbiased estimates of the turbulence statistics of the velocity field can be produced from structure functions of the Doppler lidar radial velocities if: 1) a universal description of the structure functions are known a priori, e.g., a von-Kármán model, 2) the effects of the spatial averaging by the finite extent of the lidar pulse and processing algorithms are included in the analysis, and 3) the contribution from the estimation error in the radial velocities is removed which is essential for the low-turbulence regimes ([12, 14]).

Estimates of the radial velocity statistics (longitudinal) are produced from the structure function estimates $\hat{D}_{wgt}(s)$ calculated in the direction of the lidar beam by correcting for the spatial averaging of the lidar pulse and the contribution from the estimation error of the velocity estimates [14]. The CTI lidar is well suited for turbulence measurements because of a stable Gaussian lidar pulse [24]. For a Gaussian lidar pulse with width Δr and a von Kármán model for the turbulent wind field [10]

$$D_{wgt}(s, \sigma_u, L_{0u}) = 2\sigma_u^2 G(s/\Delta p, \chi, \mu) \tag{7}$$

where $\chi = \Delta p / L_{0u}$ and $\mu = \sqrt{2 \ln 2} \Delta p / \Delta r$. The parameters σ_u, ϵ_u, and L_{0u} are determined by minimizing the weighted error between the structure function estimates $\hat{D}_{wgt}(s)$ and the model $D_{wgt}(s, \sigma_u, L_{0u})$

Improved vertical resolution of Doppler lidar estimates of the turbulence parameters is produced using structure functions of the radial velocity as a function of the spatial separation in the azimuth direction (ϕ) for each range gate distance R and elevation angle θ under the assumption that the temporal evolution of the velocity field is small during the PPI sector scan. The turbulent fluctuations are defined as the deviations of the Doppler lidar estimates from the best fit windspeed and direction as shown in Fig. 3.

The Doppler lidar azimuth structure function is given by [14]

$$D_{wgt}(s, \sigma_v, L_{0v}) \quad = \quad 2\sigma_v^2 G_\phi(s/\Delta p, \chi, \mu) \tag{8}$$

where s denotes arc distance along the azimuth scan, $\chi = \Delta p / L_{0v}$, and σ_v is the standard deviation of the transverse velocity fluctuations. The parameters σ_v, ϵ_v, and L_{0v} are determined by minimizing the chi-squared error between the noise-corrected structure function estimates $\hat{D}_{wgt}(s)$ and the model predictions $D_{wgt}(s, \sigma_v, L_{0v})$.

An example comparison of the two different turbulence estimation techniques for the data of Fig. 2 is shown in Fig. 4. Note that the effects of the spatial filtering by the lidar pulse are much less pronounced for the azimuth structure function technique. All the turbulence parameters are in good agreement indicating that the turbulence field is well described by the homogeneous and isotropic von Kármán model. This is consistent with previous measurements in convective boundary layers [27]-[29].

Profiles of the turbulence statistics are produced from radial and azimuth structure functions calculated from all the radial velocity data in given altitude intervals. To maintain similar vertical resolution, only the lower elevation angles are used for the structure functions in the radial direction. An example comparison of the two different turbulence estimation techniques is shown in Fig. 5 for weak turbulence conditions at an altitude interval of 70-90 m. The ϵ estimates are in good agreement but L_0 and σ do not agree. It is difficult to evaluate this disagreement because of various competing artifacts such as the effects of the long pulse length, the functional form of the true atmospheric turbulence model, and inhomogeneities in the turbulence field over the processing domain.

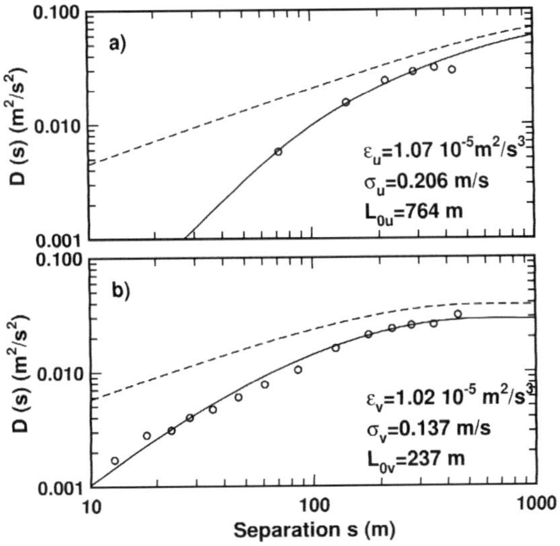

Figure 5. Structure function estimates of turbulence for a low turbulence condition with a) the longitudinal structure function technique and b) the azimuth structure function technique. The average Doppler lidar structure functions are shown as circles and the best-fit theoretical models are lines [Eq. (7) and (8)], and the corresponding best-fit von Kármán model are dashed lines.

Comparison of the in situ profiles and lidar derived profiles for a stable night time boundary layer (11.57-11.95 LT on June 6 2006, positive gradient in potential temperature indicates stable conditions) are shown in Fig. 2. The large variability in the in situ turbulence estimates (C_T^2 and ϵ) are intrinsic atmospheric variability (intermittency) and not measurement error [30]. The boundary layer height or mixing height H is defined as the sharp drop in turbulence near the first minimum in the gradients of temperature, wind speed, and relative humidity, i.e., $H \approx 100$ m. The lidar derived estimates of ϵ agree above 70 m indicating isotropy of the small scale turbulence and all three turbulence parameters agree in the height interval 100-120 m. However, there are some noticeable differences between the in situ profiles of ϵ indicating spatial variability of the turbulent field. The large scatter in L_0 above 140 m altitude is a common occurrence in the night time residual layer, especially under low turbulence conditions where large scale features such as inertial gravity waves are superimposed on the weaker small scale turbulence.

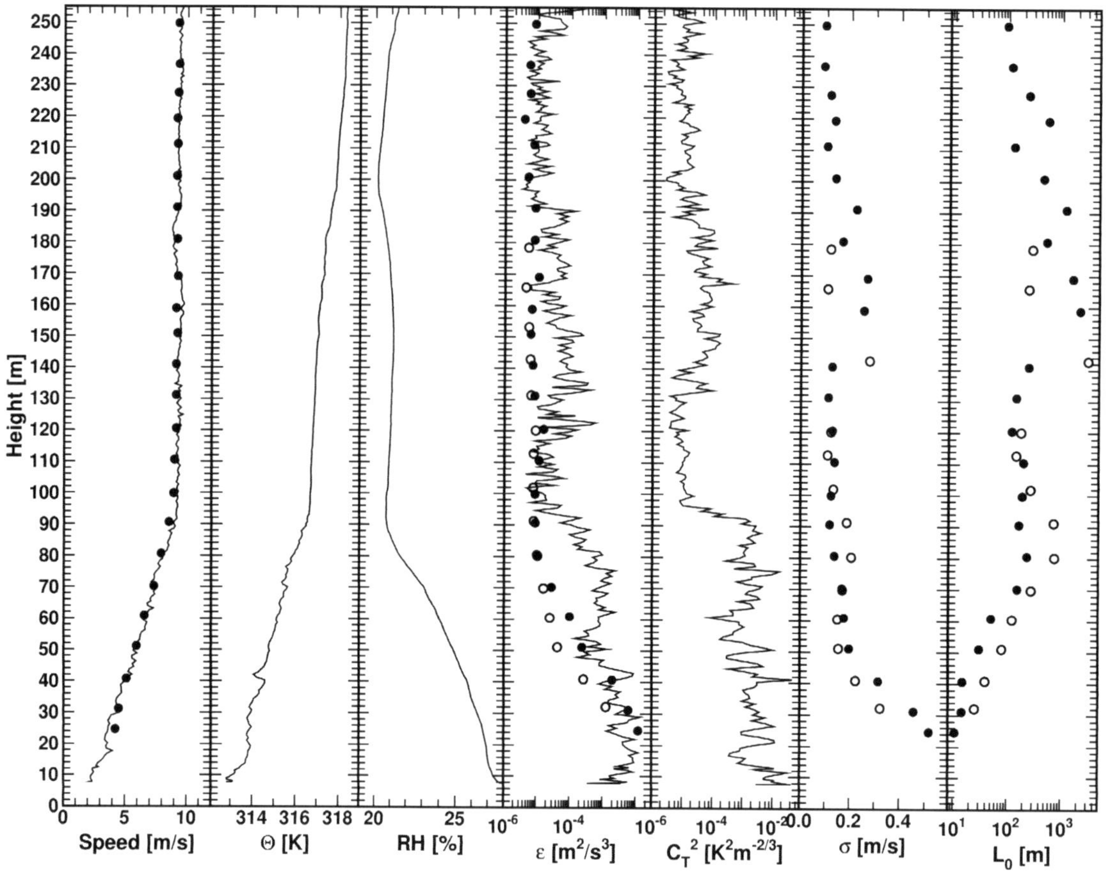

Figure 6. Comparison of in situ measurements (solid line), lidar derived measurements with the azimuth structure functions (bullets), and the radial structure functions (open circles) vs height. The panels are wind speed, potential temperature Θ, relative humidity RH, energy dissipation rate ϵ, temperature structure constant C_T^2, velocity standard deviation σ of one horizontal component, and outer scale L_0.

3. Wind Energy Applications

The mean and turbulent structure of the atmospheric boundary layer, especially for stable night time conditions, is important for wind energy [31]-[35]. Accurate profiles of the mean wind and turbulent statistics over short enough time intervals to monitor a rapid change in conditions is important for optimal operation of wind farms.

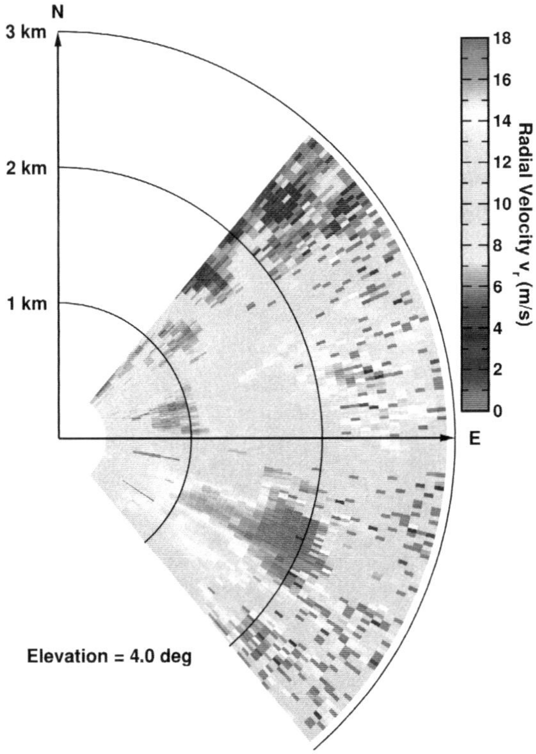

Figure 7. Doppler lidar radial velocity v_r vs azimuth angle ϕ and range for and elevation angle $\theta = 4^o$.

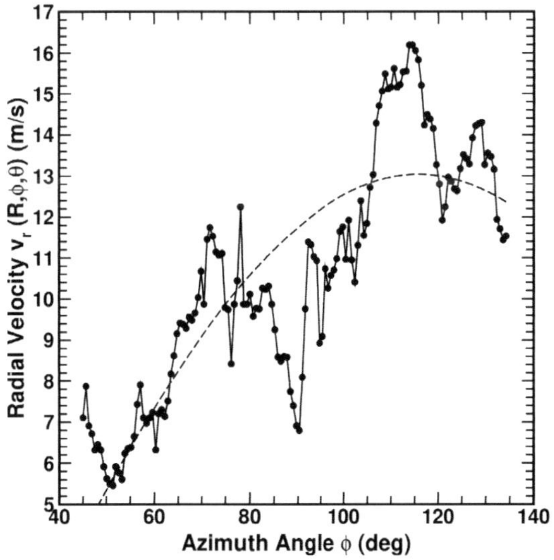

Figure 8. Radial velocity v_r vs azimuth angle ϕ from the data shown in Fig. 7 for a range-gate distance R = 1248 m, altitude H = 144 m, and best fit wind speed and direction (dashed line) are 13.1 m/s and 295°, respectively.

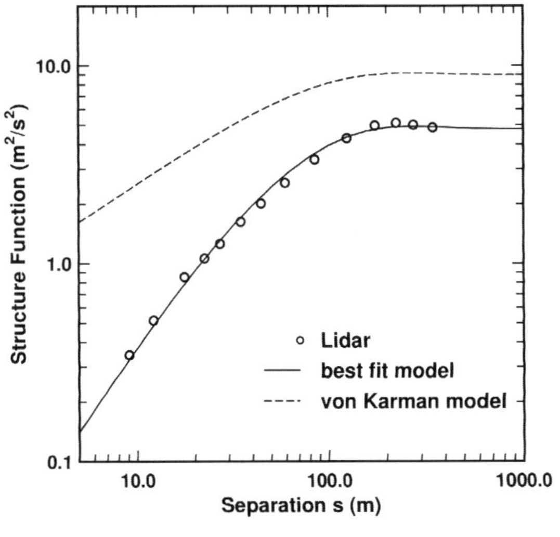

Figure 9. Lidar azimuth structure function (o), best-fit model including spatial filtering by the lidar pulse (solid line) [Eq. (7) and (8)], and the corresponding von-Kármán model (dashed line) for an altitude H=69 m, wind speed U=11.0 m/s, wind direction is 290°, ϵ=0.089 m^2/s^3, L_0=137 m, and σ=2.36 m/s.

Fig. 7 shows an example of Doppler lidar radial velocity data for a single azimuth scan at an elevation of 4° from a higher wind period of interest for wind energy applications. The large turbulent eddies are clearly visible as orange to red colored features. There are some random outliers beyond a range of approximately 1.8 km because of the low aerosol content. Note that this is night time data in a stable boundary layer. An example of the radial velocity as a function of azimuth angle for a fixed range-gate in Fig. 7 is shown in Fig. 8 as well as the best-fit mean velocity assuming a constant wind speed and direction [14, 26]. The large random fluctuations are produced from the atmospheric variability (turbulence) because the estimation error of each measurement is typically 0.1 m/s. These large velocity variations can produce large estimation errors in all the velocity statistics using tower data if short averaging times are imposed by the time scale of local stationarity [25]. However, the 3D sampling of the scanning Doppler lidar provides a larger number of independent samples of the atmospheric processes compared with tower data, and therefore, the lidar produces the most accurate statistics, when the turbulent field is approximately homogeneous and stationary over the analysis space-time domain.

Fig. 9 shows an average Doppler lidar azimuth structure function with the best-fit von-Kármán model for the case of a well resolved turbulent eddy size L_0 based on the data shown in Fig. 7. Fig. 10 shows the wind speed and turbulence profiles for the time interval 21:37-22:00 LT on March 9, 2007. The wind speed and direction estimated from the single range-gate shown in Fig. 8 is in good agreement with the average profile of Fig. 10, even with the large variations in the radial velocity with azimuth ϕ. Most of the scatter in the profile of wind direction is estimation error produced by the high turbulence level σ and large outer scale L_0. Note that the size of the eddies L_0 increase with height and above the 180 m altitude, it is difficult to estimate L_0. The standard metric of turbulence $\sigma/U \approx 0.2$ at a hub height of 90 m exceeds the IEC Class A value of 0.16 for high turbulence [36].

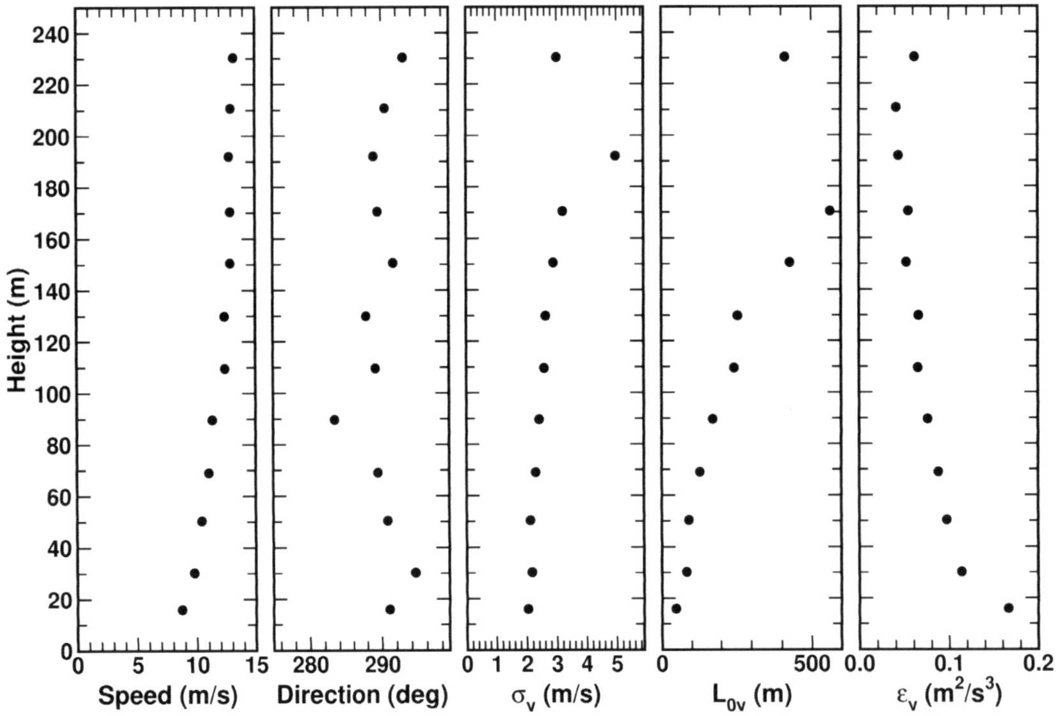

Figure 10. Profiles of wind speed U, direction, σ_v, L_{0v}, and ϵ_v from the azimuth structure function analysis (bullets).

4. Conclusion

Scanning coherent Doppler lidar is ideally suited for boundary layer measurements of the mean and turbulent velocity statistics. A three dimensional volume scan using multiple PPI scans provides timely updates over a region of interest that is essential for producing a large number of independent samples of the atmospheric processes which is difficult for other measurement systems such as towers and aircraft. However, to accurately estimate turbulent statistics, especially in regions of low turbulence, careful removal of the estimation error is important [12, 14]. There are critical lidar design issues since estimation error is smaller for a long lidar pulse but better spatial resolution is produced with a shorter pulse. For the CTI WindTracer system, the 66 m pulse width and 72 m range gate typically produces a radial velocity estimation error $\sigma_e \approx 0.05 - 0.1$ m/s which results in a correction to structure function estimates of $2\sigma_e^2 \approx 0.005 - 0.02$ m^2/s^2 which is feasible with current algorithms and 20 minutes of data [12]. A shorter pulse system may require more accurate estimation error correction for low turbulence conditions or a modification of the lidar operating parameters to reduce the estimation error with pulse accumulation [8].

The two lidar techniques for estimating the turbulent statistics of the radial (longitudinal) velocity and transverse velocity (structure function in the azimuth direction) can be used simultaneously to provide a more complete statistical description of the boundary layer processes. However, the radial technique is limited to low elevation angles and does not have the vertical resolution or the sensitivity to the small scales shown by the azimuth technique (see Fig. 6 and [14]). Even with these limitations, good agreement is produced for the small-scale statistic ϵ and in regions of horizontal isotropy, good agreement is produced in all the turbulent statistics (Fig. 6, 90-110 m height). The lidar derived turbulence profiles have an added advantage of improving the statistical accuracy compared with the in situ profiles.

The optimal design of scanning coherent Doppler lidar for wind energy applications is focused on higher wind and higher turbulence conditions. A shorter lidar pulse width and a smaller range-gate would provide more accurate estimates of the turbulence statistics and produce better profiles of the longitudinal velocity statistics, i.e., the statistics of the radial velocity in the direction of the lidar beam. The assumption of a universal model for the turbulence statistics requires more investigation and realistic turbulence descriptions [29, 37, 38] should improve the estimation of the important turbulence parameters. More research is required to determine the optimal update rate, lidar design, analysis algorithms, and turbine control algorithms for wind energy applications. Long term observations of winds and turbulence at an operational wind farm are needed to better understand the affect of atmospheric conditions on power production and maintenance issues such as excessive loading and vibration.

Acknowledgments. This work was supported by the National Science Foundation (NSF) and the Army Research Office (ARO). The authors acknowledge useful discussions with Neil Kelley, Jacob Mann, Jerry Crescenti, Justin Sharp, Andrew Oliver, Greg Poulos, Robert Sharman, and Steve Hannon. Keith Barr, Steve Hannon, and Gerard Pelk of Lockheed Martin Coherent Technologies, Inc. provided the Doppler lidar data.

References

[1] Menzies R T and Hardesty R M 1989 Coherent Doppler lidar for measurements of wind fields. *Proc. IEEE,* **77** 449-462

[2] Huffaker M R and Hardesty R M 1996 Remote sensing of atmospheric wind velocities using solid-state and CO_2 coherent laser systems. *Proc. IEEE,* **84** 181-204

[3] Henderson S W, Hale C P, Magee J R, Kavaya M J, and Huffaker A V 1991 Eye-safe coherent laser radar system at 2.1 μm using Tm,Ho:YAG lasers. *Opt. Lett.,* **16** 773-775

[4] Henderson S W, Suni P J M, Hale C P, Hannon S M, Magee J R, Bruns D L, and Yuen E H 1993 Coherent laser radar at 2 μm using solid-state lasers," *IEEE Trans. Geo. Remote Sensing* **31** 4-15

[5] Grund C J, Banta R M, George J L, Howell J N, Post M J, Richter R A, and Weickmann A M 2001 High-resolution Doppler lidar for boundary layer and cloud research. *J. Atmos. Oceanic Tech.* **18** 376-393
[6] Banakh V A and Smalikho I N 1997 Estimation of the turbulence energy dissipation rate from pulsed Doppler lidar data. *Atmos. Oceanic Opt.* **10** 957-965
[7] Frehlich, R G 1997 Effects of wind turbulence on coherent Doppler lidar performance. *J. Atmos. Ocean. Tech.* **14** 54-75
[8] Frehlich R 2004 Velocity error for coherent Doppler lidar with pulse accumulation. *J. Atmos. Ocean. Tech.* **21** 905-920
[9] Frehlich R, Hannon S, and Henderson S 1994 Performance of a 2μm coherent Doppler lidar for wind measurements. *J. Atmos. Ocean. Tech.* **11** 1517-1528
[10] Frehlich R, Hannon S, and Henderson S 1998 Coherent Doppler lidar measurements of wind field statistics. *Boundary-Layer Meteor.* **86** 233-256
[11] Davies F, Collier C G, Pearson G N, and Bozier K E 2004 Doppler lidar measurements of turbulent structure function over an urban area. *J. Atmos. Oceanic Tech.* **21** 753-761
[12] Frehlich R and Cornman L 2002 Estimating spatial velocity statistics with coherent Doppler lidar. *J. Atmos. Oceanic Technol.* **19** 355-366
[13] Smalikho I, Kopp F, and Rahm S 2005 Measurement of atmospheric turbulence by 2-μ Doppler lidar. *J. Atmos. Oceanic Technol.* **22** 1733-1747
[14] Frehlich R, Meillier Y, Jensen M L, Balsley B, and Sharman R 2006 Measurements of boundary layer profiles in an urban environment *J. Appl. Meteor. Clim.* **45** 821-837
[15] Zrnic D S 1979 Estimation of spectral moments of weather echoes. *IEEE Trans. Geosci. Electronics* **GE-17** 113-128
[16] Rye B J and Hardesty R M 1993 Discrete spectral peak estimation in incoherent backscatter heterodyne lidar. I. Spectral accumulation and the Cramer-Rao lower bound," *IEEE Trans. Geo. Sci. Remote Sensing* **31** 16-27
[17] Rye B J and Hardesty R M 1993 Discrete spectral peak estimation in incoherent backscatter heterodyne lidar. II. Correlogram accumulation," *IEEE Trans. Geo. Sci. Remote Sensing* **31** 28-35
[18] Frehlich R G and Yadlowsky M J 1994 Performance of mean frequency estimators for Doppler radar and lidar. *J. Atmos. Oceanic Technol.* **11** 1217-1230 Errata: *J. Atmos. Oceanic Technol.* **12** 445-446
[19] Frehlich R G 1996 Simulation of coherent Doppler lidar performance in the weak signal regime. *J. Atmos. Oceanic Technol.* **13** 646-658
[20] Dabas A 1999 Semiempirical model for the reliability of a matched filter frequency estimator for Doppler lidar. *J. Atmos. Ocean. Tech.* **16** 19-28
[21] Zarader J.-L, Dabas A, Flamant P H, Gas B, and Adam O 1999 Adaptive parametric algorithms for processing coherent Doppler-lidar signals. *IEEE Trans. Geosci. Remote Sensing* **37** 2678-2691
[22] Frehlich R G 2001 Estimation of velocity error for Doppler lidar measurements. *J. Atmos. Oceanic Technol.* **18** 1628-1639
[23] Banakh V A and Werner C 2005 Computer simulation of coherent Doppler lidar measurements of wind velocity and retrieval of turbulent wind statistics. *Opt. Eng.* **44** 1-19
[24] Frehlich R G, Hannon S, and Henderson S 1997 Coherent Doppler lidar measurements of winds in the weak signal regime. *Appl. Opt.* **36** 3491-3499
[25] Lenschow D H, Mann J, and Kristensen L 1994 How long is long enough when measuring fluxes and other turbulence statistics. *J. Atmos. Oceanic Technol.* **11** 661-673
[26] Browning K A and Wexler R 1968 The determination of kinematic properties of a wind field using Doppler Radar. *J. Appl. Meteor.* **7** 105-113
[27] Caughey S. J. and Palmer S G 1979 Some aspects of turbulence structure through the the depth of the convective boundary layer. *Quart. J. Roy. Meteor. Soc.* **105** 811-827
[28] Panofsky H. A., Tennekes H, Lenschow D H, and Wyngaard J C 1977 The characteristics of turbulent velocity components in the surface layer under convective conditions. *Bound.-Layer Meteor.* **11** 355-361
[29] Lenschow D H and Kristensen L 1988 Applications of dual aircraft formation flights. *J. Atmos. Oceanic Technol.* **5** 715-726
[30] Frehlich R, Meillier Y, Jensen M L, and Balsley B 2004 Statistical description of small scale turbulence in the low-level nocturnal jet. *J. Atmos. Sci.* **61** 1079-1085
[31] Blumen W, Banta R, Burns S P, Fritts D C, Newsom R, Poulos G S, and Sun J 2001 Turbulence statistics of a Kelvin-Helmholtz billow event observed in the night-time boundary layer during the Cooperative Atmosphere-Surface Exchange Study field program. *Dyn. Atmos. Oceans* **34** 189-204
[32] Banta R M, Pichugina Y L, and Newsom R K 2003 Relationship between low-level jet properties and turbulence kinetic energy in the nocturnal stable boundary layer. *J. Atmos. Sci.* **60** 2549-2555
[33] Fritts D C 2005 Conditions contributing to adverse loading of wind turbines in the nocturnal boundary layer:

Final report. *NREL/SR-500-37809*

[34] Banta R M, Pichugina Y L, and Brewer W A 2006 Turbulent velocity-variance in the stable boundary layer generated by a nocturnal low-level jet. *J. Atmos. Sci.* **63** 2700-2719

[35] Kelley N D, Jonkman B J, Scott G N 2006 The Great Plains turbulence environment: Its origins, impact, and simulation. *NREL/CP-500-40176*

[36] International Electrotechnical Commission, IEC, *Wind turbines - Part 1: Design requirements*, IEC 61400-1 Ed. 3., 2005.

[37] Kristensen L, Lenschow D H, Kirkegaard P, and Courtney M 1989 The spectral velocity tensor for homogeneous boundary-layer turbulence. *Bound.-Layer Meteor.* **47** 149-193

[38] Mann J 1994 The spatial structure of neutral atmospheric surface-layer turbulence. *J. Fluid Mech.* **273** 141-168

IOP Conf. Series: Earth and Environmental Science **1** (2008) 012018 doi:10.1088/1755-1307/1/1/012018

Windscanner: 3-D wind and turbulence measurements from three steerable Doppler Lidars

Torben Mikkelsen, Jakob Mann, Mike Courtney and Mikael Sjöholm

Wind Energy Department, Risø National Laboratory for Sustainable Energy, Technical University of Denmark, P.O. 49, DK-4000 Roskilde, Denmark.

Corresponding author: torben.mikkelsen@risoe.dk

Abstract. At RISØ DTU we has started to build a new-designed laser-based lidar scanning facility for detailed remote measurements of the wind fields engulfing the huge wind turbines of today. Our aim is to measure in real-time 3D wind vector data at several hundred points every second: 1) upstream of the turbine, 2) near the turbine, and 3) in the wakes of the turbine rotors. Our first proto-type Windscanner is now being built from three commercially available Continuous Wave (CW) wind lidars modified with fast adjustable focus length and equipped with 2-D prism-based scan heads, in conjunction with a commercially available pulsed wind lidar for extended vertical profiling range. Design, construction and initial testing of the new 3-D wind lidar scanning facility are described and the functionality of the Windscanner and its potential as a new research facility within the wind energy community is discussed.

1. Introduction

The objective of this paper is to describe our vision to establish a lidar scanning remote sensing facility for wind and turbulence measurements and turbine tests and to describe the technological challenges towards the realization of this vision, which is illustrated in figure 1. .

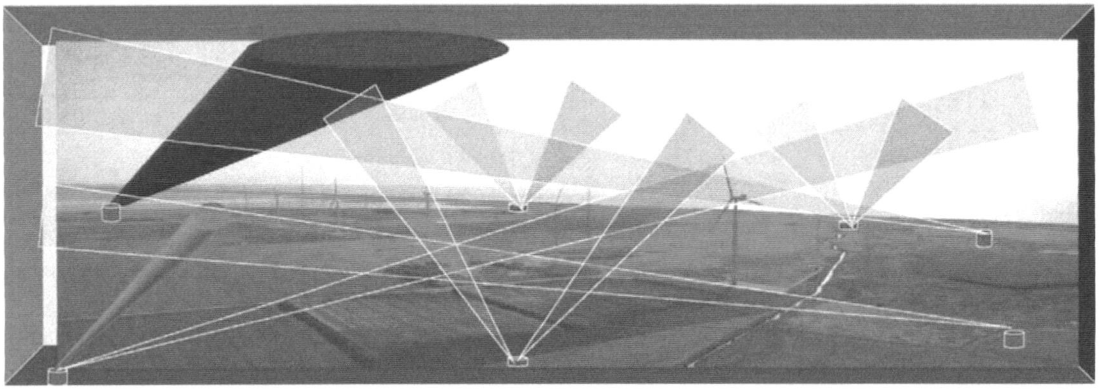

Figure 1. An "artists view" of the new mobile Windscanner facility engaged at the test station for large wind turbines, Høvsøre. Westerm Jutland, Denmark.

At Risø DTU's National Laboratory for Sustainable Energy we are currently engaged with the design, construction and testing of a new ground-based "Windscanner" research facility for true real-time measurements of the four-dimensional atmospheric velocity field, within the volumes that fully surround the huge wind turbines of today and tomorrow. With existing technology wind vectors can be measured (scanned) at a proven rate of up to 500 hundred measurement points in space per second. Our belief is that a successful implementation of the new envisioned "Windscanner Research Facility" can boost our understanding and comprehension of the turbulent atmospheric flow that surrounds the giant wind turbines. This new knowledge, as we envision, will accelerate the design and development of ever-larger wind turbines and thus provide new insight regarding external wind and turbulence conditions in the rapidly growing world market for wind energy.

Till now, our current understanding of flows around wind turbines derives largely from point measurements at a few scattered towers. But knowledge of wind speeds and turbulence statistics at a few nearby points is not sufficient. Further, even for current wind turbines, standard measurement methods that involve dense arrays of upwind wind masts would not only distort the ambient flow but also be prohibitively expensive. To fully determine anticipated power production levels and structural loads, wind researchers must develop a much more detailed understanding of the turbulent flows approaching, hitting, and propagating through and downwind to impact on further wind turbines there.

We therefore pursue a far more contemporary and advanced venue, namely a mobile, full 3-dimensional in space and high temporal resolving wind and turbulence Remote Sensing (RS) facility which we have named "The Remote Sensing Full Scale Laser Windscanner Facility", viz.:

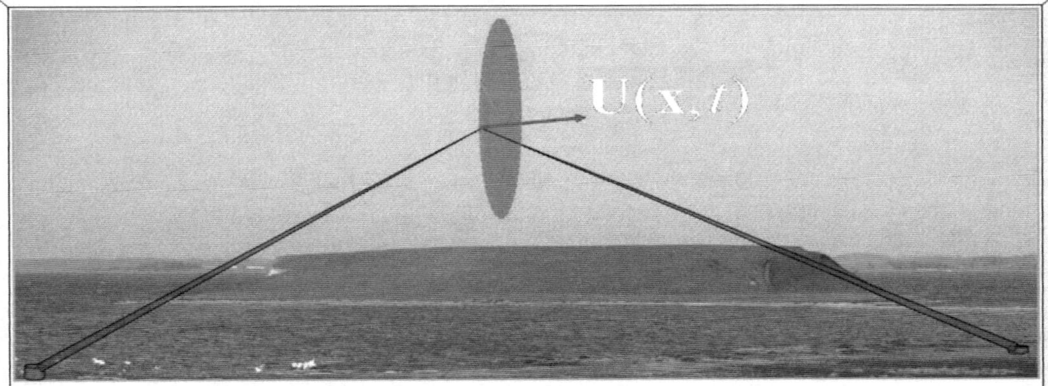

Full Scale Laser Wind Scanner

The recent advent of cheaper, eye safe and stable coherent lasers has via recent advances in telecommunications spurred the development of remote wind profiling lidar systems [1]. Our group at Risø-DTU has tested several such systems, comparing them to mast-based measurements both on and off-shore. Impressively high correlations have been obtained with cup and sonic anemometers, up to 160 m above the surface. Deployed as single units, these new lidars can measure the mean, horizontal wind within a flexibly focused volume. Thus, our Windscanner vision require the deployment of multiple lidar units, a significantly more ambitious and challenging mobile research facility, to assess the four-dimensional velocity fields as the driving external conditions during wind turbine operation.

A fully operational Windscanner facility is furthermore attractive for future wind energy research laboratories, as it would constitute the natural, open-air equivalent of an enormous wind tunnel, albeit one with negligible disturbance from surface clutter and completely unhampered by the distorting effects of walls and roofs of a wind tunnel. I.e., these mobile even hand-carried lidar units are light enough to be relocated quite easily. Thus, such a mobile facility can operate routinely with negligible disturbance to or distortion of the local terrain – whether it be over water, sloped, complex, or covered by bare soil, vegetative, forest, or other types of surface canopies. Following this our first

IOP Conf. Series: Earth and Environmental Science 1 (2008) 012018 doi:10.1088/1755-1307/1/1/012018

demonstration project, we envision to assist the dissemination of such full-scale 3-D Windscanners at the European level and at Transatlantic wind energy test sites. We therefore envision one day to be able to deploy mobile versions of the Windscanner system, deployable anywhere in close collaboration with our European and American wind energy research and WT testing collaborators and partners.

2. Objectives of the Windscanner facility

The remote sensing devices in use today measure a wind speed profile, i.e., how wind speed varies with height along a vertical line extending from the surface [2],[3],[4]. Such measurements are, in this sense, one-dimensional. The Windscanner extend this 1-D profiling capability to multiple snapshots in time of the 3-dimensional field of atmospheric winds and turbulence, i.e., a complete, 4-D scanning, turbine test facility with open boundary conditions. Some of the motives for our ambitions are given in the following subsections.

2.1. To advance the understanding of atmospheric flow around wind turbines

Though we in the international research community can now predict wind profiles in a variety of atmospheric conditions, we are only beginning to understand the full, three-dimensional structure of atmospheric turbulence, even for relatively simple terrain. Thus, the ability to 'see' full-scale three-dimensional turbulence should spur a major advance in understanding flows around wind turbines, within the open and natural context of the atmospheric boundary layer in-situ. For example, the contemporary model of three-dimensional turbulence applies strictly only when there are no temperature gradients or thermal energy fluxes (neutral stratification, as modeled in wind tunnels). In practice the atmospheric boundary layer is hardly ever in this condition.

Thus, we anticipate that the Windscanner Facility will make significant contributions in the following areas:

- Mean winds and four-dimensional profiles of turbulence under a wide range of atmospheric conditions
- Flow and turbulence over complex terrain
- Wakes behind wind turbines
- Visualization and time evolution of full-scale extreme gust events and coherent structures
- Detailed evaluation of CFD (computational fluid dynamics) codes, such as e.g. DTU/Risø's EllipSys, along with simpler models

2.2. Complex terrain and Upstream Rotor effects

As a sample problem, which the Windscanner Facility may address, wind turbines in complex terrain suffer from excessive loads and sometimes lower than expected power production. The term "complex terrain" includes both forested areas where most turbines in Northern Europe are erected, and the steep topography typical of Southern Europe. This often leads to poor financial performance for wind energy projects and in extreme cases structural failure, e.g., by blades hitting the tower.

The root of such problems is the complex flow surrounding wind turbines, induced both naturally and also by neighboring upstream turbines. Thus, a much more detailed representation of the turbulent flow is required, upwind, at the turbine site, and downstream of the rotors. I.e., to assess energy production and structural loads accurately, there is a need to measure and comprehend the instantaneous, three-dimensional wind fields at high spatial and temporal resolution, as the wind comes toward, through, and behind the rotor discs, hubs, and supporting tower structures.

Ideally, such comprehensive wind and turbulence field measurements could be obtained during a representative period of time at each future potential wind turbine site or array, and before the proposed turbines have been erected.

In-situ assessments may eventually become routine. However, meanwhile, it also remains important to improve the parallel developing capabilities of CFD (computational fluid dynamics) models. CFD is used for ultra-high resolution, complex atmospheric flow calculations over vegetated surfaces, steep terrain, and flow in and around the wakes of turbines. Thus, an on-going program of data comparison and verification with true measured 4-D wind field scanning data will inevitably spur significant advances in the initialization and further developments of CFD models for wind energy.

Continued and enhanced collaboration between the wind measurement and the modeling community is also envisioned, e.g. the Wind Energy Department at Risø DTU and DTU MEK have already collaborated closely for decades on the development of EllipSys, a leading CDF code used in the wind energy industry.

2.3. To both assess the 4-D flow and to mitigate structural loads in real-time

Since turbulence is the dominant contributor to rapidly fluctuating loads on a wind turbine, it is important for the Windscanner to map the wind fields quite quickly. We envision also a simpler, turbine-mounted unit to assess the flow immediately approaching the rotor, in tandem with ground-based units, where several laser beams scan a volume for all three components of the velocity. The former unit could be used to pitch the blades in anticipation of arriving wind gusts, thus reducing structural loads in real-time. In fact, as the blades become longer and more flexible, the more significant such control mechanisms become - the inevitable result of ever-larger turbines.

Figure 2. Two-dimensional scan patterns obtainable from a spinner-based wind lidar equipped with a dual-prism scan head. Left figure: An example of "Rosette scans" obtainable with two counter-rotating prisms with different rotating speed. Right: Spiral scans obtainable with two co-rotating prisms of different angular speed. These scan patterns are envisioned for real-time measurements of the incoming wind and gusts from a single spinner-mounted wind scanner.

3. The system design and technical specifications of the Windscanner

Risø DTU is now in the process of building and testing a first 3-D proto-type Windscanner constructed from modified commercially available fiber-based wind lidar profilers. Risø DTU has purchased 3 continuous wave (CW) "ZephIR" wind lidars from National Power, and one "Wind cube" pulsed lidar form the French company Leosphere, in order to build a first proto-type 3-D Windscanner.

Risø DTU, in collaboration with MEK DTU and COM DTU are furthermore in collaboration with QinetiQ Ltd. in the process of modifying the three CW lidars with respect to external fast steerable focus control, and fast full bandwidth throughput (up to 500 Doppler spectral measurements per second per device.) At QinetiQ Ltd. progress with modification of the optics is on-going within the three ZephIRs so they in the future will be able to discriminate the sign ambiguity of the measured Doppler shifts (i.e., QinetiQ is converting the systems into true heterodyning devices, by implementing an acoustic optical modulator. At DTU Lyngby, Denmark, we are furthermore collaborating with the Mechanical Department DTU MEK and also with an on-campus industrial mechanical design company (IPU), in order to design and construct a 2-D scan head ("Pi- scanner") for fast elevation and azimuth steering control of the lidar beams. Our design criteria is to be able to steer the beams through specified periodical two-dimensional oscillation patterns covering a full steradian (one Pi steradian) at a rate of ~ 10 full revolutions per second.

In addition we have purchased a single pulsed Windcube lidar from Leosphere for extended wind profile measurements at high vertical ranges from ~100 to 300 meters above the ground. The state–of-the-art technical specifications of the available commercial lidars profilers are shortly described in the following subsections.

3.1. The standard QinetiQ ZephIR CW lidar

A first fiber-based prototype CW lidar was developed around 2003-2005 by QinetiQ Ltd. and amended and improved in collaboration with Risø DTU [4]. QinetiQ subsequently presented the first commercial lidar profiler at the beginning of 2006. Known as the 'ZephIR', this is a continuous wave system with height discrimination achieved by varying focus.

The laser radiation is transmitted through a constantly rotating prism giving a beam deflection of 30° from Zenith and making a single complete rotation per second. Backscattered light mixed with the local oscillator is sampled at ~100 MHz. The unit is able to transform blocks of 512 time series scans to power spectra in real time. These 200.000 spectra per second are then block-averaged 4.000 at a time to give highly noise-suppressed Doppler shifted spectra at 50 Hz. A centroid method is used to obtain the frequency of the 'peak', the radial speed (i.e. in the laser beam direction) being directly proportional to this frequency. Thus 50 radial velocities, one for every 360°/50=7.2° are available from each rotation.

Unlike pulsed systems, a CW lidar does not inherently 'know' the height from which backscatter is being received. We are obliged to assume, or secure by measurements, e.g. with a ceilometer, a sensibly flat vertical profile of aerosol concentration, in which case the backscattered energy is from the focused volume, weighted according to the appropriate spatial weighting Lorentzian function centered at the focus distance [5]. The obtained radial wind speed distribution in this case is dominated by the signal from the set focus distance.

3.2. The standard Leosphere Windcube pulsed lidar

The French company Leosphere introduced in December 2006 a pulsed wind lidar named "Windcube" aimed at the wind energy market. The Windcube is a pulsed lidar with a fixed focus. Like the ZephIR, it has a 30° prism to deflect the beam from Zenith, but here the prism does not rotate continuously. Instead, the prism holds still whilst the lidar sends a stream of pulses (5000-10000) in a given direction, recording the backscatter in a number of range gates (fixed time delays) triggered by the end of each pulse. Having sent the required number of pulses, the prism rotates to the next azimuth angle to be scanned, each separated by 90°. A full rotation takes about 6 seconds.

During the rotation and before the next stream of pulses can be sent, the recorded data are processed. For each range gate, the time series from each pulse are Fourier transformed to power spectra which are block-averaged. Due to the short recording duration (200 ns), the resulting spectra have poor frequency resolution due to the well-known time x bandwidth uncertainty principle. Instead of using a centroid to obtain the frequency of the peak, Leosphere have implemented a mathematical model [6] including the most important parameters affecting the shape of the expected Doppler spectrum. This model is fitted to each block-averaged power spectrum in order to obtain the Doppler shift to a much higher resolution than could otherwise be expected. At each direction step (1.5 s), the Windcube combines the four most recent radial speeds at each height in order to obtain the horizontal and vertical speed and wind direction.

Windcube applies as standard heterodyning acousto-optic modulator adding a precise frequency offset to the local oscillator which is mixed to and beats with the returning Doppler shifted backscatter.

3.3. Wind lidars and their range resolutions:

The two lidar systems have fundamentally different range resolution properties. The ZephIR is a variable-focus, CW system and as such, the vertical probe length is determined by the depth of the focus volume. More specifically this can be defined as the FWHM (Full Width Half Maximum) of the Lorentzian function defining the collection efficiency. This probe length increases quadratically with sensing range. When measuring at short ranges, the probe length is only a few decimeters. At 100 m range, however, the FWHM of the ZephIR lidar is estimated to be close to 17 m [7], whilst at 138 m (cf. the MusketeerEx section below) the FWHM of the ZephIR's Lorentzian weighting function was estimated to be ~31m.

For the Windcube, the spatial measuring resolution depends on the pulse length and gate-open time and is constant with sensing range. The conventional estimate for a pulsed lidar's range resolution with identical pulse and gate-open times (200 ns) yields a spatial resolution (FWHM) equal to 30 m in this case.

With these parameters the estimated "break-even" range with respect to range resolution between the CW lidar (ZephIR) and the pulsed lidar (Windcube) is near ~140 m range. At longer ranges (> 140 m) the pulsed lidar will repressible exhibit a more detailed range resolution, while at shorter ranges (< ~140 m) the ZephIR CW lidar will theoretically exhibit the better range resolution. These predictions of the range resolutions of the two lidar type instruments are at the moment subject to experimental evaluation at Risø DTU, cf . [8, 9].

With the present first Windscanner configuration based on three modified ZephIR CW lidars we expect to be able to capture measurements within a 3-D volume based on the three CW lidar's range maximums engulfing a volume of the order of: 150 m x 150m x 150m, cf. figure 3 below.

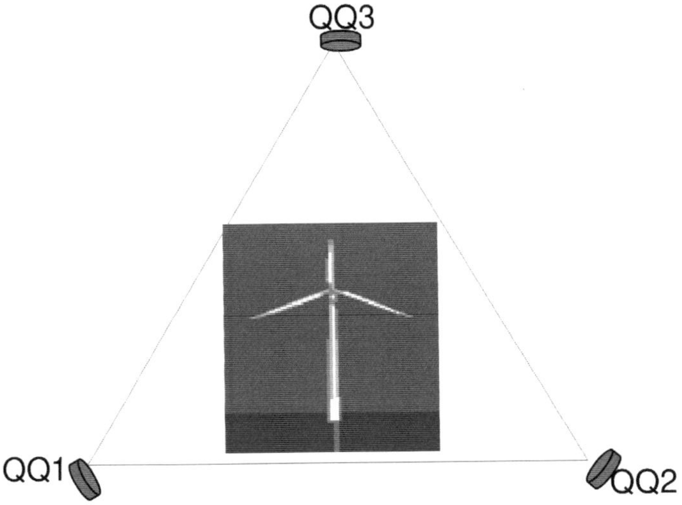

Figure 3.

Three-Dimensional Windscanner - Basis Configuration measurement volume: ~150m x 150 m x 150 m. Three modified QQ ZephIRs equipped with 2-D +/- 60 degree steerable " One-Pi" scan heads.

4. The Musketeer Experiment: 3-D Lidar vs. 3-D sonic wind measurements at 78 m height

Figure 4. Our first 3-D Lidar vs. Sonic Intercomparison field test: 3-D wind components measured at 78 m above Høvsøre during the December 2007 "MusketeerEx lidar vs. sonic field test". Three pulsed lidars and one CW lidar beam "crossed beams" near a sonic on top of the 78 m mast.

In December 2007 a first test of 3-D remote sensed wind components were performed at the Risø DTU's test station "Høvsøre" for large wind turbines, Western Jutland.

Measurements of the winds three components were recorded at the 78 meter level from an in-situ meteorological tower, and several lidars, including a single QinetiQ CW prototype lidar (lower left corner) and three pulsed Windcubes (the three white boxes in figure 4) were pointed to a measurement volume near the top-mounted sonic at 78 meters above ground, cf. figure 4.

Lidar measurement ranges were preset to ~138 meters for the two slant-angle pointing lidars (both pulsed and CW), while a third (pulsed Windcube) was positioned at the base of the met-tower and pointed vertical towards the test measurement volume at ~78 meters height.

For wind energy research in particular and for micro-meteorology in general we are interested in measuring turbulence, not just its statistics, but rather its gusty coherent time series of the fluctuating three-dimensional velocities at high resolution spatially and temporally. During the MusketeerEx 2007 we succeeded to point three pulsed lidars and one CW lidar towards the same small measurement volume to measure 3-D turbulence from a single point 78 meters above the ground. Then the Windscanner becomes fully operational with respect to two angular and one focal ranging each, it is our intention to steer the beams around the wind turbines engulfing volumes as depicted by a graphical designer (Lynge Grafik) in figure 5.

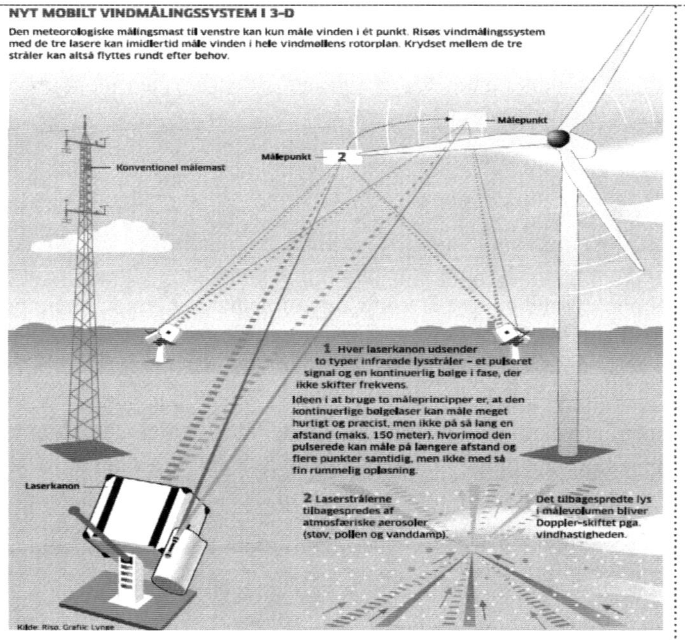

Figure 5. Danish artists view on the 3-D Windscanner in operation at the Høvsøre Wind Turbine Test site. The green rays represent the pulsed lidar's while the red beams depict the CW pulsed laser beams. The laser beams are seen to be steered towards the same measurement volumes, hence measuring the full 3-D wind vectors in several points (up to 500 measurement points) per second. The graphical view of the set-up has been published in Ingenjøren, December 12th, 2007.

During MusketeerEx 2007 we recorded wind speeds up to 21 m/s. Details about the experimental setup is given elsewhere[9]. A small fraction (10-min) of the lidar vs. in-situ measurements are shown in figure 6, where time series of wind speeds measured by three Windcubes along with simultaneous in-situ sonic measurements at 78 m height are shown. If n is a unit vector in the direction of a lidar beam and u is the wind vector measured by the sonic, then shown is n · u. The sonic speeds have been block averaged over 1.5 s corresponding to the sampling rate of the lidar. Each wind speed determination by the lidar is based on Doppler spectra averaged over 0.5 s. The agreement between the sonic and the lidar wind components are seen to be astonishing identical! Detailed analysis of the wind traces coherence and spectral contents are ongoing [9].

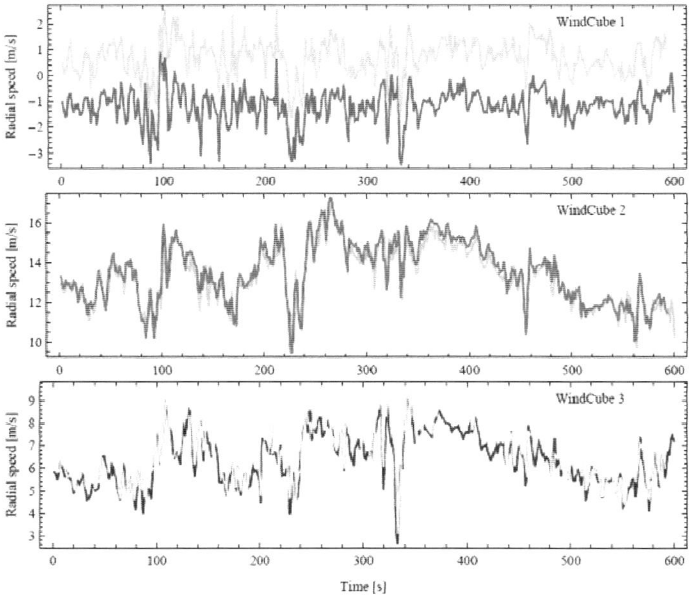

Figure 6. First 3-D Lidar Windscanner measurements compared with in-situ (Sonic) measurements at 78 meters height during the Musketeer Experiment at Høvsøre, Dec. 2007 ten minute period of wind data (Dec 8. 02:50 – 03:00). The red curve is the three-dimensional sonic data projected onto the direction of the beam of Windcube 1. The data from this lidar are shown in cyan. Likewise, the projected sonic speed and the corresponding Windcube 2 speed are shown in green and magenta. For Windcube 3 it is blue and [9].

5. Conclusions

We have designed and are well under construction with the first 3-D full-scale lidar-based mobile Windscanner facility for wind energy applications. We have based our design on existing and modified newly emerged wind profiler lidars. Our first lidar vs. in-situ 3-D measurement inter-comparison seems very promising, cf. the data in figure 6.

We already now see many interesting applications for the Windscanner under construction. E.g., for direct measurements of the external conditions within the rotor plane and correspondingly power curve performance studies, over homogeneous terrain, and as the wind approaches the turbines over various terrain in particular. We also see the Windscanner facility as an attractive tool for evaluation of flow and turbulence modes used in wind energy. Windscanner measurements can become a central part in evaluation of CFD code calculations; cf. the recent Bolund experiment near Risø December 2007/January 2008.

Further developed, the Windscanner technique may spin off and become directly integrate within a spinner-based lidar used during normal operation for active pitch control within the turbines control system to mitigate the effect of, e.g., strong gusts and possibly also to assist improve the turbines energy performances under sub-rated wind speed operation. Further testing of the Windscanner in full scale at Høvsøre will continuously play a significant role in the development of the test facility.

6. Acknowledgements

The Danish Agency for Science, Technology and Innovation is acknowledged for supporting this work with a "Pre-project maturation study of a new national infrastructure facility centered around a mobile: "Full Scale Laser Windscanner".

References

[1] Courtney M, Wagner R and Lindelöw P 2008 Commercial lidar profilers for wind energy. A comparative guide, *Conference Proceedings of EWEC 2008*

[2] Smith D A, Harris M, Coffey A S, Mikkelsen T, Jørgensen H E, Mann J and Danielian R 2006 Wind Lidar Evaluation at the Danish Wind Test Site in Høvsøre *Wind Energy* **9** 87–93

[3] Antoniou I, Jørgensen H E, von Hünerbein S, Harris M, Martí I, Aristu M, Foussekis D, Christensen L. C. 2007 Remote sensing the wind using Lidars and sodars *Conference Proceedings of EWEC (Milan, 2007)* paper BL3.290.

[4] Antoniou I, Courtney M, Jørgensen H E, Mikkelsen T, von Hunerbein S, Bradley S, Piper B, Harris M, Marti I, Aristu M, Foussekis D, Nielsen M P 2007 Remote sensing the wind using lidars and sodars (Invited Paper and Presentation). *Conference proceedings (online). 2007 European Wind Energy Conference and Exhibition, Milan (IT), 7-10 May 2007 (EWEA, Brussels, 2007)*

[5] Courtney M, Wagner R 2008 Testing and comparison of lidars for profile and turbulence measurements in wind energy *Conference Proceedings of ISARS 2008*

[6] Valla M 2005 *Estimateur du Maximum de Vraisemblance Etude d'un lidar doppler impulsionnel à laser Erbium fibré pour des mesures de champ de vent dans la couche limite de l'atmosphère Thesis from École Doctorale d'Informatique,Télécommunications et Électronique de Paris.* p 95-103.

[7] Antoniou I (Editor), Jørgensen H E, Courtney M, Mikkelsen T, Mann J, Harris M, Coffey A VonHünerbein S, Marti I, Aristu M, Nielsen M P, Foussekis D 2006 *Remote Sensing (EU FP7 UPWIND WP6, Six-month progress report; Nov 2006; Risø-I 2527(EN).*

[8] Sjöholm M, Mikkelsen T, Mann J, Enevoldsen K and Courtney M 2008 Time series analysis of continuous-wave coherent Doppler Lidar wind measurements *Conference Proceedings of ISARS 2008*

[9] Mann J, Cariou J-P, Courtney M, Parmentier R, Mikkelsen T, Wagner R, Lindelöw P, Sjöholm M, Enevoldsen K 2008 Comparison of 3D turbulence measurements using three staring wind lidars and a sonic anemometer *Conference Proceedings of ISARS 2008*

IOP Conf. Series: Earth and Environmental Science **1** (2008) 012019 doi:10.1088/1755-1307/1/1/012019

The Myres Hill remote sensing intercomparison study: preliminary results

P J M Clive[1], I Chindurza[1], I Ravey[2], J Bass[2], R J Boyle[3], P Jones[3], S J Lang[4,5], S Bradley[6,7,8], L Hay[9], A Oldroyd[10], M Stickland[11]

[1]SgurrEnergy Ltd, 79 Coplaw Street, Glasgow G42 7JG, Scotland
[2]RES Group Ltd, James Blyth House, 7000 Academy Park, Glasgow, Scotland
[3]TUV NEL Ltd, East Kilbride, Glasgow G75 0QF, Scotland
[4]Sustainable Energy Research Group, University College Cork, Ireland
[5]Westwind, 148 Blarney Street, Cork, Ireland
[6]Mighty River Power, Level 14, 23-29 Albert Street, Auckland, New Zealand
[7]Department of Physics, University of Auckland, New Zealand
[8]University of Salford, Salford, UK
[9]Garrad Hassan & Partners Ltd, 2064 Maryhill Road, Glasgow G20 0AB, Scotland
[10]Oldbaum Services Ltd, Schoolhouse, Brig o' Turk, Callander, Scotland
[11]University of Strathclyde, 16 Richmond Street, Glasgow G1 1XQ, Scotland

Email: peter.clive@sgurrenergy.com

Abstract. Two remote sensing techniques (SODAR and LIDAR) have been developed for measuring wind speed and turbulence from ground level up to altitudes of 300 m or higher. Although originally developed in the defence sector, these techniques are now generating considerable interest in the renewable energy and meteorological sectors. Despite the benefits of these instruments they are not yet generally accepted for due diligence measurements by wind energy developers and financial institutions. There is a requirement for a series of independent assessments of these new metrology techniques, comparing their measurements with the approved cup-type anemometer readings. This is being addressed at TUV NEL's Myres Hill wind turbine test site in a measurement programme supported by the DIUS National Measurement Systems Measurement for Innovators scheme and a consortium of 21 industrial collaborators. Data from SODAR and LIDAR systems are being compared with results from cup-type anemometers mounted at different heights on an 80m meteorological mast. An ultrasonic sensor is also mounted on the mast. The objective of the test programme is to assess the effectiveness of SODAR and LIDAR wind speed measurement techniques under different operating regimes and atmospheric conditions. Results from the measurements will provide definitive data on the performance of the remote wind speed sensing techniques under test on complex terrain typical of many wind farm sites. Preliminary measurements based on data acquired during the initial measurement campaign are presented.

1. Introduction

Mature portable remote sensing technologies are now successfully being used for wind power applications [1, 2]. An intercomparison study of remote sensing devices has been undertaken at the TUV NEL Myres Hill test facility near East Kilbride, Scotland. The data acquired using an AQ

Systems AQ500 SODAR and a NPC/QinetiQ ZephIR LIDAR are being compared to the measurements obtained using conventional mast mounted cup anemometry and ultrasonic anemometry on an 80m tall guyed tubular tower. This study is distinguished by two features: 1) it is industry-led and is supported by a consortium consisting of 21 organizations and 2) the test site is more typical of the upland terrain where wind farms are installed than other test sites that have been used in the past characterized by flat featureless topography [3 - 8]. The study was undertaken to assess the suitability of remote sensing techniques for wind resource assessment applications in a manner that would inform the degree of confidence in the technologies of the industrial partners in the consortium. The process of adoption of remote sensing by the wind power industry is underway but is still at an intermediate stage. The accuracy of these devices, particularly in locations typical of real wind farms, requires further demonstration to support sufficient confidence for remote sensing data to be acceptable for due diligence purposes with a less strenuous requirement for continuous instrument validation than is presently the case [9]. In addition, one of the main objectives of the test is to assess these systems in their off-the-shelf configuration.

At the time of writing the measurement campaign was still underway and as a consequence of this preliminary measurements are presented here using data acquired between the dates 2008-02-05 and 2008-02-29. The ZephIR was co-located with the mast at E 256910, N 646490 whereas the AQ500 was deployed to E 256910, N 646270 some 300m distant to the South of the mast location. The initial test setup had placed the AQ500 system 80m due West of the intended mast location prior to mast erection. However, upon completion of mast installation, a clear fixed echo could be seen in both the shear profile and the system spectra. After consultation with the manufacturer, AQ Systems, the system was redeployed 300m from the reference mast in a South Westerly direction. Care was taken to ensure the AQ500 was located upwind of the reference mast in the predominant site wind direction. The configuration of the mast is illustrated in Figure 1 and the instruments installed on it are listed in Table 1. The locations of the devices relative to each other are shown in Figure 2. Figures 3 and 4 show the ZephIR and AQ500 respectively.

Two NEG Micon NEM 900/54 wind turbines are located at E 256959, N 646765 and E 256995, N 646490. These have a hub height of 60m and a rotor diameter of 54m and present obstacles that perturbed the flow in certain direction sectors relative to the mast/ZephIR and AQ500 locations. The free stream sectors at each location were calculated in compliance with IEC 61400-12 and are detailed in Table 2. Data acquired when the wind was blowing from these free stream directions at the relevant remote sensing device locations were included in the results presented here.

Table 1. Instruments installed on the mast.

Reference	Instrument			
	Type		Height (m)	Orientation (degrees)
WS1	Anemometer	Risø P2546A	80.35	306
WS2	Anemometer	A100L2	80.35	126
WS3	Anemometer	A100L2	65.00	182
WS4	Anemometer	A100L2	50.30	270
WS5	Anemometer	A100L2	50.05	182
WS6	Anemometer	A100L2	30.25	266
WS7	Anemometer	A100L2	30.00	182
WS8	Anemometer	A100L2	20.00	183
WS9	Anemometer	A100L2	10.00	183
DIR1	Wind Vane	W200P	77.70	222
DIR2	Wind Vane	W200P	45.00	221
DIR3	Wind Vane	W200P	25.00	222
TEMP_1	Temperature		76.90	-
TEMP_2	Temperature		1.00	-

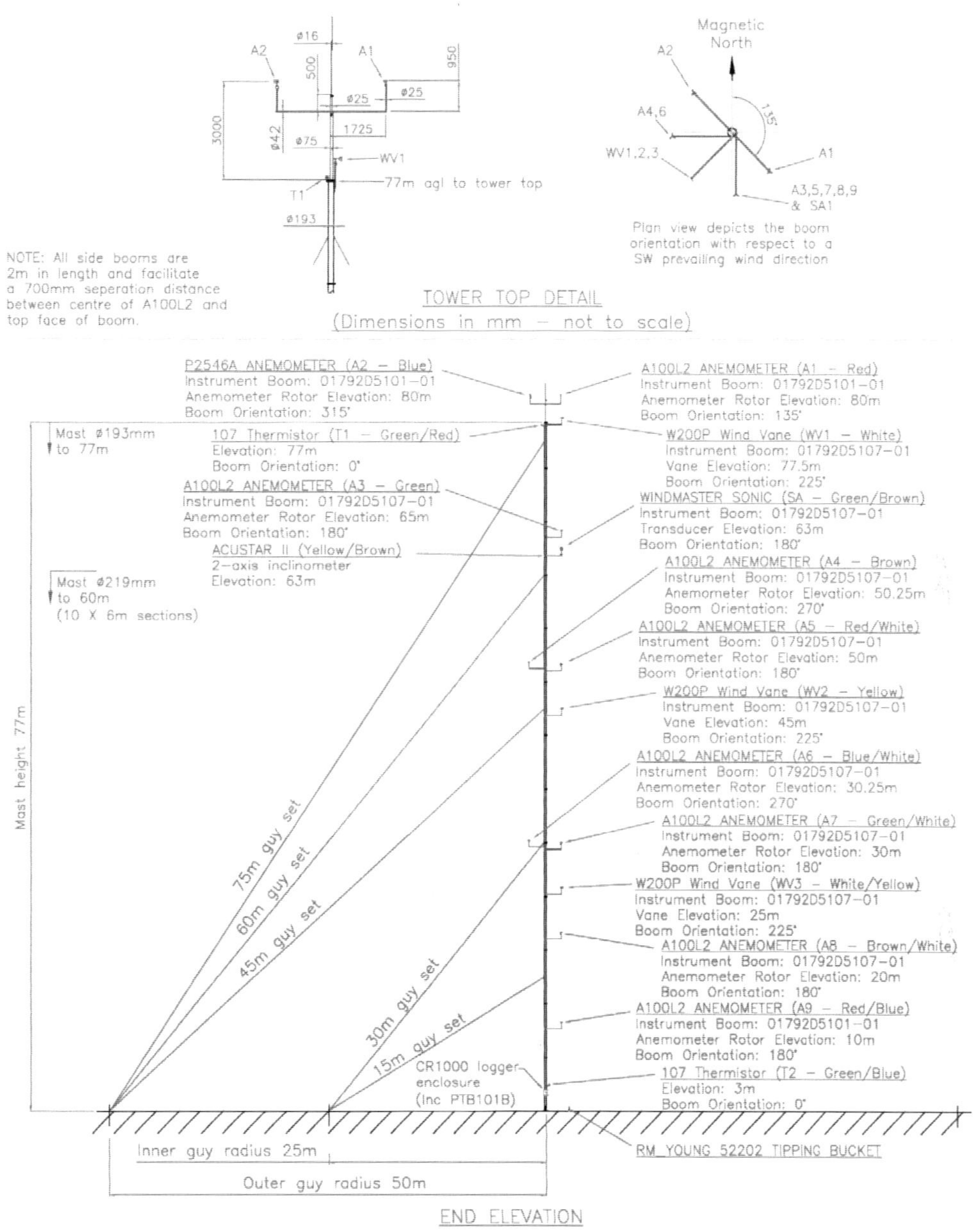

Figure 1. Mast configuration.

IOP Conf. Series: Earth and Environmental Science **1** (2008) 012019 doi:10.1088/1755-1307/1/1/012019

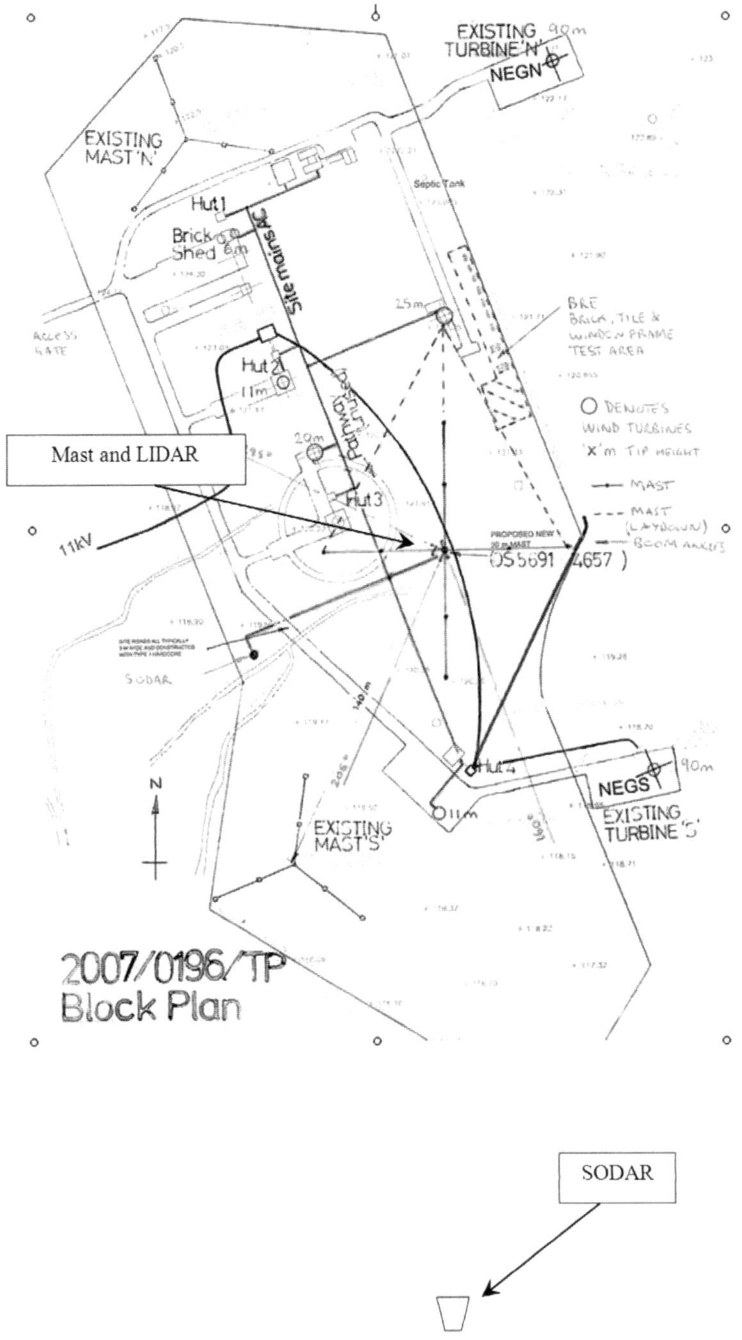

Figure 2. Instrument locations.

Table 2. Free stream sectors, clockwise from North.

Location	From (degrees)	To (degrees)
North turbine: E 256959, N 646765	205.14	139.95
South turbine: E 256995, N 646490	25.14	319.95
Mast/ZephIR: E 256910, N 646490	52.31	83.66
	182.87	335.90
AQ500: E 256910, N 646270	56.42	340.62

Figure 3. NPC/QinetiQ ZephIR Lidar co-located with mast.

2. ZephIR Measurements

Figure 5 shows a time series trace comparing the 80m ZephIR data to the 80m mast data. Figure 6 shows a time series trace of the difference between the mast and ZephIR measurements at this height. It is clear that there are periods when the ZephIR was not operating satisfactorily. Figure 7 shows the diurnal variation in the difference between the mast and the ZephIR measurements. There is some indication that the deterioration occurs mainly between the hours of 20:00 in the evening and 10:00 in the morning. Figure 8 shows the linear regression of the ZephIR data on the mast data.

Figure 4. AQ Systems AQ500 SODAR.

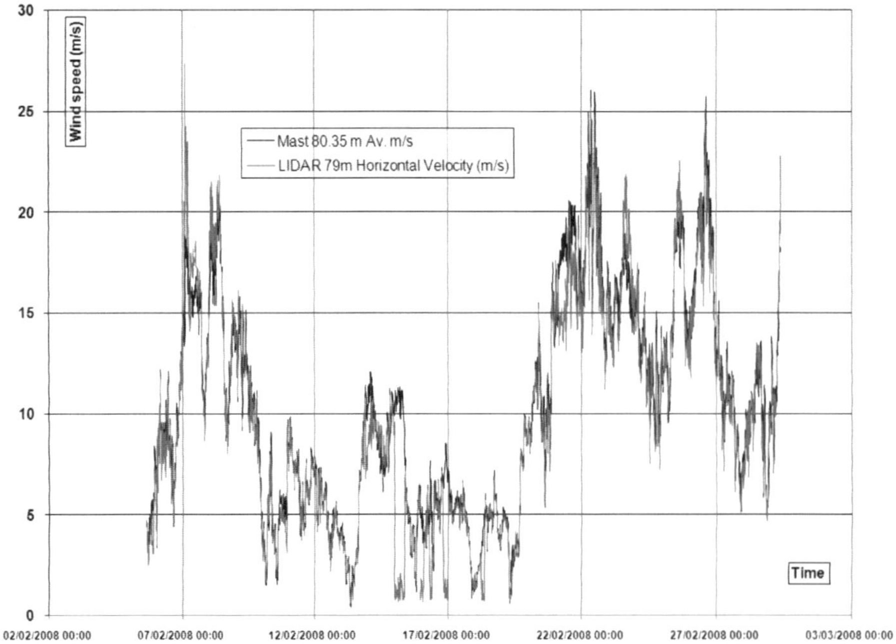

Figure 5. ZephIR and mast time series, 80m, 2008-02-05 to 2008-02-29.

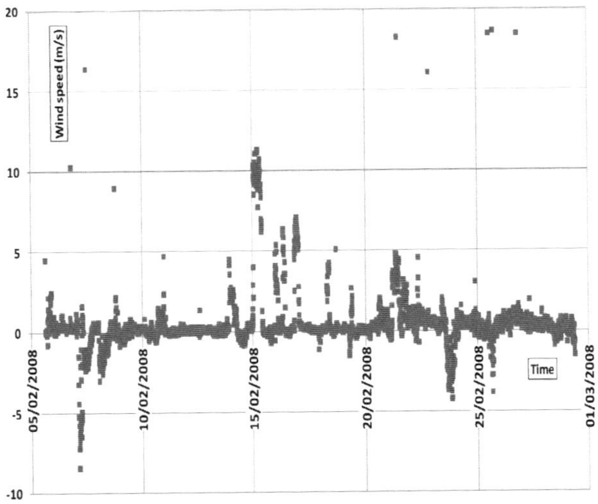

Figure 6. Time series of the difference between mast and ZephIR 80m data.

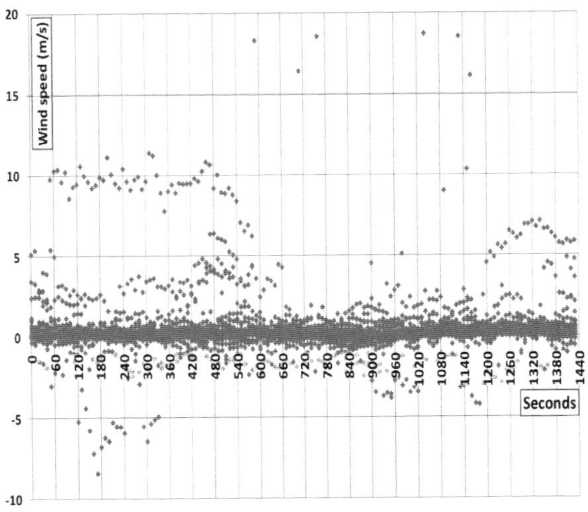

Figure 7. Diurnal variation in difference between mast and ZephIR 80m data.

It is clear that further investigation is required to ascertain the reasons for the anomalous behavior during the periods highlighted in Figure 6. Similar results are observed when comparing the data pertaining to the other measurement heights, which will be fully reported once the measurement campaign is concluded. The linear regression shown in Figure 8 has been constrained to pass through the origin to aid comparison with other, similar studies.

3. AQ500 Measurements

Figure 9 shows the time series trace of the 80m measurements from the mast and the AQ500. Figure 10 shows the linear regression of the free stream AQ500 data on the mast data at this height. This fit is not constrained to pass through the origin on the grounds that the measurement devices being compared are not co-located.

The agreement between the AQ500 and the mast was encouraging, given the distance of 300m between them. Again, a slope $m < 1$ is observed, however due to the fact that the devices are not co-located no conclusions can be drawn about possible volume averaging effects. However, the under-estimation of winds at low wind speeds is typical of the influence of echoes from fixed objects such as a mast or turbine. At low wind speeds the spectral peak from such echoes is close to that from the atmosphere and often a combined peak results, giving bias toward lower estimated wind speeds. Further analysis will be done on this.

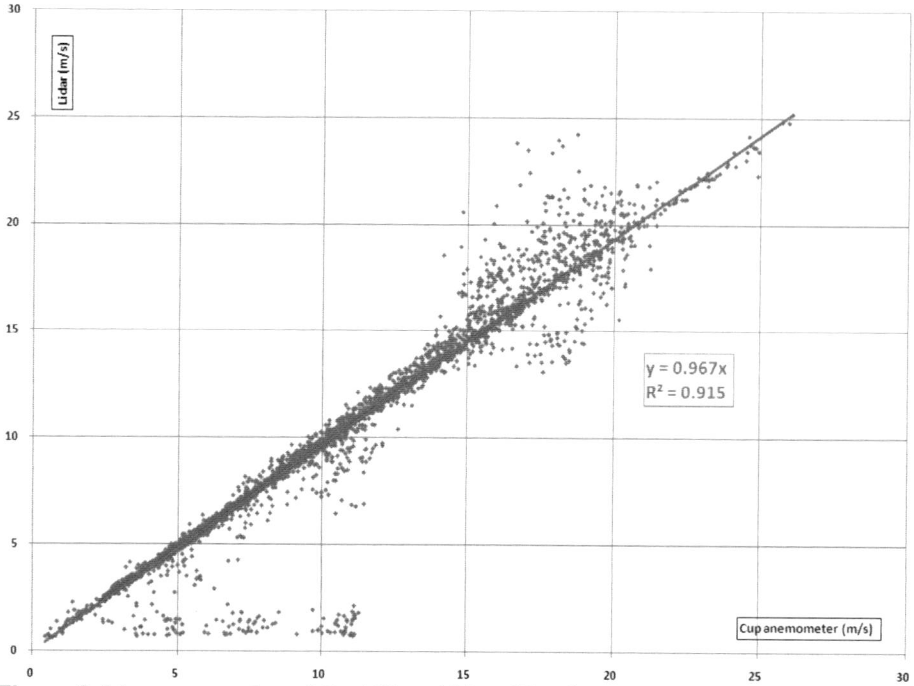

Figure 8. Linear regression of ZephIR and Mast 80m free stream measurements.

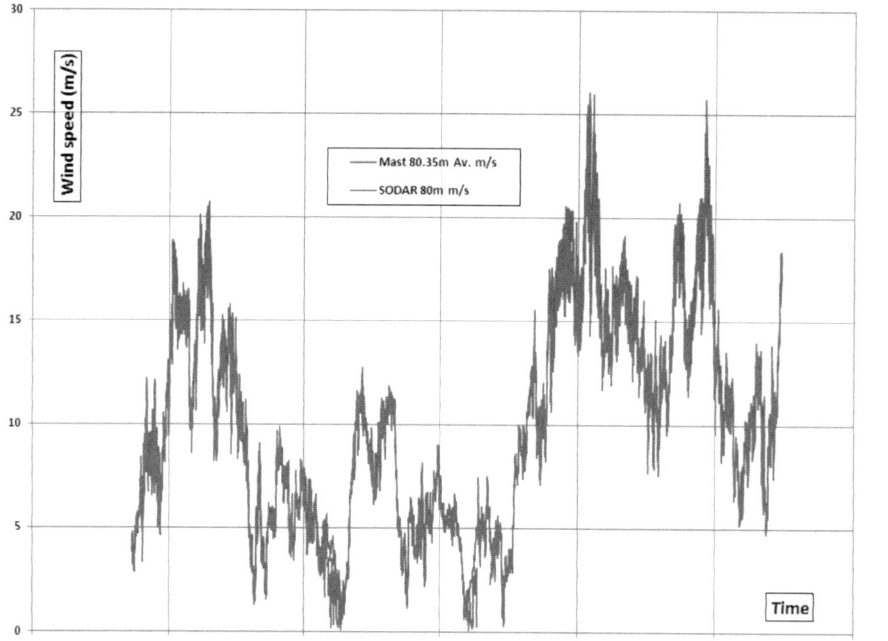

Figure 9. AQ500 and mast time series, 80m, 2008-02-05 to 2008-02-29.

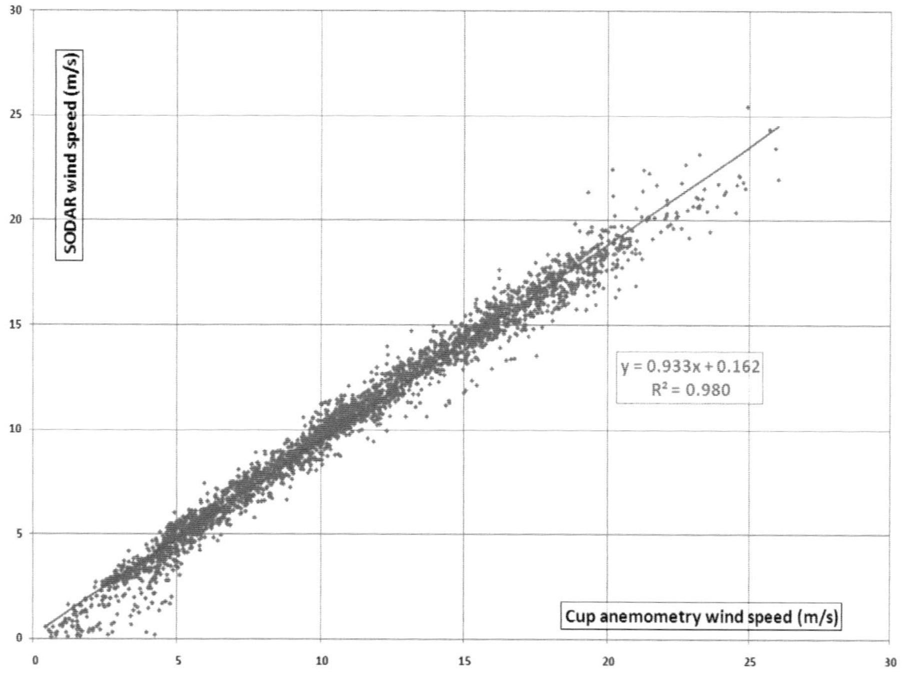

Figure 10. Linear regression of AQ500 and Mast 80m free stream measurements.

4. Conclusions

The data presented here are preliminary and the measurement campaign that produced the data has not yet concluded. The campaign is scheduled to conclude during April/May 2008 and the final analysis and results of this measurement campaign will be reported more fully in other publications in the near future, in particular at ISARS 2008 towards the end of June 2008.

Periods of anomalous behavior were observed in the ZephIR data, necessitating further investigation. It is likely that this behavior is attributable to unresolved issues around contamination of ZephIR data by higher-altitude cloud. This ZephIR is known to have a very extended range of sensitivity which, when coupled with the very high reflectance from overlying cloud, can give wind speeds from the cloud level rather than from the focal point of the laser beam. Since winds are frequently observed to increase in strength with height, this will tend to give overestimation of wind speeds. The ZephIR suppliers have developed, since these measurements were made, a new algorithm for largely removing this problem, and this algorithm will replace the current version in further comparisons.

Encouraging agreement between the AQ500 and mast data was observed.

The ongoing test program includes the co-location of the ZephIR and AQ500 to help mitigate terrain influence on the results. It is planned to use a second ZephIR unit in parallel with the current ZephIR unit. Initially, the second ZephIR will be co-located with the current unit for two weeks after which the current ZephIR will be co-located with the AQ500 for two weeks. This will allow direct comparison of the two sets of ZephIR measurements when co-located and when 300m apart. It will also provide a secondary set of ZephIR measurements that can be correlated against the mast when the AQ500 and primary ZephIR are co-located as well as a direct correlation between AQ500 and primary ZephIR at the same location.

Acknowledgements
The authors would like to thank the consortium members for their support, including:

Airtricity	NaREC	ScottishPower Renewables
AMEC Wind Energy	Natural Power	Shell WindEnergy
BP Alternative Energy	npower renewables	SgurrEnergy
E.ON UK	Oldbaum Services	TUV NEL
Garrad Hassan	REpower	University College Cork
Gamesa Energy UK	RES Group Ltd	University of Strathclyde
Mighty River Power	Scottish Enterprise	Vestas Northern Europe

References

[1] Crescendi, H.G., 1997, "A look back of two decades of Doppler sodar comparison studies", Bulletin of the American Meteorological Society, 78, 651-673

[2] Banakh VA, Smalikho IN, Köpp F, Werner C. "Representativeness of wind measurements with a CW Doppler lidar in the atmospheric boundary layer", Applied Optics 1993 **34:** 2055-2067

[3] Kindler, D; Oldroyd, A; Macaskill, A; Finch, D. "An 8-month test campaign of the QinetiQ ZephIR system: Preliminary Results. Meterologische Zeitschrift, Volume 16 No. 5, October 2007. pp 479 –489

[4] Antoniou, I; Jorgensen, H.E.; Ormel, F; Bradley, S.G.; Von Hünerbein, S.; Emeis, S; Warmbier, G. "On the Theory of SODAR Measurement Techniques". RISOE R-1410, 2003.

[5] Courtney, M; Antoniou, I. "Preliminary comparitive ZephIR LiDAR results to cup anemometer measurements. State of the art of Remote Wind Speed Sensing Techniques using SODAR, LiDAR and Satellites. Proceedings of 51st IEA Topical Experts Meeting January 2007.

[6] Antoniou, I.; Jørgensen, H.E.; Mikkelsen, T.; Pedersen, T.F.; Warmbier, G.; Smith, D., Comparison of wind speed and power curve measurements using a cup anemometer, a LIDAR and a SODAR. In: Scientific proceedings. 2004 European Wind Energy Conference and Exhibition, London (GB), 22-25 Nov 2004. (European Wind Energy Association, London, 2004) p. 47-51

[7] Bradley, S. G. (Ed), Antoniou, I., Von Hünerbein, S., Kindler, D., De Noord, M., Jorgensen, H. E. SODAR calibration for wind energy applications. Manchester, UK, University of Salford. 69pp. 2005.

[8] Antoniou, I., Jorgensen, H., Von Hünerbein, S., Bradley, S.G., Kindler, D., Warmbier, G., De Noord, M. 'The profiler intercomparison experiment (PIE)', European Wind Energy Conference, Proceedings of EWEC, 2004, London, November, 2004

[9] Tindal, A.J. "Remote Sensing", WNDGP051, Garrad Hassan, 2007

IOP Conf. Series: Earth and Environmental Science **1** (2008) 012020 doi:10.1088/1755-1307/1/1/012020

Doppler lidar measurements of the Great Plains low-level jet: Applications to wind energy

R M Banta[1], Y L Pichugina[2,1], N D Kelley[3], B Jonkman[3], and W A Brewer[1]

1. Earth System Research Laboratory, National Oceanic and Atmospheric Administration, 325 Broadway, Boulder CO 80305, U.S.A.
2. Cooperative Institute for Research in the Environmental Sciences, Boulder, CO, U.S.A.
3. National Renewable Energy Laboratory, Golden Colorado, U.S.A.

E-mail: robert.banta@noaa.gov

Abstract. The southerly low-level jet (LLJ) of the Great Plains of the United States is a recurrent flow feature of the nighttime boundary layer of the region, which has been identified as a region of high potential for wind energy. The acceleration of the LLJ after sunset produces an enhancement of the wind speed over daytime values, and provides a dependable resource for wind energy. On the negative side, occasional bursts of strong turbulence may be generated that can be of just the right frequency to excite strong oscillatory response in the turbine rotors, thereby accelerating the fatigue of the rotor parts. High resolution Doppler lidar has been used in two studies of the LLJ over the U.S. Great Plains. In this paper we show the usefulness of this remote sensing tool in documenting the mean and turbulent vertical structure, and the evolution of these vertical structures through entire nights. This leads to implications about potential usefulness of Doppler lidar in monitoring mean winds and turbulence in real time to aid in turbine operations.

One of the major wind-energy resource regions in the United States (U.S.) is the Great Plains, extending in a north-south band through the central part of the country, just to the east of the Rocky Mountains. During the warm season when winds in the well mixed daytime boundary layer (BL) can often be 5 m s^{-1} or less, a significant contributor to the wind resource is the nocturnal low-level jet (LLJ). The acceleration of the LLJ after sunset, a dynamic response in the lower BL to surface cooling (Blackadar 1957), can result in significant accelerations of the wind speed at the height of wind-turbine rotors (figure 1). The evolution of the mean wind profile after sunset is therefore an important forecast for wind energy applications, and interaction between LLJ-induced winds and the landscape below is important for wind-

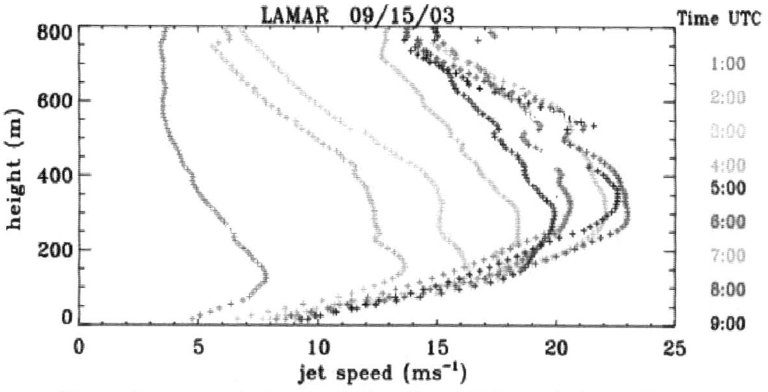

Figure 1: Hourly profiles of mean wind speed showing LLJ evolution after sunset on 15 Sept 2003 (Pichugina et al. 2005).

farm siting. On the other hand, turbulence and wave activity in the strongly sheared wind layer below the LLJ can adversely affect turbine performance by producing vibrations in the rotor blades, limiting the lifetimes of turbine hardware (Kelley et al. 2004).

Two field projects in the past decade have studied mean and turbulent properties of the LLJ over the Great Plains in detail, the Cooperative Atmosphere-Surface Exchange Study campaign in October 1999 (CASES-99; Poulos et al. 2002) and the Lamar Low-Level Jet Project of September 2003 (LLLJP-03; Kelley et al. 2004; Pichugina et al. 2004. 2006). In addition to meteorological towers instrumented with sonic anemometers, both projects featured the High-Resolution Doppler Lidar (HRDL) developed by the National Oceanic and Atmospheric Administration's Earth System Research Laboratory (NOAA/ESRL), in collaboration with the National Center for Atmospheric Research (NCAR), and with initial funding by the U.S. Army Research Office (ARO). HRDL is well suited to studying the LLJ and the accompanying stable boundary layer (SBL), because of its 30-m range resolution, ~10-20 cm s^{-1} velocity precision, narrow beam, and scanning capability. Analysis of scan data involving low elevation angles can provide profiles with vertical resolutions down to a meter or less. Elevation scans aligned with the mean wind, which provide a vertical slice of data through the atmosphere, have been analyzed by binning the velocity data in the vertical (e.g., figure 2) and calculating means and turbulent variances for each bin (Banta et al. 2002, 2006; Pichugina et al. 2008). This procedure provides vertical profiles of the mean wind speed U(z) and streamwise variance $\sigma_u^2(z)$ for each scan, generally repeated every 20-30 s.

An example of how vertical profiles of the mean horizontal wind derived from such scans can be used to document the evolution of the LLJ is shown in figure 1. These hourly profiles show a nascent LLJ at sunset (~0100 UTC) developing into a 23 m s^{-1} jet by midnight local time (0700 UTC). Accelerations at the level of the turbine rotors (up to heights of 150 m in the near future) were not well reflected at the surface, illustrating limitations of near-surface data in evaluating the wind resource. Availability of such data at 20-30 s intervals allows time-height cross sections, such as figure 3, to be generated for each night, to illustrate variations in mean winds that can occur over periods of a few minutes or more in the SBL. Velocity profile data can be further analyzed to determine the height, speed, and direction of the LLJ maximum or nose as a function of time (figure 4) for each night (e.g., Banta et al. 2002).

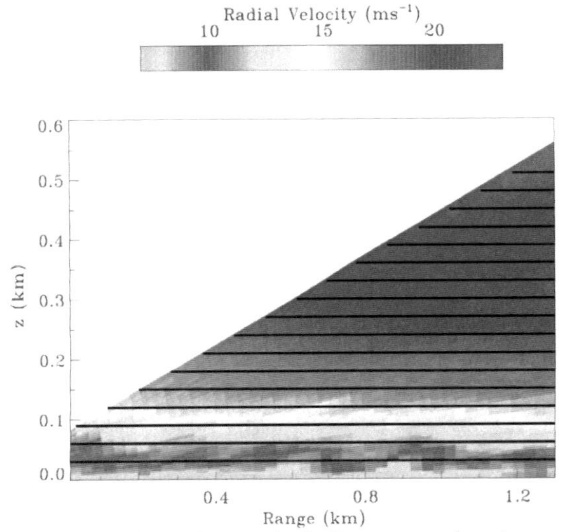

Figure 2: Sample vertical-slice scan showing binning procedure (Banta et al. 2006).

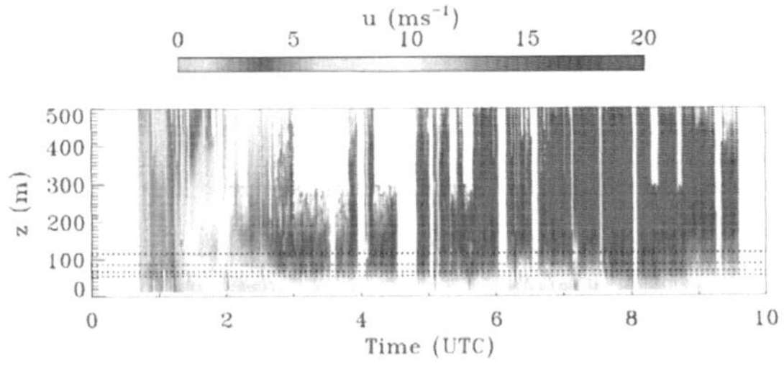

Figure 3: Time-height cross section of streamwise velocity for 15 September 2003. Abscissa is time (UTC), with sunset at 0100 UTC and midnight local time at 0700 UTC. Ordinate is height in meters (up to 500 m), and velocity scale on color bar is in m s^{-1}, from 0 to 20 m s^{-1} (Pichugina et al. 2005, 2007).

Figure 4: LLJ speed (red, scale at left from 0 to 25 m s^{-1}, and height (blue, scale at right from 0 to 500 m) vs. hour UTC for 5 nights of LLLJP.

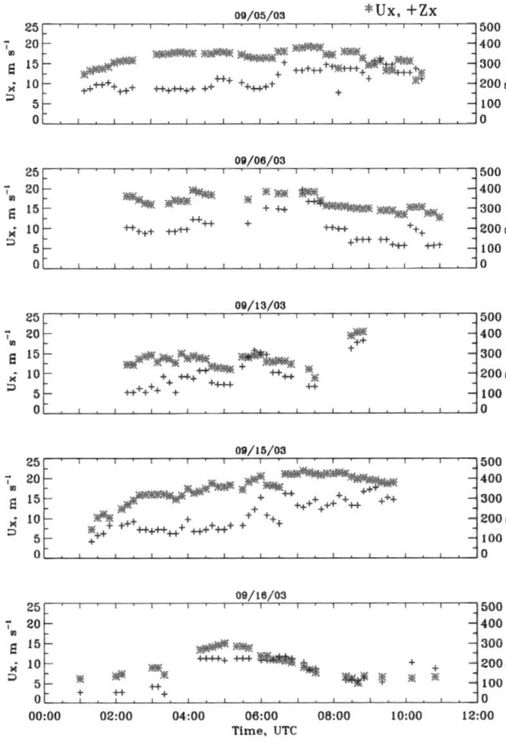

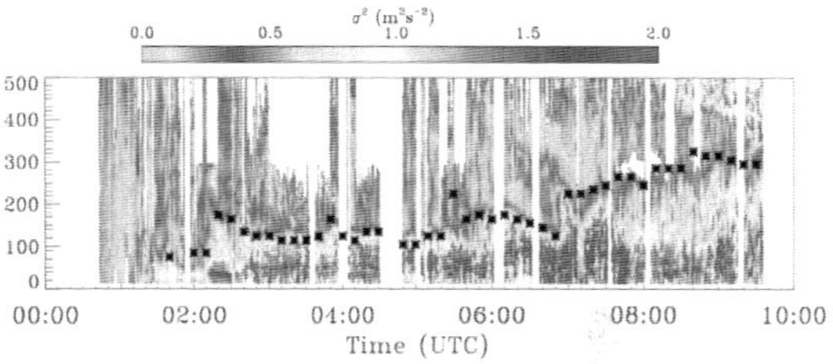

Figure 5: Time-height cross section of streamwise velocity variance for 15 September 2003. Abscissa is time (UTC), with sunset at 0100 UTC and midnight local time at 0700 UTC. Ordinate is height in meters (up to 500 m), and velocity scale on color bar is in m^2 s^{-2}, running from 0 to 2.0 m^2 s^{-2} (Pichugina et al. 2005, 2007).

Turbulent fluctuations—also of significance to wind-energy applications—are generated in the strong shear layer between the surface and the LLJ nose. The intensity and temporal variability of the turbulence is illustrated by time-height cross sections of the streamwise variance. Figure 5 shows the evolution of turbulence in the subjet layer with temporal variations of minutes or more. Higher-frequency fluctuations also occur, as revealed by individual HRDL vertical-slice scans (figure 6). Although strictly turbulent fluctuations tend to be random in time, the fluctuations in Fig. 6 are periodic shear-instability waves that have been analyzed in detail by Newsom and Banta (2002). Waves such as these or other types appearing in the atmosphere have the potential to produce significant fatigue or damage to rotors and turbine hardware, if the frequency is in resonance with the natural frequency of the rotor blades (Kelley et al. 2004).

The ability to adjust the rotors in real time to avoid these disturbances would seem to be advantageous for extended hardware lifetimes. Since numerical weather prediction (NWP) models are not useful for prediction of these kinds of disturbance in any sort of useful way for this application, the alternative would be to detect these bursts in real time upwind of the turbines. Figure 7 shows examples of using HRDL in staring mode at low elevation angles to detect oncoming turbulence. In these examples, the elevation angle was 10°, chosen to show the turbulence structure up to the height of the LLJ. Turbulence is detected at hub height 600 m upstream, providing 1 min of lead time for a 10 m s^{-1} flow speed.

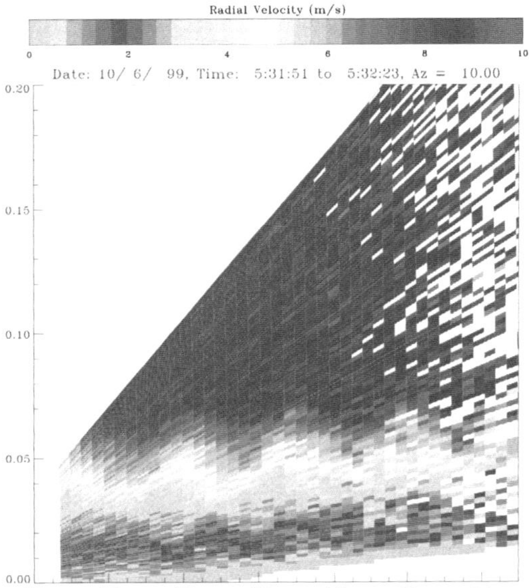

in

Figure 6: Shear-instability waves observed on 6 Oct 1999 (Newsom and Banta 2003).

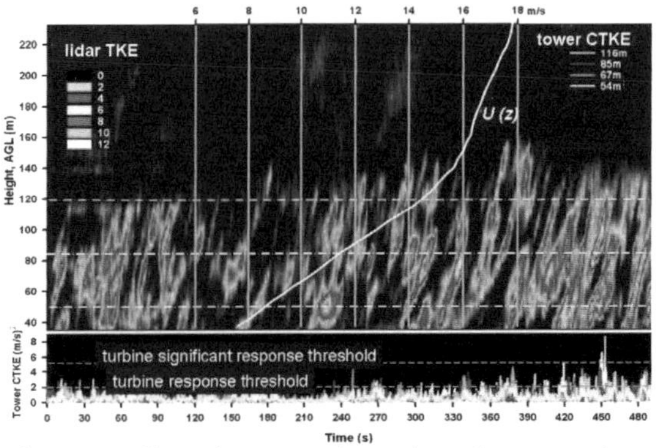

Figure 7: Sample range-time plot of streamwise velocity variance calculated from HRDL in staring mode at 10° elevation angle for the night of 9 Sept 2003, 0729-0739 UTC, where range has been scaled to match the height of the beam. Bottom strip: corresponding tower measured coherent TKE. Abscissa for both plots given in seconds.

Because of the poor representation of stable mixing processes in NWP models, such mesoscale models, even those optimized for research purposes, have not demonstrated quantitative skill in predicting LLJ properties including speed, shear, height, turbulence properties, and their temporal evolution. An important aspect of SBL research, therefore, is to produce better model physics so that such predictions can be more accurate for forecasting, siting, and other planning activities, and to produce accurate case studies for model evaluation.

Acknowledgments. Field data acquisition and much of the analysis for this research were funded by the National Renewable Energy Research Laboratory (NREL) of the U.S. Department of Energy (DOE) under Interagency Agreement DOE-AI36-03GO13094 (Lamar), and the U.S. Army Research Office (Dr. Walter Bach) of the Army Research Laboratory under Proposal No. 43711-EV (CASES-99) and CIRA Center for Geoscience/ Atmosphere Research. Analysis and manuscript preparation were also supported by the NOAA Health of the Atmosphere Program. We thank our colleagues Raul Alvarez, Lisa Darby, Joanne George, Jennifer Keane, Janet Machol, Brandi McCarty, Andreas Muschinski, Ron Richter, Scott Sandberg, and Ann Weickmann from ESRL, and the following from NREL: J. Adams, Dave Jager, Mari Shirazi and S. Wilde.

REFERENCES

Banta, R.M., Y.L. Pichugina, and W.A. Brewer, 2006: Turbulent velocity-variance profiles in the stable boundary layer generated by a nocturnal low-level jet. *J. Atmos. Sci.*, **62**, 2700-2719.

_____, R.K. Newsom, J.K. Lundquist, Y.L. Pichugina, R.L. Coulter, and L. Mahrt, 2002: Nocturnal low-level jet characteristics over Kansas during CASES-99. *Boundary-Layer Meteor.*, **105**, 221-252.

_____, Y.L. Pichugina, and R.K. Newsom, 2003, Relationship between low-level jet properties and turbulence kinetic energy in the nocturnal stable boundary layer. *J. Atmos. Sci.*, **60**, 2549-2555.

Blackadar, A.K., 1957: Boundary layer wind maxima and their significance for the growth of nocturnal inversions. *Bull Amer. Meteor. Soc.*, **38**, 283-290.

Emeis, S., M. Harris, and R.M. Banta, 2007: Boundary-layer anemometry by optical remote sensing for wind energy applications. *Meteorologische Zeitschrift*, **16**, 337-347.

Grund, C.J., R.M. Banta, J.L. George, J.N. Howell, M.J. Post, R.A. Richter, and A.M. Weickmann, 2001: High-resolution Doppler lidar for boundary layer and cloud research. *J. Atmos. Oceanic Technol.*, **18**, 376-393.

Kelley, N.D., B.J. Jonkman, G.N. Scott, and Y.L. Pichugina, 2007: Comparing pulsed Doppler LIDAR with SODAR and direct measurements for wind assessment. AWEA 2007 Windpower Conf. Los Angeles, California. NREL/CP-500-41792.

_____, M. Shirazi, D. Jager, S. Wilde, J. Adams, M. Buhl, P. Sullivan, and E. Patton, 2004: Lamar Low-Level Jet Program interim report. NREL/TP-500-34593.

Newsom, R.K., and R.M. Banta, 2003: Shear-flow instability in the stable nocturnal boundary layer as observed by Doppler lidar during CASES-99. *J. Atmos. Sci.*, **30**, 16-33.

Pichugina, Y.L., R. M. Banta, N. D. Kelley, S. P. Sandberg, J. L. Machol, and W. A. Brewer, 2004: Nocturnal Low-Level Jet characteristics over southern Colorado. *Preprints, 16th Symposium on Boundary Layer and Turbulence*, Portland ME.

_____, R. M. Banta, N. D. Kelley, 2005: Application of high-resolution Doppler lidar data for wind energy assessment. *Preprints, 2nd Symposium on Lidar Atmospheric Applications*, San Diego CA.

_____, R. M. Banta, and W. A. Brewer, 2006: Vertical profiles of velocity variances and TKE using Doppler-lidar scan data. *17th Symposium on Boundary Layers and Turbulence*, San-Diego, CA

_____., R. M. Banta, N. D. Kelley, 2007: Analysis of the southern Colorado low-level jet by high-resolution Doppler lidar data. Comparison to the Great Plains LLJ climatologies. *3d Symposium on Lidar Atmospheric Applications*, San Antonio, TX.

Pichugina, Y.L., R. M. Banta, N. D. Kelley, B.J. Jonkman, W. A. Brewer, S. P. Sandberg, and J. L. Machol, 2007: Nocturnal low-level jet characteristics over southern Colorado. *Preprints, 14th Coherent Laser Radar Conf.*, Snowmass CO.

Poulos, G.S., W. Blumen, D.C. Fritts, J.K. Lundquist , J. Sun, S. Burns, C. Nappo, R.M. Banta, R.K. Newsom, J. Cuxart, E. Terradellas, B. Balsley, M. Jensen, 2002: CASES-99: A comprehensive investigation of the stable nocturnal boundary layer. *Bull. Amer. Meteor. Soc.*, **83**, 555-581.

IOP Conf. Series: Earth and Environmental Science **1** (2008) 012021 doi:10.1088/1755-1307/1/1/012021

Testing and comparison of lidars for profile and turbulence measurements in wind energy

Michael Courtney, Rozenn Wagner and Petter Lindelöw

Wind Energy Department, Risø National Laboratory for Sustainable Energy, Technical University of Denmark, P.O. 49, DK-4000 Roskilde, Denmark.

E-mail: michael.s.courtney@risoe.dk

Abstract. Lidar profilers are beginning to gain a foothold in wind energy. Both of the currently available commercially systems have been extensively tested at the Høvsøre facility in Denmark and valuable insights have been gained. The extensively instrumented facility will be described and some examples of the results given, illustrating the strength and weaknesses of the two contrasting profilers.

1. Introduction

Wind measurements are fundamental to the wind energy industry. No wind project can be started before the available wind resource has been documented and the case for a successful return on the investment has been scrutinized and accepted. Wind turbine power performance testing (measuring electrical energy as a function of wind speed) is usually an important element in the acceptance and delivery process, often with hard cash at stake.

For both wind resource and power performance testing, the mast-mounted cup anemometer is the dominant and only formally accepted sensor for wind speed. This robust, simple and accurate instrument has served the industry well. However, the rapid increase in wind turbine size over the past few years has exposed the cup anemometer's Achilles heal – the mast structure needed to locate the cup anemometer at the hub-height of the wind turbine. With typical hub-heights now often exceeding both 80 and 90m, traditional mast-mounted cup anemometry has become an expensive and logistically complex affair. At the same time, the current practice of basing power performance measurements on a single, hub-height wind speed measurement becomes intuitively more and more suspect for such huge rotor swept areas. Recent numerical studies [1] indicate that the correlation between measured electrical power and wind speed increases if the wind speed is based on a weighted average of the wind speed profile over the entire rotor rather than on a single point measurement. Here, remote sensing is set to play a major role.

An alternative to mast-mounted anemometry is ground-based remote sensing systems. Sodar systems have been available for many years and have achieved some acceptance within the wind energy community, especially in North America. The measurement accuracy of sodar systems can not match that of cup anemometry and unless high acoustical powers are used, their availability falls in high wind speeds. A hybrid system comprising a moderately tall mast (say 40m) and a relatively low power (30-100W electrical) sodar has many attractive features - the high absolute accuracy and high availability of the cup anemometer complements the less accurate but highly relevant vertical resolution obtained from the sodar.

Since light can be much more precisely focused and spreads in the atmosphere much less than sound, lidar systems have an inherently higher accuracy and higher signal to noise ratio than a sodar. They have the potential to replace mast measurements and in time, to perform with comparable accuracy to cup anemometers.

Until recently, making wind speed measurement using lidars was prohibitively expensive and essentially limited to the aerospace domain [2]. Suitable lasers were costly, large and required elaborate cooling systems. The optical systems were built on traditional optical-benches and were inherently difficult to keep aligned. In addition it was difficult to obtain a sufficient acquisition rate while keeping these early systems eye-safe and they were therefore unsuitable for unmanned operation. All these limitations were swept aside by the emergence of sufficiently coherent lasers at wavelengths compliant with fiber optic components (so-called 'fiber lasers'). Lidar systems can now be built using quality tested, standard telecommunications components, available off-the-shelf and connected together by standard terminated optical fibers. The lasers consume relatively low electrical power and do not require elaborate cooling systems. Since precise alignment is no longer generally an issue, robust portable systems can be realized. Eye-safety concerns can also be overcome since the majority of fiber lasers emit at the eye-safe wavelength of 1.5μm.

Two commercial lidar systems are now available. Both systems have been extensively tested both in prototype and production forms, by Risø DTU at the Høvsøre Test Station in Jutland, Denmark. It is our belief that the insight gained in this testing has played a significant role in the development of these systems. It is also our hope that the Høvsøre facility can continue to contribute to the development of lidar profilers and in particular, to their acceptance within the wind energy community. The aim of this paper is to describe, contrast and compare these two commercial profilers within the context of the testing that we have carried out at Høvsøre.

2. The Høvsøre test facility

In order to support the wind turbine industry and as a research venue for wind energy, Risø DTU has established the Test Station for Large Wind Turbines at Høvsøre in western Jutland, Denmark (Figures 1 and 2). This facility comprises a line of five test stands for MW class wind turbines, each stand with its dedicated upstream (westerly) measuring mast. Two technicians are permanently stationed at Høvsøre in order to service the copious instrumentation and to assist with other research and development activities. There is excellent infrastructure including network over the entire site, offices, a workshop and meeting rooms.

As seen in Figures 1 and 2, the line of test stands is orientated N-S, parallel to the coastline and about 1.5km from the sea. At the southern end of the line, 200m from the closest wind turbine, there is an intensively instrumented meteorological mast (it can be seen in Google Earth at position 56.441045N, 8.150787E). It has a top-mounted anemometer at 116.5m and measuring stations with both cup and 3-D ultrasonic anemometers at 100m, 80m, 60m and 40m. The purpose of this mast is to supplement the wind measurements at the turbine test stands, to provide additional information about the climatology at Høvsøre and as a source of meteorological data for boundary layer research. The quality of the instrumentation is high both in terms of the mounting ($<$ 1% flow distortion for dominant wind directions) and in respect of the sensor maintenance. Cup anemometers (Risø class 1 classification) are regularly exchanged for re-calibrated units and a rigorous QC is implemented.

IOP Conf. Series: Earth and Environmental Science 1 (2008) 012021 doi:10.1088/1755-1307/1/1/012021

Due to its height, intensity of instrumentation and high data quality, this mast is well suited for testing lidar profilers. Since there is no problem with backscatter from the mast structure (unlike fixed-echo problems with sodar testing), it is possible and indeed advantageous to place the profilers close to or directly beside the mast and thus improve the accuracy of the comparison. Wind speed measurements at 160m height (both cup and ultrasonic anemometers) are also available from a light beacon mast, some 300m north of the meteorological mast.

Høvsøre is located in very flat terrain but with somewhat inhomogeneous roughness (see Figure 1). To the south is a lagoon, at the closest point, only 900m from the meteorological mast and 1.8 km to the west, the North Sea, separated from the land by a strip of sand dunes 10-20m high. The most homogeneous fetch is from the east – mostly open farmland. The presence of the five wind turbines to the north of the meteorological mast precludes testing with wind from the northerly sectors. Indeed, it is important to be constantly aware of the measuring volumes of the profilers and how these might be influenced by the inflow or wakes of the wind turbines or the mast itself. Most lidar testing makes use of wind from the sector 200-300°. Wind from these directions is unaffected by the wind turbines, can be measured by the mast-mounted cup anemometers with less than 1% flow distortion and importantly, are very common.

In order to test remote sensing profilers in the context of power performance measurements, a test pad has been established between and upstream of two of the wind turbines at the test station. Here it is possible to place lidar and sodar profilers at distances from the wind turbines typical for power performance testing. Measurements from the profilers can be related both to the measured turbine electrical power and to the conventional mast-mounted cup and sonic anemometer measurements.

3. Commercial lidar profilers

3.1. The QinetiQ ZephIR

After several prototype versions, one of which had previously been tested at Høvsøre [3], the English company QinetiQ presented the first commercial lidar profiler at the beginning of 2006. Known as the 'ZephIR', it is a continuous wave system operating at 1.5μm wavelength, with height discrimination achieved by varying focus. Five different sensing heights can be chosen. Each height is scanned for 3s before the lidar re-focuses to the next height in the sequence. Figure 3 shows a ZephIR on test at the Høvsøre test facility.

The laser light is emitted through a constantly rotating prism giving a deflection of 30° from the vertical and making one complete rotation per second. Backscattered light mixed with the local oscillator is sampled at 100MHz. The unit is able to transform blocks of 512 sample points to power spectra in real time. The 200000 spectra per second are block-averaged 4000 at a time to give highly noise-suppressed Doppler shifted spectra at 50Hz. A centroid method is used to obtain the frequency of the 'peak', the radial speed (i.e. in the laser beam direction) being directly proportional to this frequency. Thus 50 radial velocities, one for every 360°/50=7.2° are available from each rotation.

Unlike pulsed systems, continuous-wave lidar systems do not inherently 'know' the height from which backscatter is being received. We are obliged to assume (or rather, hope for) a sensibly flat vertical profile of aerosol concentration, in which case the backscattered energy is mainly from the focused volume, weighted according to the appropriate Lorentzian function centered at the focus distance [4]. The obtained radial wind speed distribution in this case is dominated by the signal from the set focus distance.

The assumption of vertical aerosol homogeneity unfortunately fails completely in the fairly common case of low level clouds (under 1500m). Here, the relatively huge backscatter from the cloud base gives significant contribution even though the cloud is far above the focus distance. The resulting Doppler spectrum has two distributions – one corresponding to the radial speed at the focused height and a second corresponding to the (usually) higher speed of the cloud base. Unless corrected for, this will introduce a bias to the wind speed measurement. For this reason, the ZephIR has a cloud-correction algorithm. An extra three second scan at 300m is inserted into the height cycle. The spectra measured at 300m are used in an attempt to remove the influence of the clouds at the desired measuring heights.

Since the ZephIR does not apply a frequency offset to the Doppler shift, it is not possible to determine the direction of the radial velocities. This fact introduces an ambiguity in the polarity of the parameter pair wind direction and vertical velocity. For this reason, the ZephIR independently senses the wind direction using a sensor mounted on a short mast fixed to the top of the lidar. Assuming that the lowest remotely-sensed wind direction will be closest to the directly-sensed value immediately above the lidar, the direction ambiguity is resolved and hence the true sign of the vertical speed is established.

3.2. The Leosphere Windcube pulsed lidar.

With the recent (December 2006) introduction of the French company Leosphere's 'Windcube', there are now two contrasting lidar profilers on the market. The Windcube, shown in Figure 4, is a pulsed lidar with a fixed focus, also operating at the 1.5μm wavelength. Like the ZephIR, it has a 30° prism to deflect the beam from the vertical but here the prism does not rotate continuously. Instead, the prism holds still whilst the lidar sends a stream of pulses (5000-10000) in a given direction, recording the backscatter in a number of range gates (fixed time delays) triggered by the end of each pulse. Having sent the required number of pulses, the prism rotates to the next azimuth angle to be scanned, each separated by 90°. A full rotation takes about 6s.

During the rotation and before the next stream of pulses can be sent, the recorded data are processed. For each range gate, the time series from each pulse are Fourier transformed to power spectra which are block-averaged. Due to the short recording duration (200 ns), the resulting spectra have poor frequency resolution. Instead of using a centroid to obtain the frequency of the peak, Leosphere have developed a mathematical model [5] including the most important parameters affecting the shape of the expected Doppler spectrum. This model is fitted to each block-averaged power spectrum in order to obtain the Doppler shift to a much higher resolution than could otherwise be expected. At each direction step, the Windcube combines the four most recent radial speeds at each height in order to obtain the horizontal and vertical speed and wind direction.

In contrast to the ZephIR, the Windcube uses an acousto-optic modulator (AOM) to add a precise frequency offset to the local oscillator which is mixed to and beats with the returning Doppler shifted backscatter. Backscatter from a fixed target, introducing no Doppler shift, appears shifted in the resulting power spectrum. Thus the polarity of the radial velocity is available and there is no ambiguity regarding the wind direction.

4. Testing lidar profilers

Since the wind industry is searching for a replacement for the mast-mounted cup anemometer, our testing has very much centred on comparing the profiler's measurements at a number of preset heights with the cup anemometer measurements made at these same heights. This is very much how we would compare, for example an ultrasonic anemometer with a cup anemometer. With a little more thought, we can see that we are actually comparing a measurement made over a volume with significant vertical and horizontal extent, with what is essentially a point measurement (the measurement volume of the cup anemometer being insignificant in comparison). Only in very exceptional conditions can we expect to measure exactly the same mean speeds – in general we will not. Whilst over flat terrain the differences are not usually very large, they can be significant especially within the context of power curve measurements. It is therefore important to understand and quantify these effects, especially when significant biasing can be expected.

When we extend the comparison to include some measure of the turbulence (e.g. the standard deviation of the horizontal speed), the situation becomes even more complicated because now we have not only different volumes but also different averaging times to consider.

4.1. Coping with the horizontal extent of the scan path

Here we must address the greatest weakness of (monostatic) lidar profilers – their inherent assumption that the wind flow over the scan path and within the scan duration is sufficiently homogeneous to allow a successful wind vector construction from the sensed radial velocities. The benefit of this assumption is that we can with one instrument, consecutively sense in 3 (or more) different directions, assume that we are seeing different projections of the same wind speed vector, solve the equations and arrive at the 'homogeneous' speed and direction. The cost of the assumption of homogeneity is that measurement errors will be introduced according to the extent to which the assumption is actually breached.

Luckily at Høvsøre, the terrain is very flat. This seems to be a very important requirement for a lidar profiler test site. Any terrain features causing curvature of the flow over the horizontal extent of the measuring volume, will introduce quite significant errors in the measured speeds.

Unluckily (with respect to homogeneity), we have several large wind turbines at Høvsøre, the closest only 200m from the measuring mast. Wind turbines produce large and meandering wakes that are truly and dramatically inhomogeneous. It is essential that the scan paths of the lidar profilers do not come in their vicinity. This requirement imposes the greatest restrictions on the wind directions we can employ. The measurement mast itself also generates a small but significant wake which must also be taken into account.

Other sources of flow inhomogeneity must also be considered. As described above, the upstream roughness at Høvsøre is quite variable. For our preferred wind directions, 200-300°, the dominating upstream feature is the sea-land roughness change 1 to 2 km upstream. Our belief is that this does not result in any significant horizontal inhomogeneities around the measuring mast but this remains to be adequately documented.

4.2. Coping with the vertical extent of the measuring volume

By their very nature, ground-based profilers measure not at one height but actually over a height range with a non-trivial (and certainly non-uniform) weighting within this range. In the details and as described below, the two commercially available lidars are very different. Nevertheless, for both types, it is important to consider the implications of the vertical measuring volume when analyzing and attempting to understand the measurement results.

A consequence of the complicated vertical weighting is that Lidars have a dependency on how much the horizontal speed changes with height, i.e. the local wind shear. It is therefore important to include some measure of this relation in any analysis of the data. In a traditional linear regression of lidar and cup anemometer wind speed (see Figure 5), these shear proportional errors will be 'smudged', influencing the estimation of both slope and offset. If instead, a two parameter regression is performed (see Figure 6), the wind speed proportional errors and the wind shear proportional errors can be separated to some degree [6].

5. Strengths and weaknesses

In this section we will discuss the relative merits of the two lidar profilers, with respect to a number of different aspects. It is not our intention to make a recommendation, but rather to present the differences as we see them.

5.1. Height resolution

Here the two systems have fundamentally different properties. The ZephIR is a variable-focus, continuous wave system and as such, the height range is determined by the depth of the focus volume. More specifically this can be defined as the FWHM (Full Width Half Maximum) of the Lorentzian function defining the collection efficiency. This increases quadratically with sensing range. At low heights, the probe length is only a few decimeters. At 100m height, it is about 20m whilst at 150m it is close to 50m.

For the Windcube, the vertical measuring depth depends mainly on the pulse length and is approximately constant with sensing range. A classical measure is the FWHM of the function describing how the collected energy varies with altitude. This FWHM is about 26m (vertically) for the Windcube and is sensibly constant with altitude. However, it is important to consider the spectral density contributed from different parts of the range gate [7] and this gives an effective probe length that is shorter than the classical value. Based on these new considerations, the vertical probe length is about 16m. Which of these two contrasting definitions best describes the actual performance of the Windcube is not yet entirely clear. We are actively pursuing this issue both theoretically and experimentally.

The weighting functions of the sample volumes for the two systems are compared in figure 7. We have chosen to plot the narrower definition based on spectral density for the Windcube. The influence of the Windcube's fixed focus has not been accounted for. At low heights, the ZephIR clearly has the smaller probe length (beneath 40m the Windcube can not measure at all). Between about 100 and 120m, the probe lengths are approximately equal. Higher up, the Windcube has the edge and at heights of over 150m, only the Windcube can measure with an acceptable vertical resolution.

5.2. Low speeds

The most significant noise source in coherent lidar is known as RIN (Relative Intensity Noise) and is caused by rapid variations in the emitted power. RIN is typically 'pink' noise, meaning that it falls off with increasing frequency. It is so severe as to essentially blind the lidar to Doppler shifts occurring at the low frequency end of the spectrum. For the ZephIR system, this affects Doppler shifts related to low wind speeds. As the actual wind speed falls below about 4 m/s, the wind speed measured becomes more and more positively biased. This is because more and more of the low frequency side of the peak in the spectrum falls into the region that has been blanked in order to prevent corruption by the RIN noise.

In contrast, the Windcube effectively shifts the range of radial speeds affected by RIN to large negative values, corresponding to implausible wind speeds. Thus the Windcube can measure in complete wind still.

IOP Conf. Series: Earth and Environmental Science 1 (2008) 012021 doi:10.1088/1755-1307/1/1/012021

5.3. Clouds

As mentioned in section 3.1, the ZephIR is sensitive to backscatter from low level clouds. When it became clear that this would be an important issue, a Vaisala ceilometer was added to the Høvsøre instrumentation. This instrument is commonly used, especially at airports, to measure the height of the cloud base. The cloud base data from the instrument are available in the database and can be used to classify lidar results according to cloud conditions. A number of important results have been achieved:

Firstly it was confirmed that the ZephIR running without the cloud correction algorithm applied, was indeed sensitive to cloud backscatter. In addition, the degree of corruption varied between different ZephIR units. It is suspected that greater photo-detector sensitivity gives a greater sensitivity to unwanted cloud base backscatter.

Secondly, we determined that using the internal cloud correction algorithm did reduce the sensitivity to cloud. However, the cost of this was general negative biasing of the data in conditions where clouds were not present. More disturbingly, the standard deviation of the horizontal speed normalized by that measured from the cup anemometer, increased with the cloud correction algorithm active. We interpret this increase as an indication of noise introduced by the cloud correction algorithm.

Recently we have been working with QinetiQ on an improvement to the cloud-correction algorithm. With the cloud correction activated, two additional heights are inserted into the measuring cycle, 800m (replacing the previous 300m) and 38m. One major improvement is that the cloud correction is now conditional, being turned on and off according to the relative backscatter at these heights. A second improvement is that with the 800m cloud detection height (essentially a collimated beam), the spectral densities are very low (and very broad) unless cloud is actually present. This aids the discrimination process and helps to prevent corruption of the lower level Doppler spectra that was previously experienced in cloud-free conditions. A further finesse is that only a proportion of the cloud level spectrum is subtracted from the lower level, this depending on the backscatter ratios between the highest and lowest levels. Having tested the new algorithm with several periods of previously recorded raw data, we are currently testing an embedded version of the new algorithm in a ZephIR lidar at Høvsøre.

In comparison to the ZephIR, clouds play only a minor role in the performance of the Windcube. Since this is a pulsed system with the backscatter collected in finite range-gates, only clouds actually present in the measuring range will have any affect. This can be considered as a discontinuity in the aerosol concentration within the measuring volume and the wind speed at the step will dominate the Doppler spectrum. Probably more significant is how mist and clouds affect the available measuring heights. Very low cloud will cause significant propagation losses and the highest measuring height may be quite restricted. On the other hand, light clouds at say 250m can increase the backscatter from these heights without extinguishing the beam and this can help the lidar to reach higher heights than would otherwise be available.

5.4. Heights - number, ranges and synchronicity

Since the Windcube is a pulsed system, the radial velocity at all heights are effectively measured simultaneously. Wind vectors at as many heights as required are produced every 1.5 s. Usually between 10 and 15 simultaneous heights are used. The lower limit of height is around 40m, the upper depends on the background aerosol level. In clean (e.g. mountain) air it may be impossible to measure regularly over 150m. In other locations the background aerosol may be significantly higher and sensible measurements can be made over a greater height range, if required. At Høvsøre, it is possible to measure up to 300m with availability of about 50%, up to 200m with 75% availability and up to 100m with 95% availability.

The ZephIR is more restricted in the number of heights that can be chosen – no more than 5. Measuring heights are scanned each 3 s at a time, with some extra time required for re-focusing between heights. A typical cycle takes about 18 s to complete. Thus at any one height, a 3 s average is available once in every 18 s period. There will be a minimum of slightly more than 3 s between measurements at consecutive heights. This has some implications for studies involving determining instantaneous wind shear or other form of extreme analysis. Heights can be chosen from 10m and above. The upper limit is largely determined by the quadratically increasing vertical probe length. At the height of 150m, the probe length is about 50m and it is usually not advisable to measure any higher than this.

5.5. Complex terrain

Both instruments rely on the assumption of horizontally homogeneous flow. In complex terrain this assumption is violated, increasingly so as the terrain complexity increases. This is not an issue that can be tested at Høvsøre but indications from other sites indicate that errors of 5-10% in the mean speed are not uncommon. With the standard configuration, both lidar profilers use the same 30° cone angle. The error for both systems will be similar at a given location.

An option for the Windcube is for the user to replace the standard 30° prism with one of 15°. This will reduce the error due the flow complexity (actually the curvature of the flow) but at the cost of reduced resolution of the horizontal speed. Almost certainly, this is a sensible choice for complex terrain measurements.

5.6. Ease of installation

One of the major advantages of lidar profilers in comparison with conventional measurement masts is that the lidar systems are easily and quickly deployed. Whilst this is true for both the ZephIR and the Windcube, installation of the latter is much simpler. The ZephIR is constructed as 3 'pods' (battery, electronics and optics), mounted one on top of the other. The assembled lidar is far too large and heavy to be moved more then a short distance so any redeployment requires a disassembly and a subsequent assembly operation at the new site. For skilled operators, each operation can easily be accomplished within one hour but there are a large number of bolts and cables to manipulate with. In adverse weather conditions, this is not a particularly amusing experience!

In contrast, the Windcube is one rectangular box with a convenient handle in each corner. Deployment is accomplished by lifting it to the required position, applying power and connecting a computer. Levelling is achieved by adjusting the legs of the Windcube until the tilt and roll indicated by the Windcube software is close to zero.

5.7. User interface

Very different approaches have been taken to the user interface and to the means of removing data from the systems. The ZephIR uses the internal processing board for all processing and data storage. In order to connect with this system, it is necessary to connect using a pc running a client program. The connection can be made using either an Ethernet or a wireless (GSM/GPRS/3G) radio link. Using the client program, it is possible to log into the ZephIR and perform a variety of tasks – configuration, data extraction and status monitoring being the most important. Both the Ethernet and wireless links are slow and only 10 minute statistics can be sensibly downloaded in this way. Data are stored in the ZephIR on a removable compact flash card. This is an alternative method for removing the data and is especially useful if it is required to store the raw Doppler spectra or the processed 3s data. A third possibility is to have a pc permanently attached to the ZephIR and use the data streaming facility in the pc client program to store continuously, raw Doppler spectra on the pc's hard disk. This requires a local, almost direct (e.g. one switch) Ethernet connection between the pc and ZephIR. It will not work reliably over a larger network with significant traffic.

All the processing in the Windcube is carried out using an internal Windows pc. This is accessed over an Ethernet connection by using the Window's 'Remote Desktop' facility. Data are stored on an internal hard disk. This is accessible over the network through an FTP server running on the Windcube pc. Also provided is a utility for periodically sending 10 minute statistics as e-mail. Network connection can be realised in remote locations by using an internal GSM wireless modem. The choice of architecture and operating system means that newer data transfer protocols can be easily implemented.

5.8. How well they measure

At Høvsøre we have tested over 10 different lidar systems, a mix of ZephIRs and Windcubes. Whilst it would be inappropriate to give individual results, we are able to give an indication of the range of the test results. Here again it is important to emphasise that a lidar profiler measurement is not the same as a point measurement on a mast. Discrepancies between the point and the lidar measurements are not necessarily because the lidar is wrong, but because it measures in a different way that we are not yet fully able to interpret. Conceivably, the volume averaged lidar measurement may be more appropriate for wind energy applications than the traditional point measurement.

Comparing cup anemometer and lidar measurements, we have various measures of the 'error'. Traditional linear regressions of measured lidar and cup speeds give a slope and, if not forced through the origin, an offset. Whilst there are statistical tests for the uncertainty of the offset, our more pragmatic approach is to perform the regression in both ways. An offset of more than a few centimetres per second is almost certainly an indication that there are other error sources that are not purely or linearly wind speed dependent – for example, a wind shear dependency as previously discussed. Comparing slopes is usually only meaningful for the regression performed with an offset forced to zero. Here typical values are between 0.98 and 1.01 and for the 10 systems tested, we have seen values in the range 0.95 to 1.02.

Another way of characterising the discrepancy between the cup anemometer and the lidar (at a given height) is to calculate the mean and standard deviation of the 'error'. Here typical values are 0.20 m/s for the standard deviation and mean discrepancies in the range -0.2 to 0.2 m/s. Whilst these figures are typical, the actual values will depend on the distribution on the wind speeds during the test campaign.

We are not able to identify the lidar that performs best in respect of measurement accuracy. Both the Windcube and the ZephIR are comparable and, in our opinion, perform remarkably well for remote sensing instruments.

6. Conclusions

Lidar profilers are beginning to play a role in the wind energy industry. We believe that these instruments can play a major role in replacing traditional measurement masts and providing detailed wind speed profiles for use in more accurate and repeatable power performance testing.

Two commercial lidar profilers are now available, the continuous wave, variable-focus Qinetiq ZephIR and the pulsed Leosphere Windcube. At the Høvsøre facility in Denmark, Risø DTU has been testing a number of examples of both these systems. Both lidars measure to a higher degree of accuracy and with higher availability than we have previously seen from ground based remote sensing systems.

When testing lidars and comparing with point measurements from, for example, mast mounted cup anemometers, it is important to remember that the lidar measures in a volume with both vertical and horizontal extent and with non-linear weighting. We do not yet fully understand how to interpret the difference between these two forms of measurements.

The main differences between the ZephIR and the Windcube have been presented. Probably most significant is how the ZephIR's vertical probe length increases with the square of the height but for the Windcube remains constant. This makes the ZephIR clearly better suited for measurements beneath about 80m and the Windcube clearly better suited for measurements above about 130m. In conditions where low cloud and mist are not uncommon, the Windcube has a significant advantage in that it is not affected by backscatter from the clouds.

Testing at Høvsøre has, we believe, played a significant role in the development of both the ZephIR and Windcube. Several minor defects have been detected and much feedback regarding both the physical and software design has been given. In particular, both the deficiencies of and the improvements to the ZephIR's cloud-correction algorithm have been based on the analysis of data measured at Høvsøre.

Acknowledgements
Our former colleague Ioannis Antoniou, now at Siemens Wind Power, has played a major role in establishing lidar testing at Høvsøre and for this he is gratefully acknowledged. The authors would also like to thank the permanent staff at Høvsøre, Bjarne Sønderskov and Anders Vestergaard for their untiring and highly professional support.

Much of the lidar testing and analysis has been carried out as part of Remote Sensing work package (WP6) of the EU Upwind project. ZephIR is a trademark of QinetiQ Ltd., UK. Windcube is a trademark of Leosphere, France.

References

[1] Wagner R., Antoniou I., Pedersen S.M.., Jørgensen H.E., *Wind profile influence on turbine performance*, submitted to the Wind Energy Journal in July 2007
[2] Huffaker R. M., Jelalian A. V. and Thomson J. A. 1970 Laser-Doppler system for detection of aircraft trailing vortices *Proc. IEEE* **58**, 322-326
[3] Smith D. Harris M. Coffey A. Mikkelsen T. Jørgensen H. Mann. J and Danelian R. 2006 Wind lidar evaluation at the Danish wind test site in Høvsøre *Wind Energy* **9**, p 87-93
[4] Lindelöw P 2008 Range dependent collection efficiency of fiber based coherent lidars *Fiber Based Coherent Lidar for Remote Sensing of Wind* Thesis from Technical University of Denmark p 56-62
[5] Valla M 2005 Estimateur du Maximum de Vraisemblance *Etude d'un lidar doppler impulsionnel à laser Erbium fibré pour des mesures de champ de vent dans la couche limite de l'atmosphère* Thesis from École Doctorale d'Informatique, Télécommunications et Électronique de Paris. p 95-103
[6] Lindelöw P A wind shear dependent error in the horizontal wind velocity sensed by a focused range gated lidar 2007 *IOP conf. Ser.: Earth Environ. Sci.* to be printed
[7] Lindelöw P 2008 Effective sample volume of range gated lidars *Fiber Based Coherent Lidar for Remote Sensing of Wind* Thesis from Technical University of Denmark p 65-69

Figure 1 Høvsøre Test Station as seen from Google Earth. The distance from the met mast to the coast to the west is 1.8 km. The met mast is at position 56.441045 N, 8.150787 E.

Figure 2 The row of wind turbines at Høvsøre with the 116.5m meteorological tower in the foreground and the 160m beacon mast between the first and second wind turbine. The picture is taken from a low flying aircraft to the south-west of the mast.

IOP Conf. Series: Earth and Environmental Science **1** (2008) 012021 doi:10.1088/1755-1307/1/1/012021

Figure 3 A ZephIR lidar on test 40m from the meteorological mast.

Figure 4 Four Leosphere Windcubes on simultaneous test at the base of the meteorological mast.

IOP Conf. Series: Earth and Environmental Science **1** (2008) 012021 doi:10.1088/1755-1307/1/1/012021

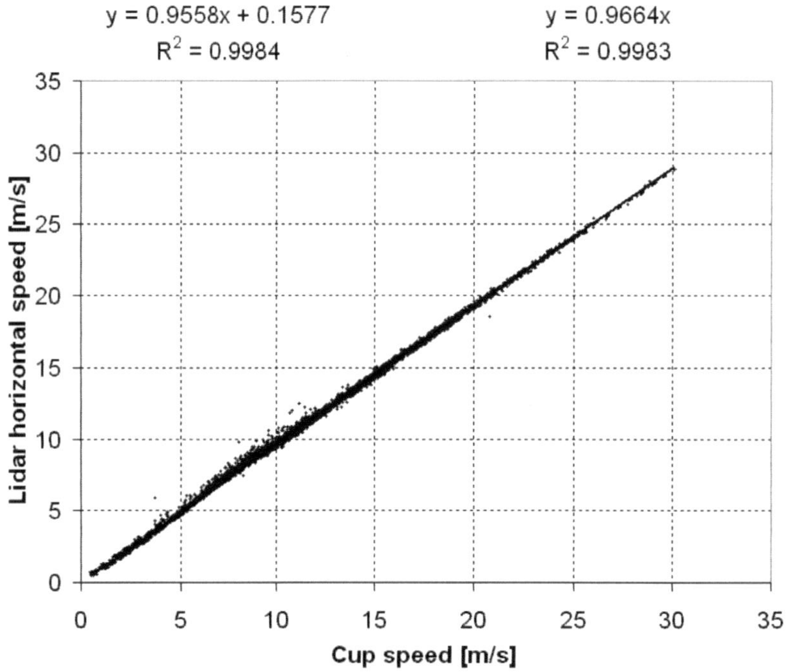

Figure 5 Example of a traditional 1-parametric regression of lidar speed on cup anemometer speed. The example shown is for a Windcube measuring at 116m with the cup anemometer at this height as the reference. The rather low value of the slope is partly attributable to an error in the cone angle of the Windcube. A correction to this has since added 1.5% to the slope.

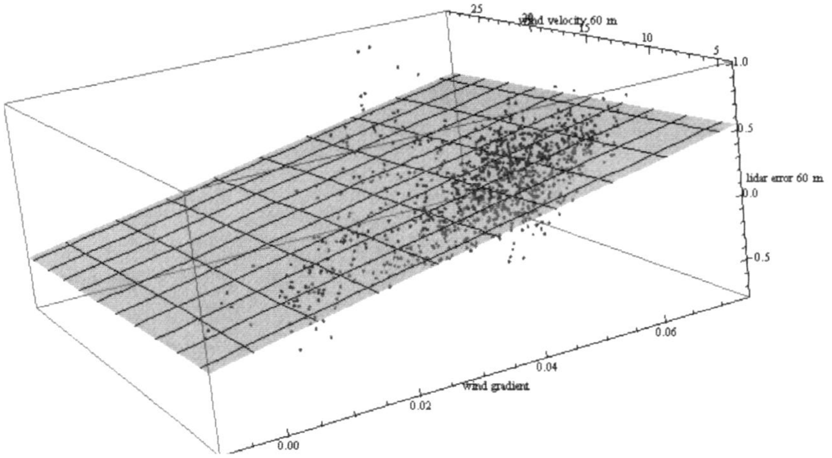

Figure 6 Example of a 2-parametric regression of lidar error as function of the cup anemometer speed and wind shear. The example shown is for a Windcube measuring at 60m with the cup anemometer at this height as the reference. The analysis estimates a horizontal wind velocity loss of 0.5% due to a cone angle error and an underestimation of 9 m in altitude.

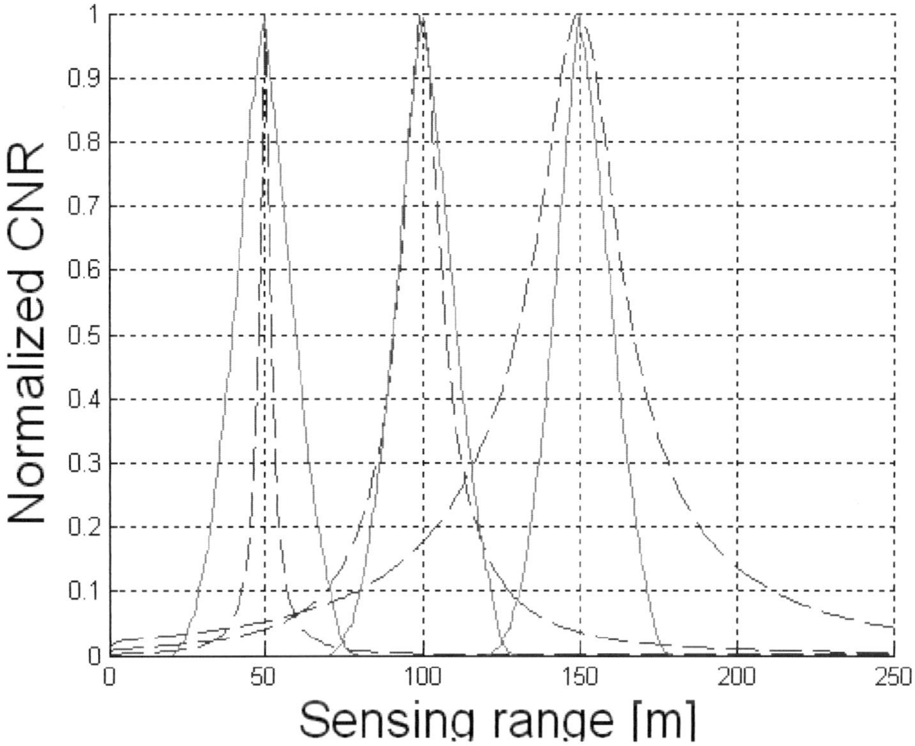

Figure 7: Comparison of the weighting functions for the Windcube and the Zephir at three sensing ranges centred on 50, 100 and 150 meters range, corresponding to 43, 87 and 130 m height. The weighting function for the Windcube, red, is represented by the spectral density of the contribution, scattered from distance z, to the wind spectra [7]. The theoretical Lorentzian profiles obtained for the Zephir are given in dashed blue [4].

IOP Conf. Series: Earth and Environmental Science **1** (2008) 012022 doi:10.1088/1755-1307/1/1/012022

Fast Wake Measurements with LiDAR at Risø Test Field

Ferhat Bingöl[1], Juan José Trujillo[2], Jakob Mann[1] and Gunner C Larsen[1]

[1] Wind Energy Department, Risø National Laboratory for Sustainable Energy, Technical University of Denmark, P.O. Box 49, DK-4000 Roskilde DENMARK
[2] University of Stuttgart, GERMANY

gunner.larsen@risoe.dk

Abstract. The vast majority of wind turbines are today erected in wind farms. As a consequence, wake generated loads are becoming more and more important. We present a new and successful experimental technique, based on remote sensing, to measure instantaneously the flow in the wake of wind turbines. Downstream wind speed can be quantified spatially in one and two dimensions. Data analysis allows us to identify the wake transversal position, thus enabling us to quantify the wake meandering as well as the instantaneous wake expansion expressed in a meandering frame of reference. The experimental results are subsequently used in a preliminary verification of the basic conjecture of a wake meandering model that essentially considers the wake as a passive tracer.

1. Introduction

The vast majority of wind turbines are today erected in wind farms. As a consequence, wake generated loads are becoming more and more important. One relevant problem is the large-scale dynamic characteristics of wind turbine wakes. At Risoe a model to describe this behaviour has been proposed. The Dynamic Wake Meandering (DWM) model [1] has as main assumption that the lateral- and vertical wake deficit transport (i.e. meandering) basically act as if the wake (deficit) was a passive tracer driven by the large-scale turbulent eddies in the atmospheric boundary layer.

The reliable application of this assumption is dependent on a good verification which is not possible with available measurements. Field studies on the deterministic wind turbine wake, which is relevant for energy production; have been performed extensively in the past. However, studies of the dynamics of the wake, which are of crucial importance for turbine loading, are scarce if non existent. This has been historically related to a lack of prioritization but also to great experimental difficulties. Dynamic wake flow measurements on-site, performed with traditional anemometry on meteorological masts, are costly and difficult in their realization. In this paper we present a new experimental technique, based on remote sensing, to measure instantaneously the flow in the wake of wind turbines. This presents enormous advantages with respect to flexibility and amount of information that can be obtained from the flow. The instantaneous wake position can be identified, thus allowing us to quantify the wake meandering as well as the wake expansion and wake turbulence characteristics expressed in a moving (meandering) frame of reference. Additionally, other detailed wake characteristics as transversal wind profile and attenuation of the wake deficit are available.

The experiment related here is designed and conducted with the primary aim of verifying the basic conjecture in the DWM model. In order to achieve this goal, a LiDAR (Light Detecting and Ranging) system has been adapted for measurements of line of sight wind speed from the nacelle of a wind turbine. The LiDAR working principle is based on the Doppler shift between emitted- and reflected laser signals from airborne aerosols assumed to follow the wind passively (Figure 1). This non-intrusive instrument enables instantaneous recording of the wake deficit at a specified down stream position and tranversal position. In the following the setup of the used LiDAR system is explained along with the results of a measurement campaign performed at Risø test field in Roskilde.

2. Test Site

The test site is a relatively flat and homogeneous terrain located near Risø National Laboratory/DTU at Roskilde in Denmark. The test field is used simultaneously for experiments on neighboring wind turbines. The site presents a yearly average wind direction of 289°. The experimental setup consists of the LiDAR system mounted at the nacelle of a test turbine along with two reference meteorological masts dedicated to wind field measurements.

Figure 1. ZephIR Prototype is being used at Risø since 2004. (a) Original QinetiQ design conical scanning mode ZephIR, at operation at Risø Høvsøre Test centre, a homogenous flat terrain. (b) Risø adaptation of ZephIR, operational at Risø Roskilde Test Centre for fast laser wake

The test turbine is a Tellus of 95 kW rated power with an hub height of 29.3 m and a rotor diameter of 19 m. The primary reference mast (MM2) is situated 36 m from the test turbine to the west. This mast is 33m high and equipped with wind speed and wind direction sensors at various heights to supply mean wind speed as well as wind direction data that can be suitably averaged to filter away that fast fluctuating part of the wind direction signal associated with the small scale turbulence. The secondary reference mast (MM1) is situated 114 m from the test turbine to the north-west. This reference mast is likewise equipped with wind speed and wind direction sensors in various heights and additionally with temperature, pressure and rain sensors. A second turbine, a Nordtank of 500 kW rated power, is erected 70 m approximately north of the test turbine. Its hub height is 36m and its rotor diameter is 41m. This turbine is not part of this experiment; however, it affects certain wind sector which has to be excluded from the analysis. The positions of the turbines and meteorological masts at the site are shown in Figure 2. The details of the available sensor signals appear in Table 1.

Table 1. Available data in the database. WS: Wind Speed, WD: Wind Direction, WSN: Wind Speed Nacelle, WDN: Wind Direction Nacelle, YM: Yaw Misalignment

T/M	**Channel**	**Height**
MM1	WS	35 m
	WD	36 m
	Temperature	3 m
	Pressure	3 m
	Rain	-
MM2	WS	29.3 m
	WD	19.7 m
	WD	29.3 m
	WD	35 m
Tellus	WSN	35 m
	WDN	35 m
	Power	-
	YM	-
Nordtank	WSN	29 m
	Power	-
	RPM	-
	Status	-

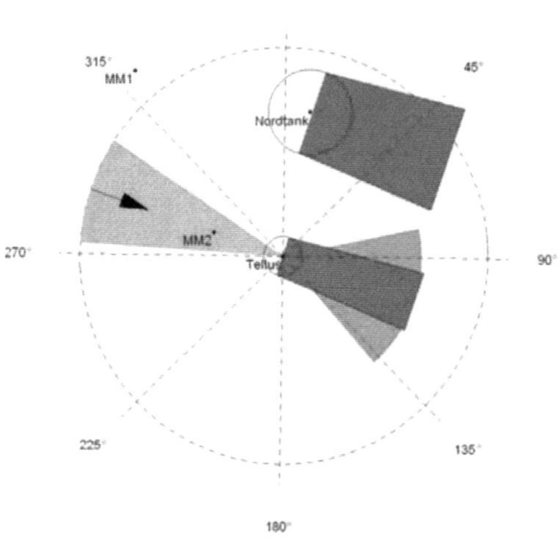

Figure 2. Shown are the positions of the two masts and two wind turbines. Light blue area denotes the selected wind direction sector. All represented data in this paper are from this sector while the Tellus nacelle position is in the same direction. The sector is 30° wide between from 274 ° to 304 °; thus centering the primary reference mast, MM2. The red area behind the Tellus is the possible wake width up to 70 m downstream according to the N.O.Jensen wake model [2] likewise for the Nordtank wake. Green is the example swept area of the LiDAR Line Scan measurement.

3. ZephIR Prototype

Risø possesses since 2004 a prototype model of the ZephIR LiDAR produced at that time by the company QinetiQ. The original design of the ZephIR Prototype was as a vertical profiling tool facilitating focusing of the laser beam at different focus distances (heights). At a given altitude it scans a horizontal circular path with a cone half-angle of 30.6° and measures line of sight wind speed at multiple points. The conning angle is obtained by deflecting the vertical laser beam optically with a rotating vertical-axis wedge. A unique value of horizontal wind speed is derived from all the measured points. A thorough explanation of working principles of ZephIR can be found in reference [3].

The first tests to assess the quality of the instrument have been performed at Risø Høvsøre Test Centre in Denmark at the end of 2004 and beginning of 2005 [4]. These and other studies show very good correlation with Risø cup anemometer [5] and SODAR measurements [6]. Since then the instrument has been used in various experiments at Risø and will be used in the future with new adaptations for different application to wind energy.

3.1. Adaptation

The initial standard configuration of the LiDAR system showed up to have some shortcomings in respect for the application to wake flow measurements from the nacelle. First of all, the data processing time in the supplied software showed up to have an unacceptable performance. Rewriting the data processing software in C and replacing the existing FFT algorithm with the Fastest Fourier

Transform in the West (FFTW - www.fftw.org) has increased the processing speed significantly thus increasing the scanning rate to 136 measurements per second. This gives the possibility of scanning the wake behind the turbine fast enough before the wake changes its form or position significantly. The modified software is developed at Risø in cooperation with QinetiQ.

Secondly, the conventional conical scanning mode is designed for measurements of spatial averaged wind characteristics, whereas the desire here is to record instantaneous wind speeds in many points. To solve this problem, the wedge has been taken off. This makes the LiDAR device a straight shooter which measures the wind speed in the direction it is pointed to. The modified ZephIR has been mounted on a pan and tilt head, which is originally designed for security cameras. In the present experiment only the pan movement (with pan range ±35°) is let free, and tilt is fixed and adjusted to align the laser beam with the horizontal plane. The recording process is controlled by a computer assuring that the pan signal is recorded simultaneously with the velocity measurements. Time stamp for each measurement is also recorded.

The optical head of the modified ZephIR prototype has been mounted on the Tellus wind turbine looking downwind (Figure 3). It is located nearly at 29 m a.g.l. which is the hub height. The laser source and process computer of the ZephIR is located in a shelter at the turbine base and can be remotely controlled over a local area network. Various scanning modes have been developed and will be explained in the next subsection.

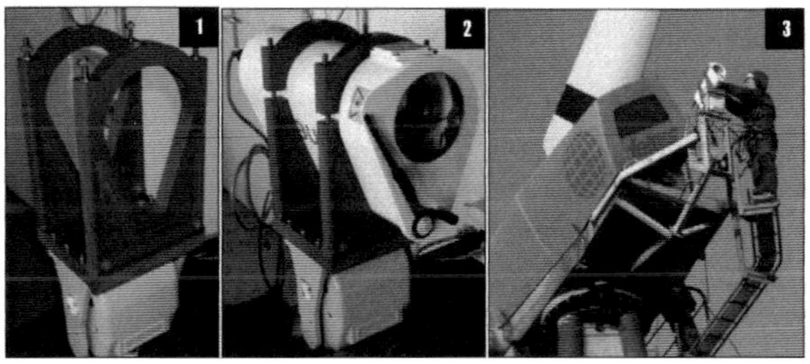

Figure 3. Adaptation in three steps; 1) Tilt and pan head is modified with a holder. 2) The LiDAR head is mounted on it. 3) Mounted behind the Tellus turbine looking downstream.

3.2. Working Modes

After adaptation of the prototype three working modes have been developed including configurations with as well as without the wedge (Figure 4). The modes are named as follows:

3.2.1. Line Scan Mode

In this mode the LiDAR focuses to a single distance, and pans between ±30° generating an arc scan behind the wind turbine. Since the measurement focus distance is constant we name the mode "Line Scan". Each line scan takes 1.3 seconds, while the wind speed is measured nearly 180 times on different positions along the arc. See Figure 4a.

3.2.2. Deep Line Scan Mode

As an additional operation to the previous mode, the focus distance is changed every other line scan. Each focus distance thus includes nearly 360 wind measurement points. The focus is changed so that the instrument roughly follows the same air parcel downstream. The focus limit of the ZephIR is 200 m, so when it reaches this limit, focus distance is changed to 20 m to start the operation all over again. We have tried the Deep Line Scan Mode in the range from 1 diameter to 10 diameter distance with 4-5 stations which takes between 18 and 24 seconds for one cycle (Figure 4b).

3.2.3. Sphere Scan Mode

Unlike the two modes described above the "Sphere Scan" mode does includes a vertical movement of the focus point. The horizontal movement is made like explained in the last scanning modes and the vertical is performed through optics. We have introduced the standard wedge but have constrained its total rotation to a back-and-forth movement. The special mechanism drives the wedge in a similar way to that on a car wiper. With the wedge properly mounted the focus point path follows a vertically oriented arc. A combination of this movement and horizontal panning of the head, results in a scanning pattern similar to that shown in Figure 4c. Since the LiDAR focus distance is kept constant the surface covered is a spherical patch. In the following this type of scanning is going to be referenced as 2D due to its surface character. One pan movement is made in slightly more than one second, which means that we can cover the surface at reasonably fast rate and therefore we call this a "snapshot".

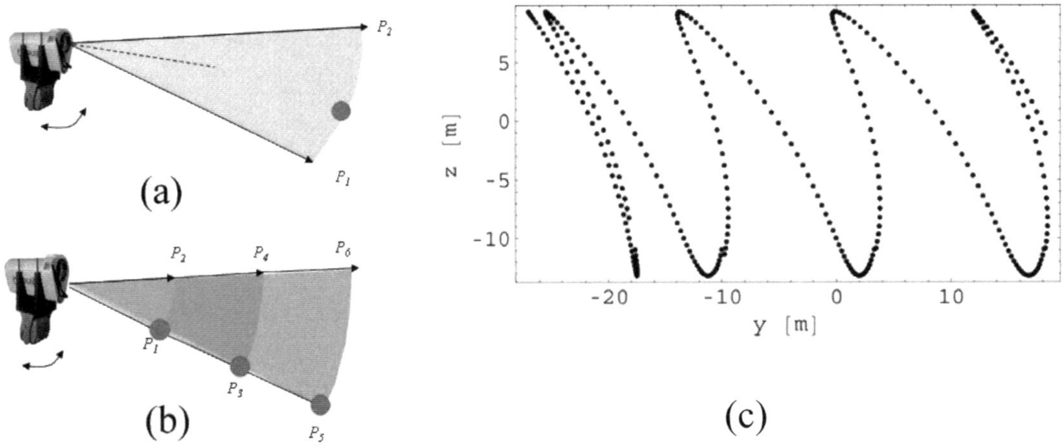

Figure 4. (a) Line Scan Mode focuses to a single distance, the red dot, while panning between points p1and p2. (b) Deep Line Scan Mode focuses to programmed distances while panning. (c) A sample of the Sphere Scan Mode pattern. Focus distance is 58m. Horizontal axis is the lateral deflection (in meters) associated with the pan position (±25°) and vertical axis is the vertical deflection (in meters) associated with the wedge rotation (45°). Each full 2D pattern scan might end up with different point positions, but they will always be within the same boundaries.

3.3. Data Processing and Database

The data from turbines and meteorological masts are collected into one single synchronized MySQL database to facilitate data analysis. The combined database includes 17 channels of data originating from three different measuring campaigns in May 2005, November 2005 and February 2006, respectively. The signals are recorded using up to four synchronized measurement PC's.

Table 2 Measurement Calendar

Setup	Scanning Mode	Start	End
1	Line Scan & Deep Line Scan	2004-09-15	2005-05-01
2	Deep Line Scan	2005-10-14	2005-10-29
2	Line Scan	2005-10-29	2005-11-21
3	Sphere Line Scan	2006-04-26	2006-05-11

3	Sphere Line Scan	2006-06-06	2006-07-12

4. Results

The first data of Line and Deep Line Scanning mode have been collected in 2004. At the end of 2005, mast MM2 is erected and measurements are repeated.

Horizontal profiles of the wind turbine wake are observed in Figure 5. Four focus distances have been selected, and from each station nearly 300 data points are collected. In the closest focus distance all measurements are in the wake, but further downstream the wake as well as undisturbed wind on the sides of the wake can be seen.

As for the line scan measurements, blocks consisting of 10 minutes of data are analyzed. Each scanned arc of velocities are included in a plot side-by-side with a gray scale indicating the measured wind speed (cf. Figure 6), thus resulting in a depiction of the wake deficit displacement at the particular focus distance as a function of time. The light areas in the figure symbolize high wind speeds, whereas dark areas symbolize low wind speeds. The red symbolizes model predictions of the wake deficit positions (meandering) based on the spatially averaged (instantaneous) direction evaluated from the three wind vanes on the reference MM2 mast, corrected for the measured yaw position of the Tellus nacelle as well as for the advection time (MM2 to focus distance).

With the selected representation of data and model, it is easy to see the measured wake width and wake movements as well as the correlation between the predicted instantaneous wake position and measured instantaneous wake position. The model of the wake dynamics assumes the wake deficit to be advected in the mean wind direction with the mean wind speed and advected in the lateral direction with the large scale lateral turbulence component [1]. Assuming Taylor's hypothesis, the "driving" lateral velocities are perfectly correlated with the filtered wind direction measurements at the reference meteorological mast, however, delayed with the transportation time from the mast to the downstream point of interest. Thus for the present case, where the meteorological mast is 36 m upstream the turbine, the requested delay is the time for an air particle to move 36 m plus the focus distance with the mean wind speed. In general, the predicted wake movements show a convincing correlation with the measured wake movements, thus supporting the wake meandering hypothesis stated in the introduction.

One of the problems in interpreting the data in Figure 6 is that the wake width is often smaller than anticipated. The Tellus turbine has rotor diameter of 19 m, and focus points are 3 diameters downstream. The wake width can be estimated to approximately 12° to 18° in pan, corresponding to a wake diameter of the order 12 m to 18 m, which is even smaller than the rotor diameter. According to the currently available wake models, the width should be larger. Currently we do not fully understand this discrepancy, which also contradicts the measurements from the first measuring campaign. In second quarter of 2006, the LiDAR is physically updated with a new wedge mechanism. However, most likely we cut the wake horizontally not in the middle but closer to the top or the bottom.

The problem with interpretation of the observed wake widths from the line scan measurements can be studied more in detail if based on Sphere Scan recordings, which resolves both the lateral- and the vertical wake position at a prescribed downstream distance.

Sphere scans

Measurements based on the sphere scan mode (2D) are processed and analyzed in a different manner. We estimate the "line of sight" wind field in the entire scanned surface through linear interpolation of the measured wind speed for each snapshot. The process is done mainly in three steps performed as follows. First, the quality of the line of sight wind speed is checked and corrected if necessary at each focus point. Second, the focus points for a particular snapshot are selected, and third the wind speed is calculated at the nodes of a regular grid by means of interpolation using a Delaunay triangulation. This means that no projection for wake-axis alignment of the measured wind speed

vector is performed. An exemplary snapshot can be seen in Figure 7 where an instantaneous wake deficit can be observed compared to the rotor diameter (cut circle). The measurement was performed during stable atmospheric conditions with an inflow average wind speed of 5.4m/s and at a distance equivalent to four rotor diameters. The wind field characteristics represented in this plot coincides qualitatively with our expectations. Appreciable is a shift of the wake with respect to the wake-axis of that measurement period. Moreover, the undisturbed wind shear is clearly visible. This is used in a posterior analysis to "extract" the deficit from the whole wind field and automate a procedure to estimate the center position of the deficit. In a further study, at the moment in preparation for publication, the analysis is performed for every snapshot for time series of 10 min. There we have the capability of determining the position of a reference frame moving simultaneously with the wake center. With this information we can measure the averaged instantaneous wake. In other words, this scanning mode expands our capabilities to validate not only wake meandering models but also models including details like the wake radius and wake shape.
Observation of the average of the wake center vertical position has shown a slight upward shift with respect to hub level. This result supports that during the line scan campaign the wake center was out of scanned arc. The shifted wake combined with a possible non-perfect alignment of the LiDAR with the horizontal could explain it.

During a posterior campaign the 2D measurements are also performed varying the focus distance as in the deep line scan. This enables observation of wake decay and wake radius expansion with more detail. However, these data are still untouched and further analysis is still to come.

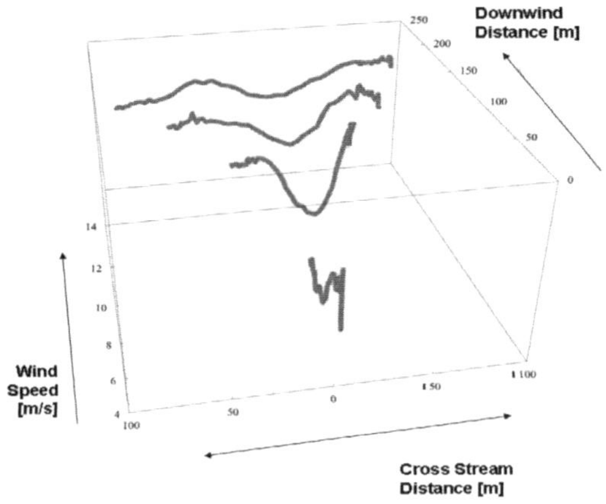

Figure 5. Deep Line Scan; data is collected from four different focus distances from 1 diameter to 9 diameter (19 m, 76 m, 130 m and 176 m). The initial wake is of the same size as the rotor and widens gradually downstream.

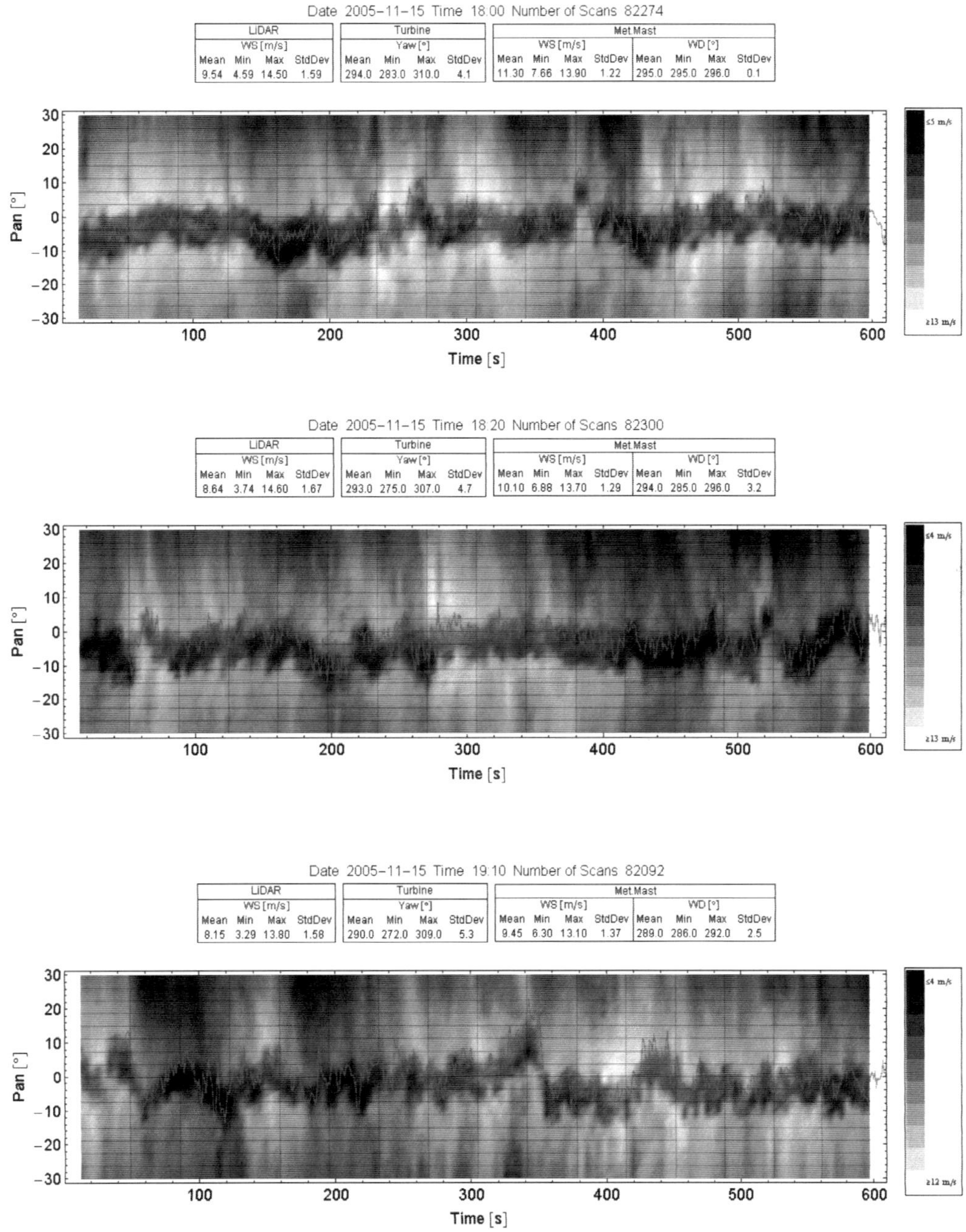

Figure 6. Line Scan data plotted in gray scale as function of time. Low wind speed is shown as black and higher wind speed is shown as white. Focus distance is 3 diameters (58 m).

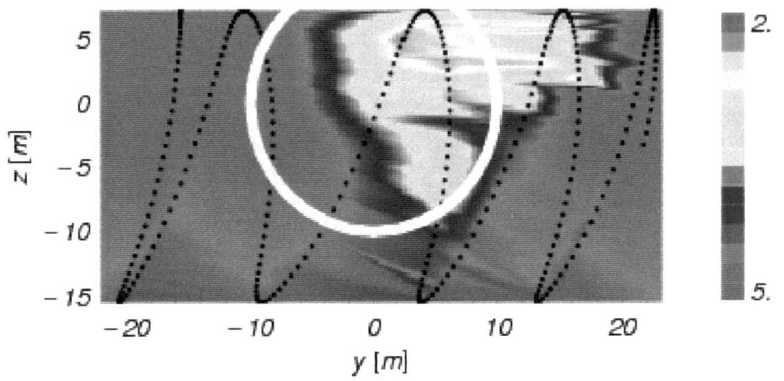

Figure 7. [Color] Snapshot (1.1seconds) of line of sight wind speed in the wake of a wind turbine at approx. four diameters downstream. Measurement performed with LiDAR in an sphere scanning mode. Cutted circle represents the turbine rotor. Dots represent measurement points.

5. Conclusion

First time ever full-scale instantaneous wakes deficits of wind turbines have been measured directly in one and two dimensions. This feature has enabled a detailed experimental investigation of the wake meandering phenomenon as well of the shape, widening and attenuation of the wake deficit.
Resolving the wake meandering dynamics has allowed us to investigate the basic assumption in the Dynamic Wake Meandering model - i.e. that the wake deficit is advected passively by the larger than rotor size eddies in the inflow. The performed experimental analysis seems to support fully this hypothesis for the investigated full scale turbine located in the atmospheric boundary layer.

Acknowledgment

The project is supported by internal Risø funding as well as by a Danish Energy Agency grant under the project "Dynamic Wake Model for detailed aeroelastic Simulations of Wind Turbines in Farms" (33030-004)

References

[1] G.C Larsen, H.Aa. Madsen, K Thomsen and T.J. Larsen. Wake meandering - a pragmatic approach. Submitted for publication in Wind Energy.
[2] N.O. Jensen. A note on wind generator interaction. Technical Report Risø-M-2411, Risø National Laboratory, 1984.
[3] Ferhat Bingöl. Adapting a Doppler laser anemometer to wind energy. Master's thesis, Technical University of Denmark, 2005.
[4] David A. Smith, Michael Harris, Adrian S. Coffey, Torben Mikkelsen, Hans E. Jørgensen, Jakob Mann, and Régis Danielian. Wind Lidar evaluation at the Danish wind test site Høvsøre. Wind Energy, 9(1–2):87–93, 2006.
[5] I. Antoniou, H. E. Jørgensen, T. Mikkelsen, T. F. Pedersen, G. Warmbier, and D. Smith. Comparison of wind speed and power curve measurements using a cup anemometer, a lidar and a sodar. European Wind Energy Conference and Exhibition, pages 47–51, November 2004.
[6] I. Antoniou, H. E. Jørgensen, T. Mikkelsen, S. Frandsen, R. J. Barthelmie, C. Perstrup, and M. Hurtig. Offshore wind profile measurements from remote sensing instruments. European Wind Energy Conference and Exhibition, February 2006.

IOP Conf. Series: Earth and Environmental Science 1 (2008) 012023 doi:10.1088/1755-1307/1/1/012023

Wind shear proportional errors in the horizontal wind speed sensed by focused, range gated lidars

P Lindelöw[1], M Courtney[1], R Parmentier[2] and J P Cariou[2]
[1] Wind Energy Department, Risø National Laboratory for Sustainable Energy, Technical University of Denmark, P.O. 49, DK-4000 Roskilde, Denmark
[2] Leosphere, Orsay, France

E-mail: petter.lindeloew@risoe.dk

Abstract. The 10-minute average horizontal wind speeds sensed with lidar and mast mounted cup anemometers, at 60 to 116 meters altitude at Høvsøre, are compared. The lidar deviation from the cup value as a function of wind velocity and wind shear is studied in a 2-parametric regression analysis which reveals an altitude dependent relation between the lidar error and the wind shear. A likely explanation for this relation is an error in the intended sensing altitude. At most this error is estimated to 9 m which induced errors in the horizontal wind velocity of up to 0.5 m/s as compared to a cup at the intended altitude. The altitude errors of focused range gated lidars are likely to arise partly from an unaccounted shift of the weighting functions, describing the sample volume, due to the range dependent collection efficiency of the focused telescope. Possibilities of correcting the lidar measurements both for wind velocity and wind shear dependent errors are discussed. The 2-parametric regression analysis described in this paper is proven to be a better approach when acceptance testing and calibrating lidars.

1. Introduction

The rapid increase of wind turbine size makes wind assessment with mast-mounted cup anemometers, the only technique for site evaluation and assessment of turbine power performance which is currently accepted by the wind energy community, expensive and logistically complex. Commercially available conically scanning lidars have potential to replace mast-mounted sensor when sensing wind over flat terrain at altitudes corresponding to modern turbine rotor heights, currently 60-120 m. Direct sensing of the wind field over the rotor area is likely to improve the accuracy of potential output predictions by improving the accuracy of site evaluations and power curve calibrations. For a successful introduction of lidars to the wind energy industry, it is important to describe and precisely estimate, experimentally as well as theoretically, the deviations of lidars as compared to cup anemometers.

2. Wind shear proportional lidar errors

Lidars have previously been compared to cup anemometers by regression analysis of the 10-minute average horizontal wind speed [1, 2]. These studies aimed at estimating errors which are linearly proportional to the wind velocity.

A likely error of conically scanning lidars is an error in the intended cone angle, φ, see figure 1. The error in cone angle, $\Delta\varphi$, is transferred to an error in the sensed horizontal wind speed, u_{meas}, which is linearly proportional to the wind velocity, u. The lidar error, ε_{lidar}, exclusively due to a cone angle error can be expressed as

$$\varepsilon_{lidar} = u_{meas} - u = \left(\frac{\sin(\varphi - \Delta\varphi)}{\sin(\varphi)} - 1 \right) \cdot u = k_u \cdot u \tag{1}$$

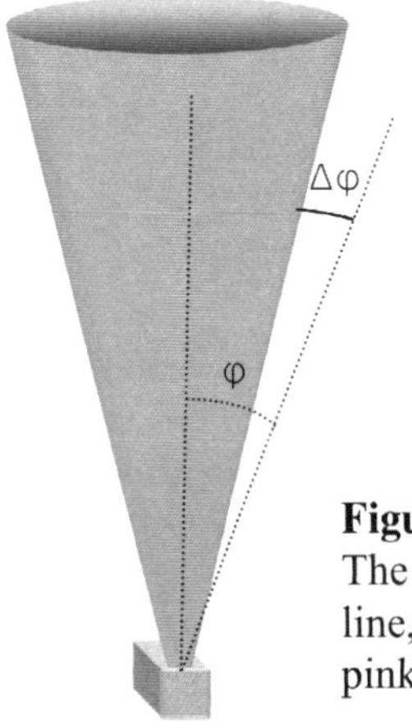

Figure 1: Illustration of the cone angle error. The intended cone angle is shown as a dotted line, the actual cone angle is illustrated by the pink cone.

Typical lidars use a cone angle of 30°. An error in the cone angle of a mere 0.5° would induce a k_u of -1.5%, see figure 2. A lidar suffering uniquely from a cone angle error should have a k_u which is constant with sensing range, nor should it have an offset to the linear fit of the lidar-cup regression. However, this has typically not been the case from the conventional 1-parametric regression analyses, e.g. as for Risø DTU's commercial Windcube unit in table 4.

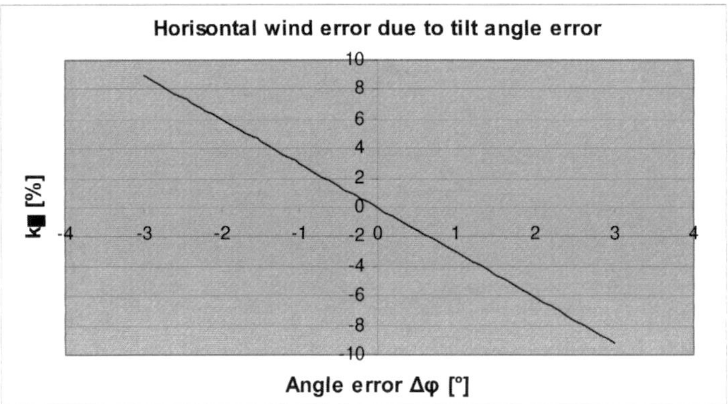

Figure 2: The error in horizontal wind due to cone angle error, Δφ, of a lidar with an intended cone angle of 30 °.

However, an error in cone angle will also introduce an error in sensing altitude since the lidar will measure at a sensing altitude which corresponds to the intended height and the actual cone angle. This altitude error, Δh, increases with sensing altitude according to

$$\Delta h = \left(\frac{\cos(\varphi - \Delta\varphi)}{\cos(\varphi)} - 1 \right) h_i \tag{2}$$

An altitude error will introduce errors on the estimated radial wind speed which in a first approximation are linearly proportional to the wind shear at that altitude. For typical wind shears the error in altitude, due to the cone angle error, will to a small degree balance the wind velocity error. For example, a cone angle error of -0.5°, on an intended 30°, will induce a k_u of 1.5% but also an overestimation of the actual sensing altitude of 0.5 m when intending to sense at 100 m. A 10 m/s wind with a wind shear increasing with 0.1 m/s per meter altitude would thus be sensed with an effective error of $10 \cdot 0.015 - 0.5 \cdot 0.1 = 0.1$ m/s.

A wind shear dependent error will typically influence the traditional 1-parametric regression analysis since the wind velocity and the wind shear are usually not uncorrelated [3].

Other altitude errors could result from inaccurate focusing of continuous wave variable focus lidars [1] or trigger errors in range gated lidars [4]. These errors could be classified as fabrication errors and would typically differ from system to system.

However, lidars can also suffer from inherent wind shear dependencies since they, in contrast to cup anemometers, measure the wind distribution from a sample volume of some length. The dominant line-of-sight, or radial, wind speed is estimated from the measured wind spectra, which corresponds to the wind distribution within the sample volume weighted with the appropriate function [5]. At least three dominant radial wind speeds are sensed in different directions and the horizontal and vertical wind speeds and direction are constructed from these vectors under the assumption that the wind is homogeneous over the scan path.

Depending on the estimator it is assumed that the dominant line-of-sight wind velocity is representative of the wind velocity at the altitude which corresponds e.g. to the peak or the centroid of the weighting function in the sample volume. It is not unlikely that the estimation of the dominant wind signal slightly will depend on the wind shear in the sample volume. Although this effect is not a direct altitude error it can in a first approximation be treated with the same analysis. Worse is that this wind shear dependency could be non-linear and thus not apt to be studied by a linear regression analyses. Nevertheless, the combined wind shear dependencies observed in this study are dominantly linear.

Another inherent error could be due to unadjusted distortion of the range gates. Fiber based lidars, even pulsed, typically use a focused telescope to collect more backscatter from the intermediate range. The range dependent collection efficiency of the focused telescope will shift the classical weighting function in the range gate [6], for example according to figure 3.

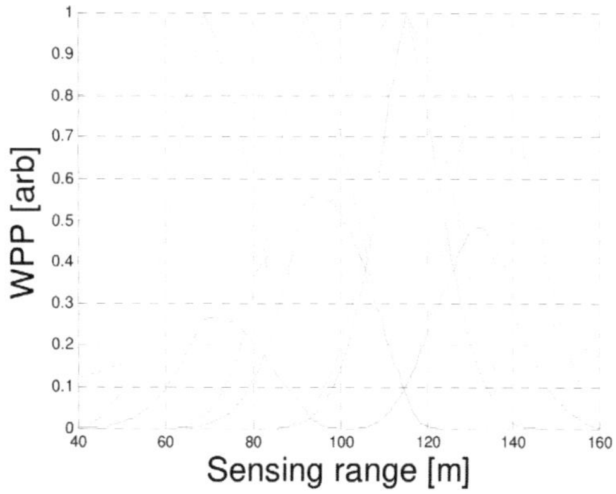

Figure 3: Illustration of a possible case of range gate distortion of a focused range gated lidar. The green dashed graph is a normalised theoretical collection efficiency function of the Windcube prototype telescope focused at 110 m distance. The red graphs show the normalised experimentally measured narrow bandwidth contribution of 4 unadjusted range gates from the Windcube prototype. The blue curves show the distorted range gates, note how the center of the range gates shift towards the focus distance.

Inhomogeneous aerosol distributions, either regarding backscatter coefficients or coherent scatter duration, could also influence the center of the weighting function, especially in continuous wave systems, giving rise to wind shear proportionalities which would alter with e.g. meteorology and wind direction. Even a system tilt could introduce an altitude error in the order of 1-2% of the sensing range. These altitude errors will differ for different scanning directions which introduces a complicated dependency on the horizontal wind velocity error. Neither of these two errors is further considered in this paper. The assumption in this paper is that the altitude errors are exclusively linearly proportional to the wind shear.

The errors in sensing altitude could be constant with sensing range, as in the case of a trigger offset, or change with sensing range, as in the case of an error in focus setting or a distorted range gate. A number of possible lidar errors and their implications are listed in table 1. None of these altitude errors will be clearly revealed in the traditional lidar vs cup regression analysis but they will influence the estimate of the cone angle error and could introduce offsets in the traditional fit.

Table 1: Possible lidar errors and their implications.

Lidar error	Mainly occurring in	Implication	Change with sensing range
Cone angle	Any lidar	• Wind speed proportional • Altitude error	• Constant with altitude • Magnitude increases with height
Trigger error	Range gated	Altitude error	Constant with altitude
Focus error	Continuous wave	Altitude error	Magnitude typically increases with altitude
Range gate distortion	Fixed focus range gated	Altitude error	Changes sign around focus distance
Estimator error	Any lidar	Wind shear dependency	Depends on weighting function and thus on sensing range. Maybe non-linear.
Inhomogeneous aerosol distribution	Continuous wave	Altitude error	Depends on meteorology and wind direction
System tilt	Any lidar	Altitude error	Depends on wind direction and increases with altitude

3. Experimental set up

In this study we compare 10-minute average wind speeds collected at the wind turbine test station in Høvsøre by a commercial Windcube and a prototype Windcube with those sensed with mast mounted cup anemometers. The test site is flat, surrounded by farmland with no major obstacles in the immediate neighborhood and at a distance of 1.7 km from the coast of the North Sea in the northwest of Denmark. The turbines are aligned in a north–south direction, separated by 300 m from each other. A 120 m met tower is positioned 200 m south of the southernmost turbine. Additional met masts are located 240 m to the west of each turbine. The met tower is equipped at 10, 40, 60, 80, 100, 116.5 meters with boom mounted Risø P2546 cup anemometers [7].

The Windcube prototype was installed 40 m north of the 120 m met tower during the winter of 2006-07 while the commercial Windcube was installed right next to the same met tower in the winter of 2007-08, see figure 4. Next to the met mast tower is a ceilometer which senses the cloud base height.

Figure 4: Turbine test station in Høvsøre. The Windcube prototype position is indicated by the blue dot, while the commercial unit position next to the met mast is indicated in red.

The Windcubes are pulsed lidars which have telescopes focused at a fix distance to improve the collection efficiency from the first few hundred meters. The specified FWHM of the vertically projected weighting function of the sample volume, is about 20 m.

The collected 10-minute averages have been screened since cup and lidar measurements are sensitive to several factors. Only wind from the sector 40° to 60° and 220° to 290°, counting clockwise from north, have been included in this conservative analysis to avoid measurements which are influenced by turbulence from the tower and from the wind turbines. The most northerly sensing direction of the windcube prototype is suspected to be influenced by turbine wakes from the east, thus only westerly wind directions are included for that study. Furthermore, only wind velocities faster than 4 m/s are included to avoid cup overspinning, air temperatures over 2° C assure unfrozen cups and the data have been screened for rain since this might affect the sensors. Finally, only Windcube measurements with a signal-to-noise ratio higher than -20 dB have been accepted.

The horizontal wind velocity measured by the Windcube prototype has been post corrected with 2.1% since the cone angle was measured to be 0.7° wider than intended. The altitude error due to the cone angle error has not been accounted for in the data set.

4. 2-parametric regression analysis of the lidar error

A 2-parametric regression analysis of the lidar error, $\varepsilon_{lidar}(z)$, as a function of the horizontal wind velocity, u(z), and the wind shear, g(z), is a method which can simultaneously reveal cone angle errors and altitude errors.

The wind shear, g(z), is calculated as the derivate of the wind profile generated from a $a+b\ln z+cz+dz^2+ez^3$ fit to the wind speed sensed with the cup anemometers at 10, 40, 60, 80, 100 and 116.5 m. The fitting function used diverges from more standard wind profile models, e.g. $U_0+A\ln z+B\ln z^2$ [8], but fits better to some wind profiles measured in Høvsøre. Note that the accuracy of the shear fit is lower at 116.5 m altitude since the wind profile fit extrapolates above the last cup value.

The lidar error is described in the 2-parametric regression analysis as $\varepsilon_{lidar}(z) = C + k_u(z)\cdot u(z) + k_g(z)\cdot g(z)$, where k_g is the wind shear regression coefficient describing the lidar error in m/s per m/s wind velocity increase per meter altitude, i.e. in meters of altitude. The wind shear regression coefficient, k_g, gives directly an estimation of the altitude error in meters if the altitude error is limited so that the sensed wind velocity at the actual sensing altitude, $u(z + \Delta z)$, is well described by $u(z) + g(z)\cdot k_g(z)$, i.e. as the wind velocity at the intended altitude plus the wind change over the erroneous altitude estimated from the wind shear at the intended altitude, g(z). Note that a system can suffer from several altitude errors simultaneously. This 2-parametric regression analysis can not differentiate between them and will only reveal the combined effect.

To illustrate the effectiveness of the 2-parametric regression analysis we compare the wind measured by the well calibrated and accurately placed cup at 80 m with the cup at 60 m. We expect the wind velocity dependency k_u and C to be negligible while we expect an "altitude error" k_g close to 20 m. Although the error in altitude is relatively high, and the assumption that the derivative of the profile can be used to extrapolate the wind velocity at the higher cup we arrive with a very accurate estimation since the cup error can be described from $\varepsilon = -0.06 - 0.002\cdot u + 19.7\cdot g$. The 95%-confidence interval for k_u and k_g is from -0.0027 to -0.0012 and 19.5 to 19.8 m respectively.

5. Measurements of altitude and cone angle errors

The predicted prototype Windcube error at 60 m, estimated from a least-square fit 2-parametric linear regression analysis, is shown as a plane together with the data set in figure 5. In figure 6 the graph is displayed in parallel to the plane along the wind velocity respectively wind shear axes to give a good illustration of the lidar error dependency on u(60) and g(60).

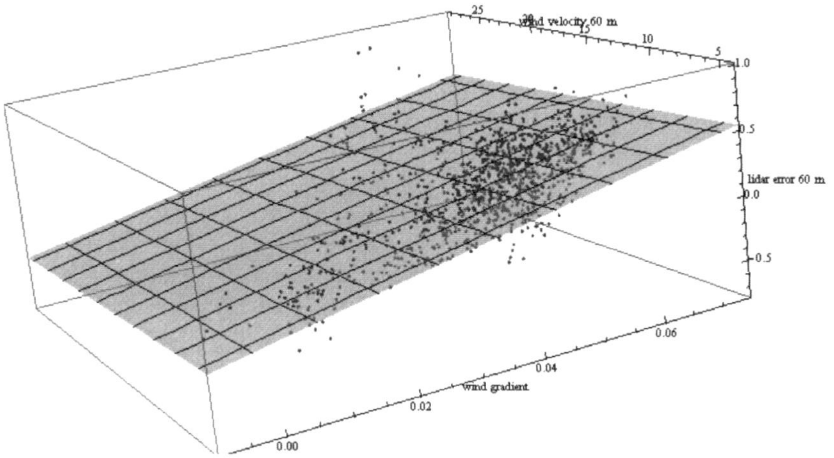

Figure 5: Predicted and actual lidar error at 60 m as a function of wind shear and velocity.

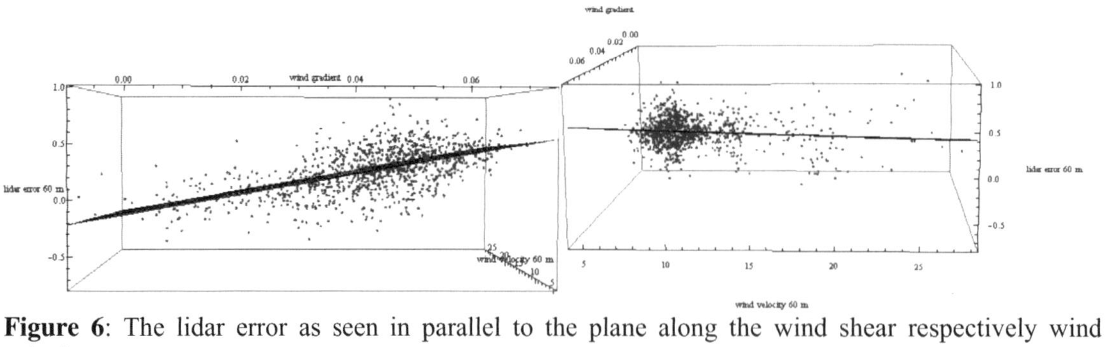

Figure 6: The lidar error as seen in parallel to the plane along the wind shear respectively wind velocity axes.

The regression parameters with their 95% confidence interval, estimated from the traditional 1-parametric and the proposed 2-parametric regression analysis, for the windcube prototype are gathered in table 2 and 3. The analysis is made over 1071 10-minute average horizontal wind speeds.

Table 2: Lidar error parameters and their 95% confidence interval from the traditional 1-parametric regression analysis of the Windcube prototype.

Intended Altitude [m]	Offset [m/s]	k_u [%]
60	0.17 ± 0.04	0.6 ± 0.4
80	0.07 ± 0.03	-0.5 ± 0.3
100	0.03 ± 0.03	0.1 ± 0.2
116.5	0.00 ± 0.04	-0.9 ± 0.3

Table 3: Lidar error parameters and their 95% confidence interval from the 2-parametric regression analysis of the Windcube prototype.

Intended Altitude [m]	Offset [m/s]	k_u [%]	k_g [m]
60	-0.11 ± 0.04	-0.5 ± 0.3	9.0 ± 0.8
80	-0.04 ± 0.03	-1.0 ± 0.2	4.3 ± 0.7
100	0.02 ± 0.02	-0.4 ± 0.2	3.1 ± 0.6
116.5	-0.02 ± 0.04	-0.7 ± 0.4	-0.4 ± 0.5

The offsets, which should approach zero are factors 1.5-1.75 smaller in the 2-parametric analysis, however, even the 95% confidence interval does not stretch over 0 for all altitudes. It is possible that the relatively large offset at 60 m, -0.11 m/s, is due to the quite large altitude error which gives less value to the assumption that the wind profile is linear over the altitude error. Note that the magnitude of the offsets decrease with a decreasing altitude error. Also the confidence intervals are 0-30 % smaller than for the 1-parametric regression.

The $k_u(z)$:s are more similar for the different altitudes when estimated from the 2-parametric regression, differs with 0.6%, as compared to 1.5% from the 1-parametric. This is because the altitude error influence is taken up by k_g in the 2-parametric regression. It can for example be seen that k_u(116.5 m), an altitude with a relatively small altitude error according to table 3, from the 1-parametric analysis is similar to the k_us in the 2-parametric regression analysis. Nevertheless, it should be noted that e.g. k_u(80) and k_u(100) still differs with more than the 95% confidence interval also in the 2-parametric analysis. However, this irregularity can easily be explained by cup-to-cup variations. From the most trustworthy analysis, i.e. the 2-parametric at 60, 80 and 100 m, it is possible to estimate an average k_u of -0.6%, whereas the 1-parametric analysis would estimate an average k_u of -0.2%. The k_u of -0.6% corresponds to a residual cone angle error of 0.2° after the initial post correction of -0.7°.

Most importantly large altitude errors are deduced from the 2-parametric regression analysis. Wind shears are typically 0.05 m/s/m in Høvsøre and the altitude errors thus introduces typical horizontal wind velocity errors from a negligible -0.02 m/s at 116.5 m altitude to an important overestimation of 0.45 m/s at 60 m.

The altitude error alters with 9.4 meters from 60 to 116.5 m and it is likely that range gate distortion plays a significant role in this. Range gate distortion effects should change sign at the focus distance which therefore could be concluded to lie at slightly lower than 116.5 m for the windcube prototype. However, nothing rules out other constant or increasing altitude errors which would translate the sign shift, due to the range gate distortion, in altitude.

The commercial Windcube has better control of the focus position which is estimated to lie at approximately 100 m. The focus position concurs with on-site CNR vs. height measurements. This data set was collected from the 12th of December 2007 to the 3rd of January 2008. Out of 3312 10-minute average horizontal wind velocities 1776 remained after screening. The regression parameters can be found in table 4 and 5.

Table 4: Lidar error parameters and their 95% confidence interval from the traditional 1-parametric regression analysis of the commercial Windcube.

Intended Altitude [m]	Offset [m/s]	k_u [%]
60	0.18 ± 0.05	-2.7 ± 0.3
80	-0.01 ± 0.04	-2.5 ± 0.3
100	0.13 ± 0.03	-2.7 ± 0.2
116.5	0.18 ± 0.03	-3.6 ± 0.2

Table 5: Lidar error parameters and their 95% confidence interval from the 2-parametric regression analysis of the commercial Windcube.

Intended Altitude [m]	Offset [m/s]	k_u [%]	k_g [m]
60	0.16 ± 0.05	-2.9 ± 0.4	0.9 ± 1.0
80	-0.02 ± 0.05	-2.6 ± 0.4	0.5 ± 1.0
100	0.13 ± 0.03	-2.5 ± 0.2	-0.7 ± 0.7
116.5	0.15 ± 0.03	-2.7 ± 0.3	-3.7 ± 0.7

The altitude errors are less significant for this production unit. The offsets, confidence intervals and agreement of the k_u estimations are therefore less different between the two regression analyses. However there is a significant wind velocity proportional error. On average it could be estimated to -2.7%. The most common wind speed in this data set is 12 m/s giving a typical error of about 0.35 m/s related to the estimated cone angle error of 0.9°. The cone angle has been verified afterwards by direct measurement of the beam positions in the 4 sensing directions as compared to the initial wedge angle verification from a direct geometrical measurement of the wedge. This exercise revealed a cone angle error of 0.45° on the intended 28.26°, explaining approximately -1.5% of the wind velocity proportional error. Nevertheless, 1% is still left unexplained.

The altitude errors on the production unit alter with 4.6 meters from 60 to 116.5 m. These errors are smaller than for the prototype unit, but the combined altitude errors still introduced a non-negligible error of about 0.2 m/s at 116.5 m. If the altitude errors were exclusively due to range gate distortion we would expect a focus corresponding to an altitude between 80 and 100 meters. If the altitude errors due to the suspected cone angle error of 0.9° in the production unit, 0.5, 0.7, 0.8 and 1.0 m for 60, 80, 100 and 116.5 m respectively, are taken apart the sign shift would occur between 60 and 80 m which does not concur with the focus setting.

The altitude errors of the two systems are expected to differ since they are focused at different ranges and have different cone angle errors. If the altitude errors due to the suspected cone angle error are taken apart the spread of the residual $k_g(z)$:s increases with 0.5 m for the production unit. The altitude errors, -0.3, -0.4, -0.5 and -0.6 m, due to the best guess of unadjusted cone angle error on the prototype, 0.2°-0.7° = -0.5°, instead reduces the spread with 0.3 m. Since the difference is small between the prototype and the commercial unit it is difficult, from these two surveys, to estimate which altitude error behaviour is most typical.

6. Correcting and calibrating the lidar errors after a verification campaign

Cone angle errors can easily be corrected by using the actual cone angle, estimated from a verification campaign, in the calculations of the sensing distance and radial wind projections. Errors in the variable focus distances can also be corrected for with ease. However, direct corrections of e.g. range gate distortion and estimator dependency could be difficult. Corrections of these lidar measurements can be done in several ways. A calibration approach could be to use the estimations from the 2-parametric regression analysis from each altitude independently. The wind gradient, g(z) would in the general case, where masts are unavailable, have to be estimated from uncorrected lidar data, it is likely that an iteration procedure could improve these corrections. The k_u could alternatively be the average of the estimations at each height while the estimated k_g:s could be fitted to a model describing the altitude error. Another approach would be to adjust the range gate positions with the appropriate k_g. Further investigation is necessary to see the merits of the different approaches

Without further theoretical studies of the origin of the altitude error it will be difficult to make proper interpolations of the estimated k_g:s so that the lidar can be corrected for any altitude. It will be important, for other wind lidar applications, to describe how altitude errors will develop for very long sensing ranges, up to 2 km, where cup comparisons will be impossible to execute with the lidar in vertical set up.

7. Conclusions

Although commercial lidars have shown impressive precision, several residual errors and deviations can still be detected in comparisons to mast mounted cup anemometers. The consistency of these first generation lidars is very high and allows for high precision analyses of wind velocity and wind shear dependent errors. The prevailing part of the errors observed in these two campaigns can with high likelihood be connected to a cone angle error and from a multitude of altitude errors which can be estimated from 2-parametric regression analyses. In this report two lidar systems have been tested and suspected cone angle and altitude errors, inducing horizontal wind velocity errors of up to 0.5 m/s individually, have been revealed. However, additional error sources with different dependencies can not be excluded.

These errors are too high to be ignored in wind assessment for the wind power industry. The 2-parametric regression analysis described in this paper is a better approach to provide calibrations for lidar systems than traditional 1-parametric analyses. Some of the errors discussed, are probably more effectively detected and corrected by hardware measurements of, e.g. the actual cone angle, while other errors like the estimator dependency on wind shear and range gate distortion has to be theoretically calculated from high precision measurements of the true collection efficiency, which might be non-trivial and would need verifications in studies like those presented.

Acknowledgements

The authors would like to thank the permanent staff at Høvsøre, Bjarne Sønderskov and Anders Vestergaard for their untiring and highly professional support.

Much of the lidar testing and analysis has been carried out as part of Remote Sensing work package (WP6) of the EU Upwind project. Windcube is a trademark of Leosphere, France.

References

[1] Smith D, Harris M, Coffey A, Mikkelsen T, Jørgensen H, Mann J and Danelian R 2006 Wind lidar evaluation at the Danish wind test site in Høvsøre *Wind Energy* **9**, p 87-93

[2] Albers A 2006 Evaluation of Zephir *Internal report*, VC05050, PWG 06005, Deutsche WindGuard

[3] Panofsky H and Dutton J A 1984 *Atmospheric Turbulence* (New York: John Wiley & Sons)

[4] Cariou J P, Parmentier R, Valla M, Sauvage L, Antoniou I and Courtney M 2007 An innovative and autonomous 1.5 µm coherent lidar for PBL wind profiling, *Proceedings of 14th Coherent Laser Radar Conference*

[5] Lindelöw P 2008 Effective sample volume *Fiber Based Coherent Lidars for Remote Sensing of Wind* Thesis from Technical University of Denmark p 63-76

[6] Lindelöw P 2008 Range dependent collection efficiency of fiber based coherent lidars *Fiber Based Coherent Lidars for Remote Sensing of Wind* Thesis from Technical University of Denmark p 56-62

[7] Kristensen L, Frost Hansen O 2002 Distance constant of the Risø cup anemometer *Risø-R-1320(EN)*, Risø National Laboratory, Roskilde G X 2001

[8] Högström U 1987 Non-dimensional wind and temperature profiles in the atmospheric surface layer : A re-evaluation *Boundary-Layer Meteorology* **42**, p 55-78

14th International Symposium for the Advancement of Boundary Layer Remote Sensing IOP Publishing
IOP Conf. Series: Earth and Environmental Science **1** (2008) 012024 doi:10.1088/1755-1307/1/1/012024

Sodar detection of mixing height in flat and mountainous terrain

Barbara Hennemuth [1)] **and Hans-Jürgen Kirtzel** [2)]

[1)] Consulting Meteorologist, Hamburg, Germany, [2)] METEK GmbH, Elmshorn, Germany

E-mail: barbara.hennemuth@zmaw.de

Abstract. The atmospheric boundary layer plays an important role in air pollution and dispersion problems because the transport processes are managed within this layer and its top limits the vertical exchange of pollutants. A method for the derivation of the mixing height from measurements of sodar, RASS and sonic anemometer-thermometer is presented for flat terrain. It does not only use vertical profiles of measured parameters but also bulk information like histograms and time evolution. Results from a two-years period are verified by radiosonde-derived mixing height values and show the potential of the combination of the three systems to monitor the mixing height. Difficulties arise at locations in mountainous terrain where thermal wind regimes dominate which are highly non-local. An additional problem is a strong local heat source at an industrial site where even the definition of the mixing height is unclear.

1. Introduction

Air pollution problems and efforts in avoiding high concentration of harmful constituents in the living space of man are still an actual topic. An important parameter controlling the concentration of pollutants in the atmosphere is the depth of the mixing layer (ML). According to [2] the mixing height (MH) is *'the height of the layer adjacent to the ground over which pollutants or any constituents emitted within this layer or entrained into it become mixed by convection or mechanical turbulence within a time scale of about one hour or less'*.

Dispersion models, which are often prescribed by the local or national administration, mostly derive the mixing height in a simplified manner from near-surface parameters. In flat and homogeneous terrain this assumption may hold although e.g. the German administrative regulation TA-Luft [6] describes a formalism which does not always give realistic results. In mountainous terrain with elevated surfaces the derivation of MH from near-surface data fails because three-dimensional wind regimes develop which influence the dispersion of constituents.

A method for the derivation of boundary layer height from measurements of sodar, RASS and sonic anemometer-thermometer data is presented. Results for a location in flat terrain are presented and validated against radiosonde data. Problems arise at locations which are influenced by thermal wind systems in mountainous terrain. There the structure of the atmosphere is strongly influenced by non-local parameters and the orography. Moreover, the definition of the mixing height is not clear, as discussed e.g. by [7].

2. Instrumentation

SODAR technique has established as a reliable and cost efficient tool in the field of meteorological instrumentation for monitoring the atmospheric bounday layer. A large number of valuable parameters can be derived operationally, such as the mean profiles of wind directions and wind speed, and the standard deviation of the wind components, in particular of vertical wind, or vertical profiles of reflectivity as a measure for turbulence. This data set allows to identify with high reliability wind shear, low level jets, built up and decay of ground based or elevated inversion, stability classes, etc. But the mixing height as one of the most important parameter in air pollution modelling can be derived from sodar measurements alone only with some uncertainties. With the extension of simultaneously operated RASS systems for vertical temperature profiling the vertical stratification structure can be observed directly, which is informative in situations with ground based inversion. The RASS does not only identify the height of specific temperature gradients, but also their strength as a quantitative measure. Furthermore, a sonic anemometer which measures the turbulent fluxes of heat and momentum delivers the sources for the further time development of the present stratification.

Sets of the three measuring devices were installed for operational instrumentations at three sites which differ strongly in their orographic structure. Two sites in mountainous terrain were situated within industrial areas while one site was located in an open field at an operational measuring site used by the German Meteorological Service (DWD).

For the sodar part the long range Doppler sodar of type Metek PCS.2000-64 was used. Its acoustic part consists of an 8x8 phase array antenna housed by an efficient acoustic shield to reduce noisy signals which are often present at industrial sites. The sodar uses a WINDOWS based PC-unit for signal generation, signal analysis, data evaluation and links to host stations and to external data servers. The typical height range in multiple frequency mode (typically five frequencies) under fair measuring conditions is about 500-700 m for the Doppler components and about 700 m for the reflectivity profiles. The RASS extension uses a CW radar transmitter at 1290 MHz (according to national rules) and independant grid antennas for transmission and reception, each of 2.4 m diameter. For measurements of the surface sensible heat flux at 2 m height the USA-1 system of Metek was used which samples instantaneous data of the three wind components and the sonic temperature at a rate 10-40 Hz. The USA-1 computes on-line the means of the wind field and the sonic temperature and the turbulent quantities, such as variances, covariances and fluxes of heat and momentum.

IOP Conf. Series: Earth and Environmental Science **1** (2008) 012024 doi:10.1088/1755-1307/1/1/012024

The instruments are synchronized, time intervals of 15 min are used.

3. Determination of mixing height

The method has been developed for flat terrain and follows the suggestions of [4]. Datasets from the Meteorological Observatory Lindenberg (Germany) have been used. The method does not only use single vertical profiles of sodar backscatter R_3 or of derived parameters but also uses bulk information, e.g. histograms and boundary layer evolution over time. The algorithm includes the following steps:

- ABL height exists if second moment and fourth moment of R_3 exceed certain threshold values, otherwise no ABL height can be derived from sodar data. Figure 1 shows examples of different cases with the upper figure illustrating a situation where a boundary layer cannot be derived.
- Determine local minima in the frequency distribution of R_3 and discriminate a 'gradient type' (double peak in the pdf, see mid figure) or a 'maximum type' (secondary maximum at high backscatter values, see bottom figure).
- Three criteria for ABL height are considered:
  - Find the lowest height z_i for which the condition holds: $R_3(z_i+2) < S-2$ and $R_3(z_i+1) < S$and $R_3(z_i) < S+2$. (with S = local minimum of the histogram or predefined).
  - Find the height of the largest local maximum in the R_3-profile.
  - Find the lowest height of a prescribed gradient of horizontal wind speed.

  Height values might not be detectable for each time step.
- Derive a unique ABL height from the three height values under consideration of the backscatter type. This curve may also contain gaps.
- If this timeseries is strongly variable: Search for a growing convective ABL height between 04:30 UTC and 11:30 UTC.
- Decide from the following profiles whether the ABL may be higher than the sodar range of 700 m or whether an ABL height within the sodar range can be traced.
- Find the top of a shallow stable ABL in the late afternoon or evening which is the successor of the inactive deep ABL of the afternoon.

If an inversion height can be derived from RASS temperature profiles, these values are used to correct sodar-derived mixing heights. Sonic heat flux values force the mixing height to larger than 700 m if a growing CBL is found and the heat flux still increases. In case of a negative heat flux the lowest available mixing height - whether from RASS or sodar - is taken. For further details see [5].

Results are shown in Figure 2 where also radiosonde-derived mixing height values are plotted. Both figures show a shallow stable ABL during night-time and a growing CBL in the morning which final depth is within the sodar range on 6 January (left figure) or reaches beyond the sodar range on 27 August (right figure). The formation of

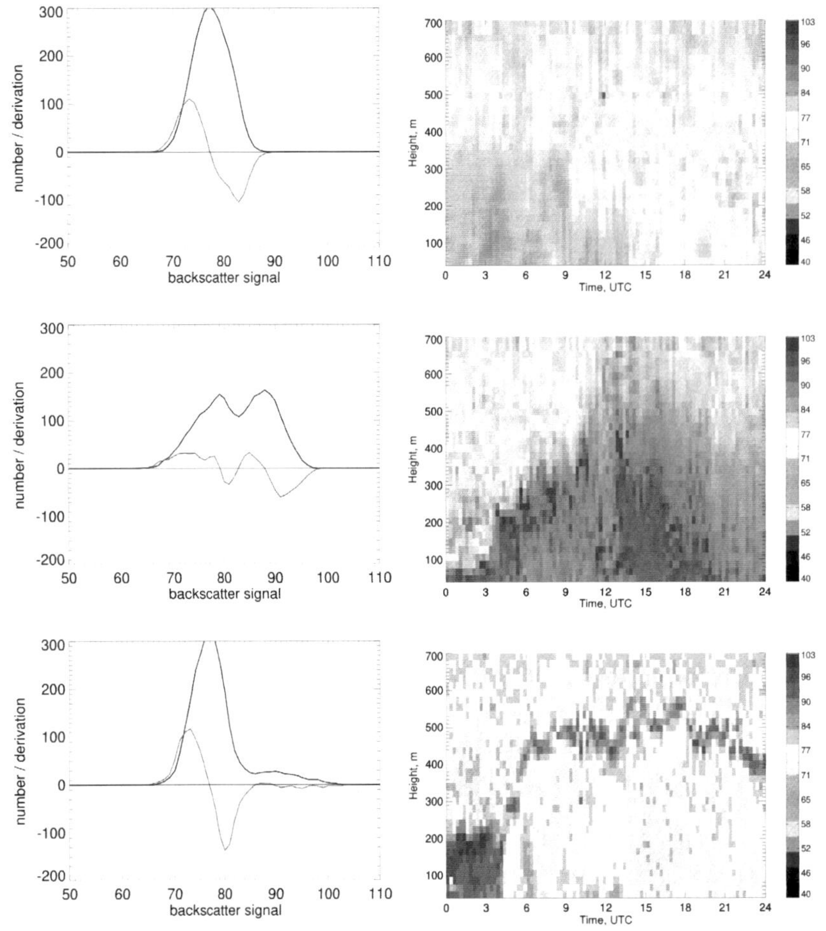

Figure 1. Histogram (left) and time-height section of backscatter R_3 on 17 February 2005 (upper panels), 12 February 2005 (middle panels), and 11 November 2004 (lower panels). The left figures additionally show the first derivative of the histogram (red lines) to illustrate the local minima.

a surface inversion in the evening may pose problems to the RASS which is evident in Figure 2, left panel at 23 UTC. In general, problems arise from very low mixing height below the lower range of sodar or RASS (right figure, 23 UTC), and during times when the mixing height is not well defined (both figures, afternoon).

4. Results

4.1. Flat terrain

Measurements over two years from the rather flat site of Lindenberg were analysed. The measurements took place at Falkenberg where the boundary layer field site of the Meteorological Observatory Lindenberg - Richard Aßmann Observatory (German Meteorological Service) is located. The landscape around Falkenberg is rather flat and

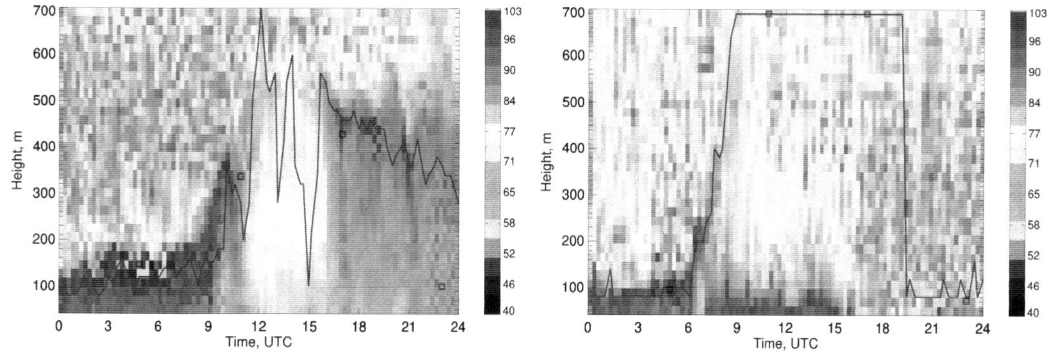

Figure 2. Time-height section of sodar backscatter signal on 6 January 2004 (left) and on 27 August 2005. The blue line is the sodar/RASS/sonic-derived ABL height, the black squares mark the radiosonde ABL height. The radiosonde-derived ABL height on 27 August 2005 is larger than 700 m at 12 an 18 UTC, but for convenience plotted here at 700 m.

dominated by a mixture of agricultural landuse, forests and small villages, the measuring site is a flat meadow with open fetch.

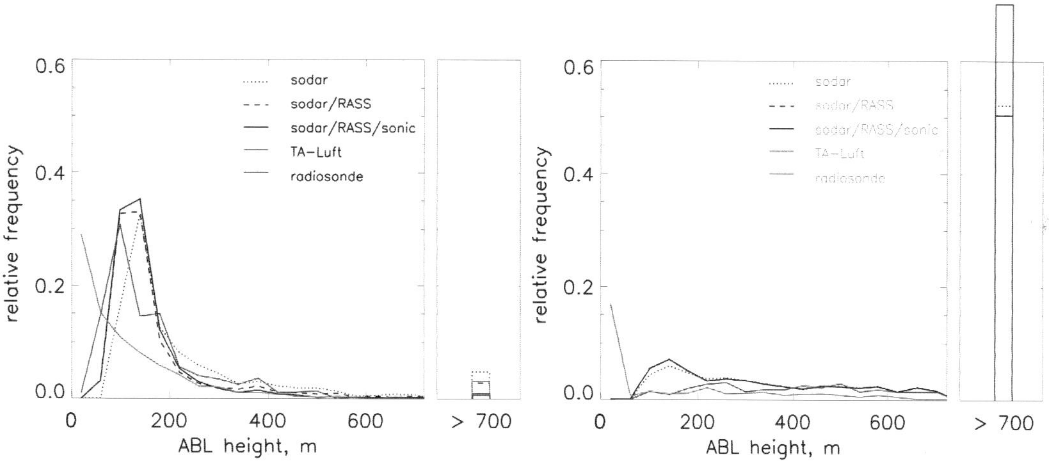

Figure 3. Histograms of ABL height derived from sodar (dotted), from sodar and RASS (dashed), and from sodar, RASS and sonic (solid) for 2004 and 2005 at Lindenberg. The results are shown for 00 UTC (left), and 12 UTC (right). Red lines shows the results for the administrative regulation TA-Luft. The blue lines show the results derived from radiosonde data.

The results are presented as histograms of the mixing height at 00 UTC and 12 UTC (Figure 3). There is a large frequency of shallow boundary layers during night-time, early morning, and evening (the latter two time periods are not shown here). The frequency decreases strongly with increasing mixing heights because these height ranges preferably occur during the growing phase of the CBL and each height value only lasts for a short time. Height values larger than 700 m occur rarely at night, they are set to '> 700 m'. The frequency of ABL heights of '> 700 m' is very large at 12 UTC (approximately 50%), and at 18 UTC still 20% are counted. The histograms for 00 UTC and 18 UTC show few ABL height values in the range of 300 m and 700 m whereas particularly at 12 UTC this range is nearly equally filled. The addition of RASS temperature profiles enhances the number of shallow MH (< 200 m) during all times of the day whereas the addition of heat flux values from the sonic changes little. Only in the afternoon (not shown here) the frequency of ABL height values between 40 m and 300 m is enhanced.

For comparison the histograms of the MH which is calculated according to [6] are plotted in red. It is obvious that there is an overestimation of very shallow MH and a lack of MH between 100 m and 300 m, depending on the time of the day.

4.2. Mountainous terrain

Sodar / RASS measurements were performed at Jesenice (Slovenia) near an industrial plant from 14 September to 14 December 2006. The measuring location is situated on the bottom of a large alpine valley directed 120° - 300° and with crest heights of at least 1000 m above the valley floor. At the measuring site a small north to south oriented tributary valley leads into the main valley. The orography entirely dominates the flow charachteristics, either for thermally driven mountain - valley wind systems or for dynamical channeled flow.

Figure 4 shows an example of a mountain - valley wind regime out of a period of several days. The vertical profiles of wind direction in the mountain wind and valley wind periods are nearly constant up to at least 300 m (the upper range of the sodar). The northern wind up to 200 m height is the outflow out of the tributary valley and is later replaced by the mountain wind of the main valley. The horizontal wind is moderate, in the valley wind regime larger than in the mountain wind regime and zero during the change of regimes. The vertical wind is upward during valley wind periods. The temperature profiles do not show a strong inversion topping the night-time boundary layer. The morning transition starts aloft and reaches the ground one or two hours later, while the evening transition starts at the ground. This can be seen in the temperature and in the wind profiles.

This example shows that - excluding the transition times and possible effects of a tributary valley - the wind regimes extend over deep layers. In this particular case the low upper sodar range cannot detect the upper range of the wind systems. But other studies show that the daytime wind regime covers the whole valley atmosphere and the height of the night-time inversion depends on the valley geometry and on the position within the valley, i.e. near the beginning or near the mouth of the valley (see e.g. [1, 3]). The crucial periods for the evolution of pollutant concentrations are the transition periods and in these periods turbulent parameters play a minor role compared to vertical wind. An important role in the dispersion of pollutants play counter flows

IOP Conf. Series: Earth and Environmental Science **1** (2008) 012024 doi:10.1088/1755-1307/1/1/012024

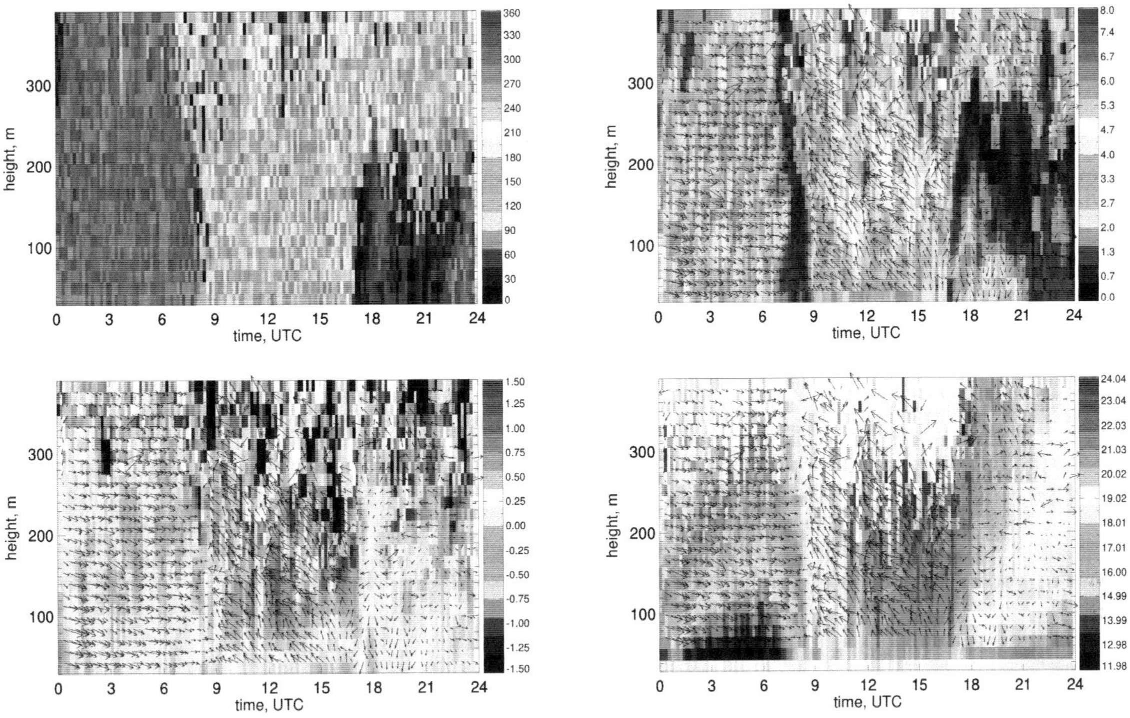

Figure 4. Time-height section of wind direction (in *circ*, top left), wind speed (in m/s, top right), vertical wind (in m/s, bottom left), and potential virtual temperature (in °*C*, bottom right) at Jesenice (Slovenia) on 21 September 2006 .

at elevated layers as e.g. depicted in Figure 4 after 18 UTC. At this particular location such flows are often observed connected with a flow into or out of the tributary valley. Another measuring campaign took place in the broad Danube valley at the location of a large industrial plant. The enormous heat emission during the production process influences the local valley atmosphere. Measurements were performed with a sodar / RASS and a sonic anemometer-thermometer placed right within the industrial area. Figure 5 shows time-height sections of sodar backscatter and wind direction over a three-days period.

The sodar backscatter plots show a quite strange time evolution with increasing backscatter during night-time and decreasing backscatter during daytime. It is evident, that two main wind direction sections dominate (270° and 90°) according to the valley axis. Preferably with westerly wind air masses from the artificial heat source are transported over the sodar and sonic. This results in larger backscatter values as observed during night-time. At the same time a large heat flux is measured by the sonic. Easterly winds are frequently associated with low backscatter. The structure of the lower atmosphere is thus strongly determined by the local heat source which influences the measurements when the instrumentation is situated in the lee of the source.

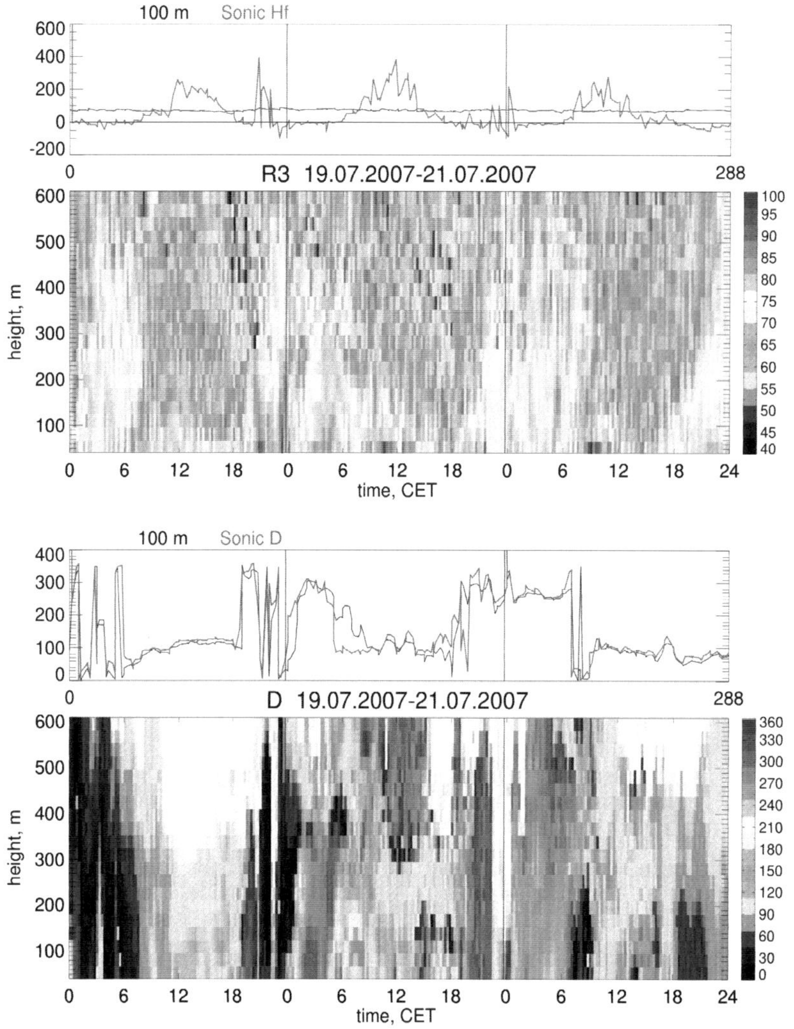

Figure 5. Time-height section of sodar backscatter (top) and wind direction (bottom) at an industrial location from 19 July to 21 July 2007. The time series above these figures show surface heat flux (top) and wind direction at 2 m and 100 m.

On the other hand, a lot of observed phenomena at this location can not be easily explained. Not only the orography of the surroundings is rather complicated with a bending main valley and a tributary valley just at the site of the plant, also the large area of the industrial plant is structured and shows difference in altitude of some 100 m. In an area like this the mixing height is suspected to be inhomogeneous, and it strongly depends on the overall wind field in the valley and the locally induced thermal structure of the lower atmosphere. A simple determination of the mixing height - as suggested by regulations - fails but also the mixing height which could be determined e.g. by a sodar / RASS is not representative for the dispersion of constituents.

5. Conclusions and outlook

The features presented above illustrate the difficulties in defining and determining the mixing height at orographically complicated sites. At flat and nearly homogeneous sites the development of the mixing height during undisturbed days is well known and it can be successfully derived from sodar, RASS and sonic data. Problems arise when the radiative forcing vanishes or when the synoptic situation is changing and accompanied by events like storms or frontal passages. At those times a boundary layer may not be defined. But in mountainous regions a lot of other factors play a role and influence the development of the mixing height. The wind regime - either dynamically forced or thermally induced - together with non-vanishing average vertical wind strongly influence the transport of energy and matter as well as of pollutants.

Simple prescriptions to determine the mixing height can be shown to have shortcomings particularly in the range which is of most interest, i.e. between 100 m and 300 m. In mountainous regions such methods which use local near-surface parameters nearly completely fail. The wind regimes are influenced by upstream conditions and conditions at the slopes which are not captured by a single surface measurement at the valley floor.

The operation of remote sensing instrumentation at a site which is highly complicated - either because of its complicated orographic structure or because of artificial heat sources - is only recommended for a simple survey and cannot be used to determine the representative mixing height. In such situations the application of a mesoscale model is recommended and the observation-derived mixing height should be used for model verification.

References

[1] Baumann-Stanzer, K. and Groehn, I. (2004). Alpine radiosoundings - feasible for mixing height determination? *Meteorol. Z.*, 13:131–142.

[2] Beyrich, F. (1997). Mixing height estimation from sodar data - A critical discussion. *Atmosph. Environm.*, 31:3941–3953.

[3] Chemel, C. and Chollet, J.-P. (2006). Observations of the daytime boundary layer in deep alpine valley. *Boundary-Layer Meteorol.*, 119:239–262.

[4] Emeis, S. and Türk, M. (2004). Frequency distributions of the mixing height over an urban area from SODAR data. *Meteorol. Z.*, 13:361–367.

[5] Hennemuth, B. and Kirtzel, H.-J. (2008). Towards operational determination of boundary layer height using sodar/rass soundings and surface heat flux data. *Meteorol. Z.*, (accepted for publication).

[6] TA Luft (2002). *Erste allgemeine Verwaltungsvorschrift zum Bundes-Immissionsschutzgesetz (Technische Anleitung zur Reinhaltung der Luft - TA Luft).* Bundesministerium für Umwelt, GMBl 25-29, Bonn, Germany.

[7] Wekker, S. D., Steyn, D., and Nyeki, S. (2004). A comparison of aerosol-layer and convective boundary-layer structure over a mountain range during STAAARTE '97. *Boundary-Layer Meteorol.*, 113:249–271.

Temperature profiles by ground-based remote sensing and in situ measurements

S Argentini[1], **I Pietroni**[1,2], **C Gariazzo**[3], **A Conidi**[1], **G Mastrantonio**[1]
A Pelliccioni[3], **I Petenko**[1, 4], **A Viola**[1] and **A Amicarelli**[3]
[1] ISAC-CNR, Via del Fosso del Cavaliere, 100, 00133 Roma, Italy

[2] Università di Siena, Via del Laterino, 8, 53100 Siena, Italy

[3] ISPESL Dipartimento Insediamenti Produttivi e Interazione con l'Ambiente, Via Fontana Candida, 1, 00040 Monteporzio Catone (RM), Italy

[4] IAPh RAS, Pyzhevskiy, 3, Moscow, Russia

E-mail: s.argentini@isac.cnr.it

Abstract. This study focuses on the accuracy of the temperature profiles measured with a Doppler Radio-Acoustic Sounding System and a Microwave Temperature Profiler during a period of about 3 months in winter 2007-2008. The experiment was carried on at the experimental facility of the Institute of Atmospheric Sciences and Climate (ISAC) of the Italian National Research Council (CNR). The temperature data measured with remote sensors were verified with *in situ* measurements on a mast as well as with tethered balloon data. The facsimile echograms obtained with the ISAC Doppler SODAR were analysed to understand to which extent the RASS and Radiometer temperature profiles behaviour can represent the real thermal structure of the atmosphere.

1. Introduction

Although the Radio Acoustic Sounding System (RASS) suggested by Atlas [1] is very clear in its physical base and was easily realised [2], the practical use for operational measurements needs a correct understanding of the received frequency and power variations with height, and their dependence on various system parameters such as transmitted power, antenna gain, wavelength, etc. The main difficulty in the correct description of the received signal arises from the fact that the echo signal parameters are strongly influenced by a variety of environmental variables existing in the real atmosphere. The environmental variables which influence the RASS measurements are: the vertical and horizontal winds, the atmospheric turbulence, humidity, the vertical temperature gradients, and the absorption of acoustic wave. These errors were discussed in many works (see, for example, [3], [4], [5]). To accurately quantify the effect of these factors on the received signal is difficult. A rigorous approach to the description of the RASS theory, which takes into account the diffraction effects in the Fresnel approximation, has been developed by Kon and Tatarskii [6]. Both experimental and theoretical results are summarised by Kallistratova and Kon [7] which mostly consider a bistatic configuration with separated transmitting and receiving radar antennas. Later, Petenko [8] used the same theoretical approach to improve the estimate of the errors due to the antenna geometry for the Bragg-RASS based on a monostatic radar wind profiler. Görsdorf and Lehmann [9] improved the algorithm for the range correction of the Bragg RASS, which currently is used in the network of wind profiler/RASS measurements.

Using a RASS with a continuous wave Doppler radar and a pulsed acoustic source (the so called Doppler RASS) the violation of the Bragg condition influences not only the echo power, but also the frequency of the received signal. This effect can reduce the accuracy of the temperature measured by the RASS, especially when estimating the temperature gradients. The origin of this effect was explained by Petenko and Bedulin [10]. The same authors proposed a method for compensating this error by using acoustic pulses with a linearly modulated frequency.

The RASS temperature profiles were compared in the past versus *in situ* sensors located over meteorological towers, the thersondes and radiosondes. Engelbart et al. [11] made a comparison of the SODAR/RASS data with the AIR TMT-5A thethersondes and the 99 m Lindenberg tower data. The comparisons of the SODAR/RASS-derived profiles with in-situ measurements showed an accuracy within 0.1 K.

The radiometric techniques and the comparison of the performance of these techniques with the RASS were discussed by Westwater et al. [12]. The results of a comparison study of the temperature profiles by a continuously and a discrete scanning radiometer, a RASS and the data of the BAO tower in Colorado were shown. The temperatures of the microwave radiometer and the tower showed an agreement within one degree while the results of the comparison with the RASS data, which are shown only at 200 m present a lower level of agreement as the measurements experienced severe radio frequencies interference.

Taking advantage of the ISAC-CNR experimental facility a scanning microwave radiometer (MTP5-P) of the same type described by Kadygrov et al. [13], was used and the profiles compared with those of a METEK 1290 MHz SODAR/RASS. Tethered balloon measurements have been carried out in a few cases for comparison in the first 150 m a.g.l. To study the capability of RASS and the MTP5-P to catch the layer structure of the temperature field we compared the temperature profiles with the thermal structure of the atmosphere obtained by the backscattered echo of ISAC-Doppler SODAR.

2. Description of the experiment

A field campaign is in progress at the experimental facility of the ISAC-CNR located at Tor Vergata (RTV), in the suburbs of Roma. The purpose of this field experiment is to compare, under a wide range of meteorological conditions, the remote measure of the temperature profiles given by a SODAR/RASS and a scanning microwave radiometer. The results presented in this study refer (unless different specification) to the period 2 November, 2007 - 7 February, 2008.

2.1. The site

The RTV site (41°50' N, 12°38' E) is located 100 m a.s.l. and 25 km from the Thyrrenian sea (at SW). At 10 km, in the SE direction, the Albani hills lie, of 600 m of mean altitude (the top of the hills reaches 1000 m). At 20 km in NW direction the great urban centre of Rome, 20 km of diameter, lies.

2.2. The instrumentation

The sensors utilised in our study consist in a microwave radiometer and a SODAR/RASS for the remote measure of the temperature. In a few days the temperature measured with a sonde on a tethered balloon was used for comparison with the remote sensors profiles. Thermometers at 1.6 m, 2 m, 10 m, and radiometers for the estimate of radiative budget are located in the same site. A Doppler SODAR is used to record the thermal structure of the atmosphere. In the following a brief description of the instrumentation is given.

Microwave Radiometer (MTP5-P, Meteorological Temperature Profiler- Polar). The MTP5-P is an instrument for the remote measure of the air temperature profile from the ground level to 600 m. The MTP5-P was designed for the investigation of the atmospheric boundary layer in polar regions

[13] [14]. Unlike any other ground-based microwave temperature profiler the MTP-5P has a larger antenna (diameter = 0.6 m and a beam width of about 0.5°) and consequently a higher vertical resolution (10 m) in the lowest 100 m. The MTP5-P is an angular scanning single-channel microwave radiometer with the working frequency in the center of molecular oxygen absorption band at about 60 GHz which was designed to provide continuous, unattended observations in extreme meteorological conditions (very low temperature, strong wind, etc). It measures the thermal emission of the atmosphere with a high sensitivity (0.04 °C at 1 s of integration time) at different zenith angles. On the base of these measurements it is possible to retrieve the temperature profile with an accuracy of 0.5 °C. The angular-scanning system has the sharper height resolution in the lower 300 m and is used for the investigation of atmospheric boundary layer (ABL) [13].

Metek SODAR/RASS. Virtual temperature profiles measurements were performed with a commercial 1290 MHz Metek SODAR/RASS system. Narrow acoustic beams are obtained by means of a phased array of 16 loudspeakers. The system is able to derive the wind components and virtual temperature data up to 400 m a.g.l. with a vertical resolution of 28 m and 38 m as lowest measuring level. The 10 minutes averaged data were used in further analyses only if the acceptance test was satisfied. The latter is based on different parameters such as signal to noise ratio and statistical significance of measured signal and ambient noise. Signal to noise ratios were used for quality assurance, selecting only those data with positive values. The radial sound velocity was not corrected for the vertical wind component which in 78 % of cases was $\pm$ 0.2 m/s. Although the RASS measures the virtual temperature data were directly compared with the absolute temperature retrieved by the radiometer because the humidity and pressure vertical profiles were not routinely collected. However, we estimated that the difference between absolute and virtual temperature at ground level with the local meteorological conditions (1000 hPa, 10 °C and 60 % of relative humidity) is about 0.05 °C. The only correction was made to take into account the geometry of bistatic antenna configuration. The formula obtained by Kallistratova and Kon [7] and presented in [8] has been used for this kind of correction taking into account the diffraction effects in the Fresnel approximation. For the first range gate (38 m) this correction is $\approx$ 0.56 °C, for the second range gate (66 m) $\approx$ 0.36 °C, but these are an approximate estimate. This correction should be refined basing on the comparison with in situ measurements.

ISAC Doppler SODAR. This Doppler SODAR is an upgrade version [15] of the SODAR described in Mastrantonio and Fiocco [16] and Mastrantonio and Argentini [17]. It uses three antennas to radiate acoustic tones every 6 s, each at a different frequency: 1750, 2000, and 2250 Hz. The pulse repetition frequency allows for an instrumental range of approximately 1000 m. The echoes recorded by the three antennas, after suitable filtering, are added, sampled at a frequency of 1600 Hz, and analysed in real time. A fast Fourier transform is applied to the digitised echo, 256 points at a time, corresponding to a height interval of approximately 27 m, the first being centred at 40 m. A two step procedure [16] is used to minimise the influence of the ambient noise on the measures and to reject not reliable data.

Surface measurements. Meteorological surface parameters, such as wind speed and direction at two different heights (1.6 and 10 m a.g.l.), absolute air temperature and its surface gradient (1.6 and 10 m a.g.l.), soil temperature, net and global solar radiation, atmospheric pressure and relative humidity were measured at 1 Hz and averaged over 10 minutes. Air temperature measurements were conducted by means of PT1000 thermometers with 0.2 °C of accuracy. In addition data from an Automatic Weather Station (AWS) were available. The AWS is a Vaisala MILOS 520 equipped to measure the meteorological parameters: temperature, pressure wind speed and direction within the WMO standard. The anemometer and the wind vane are positioned at the top of a 10 m high pole. The thermo igrometer and radiation sensors are fixed on the same arm at 2 m from the ground. These data are used as reference for the other sensors of the experiment as well as to initialize the tethersonde before the launch. The meteorological data are recorded on a 10 minutes base average.

Tethered balloon meteorological sounding. The system consists of a receiver operating in the range of meteorological frequencies 395 - 410 MHz, a tethersonde, a helium inflated balloon and a winch with a 1000 m tetherline. The balloon, properly shaped to facilitate orientation upwind retrievable by the winch is used to fly the tethersonde. The tethersonde package sensors include a three cup anemometer, a magnetic compass, an aneroid barometer, a thermo-resistance and a humicap sensor for wind speed and wind direction, pressure, temperature and relative humidity measurements respectively. The temperature sensor has an accuracy of ± 0.5 °C, the wind speed has an accuracy of ± 0.5 m/s, the wind direction measured with the magnetic compass that relies on the orientation of the balloon within an accuracy of ± 10°, when the balloon is suitably stable. The pressure measurement has an accuracy of ± 1 hPa.

The altitude of the sensor package above the surface is determined from the pressure and temperature. During flight, the electronics of the tethersonde sequentially scans the sensors and transmits every 6 s the raw data to the ground station. After the acquisition the data are processed and plotted on line by the PC connected to the receiver. The correct data acquisition can be checked on line so that possible modification of the flight can be done.

3. Results

3.1 Comparison of MTP5-P and SODAR/RASS temperature time series and profiles

Figures 1a,b show the temperature time series of the MPT5-P and the RASS for one week of December 2007 at 40 m and 200 m, respectively. In each graph is also plotted the difference $T_{MTP5} - T_{RASS}$ between the two temperatures. This behaviour is the same at both levels and was observed during the whole field experiment. The larger disagreement is observed during the night when $T_{MTP5} - T_{RASS}$ is mostly negative. During the warmer hours of the day $T_{MTP5} - T_{RASS}$ is positive in most of the cases.

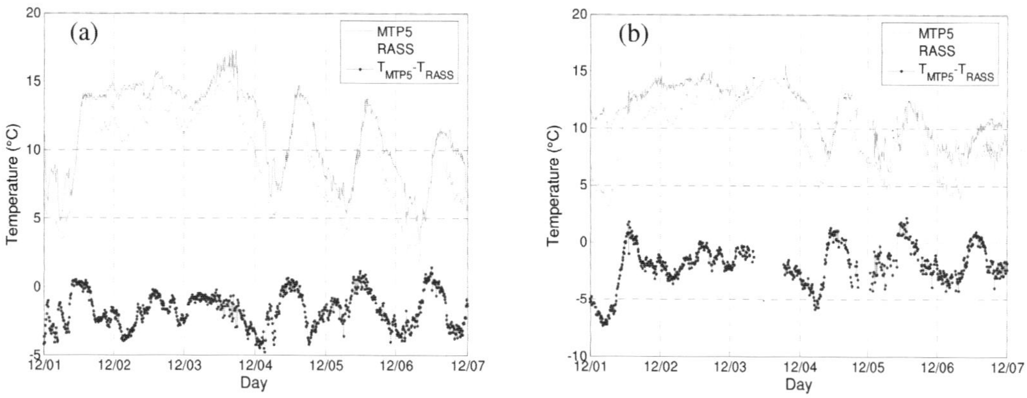

Figure 1. Time series MTP5-P (grey) and RASS (black) at 40 m (a) and 200 m (b) for one week of December. $T_{MTP5} - T_{RASS}$ is also given in the same graph.

At 40 m and 200 m the bias are -1.2 ± 1.0 and -1.8 ± 1.3 respectively. As the temperature difference varies during the day we studied the behaviour of $T_{MTP5} - T_{RASS}$ for different incident solar radiation conditions. We considered the net radiation time series. Negative values of the net radiation at the surface correspond, in almost all the cases, to nocturnal (up to sunrise) cases. The net radiation over the period 2 November – 7 February has been calculated using the data from radiometers located in the same area. The scatter plot of $T_{MTP5} - T_{RASS}$ at 40 m and 200 m versus the net radiation is shown in the Figures 2a and 2b respectively. These figures confirm the results of Figures 1a and 1b which

IOP Conf. Series: Earth and Environmental Science **1** (2008) 012025 doi:10.1088/1755-1307/1/1/012025

indicate that for negative values of the net radiation $T_{MTP5} - T_{RASS}$ is negative and large (in most of the cases the difference ranges between 0 °C and – 4.5 °C) in some cases, especially at 200 m may also reach -6 °C. For positive values of the net radiation $T_{MTP5} - T_{RASS}$ is generally positive and less than 1°C, in a few cases is negative, but the scatter reduces to -1.5 °C - 2 °C. Similar results are obtained at 200 m, but $T_{MTP5} - T_{RASS}$ is mostly negative and the values are more scattered.

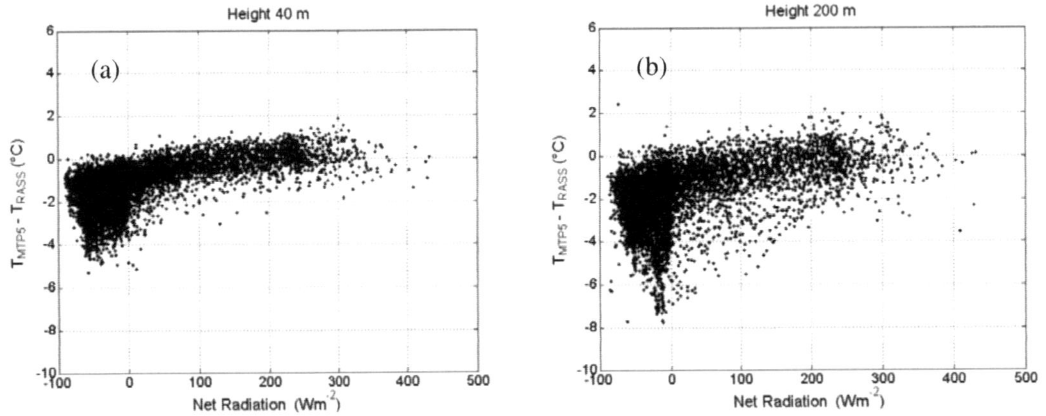

Figure 2. Scatter plot of $T_{MTP5} - T_{RASS}$ versus net radiation at 40 m (a) and 200 m (b) for the period 2 November, 2007– 7 February, 2008.

To study the height behaviour of $T_{MTP5} - T_{RASS}$ the probability distribution of the cases giving a certain $T_{MTP5} - T_{RASS}$ at a given height has been computed (Figures 3a,b). On the right side of the graph is also shown the behaviour of the percentage of data used at each level in computing the probability distribution. Considering the results of Figures 1 and 2 the diurnal (1000-1500 LT) (Figure 3a) and nocturnal (2200-0300 LT) (Figure 3b) periods have been analysed separately. During the day hours (Figure 3a) -below 150 m- the 50 % of cases shows a difference which varies between 0 and 1°C, this difference reduces to 0.5 °C above 150 m. A larger difference - even greater than 2 °C - but with a lower occurrence (10 % to 20 %) - is observed at all the heights. During the night the difference is greater. Only in the10 %- 20 % of the cases $T_{MTP5} - T_{RASS}$ is below 1 °C while in the 50 % of the cases is about 2 °C. In the 20 % to 30 % of the cases the difference may also reach 4 °C.

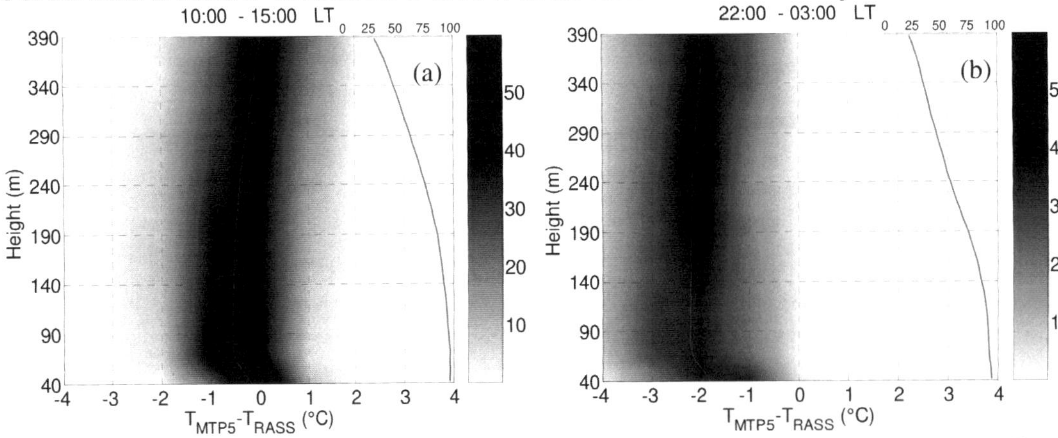

Figure 3. Probability distribution at 1000-1500 LT (a) and 2200-0300 LT (b) of temperature difference between MTP5-P and RASS data at different heights

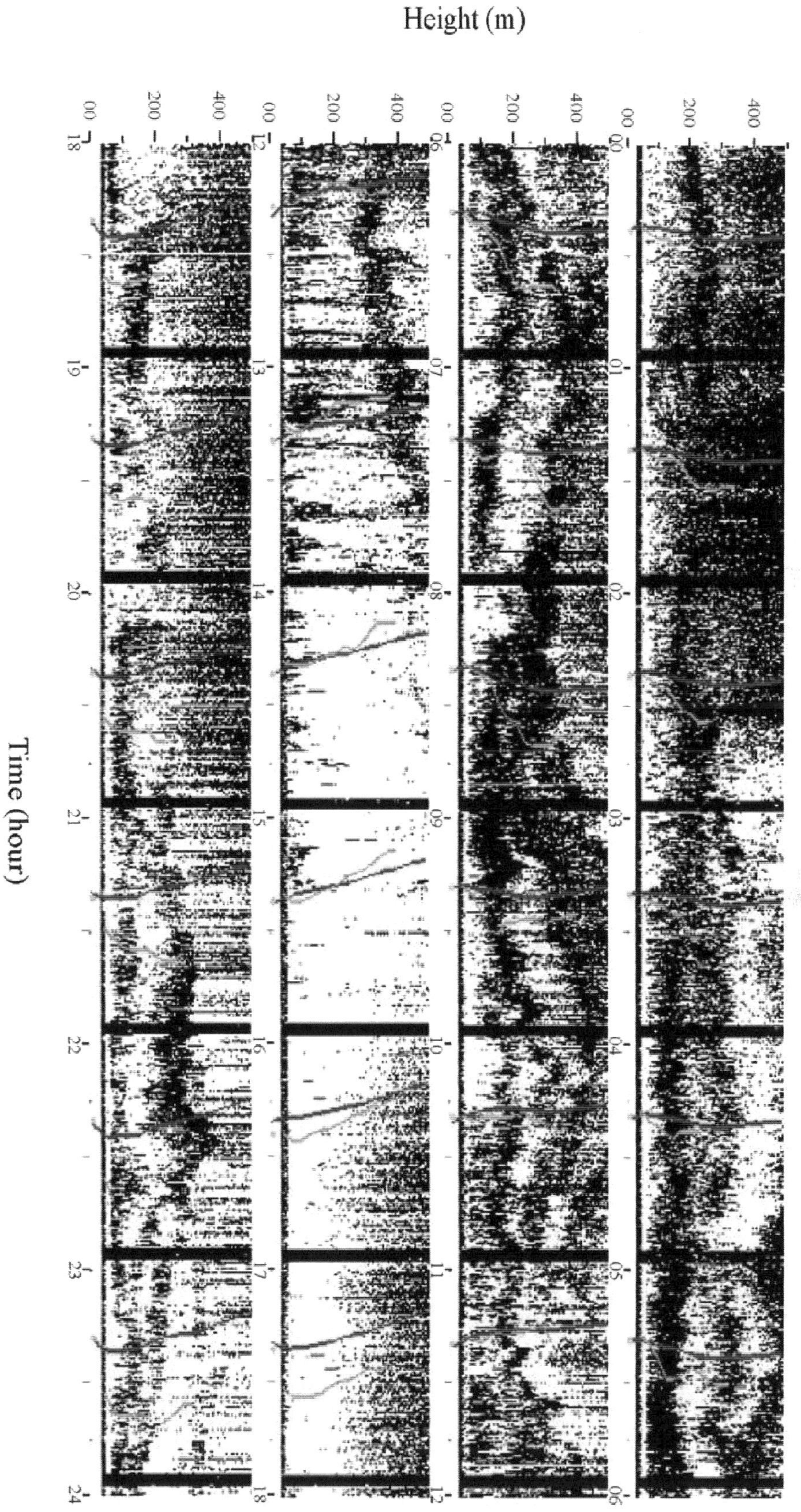

Figure 4. Facsimile representation for 30 November 2007 given by ISAC SODAR with superimposed MTP5-P (red) and RASS (green) profiles.

3.2 Vertical structure of the temperature field

It is well known that the layer structure of the temperature fluctuation field under stable stratification detected by the SODAR indicate the presence of strong temperature gradients located in narrow layers with a thickness of a few ten meters. The correlation of such thermal structure pattern with the gradients of the relevant meteorological parameters was studied by Gossard et al. [18].

Figure 4 shows the thermal structure of the atmosphere for 30 November, 2007 given by the Doppler SODAR, the level of grey in the figure is proportional to the intensity of the backscattered signal. At each hour, starting from 0030 LT in the morning, we superimpose the instantaneous temperature profiles. Red and green lines represent the MTP5-P and RASS profiles respectively reported on the same scale (we do not give values for keeping the figure more clear). The vertical scale is the same of the SODAR echogram. In general the RASS gives a better representation of the thermal structure of the atmosphere than the MTP-5P although the RASS vertical range is always lower than the theoretical range of sounding (400 m). In most of the cases the MTP5-P it is not able to catch the strong gradients occurring in a narrow range. This may explain in part the reason for the strong negative differences observed during the night.

3.3 Comparison of MTP5-P, RASS and tethersonde profiles

Due to logistic problems only a few tethersonde soundings could be done. The result of the comparison of these sounding with MPT5-P and the RASS profiles are shown in the Figures 5a,b,c, for 13 February 2007 at 1100 LT, 13 February 2007 at 1210 LT, and 14 February 2007 at 0910 LT respectively. Temperature at 1.6 m a.g.l. (star) is also plotted in the same graph. The three profiles of Figure 5a are in good agreement. However about one hour later (Figure 5b) the MTP5-P and the tethersonde profiles have mostly the same behaviour while the RASS seems to overestimate the tethersonde data of about 1.0 °C to 1.5 °C (depending on the height).

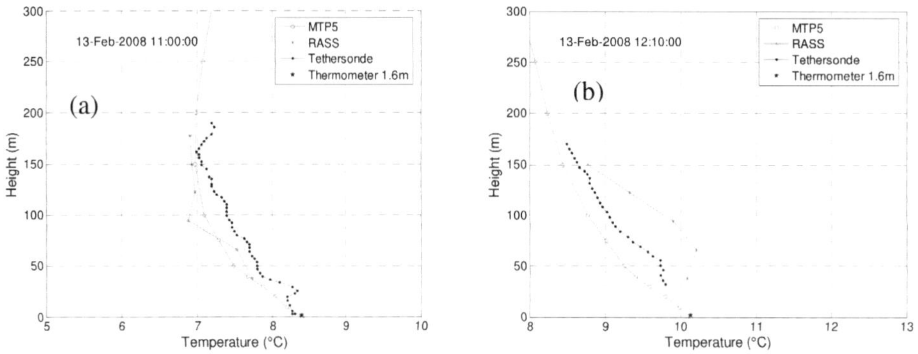

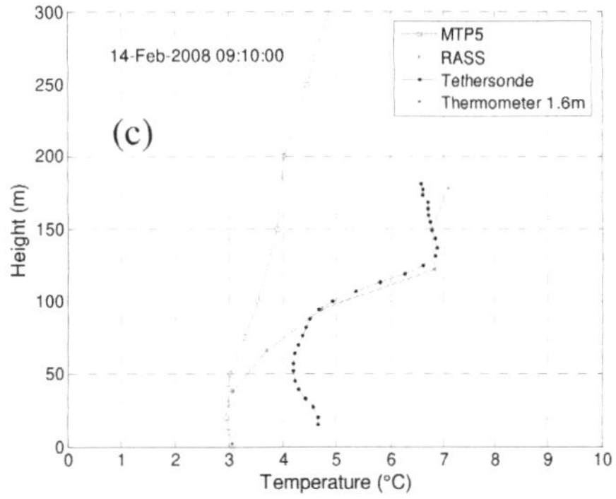

Figure 5. MTP5-P (circle), RASS (star), tethersonde (dot) for day 13 February, 2008 at 1100 LT (a), 13 February, 2008 at 1210 LT (b), 14 February, 2008 at 0910 LT (c).

Figure 5c shows that the RASS follows properly the strong temperature jump observed by the tethersonde (the temperature decreases more than 2 °C in about 50 m) while the radiometer temperature only evidences a weak tendency to decrease. The conclusion which can be given from the observation of these three cases is that both the MTP5-P and RASS have some limitations, limitations which occurs in different circumstances. A larger data set, obtained in different meteorological conditions, is then needed to fully understand which are the parameters influencing more the

Conclusions and future work

A difference greater than 1°C has been observed between the SODAR/RASS and MTP5-P temperature data. We can summarize the main results as follows:

(1) The bias between the temperature data at 40 m is -1.2 ± 1.0 while at 200 m is -1.8 ± 1.3.
(2) The profiles reach a better level of agreement in the warmest hours of the day
(3) The vertical range of the SODAR/RASS was limited in most of the cases at the first 200-meters.
(4) The RASS seems to better represent the multi-layer thermal structure of the atmosphere, the radiometer data being too smoothed
(5) The comparison with the tethersonde profiles did not help in giving a definite conclusion due to the limited number of profiles available for comparisons.

A definitive assessment of the profile data quality was not achieved because we could not include in our study a sufficient number of tethersonde profiles covering a wide range of meteorological conditions. In the next future we plan to make systematic, and all weather conditions measurements with the tethersonde, in order to achieve a full understanding on the RASS and MTP5 profiles data quality.

Acknowledgments

The authors like to thank Eugene-Miller of Attex for helping in improving the MTP5-P data quality and Dr. Uli Görsdorf of the Lindenberg Meteorological Observatory and the staff of Metek for useful discussion on RASS data quality. This research was partly supported by the Piano Nazionale Ricerche in Antartide in the frame of French-Italian "Dome Concordia" project.

References

[1] Atlas D 1962 Indirect probing techniques *Bull. Amer. Meteor. Soc.* **43** 457-466
[2] Peters G 2002 History and development of radio acoustic sounding *Proc. 11th Inter. Symp. on Acoustic Remote Sensing and Associated Techniques of the Atmos. and Oce., ISARS'02* 75-81
[3] Angevine W M and Ecklund W 1994 Errors in radio acoustic sounding of temperature *J. Atmos. Oceanic Technol.* **11** 837–842
[4] Görsdorf U and Petenko I 1998 Some results of experimental investigation of the RASS-measured temperature accuracy *Proc. 9th Inter. Symp. on Acoustic Remote Sensing and Associated Techniques of the Atmos. and Oce. (ISARS'98),* E. Mursch-Radlgruber and P. Seibert, Eds., 160–163
[5] Görsdorf U 1998 About the accuracy of temperature measurements with RASS. *Meteor. Z.,* **7,** 241–247
[6] Kon A I and Tatarskii V I 1986 The power of a signal under conditions of radioacoustic sounding of the atmosphere *Radioteknica i Electronica* **10** 1903-1908
[7] Kallistratova M A and Kon A I 1985 *Radio-Acoustic Sonding of the Atmosphere.* Moscow, Nauka, 198 pp. (In Russian)
[8] Petenko I V 1999 Improved estimation of errrors due to antenna geometry in RASS based on a radar wind profiler *Meteorol. Atmos. Phys.* **71** 69-79

[9] Görsdorf U and Lehmann V 2000 Enhanced accuracy of RASS measured temperatures due to an improved range correction *J. Atmos. Oceanic. Technol.* **17** 406–416

[10] Petenko I V and Bedulin A N 1994 Accuracy of RASS measurements and the Bragg condition *Int. J. Remote Sensing* **15** (2) 369-374

[11] Engelbart D, Steinhagen H, Gorrsdorf U, Neisser J, Kirtzel H J, Peters G 1999 First results of measurements with a newly-designed phased-array SODAR with RASS *Meteorol. and Atmos. Phys.* **71** 61-68

[12] Westwater E R, Han Y, Irisov V G, Leuskiy V, Kadygrov E N, Viazankin S A 1999 Remote sensing of boundary layer temperature profiles by a scanning 5-mm microwave radiometer and RASS: comparison experiments *J. Atmos. Oce. Tech.* **16** 805-818

[13] Kadygrov E N, Kodaev A V, Viazankin A S, Argentini S, Conidi A 2003 New Microwave temperature profiler - first results of measurements in polar regions. Atmospheric Radiation Measurement (ARM) Program Science Team Meeting. Broomfield, Colorado, USA (31-marzo – 4 aprile 4)

[14] Argentini S, Conidi A, Viola A, Mastrantonio G, Ferrara N, Petenko I, Kadygrov E N, Koldaev A V, Viazankin A. S 2005 Temperature measurements at Dome C using a new microwave temperature profiler. *X Workshop sull'atmosfera antartica*, Roma (22-24 ottobre, 2003),

[15] Contini D, Mastrantonio G, Viola A and Argentini S 2004 Mean vertical motions in the PBL measured by Doppler SODAR: accuracy, ambiguities, aossible improvements. *J. Atmos. Oce. Tech.* **21** 1532-1544

[16] Mastrantonio G and Fiocco G 1982 Accuracy of wind velocity determinations with Doppler sodar *J. Appl. Meteorol.*, **21**, 820– 830

[17] Mastrantonio G and Argentini S 1997 A modular PC-based multiband sodar system *Acoustic Sounding and Applications*, edited by S. P. Singal, , Narosa, New Delhi, 105– 116

[18] Gossard E E, Gaynor J E, Zamora R J and Neff W D1985 Finestructure of elevated stable layers by sounder and in situ tower sensors *J. Atmos. Sci.* **42** 2156-2169.

IOP Conf. Series: Earth and Environmental Science 1 (2008) 012026 doi:10.1088/1755-1307/1/1/012026

Ten years of operational boundary-layer measurements at the Richard – Aßmann Observatory Lindenberg: The role of remote sensing

F Beyrich and D A M Engelbart
Deutscher Wetterdienst
Meteorologisches Observatorium Lindenberg / Richard– Aßmann Observatorium
Am Observatorium 12
D – 15864 Tauche – OT Lindenberg

Phone: +49 33677 60228
Email: frank.beyrich@dwd.de

1. Introduction

The German Meteorological Service (DWD) performs a comprehensive operational measurement program at the Meteorological Observatory Lindenberg / Richard– Aßmann Observatory (MOL-RAO) in order to provide a reference data set on the physical structure and processes in the atmosphere – the so-called "Lindenberg Column". Boundary-layer measurements form an essential part of the observatory's activities, and a special boundary-layer field site has been established in the 1990ies at an open field 5 km to the south of the observatory site [1](Figure 1).

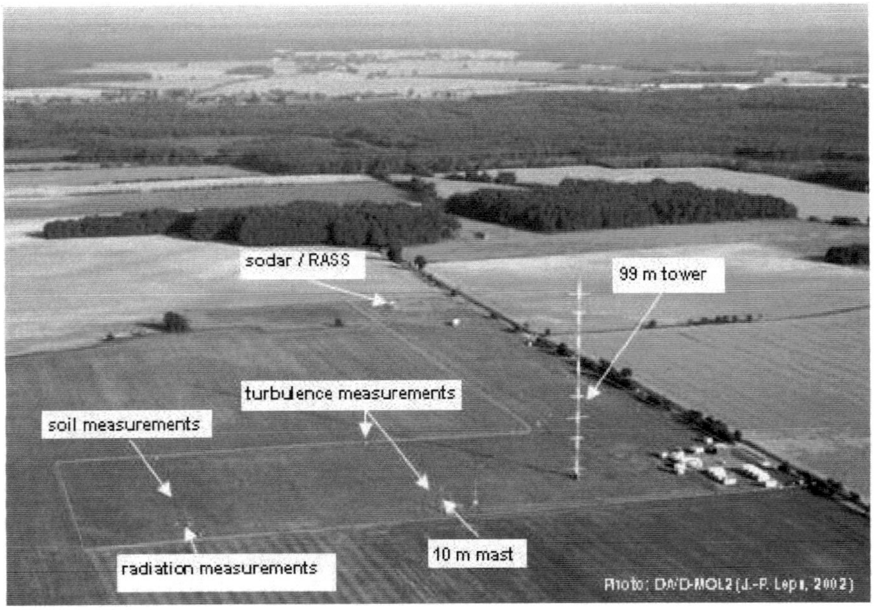

Figure 1. The MOL-RAO Boundary-Layer Field Site (GM) Falkenberg

Measurements at this boundary-layer field site (the "Grenzschichtmessfeld" – GM – Falkenberg) became operational in May, 1998. They comprise in-situ profile measurements of wind, temperature and humidity at towers of 10m and 99m height, micrometeorological measurements (soil parameters, radiation, turbulent energy fluxes) close to the ground and the operation of a sodar/RASS for temperature and wind profiling of the lower atmospheric boundary layer [2].

2. Milestones of the Boundary-Layer Measurement Programme at MOL-RAO

The establishment and evolution of a boundary-layer measurement programme at MOL-RAO was essentially stimulated by the LITFASS project (**L**indenberg **I**nhomogeneous **T**errain: **F**luxes between the **A**tmosphere and the **S**urface – a long-term **S**tudy), a research programme carried out at MOL-RAO between 1995 and 2001, in order to set up and to test an operational system for the investigation of land surface – atmosphere interaction processes over a heterogeneous landscape typical for the Northern part of Germany (and also for large regions of Northern Central Europe south of the Baltic Sea, e.g., [3]. The boundary-layer measurement programme has a strong operational component (Figure 2) allowing for a comprehensive and detailed characterisation of the boundary-layer structure and processes under a wide variety of synoptic conditions.

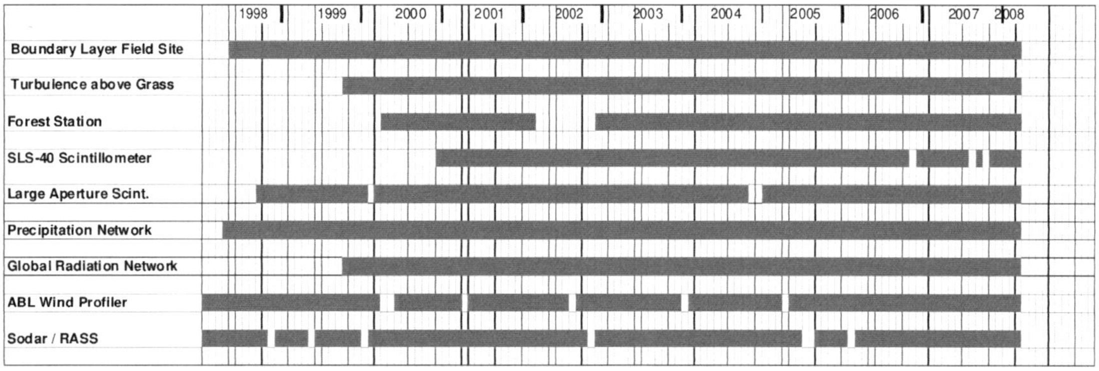

Figure 2. Elements of the boundary-layer measurement programme at MOL-RAO and their operational status (as of March 2008)

In DWD, data from these operational measurements are used for a monitoring of the performance and for the testing of modifications of the soil and boundary-layer parameterization schemes of the operational numerical weather-prediction models (e.g., [4],[5]). In addition, data sets are regularly transferred to international data bases being part of the world climate research programme activities (e.g., [6]).

Besides the operational measurements, several internal (DWD), national, and international field experiments with a strong focus on the atmospheric boundary layer have been organised and performed at the GM Falkenberg and in the surrounding area during the last 10-12 years. An overview is given in Table 1.

Table 1. Boundary Layer Experiments at the Meteorological Observatory Lindenberg

Name	Time Period	Part. Groups	Scientific Focus	Reference
LINEX-96/2	Jun 1996	4	Characterisation of the Falkenberg field site, test of turbulence instruments	Foken (Ed.): *DWD Offenbach - Arbeitsergeb-nisse FE* **No. 48** (1997)
LINEX-97/1	Jun 1997	4	Test of measurement systems and measurement strategy for LITFASS-98	Foken (Ed.): *DWD Offenbach - Arbeitsergeb-nisse FE* **No. 53** (1998)
LITFASS-98	May / Jun 1998	15	Turbulent fluxes over a heterogeneous landscape and radiation / cloud coupling	Beyrich et al. (Eds.): *Theor. Appl. Climatol.* **73**, 1-105 (2002)
LINEX-2000	Aug / Sep 2000	4	Variability of the ABL structure over the LITFASS area, turbulent fluxes from sodar measurements	Engelbart et al.: *Proc. 11th ISARS*, pp. 338-341 (2002)
STINHO-2	Jul 2002	5	Turbulent exchange over a heterogeneous land surface at the micro-α Scale	Raabe et al.: *Meteorol. Z. (N.F.)* **14**, 315-327
LITFASS-2003	May / Jun 2003	15	Evaporation over a heterogeneous land surface, area-averaged surface fluxes	Beyrich et al. (Eds.): *Boundary-Layer Meteorol.* **121**, 1-220 (2006)

3. The Role of Remote Sensing

The use of remote sensing systems has played a central role in the boundary-layer measurement programme of the MOL-RAO from the beginning. Operation of a sodar/RASS system has become an element of the GM Falkenberg boundary-layer field site in the design phase already, and the coupling of the measurements at GM Falkenberg with the wind profiler measurements at the observatory site has always represented the interface between the two sites in creating the"Lindenberg Column". Moreover, numerous methodical and case studies have been performed dealing with various aspects of the atmospheric boundary-layer structure and processes. An overview on some of these studies is given in Table 2.

An assessment of techniques and methods for the determination of turbulent fluxes from ground-based remote sensing instruments has been given recently in [7].

It should be noticed that the GM Falkenberg as a well-equipped and logistically flexible and powerful measurement and reference site has also been used for boundary-layer studies with remote-sensing instruments by research groups from other institutes together with the scientists at MOL-RAO, a few examples of studies referring to different issues using remote-sensing systems are:

- Boundary-layer height determination (e.g., Hennemuth and Lammert, 2006: *Boundary-Layer Meteorol.* **120**, 181-200)
- Water-vapour profiles and fluxes (e.g. Hennemuth et al., 2008, submitted to *Atmos. Chem. Phys.*)
- Acoustic tomography (Raabe et al., 2001, *J. Atmos. Ocean. Technol.* **19**, 1225-1230)
- Sodar evaluation (Pietschmann, 2007, *Wiss. Mitt. Meteorol. Inst. Univ. Leipzig* **Bd. 41**, 123-138).

Table 2. Lindenberg studies on ABL structure and processes with focus on remote sensing

Study Area	Study Focus	Reference
sensor characterisation	sodar – wind profiler	Beyrich et al., 1997: *Meteorol Z. (N.F.)* **7**, 220-225
sensor synergy	composite profiles	Görsdorf et al., 2002: *Theor. Appl. Climatol.* **73**, 97-105
boundary-layer height	sodar + wind profiler	Beyrich and Görsdorf, 1995: *Boundary-Layer Meteorol.* **76**, 387-394
	wind profiler	Engelbart, 1997: *Proc. COST-76 Profiler Workshop Engelberg*, 286-289
turbulent fluxes	heat flux profiles	Engelbart and Bange, 2002: *Theor. Appl. Climatol.* **73**, 53-65, Vogt and Engelbart, 2002: *Proc. 11th ISARS*, 343-345
	momentum flux profiles	Engelbart et al., 2002: *Proc. 11th ISARS Rome*, 339-342, Kouznetsov et al., 2004: *Meteorol. Atmos. Phys.* **85**, 93-99
boundary-layer processes	low-level jet climatology	Görsdorf et al., 2004: *Proc. 12th ISARS*
spatial variability	wind and temperature profiles	Petenko et al., 2000: *Proc. 10th ISARS*, 318-321

4. Scintillometry

Scintillometers measure intensity fluctuations of (optical and / or microwave) electromagnetic radiation propagating across the turbulent atmosphere. These intensity fluctuations are caused by atmospheric turbulence, suitably characterised by the refractive-index structure parameter (or equivalently the temperature and humidity structure parameters). Using similarity theory, the turbulent fluxes of sensible (and latent) heat can be derived from the structure-parameter information.

Scintillometer systems have been increasingly used over the last 10-15 years for the estimation of path-integrated sensible heat fluxes at the field scale and over distances of several kilometers, respectively, thus bridging the gap between local micrometeorological measurements and the typical spatial resolution of regional-scale meteorological models and of satellite images (e.g., [8], for a summary). Heat fluxes derived from scintillometer measurements have been shown to compare well with eddy-covariance measurements or with independent estimates of area-averaged fluxes for a footprint area of several square kilometers (e.g. [9], [10]). The potential of so-called Large-Aperture Scintillometers (LAS) for the determination of regional-scale sensible heat fluxes has been demonstrated for a variety of surface types from quasi-homogeneous to heterogeneous agricultural, different types of forest, and mixed farmland– forest landscapes. Moreover, the LAS has been shown to have the potential for continuous, unmanned long-term operation making it ideally suited for monitoring applications.

MOL-RAO has been operating a LAS over a nearly 5km path between the GM Falkenberg and the Lindenberg observatory sites for about ten years in co-operation with the Meteorology and Air Quality Group from Wageningen University. These continuous LAS measurements became a part of the operational measurement program at MOL-RAO and represent an important contribution to LITFASS [11]. Long-term operation of scintillometers has been proven possible with relatively little maintenance, and the flux data availability from scintillometers is comparable to other turbulence measurement techniques for daytime conditions. The heat fluxes derived from the LAS measurements appear to represent a suitable average over the mixture of land-use types found in the footprint area of the scintillometer path (Figure 3).

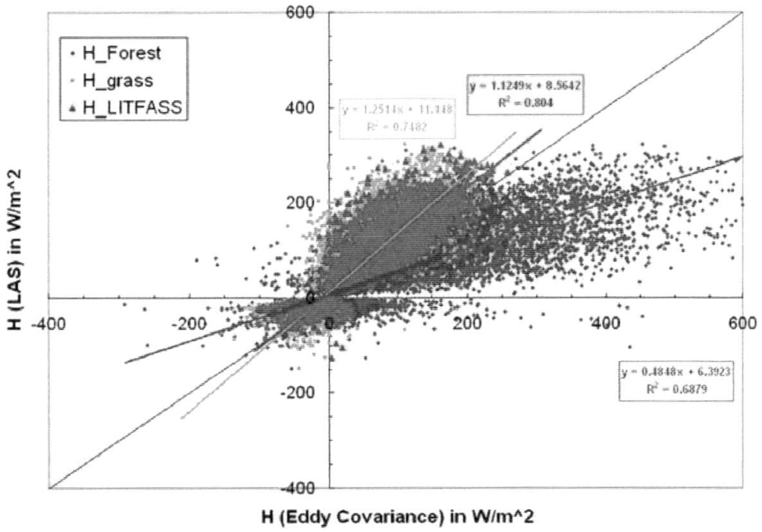

Figure 3. Sensible heat fluxes as derived from LAS measurements in comparison with those from eddy-covariance measurements over different types of land use and as a land-use weighted average (data from 2006).

5. Summary and Outlook

Boundary layer measurements using remote-sensing techniques have become an essential part of the operational measurement programme at the MOL-RAO. They provide a significant contribution to the "Lindenberg Column".

Profile measurements of atmospheric state variables form the basic application of remote-sensing systems at MOL-RAO. Several methods for the derivation of process parameters have been tested, but their operationalisation is still an unsolved issue.

Scintillometry has proven to be a reliable technique for the quasi-operational determination of structure parameters (and for the derivation of turbulent heat fluxes from these structure parameters) which are representative for a characteristic length scale of several kilometres corresponding to the resolution of current regional numerical weather-prediction and climate models and also to the pixel resolution of the images from contemporary earth-observing satellites.

Interested researchers are invited to use the facilities available at the MOL-RAO and its boundary-layer field site GM Falkenberg for their own experiments and / or to use the comprehensive data sets collected operationally for scientific studies.

References

[1] Neisser J, Adam W, Beyrich F, Leiterer U, and Steinhagen H 2002 Atmospheric boundary layer monitoring at the Meteorological Observatory Lindenberg as a part of the „Lindenberg Column". Facilities and selected results *Meteorol. Z.* **11** 241-253

[2] Engelbart D, Steinhagen H, Görsdorf U, Neisser J, Kirtzel H-J, and Peters G 1999 First results of measurements with a newly-designed phased-array sodar with RASS *Meteorol. Atmos. Phys.* **71** 61-68

[3] Beyrich F, Herzog H-J and Neisser J 2002 The LITFASS project of the DWD and the LITFASS-98 experiment: The project strategy and the experimental setup *Theor. Appl. Climatol.* **73** 3-18

[4] Heret C, Tittebrand A and Berger F H 2006 Latent heat fluxes simulated with a non-hydrostatic weather forecast model using actual surface properties from measurements and remote sensing *Boundary-Layer Meteorology.* **121** 175-194

[5] Beyrich, F, Heret C, Leps J-P, Richter S H, Rummel U, Vogel G and Weisensee U 2007 Komponenten des Wasserkreislaufes über einer heterogenen Landoberfläche: Messungen und NWV-Modellergebnisse für das LITFASS-Gebiet
(available at: http://meetings.copernicus.org/dach2007/download/DACH2007_A_00103.pdf)

[6] Beyrich F, Adam W 2007 Site and Data Report for the Lindenberg Reference Site in CEOP – Phase I. Offenbach a.M. - Selbstverlag des Deutschen Wetterdienstes *Berichte des Deutschen Wetterdienstes.* **Nr. 230** 55 pp (ISSN 0072-4130)

[7] Engelbart D A M, Kallistratova M A and Kouznetsov R 2007 Determination of the turbulent fluxes of heat and momentum in the ABL by ground-based remote-sensing techniques (a review) *Meteorol. Z.* **16** 325-335

[8] De Bruin H A R (Ed) 2002 Recent developments in scintillometry research *Boundary-Layer Meteorol.* **105** 1-193

[9] Meijninger W M L, Hartogensis O K, Kohsiek W, Hoedjes J C B, Zuurbier R M and De Bruin H A R 2002 Determination of Area-Averaged Sensible Heat Fluxes with a Large Aperture Scintillometer over a Heterogeneous Surface - Flevoland Field Experiment *Boundary-Layer Meteorology.* **105** 37-62

[10] Meijninger W M L, Beyrich F, Lüdi A, Kohsiek W, de Bruin H A R 2006 Scintillometer-Based Turbulent Fluxes of Sensible and Latent Heat over a Heterogeneous Land Surface - A Contribution to LITFASS-2003 *Boundary-Layer Meteorol.* **121** 89-110

[11] Beyrich F, de Bruin H A R, Meijninger W M L, Schipper J W and Lohse H 2002 Results from one-year continuous operation of a large-aperture scintillometer over a heterogeneous land surface *Boundary-Layer Meteorol.* **105** 85-97

IOP Conf. Series: Earth and Environmental Science 1 (2008) 012027 doi:10.1088/1755-1307/1/1/012027

Long-term observations of the urban mixing-layer height with ceilometers

S Emeis[1,3], K Schäfer[1], C Münkel[2]

[1] Institut für Meteorologie und Klimaforschung – Atmosphärische Umweltforschung, Forschungszentrum Karlsruhe, Garmisch-Partenkirchen, Germany

[2] Vaisala GmbH, Hamburg, Germany

stefan.emeis@imk.fzk.de

Abstract. Urban air quality assessment requires the knowledge of the temporal and spatial structure of the mixing-layer height (MLH), because this variable controls the vertical space for rapid mixing of near-surface pollutants. Because MLH is a consequence of vertical temperature and moisture profiles in the lower atmosphere, remote sensing is a suitable tool to monitor MLH. Two ceilometers, a Vaisala LD40 and a Vaisala CL31, have been run for many months in the German city of Augsburg to observe the vertical aerosol distribution. Wind and temperature profile information have been obtained for the same period from sodar observations. This paper compares the MLH determined from the optical backscatter intensity received by the two ceilometers among each other and with the MLH derived from the acoustic backscatter intensity and the variance of the vertical wind component from sodar measurements.

1. Introduction

The mixing layer height (MLH) is assumed to be a key parameter for the characterisation of air pollution, because this variable controls the vertical space for rapid mixing of near-surface pollutants. The determination and modelling of the MLH has therefore found considerable interest in the recent decade [1], especially in urban meteorology. With today's availability of ground-based remote sensing devices for monitoring the structure of the atmospheric boundary layer (ABL) it has been shown that the ABL sometimes exhibits multiple layering (e.g. internal boundary layers, near-surface inversions and residual layers at night-time and in the morning hours)[2]. It was demonstrated that the lowest stable layer or inversion limits the vertical exchange of primary pollutants emitted at or near the surface [3] and thus controls the near-surface pollutant concentrations. Acoustic remote sensing has also been used to derive climatologic statistics of MLH for towns ([4], [5]) and wind and turbulence profiles over towns [6]. Multiple inversions within a few hundred meters above ground have been detected both with optical and acoustic remote sensing in an Alpine valley [7]. MLH information is also necessary for special kinds of satellite data interpretation, e. g. the retrievals of optical depths for the particle concentration near the surface [8].

[3] To whom any correspondence should be addressed

MLH is supposed to vary on the regional scale due to the weather conditions with smaller variations due to land surface properties types. Therefore it is interesting to analyze whether a town like Augsburg (about 265 000 inhabitants in 2007) influences and modifies MLH. Augsburg is situated about 60 km west-north-west of Munich in a rural area on the river Lech. The Lech is flowing north from the Alps (about 100 km south of Augsburg) towards the Danube in a shallow valley, which is about 10 km wide and about 100 m deep. Under synoptically calm conditions with weak pressure gradients we observe light winds from the south at night and from the north to north east at daytime. At the presence of stronger pressure gradients the winds do not deviate much from the large-scale synoptic winds. Prevailing wind direction of such cases is from the south west.

For the investigation two ceilometers were placed at the northern edge of the town, in the town centre, and the sodar at the southern edge of the town. The investigation is complicated by the fact that two different ceilometers have been used in this measurement experiment. Therefore, in a comparison experiment, all three remote sensing instruments have been situated side by side for four weeks. The enhancement of the methodology for the determination of MLH from an integration of these different remote sensing methods is a further objective of this experiment.

The paper presents results and conclusions from this intercomparison exercise.

2. Methodology

In the following subsections we will describe the algorithms with which the MLH is derived from ground-based remote sensing data. We will denote with the letter H and an attached number certain derived heights which are related to inversions and the MLH; while we will use the variable z to denote the normal vertical coordinate.

2.1. Acoustic detection of MLH

Beyrich [9] has listed possible ways to analyze MLH from acoustic backscatter intensities measured by a sodar. This algorithm has been enhanced in [4] by simultaneously using acoustic backscatter intensity and the variance of the vertical velocity component σ_w. In [7] the algorithm has been extended for the detection of multiple inversions present simultaneously. The algorithm determines (Figure 1):

- the height of a turbulent layer (H1) characterised by high acoustic backscatter intensities R(z) due to thermal fluctuations (thus having a high σ_w),
- several lifted inversions (H2_n) characterized by secondary maxima of acoustic backscatter due to a sharp increase of temperature with height and simultaneously low σ_w, and
- the height of a surface-based stable layer (H3) characterised by high backscatter intensities due to a large mean vertical temperature gradient starting directly at the ground and having a low σ_w.

The mixing layer height (MLH) is defined as the minimum of the heights H1, the lowest H2 (i.e. H2_1), and H3. Acoustic remote sensing by a SODAR also yields wind fields that in principle could be used for determination of the mixing layer height, too. Because the wind information above the inversion is not regularly available from SODAR measurements, wind data have not been included into the scheme for the determination of the mixing layer height.

The height H1 corresponds to a sharp decrease of the acoustic backscatter intensity R(z) with height z usually indicating the top of a turbulent layer:

$$H1 = z, \text{ if } (R(z) < 88 \text{ dB and } R(z+1) < 86 \text{ dB and } R(z+2) < 84 \text{ dB}). \quad (1)$$

The dB values are derived from an arbitrary scale because the received acoustic backscatter intensities from a SODAR cannot be calibrated. Therefore, the R(z) values are specific for the type of SODAR used. Therefore, the numbers in (1) can be taken for illustration purposes only.

Elevated inversions are diagnosed from secondary maxima of the backscatter intensity that are not related to high turbulence intensities. For elevated inversions an increase in backscatter intensity below a certain height z = H2 and a decrease above is stipulated while the turbulence intensity is low:

$$H2\_n = z, \text{ if } (\partial R/\partial z|_{z+1} < 0 \text{ and } \partial R/\partial z|_{z-1} > 0 \text{ and } \sigma_w < 0.70 \text{ ms}^{-1}) \quad (2)$$

for n = 1, …, N. In [7] N was chosen to be five. The determination of the height of the stable surface layer H3 is started if the backscatter intensity in the lowest range gates is above 105 dB while σ_w is smaller than 0.3 ms^{-1}. The top of the stable layer H3 is at the height where either the backscatter intensity sinks below 105 dB or σ_w increases above 0.3 ms^{-1}.

$$H3 = z, \text{ if } (R(z) > 105 \text{ dB and } R(z+1) < 105 \text{ dB and } \sigma_w(z) < 0.3 \text{ ms}^{-1}) \text{ or if } (\sigma_w(z) < 0.3 \text{ ms}^{-1} \text{ and } \sigma_w(z+1) > 0.3 \text{ ms}^{-1} \text{ and } R(z) > 105 \text{ dB}) \quad (3)$$

The dB values in (3) are given for the same illustrative purpose as in (1).

The search for H2 to H3 can be integrated into one search algorithm. If a surface-based stable layer with height H3 has been detected, the search for the first height H2_1 is started at the first range gate above H3, otherwise it starts at the lowest range gate. The search for elevated inversion is stopped when N elevated inversions have been found or the highest range gate of the available data has been reached.

The threshold values for σ_w used in (2) and (3) have been determined by optimizing the automatic application of the detection algorithm. In doing so it turned out that no lifted inversions occurred while the variance σ_w was lower than 0.7 ms^{-1}. Thus this threshold made it possible to distinguish between inversions and elevated layers of enhanced turbulence. Further, the variance σ_w in nocturnal stable surface layers was always below 0.3 ms^{-1}, so that this threshold made it possible to differentiate between nocturnal stable surface layers and daytime super-adiabatic surface layers although both types of surface layers yield more or less the same level of backscatter intensity. Finally MLH from the acoustic remote sensing is determined as the minimum of H1, H2_1, and H3:

$$MLH_{ac} = \min (H1, H2\_1, H3) \quad (4)$$

2.2. Optical detection of MLH

The heights of the near surface aerosol layers (H4_n) can be analysed from the optical vertical backscatter profiles. Here the gradient method is used. In this method the heights H4_n are characterized by minima of the vertical gradient (the term 'gradient minimum' is used here to denote the most negative value of the gradient, see also Fig. 1) of the optical attenuated backscatter intensity (B(z)). [10], [11] and [12] discussed the principle of the retrieval of this height. In [7] this approach has been refined and extended to enable the calculation of up to n=5 lifted inversions.

Prior to the determination of gradient minima the overlap and range corrected attenuated backscatter profiles have to be averaged over time and height to suppress noise generated artefacts (see also section 3.1). Therefore the H4 values are determined in a two-step procedure. Between 140 m and 500 m height sliding averaging is done over 15 min and a height interval Δh of 80 m. In the layer between 500 and 2000 m Δh for vertical averaging is extended to 160 m. Two additional parameters have been introduced to further reduce the number of false hits. The minimum accepted attenuated backscatter intensity B_{min} right below a lifted inversion is set to $200*10^{-9}$ $m^{-1}sr^{-1}$ in the lower layer and $250*10^{-9}$ $m^{-1}sr^{-1}$ in the upper layer. Additionally the vertical gradient value $\partial B/\partial z_{max}$ of a lifted inversion must be more negative than $-0.30*10^{-9}$ $m^{-2}sr^{-1}$ in the lower layer and more negative than $-0.30*10^{-9}$ $m^{-2}sr^{-1}$ in the upper layer.

If B(z) denotes the measured attenuated backscatter intensity in the height z above ground averaged over time and height and Δh is the height averaging interval, then the gradient $\partial B/\partial z$ in the height z is calculated as

$$\partial B/\partial z|_z = (B(z+\Delta h/2) - B(z-\Delta h/2)) / \Delta h \quad (5)$$

A gradient minimum is characterized by a change of sign from minus to plus of the second derivative of B(z). The height interval under examination is searched from bottom to top for these gradient minima H4_n. The second derivative of B(z) in the height z is

$$\partial^2 B/\partial z^2|_z = (\partial B/\partial z|_{z+\Delta h/2} - \partial B/\partial z|_{z-\Delta h/2}) / \Delta h. \quad (6)$$

Thus H4_n is detected in a height z if the second derivative of B(z) one range gate below z is not positive, if the second derivative of B(z) in the height z is positive, and if the abovementioned false hit conditions are fulfilled:

$$H4\_n = z, \text{ if } \partial^2 B/\partial z^2|_{z-1} \leq 0 \text{ and } \partial^2 B/\partial z^2|_z > 0 \text{ and } B(z-\Delta h/2) \geq B_{min} \text{ and } \partial B/\partial z|_z \leq \partial B/\partial z_{max} \quad (7)$$
for n = 1, …, 5.

The MLH from optical remote sensing is taken as the lowest height H4 (see Figure 1):

$$MLH_{op} = H4\_1 \quad (8)$$

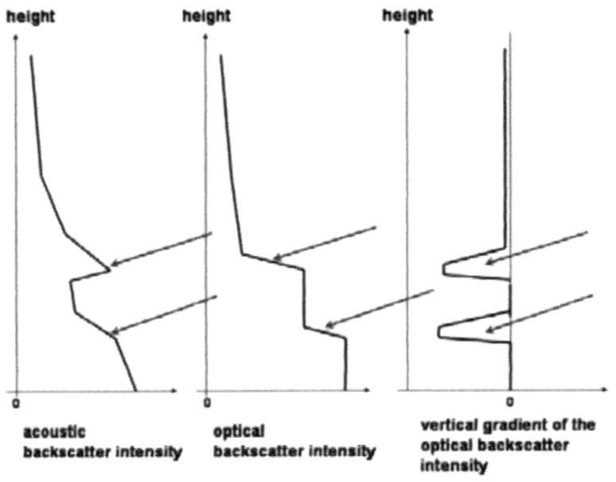

Figure 1. Schematic comparison of MLH determination algorithms in the case of a nocturnal stable surface layer and an elevated inversion above it. Left: vertical profile of acoustic backscatter intensity, middle: vertical profile of optical backscatter intensity, right: vertical profile of the vertical gradient of the optical backscatter intensity. Lower arrows mark the position of the height H3 (left) and H4_1 (middle and right). Upper arrows mark the heights H2_1 (left) and H4_2 (right). From [7].

The comparison between the two schemes for the determination of MLH from acoustic and optical backscatter intensities reveals an interesting difference which should be noted carefully (Figure 1): While we detect H1 and H3 from the acoustic backscatter intensity itself and H2 from the first derivative of this backscatter intensity, we need the first and the second derivative of the optical backscatter intensity (but not the optical backscatter intensity itself) to determine H4. This difference comes from unequal scattering processes for acoustic and optical waves: Acoustic waves are scattered at atmospheric refractivity gradients and thus at temperature gradients ([13]) while optical waves are scattered at small particles. Therefore the optical backscatter intensity is proportional to the aerosol concentration itself. The MLH on the other hand, which we desire to derive from these backscatter intensities, is in both cases found in heights where we have vertical gradients of the temperature and of the aerosol concentration. Therefore, in principle, the vertical distribution of the acoustic backscatter intensity should look very much alike the vertical distribution of the vertical gradient of the optical backscatter intensity.

The analysis of the sodar data and the ceilometer data can be combined to one single MLH information by forming the minimum from (4) and (8):

$$MLH = \min (MLH_{ac}, MLH_{op}) \tag{9}$$

3. Instrumentation

3.1. Ceilometer

Two different ceilometers [10] have been used. One is a double-lens Vaisala LD40 (Figure 2, right) using infrared light with a wavelength of 855 nm, a height resolution of 7.5 m, and a maximum range of 13000 m. The other is the one-lens CL31 (Figure 2, left) using infrared light of 910 nm, a height resolution of 5 m, and a maximum range of 7500 m. The slightly different wavelengths and the unequal optical construction of both instruments lead to differences in the recorded optical backscatter intensities (see next section and Figure 4). Both ceilometers were sounding at zero zenith angles.

For both ceilometers averaging over 2 min and 60 m is not sufficient to suppress the detection of noise generated artefacts as MLH; 4 min and 120 m averaging works well for convective layers below 800 m but more reliable results are achieved with a time averaging interval of at least 10 min.

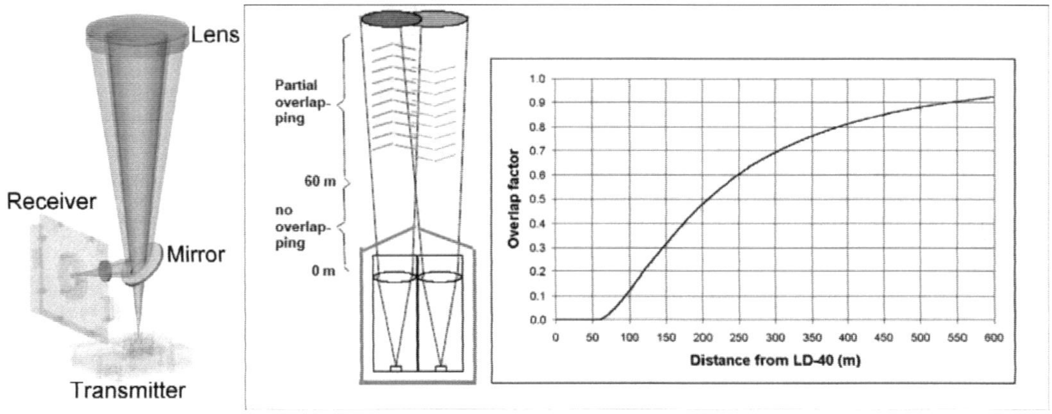

Figure 2. CL31 and LD40 ceilometer schematics

Figure 3. METEK DSDR3x7 sodar.

3.2. Sodar

The sodar (Figure 3) used in this study is a three-antenna METEK DSDR3x7 sodar [14] using an acoustic frequency of 1670 Hz, a height resolution of 30 m, and a maximum range of 1255 m. Two antennas were tilted at 20 degrees zenith angle, the third antenna was looking vertical.

4. Results

Some results from the intercomparison period in May 2007 will be displayed here to show the ability of the three remote sensing instruments. Figure 4 presents a comparison of the optical backscatter intensities from the two ceilometers. The largest differences are found in the lower 200 m close to the ground due to the different optical paths in the two instruments (one optical axis in the CL31 versus two optical axes in the LD40). It turns out, as expected, that the one-lens system (the CL31) is much better suited to investigate the aerosol distribution in the lower layers close to the ground. But also in greater heights some differences can be found, which probably have to be credited to the slightly different wavelengths of the two instruments.

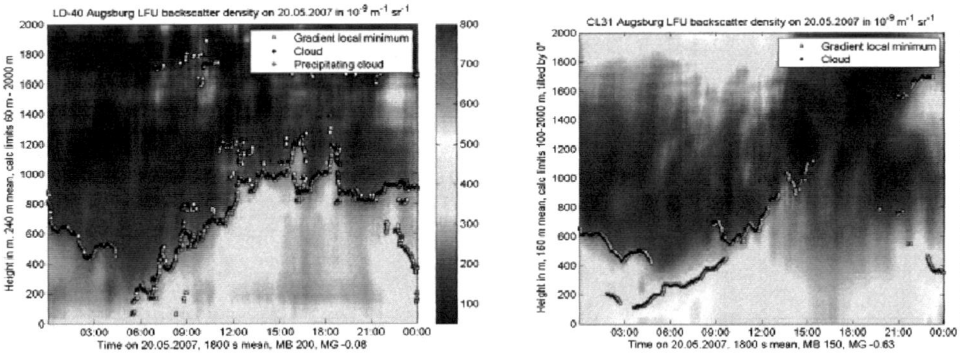

Figure 4. Comparison of optical backscatter intensity for heights up to 2000 m above ground from the Vaisala LD40 (left) and from the Vaisala CL31 (right) during the intercomparison exercise on 20 May 2007 in Augsburg, Germany.

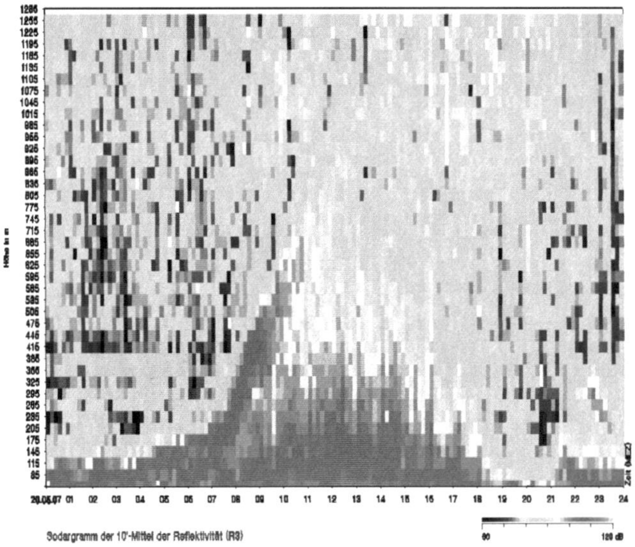

Figure 5. Acoustic backscatter intensity from the METEK DSDR3x7 sodar for heights up to 1270 m above ground during the during the intercomparison exercise on 20 May 2007 in Augsburg, Germany.

Figure 5 displays the acoustic backscatter intensity recorded by the vertical antenna of the sodar. The other two antennas are used together with the vertical antenna in order to measure the three-dimensional wind vector from the radial Doppler shifts of the backscattered sound pulses.

In Figure 6 the mixing layer heights derived from the optical data of the two ceilometers are plotted. For large parts of the day both instruments give nearly the same results. In the morning hours it can clearly be seen that the CL31 is much better able to detect the shallow mixing layer height below 200 m. On the other hand, the LD 40 is a little bit better in following the top of the convective boundary layer in the later afternoon.

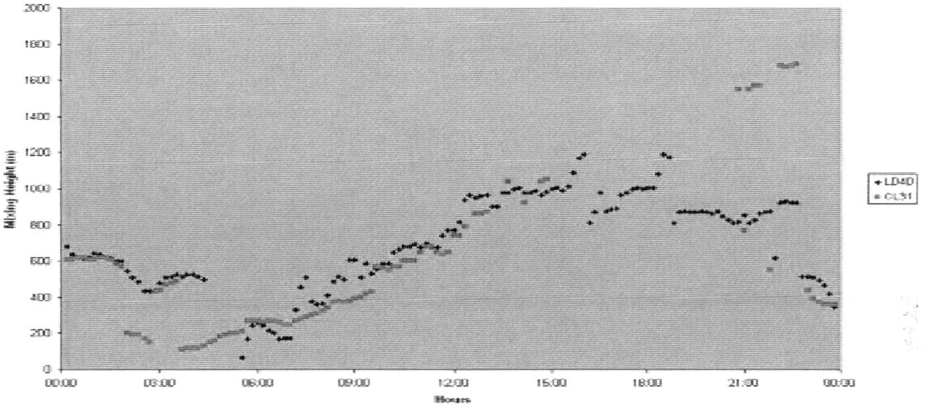

Figure 6. MLH derived from the Vaisala LD40 ceilometer (blue dots) and from the Vaisala CL31 ceilometer (purple dots) during the during the intercomparison exercise on 20 May 2007 in Augsburg, Germany. Sunrise was at 03:31, sunset at 18:56.

In Figure 7 the MLH from the one-lens ceilometer (CL31) and sodar data are compared. Among the plotted sodar data 'top turb.' denotes H1 from eq. (1), 'stable layer' H3 from eq. (3), and 'MLH' MLH_{ac} from eq. (4). Among the ceilometer data 'MLH 1' and 'MLH 2' denote the first two H4_n from eq. (7).

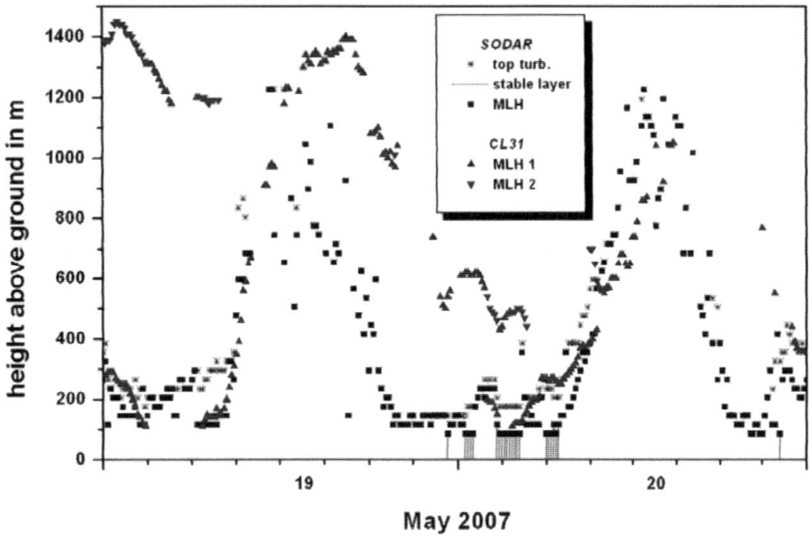

Figure 7. Mixing-layer height in Augsburg, Germany, on 19 and 20 May 2007 from CL31 (triangles) and sodar (Squares, asterisks, and vertical lines) measurements. See text for further details.

We see that both types of instruments (the acoustic and the optical one) generally give the same results. But there are a few differences which are worth to be looked at in a little more detail.

- On the morning of the first day plotted in Figure 7 two heights are detected by the ceilometer (between 100 and 300 m and between 1200 and 1400 m) while the sodar, due to its maximum vertical range, can only detect the lower one. The layer between these two heights is the residual layer left over from the convective boundary layer of the day before. The same partially happens in the next night (in the centre of the Figure).
- On the second day the rise of the acoustically detected MLH is much faster than the rise of the optically detected MLH. This indicates that the vertical aerosol distribution needs some time to follow the thermal structure of the ABL.

Both facts have already been documented in [2].

5. Discussion

Details of the optical construction (one lens versus two lenses) of ceilometers and the choice of the wavelength have some influence on the received optical backscatter intensity. Especially the CL31 ceilometer used in this study has advantages when monitoring shallow nocturnal surface layers and mixing layer heights.

Acoustic remote sensing with sodars allows a good detection of shallow nocturnal mixing layers but sodars have the decisive disadvantage that their vertical range is rather limited. On the other hand, a sodar permits the distinction between stable surface layers, convective boundary layers and the top of turbulent layers.

Acoustic and optical remote sensing mostly deliver complementing results. When they both show the same process, e.g. the start of the development of the convective boundary layer in the morning, they allow the analysis of how fast the vertical aerosol distribution is able to follow the developing thermal structure of the boundary layer.

The intercomparison experiment showed that the intended derivation of spatial structures in the urban mixing layer height over Augsburg will remain difficult from these three instruments used here. For this purpose identical instruments should be used at the different sites. This has been realized in the meantime by substituting the LD40 with a CL31 ceilometer. Results are forthcoming.

6. Conclusions and Outlook

Remote sensing with ceilometers and sodars is a good way to monitor the vertical structure of the ABL. These instruments deliver data with high temporal and vertical resolution.

The combined application of acoustic and optical remote sensing allows an assessment of the overall vertical structure of the ABL by indicating the occurrences of nocturnal stable surface layers, convective boundary layers and residual layers.

The comparison of the increase of the height of the convective boundary layer in the morning from acoustic and from optical remote sensing gives some insight in the dynamics of vertical turbulent diffusion in the developing convective boundary layer.

The delay in the development of the vertical aerosol distribution with respect to the development of the thermal structure of the boundary layer is an important result which has to be kept in mind if only ceilometers are used to monitor the mixing layer height.

The deployment of a radio-acoustic sounding system (RASS) could lead to an even more complete monitoring of the structure of the boundary layer since a RASS delivers the temperature profile itself. The then remaining task would be a simultaneous vertical profiling of the moisture content of the boundary layer. But the existing instruments for the measurement of the moisture profile in the atmosphere (Raman-Lidar, water vapour lidar, micro-wave radiometer) still do not have the same vertical resolution and lower range gate which ceilometers and sodar have.

7. Acknowledgements
The measurements in Augsburg were made in cooperation with Jürgen Schnelle-Kreis of the Bayerisches Institut für Angewandte Umweltforschung und –technik (BIFA), Augsburg and Jan Bernkopf of the Bayerisches Landesamt für Umwelt (LfU), Augsburg.

References
[1] Piringer M, Joffre S, Baklanov A, Christen A, Deserti M, De Ridder K, Emeis S, Mestayer P, Tombrou M, Middleton D, Baumann-Stanzer K, Dandou A, Karppinen A and Burzynski J 2007 The surface energy balance and the mixing height in urban areas – activities and recommendations of COST-Action 715 *Bound.-Lay. Meteorol.* **124** 3-24

[2] Emeis S and Schäfer K 2006 Remote sensing methods to investigate boundary-layer structures relevant to air pollution in cities *Bound-Lay. Meteorol.* **121** 377-85

[3] Schäfer K, Emeis S, Hoffmann H and Jahn C 2006 Influence of mixing layer height upon air pollution in urban and sub-urban areas *Meteorol. Z.* **15** 647-58

[4] Emeis S and Türk M 2004 Frequency distributions of the mixing height over an urban area from SODAR data *Meteorol. Z.* **13** 361-67

[5] Lokoshchenko M A 2002 Long-Term Sodar Observations in Moscow and a New Approach to Potential Mixing Determination by Radiosonde Data *J. Atmos. Oceanic Technol.* **19** 1151-62

[6] Emeis S, Baumann-Stanzer K, Piringer M, Kallistratova M, Kouznetsov R and Yushkov V 2007 Wind and turbulence in the urban boundary layer – analysis from acoustic remote sensing data and fit to analytical relations *Meteorol. Z.* **16** 393-406

[7] Emeis S, Jahn C, Münkel C, Münsterer C and Schäfer K 2007 Multiple atmospheric layering and mixing-layer height in the Inn valley observed by remote sensing *Meteorol. Z.* **16** 415-24

[8] Sarigiannis D A., Soulakellis N, Schäfer K, Tombrou M, Sifakis N I, Assimakopoulos D, Lointier M, Dandou A and Saisana M 2002 ICAROS: An Integrated Computational Environment for the Assimilation of Environmental Data and Models for Urban and Regional Air Quality *Int. J. Water, Air, and Soil Pollution: Focus* **2** 641-54

[9] Beyrich F 1997 Mixing height estimation from sodar data - a critical discussion *Atmos. Environ.* **31** 3941-53

[10] Münkel, C., J. Räsänen, 2004: New optical concept for commercial lidar ceilometers scanning the boundary layer, Proceedings of SPIE, Vol. 5571, 364–74

[11] Schäfer K, Emeis S M, Rauch A, Münkel C and Vogt S 2004 Determination of mixing-layer heights from ceilometer data. In: *Remote Sensing of Clouds and the Atmosphere IX*, Klaus Schäfer, Adolfo Comeron, Michel Carleer, Richard H. Picard, Nicolas Sifakis (eds.):, Proc. SPIE, Bellingham, WA, USA, **5571** 248-259

[12] Schäfer K, Emeis S, Junkermann W and Münkel C 2005 Evaluation of mixing layer height monitoring by ceilometer with SODAR and microlight aircraft measurements. In: *Remote Sensing of Clouds and the Atmosphere X*, Klaus Schäfer, Adolfo T. Comeron, James R. Slusser, Richard H. Picard, Michel R. Carleer, Nicolaos Sifakis (eds.), Proc. SPIE, Bellingham, WA, USA, **5979**, 59791I-1 – 59791I-11

[13] Neff, W.D., R.L. Coulter, 1986: Acoustic remote sensing. In: Lenschow D H (ed.) *Probing the Atmospheric Boundary Layer*, Amer. Meteor. Soc., Boston, MA, 201–39

[14] Reitebuch O und Emeis S 1998 SODAR-measurements for atmospheric research and environmental monitoring *Meteorol. Z.* **7** 11-14

IOP Conf. Series: Earth and Environmental Science 1 (2008) 012028 doi:10.1088/1755-1307/1/1/012028

Use of Sodar Data for Analysis of Relations between Concentrations of Minor Atmospheric Gases

Lokoshchenko M.A.[1], Elansky N.F.[2]

[1]Department of Meteorology and Climatology, Geography Faculty, Lomonosov Moscow State University. 119991, Lengory, Moscow, Russia
Phone: +7-495-9394284; Fax: +7-495-9392479; E-mail: loko@geogr.msu.su
[2]Obukhov Institute of Atmospheric Physics of the Russian Academy of Sciences. 119017, 3, Pyzhevsky lane, Moscow, Russia.

Abstract. The statistical relations between surface concentrations of minor atmospheric gases have been studied. As it was found a relation between concentrations of gases having the same sources (e.g., fuel burning for NO and CO) is close to linear trend. On the other hand, two species which are both a source and a sink to each other (e.g., O_3 and NO) demonstrate hyperbolical regression line. Finally, concentrations of two species one of which is partially a source for another (e.g., NO and NO_2) demonstrate exponential (logarithmic) type of trend. However, real relations between concentrations of two minor gases for any day represent non-homogeneous and complicated closed graphs like Carno cycle. As it was found a structure of these graphs during a day may be successfully explained by dynamics of the thermal stratification with the use of sodar data. Two examples of this analysis have been demonstrated.

1. Introduction

A sodar record ('echogram') represents the oldest kind of the acoustic remote sensing data. Since McAllister [1] a lot of works was devoted to studying of the thermal stratification's dynamics on a base of analysis of echo-signal intensity in the 'time-height' co-ordinates. Some of them have been collected and discussed in [2-4]. Now sodars are used mainly for profiling of wind speed, wind direction and some turbulent parameters in the Atmospheric Boundary Layer (ABL). However, classic sodar echogram still remains a useful tool for some applications, e.g. – for the air pollution monitoring.

Studying of statistical relations between surface concentrations of minor atmospheric gases is an important task. It leads to better understanding of their dynamics and of its influence on human health. It should be noted that a total effect of various air pollutants to a health sometimes isn't equal to summarized separate actions of each ones. For instance, combined damage to human lungs from sulfur dioxide and smoke particles is much more than separate effects from both these pollutants. Other words, a synergism between SO_2 and a smoke takes place when they present in the air simultaneously (e.g., [5]). Thus, an importance of this studying is evident. However, relations between concentrations of atmospheric pollutants from one minute to another are seem to be almost not investigated before.

In Moscow the measurements of minor atmospheric species (O_3, NO, NO_2, CO, SO_2, aerosol particles, etc.) on the level of 4 m above ground have been started in 2002 at Meteorological Observatory of Lomonosov Moscow State University (MSU). Special ecological station has been created there by common efforts of Obukhov Institute of Atmospheric Physics and Geography Faculty of MSU. Since that time measurements of surface concentrations of air pollutants are carried out continuously with one minute temporal resolution. This station is situated at the south-western periphery of Moscow city and operates accordingly to standards of the Global Atmospheric Watch Service. The continuous monitoring of the ABL structure is carried out at this site by use of two sodars: vertical sodar "ECHO-1" of GDR production, and Doppler sodar "MODOS" produced by METEK firm (Germany). Both sodars operate continuously in close vicinity of the station – at a distance of 70 m from chemical sensors. Both wind profiles, and the thermal stratification are detected by them synchronously. The sodar data there have been successfully used for explanation of dynamics of various trace gases [6-9, etc.]

2. General particularities of pair relations between minor gases.

Let us discuss now main features of pair statistical relations between minor atmospheric gases – ozone, nitrogen oxides and carbon oxide. A total kind of regression line between two species mainly depends on interrelations between them as a source and a sink to each other.

One possible case is a pair of species which have the same source (e.g., fuel burning). Typical example of such gases are NO and CO. Both of them are primary pollutants produced mainly by internal-combustion engines of automobiles. As it is seen in Fig.1, these pollutants demonstrate clear linear relation during one full day. The linear correlation coefficient is 0.96. This day as an example is quite representative. At any other day the statistical relation between NO and CO as a rule is close to linear.

It should be noted as well that linear trend intersects the vertical axis in Figure 1 at the value close to 1.1 ppm whereas the least one minute CO concentration during this day was only a bit below 1.0 ppm. The minimal measurable value of this gas is much less – 0.05 ppm.

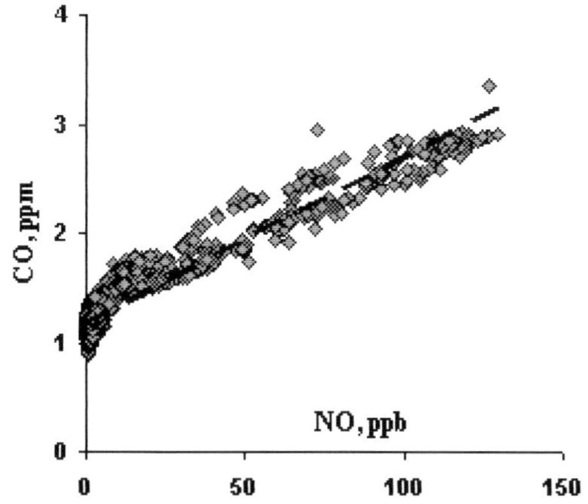

Figure 1. Relation between NO and CO on August 13th, 2002. Dashed line is linear trend. Points indicate surface concentrations per every minute.

Thus, concentrations nearly of 1.0 ppm represent mean conditions of a big city. They are close to upper limit of background levels of CO in European Russia – 0.7 ppm [10].

Air pollutants which can be both sources and sinks to each other demonstrate principally another kind of statistical relation. For example, O_3 and NO take part in complicated system of photochemical reactions (e.g., [11]). As it is known ozone and nitrogen oxides generate together photochemical equilibrium in the atmosphere. As a consequence, a feedback between O_3 and NO is clear in Fig.2 so that the more is one the less is another. Evidently, this type of relation may be successfully described by hyperbolic trend.

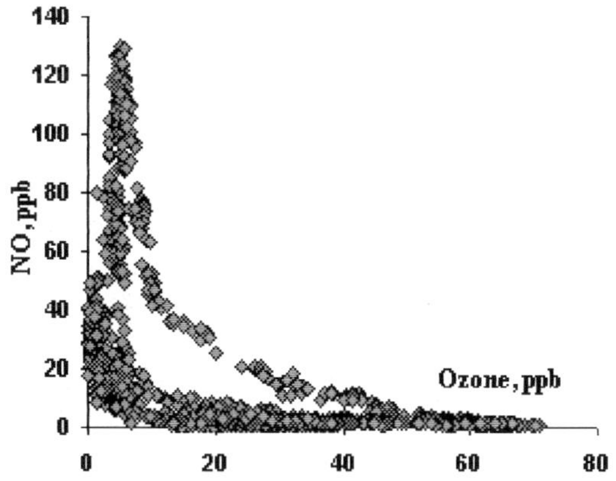

Figure 2. Relation between O_3 and NO on August 13th, 2002. Points indicate surface concentrations per every minute.

Finally, let us consider one more kind of relation – between gases one of which is a partial source of another. A clear example of it is NO and NO_2. As it is known the most part of nitrogen dioxide in the air is generated from NO oxidation and only a few of it is produced by automobile engines directly. As one can see in Fig.3 the statistical relation at this case is direct but complicated and non-linear. Points create two separate accumulations divided by blank space in which there are no data. Each of both accumulation zones can be successfully derived by logarithmical (exponential) kind of trend.

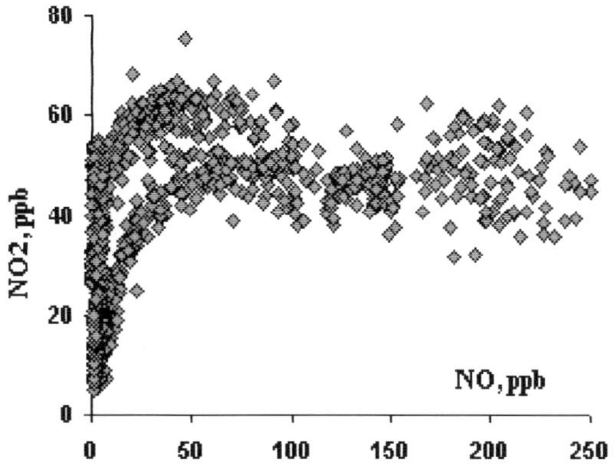

Figure 3. Relation between NO and NO_2 on May 08th, 2002. Points indicate surface concentrations per every minute.

One of them corresponds to more values of NO_2, and another – to less ones. Non-linear relation between these gases may be explained by specific of their interaction.
For example, in time of NO reducing in the ground air layer one part of it disperses into upper air layers whereas another part is spent to oxidation from NO into NO_2. Thus, the surface concentration of NO_2 reduces due to dispersion more slowly than the one of NO because of its additional source – generation from NO. As a consequence, the statistical relation is nonlinear.
Although a total tendency of statistical relation is clear a structure of relation is complicated everywhere. In all three figures points are accumulated into separate zones (each point indicates mean value during one minute). These zones are divided from each other by blank space where data are absent. Prima facie, a structure of any relation seems to be chaotic. So, let us apply the acoustic remote sensing data about the ABL thermal stratification to explain it in details.

3. Analysis of the relation's structure with the use of sodar data.
As it is known classic sodar record of echo-signal indicates dynamics of the ABL thermal stratification. It allows to recognize a type of stratification with an accuracy of the vertical thermal gradient sign and to detect layers of turbulent structures connected with inversions [3, 4, etc.].
Sodar "ECHO-1" operates at the Meteorological observatory of Lomonosov Moscow State University since 1988. The operation frequency is 1666.7 Hz, the vertical range of sounding is in limits since 25 till 800 m, and spatial resolution is 12.5 m.
Let us discuss thermal stratification on August 13th, 2002. In Fig.4 the sodar record during period from 6:00 a.m. to 04:00 p.m. at this day is presented.

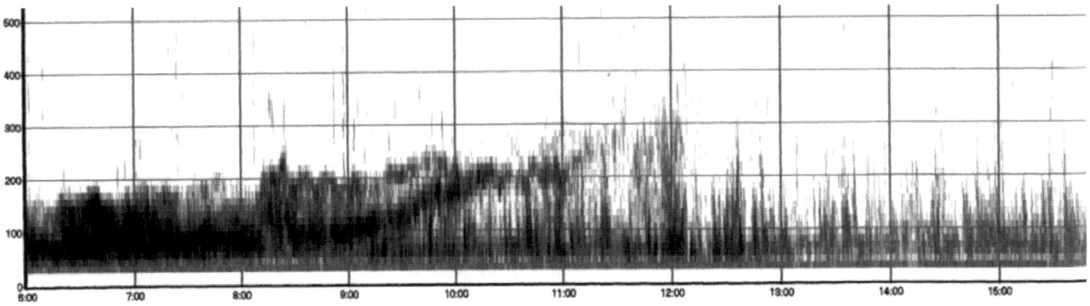

Figure 4. Sodar record for August 13th, 2002. Moscow University, Moscow.

As one can see the dynamics of thermal stratification was typical for summer day. The surface inversion after sunrise has been lifted above ground and transformed into the elevated one not earlier than at 07 a.m. Then it existed above a layer with unstable stratification until 12:05 a.m. After noon the unstable stratification is detected by sodar data in all the range – up to 800 m. After 05:30 p.m. thermal convection connected with vertical 'plumes' (typically convective structures) goes to nothing, and slightly-stable stratification appears.
So, results of measurements of trace gases at this day have been analyzed separately for four temporal intervals: from 00:00 to 07:00 a.m. (surface inversion); from 07:01 to 12:05 a.m. (elevated inversion); from 00:05 to 05:30 p.m. (unstable stratification) and from 05:31 to 11:59 p.m. (slightly-stable stratification).
In Fig.5 the daily course of NO and O_3 for this day is shown. As one can see the ozone surface concentration has two maxima – the main in the afternoon and the secondary at night which is specific feature of a big city. On the contrary, maximal concentrations of NO occur in morning and in evening. So, daily courses of both minor gases are quite typical for summertime [8]. The dynamics of NO is shown by various markers depending on a type of thermal stratification. Additionally, a time of elevated inversion's existence were divided there into two periods: the early morning from 07:01 to 09:47 a.m. and the late morning from 09:48 to 12:05 a.m. The temporal margin between them corresponds to a moment when the surface concentration of NO was maximal (129 ppb at 09.47) during morning rush-hour.

This additional margin may be connected both with dynamics of thermal stratification as well (because elevated inversion becomes weaker and gradually goes to nothing in late morning), and with dynamics of traffic intensity in the city (which reduces at the end of rush-hour).

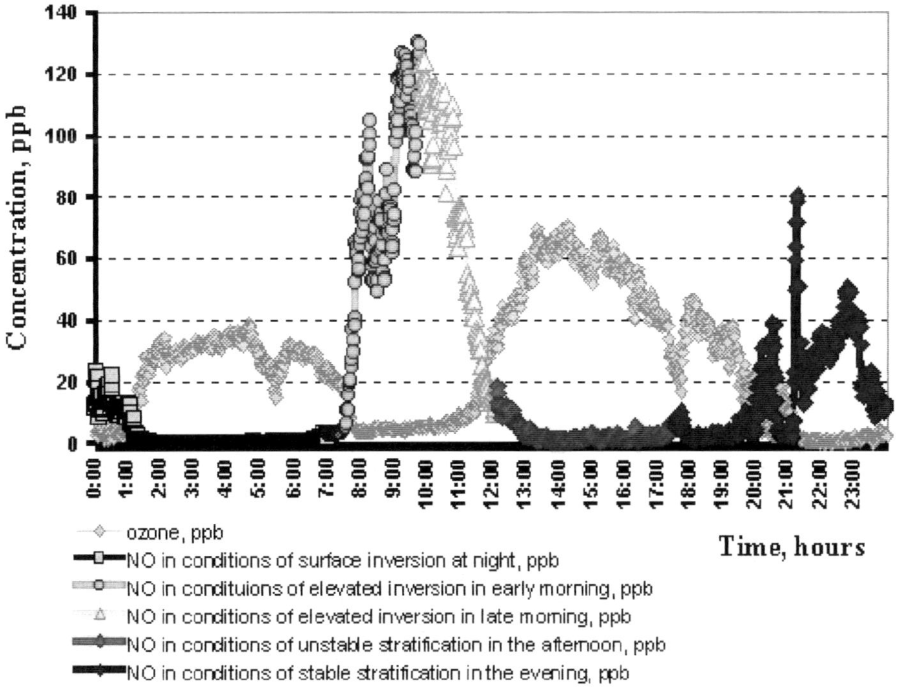

Figure 5. The daily course of O_3 and NO surface concentrations in August 13th, 2002. The dynamics of nitrogen oxide is shown by different markers which indicate a type of the thermal stratification.

So, let us return now to Figure 2 and analyze this relation with the account of the thermal stratification type. As it is seen in Figure 6, a structure of this relation strongly depends both on traffic intensity, and on thermal stratification which determines an intensity of vertical mixing and, hence, an intensity of air pollution dispersion into upper air layers. Nocturnal data about NO (squares) indicate its extremely low contents because traffic is almost absent at this time. Then, in morning, values of NO become much more because of high-intense traffic and elevated inversion which prevents vertical mixing and upward dispersion of this gas. At the same time concentrations of O_3 fall due to its absorption by ground. As a result one hyperbolic branch (circles) is clear.

Then, a reducing of ozone goes to nothing because ultraviolet radiation grows in morning and the additional source of ozone appears, that it photo-dissociation of NO_2 molecules. This source compensates the ozone consumption to absorption by ground, so that hyperbolical line asymptotically approaches to extremely small value of O_3 – nearly of 3.5 ppb.

In late morning when traffic intensity reduces and elevated inversion begins to weaken points in Figure 6 go back. Surface concentration of NO falls and the one of O_3, vice versa, grows. As a result, the second hyperbolical line (triangles) appears. It should be noted that it replaced in these co-ordinates to the right as comparable as the first one (circles) due to more background of ozone in late morning. At noon, when elevated inversion finally disappears, the ozone concentration reaches 35 ppb. As it was seen in Figure 4, the sodar record indicates strong thermal convection (so-called 'plumes') in the afternoon. At this time downward flow of ozone becomes intensive and its concentration near the ground grows up to 70 ppb. Simultaneously the NO concentration reduces till extremely low values due to intensive vertical mixing.

Finally, slightly-stable stratification in the evening leads to weakening of the vertical mixing. Besides, the ozone generation from NO_2 falls at the same time because short-wave ultraviolet radiation reduces.

As a result of both effects, the third quasi-hyperbolical line (dark rhombuses) appears in Figure 6. The surface concentration of NO grows again, and the one of O_3, vice versa, falls up to the lowest values during a day (less than 1 ppb).

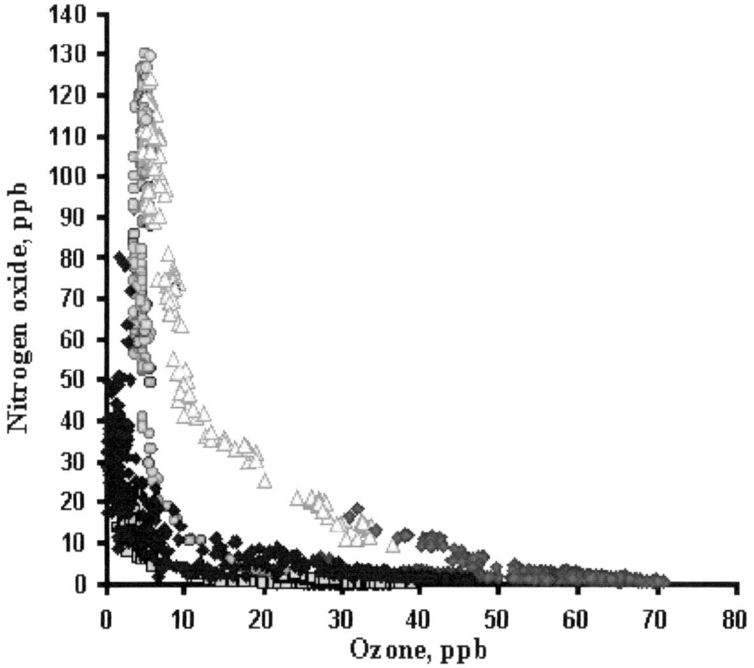

Figure 6. Statistical relation between ozone and NO in August 13th 2002 with the account of the ABL thermal stratification. Markers are the same as in Fig.5.

Thus, a structure of the relation between O_3 and NO is not disorderly. A full cycle of changes of both gases during a day represents closed plane figure – like Carno cycle in thermodynamics. Besides a total tendency, one can see in Fig.6 a scatter of separate points around hyperbolical lines and partial superposition of different markers to each other that means back motions, opposite to total tendency. Probably, it is a result of the atmospheric turbulence and of irregular intensity of traffic in the city.

As we can see the dynamics of surface concentrations of both minor gases during a day may be explained with the account of three factors which influence on it: thermal stratification, photochemical generation of ozone and intensity of automobile traffic (which is the main source of NO). However, an influence of the thermal stratification on dynamics of O_3 and NO seems to be deciding among others. It is confirmed by evident fact that the most of separate values for every minute are concentrated in Fig.6 into compact zones indicating a type of the thermal stratification. It should be noted that in the afternoon the NO values (light rhombuses) are as low as at night (squares), in spite of much more intensity of traffic in a day than at night. The reason of it is strong vertical mixing under unstable stratification in the afternoon that leads to quick dispersion of pollutants.

Let us analyze in details one more statistical relation at the same day – between surface concentrations of NO and CO (Fig.1). The maximal value of CO took place nearly at the same moment as NO.

A structure of statistical relation between these gases with the account of a type of the ABL thermal stratification and of their maximal values has been presented in Fig.7. Using there markers are the same as in Fig.6. As it is seen the most values have been observed in the morning in conditions of the elevated inversion and rush-hour of traffic intensity. It should be noted that in early morning the NO contents grows under smaller values of CO than reduce later, after 09:47. This effect demonstrates gradual accumulation of CO in the ground air layer during a morning because it oxidizes into CO_2 slower than NO into NO_2. Thus, a lack of coincidence between circles and triangles in Fig.7 indicates more inertia of CO dynamics in time. The linear correlation co-efficient of partial samples R from 07:01 to 09:47 and from 09:48 to 12:05 a.m. consisted of 0.98 and 0.97 correspondingly.

At night, in the afternoon and in early evening the surface concentrations of both gases are few due to absence of intense sources at night and due to strong vertical mixing during a day.
It is interesting to note non-linearity of nocturnal, daily and evening extremely low values of NO. For example, a partial sample of evening values (dark rhombuses) may be approximated by logarithmical trend with high statistical validity: the non-linear correlation co-efficient for this sample R = 0.97. As a result of this local non-linearity as well as of non-coincidence of values in the morning a total R for whole sample for a day is a bit less then R for partial samples. The cause of non-linearity near the vertical axis is the same, i.e. a great inertia of CO in time. One more consequence of it is comparatively high background level of this gas.

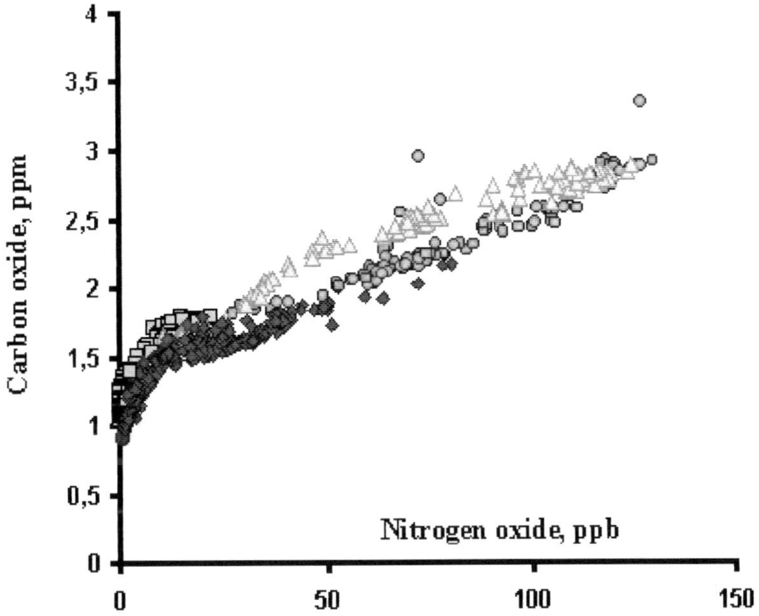

Figure 7. Statistical relation between NO and CO in August 13th 2002 with the account of the ABL thermal stratification. Markers are the same as in Figures 5 and 6.

Thus, we can see that the relation between NO and CO demonstrates too an evident influence of the ABL thermal stratification. The intensity of automobile traffic during a day is only a bit less than in morning [7]. However, an existence of elevated inversion which prevents vertical mixing leads to much more level of both pollutants in morning: on 6-7 times more than in the afternoon in conditions of unstable stratification. Besides thermal stratification, one more factor creates this effect too. It is wind speed in the ABL which in average is more in the afternoon than in morning time [7] that leads, as well as the unstable stratification, to more intense dispersion.
A structure of the third relation on Fig.3 has been analyzed by authors as well and it was found to be clearly dependent on the thermal stratification too.

4. Conclusions.

1. Statistical relation between surface concentrations of minor gases having common source is direct (the more is one the more is another) and close to linear.
2. Statistical relation between surface concentrations of minor gases one of which is a source of another is direct as well, but non-linear. It may be approximated by logarithmical trend.
3. Statistical relation between surface concentrations of minor gases which can play a role of both source and sink to each other is opposite and may be approximated by hyperbolical trend.
4. A structure of any relation for a day is complicated. Usually it may be successfully explained by dynamics of the ABL thermal stratification with the help of sodar data and, besides, with the account of the daily course of minor gases. In morning as a rule there is a closed loop in

co-ordinates of concentrations of two gases – similar to well-known Carno cycle. In case of O_3 and NO comparison this loop is created by photochemical generation of ozone in morning time; in case of NO and CO it indicates different life times of NO and CO in the atmosphere and, hence, different inertia of their accumulation in the ABL.

Acknowledgements.

This work has been supported by Russian Basic Research Foundation (projects 07-05-00874, 07-05-00428 and 06-05-08089).

References.

1. McAllister L.G., Pollard J.R., Mahoney A.R. and Shaw P.J.R. 1969 Acoustic sounding – a new approach to the study of atmospheric structure: *Proc. IEEE*, **57**, No.4, pp.579-587.
2. Brown E.H. and Hall F.F. 1978 Advances in atmospheric acoustics. *Reviews of Geophysics and Space Physics,* **16**, No.1, pp.47-110.
3. Krasnenko N.P. 2001 *Acoustic Sounding of the Atmospheric Boundary Layer,* in Russian. (Tomsk: IOM, Russian Academy of Sciences).
4. Lokoshchenko M.A. 1996 *Atmos. Oceanic Optics*, Tomsk, **9**, No.7, pp.616-628.
5. Brimblecombe P. 1986 *Air Composition and Chemistry.* (Cambridge: Univ. Press, Cambridge, United Kingdom).
6. Lokoshchenko M.A. and Elansky N.F. 2006 *Izvestiya, Atmospheric and Oceanic Physics* **42,** No.2, pp.148-155.
7. Lokoshchenko M.A., Elansky N.F., Belikov I.B., Skorokhod A.I., Semenova N.V., Isaev A.A. and Zvyagintsev A.M. 2006 *Preprints of the Sixth International Conference on Urban Climate ICUC-6*, Göteborg, Sweden, pp.104-107.
8. Elansky N.F., Lokoshchenko M.A., Belikov I.B., Skorokhod A.I. and Shumskii R.A. 2007 *Izvestiya, Atmospheric and Oceanic Physics,* **43**, No.2, pp.219-231.
9. Lokoshchenko M.A., Elansky N.F., Malyashova V.P. and Trifanova A.V. 2008 *Atmos. Oceanic Optics*, Tomsk, **21**, No.5.
10. Elansky N.F. et al. 2006 *Mobile Observatory TROICA and Observations of Atmospheric Composition over Russia,* in Russian. (Moscow: Saturn), 46 p.
11. Graedel T.E. and Crutzen P.J. 1993 *Atmospheric Change: an Earth System Perspective.* (New York, USA: W.H.Freeman and Company).

IOP Conf. Series: Earth and Environmental Science **1** (2008) 012029 doi:10.1088/1755-1307/1/1/012029

Doppler lidar measurements of boundary layer winds and sensible heat flux

J C Davis, F Davies, C G Collier and G N Pearson
The University of Salford, Salford, Greater Manchester, M5 4WT, UK

Email: j.c.davis@salford.ac.uk

Abstract. During the Convective and Orographically induced Precipitation Study the Salford Universities' Facility for Atmospheric Measurement 1.5 micron scanning Doppler lidar was deployed at Achern, Bavaria, Germany from 13th June to 16th August 2007. Vertical velocity profiles through the boundary layer were measured every 3 minutes with vertical profiles of horizontal wind velocity being derived from performing azimuth scans approximately every 30 minutes. During Intense Observation Periods radiosondes were launched from the site. In this paper, a case of convective development on 15th July 2007 is investigated. Estimates of eddy dissipation rate are made from the vertically pointing lidar data and used as one input to the velocity temperature co-variance equation to estimate sensible heat flux. The assumptions made in this analysis are tested as far as possible.

1. Introduction

The Convective and Orographically-Induced Precipitation Study (COPS), was conducted in the Black Forest region of Germany during the summer of 2007. Its aim was to advance the quality of forecasts of orographically-induced convective precipitation using extensive observations and modelling. Universities' Facility for Atmospheric Measurement (UFAM) is funded through the UK Natural Environment Research Council to participate in the project. The University of Salford instrumentation included a 1.5 micron Doppler lidar system, mounted in a mobile laboratory with a full hemispheric scanning capability using a dual-mirror scanning system [1], a 14 channel microwave radiometer and an automatic weather station (AWS). All Salford instruments were set up to run continuously from 13th June to 16th August 2007. The lidar, built by Halo Photonics, was fully controllable over the internet and had been previously deployed during the Helsinki Testbed [2].

The Salford Autonomous Lidar System, operating at a wavelength of 1.5 microns, employs novel optical technology and the design approach has led to a new type of eye-safe Doppler lidar providing a high level of performance and exhibiting exceptional stability [1]. In Figure 1 the system is illustrated *in situ* at the COPS field site in Achern, Germany.

IOP Conf. Series: Earth and Environmental Science 1 (2008) 012029 doi:10.1088/1755-1307/1/1/012029

Figure 1: The Salford University Doppler lidar with antenna (inset).

The lidar works by transmitting a short laser pulse and collects the backscattered signal from the illuminated aerosol targets along the path of the laser beam. The primary scatterers are small atmospheric aerosol particles whose diameters are within an order of magnitude of the lidar wavelength. At optical wavelengths, scattering within the lower atmosphere is primarily by particles with diameters less than 3 μm, which are sufficiently small to be advected by the wind, and serve as an effective tracer of wind velocity. Accurate estimates of the radial component of the velocity (along the line of sight of the laser beam) are produced as a spatial average over the sensing volume of the transmitted pulse.

The lidar system has a modular design arranged in three separate units; the optical base unit, the weather-proof antenna and the signal processing and data acquisition unit. The base unit has approximate dimensions 56 cm x 54 cm x 18 cm and contains the optical source, interferometer, receiver and the electronics. The weather-proof antenna is attached to the base unit via an umbilical. The antenna can be deployed permanently outside whilst the base unit and data acquisition system are housed within a laboratory environment. The signal processing has been developed with a view to providing a high level of flexibility with respect to the data acquisition parameters. Users are able to set parameters such as the range gate length, maximum range and number of pulses accumulated for each measurement, as detailed in Table 1.

Table 1. Autonomous Doppler lidar system parameters.

Parameter	Value
Operating Wavelength	1.5 μm
Range gate length	30 m
Minimum range	100 m
Maximum Range	7 km
Temporal resolution	0.1 – 30 s
Pulse Repetition Frequency	20 kHz

The lidar was set up to run continuously throughout the duration of the field trial, from 13th June to 16th August 2007. Measurements were limited only by atmospheric moisture (rain / fog), a lightning strike and a mains power failure.

The Doppler lidar data was selected to test a methodology outlined in [5] for calculating sensible heat flux (Q_H), in an urban area. During the COPS field trial, there were few instruments measuring Q_H, and none of these were located at the Achern site. However, the conditions were suitable for testing this methodology in a non-urban area and since Q_H is an important variable when investigating the development of thermals it was considered an ideal opportunity to test this methodology.

2. Measurements

During much of the field campaign, the lidar system performed a series of scan patterns pre-programmed into the system software. Generally, the system carried out:

- 5 minute azimuth scan;
- 25 minute vertically pointing scan

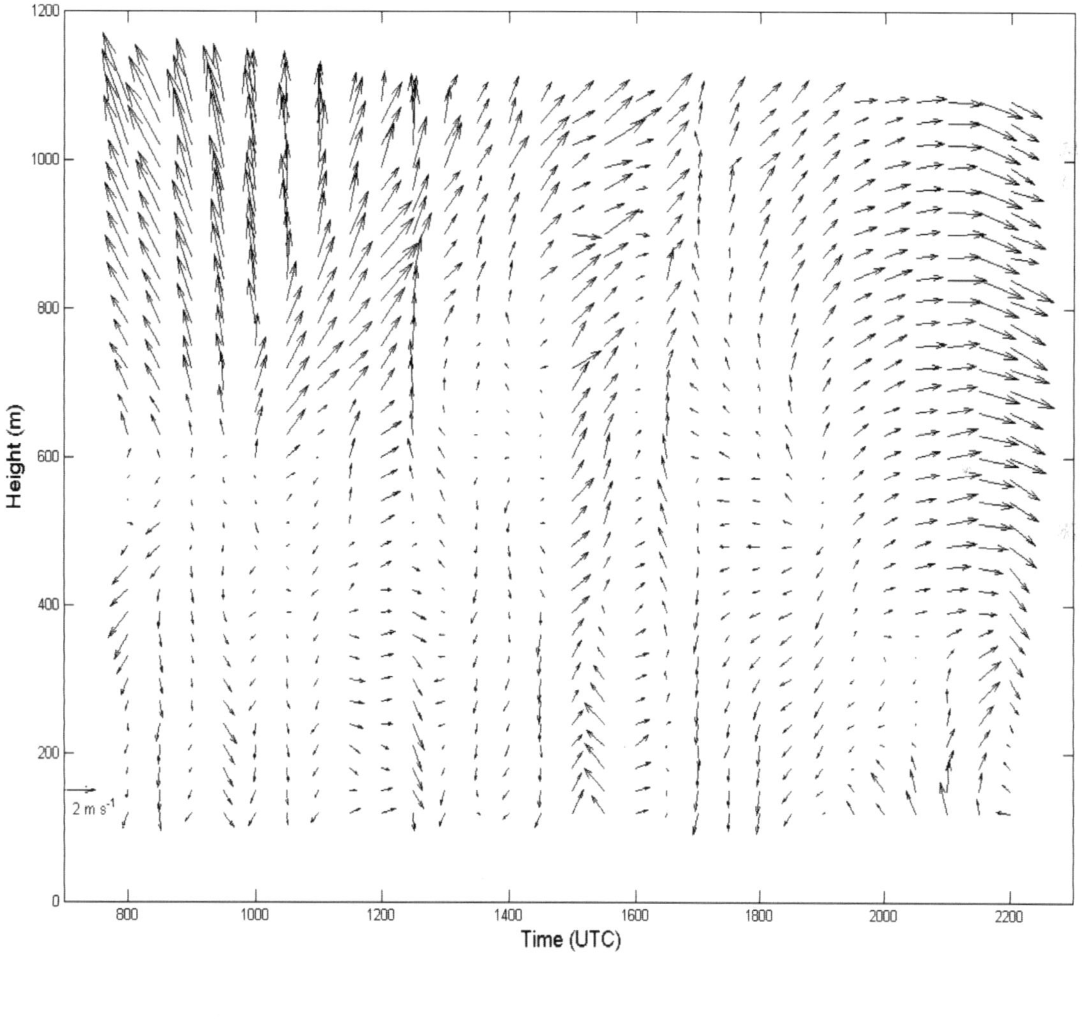

Figure 2: Wind vector plot for 15/07/07, 0800 - 2200 h UTC, arrows represent relative horizontal wind speed and direction. Due to some wind velocities being very small, the arrows appear as dots. The legend (bottom left) shows an arrow representing a 2 m s^{-1} wind speed, with westerly direction.

The azimuth scan was performed to determine wind velocity profiles and these data were concatenated enabling a continuous day-long vector plot, as illustrated in Figure 2. The vertically pointing scans were concatenated to produce a complete overview of vertical velocities measured by the Doppler lidar each day, as illustrated in Figure 3. Both Figure 2 and Figure 3 show data up to a range of 1100 m. Beyond this range, on this day, the lidar 'sees' only noise. Data such as those illustrated can be used to calculate a variety of products [3], including mixed layer height [4], and turbulence statistics [5].

The data from the 15th July 2007 show that thermals were building throughout the day, suggesting a suitable day to use lidar data to calculate sensible heat flux (Q_H), a parameter important in monitoring the development of thermals. A novel method outlined here was put to use and has been discussed in more detail in [6].

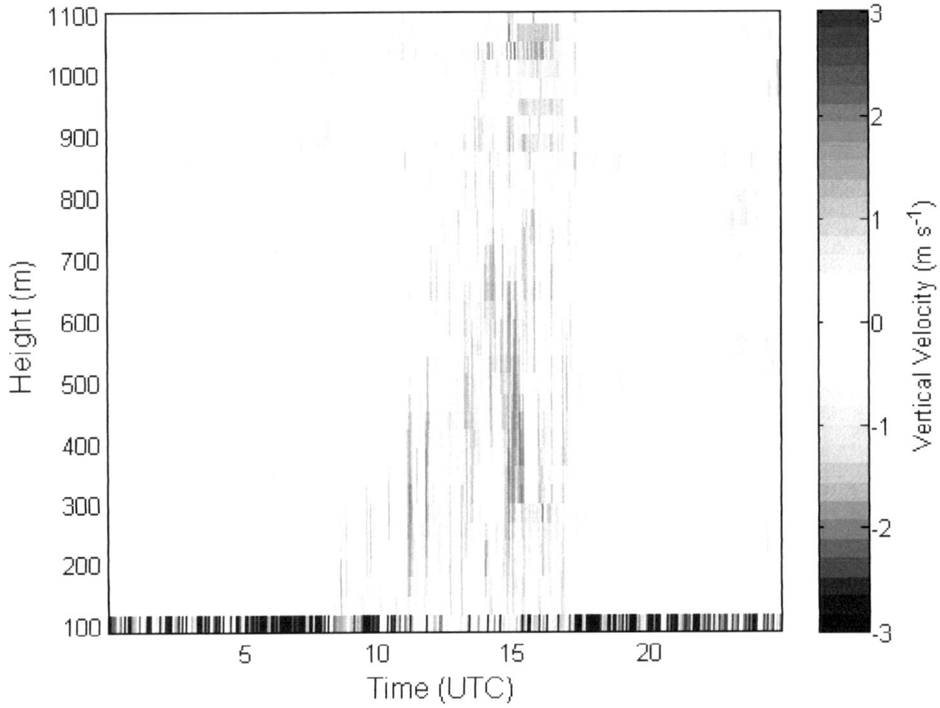

Figure 3: Full day view, 15/07/07, showing thermals building throughout the day.

3. Methodology

The methodology for calculating Q_H is summarised, with details to follow. Firstly, since the methodology requires unstable conditions, these must be ascertained. Then, the mean vertical velocity, $\overline{w}$ and its variance w'^2, can be deduced from different scanning strategies, either elevation scans or simply by pointing the beam as a stationary 'stare' scan. The turbulence kinetic energy dissipation rate (ε) can be deduced from stare scans. Once these data have been obtained, using the equations detailed, Q_H can be calculated.

The method requires unstable conditions, so to confirm whether this was the situation for the 15th July 2007 case, data from two radiosonde ascents carried out at the Achern site were collected and are illustrated in Figure 4.

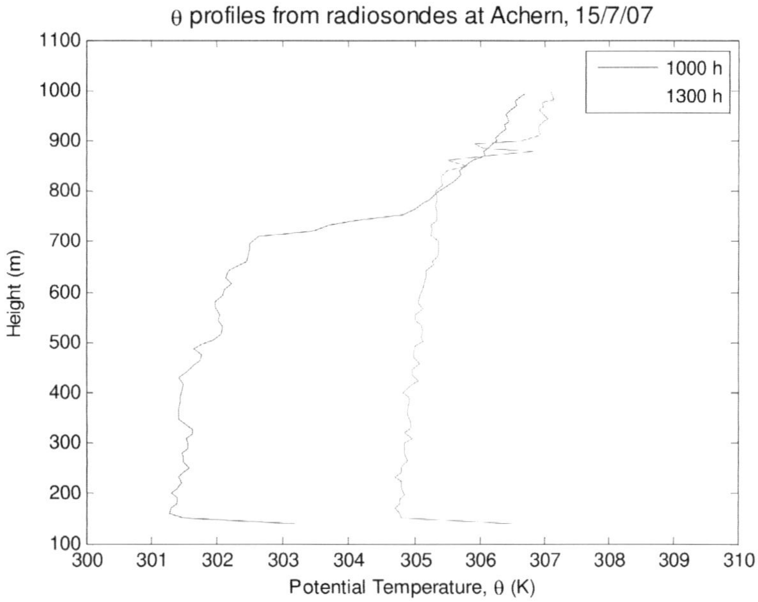

Figure 4: Potential Temperature profiles on 15/07/07 suggesting unstable conditions are present.

Given these conditions, Q_H can be calculated from:

$$Q_H = \rho_0 C_P \overline{w'\theta'} \tag{1}$$

taking C_P, the heat capacity of air at constant pressure, as 1004 J deg^{-1} kg^{-1} , ρ_0, the air density, as 1.275 kg m^{-3} and $\overline{w'\theta'}$, the potential temperature-vertical velocity covariance. Under unstable conditions, as suggested by the Doppler lidar scans and radiosondes (Figure 2, Figure 3 and Figure 4), $\overline{w'\theta'}$ can be calculated from [7], [8]:

$$\overline{w'\theta'} = \frac{\theta_0}{g}\left[\left(\frac{3}{2}\overline{w'^2}\frac{\partial \overline{w'}}{\partial z}\right) + \frac{\varepsilon}{3}\right] \tag{2}$$

A traditional way to estimate ε is by examining the line spectra of the longitudinal velocity correlation. In the inertial subrange, the expected relationship is [9]:

$$\overline{f}(\kappa) = \alpha \varepsilon^{2/3} \kappa^{-5/3} \tag{3}$$

where κ is the wave number, α is a universal constant (0.5) and $\overline{f}(\kappa)$ is the Fourier transform of the longitudinal velocity correlation

IOP Conf. Series: Earth and Environmental Science **1** (2008) 012029 doi:10.1088/1755-1307/1/1/012029

With the data collected from the lidar and radiosondes, it was possible to calculate ε and $\overline{w'\theta'}$, both of which are required in equation 2 to calculate sensible heat flux (Q_H).

4. Results

Figure 5 shows one of the spectra produced from a stare scan in order to obtain a single value of ε.

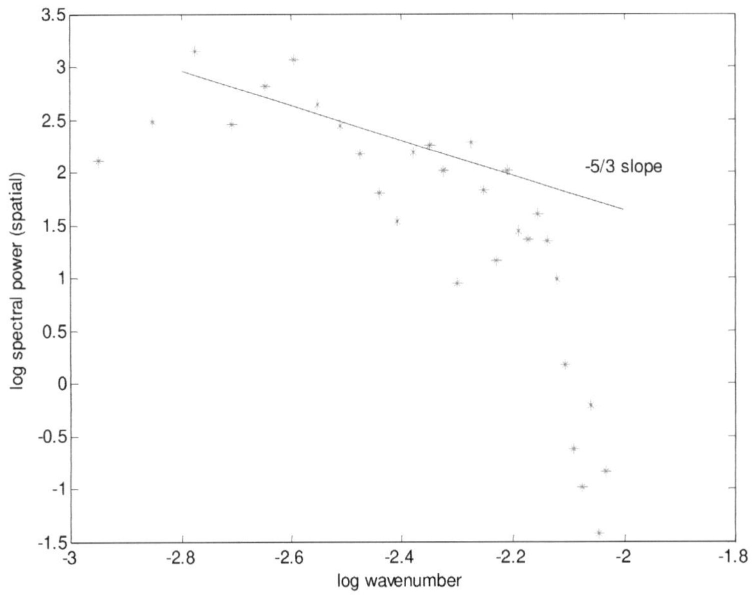

Figure 5: Longitudinal spectra with a -5/3 line in black to highlight the inertial subrange.

With data collected on the 15/07/07, a time series has been produced and shown in Figure 6 along with associated error margins. Errors can be caused by missing data points, the presence of birds and insects and other horizontal inhomogeneities. There are also errors inherent in the Doppler lidar system itself, as outlined in Table 2.

Table 2. Errors associated with Doppler lidar calculations of $\overline{w'^2}$ and ε

Error type and extent		
Vertical velocity variance, $\overline{w'^2}$	Bias = 0.02 $m^2 s^{-2}$ Random = ±0.68 $m^2 s^{-2}$	[10]
Turbulent kinetic energy dissipation rate (ε)	Error = 10%	[11]

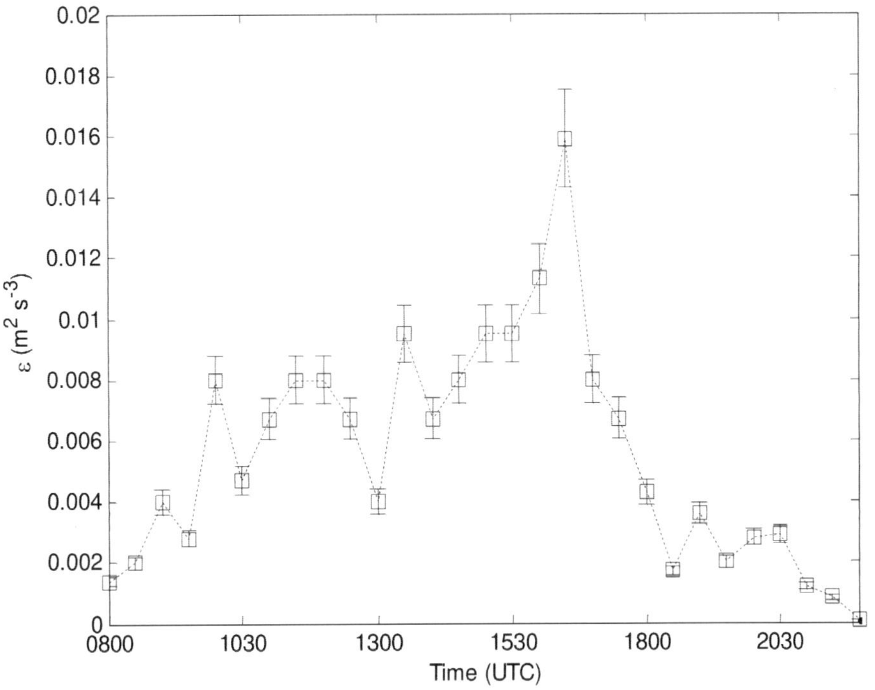

Figure 6: Variation of the turbulence kinetic energy dissipation rate during 15/07/07 as estimated from Doppler lidar data.

The time series in Figure 6 illustrates ε changing throughout 15/07/07 as would be expected – increasing towards noon, then decreasing towards the evening. With knowledge of ε, a value for Q_H was calculated from each ε and this is shown in Figure 7.

As with ε, Q_H values tend to increase during the day, reaching an approximate peak at noon. They then begin to fall at around 15.30 h UTC, consistent with Q_H values measured by an energy balance station within the COPS study area, some 10 km from the Achern site. [12].

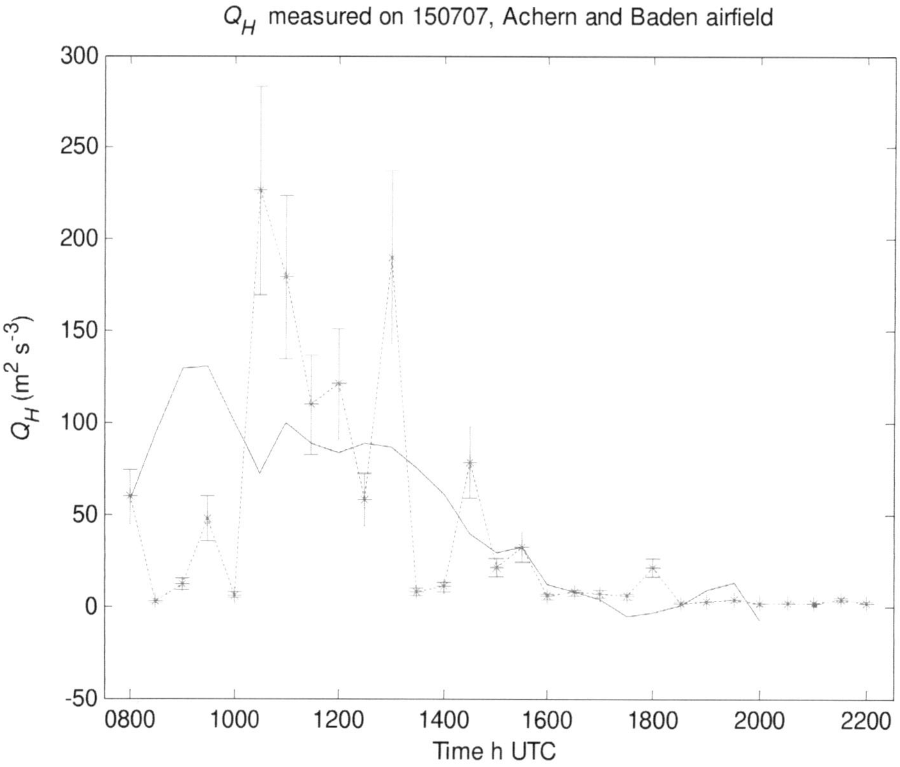

Figure 7: Q_H timeseries calculated for 15/07/07, showing an estimation of the errors for that day is shown in red. Half-hourly averages of Q_H, as measured by the IMK-FZK energy balance station for that day, is shown in blue.

5. Conclusions and Future Work

It has been shown that by using a method outlined in [6], it is possible to calculate values of Q_H and it is considered that these values fall within an acceptable range, comparable with other Q_H values measured that day at another site within the COPS region [12]. However, Q_H calculated from Doppler lidar data using this method would differ from that calculated using a 'point' sensor such as an energy balance instrument or a sonic anemometer since the data collected with a Doppler lidar is integrated over a 'volume' of atmosphere advecting through the laser beam. Such measurements are a useful addition to the other measurements provided by the lidar, requiring no extra equipment.

It is considered that further investigations of employing this method to calculate Q_H are needed to establish the overall applicability of the method and build on the preliminary calculations. It is expected that such investigations will include:

- Calculation of ε using alternative methods;
- investigation of ε at different heights and atmospheric conditions; and
- investigation over different surface types.

6. Acknowledgements

The authors would like to acknowledge Dr N Kalthoff at IMK FZK Karlsruhe for supplying Q_H data from their energy balance station located at Baden Airfield, Germany, collected as part of COPS, July 2007.

7.References

[1] The Halo Photonics pulsed Doppler LiDAR system available on the world wide web at <http://halo-photonics.com/>
[2] Bozier K E, Pearson G N, Davies F and Collier C G 2007 Doppler lidar observations of Russian forest fire plumes over Helsinki W*eather* ***62*** 203-208
[3] Collier C G, Davies F, Bozier K E, Holt A R, Middleton D R, Pearson G N, Siemen S, Willetts D V, Upton G J G and Young R I 2005 Dual Doppler lidar measurements for improving dispersion models *Bull. Amer. Meteorol. Soc.* **86** 825-838
[4] Davies F, Middleton D R and Bozier K E 2007 Urban air pollution modelling and measurements of boundary layer height *Atmos. Env.* **41** 4040-4049
[5] Davies F, Collier C G, Pearson G N and Bozier K E 2004 Doppler lidar measurements of turbulent structure function over an urban area *J. Atmos. Oceanic Technol.* **21** 753-761
[6] Davis J C, Bozier K E, Collier C G and Davies F 2008 Spatial variations of sensible heat flux over an urban area (*Preprint Met. Apps. MET-08-0010.R3*)
[7] Wyngaard J C and Coté O R 1971 The budgets of turbulent kinetic energy and temperature variance in the atmospheric surface layer *Journal of Atmospheric Science* **28** 190-201
[8] Gal-Chen T, Xu M and Eberhard W L 1992 *J. Geophys Res.* **97** D17 18409-18423
[9] Batchelor G K, 1967 *The Theory of Homogeneous Turbulence* 5th ed. Cambridge University Press, London
[10] Davies F, Collier C G, Pearson G N, Bozier K E 2003 On the accuracy of retrieved wind information from Doppler lidar observations *Q.J.R. Meteorol. Soc.* **129** 321-334
[11] Frelich R and Cornman L 2002 Estimating spatial velocity statistics with coherent Doppler lidar *J. Atmos. Oceanic Technol.* **19** 355-366
[12] Kalthoff N, 2008 *personal communication* IMK-FZK Karlsruhe, Germany

IOP Conf. Series: Earth and Environmental Science 1 (2008) 012030 doi:10.1088/1755-1307/1/1/012030

Research of the Arid Aerosol Carrying Out With a Help of a Set of Sodars

Kramar Valery, Granberg Igor

A.M. Obukhov Institute of Atmospheric Physics, Russian Academy of Sciences, 3 Pyzhevskii lane, Moscow 109017, Russia

vkramar@ifaran.ru, igran@ifaran.ru

Abstract. Study of spatial and vertical distribution of vertical wind velocity was carried out in the "Har-Gzyr 2007" field program in the "Black Sands" region, Russia, with the help of four sodars. Continuous (about half an hour) and stable anabatic motions of the air were revealed for altitudes 10m to 800m over the area of some square kilometres. The measured vertical speed reached 2m/s. Permanent and quick transfer of sub-micron aerosol particles from the surface layer to the troposphere can be explained by this phenomenon. The scheme is proposed to design a mass balance equation to calculate aerosol concentration in the upper layers of the ABL.

1. Introduction

Research conducted in 1991-2004 by the Obukhov Institute of Atmospheric Physics (IAPh) and the Karpov Institute of Physical Chemistry (with participation of research officers of other scientific institutions) showed that the atmosphere at the height of about 1km contains high concentration of dust and salt particles even at the periods when there are no dust storms [1]. One of the reasons of the phenomenon are convective updraft flows rising due to overheating of the base surface. It was established in the experiments that during a hot season in the absence of sand storms convective processes raise out of the Kalmykiya's sand areas into the air a big amount of aerosol particles of less than 5μm in dimension, including very low dimension particles of 0.01 to 0.1μm.

Conventionally dust aerosol transfer may be divided onto three stages: a) raise of the dust and salt particles from the surface and formation of a near-surface particle layer, b) transfer of the particles in the boundary layer of atmosphere, c) far distance transfer of the particles in the troposphere. Present research is devoted to the determination of the mechanism of aerosol raising from the near surface layer to the upper levels of atmospheric boundary layer (ABL), i.e. stage b).

A number of observations as well as theoretical analysis reveal two main forms of existence of convective vortex coherent structures [2,3]. One of them are cells having a form close to the toroidal one, the other are roll-form structures. According to the mentioned researches typical dimensions of such large-scale vortexes are of 3 to 4km in cross-section and about 1km in height. Research [2] shows that calculated vertical velocities in the structure reach their maximum of 0.2-0.5m/s on the altitudes of 200-400m. Toroidal structures prevail when the average horizontal velocity of the wind is less than 3-5m/s; rolls prevail with velocity growth up to 10m/s. Further growth of wind velocity leads to destruction of all structures. The described phenomenon is rather typical for ABL [3].

To show how aerosol raised from the surface of the soil of a desert region can reach the altitude of about 1km it is enough to show that in the observation zone there are stable continuous vertical flows with vertical velocities providing aerosol transfer to the above altitude during the time of the flow existence.

2. Instrumentation and site

Four sodars were disposed in the region of the field experiment "Har-Gzyr 2007" that was carried out in July-August, 2007 close to the same title region of the National Park of the Republic of Kalmykia. Three devices were mini-sodars. Base carrier frequency of emitted pulse of the mini-sodars was close to 3.5kHz. The sodars gave detailed information about the ABL structure for the heights from 15m to approximately 200-300m. Long-wavelength sodar with base carrier frequency of 2kHz was installed for observation of the whole scale of the ABL, and it registered vertical profiles of parameters up to the altitudes of 700-800m in the daytime.

The points of the allocation of units are shown on Figure 1. The distances between points were chosen to simultaneously register parameters on different points of vortex coherent structure that has horizontal dimensions of 3 to 4km. The dimensions of the structures had been predicted by the theoretical forecasts [2,3] as well as by earlier observations [3].

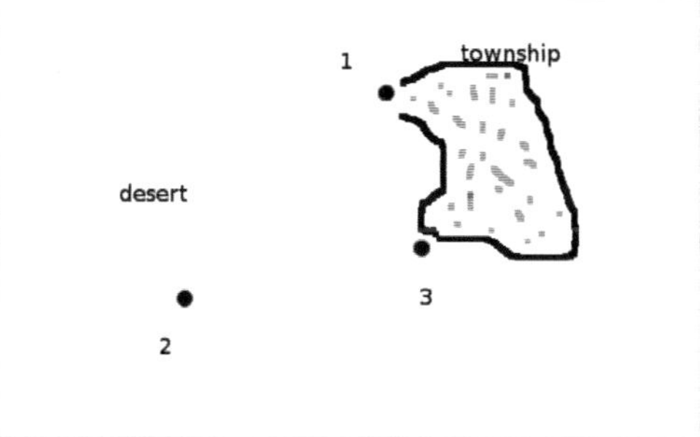

Figure 1. Scheme of points of the allocation of units in the region of observations. Sides of a triangle: 1-3 is 1.4km, 2-3 is 3.2km, 1-2 is 3.5km. One of the set of devices was disposed directly in the desert region with sand-dunes covered by weak vegetation (point 2). Northern (point 1) and southern (point 2) devices were disposed in the western periphery of Komsomolsky township of the Republic of Kalmykia.

Sodars of the next generation were used in the experiment. They were developed at the Obukhov IAPh on the basis of Latan-3 sodar. The architecture of the electronics of the unit, the set of analogue electronics, as well as software of the sodar were deeply re-engineered.

After re-engineering it became possible to use multi-frequency sounding regime of functioning of the unit as well as to emit sounding impulses by several (from 3 to 6) sodar aerials simultaneously. Other characteristics of the devices and the features of the algorithms of the received signal processing are given in the research [4].

3. Results of measurements

Figure 2 shows comparison of vertical velocities vs. time for different locations of measurement as well as for long-wavelength sodar in July 28, 2007 episode of the experiment. Here sub-figure a) the results of measurements at the point 1 (see Figure 1); sub-figure c) - the results of measurements of the unit disposed at the point 3; sub-figure b) the results of measurements at the point 2 in the desert region. d) and e) present the results of measurements with the help of long-wavelength sodar disposed at the point 1. On d) and e) parts of the figure vertical scales for all values are two times less than scales on parts a) – c). Part e) of the figure shows vertical profiles of the average horizontal wind velocity measured by long-wavelength sodar. On a) – d), in particular, one can see the repeatedness of large-scale structure with some shift and deformation for the period between 4 and 5 p.m. LMT.

According to the results of calculations presented in the research [2] values of vertical velocities in vortex coherent structures should be close to 0.2-0.3m/s at the altitudes 200-400m. Nevertheless the sodars register typical values that are many times larger, namely 1-2m/s. This large difference in the values of vertical velocities may be explained by strong overheating of the ground at the time of the experiment. Ultrasonic thermometers-anemometers registered the heat fluxes from the ground up to 150-250W/m^2. However these large values of updraft velocities support transfer of sub-micron aerosol particles from the surface layer to the altitudes of about 1km. Figure 2 shows that typical time of the lifting motion for a given location is approximately 10 to 20min. During this time period measured updraft velocities can raise aerosol particles to the upper bound of the ABL.

Besides, the figure 2 shows that frequently rather long time updraft fluxes are registered simultaneously in two locations of measurements, and for some periods – in all three locations. This fact taking into consideration the distance between points of measurement indicates the scale of the phenomenon. As a result the phenomenon can rise a big amount of aerosol particles from the large area to the upper level of ABL.

4. Using of sodar data in models of aerosol transfer

The results make it possible to formulate a variant of using of sodar data in a simple model of aerosol transfer into the upper layers of ABL. The model may be used for parametrization of the phenomenon.

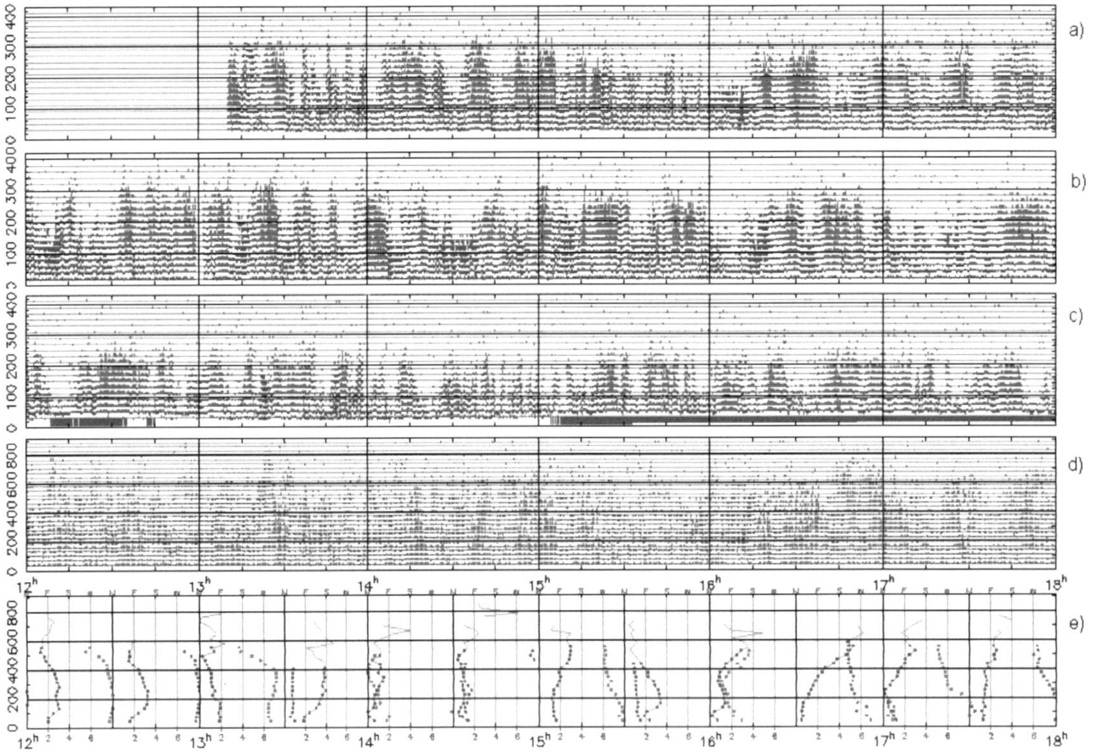

Figure 2. 08.28.2007. Vertical velocities of wind in ABL according to sodar measurements. a) – c) the measurements made by mini-sodars for the heights of 15-400m in points 1 - 3 of the region. For the mini-sodars vertical velocity is shown with a 20m step. The scale of velocity value equals to 10m/s, d) the measurements of the long-wavelength sodar disposed at the point 1 and operating for the heights of 50-800m. Red (more dark on grayscale version) marks indicate upward velocity and green ones - downward, e) velocity and direction of horizontal wind according to the long-wavelength sodar. Green points (more dark for grayscale version) show wind magnitude, other ones show wind direction.

The following assumptions can be used in a model. Two layers – the upper and the lower ones having different concentrations of aerosol particles connect between each other through up and down fluxes of large-scale coherent structure. The lower layer is a surface layer; the upper one is at the upper level of ABL. Concentration of aerosol particles in the lower layer is defined by the conditions on the surface [1]. The concentration in the upper layer can be defined with the help of a simple equation of mass balance taking into consideration that the aerosol particles carried in by the anabatic flow are distributed in the upper layer evenly. The thickness of the upper layer seems to be needed to form the equation of balance. Constant value of updraft velocity determined out of sodar measurements can be used to simplify calculations.

In the experiment along with the sodar measurements the group of researchers measured concentrations of aerosol as well as distribution of aerosol particles by their dimensions on the surface layer in the desert region 5km to the west from the sodar in the location b). The observed values of concentration were from 50 to 300μg/m^3. Earlier [1], concentrations and distribution of sub-micron aerosol on different altitudes almost in the same region had been determined. Airplane laboratory had been used for measurement of concentration, distribution and elements of aerosol particles. The layer of high sub-micron aerosol particles concentration had been discovered at this region for altitudes 800-1000m for the first time.

For example the balance equation for the upper layer in its simplest form can include one onflow of aerosol with the updraft velocity of the structure and with the C_0 concentration that is equal to the concentration of aerosol in the lower layer and two outflows. Let unknown concentration of aerosol in the upper layer be C_1. The outflow of aerosol out of the upper layer will consist of two parts. One of them is outflow of aerosol with C_1 concentration by katabatic flow of a coherent structure. The other one is outflow of aerosol of the same concentration with horizontal wind. The thickness of the upper layer (approximately 200m) can be taken from the above 'airplane' measurements; all velocities – from the sodar measurements. Linear dimensions of the structure can be found in the theories [2,3]. Aerosol concentration in the lower layer can be taken from the measurement data or calculated with a help of empirical formula [1]. After that the unknown concentration C_1 can be estimated with the help of the balance equation. Analogous scheme may be used for composition of more complex models or parametrization of the phenomenon.

As a whole, the data obtained make it possible to verify the suggested hypothesis and assumptions of an aerosol transfer models.

5. Conclusions

By using Doppler sodars, it is possible to receive information about the structure of large-scale vortex coherent structure. According to the presented measurements the updraft velocities of air masses in such structure achieve 1-2m/s and these velocities retain almost the same from the altitudes of 10m up to 700-800m. At the same location the typical time period of such updraft flow registration is 10-20min. It means that such period is enough for carrying out of sub-micron aerosol particles by the anabatic fluxes of the structures to the altitudes over 1km. The data can be used to design simple balance models as well as for parametrization of aerosol transfer process.

Acknowledgements

The work was supported by Russian Fund of Basic Researches through grants 04-05-64167, 06-05-08086, 07-05-00521, 08-05-00671, 06-05-65216. We are heartily grateful to professor M. Kallistratova for discussion and important remarks.

References

[1]Golitsyn G S, Granberg I G, Andronova F V, Zilitinkevich S S, Smirnov V V, Ponomarev V M 2003 Investigation of boundary layer fine structure in arid regions *Water, Air, and Soil Pollution: Focus 3* (Netherlands: Kluwer Academic Publishers) pp 245-257

[2]Lappen C-L and Randall D 2005 Using Idealized Coherent Structures to Parametrize Momentum Fluxes in a PBL Mass-Flux Model *Journal of the Atmospheric Sciences* **62** 2829-2846

[3]Mikhailova L A and Ordanovich A E 1991 Coherent Structures in the Atmospheric Boundary Layer (review) *Izv. Acad. Sci. USSR. Atmos. Oceanic Phys.* **27** 593-613

[4]Kouznetsov R 2008 Multi-frequency acoustic sounding of the ABL with dish antenna *This collection*

IOP Conf. Series: Earth and Environmental Science **1** (2008) 012031 doi:10.1088/1755-1307/1/1/012031

3-dimensional observations of atmospheric variables during the field campaign COPS

Sandip Pal, Andreas Behrendt, Heinz Bauer, Marcus Radlach, Andrea Riede, Max Schiller, Gerd Wagner, and Volker Wulfmeyer

Institute of Physics and Meteorology, University of Hohenheim, Garbenstrasse 30, Stuttgart, Germany, 70599.

sandip@uni-hohenheim.de

Abstract: The Convective and Orographically-induced Precipitation Study (COPS) has the aim to advance the quality of forecasts of orographically-induced precipitation in complex terrain. COPS is a Research and Development Project of the World Weather Research Program and considered to be one of the largest field campaigns on quantitative precipitation forecasting that has been performed so far. A network of state-of-the-art active and passive remote sensing systems was combined with in total 10 airborne platforms, Meteosat rapid scans and dense networks of standard meteorological instruments during the three months long field phase (June-August, 2007) in south-western Germany/eastern France to observe atmospheric variables in the three spatial dimensions and in time. By the University of Hohenheim, two novel ground-based mobile scanning lidar systems were deployed: a scanning rotational Raman lidar which provides combined measurements of the field of atmospheric temperature and aerosol optical properties, and a water-vapour differential absorption lidar (DIAL). During COPS, both lidar systems were located at Hornisgrinde, the highest peak of the northern Black Forest together with a large suite of other remote sensing systems. In this contribution we present and discuss highlights of these measurements.

1. Introduction

The main objective of the Convective and Orographically-induced Precipitation Study (COPS, www.uni-hohenheim.de/cops/) is to identify the physical processes responsible for the deficiencies in quantitative precipitation forecast over low-mountain regions with the goal to improve their model representation [1]. Within COPS, a dense network of state-of-the-art active and passive remote sensing systems was deployed to observe atmospheric key variables in as much detail as possible. A detailed description of COPS is given in the COPS Field Report [2].

The study of atmospheric boundary layer (ABL) processes in complex terrain under various meteorological conditions is a key component to reach the COPS objectives. First results of COPS related to boundary layer studies made with two novel instruments, the scanning rotational Raman lidar (RRL) [3] and the scanning water-vapour differential absorption lidar (DIAL) of University of Hohenheim, are discussed in the following. DIAL allows to profile the atmospheric water-vapour number density with higher resolution and accuracy than any other remote sensing technique (e.g., [4-6]). The combination with scanning measurements of atmospheric temperature by RRL [7] is very beneficial for a comprehensive investigation of processes in the ABL. The two instruments were deployed on top of Hornisgrinde, the highest peak in the Northern Black Forest, at an elevation of 1161 m above sea level (ASL) during the COPS campaign together with a large suite of other in-situ and remote sensing instruments forming the so-called COPS Supersite H.

2. Differential absorption lidar of University of Hohenheim

The new scanning water-vapour DIAL which has been developed within three interlinked projects funded by the German Research Foundation (Deutsche Forschungsgemeinschaft, DFG) has been deployed for the first time in the field during COPS. The instrument is affiliated to Institute of Physics and Meteorology (IPM) of University of Hohenheim (UHOH) which was responsible for the coordination of these three projects and also operates the DIAL. For brevity, we refer to this system as UHOH DIAL in the following.

Differential absorption lidar is a technique which measures the water-vapour number density with very high resolution in space and time by alternate emission of short laser pulses which are absorbed strongly by water-vapour in the atmosphere (online) and which are weakly absorbed (offline) and by time-resolved detection of the atmospheric backscatter signals [8]. For the measurements shown below, the absorption cross-section for the online signal is a factor of $\sim 10^3$ higher than for the offline signal; the two wavelengths are very close to each other with a difference of < 0.5 nm. The UHOH DIAL employs laser radiation in the near infra-red (IR) between 815 and 820 nm, a spectral region where several water-vapour absorption lines are present and can be selected according to range of interest for optimum measurements and amount of humidity present in the atmosphere. With the water-vapour absorption coefficient for the online radiation known from high-accuracy laboratory measurements, the atmospheric backscatter profiles alone provide a profile of the water-vapour number density; instrumental parameters cancel out. Consequently, DIAL is called self-calibrating as no calibration of the DIAL data need to be performed.

Worldwide, a number of water-vapour DIAL systems exist. All these systems are unique research instruments and have been developed by the institutions which also operate the instruments. During COPS, in addition to the UHOH DIAL two water-vapour DIAL systems were employed from airborne platforms, namely the DIAL of Deutsches Zentrum für Luft- und Raumfahrt (DLR, [9-10]) and the DIAL of the Centre National de la Recherche Scientifique (CNRS) [11-13]. Besides these three water-vapour DIAL systems, three other lidars provided water-vapour data but with the Raman lidar technique: the lidars of CNRS, U. Basilicata, and Institute of Tropospheric Research (IfT) at the Supersites V, R, and M, respectively. But compared to DIAL, water-vapour Raman lidar systems provide data with lower resolution in daytime. Simulations show that also space-borne DIAL is feasible and would provide data with very high performance [14].

The UHOH DIAL provides to date the highest resolution remote sensing data of water-vapour worldwide. This is achieved by high laser power (during COPS > 4 W average power and a repetition rate of 250 Hz) [15], a large telescope (80 cm diameter of the primary mirror), as well as very efficient receiver optics and detectors. The whole system is mounted on a 12-m trailer so that it can be transported. During certain COPS missions (see COPS Field Report, [2]), the UHOH DIAL was operated in scanning mode with typically 1 degree/s scan speed performing different kinds of range-height indicator (RHI) or plan-position indicator (PPI) scan patterns (for the former the elevation angle is varied while the azimuth angle is kept constant; for the latter vice versa).

During COPS, the backscatter profiles of the DIAL were stored for each laser pulse, i.e., with 250 Hz, and a range-resolution of 15 m with >20 km range (table 1). Thus maximum freedom is possible in post-processing of the data. The data products of the DIAL are profiles of water-vapour number density with typical resolutions of 15 to 150 m with a temporal resolution of 1 to 10 s. But spatial and temporal resolution can be traded off against each other. An accuracy of better than 5 % is expected [16-17]. To verify the high accuracy, intercomparison of the water-vapour data measured with ground-based and airborne remote sensing will be made within the project in cooperation with Paolo di

Girolamo of University of Basilicata by a similar approach [5-6] like with data of the International H_2O project (IHOP_2002, [18]).

3. Rotational Raman lidar of University of Hohenheim

The rotational Raman lidar of University of Hohenheim (UHOH RRL, [3]) is a new research lidar which has been developed within COSI^{TRACKS} (Convective Storms Virtual Institute, with contribution to TRACKS, Transport and Chemical Conversion in Convective Systems), a Virtual Institute of the Helmholtz Research Association.

The UHOH RRL emits ultra-violet (UV) laser pulses at 354.7 nm (during COPS: 50 Hz, ~200 mJ/pulse, [19]), collects the backscatter signal with a telescope with a 40-cm primary mirror and extracts three signals out of the backscatter spectrum: the elastic-backscatter signal of molecules and aerosols at 354.7 nm and two rotational Raman backscatter signals near 353.1 nm and 354.1 nm which show opposite temperature dependency. The ratio of the latter two provides after calibration directly and without any a-priory assumption on the state of the atmosphere a profile of the atmospheric temperature ([3],[7], [20-22]). Compared to vibrational Raman lidar, rotational Raman lidar has much larger signal intensities. The UHOH RRL filters were optimised for measurements from close to the ground to the middle troposphere. This is the reason why this system is the only RRL worldwide which provides temperature profiles with a temporal resolution of 10 – 60 s with a range resolution of ~100 m and a precision of better than 1 K also in daytime [19]; most other RRLs are restricted to night time measurements only. Like for any other lidar, the temporal and range resolution of the derived lidar data products can be traded off against each other. Therefore, the UHOH RRL data were stored with a raw data resolution of 10 s and 3.75 m up to a range of 30 km and with 37.5 m resolution up to 75 km (table 1). These data are available for detailed post-processing. The UHOH RRL contains a scanner so that the profiles can be measured in any direction of interest which is highly beneficial for boundary layer studies. During COPS, two more rotational Raman lidar instruments were operated, i.e., by U. Basilicata, and Institute of Tropospheric Research (IfT) at the Supersites R and M, respectively. The system of U. Basilicata was the first RRL operated in the UV [23].

UHOH RRL performed different scan patterns during COPS and these include

- continuous vertical measurements,
- continuous RHI scanning measurements with an increment of 5° within the full upper hemisphere,
- combination of time-synchronous RHI scan measurements and vertical measurements

Table 1. Key features of the UHOH DIAL and RRL

	UHOH DIAL	UHOH RRL
Method	Differential absorption	Rotational Raman effect
Measured parameters	Water-vapour number density, relative aerosol backscatter	Temperature, particle backscatter coefficient and particle extinction coefficient at 355 nm
Measurement time	Day and night (absence of rain)	Day and night (absence of rain)
Range resolution (raw data)	15 m	3.75 m
Time resolution (raw data)	1/250 s	10 s
Range (raw data)	>20 km.	>30 km.

IOP Conf. Series: Earth and Environmental Science 1 (2008) 012031 doi:10.1088/1755-1307/1/1/012031

4. Measurement examples

Both lidar systems collected data extensively on selected intensive observation periods (IOPs). During these days, the lidars were running continuously with the only exception of prohibitive weather conditions due to fog or rain.

4.1 Synergy of two UHOH lidar systems

Synergetic measurements of moisture, temperature and aerosol optical properties are beneficial to study boundary layer processes and their relation to the initiation of deep convection. On 20 July 2007 during IOP 9c UHOH lidars simultaneously performed the observations as shown in figure 1. Clearly visible is the stable boundary layer until an altitude of 1.2 km above ground level (AGL) with high particle backscatter coefficient (PBC) (figure 1b). This is a regime of morning stable boundary layer and in the PBC field no significant signatures of thermals are present which is the sign of a convective boundary layer (CBL). During this case, the variation of the top of the ABL is mainly due to the advection of different air masses. The PBC field is characterized here by different aerosol layers. High values of PBC coincide with the stable ABL. The PBC field also shows that clouds are moving in at ~ 0915 UTC with a cloud base of 3.0 km. Figure 1a shows the time-height cross-section of the gradient of the potential temperature field during the same time domain. The potential temperature is defined as the temperature that a parcel of dry air would have if it was compressed to a standard pressure of 1000 mb. Since potential temperature is conserved for dry-adiabatic motion in the atmosphere, θ is kept constant within well mixed layers and for descending (ascending) air parcels that are undergoing dry-adiabatic warming (cooling). The maximum of the gradient is present at the top of the stable boundary layer. The humidity field as shown in figure 1c is characterized by high values of mixing ratio (~7g/kg) in the stable ABL up to 1.2 km AGL. ABL humidity field does not change much while there is a large variability in the lower free troposphere with several aerosol layers present. Between 0800 and 0845 UTC a quite dry layer is observed first at ~ 3.0 km AGL descending down to the ABL top at ~ 1.2 km AGL. Between 0530 and 0600 UTC the PBC field shows two separated layered aerosol structures at 2.0 km and at 3.0 km AGL (figure 1b) which are highly correlated with the humidity field in the corresponding altitude. At 0830 UTC a very humid layer is approaching first in ~ 3.0 km AGL and then extending down to the ABL and apparently merging with

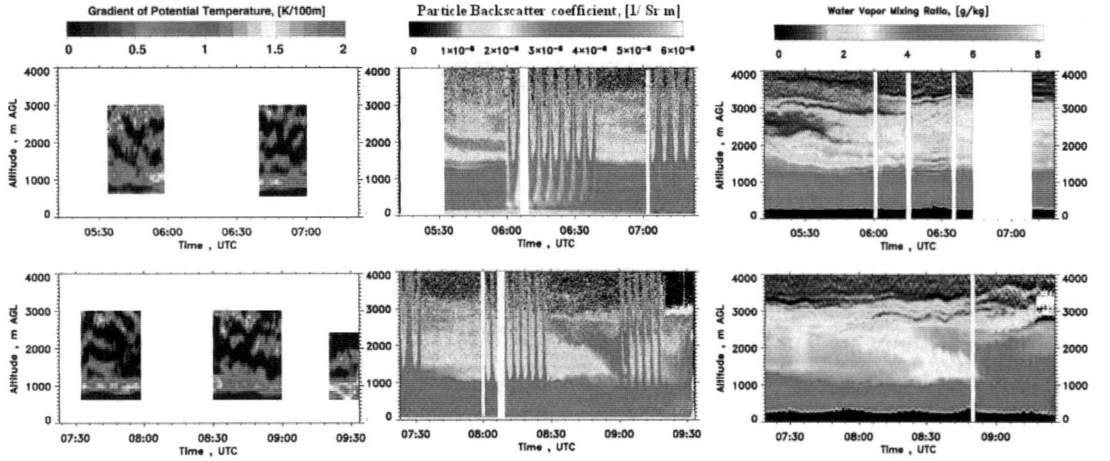

Figure 1. a: Time height cross-section of gradient of potential temperature field between 0530 and 0930 UTC on 20 July 2007 obtained with UHOH scanning RRL. Measurement during scanning measurements are not shown and blocked with white bands b: Same as (a) but for PBC field. Repetitive bow-like structure on the image are due to the RHI scanning in that time. c: Same as (a) but for humidity obtained with the UHOH DIAL

the ABL. Penetration of thin filaments of drier air mass is also present in the humidity field (figure 1c) around 0530 UTC above 2 km AGL. Free-tropospheric dry air is present at an altitude above 3.6 km AGL (figure 1c). Aerosol structures are visible (around 2.0 km AGL) between 0830 and 0900 UTC (figure 1b) and are accompanied with higher values of the water-vapour mixing ratio (WVMR) than their surroundings (figure 1c). The present case is an example for the modification of the local ABL mainly by advection. During this time forced convection initiation took place in the eastern region of Supersite H. A pre-convective gust front reached the Supersite at 0930 UTC and precipitation started at 1010 UTC.

4.2 DIAL measurements of water-vapour mixing ratio

UHOH DIAL deployment on 1-2 August 2007 (IOP 13 a-b) yields a dataset of the continuous development of the moisture profiles for more than 16 hours in the daytime and night-time boundary layer as shown in figure 2. These observations allow the study of a warm moist air mass with an embedded Saharan dust and its interaction with forced convection. Temporal and spatial resolution in the retrieved mixing ratio data are 10 s and 15 m, respectively. A gliding window averaging (150 m) is applied in height. From 2130 UTC onwards, the bottom height of the data collected is lower because we include data of a near-range telescope. Vertical and horizontal white bands on the figure are due to gaps in measurement and incomplete overlap, respectively.

A convective boundary layer is observed in the lowest altitude with increasing top height from 500 m to 800 m AGL between 0845 and 1800 UTC. A drier layer with thickness around 400 to 500 m is observed above 500 m AGL (indicated by A in figure 2) from the beginning of the measurement period until 1700 UTC. Above this, a relatively moist layer with a mixing ratio value of up to 4g/kg is

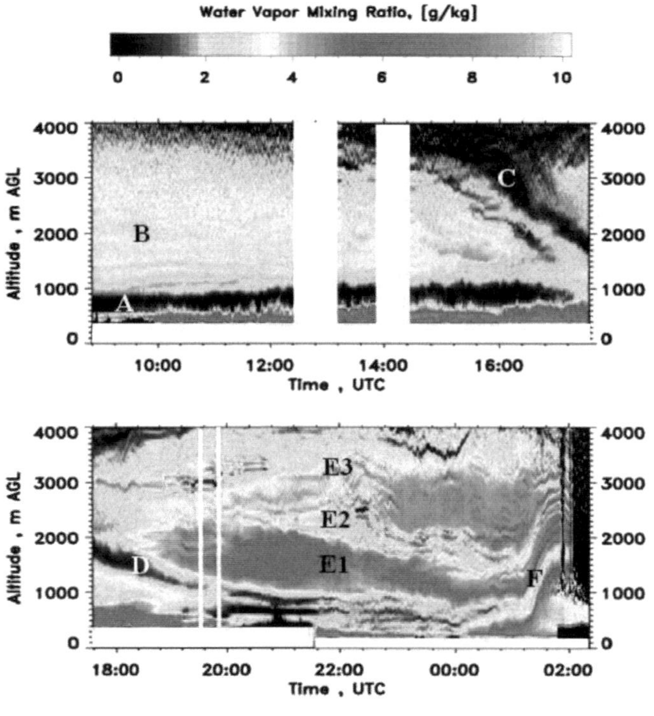

Figure 2. Time-height cross-section of mixing ratio obtained with UHOH DIAL from 1 August 0930 UTC until 2 August at 0220 UTC. Note that an out flow boundary passed over the lidar site around 0112 UTC on 2 August 2007. Time resolution and range resolutions are 10 s and 15 m, respectively with 150 m gliding average. For further details, see text.

present throughout this time (B). Penetration of drier air from above 3.5-4.5 km is visible prominently on the image (C) and this drier layer started to descend into the lower boundary layer around 1600 UTC and reaches to 1 km AGL around 1900 UTC (D). The corresponding WVMR is only < 0.5g/kg (5-6 times smaller than the columnar average in this case). Parallel to this feature, another thin and dry tongue with WVMR value of <0.3 g/kg moves downwards (c) but disappears around 1730 UTC. Warm moist air is advected from west in the lower free troposphere above the lidar site after 1800. During this time, the humidity field shows large variability and different layers are clearly visible (E1, E2, E3); e.g., a drier layer embedded between two moist air layers between 2200 and 2400 UTC (E2). A sharp rise is clearly visible in the ABL moisture regime at around 0112 UTC (F) which is an outflow boundary. Wind speed measured by the radiosonde at 2300 UTC was relatively high with ~ 12m/s near 500 m AGL.

A zoom-in-view of the humidity field between 0100 and 0220 UTC on 2 Aug 2007 is shown in figure 3. The resolution of the data is same as in figure 2. The sudden rise of the moist layer at 0112 UTC is due to the strong out-flow boundary (produced by the mesoscale convective system) that passes over the region. At the end of the measurement period precipitation reaches the lidar site around 0240 UTC. High-resolution DIAL measurements enable us to detect the exact time of the boundary passage. Some 'ripple'-like oscillatory structures in the moisture field are believed to be due to waves excited by this boundary flow.

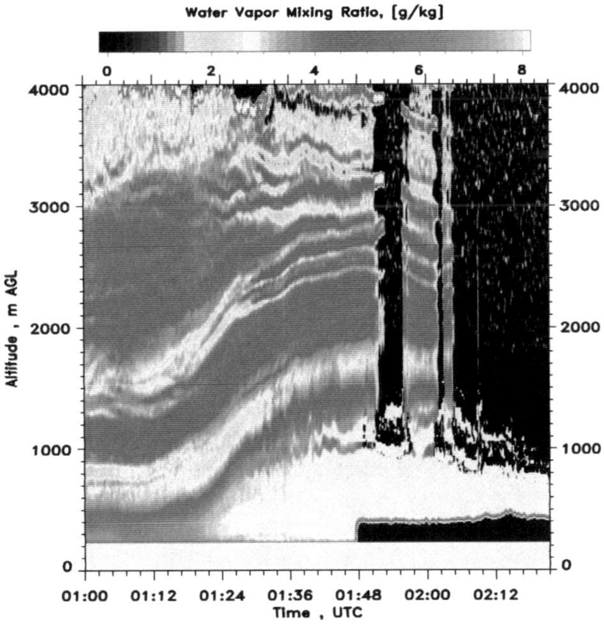

Figure 3. Zoom-in-view of figure 2. High-resolution water-vapour mixing ratio field during 0100-0220 UTC on 2 August 2007 showing an out-flow-boundary. Temporal and spatial resolutions are 10 s and 15 m, respectively, with 150 m gliding average in height.

4.3 Scanning temperature and aerosol measurements with RRL

On 25 August 2007 during IOP 18b, UHOH RRL performed a continuous RHI scan pattern as a part of Supersite cross-section scan scenario. The aim of this scan pattern is to collect atmospheric variables in 3-dimensions in such a way that the data of the neighbouring Supersites can be linked. Figure 4 shows the field of potential temperature (θ) that was measured between 1600-1700 UTC covering elevations from 0 to 180° with a step of 5°. At this time the elevation of the sun ranged between 20° to 10°. The lidar data is corrected for overlap effects at ranges below 1.3 km. As can be seen from the wind barbs in figure 4 of the radiosonde data the air flow was mainly westerly in lower heights (<2000m ASL or ~ <900m AGL) and in the free troposphere (>3000m ASL or ~ >1900m AGL). In between the wind was south-westerly orientated. Figure 4 shows some features that are consistent with ascending air motion due to overflow over the ridge as for rising air constant θ is expected. A warm θ-tongue is stretching downward just east of the mountain peak. This supports a process which is mixing down air from aloft to the ground layer. Also the PBC is slightly increased (as detailed in figure 5, see text below) within this θ-tongue.

Figure 5 shows the PBC field of six consecutive RHI scans from 1600-1630 UTC. PBC is calculated by use of the RRL technique with the aerosol data measured simultaneously to the temperature data. Each scan took ~ 5 minutes and RRL averages 650 shots in each line of sight. Temporal and spatial resolutions are 13 s and 37.5 m (10 range bins averaged), respectively. Close to the ground (< 1 km AGL), the aerosol field shows large variability with high PBC values following roughly the underlying orography. At higher altitudes, stable aerosol layers are present here which are quite constant in height independent of the orography below. In scan 1, two separate aerosol layers are seen centred in about 1.5 and 2.3 km AGL which later (scan 3 onwards) are merged to a single layer with a vertical extent of ~ 1 km. Orography induced aerosol flow dynamics is clearly visible near the altitude of lidar location.

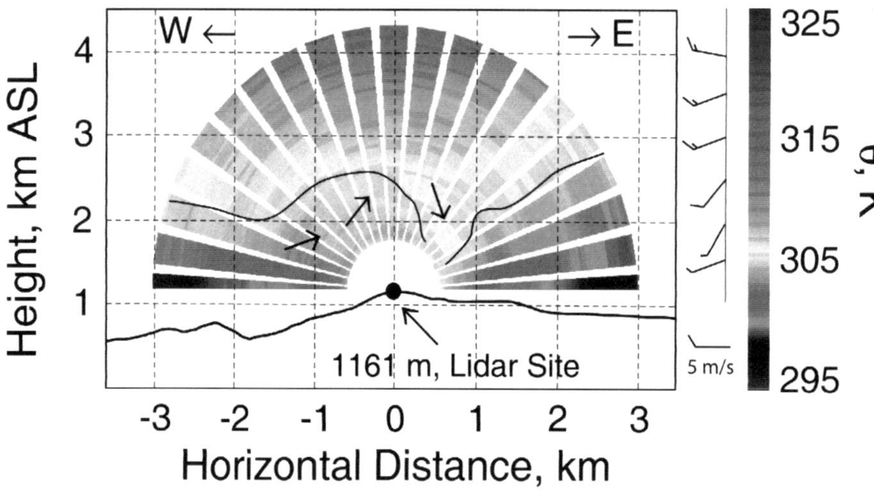

Figure 4. Field of potential temperature at 21 elevation angles measured at Hornisgrinde on 25 August 2007 between 16-17 UTC. The plane in which the RHI-scan was performed was orientated towards the neighbouring COPS Supersites (azimuth = 291°). The wind barbs show horizontal wind data measured by the radiosonde launched at the lidar site at 17 UTC. The black line marks the potential temperature of 305 K. For this field 13 consecutive raw-data scans were averaged giving an averaging time of 3 minutes for each profile. The range resolution is 37.5 m. A gliding average of 300 m was applied.

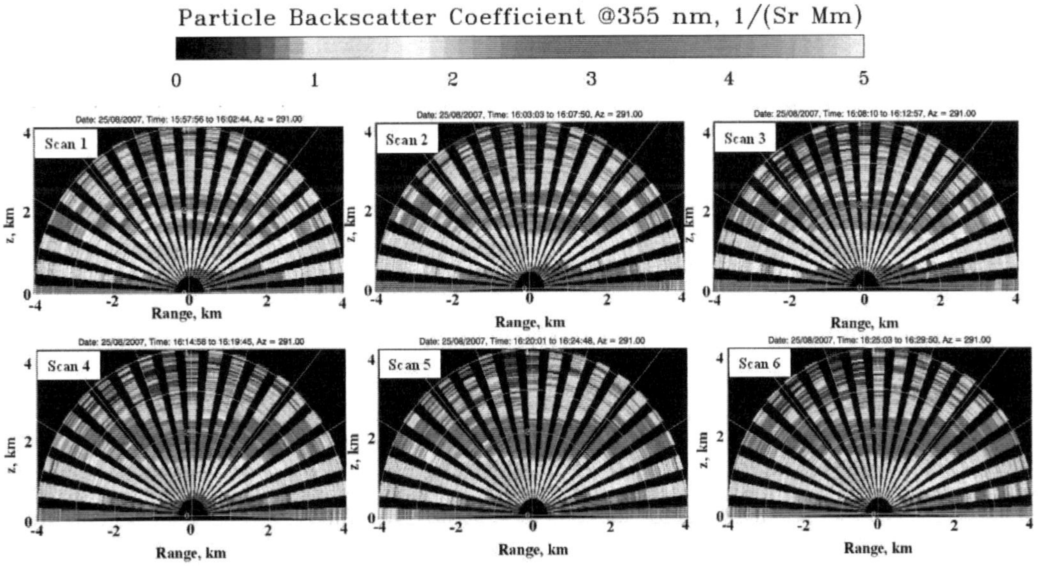

Figure 5. Sequence of six consecutive full-hemispherical RHI scans (upper left to lower right) of the particle backscatter coefficient field at 355 nm (COPS IOP 18b, 25 August 2007)

4.4 Scanning DIAL measurements

The UHOH DIAL system performed various scan patterns during the COPS IOPs. Two examples of RHI scan measurements of background-subtracted and range-square corrected offline signal intensity collected on 25 August 2007 (IOP 18 b) are shown in figure 6. An aerosol plume is observed in this region as red against the green/yellow background. This shows how fast the aerosols can be lifted up to higher altitudes (towards the CBL top) triggered by the orography-induced convective activities. Wind was blowing from west-south-west as evinced by the radiosonde profile (wind barbs in figure 4). For IR wavelengths (820 nm here), lidar received signal intensity is approximately directly proportional to the aerosol backscatter coefficient as the molecular Rayleigh scattering is relatively low in this wavelength region. Continuous RHI scans were performed in such a way that each scan took ~ 48 s both in forward and backward direction. The scan speed was set low enough (0.5°/s) in both directions to provide high angular resolution. The high laser power of the UHOH DIAL system allows capturing the aerosol structure up to a distance of more than 12 km.

Figure 6 shows diverse aerosol structures around a Cumulus mediocris (Cu med) cloud. This cloud attenuated the signals beyond it. A thick aerosol layer is present at an altitude of 2.0 km AGL throughout the measurement period as seen in the time-lapsed animation of 30 such consecutive RHI scans (available with the online version of this paper at the Institute of Physics web server). The animation reveals that two detached aerosol layers exist which are coupled as time progresses. Enhanced convective activity and the associated dynamics of the aerosol plume in the height between 0.2-0.6 km is visible. Wind driven aerosol dynamics at a height of 1.0 km AGL can also be seen in the animation. High dynamic range together with high resolution of the UHOH DIAL system offers an extraordinary tool to study the dynamics and aerosol transport in the CBL. The retrieval of the humidity field from such scanning measurements is in progress.

IOP Conf. Series: Earth and Environmental Science 1 (2008) 012031 doi:10.1088/1755-1307/1/1/012031

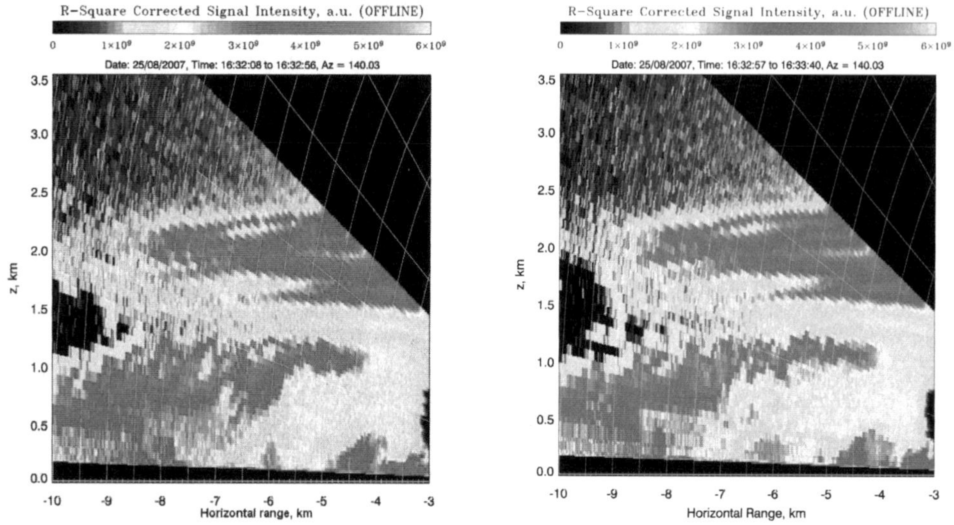

Figure 6 Example of RHI scans with the UHOH DIAL: range-corrected backscatter intensities showing the aerosol field around a Cumulus mediocris (Cu med) cloud. The scan speed was 0.5°/s, each profile is with 1-s average giving an angular resolution of 0.5°. Elevation angles of 2 to 25° are covered. The range resolution is 15 m. The horizontal scale gives the distance to the lidar in km. The scan direction is towards south-east. An animation of this event is available with the online version of this paper.

5. Summary and outlook

First results obtained from the COPS field campaign with two scanning lidar systems are presented here. High-resolution measurements of humidity and temperature during daytime are two of the most exciting features. The results show the multitude of ABL processes in a complex mountainous region. The analysis of the COPS data set is still in an early stage. Intensive research efforts have been initiated to connect all the data products from the different ground-based and airborne, active and passive remote sensing instruments to each other and to study the chain of processes leading to convective precipitation.

Acknowledgements

The new scanning water-vapour DIAL which has been developed within three interlinked projects was funded by the German Research Foundation (Deutsche Forschungsgemeinschaft, DFG). We like to acknowledge our colleagues at Institute for Tropospheric Research (IfT, Leipzig) for building the DIAL receiver and Deutsches Zentrum für Luft- und Raumfahrt (DLR Oberpfaffenhofen, Germany) for providing the seeders to stabilize the DIAL transmitter. The authors are grateful to GKSS, Geesthacht, Germany for the supply of the laser as of the scanning RRL well as to the Institute of Tropospheric Research, Leipzig, Germany for its injection-seeder .

References

[1] Wulfmeyer V, Behrendt A, Bauer H -S, Kottmeier C, Corsmeier U, Blyth A, Craig G, Schumann U, Hagen M, Crewell S, Di Girolamo P, Flamant C, Miller M, Montani A, Mobbs S, Richard E, Rotach M W, Arpagaus M, Russchenberg H, Schlüssel P, König M, Gärtner V, Steinacker R, Dorninger M, Turner D D, Weckwerth T, Hense A and Simmer C 2008 *Bull. Amer. Meteor. Soc.* (in print)
[2] Wulfmeyer V and Behrendt A (Eds.) and the COPS Investigators 2007 COPS Field Report

https://www.uni-hohenheim.de/cops/further_reading/further_reading.htm
[3] Radlach M, Behrendt A and Wulfmeyer V 2008 *Atmos. Chem. Phys.* **8** 159-169 http://www.atmos-chem-phys.net/8/159/2008/acp-8-159-2008.pdf
[4] Wulfmeyer V and Bössenberg J 1998 *Appl. Opt.* **37** 3825-44.
[5] Behrendt A, Wulfmeyer V, Di Girolamo P, Kiemle C, Bauer H -S, Schaberl T, Summa D, Whiteman D N, Demoz B B, Browell E V, Ismail S, Ferrare R, Kooi S, Ehret G, Wang J 2007 *J. Atmos. Oceanic Technol.* **24** 3-21
[6] Behrendt A, Wulfmeyer V, Kiemle C, Ehret G, Flamant C, Schaberl T, Bauer H -S, Kooi S, Ismail S, Ferrare R, Browell E V, Whiteman, D N 2007 *J. Atmos. Oceanic Technol.* **24** 22-39
[7] Behrendt A 2005 Temperature measurements with lidar In: C. Weitkamp (Ed.), Lidar: Range-Resolved Optical Remote Sensing of the Atmosphere (Springer Series in Optical Sciences 102 ISBN 0-387-40075-3 Springer New York) 273-305
[8] Schotland R M 1966 *Proc. Fourth Symp. on Remote Sensing of the Environment,* Ann Arbor, MI, Environmental Research Institute of Michigan, University of Michigan, 273–283
[9] Ehret G, Hoinka K P, Stein J, Fix A, Kiemle C, and Poberaj G 1999 *J. Geophys. Res.* **104** 31351-59.
[10] Poberaj G, Fix A, Assion A, Wirth M, Kiemle C, and Ehret G 2002 *Appl. Phys. B.* **75** 165-172
[11] Bruneau D, Cazeneuve H, Loth C and Pelon J 1991 *Appl. Opt.* **30** 3930–37
[12] Bruneau D, Quaglia P, Flamant C and Pelon J 2001 *Appl. Opt.* **40** 3462–75
[13] Bruneau D, Quaglia P, Flamant C, Meissonnier M and Pelon J, 2001 *Appl. Opt.* **40** 3450–61.
[14] Di Girolamo P, Behrendt A, Kiemle C, Wulfmeyer V, Bauer H, Summa D, Dörnbrack A and Ehret G 2008 *Remote Sensing of Environment* **112** 1552-68
[15] Schiller M, Wagner G and Wulfmeyer V 2007 CLEO conference 2007 USA
[16] Wulfmeyer V and Walther C 2001 *Appl. Opt.* **40** 5304-20
[17] Wulfmeyer V and Walther C 2001 *Appl. Opt.* **40** 5321-36
[18] Weckwerth T M, Parsons D B, Koch S E, Moore J A, Lemone M A, Demoz B B, Flamant C, Geerts B, Wang J and Feltz W 2004 *Bull. Amer. Meteorol. Soc.* **85** 253-277
[19] Radlach M, Behrendt A, Pal S, Riede A and Wulfmeyer V 2008 in 24th International Laser Radar Conference Boulder USA
[20] Arshinov Y F, Bobrovnikov S M, Zuev V E and Mitev V M 1983 *Appl. Opt.* **22** 2984
[21] Behrendt A and Reichardt J 2000 *Appl. Opt.* **39** 9 1372-78
[22] Behrendt A, Nakamura T, Onishi M, Baumgart R and Tsuda T 2002 *Appl. Opt.* **41** 7657-66
[23] Di Girolamo P, Marchese R, Whiteman D N and Demoz B B 2004 *Geophys. Res. Lett.* **31** L01106, doi:10.1029/2003GL018342 2004

IOP Conf. Series: Earth and Environmental Science 1 (2008) 012032 doi:10.1088/1755-1307/1/1/012032

Remote sensing of PBL meteorology and air quality: the outcome of the ESCOMPTE experiment

P Drobinski
Institut Pierre Simon Laplace/Laboratoire de Météorologie Dynamique, Ecole Polytechnique, Palaiseau, France

E-mail: philippe.drobinski@lmd.polytechnique.fr

Abstract. In the French Mediterranean basin, the large city of Marseille and its industrialized suburbs (oil plants in the Fos-Berre area) are major pollutant sources which cause frequent and hazardous pollution episodes especially in summer when intense solar heating enhances the photochemical activity and when sea-breeze circulation redistributes pollutants further north in the countryside. This paper summarizes the findings of five years of research on the sea-breeze in southeastern France and related mesoscale transport and dilution of pollutants within the ESCOMPTE program held in June and July 2001 (field experiment to constraint models of atmospheric pollution and emissions transport), obtained thanks to a composite observing system and a combination of remote sensing and in situ systems which produced a wealth of data. Indeed, the combination of established and novel and highly sophisticated remote sensing instruments with conventional in situ measurements (dense surface network and radiosondes) allowed to capture previously unseen details of the fine structure of the sea breeze, allowed unprecedented insight into the structure of the sea breeze flow and its contribution to ozone redistribution and allowed the validation of ultrahigh-resolution numerical research and weather prediction models as well as chemistry transport models.

1. Introduction

Summer atmospheric stagnant air episodes in coastal areas, leading to land/sea breeze flows and associated recirculations of air, have long been recognized as a major factor of development of severe photo-oxidant episodes. Those movements of air masses result from a complex combination of local and mesoscale winds that transport and redistribute pollutants across the air basins, generating exceedances of ozone regulatory thresholds that may recur for several consecutive days when ozone is retained in reservoir at high altitudes. Fast transport of high ozone concentration to the mid-troposphere even frequently occurs when these recirculations are combined with up-slope winds in the presence of mountains. Experimental evidence for such phenomena has been reported for large cities of North-America, and in Mediterranean areas where strong insolation and low wind speeds provide a favorable environment, and also on the highly polluted coasts of Asia.

The study of air quality along the coast of the Provence region (in southern France, see Fig. 1) is of particularly interest because of the sea breeze circulation and the presence of the large city of Marseille and its industrialized suburbs (oil refineries in the Fos-Berre area), which emit huge amounts of pollutants into the atmosphere. Subsequent degradation of air quality causes numerous and major pollution episodes especially during summer. This region witnesses a high frequency of photochemical pollution events compared to other French regions within large urban areas, with up to 6 days per month of surface ozone concentrations exceeding 90 ppb during one hour on average.

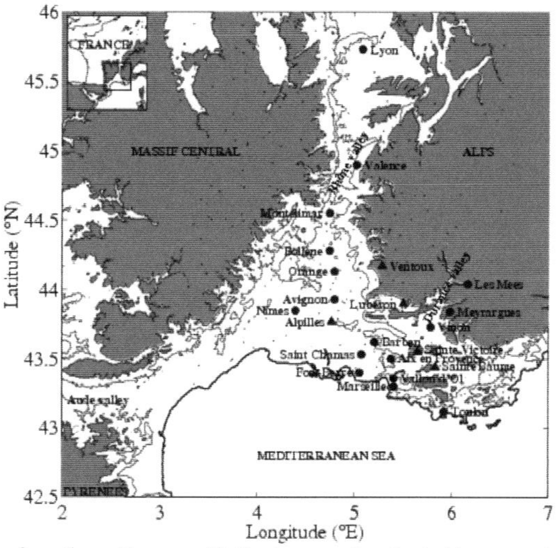

Figure 1: Map of southern France with the topography shaded in grey when higher than 500 m above sea level. 200-m height isocontour as well as the coastline are shown with solid and thick solid lines, respectively. Dots indicate the locations of the operational meteorological surface stations operated by Météo-France. The large black dots and triangles indicate the locations of cities and of mountain ranges or peaks, respectively. The small sub-panel at the upper left corner indicates with a triangle the region of France shown in the main panel.

The other peculiarity of this coastal area is the complexity of topography, combining the coast and the presence of two mountain barriers of different heights (the French Alps to the east that culminates at 4807 m and the Massif Central to the west that culminates at 1885 m), separated by the Rhône valley, a gap of 200 km length and 60 km width. On a southwestern-northeastern axis also lies the Durance valley with a fairly constant width of about 10 km and a length of about 60 km. These topographical features perturb the main sea-breeze flow, and put major limitations for numerical weather simulation models with coarse resolution (10-30 km horizontal resolution).

During summer 2001, meteorological and chemical measurements were taken in southern France in the framework of the ESCOMPTE experiment [1]. This campaign lasted nearly 6 weeks from June 7 to July 16, 2001 (hereafter and as examples June 7 will be written J07 and July 7, Ju07), and aimed at collecting a large amount of measurements from in-situ, passive and active remote sensors to document the four-dimensional state of the atmosphere over both meteorological and chemical aspects, and understand the precursors of photochemical pollution in the region. The ESCOMPTE program was designed to improve the understanding and forecasting of the life cycle of the sea-breeze, including its three-dimensional structure and associated contribution to transport and mixing of pollutants emitted from the Marseille and Fos-Berre areas, and to address the following questions:

- can a part of the temporal and spatial variability of pollutants in the ESCOMPTE target area be attributed to large-scale transport?
- how do land and sea breezes affect the diurnal redistribution of pollutants? Can the Rhône and Durance valleys modify the sea-breeze characteristics and consequently the pollutant distribution? Do slope winds (anabatic and katabatic winds) contribute significantly to pollutant ventilation?
- does the coastline shape have an impact on sea/land breeze circulation and transport? Does the urban area contribute to local-scale circulations (e.g. urban breezes) or/and disrupts the meso-scale land- and sea-breeze circulations?

This paper outlines the main results obtained from the composite observing network including in-situ and a very large number of active remote sensors (sodars, radars and lidars). More details on the challenges and results of the ESCOMPTE program, with all related references, can be found in the overview published by Drobinski et al. [2007] in the *Journal of Geophysical Research* [2] from which all the figures of this abstract are extracted.

2. Southern France as a target area for pollution studies

A 120×120 km^2 area around the city of Marseille and the Berre pond, in southern France, has been selected to host the ESCOMPTE field campaign (Fig. 1). The selection criteria were based upon strong pollutant sources (urban and industrial) within the experimental area, high occurrence of photochemical pollution events and the presence of a dense, operational pollution monitoring network. This area best fulfills these criteria.

This area is constituted by three different zones: (i) the urban area of Marseille with 1.2 million inhabitants; (ii) the Fos-Berre zone around a pond located 15 km northwest of Marseille and separated from the city by 300-m-high hills. It is characteristic of a strong industrial zone (refineries, power plants,...) surrounded by residential areas with important traffic; (iii) rural areas either agricultural or covered by Mediterranean natural landscape.

The atmospheric circulation over this area is highly influenced by orography. On large scale, the Alps and Massif Central at times reinforce in the Rhône valley a strong northerly flow called Mistral. On a smaller scale, the ranges of smaller mountains parallel to the coastline (i.e., Sainte Baume, Sainte Victoire, Luberon, Mont Ventoux) tend to channel the air masses in a west-east manner. The sea-land contrasts induce sea-breeze during daytime, with advection of marine air masses as far as 100 km inland.

The period and the area of the ESCOMPTE campaign were particularly favorable to the development of photochemical pollution events. During the studied month, no less than 11 days were affected by atmospheric pollution. Intensive observing period (IOP) 1 covered J14 and J15 which were considered as a test for the whole ESCOMPTE experimental setup. IOP 2 started on J21 and ended on J26. It was decided to split this IOP into two contiguous parts, called 2a and 2b, since meteorological conditions changed significantly on J23 at 1700 UTC. IOP2a corresponds to the end of a Mistral situation with a moderate northwesterly to westerly wind, clear skies, hot temperature (>30°C). The Marseille and Fos-Berre plumes then extended towards the east and over the sea (see Fig. 1). The highest surface ozone concentrations were observed near Toulon, in the easternmost part of the ESCOMPTE domain. IOP2b best characterizes a pollution event. During three windless days, temperature (>34°C) and surface ozone concentration reached high values (125 ppb all around Aix en Provence on J24; 100 ppb over the whole domain and up to 150 ppb in the Durance Valley on J25; 115 ppb in the Rhône valley on J26 with a light southerly wind). This episode was heavily documented. IOP 3 started on Ju02 and ended on Ju04 and was a short but interesting episode, characterized by a sea-breeze interacting with a large-scale flow veering from the northwest to the southwest in altitude and blowing from the south/southwest near the surface. On the last day, the southwesterly wind became stronger and cloudiness appeared on the domain. The last IOP (IOP 4), starting on Ju10 and ending on Ju13, was launched under hot temperature and clear sky conditions but with a westerly/northwesterly synoptic wind that prevented the accumulation of pollutants.

3. Challenges for new measurements

During ESCOMPTE, aircraft flights, together with tethered and constant volume balloons, radiosonde systems and remote sensing measurements, and associated with the continuous surface measurements, provided the most comprehensive picture of atmospheric dynamics and chemistry over a wide range of relevant spatial and temporal scales in the French Mediterranean area. An extensive description of the instrumental set-up, with the location of all instruments, can be found in [1]. The instrumental set-up was designed in order to document the main sea-breeze features and their modulation by the regional complex topography and the urban area and to relate them to ozone distribution.

Since the sea-breeze is driven by horizontal gradient of surface temperature and heat fluxes, the instrumental set-up relied considerably on a dense network of surface meteorological stations. For ESCOMPTE, the already dense network of conventional meteorological surface stations (operated by Météo-France) was completed with 14 additional stations (measuring pressure, temperature, relative humidity, wind speed and direction, short wave and/or total radiation measurements). In addition, the surface energy budget was measured in nine of these stations. Temperature and relative humidity were also measured onboard three mobile cars at rural sites, and on an network of 20 sensors spread over Marseille agglomeration. This extremely dense station network was necessary to resolve small scale features and to achieve the same resolution for observational data as for numerical weather prediction and research models, which was essential for the validation of these models. The vertical structure of the sea-breeze was documented by a tethered balloon, conventional radiosoundings and a large deployment of active remote sensors like Doppler lidars [3,4], sodars [5], UHF wind profilers [6] in the city of Marseille, in semi-urban and rural zones. The Doppler lidars were operated in the infrared spectral region, one was operated from the ground at the Vallon d'Ol in the northern suburbs of Marseille (TWL for transportable wind lidar) (Fig. 2 and Table 1), the other one was on board the German Falcon 20 (WIND for Wind Infrared Doppler lidar). The UHF wind profilers, operated at 1238 MHz (~0.3 m wavelength), were deployed in a network at different distances from the Marseille city center (Fig. 2). The surface ozone concentration was documented by the stations operated by the local air quality networks completed by 18 continental surface stations used to follow the pollutant plumes and 2 stations onboard of ships used to obtain information on the chemical conditions at the outskirts of the ESCOMPTE region and monitor atmospheric chemistry over the sea. The vertical profiles of ozone concentration were observed by ground-based ozone lidars operated in the ultra-violet spectral region [7,8,9] (Fig. 2 and Table 1). Several precise locations were defined, combining a wind profiler (Doppler lidar, radar or sodar) with an ozone lidar, in order to create a multi-scale (local and regional) remote sensing network on emission zones where photochemical processes are dominant, on intermediate zones where ozone is mainly generated by transport processes, and on rural zones in order to give information on the domain limits.

During ESCOMPTE, the use of the airborne Doppler lidar WIND on board the German Falcon 20, opened the additional spatial dimension of the aircraft's track and allowed the retrieval of the wind field in a two-dimensional plane in the clear-air or partially cloudy atmosphere. In addition to the Falcon 20 flights, the measurements of the mean and turbulent meteorological and chemical variables in the PBL were provided by three French research aircraft (Fokker 27 ARAT, Merlin IV and Piper Aztec 23) and two additional German research aircrafts (Dornier 128 and microlight research aircraft) performing classic dynamical, thermodynamical and air chemical slow and fast measurements. Constant volume balloons were also used to follow pollutant plumes, and to investigate the evolution of the plume thermodynamics and chemical properties in a "lagrangian mode".

Table 1: Specifications of remote-sensing instruments (S stands for wind speed, D for wind direction and W for the vertical velocity; IOP stands for intensive observing period).

Parameters	Method	Time resolution (s)	Spatial resolution (m)	Altitude range (m)	Accuracy	Operation
0_3, aerosols	Angular lidar	2	50-200	0 3000	2 µg m^{-3}, 0.01 km^{-1}	IOP
0_3, aerosols	Vertical lidar	30	200-1000	500-7000	2 µg m^{-3}, 0.01 km^{-1}	IOP
0_3, aerosols	Vertical lidar	3	100-200	100-3000	3 µg m^{-3}, 0.01 km^{-1}	24 h/day
Radial wind, aerosols	Wind lidar	2	300-450	0-10000	1 m s^{-1}	IOP
S, D, W	UHF radar	5	75-375	150-4000	1 m s^{-1}	24 h/day
S, D, W	VHF radar	15	375	1500-15000	1 m s^{-1}	24 h/day
S, D, W	Mini VHF radar	5	300	600-5000	1 m s^{-1}	24 h/day
S, D, W	Sodar	15-30	20-50	50-800	1 m s^{-1}	24 h/day
S, D, W	Mini sodar	15	5	10-100	1 m s^{-1}	24 h/day

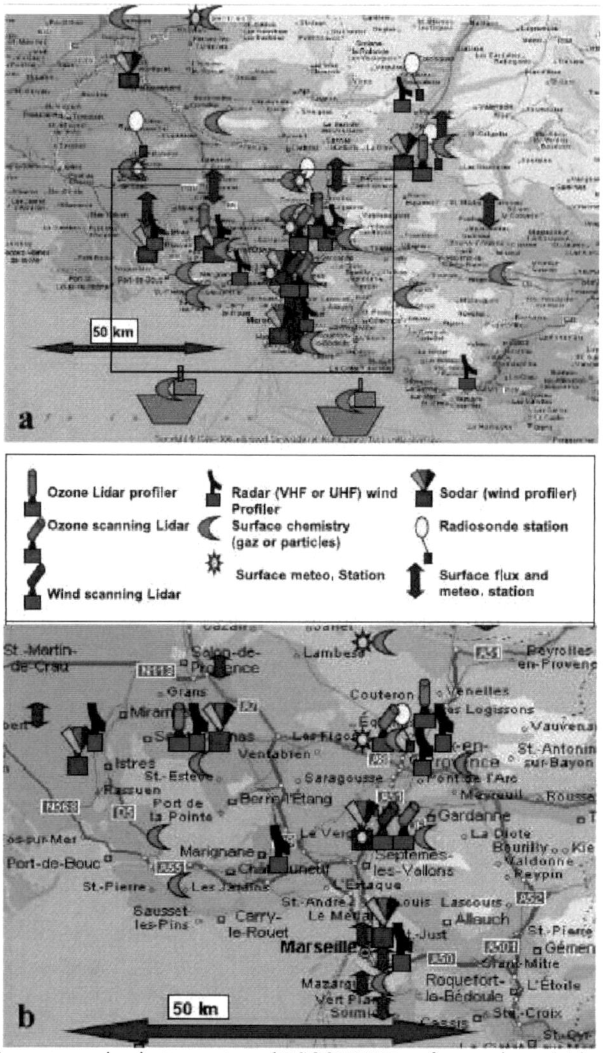

Figure 2: Locations of remote sensing instruments and ESCOMPTE surface stations: (a) in the whole domain, (b) near the emission zones during the field experiment.

4. Results

4.1. Large-scale contribution to local ozone distribution

In the PBL, the afternoon development of local pollution plumes is observed almost every day with a large spatial variability caused by the complex local atmospheric circulation and a large surface heterogeneity of the emission sources. Conversely, above the PBL, the spatial variability is much smaller, indicating the large horizontal extent of layers presenting an excess or a depletion of ozone. As an illustration, Fig. 2 shows the time evolution of the vertical profile of ozone measurements on J21 (IOP2a) and J26 (IOP2b) by two ozone lidars located 30 km apart, at Aix les Milles (near Aix en Provence) and Saint Chamas. Below about 1500 m height, small time scale variability can differ from one location to the other whereas in the free troposphere, a broad agreement is found, meaning that similar air masses are sampled by the two lidars. Thus, above the PBL, the ozone concentration variability is driven by dynamical processes at large scales (typically larger than 50 km) while smaller scale variability in the PBL is due the mesoscale circulations.

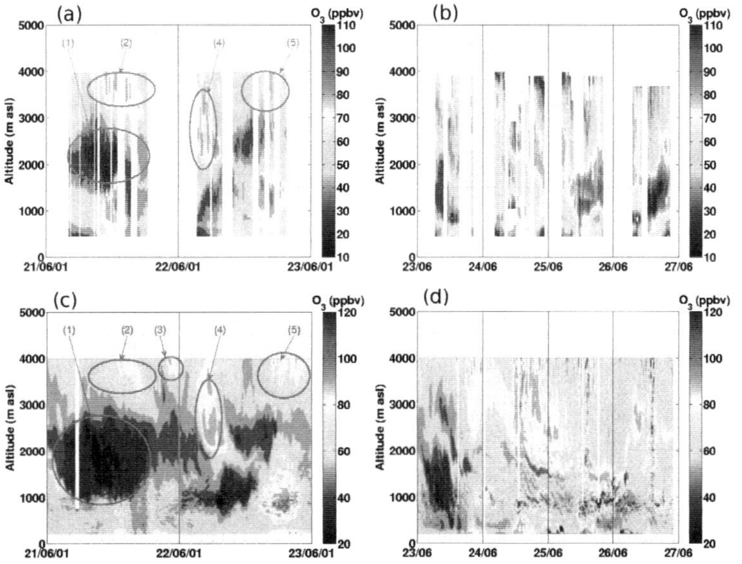

Figure 3: Vertical profiles of ozone concentration as a function of time measured by two DIAL ozone lidars during IOP 2a (a and c) and IOP 2b (b and d): the ALTO ozone lidar located at Aix les Milles (a and b) and the EPFL ozone lidar located at Saint Chamas, 30 km west of Aix les Milles (c and d).

In order to assess the dynamical processes responsible for the ozone variability in the free troposphere, numerical modelling and lagrangian studies were conducted for IOP2, during which typical types of weather were observed. These studies showed that long-range transport but also stratospheric intrusions contribute to the high variability over the whole tropospheric column. In detail, for IOP2a (J21 and J22), synoptic scale processes above the Northern Atlantic region drive the observed ozone variability (mainly from stratospheric origin, see ellipses in Fig. 3a,c) whereas during IOP2b (J23 to J26), most free tropospheric air masses come from the PBL of the Iberian Peninsula.

4.2. Horizontal mesoscale transport and vertical mixing

Horizontal transport of ozone was thoroughly analyzed by means of surface measurements, ground-based remote sensors and numerical mesoscale models that documented in detail the vertical and horizontal structure of the sea-breeze, its temporal evolution and associated mass transport. Particularly, the dataset provided by the airborne Doppler lidar allowed a unique insight on the three dimensional structure of the sea-breeze including its inland penetration, its depth and intensity and the perturbation by the complex terrain of its main features (Fig. 4). The impact of the Rhône and Durance valleys (Fig. 1) on the sea-breeze dynamics (depth, intensity and penetration) has been investigated and it was shown that the sea-breeze is generally affected by flow splitting between the Durance and the Rhône valley whenever it reaches the bifurcation zone (it is not the case during IOP2a, see Fig. 4a). If the sea-breeze flow reaches the entrance of the Durance valley (other IOPs), it is then accelerated in the Durance valley due to lateral constriction and progresses further inland. In presence of an offshore flow in the Rhône valley (J21 to J25 and Ju10), lateral constriction does not affect the sea-breeze. On J26, Ju03 and Ju04, the combination with an onshore flow leads to further penetration inland and intensification of the low-level southerly flow in the Durance but also in the Rhône valley (see Fig. 4c,d,e). If the onshore synoptic flow is too strong, the sea-breeze flow can be inhibited as on Ju04 in a large area in the Rhône valley.

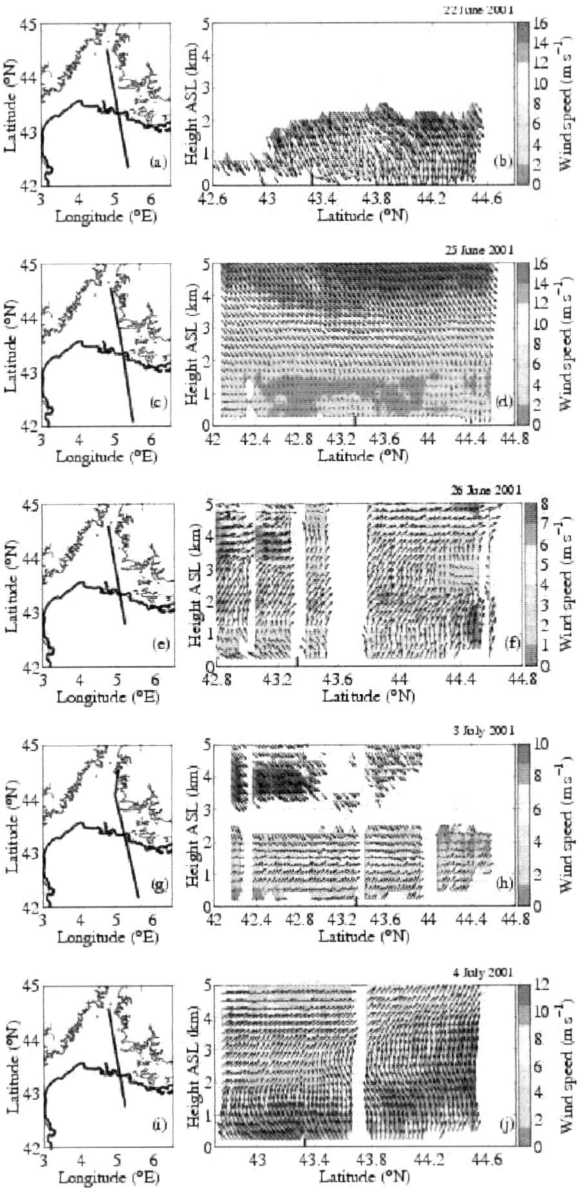

Figure 4: Panels a, c, e, g and i: Topography of the ESCOMPTE area. The thin and thick lines indicate the 500-m isoline and the coastline, respectively. The thick segment represent the flight track of the DLR Falcon 20 along which the vertical cross section of the horizontal wind field from the WIND data is extracted and shown in panels b, d, f, h and j. Panel b, d, f, h and j: Vertical cross section of the horizontal wind field from the WIND data along the leg shown in panels a, c, e, g and i on 22, 25 and 26 June 2001 and 3 and 4 July 2001. Arrows indicate the horizontal wind speed and direction as a function of height and the superimposed color indicates the wind speed. The absence of data corresponds to discarded unreliable data. The dark thick tick indicates the location of the coast.

The impact of valley channelling on ozone plume penetration has been investigated quantitatively for IOP2b. Time series of ozone concentration at different ground stations along the Durance and Rhône valleys indicate clearly the distinction between the peak of photochemistry at 1200 UTC, and the peak due to ozone production associated with sea-breeze transport around 1500-1800 UTC (Table 2). Between 1500 and 1800 UTC, photochemistry production decreases to about 25 % of the maximum found at 1200 UTC. The observed ozone increase between 1200 and about 1800 UTC, at a rate of about 7-10 ppb hr^{-1} (with a maximum of 40 ppb hr^{-1}), is thus mainly due to transport. In the Rhône valley, wider than the Durance valley, the plume of ozone precursors is more diffuse and the ozone peaks are generally lower.

IOP Conf. Series: Earth and Environmental Science **1** (2008) 012032 doi:10.1088/1755-1307/1/1/012032

Table 2: Ozone concentration in ppb at different ground stations along the Rhône and Durance valleys at about 1200 UTC, and between 1500 and 1800 UTC (left/right numbers, respectively).

	Distance from the coast	J24	J25	J26
Alpilles (Rhône valley)	35 km	72.3/85.4	100.2/111.7	80.5/57.5
Le Pontet (Rhône valley)	50 km	65.7/78.9	78.9/94.0	85.4/111.7
Bollène (Rhône valley)	90 km	63.4/71.6	75.1/78.6	65.7/91.5
Meyrargues (Durance valley)	45 km	79.3/114.1	96.7/106.4	81.2/98.6
Vinon (Durance valley)	60 km	55.8/91.7	71.9/124.0	62.6/119.7
Les Mees (Durance valley)	80 km	57.3/87.8	72.5/135.9	71.0/109.2

The leading edge of the sea-breeze forms in general a typical frontal zone with intense turbulence mixing and strong updrafts while a sharp dividing line is maintained between the sea-breeze flow and the environmental flow. It is the location of strong vertical transport and mixing of pollutants. Figure 5 shows the passage of the sea-breeze front above the UHF wind profiler on J24 (IOP2b) at Saint Chamas.

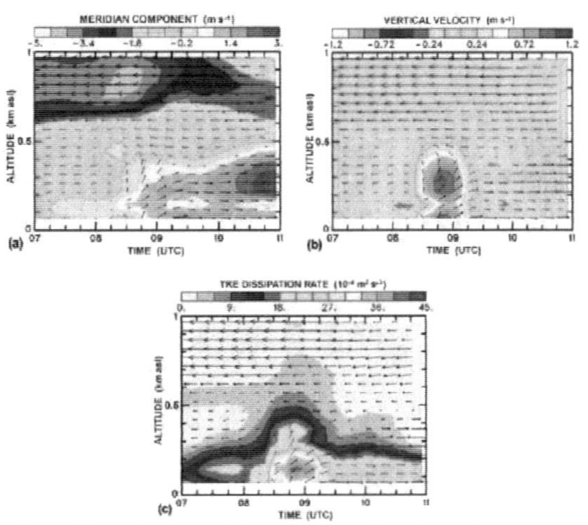

Figure 5: Time-height sections of meridian wind speed (i.e. the south/north wind component v), vertical wind components and dissipation rate measured with Saint Chamas UHF wind profiler on J24, with superimposed (v,w) wind vectors.

This unique data set perfectly illustrates the dynamics of the sea-breeze front. The sea-breeze starts blowing at Saint Chamas (very close to the shore) at about 0900 UTC when the wind takes a southerly component. The passage of the sea-breeze is materialized by strong updraft (>1 m s^{-1}) and large dissipation rate (about 45×10^{-4} m^2 s^{-3} maximum) up to about 0.5 km. Maximum turbulence is found up to about 0.3 km within the sea-breeze head due to large gradient of horizontal wind speed through the sea-breeze front generating shear induced turbulence. After the passage of the front, the sea-breeze is only 0.3 km deep with still large dissipation rate (about 30×10^{-4} m^2 s^{-3} maximum) due to near-surface shear and buoyancy production of turbulence by convection. One can note the weak downward motion following the passage of the sea-breeze front associated with detrainment in the wake of the sea-breeze front. The vertical mass fluxes can take values half as large as the sea-breeze induced horizontal air mass flux in the Rhône valley. Detrainment in the wake of the sea-breeze front does not compensate for this net upward export of air mass from the PBL to the free troposphere. This is visible in Fig. 5 with upward motion penetrating deeply in the free troposphere.

4.3. Advances in physical and chemical numerical modelling thanks to remote sensor data

The deployment of innovative remote sensors to document at high temporal and spatial resolutions the three-dimensional flow structure was the instrumental core for new methods of validation of high resolution modelling, especially for weak flow dynamics (at most 5-6 m s^{-1}) which has never been done before. The fine-scale flow was simulated with mesoscale numerical model Méso-NH using nested domains with 250-m horizontal resolution for the finest domain, similar to the radial resolution of a ground based Doppler (called TWL for transportable wind lidar) deployed at Vallon d'Ol. To do so, similar simulated radial velocity fields are computed from the simulated flow using the relation between the radial velocity and the three components of the wind:

$$v_r = u\cos\phi\sin\psi + v\cos\phi\cos\psi - w\sin\phi \qquad \text{(Eq. 1)}$$

where u is the zonal wind component, v the meridional wind component, w the vertical wind speed, ϕ the elevation angle and ψ the azimuth angle. Figure 6 shows the comparison between the radial velocities from the TWL (left row) and from the model (right row) for one vertical scan across the northern district of Marseille, at 190° from the north on 26 June 1001 at 1400 UTC. The TWL data indicate the superimposition of 4 layers (labelled A, B, C, and D in Fig. 6). In layers B, C and D, the radial velocities are globally negative. Near the surface, the TWL measures positive radial velocities up to 600-700 m ASL to a distance of 4.7 km away from the TWL. This layer corresponds to layer A. Beyond 4.7 km to the south of the TWL, negative velocities are measured near the surface. This change of sign of the radial velocity illustrates the convergence between a flow with a northerly component and a flow with a southerly component above the city core. Despite some discrepancies (location of the convergence zone, depth of layer A), the numerical model is able for this particular case to reproduce the very fine scale structure of the sea-breeze above the city of Marseille, both qualitatively and quantitatively.

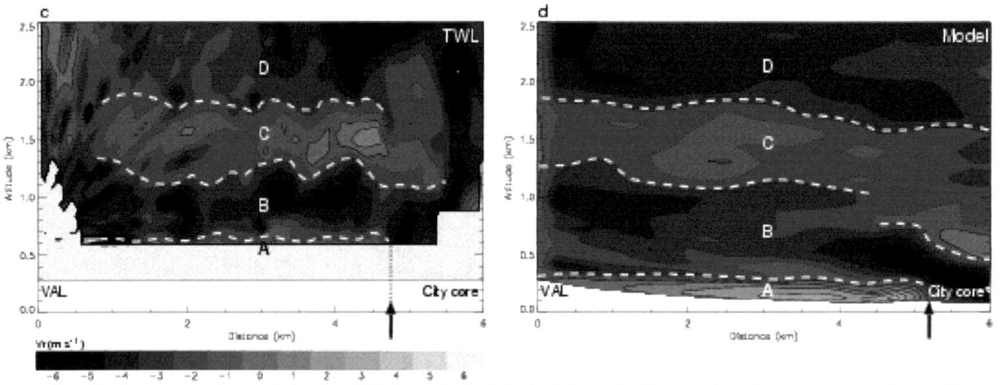

Figure 6: Vertical cross sections of observed (left) and modelled (right) radial velocities along the ground-based Doppler lidar line-of-sight on 26 June 2001 at 1400 UTC. The solid and the dashed isolines distinguish the positive and the negative radial velocities, respectively. The model outputs correspond to the 250 m domain grid size, which is similar to the radial resolution of the ground-based Doppler lidar measurements.

Ozone lidars also proved to be an extremely relevant tool to validate chemistry transport models, allowing a new insight on ozone meso-scale variability in the free troposphere. Figure 7 shows the time-series of observed ozone values during IOP2 at 1500 m and 3100 m above ground level and the corresponding simulated values obtained with the chemistry transport model MOCAGE. Part of the signal observed with the lidars is due to a tropopause fold event that occurred on the Atlantic and that travels up to the ESCOMPTE area. Small-scale spatial and temporal variabilities are related to the actual positions of the instruments regarding local emissions and re-circulations and present a diurnal variability that the models can generally quite well capture. Interestingly, differences also appear between the two lidars above 2500 m. Such differences are rather unexpected at such altitudes given the proximity of the two sites, only located 30 km apart. The model used here is not capable of capturing these differences (the chemistry transport model resolution being approximately 10 km). A much better agreement is observed just above the PBL (1500 m) between the observations at the two sites and between the observations and the model. The differences observed at the meso-scale in the free troposphere requires further investigation, as the 2500-4000 m altitude range is the one where residual pollution layers and upper tropospheric and stratospheric air masses come in contact.

IOP Conf. Series: Earth and Environmental Science 1 (2008) 012032 doi:10.1088/1755-1307/1/1/012032

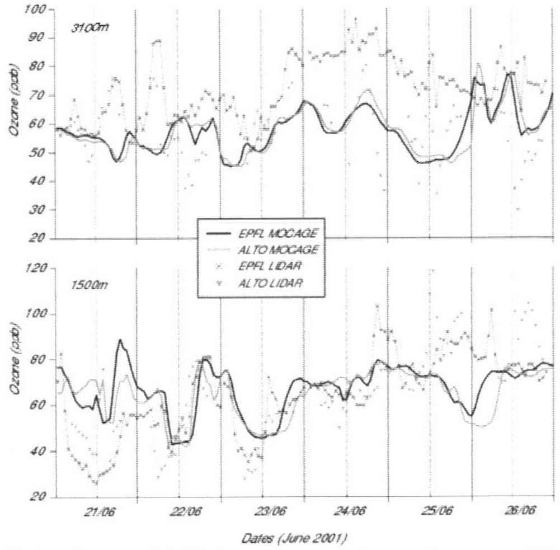

Figure 7: Lidar observations (Saint Chamas (EPFL) crosses, and Aix en Provence (ALTO) triangles) and numerical simulations (MOCAGE chemistry transport model) of ozone time-series at Saint Chamas (black) and Aix les Milles (gray), 1500 m (bottom) and 3100 m (top) above ground level.

5. Conclusion and perspectives

Looking back from some distance, the design of the composite observing network may be assessed as quite positive, even though lessons can always be learned. The composite observing system and the combination of remote sensing and in-situ systems produced a wealth of data which allows unprecedented insight into the structure of the sea-breeze flow and its contribution to ozone redistribution. The combination of established and novel and highly sophisticated remote sensing instruments with conventional in situ measurements (dense surface network and radiosondes) allowed to capture previously unseen details of the fine structure of the sea-breeze and allowed the validation of ultra-high resolution numerical research and weather prediction models as well as chemistry transport models. From this work, an operational numerical platform for air quality forecast at fine-scale, has been developed and is currently used by the local air quality network.

The main findings are that, despite the complexity of the terrain, the horizontal transport is mainly driven by the sea-breeze and its inland extent by the location of the sea-breeze front. The pollutant dilution along its horizontal transport is modulated by the PBL depth which remains shallow near the coast (and thus can contribute to high primary and secondary compound concentrations) and becomes deeper further inland. The vertical exchanges are predominantly directed from the PBL into the free troposphere, even though entrainment of rich-ozone air within the PBL can be found at sunrise and over the mountains in some occasions. Most of the vertical air mass flux occurs at the sea-breeze front, however remaining smaller than the sea-breeze induced horizontal mass flux, thus contributing to pollution accumulation just upstream of the sea-breeze front. Detrainment at the top of the following breeze within the PBL can not compensate for the upwards export. Pollution venting along the slopes is the second contribution to the net vertical export into the free troposphere. It is enhanced by the combination with the sea-breeze flow which induces a gravity wave-like oscillatory motion along the slope which affects the time evolution of ozone and nitric oxide concentration. Comparatively to the significant vertical exchanges at the sea-breeze front and along the slopes, entrainment/detrainment at the PBL top is an order of magnitude weaker especially when exchanges of ozone are considered. So, aged tropospheric polluted air masses are not substantially incorporated into the PBL.

Finally, despite the significant progress made in sea-breeze understanding, modelling and forecasting thanks to the ESCOMPTE programme, several key issues are still at best partly understood: (i) a quantitative experimental estimation of the exchanges of pollutant between the PBL and the free troposphere along the mountain slopes has not been conducted during ESCOMPTE because the instrumental set-up did not initially aimed at quantifying this contribution, especially in the countryside; (ii) the model initialization proved to be in some occasions problematic because of a lack representativity of the measurements in the valleys, at hill tops or over very heterogeneous terrain especially during anticyclonic conditions; (iii) the numerical simulation of small scale tropospheric ozone variability, well above the PBL, is currently very challenging as the accurate simulation of these features requires that both long-range transport and detailed local circulations and photochemistry are well accounted for.

Further modelling work, involving a hierarchy of coupled chemistry and transport modelling systems that allows to cover the important range of time and spatial scales involved, as well as future observing efforts in the Mediterranean area are needed to further investigate such small-scale variability of free tropospheric ozone.

6. References

[1] Cros B, Durand P, Cachier H, Drobinski P, Frejafon E, Kottmeier C, Perros PE, Peuch VH, Ponche JL, Robin D, Saïd F, Toupance G, and Wortham H 2004 The ESCOMPTE program. An overview *Atmos. Res.* **69** 241

[2] Drobinski P, Saïd F, Ancellet G, Arteta J, Augustin P, Bastin S, Brut A, Caccia JL, Campistron B, Cautenet S, Colette A, Cros B, Corsmeier U, Coll I, Dabas A, Delbarre H, Dufour A, Durand P, Guénard V, Hasel M, Kalthoff N, Kottmeier C, Laskri F, Lemonsu A, Lohou F, Masson V, Menut L, Moppert C, Peuch VH, Puygrenier V, Reitebuch O and Vautard R 2007 Regional transport and dilution during high pollution episodes in Southern France: Summary of findings from the ESCOMPTE experiment *J. Geophys. Res.* **112** D13105 doi:10.1029/2006JD007494.

[3] Bastin S, Drobinski P, Dabas AM, Delville P, Reitebuch O and Werner C 2005 Impact of the Rhône and Durance valleys on sea-breeze circulation in the Marseille area *Atmos. Res.* **74** 303-328

[4] Drobinski P, Bastin S, Dabas AM, Delville P and Reitebuch O 2006 Variability of the three-dimensional sea-breeze structure in Southern France: Observations and evaluation of empirical scaling laws *Ann. Geophys.* **24** 1783-1799

[5] Bastin S and Drobinski P 2005 Temperature and wind velocity oscillations along a gentle slope during sea-breeze events *Bound.-Layer Meteorol.* **114** 573-594

[6] Guénard V, Drobinski P, Caccia JL, Campistron B and Bénech B 2005 An observational study of the mesoscale mistral dynamics *Boundary Layer Meteorol.* **115** 263-288

[7] Ancellet G and Ravetta F 2005 Analysis and validation of ozone variability observed by lidar during the ESCOMPTE-2001 campaign *Atmos. Res.*, **74** 435- 459

[8] Simeonov P, Ristori P, Balin I, Calpini B and Van Den Bergh H 2004 A new DIAL based on a N2 Raman converter for ozone monitoring: Design and application during the ESCOMPTE Campaign, paper presented at XX Quadrennial Ozone Symposium, Int. Ozone Comm., Kos, Greece

[9] Augustin P, Delbarre H, Lohou F, Campistron B, Puygrenier V, Cachier H, and Lombardo T 2006 Investigation of local meteorological events and their relationship with ozone and aerosols during an ESCOMPTE photochemical episode *Ann. Geophys.*, in press

14th International Symposium for the Advancement of Boundary Layer Remote Sensing IOP Publishing
IOP Conf. Series: Earth and Environmental Science **1** (2008) 012033 doi:10.1088/1755-1307/1/1/012033

Evaluation of data assimilation techniques for a mesoscale meteorological model and their effects on air quality model results

[1]Amicarelli, Andrea, [1]*Gariazzo, Claudio, [2]Finardi, Sandro, [1]Pelliccioni, Armando, [2]Silibello, Camillo

[1]ISPESL – Dipartimento Insediamenti Produttivi e Interazione con l'Ambiente,

Via Fontana Candida, 1 00040 Monteporzio Catone (RM) Italy

[2]ARIANET, via Gilino 9, 20128 Milano, Italy

* claudio.gariazzo@ispesl.it

Abstract. Data assimilation techniques are methods to limit the growth of errors in a dynamical model by allowing observations distributed in space and time to force (nudge) model solutions. They have become common for meteorological model applications in recent years, especially to enhance weather forecast and to support air-quality studies. In order to investigate the influence of different data assimilation techniques on the meteorological fields produced by RAMS model, and to evaluate their effects on the ozone and PM_{10} concentrations predicted by FARM model, several numeric experiments were conducted over the urban area of Rome, Italy, during a summer episode.

1. Introduction

Atmospheric dispersion models (ADMs) are widely used for air quality assessment. It is well known that the performance of these models strongly depends on the driving meteorological data. The latter are usually provided by prognostic meteorological models. In theory, these models are capable of producing realistic results. However, models only approximate the description of physical processes through numerical solution of coupled partial differential equations. The modeling capability is therefore limited by the computational resolution, and subgrid phenomena have to be described by proper parameterizations. They consequently cannot replicate observations exactly. Agreement between model solutions and observations can be increased by data assimilation techniques that force model solutions to be more consistent with observations. However, those techniques have to be carefully applied, particularly in developing the data insertion strategy that controls when and where the observations are assimilated or how strongly they affect the solutions. A study devoted to how different data assimilation techniques can influence the meteorological fields and what are the correspondent effects on air quality predictions, has been performed over the urban area of Rome, Italy.

2. Description of the modelling system

2.1. Meteorological modelling

The meteorological fields used for this study have been obtained by means of the RAMS ver. 6 [1] prognostic model. To better reproduce the air flow within Rome urban area, the wind fields in the urban domain were calculated by means of the MINERVE [2] diagnostic mass consistent model using observed meteorological data as well as wind vertical profiles produced by RAMS at the boundaries of the urban domain. The meteorological modeling system is completed by SURFPRO [3], a diagnostic module to produce PBL scaling parameters, dry deposition velocities and turbulent diffusivities fields, on the basis of the meteorological fields and landuse maps.

2.2. Chemical Transport Model

The dispersion and the chemical evolution of the pollutants are based on the FARM model [4]. FARM is a three-dimensional Eulerian model dealing with transport, dispersion and multiphase chemistry of pollutants in the atmosphere. Photochemical reactions are described by means of SAPRC-90 chemical scheme [5]. The Models-3/CMAQ Aero-3 module [6], based on the modal approach, is employed for particulate matter. In this approach particle size distributions are represented as the superposition of three lognormal sub-distributions or modes.

3. Methodology

3.1. Models setup and input data

To consistently take into account the effect of local as well as distant sources and to describe processes dominated by scales larger than the city scale (e.g. sub-synoptic flow features, photochemical smog), a nested approach considering three domains was adopted. The background domain covers a large portion of the Italian peninsula (62x62 cells at 16 km of grid spacing), the intermediate domain considers a significant portion of Central Italy (66x58 cells at 4 km of grid spacing) and the target domain includes urban area of Rome (61x61 cells at 1 km of grid spacing). A summer episode (20-24 June 2005) was selected for this sensitivity study. It was characterized by high ozone concentrations associated with typical summer Mediterranean high pressure conditions (prevailing local sea-land breeze circulation).

To better characterize the meteorological conditions over the Rome area, the data routinely collected by the local monitoring network and by Air Force Meteorological Service/World Meteorological Organization (SMAM/WMO) stations were integrated with surface and upper-air measurements carried out at different locations during experimental field campaigns conducted during June 2005. These measurements include: wind and turbulence parameters collected by means of two sonic anemometers placed on the roof of the Istituto Superiore Prevenzione e Sicurezza sul Lavoro (ISPESL) downtown Rome and in the Villa Pamphili park (west of the city); urban wind and temperature, determined at the Ufficio Centrale di Ecologia Agraria (UCEA) Institute tower (downtown Rome) and at the Istituto Nazionale di Astrofisica (INAF) Astronomical Observatory (top of Monte Mario hill, east of Rome); wind vertical profiles carried out at Villa Pamphili park, by means of a SODAR/RASS instrument, and at the Tor Vergata science park (rural area, south of Rome), by means of the Institute of Atmospheric Sciences and Climate of the National Research Council (CNR-ISAC) Institute SODAR system. The SODAR/RASS system located at Villa Pamphili park was used for comparison with model results. It is a commercial Metek SODAR/RASS system. Narrow acoustic beams are obtained by means of a phased array of 16 loudspeakers. It provided 10' average values between 40m and 400m with a vertical spatial resolution of 20m. Radial wind components are derived from the mean spectra with an accuracy of about 0.1m/s. The 10 minutes averaged data were used in further analyses only if the acceptance test was satisfied. The latter is based on different parameters such as signal to noise ratio and statistical significance of measured signal and ambient noise. Signal to noise ratios were used for quality assurance, selecting only those data with positive values. During the studied period its data availability were 73% at 180 m height. As hourly wind data, required by the RAMS model, were not available for the SODAR systems, the last 10 minutes averaged data, available at every hour, were used instead.

SODAR data were integrated with information provided by SMAM/WMO radiosoundings collected every 6 hours at the Pratica di Mare military airport (40 km south of the city). The European Centre for Medium-Range Weather Forecasts (ECMWF) analysis at 0.5° and 6 hours resolution were used for the initial and boundary conditions on the nested computational domains.

The emission inventory is based on the Italian national inventory with special attention to the urban traffic emissions of Rome. They were directly estimated from vehicles flows on more than 9000 links making up the 700 Km of the primary road network of Rome.

Further details on the emission inventory, models set up and input data can be found e.g. [7].

3.2. Data Assimilation techniques

Three different assimilation methodologies were tested. The first was a nudging approach based on the ISentropic ANalysis (ISAN) preprocessing module available within the RAMS model. The first step of ISAN analysis procedure is to access ECMWF pressure level gridded fields and interpolate them onto the RAMS polar-stereographic grids, creating a polar-stereographic/pressure coordinate dataset over the area of interest. Meteorological fields are then interpolated vertically to both isentropic vertical coordinate and terrain-following, σ coordinate. Once the large scale data has been processed, any available surface and upper air observation may be accessed. The Barnes objective analysis scheme [8] is then applied to the wind, pressure, and relative humidity on the isentropes and the wind, temperature, and relative humidity on the σ levels. User-specified parameters control the smoothing characteristics of the objective analysis and the relative weighted background fields and observations. Time interpolation of the analysed fields are then assimilated to the prognostic fields at every time step through a nudging technique, realised by a forcing term added to the prognostic equation of pressure, wind, temperature and humidity. The forcing term acts on every grid point of the computational mesh. The RAMS simulation was carried out using 16 surface stations and three SODAR/radiosounding profiles to produce ISAN analysis (RAMSISAN hereafter).

The second approach was an Observation Data Assimilation (ODA) technique. The ODA method is a generalized observational nudging scheme. It examines each station, interpolates data in time to each timestep, then performs a "kriging" interpolation to produce three dimensional value and covariance fields. These will only nudge the model fields in the grid cells "close enough" to the monitoring stations. Different parameters can be used to control the behaviour of the ODA scheme, such as: frequency at which to update the interpolated observational value and covariance fields with the kriging scheme (600 s in our case); relative weights applied to the ODA nudging weights for each grid (0.8 and 1.0 for intermediate and target domain respectively); radii for the kriging scheme to control the smoothness of the analysis at the surface and upper air. For the latter two different values were tested: 20 and 40 km (RAMSODA20 hereafter) and 6.7 and 13.2 km (RAMSODA7 hereafter) for respectively surface and upper air radii. The distance where the influence drops by a factor of e^{-2} was set at one order of magnitude lower than the above values. The other ODA parameters were set at the suggested values.

The third assimilation procedure was obtained by coupling a diagnostic model (MINERVE) with a prognostic (RAMS) meteorological model to improve the description of urban scale winds in the target domain (MINER-RAMS hereafter). Wind profiles produced by RAMS at the boundaries of the target domain, together with data from 16 surface stations and the three SODAR/radiosounding profiles, were used to reconstruct the 3D wind fields during the selected episode by means of the MINERVE model. For temperature, humidity and pressure fields the values predicted by RAMS were used.

In order to consider the possibility to not use observations in the meteorological simulation and to better quantify the effect of the data assimilation techniques, a simulation with no data assimilation (forecasting mode) was also performed (RAMSnoass hereafter). Here only the background ECMWF meteorological fields are provided to RAMS to define initial and boundary conditions.

4. Results and discussions

For each of five simulations previously described, meteorological fields have been produced and stored at hourly resolution during the selected episode.

The meteorological fields show the typical evolution of land- and sea-breeze circulations frequently occurring in this season. According to the results obtained at the Villa Pamphili urban park station, the sea-breeze starts at about 11am (direction 250°), whereas the land-breeze between 9pm and 1am (direction 25°). The highest SODAR sea-breeze wind speed value at 36 m (app. 7 m/s) occurs at 3-5 pm, whilst the land-breeze one at 6 am (app. 3 m/s). A statistical analysis was conducted comparing the estimated wind values with SODAR observations at the Villa Pamphili urban park station. The SODAR range gates centered at 40 and 180 m heights with 20 m of thickness were used as observed values, whereas the RAMS vertical levels centered at 32 and 180 m with respectively 24 and 100 m of thickness were used as simulated values. Results are shown on Table 1 and 2 for wind speed and direction respectively.

Table 1. Statistical comparison of modelled and SODAR observed wind speed at the Villa Pamphili urban park station.

	Mean value (m/s)		RMS (m/s)		Mean Bias (m/s)		RMS Bias (m/s)		Mean Gross Err. (m/s)		R		FAC2 (%)	
	Height (m)		Height (m)		Height (m)		Height (m)		Height (m)		Height (m)		Height (m)	
	36	180	36	180	36	180	36	180	36	180	36	180	36	180
SODAR	2.7	4.9	1.5	2.3	/	/	/	/	/	/	/	/	/	/
MINER-RAMS	2.4	4.9	1.3	2.2	-0.3	-0.1	0.4	0,7	0.4	0.3	0.97	0.96	99	97
RAMSnoass	2.9	4.7	2.1	2.9	0.3	-0.3	2.1	3.1	1.6	2.4	0.36	0.30	66	66
RAMSISAN	3.2	4.9	2.3	3.1	0.5	0.0	1.7	2.9	1.5	2.5	0.63	0.44	59	60
RAMSODA20	2.5	3.4	1.6	1.9	-0.1	-1.6	1.2	2.1	0.9	2.1	0.68	0.51	82	68
RAMSODA7	2.9	4.4	1.9	2.7	0.3	-0.5	1.7	2.8	1.3	2.1	0.48	0.38	77	70

Table 2. Statistical comparison of modelled and SODAR observed wind direction at the Villa Pamphili urban park station.

	Mean Gross Error (°)		Gross error Stand. Dev. (°)	
	Height (m)		Height (m)	
	36	180	36	180
MINER-RAMS	6	7	18	17
RAMSnoass	50	48	51	49
RAMSISAN	45	39	48	41
RAMSODA20	31	31	36	36
RAMSODA7	34	36	39	38

As far as the wind speed is concerned (table1), the best performance is obtained by the diagnostic/prognostic assimilation approach (MINER-RAMS). This is an expected result due to the intrinsic nature of diagnostic models, where initial interpolation of measurements is followed by a divergence adjustment to ensure mass-consistency. Among the other assimilation procedures available in RAMS model (ISAN and ODA), the ODA with 20 km of influence radius (RAMSODA20) shows the best performance in terms of mean bias (-0.1 m/s), mean gross error (mean of the absolute values of modeled and observed differences; here 0.9 m/s), correlation coefficient R (0.68) and FAC2 (percentage of data that satisfy $0.5 <= P_m/P_o <= 2$ where P_m and P_o are the modeled and observed parameter respectively; here 82%), at least at 36 m height. At 180 m, this assimilation technique underestimates the observed value (3.4 vs. 4.9 m/s), also confirmed by the correspondent mean bias value (-1.6 m/s), but with the best values of mean gross error (2.1 m/s), R (0.51) and FAC2 (the percentage of simulated values 68%). The same assimilation procedure, but with a lower influence radius (RAMSODA7), shows worse performance than that with 20 km (higher mean gross error and lower R values), overestimating the wind speed at 32 m (0.3 m/s) and underestimating it at 180 m height (-0.5 m/s). The results obtained using RAMS in forecasting mode (RAMSnoass), where only ECMWF fields are used, provide the worst performance in terms of mean gross error and R at both heights. Although it shows a mean bias value as those obtained with the ODA technique, the highest values of RMS bias and gross error demonstrate that deviations from the observed values are larger than those obtained with assimilation techniques. The application of the isentropic analysis (RAMSISAN) improves the correlation coefficient R (0.63 and 0.44) respect to the values obtained without data assimilation (RAMSnoass), but overestimations of the observed wind speed are detected at 32 m (3.2 vs. 2.7 m/s) with the largest RMS values (2.3 and 3.1 m/s at 32 and 180 m respectively).

According to the statistical comparison results for wind direction, shown in table 2, the best performance is obtained by the diagnostic assimilation approach (MINER-RAMS), as already observed for wind speed. The ODA assimilation procedure seems to guarantee lower mean gross errors (about 31-36 degrees) than those calculated by means of both ISAN and forecasting mode. No significant variation with height has been detected.

The SODAR observed and model calculated wind speed up to 400 m were used to produce pdf functions. Results are shown in figure 1: the model results underestimate the frequencies of the highest speeds whilst those of the lowest ones are overpredicted; MINER-RAMS and RAMSODA20 show the smaller variances.

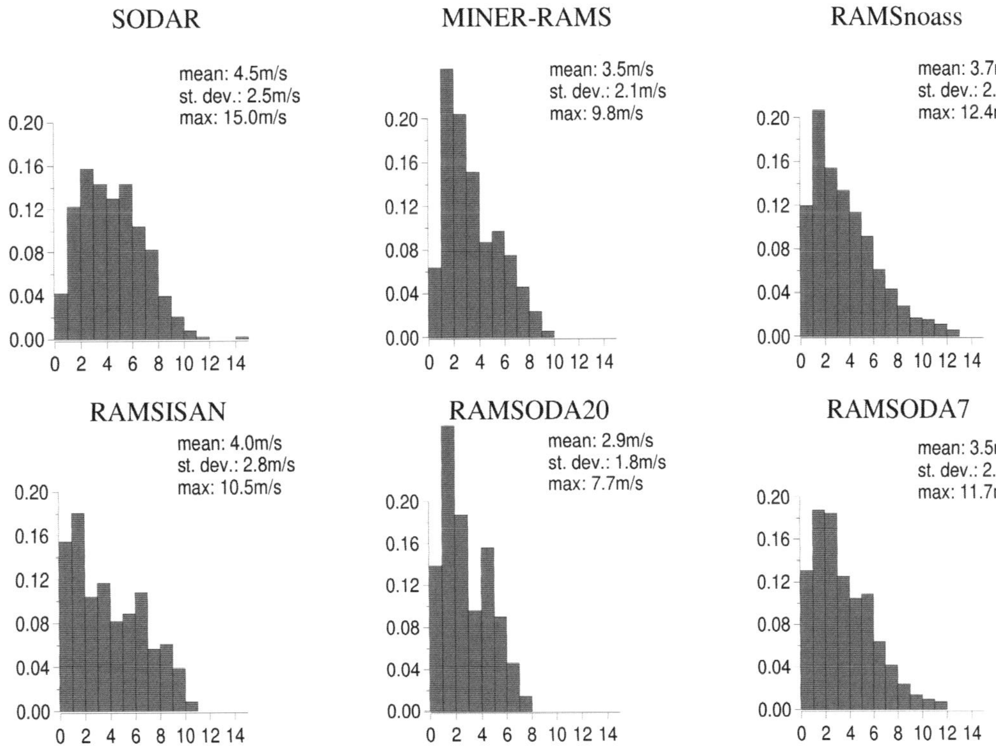

Figure 1. Frequencies of SODAR observed Vs. modelled wind speed (m/s) using data up to 400 m.

As far as the spatial and temporal distribution of wind speed is concerned, the MINER-RAMS results show a more effective reproduction of sea-breeze structure and evolution than that obtained applying the RAMSISAN assimilation procedure, followed by RAMSODA20 technique. Discrepancies in the reproduction of both land and sea breezes were sometimes detected among the different assimilation procedures. In particular, significant differences were found in reproducing the sea breeze development and its inland dynamic on 24 June, as shown in figure 2. As an example, while the MINER-RAMS and RAMSISAN assimilation techniques produce a complete penetration of sea-breeze over the urban target domain at 15:00 lst, the remaining ones show a sea-breeze which only diffuses inland over the first 40 km, with an opposite wind direction over the rest of the territory. Other discrepancies are detected for the land-breeze, particularly in terms of spatial distribution of wind speed. For the remain days (20-23 June) those discrepancies were not detected.

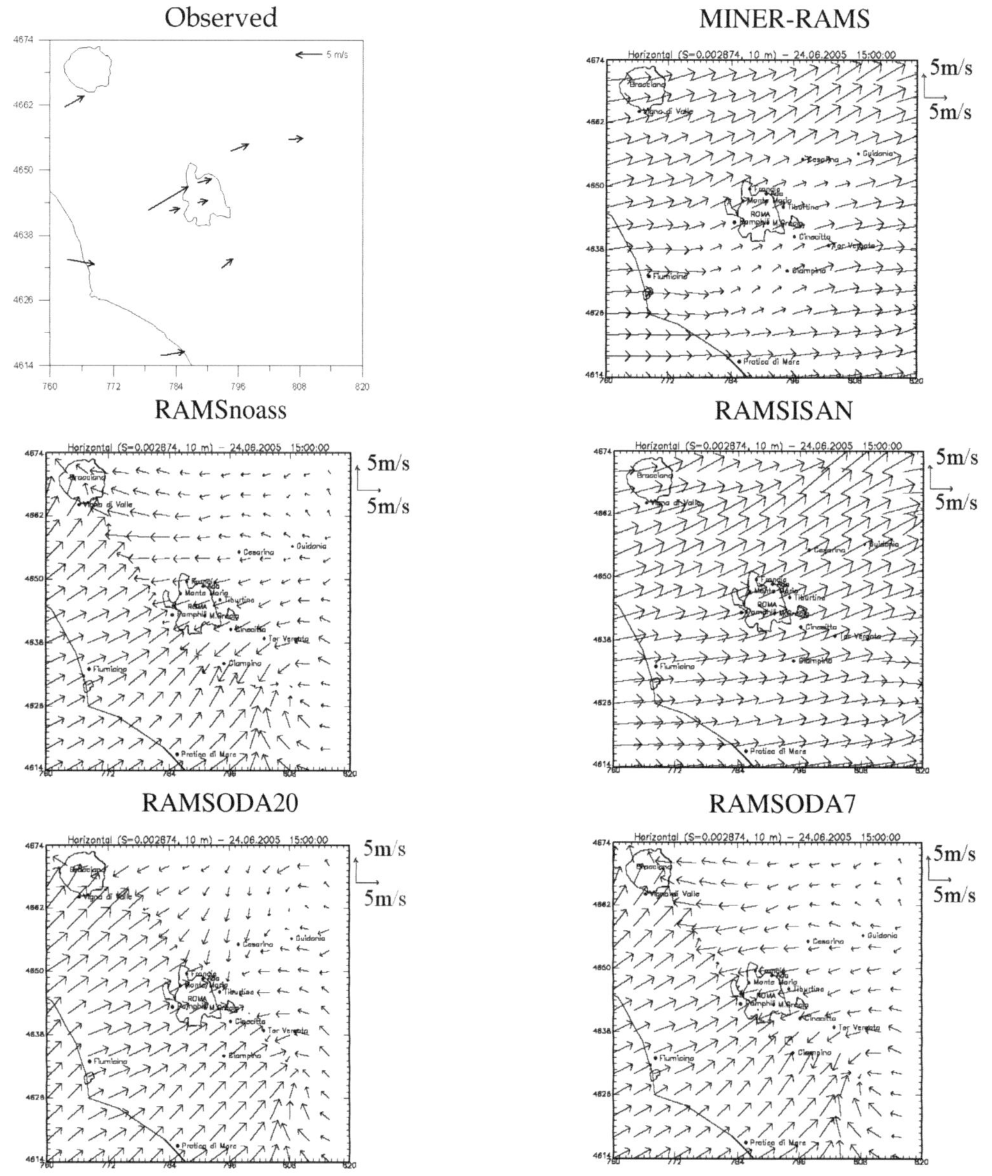

Figure 2. Surface wind fields on June 24th at 15:00 obtained with different assimilation procedures.

Table 3 shows a statistical comparison of modeled and observed surface temperature at Villa Pamphili urban park station. The forecasting simulation (RAMSnoass) usually underestimates the observed mean temperature (mean bias -1.5 °C) with the largest mean gross error (2.2 °C) and the lowest R (0.76). The ODA technique with 7 km of influence radius shows about the same performances, with a slight improvement in the mean gross error and R values. The temperature

IOP Conf. Series: Earth and Environmental Science **1** (2008) 012033 doi:10.1088/1755-1307/1/1/012033

predictions are better reproduced by using either the ODA with 20 km of influence radius or the isentropic ISAN assimilation procedures, which show similar performances. MINER-RAMS temperature results were not taken into account, as this assimilation procedure has been used only for the wind field, while temperatures were derived from RAMSISAN results.

The meteorological fields obtained using the discussed assimilation techniques were then separately provided, together with emission data and initial/boundary conditions, to the chemical transport model FARM to evaluate their possible effects on ground level predicted pollutants concentrations. Fifteen monitoring stations located in the urban area were used to compare measured with simulated pollutants concentration. Daily PM_{10} (particulate matter with characteristic length smaller than 10μm) and hourly ozone observed concentrations were available for comparisons. PM_{10} results are shown in table 4.

All simulations results are affected by underestimation of measured PM_{10} concentrations. It was mainly due to an underestimation in the organic components of the aerosol, as reported by Gariazzo et al [7], and in a lesser extent to underestimations of sulphate and sea-crustal components. Here what is more interesting to notice is the relative differences among simulations driven by the different meteorological fields. The comparison based on daily average observations likely hides some effects produced by differences among the meteorological simulations, but relevant differences among the highest daily average values remain evident.

Table 3. Statistical comparison of modelled and observed surface temperature at the Villa Pamphili urban park station.

	Mean value (°C)	RMS obs. (°C)	Mean Bias (°C)	RMS Bias (°C)	Mean Gross Error (°C)	R
Observed	24.7	3.8	/	/	/	/
RAMSnoass	23.2	2.2	-1.5	2.4	2.2	0.76
RAMSISAN	25.0	3.3	0.3	1.0	0.8	0.96
RAMSODA20	24.2	3.2	-0.5	1.2	1.0	0.95
RAMSODA7	23.3	2.4	-1.5	1.9	1.8	0.88

Table 4. Comparison of modelled and observed surface daily PM_{10} averaged over all stations.

	Mean ($\mu g/m^3$)	Max ($\mu g/m^3$)
Observed	37	43
MINER-RAMS	17	21
RAMSnoass	15	25
RAMSISAN	14	17
RAMSODA20	16	19
RAMSODA7	14	22

Table 5. Statistical comparison of modelled and observed surface ozone averaged over all stations.

	Max value (ppb)	Mean value (ppb)		RMS (ppb)		Mean Bias (ppb)		RMS Bias (ppb)		Mean Gross Error (ppb)		R		FAC2 (%)	
	Time avg	Time avg		Time avg		Time avg		Time avg		Time avg		Time avg		Time avg	
	all	all	11-18	all	11-18	all	11-18	all	11-18	all	11-18	all	11-18	all	11-18
Observed	60.5	36.7	60.7	21.0	7.8	/	/	/	/	/	/	/	/	/	/
MINER-RAMS	68.6	43.2	61.2	15.1	7.3	6.4	0.5	10.5	8.8	9.7	7.1	0.88	0.31	77	100
RAMSnoass	89.8	45.4	64.8	17.6	9.9	8.6	4.1	12.1	14.2	11.6	10.3	0.82	-0.28	75	100
RAMSISAN	69.1	48.0	64.4	15.1	6.4	11.2	3.7	9.7	7.2	12.4	6.4	0.91	0.49	76	100
RAMSODA20	80.6	44.6	64.3	16.9	7.7	7.8	3.6	10.3	10.8	10.8	8.9	0.88	0.02	77	100
RAMSODA7	88.7	45.5	65.1	17.8	9.9	8.7	4.4	12.0	14.0	11.8	10.3	0.82	-0.24	76	100

As far as the ozone is concerned, both all hourly values of the period and those occurring at the hours when ozone peaks are detected (11-18), were considered for the statistical comparison. Table 5 shows the modelled/measured ozone concentrations results for the different simulations. As already noticed for meteorology, the best results are obtained with diagnostic MINER-RAMS assimilation procedure, followed by RAMSODA20 approach, with a general overestimation. In general results are better for the daytime ozone (11-18) than for the whole analyzed period (all data). It is well known the difficulty encountered by photochemical models in correctly simulating night-time ozone concentrations. Different behaviors are observed for the different models when data are analyzed as a whole series (all data) and when the analysis is restricted to the daily ozone peaks (11-18). In particular, among the RAMS assimilation techniques, the RAMSISAN results show the worst performance when all data are considered (48 μg/m^3), but the best one when analysis is restricted to daytime peaks (64.4 μg/m^3).

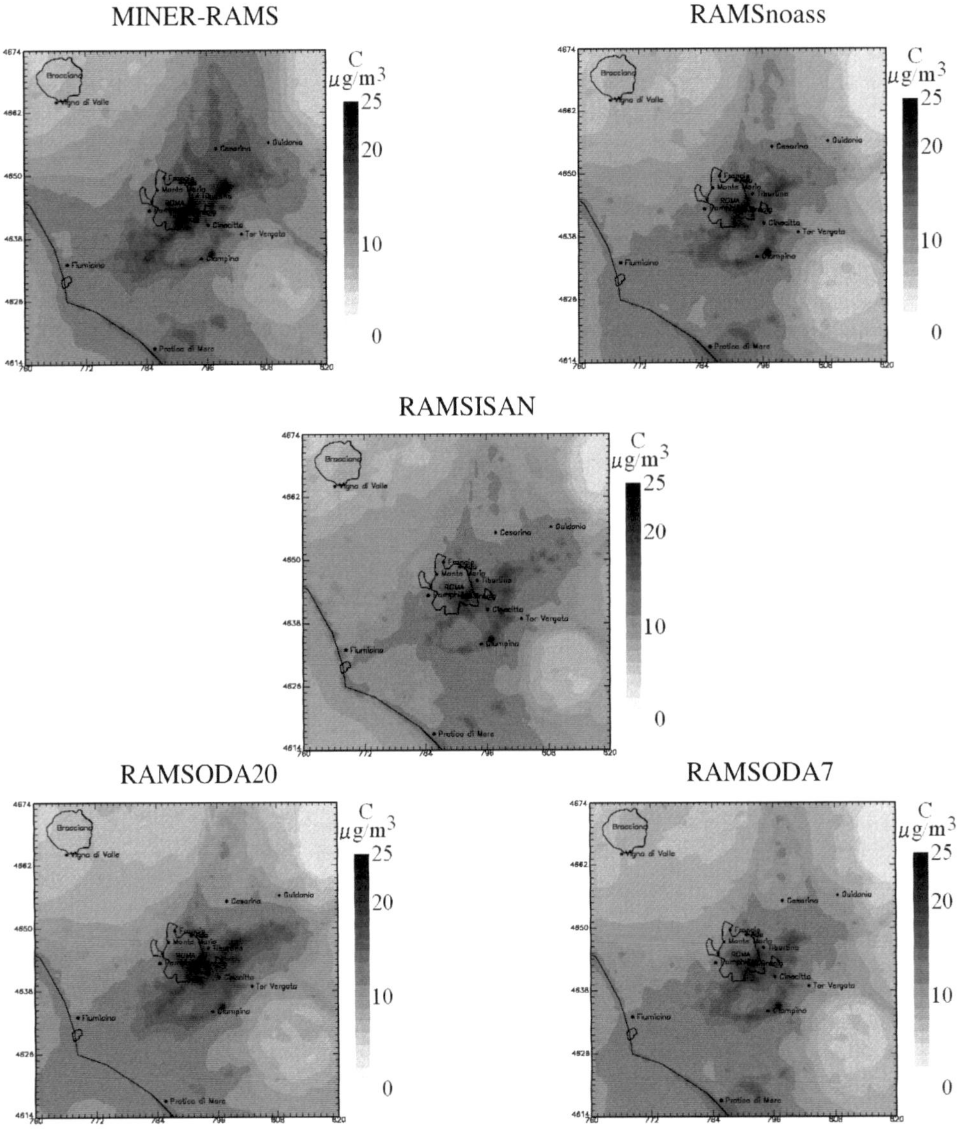

Figure 3. PM_{10} concentration maps averaged over the simulation period for different assimilation procedures (the lower limits of the colour classes are reported in legend).

It is worth noticing that with respect to the performance obtained without using data assimilation (RAMSnoass), the application of these techniques (ISAN or ODA) do not generally produce significant effects on the improvement of both mean value prediction (64.3-65.1 ppb vs. 64.8 ppb for RAMSnoass) and mean bias (3.6-4.4 ppb vs. 4.1 ppb for RAMSnoass) during the daytime (11-18) hours. A reduction in both the RMS values is obtained applying the ISAN and ODA at 20 km assimilation techniques (e.g. from 14.2 to 7.2 ppb for RMS of mean bias). Differences among performances have instead been detected at the same hours for both the mean gross error (6.4-10.3 ppb vs. 10.3 ppb for RAMSnoass) and the R values (-0.24-0.49 vs -0.28 for RAMSnoass). These statistical results suggest that the application of assimilation techniques in the RAMS model can account for up to 1% (0.5 ppb) in the daytime mean value and in reducing the mean bias of the same amount. A more significant improvement is obtained for both the mean gross error and the RMS bias during daytime.

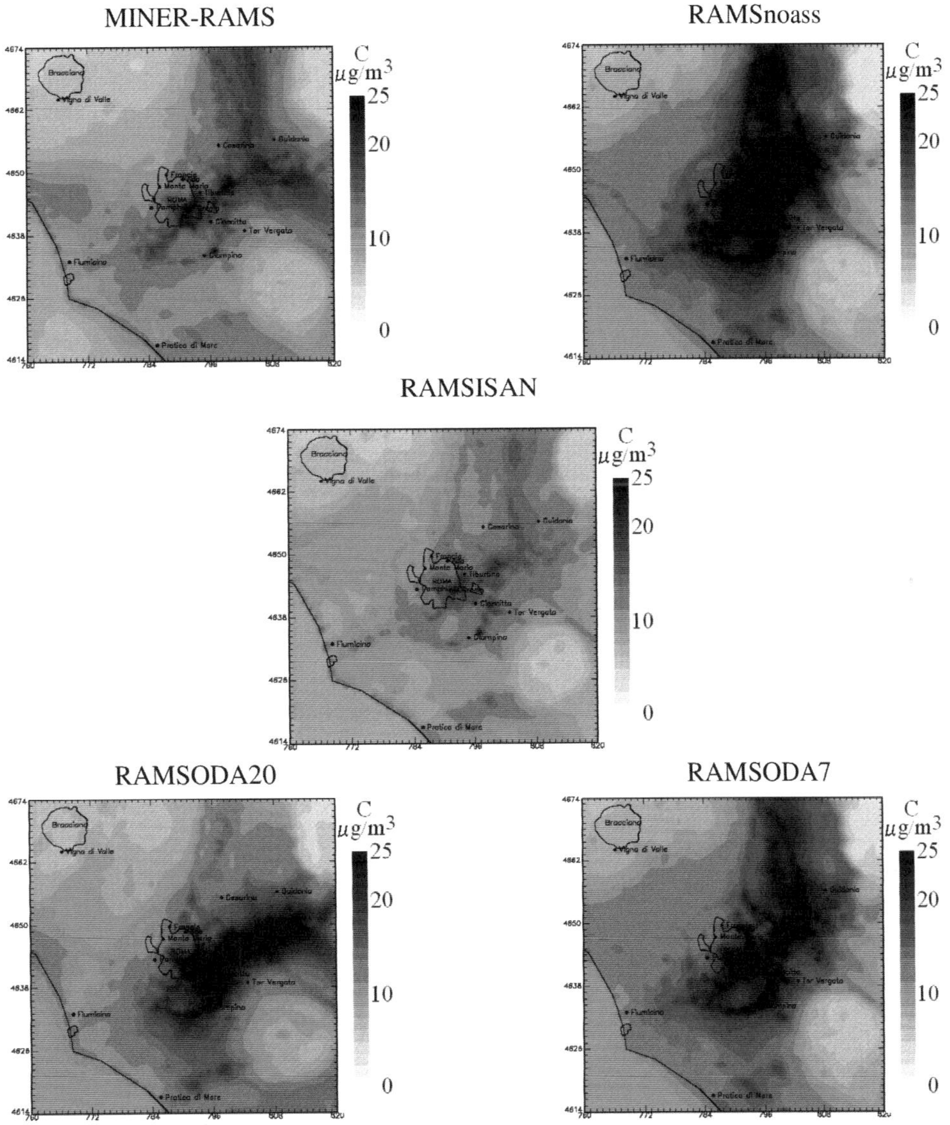

Figure 4. Daily PM_{10} concentration maps on June 24th obtained with different assimilation procedures (the lower limits values in legend).

Time series of ozone concentrations are well reproduced by different model simulations, but there is a systematic overestimation during nighttime (RAMSISAN above all) and in the last day. MINER-

RAMS, RAMSISAN and RAMSODA20 simulations better reproduce the observed ozone behavior, whilst RAMSODA7 values are very close to RAMSnoass with lower performances.

Changes in wind speed and directions influence concentration values and their fields shape. As can be seen in figure 3, PM_{10} maps averaged over the full period show the same main features, with a few differences in particular areas. Larger differences are observed for the highest concentration maps. Other differences are detected during those days when the provided meteorological fields are different, such as during June 24th. Here the differences in the sea breeze reproduction, produce significant discrepancies among the simulations, especially for PM_{10} maps, as shown in figure 4.

5. Conclusions

In order to investigate the influence of different data assimilation techniques available in the meteorological model RAMS, a study has been conducted over the urban area of Rome, Italy, during a summer episode. The effects produced by the different meteorological fields on the ozone and PM_{10} concentrations predicted by the chemical transport model FARM have been also evaluated. Three different assimilation methodologies were tested. Two of them are available within the RAMS modeling system: the ISentropic ANalysis (ISAN) preprocessing analysis and the Observation Data Assimilation (ODA) techniques. The third method was the coupling of a diagnostic (MINERVE) with a prognostic (RAMS) meteorological models. Results show the diagnostic/prognostic approach provided the best results in reproducing observed wind, PM_{10} and ozone concentrations, followed by the ODA techniques used with 20 km of influence radius. The assimilation techniques used in RAMS model was found to influence up to 1% (0.5 ppb) in the daytime ozone mean value and in reducing of the same amount its mean bias. Higher reductions were found for ozone mean gross errors (up to 6% corresponding to 3.9 ppb). Significant differences among pollutants concentration maps were observed, especially during those days when the different assimilation techniques showed significant differences in the meteorological fields.

References

[1] Cotton W.R., Pielke R. A., Walko R. L., Liston G. E., Tremback C. J., Jiang H., McAnelly R.L., Harrington J. Y., Nicholls M. E., Carrio G.G., McFadden J. P. 2003 RAMS 2001: Current status and future directions. *Meteorol. Atmos. Phys.* **82** 5-29.

[2] Aria Technologies 2001 *Minerve Wind Field Models version 7.0 General Design Manual* (ARIA Tech. Report)

[3] ARIANET 2005 *SURFPRO (SURrface-atmosphere interFace PROcessor) User's guide* (ARIANET Tech. Report).

[4] Silibello C., Calori G., Brusasca G., Giudici A., Angelino E., Fossati E., Peroni E., Buganza E., Degiarde E. 2005 Modelling of PM10 concentrations over Milan urban area: validation and sensitivity analysis of different aerosol modules. In: *Proceedings of 5th International Conference on Urban Air Quality (Valencia, Spain, 29-31 March 2005).*

[5] Carter W.P.L 1990. A detailed mechanism for the gas-phase atmospheric reactions of organic compounds. *Atmospheric Environment* **24A** 481-518.

[6] Binkowski F. S. 1999 *The aerosol portion of Models-3 CMAQ* In *Science Algorithms of the EPA Models-3 Community Multiscale Air Quality (CMAQ) Modeling System* Eds D.W. Byun, and J.K.S. Ching (EPA-600/R-99/030, National Exposure Research Laboratory, U.S. Environmental Protection Agency, Research Triangle Park, NC, 10-1-10-16) Part II: chapters 9-18.

[7] Gariazzo C., Silibello C., Finardi S., Radice P., Piersanti A., Calori G., Cecinato A., Perrino C., Nussio F., Cagnoli M., Pelliccioni A., Gobbi G. P., Di Filippo P. 2007 A gas/aerosol air pollutants study over the urban area of Rome using a comprehensive Chemical transport model *Atmospheric Environment* Vol. **41/34** pp. 7286-7303.

[8] Barnes S.L. 1973 Mesoscale Objective Map Analysis Using Weighted Time Series Observations NSSL

Boundary Layer Height Estimation by Sodar and Sonic Anemometer Measurements

D. Contini [1], D. Cava, P. Martano, A. Donateo, F.M. Grasso
CNR – Istituto di Scienze dell'Atmosfera e del Clima, U. O. di Lecce
Str. Prv. Lecce-Monteroni km 1.2, 73100, Lecce, Italy

E-mail: d.contini@isac.cnr.it

Abstract. In this paper an analysis of different methods for the calculation of the boundary layer height (BLH) using sodar and ultrasonic anemometer measurements is presented. All the methods used are based on single point surface measurements. In particular the automatic spectral routine developed for Remtech sodar is compared with the results obtained with the parameterization of the vertical velocity variance, with the calculation of a prognostic model and with a parameterization based on horizontal velocity spectra. Results indicate that in unstable conditions the different methods provide similar pattern, with BLH relatively low, even if the parameterization of the vertical velocity variance is affected by a large scatter that limits its efficiency in evaluating the BLH. In stable nocturnal conditions the performances of the Remtech routine are lower with respect to the ones in unstable conditions. The spectral method, applied to sodar or sonic anemometer data, seems to be the most promising in order to develop an efficient routine for BLH determination.

1. Introduction

In the past few decades a great effort has been made to obtain detailed descriptions of the atmospheric boundary layer (BL) properties starting from surface parameters. This is considered an important goal from both a scientific and a practical point of view [1, 2]. Measurements of the boundary layer height (BLH) are important for many applications, such as chemical and air pollution studies, mesoscale and synoptic modelling. The BLH directly affects pollution dispersion and its diurnal evolution can affect ground-level aerosol concentrations in specific conditions [3]. The geographic position and the absence of relevant orography in the Salentum peninsula (Apulia, south of Italy), where our measurement site is placed, are the main causes of advection of air masses from the sea that limit the growth of the BL and the differences between the BLH in winter and summer. Therefore the influence of the BLH on the air quality could also be stronger with respect to other sites. Radio soundings are the most common source of data for operational determination of the BLH [4] because data is continuously controlled and measurements are relatively wide distributed in Europe. However in the Apulia region it is available only one radio sounding station in the coastal area of Brindisi and data is available only three times a day at specified synoptic times (0 UTC, 6 UTC and 12 UTC). Therefore the BLH from the radio soundings could be a starting point for the analysis but they are hardly representative of the BLH over land in the internal zones of the peninsula and it is difficult to

[1] To whom any correspondence should be addressed.

reconstruct the daily evolution without using some kind of meteorological model. Remote sounding systems (lidar, sodar, rass and wind profiling radar) provide an interesting alternative for BLH estimations. The increasing availability of remote sensing device allows easier direct measurement of vertical profiles. Furthermore the progress in acoustic sounding techniques has led to the development of advanced devices for monitoring the main characteristics of the atmospheric boundary layer. The basic advantages of remote sounding systems are the continuous operation and that they do not cause any modification of the investigated flow. On the other hand, simple parameterization based on standard micrometeorological single point surface observations as well as numerical models are widely used in practice in meteorological and in environmental service. It has to be put in evidence that diagnostic models (like CALMET for example) are also widely used for BLH evaluation and they could operate on a grid interpolating the input of a surface sensors network. However very often a network of surface measurements is not available, being sensors concentrated in single sites, and in these cases the advantage of grid interpolation vanishes.
In this paper different methods to calculate the boundary layer height, starting from single point surface measurements, have been tested in a period during which two sodar systems and a micrometeorological station were operating together. A comparison between different results has been made to put in evidence the limit and potentiality of the different methods separating the stable cases from the unstable ones.

2. Measurement site, instrumentation and data processing

Measurements used in this paper were collected during the period between 7 and 25 May 2005 in the experimental field of the Istituto di Scienze dell'Atmosfera e del Clima (ISAC-CNR) located inside the University Campus in a rural area at south-east of the city of Lecce. The site is characterised by the presence of short vegetation and the surrounding area is constituted for at least 1 km by patches of trees and small two-storied buildings [5]. A 16 m high meteorological mast, equipped with a sonic anemometer (Gill Solent), a KH20 (Campbell Scientific) fast hygrometer and a Rotronic sensor for temperature and relative humidity measurements, was located close to the two acoustic sounders. The sampling rate of the anemometer was 20.8 Hz and it provided a direct method (eddy correlation) for estimating sensible and latent heat vertical turbulent fluxes as well as momentum flux. All recorded data was processed in the streamlines reference system on 30 and 60 minutes averages. A Remtech PA1 sodar was placed over the roof of our building (8 m tall) about 200 m apart from the experimental site and it was operating in order to provide 30 minutes average at 50 m vertical resolution with the first available range gate at 33 m above the ground. A second minisodar was operated with a single vertical antenna, vertically oriented. This minisodar is the PC-MTSodar developed at ISAC-CNR [6, 7] and it was operated emitting pulses at 4000 Hz (0.1 s long) with a 2 s repetition rate. For this minisodar, in the configuration used, the first range gate is about 24 m above the ground, with a vertical resolution of about 13.4 m. The Doppler shift and the Signal to Noise ratio (S/N) is calculated according to procedure described in [8] after correction of the received signals using the transfer function correction that has proven to be useful for the analysis of vertical wind velocity [9].

3. BLH calculated by the Remtech sodar

Remtech sodar software provides an internal routine for the determination of BLH based on the spectral analysis of the vertical wind velocity [10]. According to [11] and to [12] from the power spectrum of the vertical velocity, the characteristic frequency scale of the most energetic eddies is deduced. A Fourier transform is applied on each record (range gate) to obtain the power spectrum of the recorded signal and from each power spectrum a characteristic frequency scale is calculated for the largest eddies which, for their size, are correlated with the height of the boundary layer [13]. Finally the mixing height can be obtained, apart from a proportionality coefficient, as the product of the time scale and the velocity scale. With this method the boundary layer height can be obtained even if it is well above the maximum range reached by the sodar. The method based on the relationship between the scale of the most energetic eddies and the BLH was suggested for unstable boundary-layer by [14], however the Remtech software uses this approach without any respect for stability conditions [12]. In the scientific literature different point of view are reported about the efficiency of the Remtech routine. In [12] poor performances are reported for both stable and unstable boundary-layer while in [13] good performances in unstable boundary-layer have been observed during day-time but poor performances during night-time in stable conditions. This put in evidence that further research is needed for the development/improvement of automatic BLH calculation especially in stable conditions. In our measurement site the Remtech sodar gave in 2005 the average daily pattern of BLH reported in figure 1 where the cold season (October to March) has been separated from the hot season (April to September). Remtech sodar shows that the diurnal change in the BLH is, on average, around 20% between the two seasons and that the BLH is generally low considering the sensible heat fluxes involved in the summer period (often more than 300 W/m^2). This is due to the presence of relatively intense winds with consequent advection of air masses from the Mediterranean sea that contribute in keeping the BLH low. The histograms reported in figure 1 also show that the BLH estimated by the Remtech routine is rarely lower than 150m even in nocturnal stable boundary layer and this could be due to the limitations of the automatic routine and will be discussed later on.

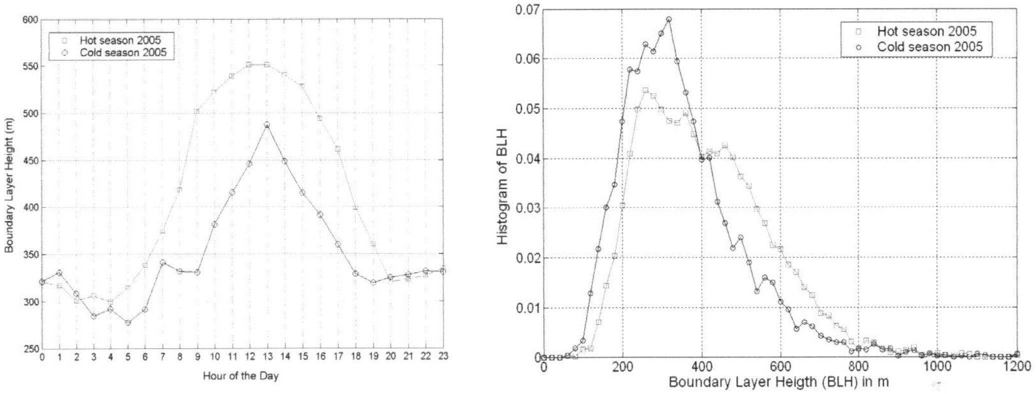

Figure 1. Typical daily pattern of BLH for hot and cold season during 2005 (left) and the relative distribution of BLH (right).

4. BLH calculated from prognostic equations based on single point surface measurements

To calculate the boundary layer height h for the diurnal cycle the algorithm described in [15] and in [16] has been used. This model is based on the growth equations of the Tennekes-Carson scheme [17], written as total (material) derivatives to take into account advection effects from coastlines, $D/dt = \partial/\partial t + U\partial/\partial x$, where U is the average wind speed in the mixing layer, (e.g.: [18]):

$$h\left(\frac{D\theta_m}{Dt}\right) = Q_s - Q_i \tag{1}$$

$$\Delta\left(\frac{D\theta_m}{Dt}\right) = -Q_i \tag{2}$$

$$\frac{D\Delta}{Dt} = \gamma\left(\frac{Dh}{Dt}\right) - \frac{D\theta_m}{Dt} \tag{3}$$

Here, Q is the potential temperature flux at the surface (subscript s) and at the upper temperature inversion (subscript i), Δ is the temperature gap above the mixing height (inversion strength), θ_m is the average potential temperature in the mixing layer and $\gamma = d\theta/dz|_{z=h}$ is the lapse rate above the thermal gap. Knowing the surface fluxes, the model can be mathematically closed through [19] expression for the heat flux Q_i in the upper inversion:

$$\frac{g}{T}Q_i = -C_K\frac{w_m^3}{h} + C_T U\frac{w_m^2}{h}\frac{Dh}{Dt} \tag{4}$$

where $w_m^3 = w_*^3 + C_N u_*^3$, u_* is the friction velocity (square root of the module of the velocity flux), $w_*^3 = ghQ_s/T$ is the convective velocity scale, g is the gravity acceleration, T the absolute temperature and C_N, C_K, C_T, are constants [20]. Inserting (4) in (2) an expression for Dh/Dt as function of h and Δ can be obtained:

$$\frac{Dh}{Dt} = \frac{\partial h}{\partial t} + U\frac{\partial h}{\partial x} = \frac{C_K w_m^3}{C_T w_m^2 + gh\Delta/T} \tag{5}$$

This equation can be transformed in an ordinary differential equation in time using Luhars stationary solution and a parameterized expression for Δ [20], and then solved by direct numerical integration in time, as described in details by [15].

At sunset the algorithm assumes an exponential collapse of the boundary layer height towards an equilibrium height depending on the local conditions [21], by a relaxation time equal to the mixed layer turbulence turnover time scale h/w_*. In night-time the evolution of h is again an exponential decay towards the instantaneous equilibrium height by a relaxation time taken as the minimum between the surface cooling time rate scale [16, 22] and the Ekman time scale (inverse of the Coriolis parameter, [21].

The required input data are time series of single point values of the surface momentum and sensible heat fluxes, wind speed and temperature. The vertical lapse rate (dθ/dz), that controls the mixed layer growth, can be either entered as external parameter or estimated within the model algorithm by a proper routine simulating the nocturnal inversion development on the base of the night-time surface cooling rate [16].

5. BLH calculated in unstable conditions by using a parameterization based on σ_w

Many of the existing similarity methods to estimate turbulence parameters and the BLH from sodar measurements as presented in [23] and discussed in review papers [2, 4]. According to similarity scaling [1, 13], the relationship between the standard deviation of vertical wind component σ_w in a convective boundary layer and its velocity scaling w_* is the following:

$$\frac{\sigma_w^2}{w_*^2} = C_1\left(\frac{z}{h}\right)^{2/3}\left(1 - C_2\frac{z}{h}\right)^2 \tag{6}$$

where $C_1 = 1.8$, $C_2 = 0.8$ and z is the height above the ground and h is the boundary layer height. The above equation is valid in the vertical region between $0.1h$ and $0.7h$ [1].In [24] an integration of Eq. (6) provided the relationship:

$$\frac{\sigma_w^2}{w_*^2} = b \tag{7}$$

where b is a parameter with an average value of 0.4-0.45 [1].

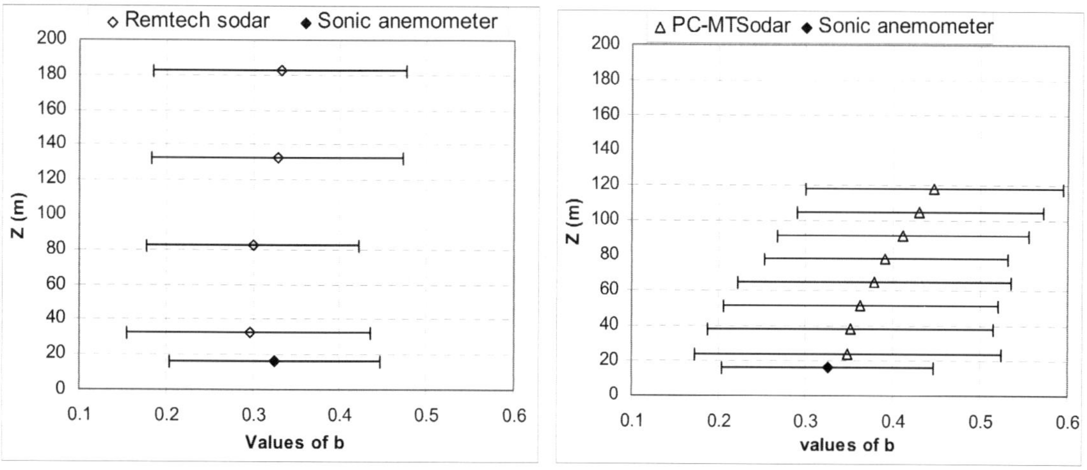

Figure 2. Calculated values of b with their standard deviation.

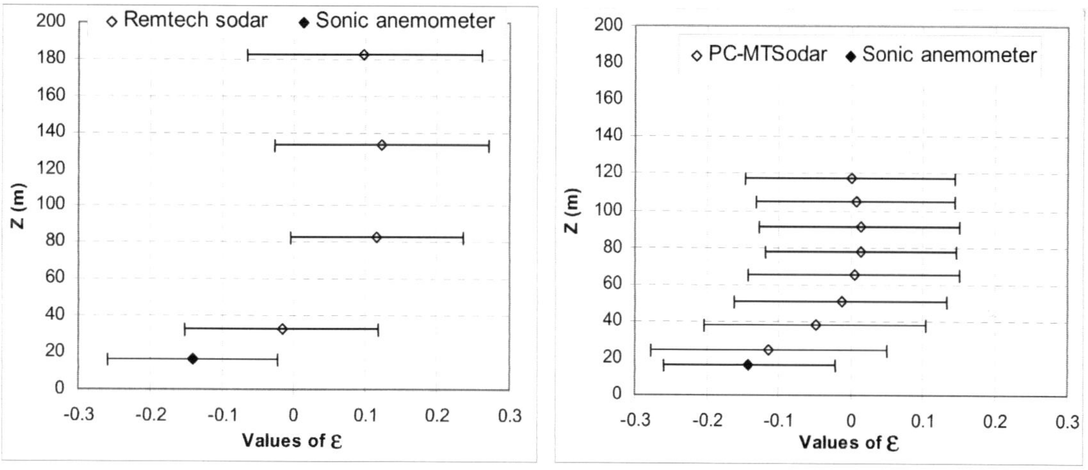

Figure 3. Calculated values of ε with their standard deviation.

Assuming a constant value for b it is possible to use equation (7) for the determination of the BLH using as input parameter the σ_w measured by the sodar in a specific range gate and the heat flux measured by the sonic anemometer. With our data-set we evaluated the parameter b using a scaling velocity w_* calculated through the BLH provided by the Remtech sodar (h_R) and the heat flux measured by the sonic anemometer. The variance of vertical wind velocity measured by the Remtech sodar, by the PC-MTSodar and by the sonic anemometer was used for comparison.

In figure 2 a plot of b at different heights is shown as obtained by the σ_w measured with the two sodar systems. Also the values of b obtained by using the σ_w measured by the sonic anemometer are reported. The analysis is limited to unstable cases in which $h_R/L < -2$ and the standard deviation of b furnishes an indication of the variability of this parameter. The average values observed as well as the wide variability of b are comparable with the ones reported in the scientific literature [25, 26]. Also the values of $\varepsilon = \frac{\sigma_w^2}{w_*^2} - C_1 \left(\frac{z}{h_R}\right)^{2/3} \left(1 - C_2 \frac{z}{h_R}\right)^2$ have been calculated and shown in figure 3 and results indicate that the values obtained with the PC-MTSodar are near zero (especially at z around 100 m) even if a large standard deviation is present as it was for b. The values of ε calculated using σ_w measured by the Remtech sodar are slightly larger than the ones obtained with the PC-MTSodar. This could be due to the underestimation of σ_w in convective conditions caused by the incomplete coverage of the turbulence spectrum. This effect is related to the sodar measurement principle and it is also present in the PC-MTSodar data even if to a lower extent [7]. Therefore the evaluation of BLH through this parameterization has been performed only using the PC-MTSodar data collected in the fifth range gate (78 m) and no significant differences have been observed for the different range gates. It has to be put in evidence that a parameterization based on constant b (in our case 0.39) allows to evaluate the BLH with a relatively large error due to the large standard deviation of b therefore giving only an indication of the BLH in relatively strong convective cases as will be shown in the next section.

6. BLH calculated from spectral analysis of the wind velocity

In order to have a further estimate of BLH a methodology that requires only the peak frequency of horizontal velocity spectra in the surface layer has been used. This approach has the advantage of being convenient in field experiments [27, 28] and it could be applied, in theory, to both sodar and sonic anemometer data. The procedure used infers an estimate of the BLH by using the empirical relationship firstly proposed in [14]:

$$\Lambda_m = 1.5h \tag{8}$$

that relates the wavelength of maximum energy (Λ_m) in the u- and w- spectrum to the BLH. Equation (8) holds for w-spectrum just in the limit $0.1\ h < z < h$ therefore high level data from the sodar should be used. However given the low temporal resolution of the sodar data and the frequency of data missing (at the high range gates) due to the selection on signal-to-noise ratio the w-spectra were noisy. Therefore we applied this methodology only to u-spectrum computed by the sonic anemometer data. Spectra of longitudinal wind component have been computed by using hourly time series and equation (8) has been applied to obtain BLH independently from the atmospheric stratification. As remarked in section 3, the spectral method was proposed in [14] for unstable stratification, however given the need to compare our results with the ones of the Remtech routine, and also the lack of a similar approach for stable atmospheric stratification we used equation (8) for all stability conditions. Therefore the proposed analysis is only a first approach to obtain a spectral parameterization that can be extended also in nocturnal conditions through the comparison with the other described. Further work will include the development of a spectral method specific for stable cases, as matter of fact other spectral relationships have been proposed in literature but they hold only for convective conditions (see for example [27]).

7. Discussion of results

In figure 4 the BLH evaluated with the different methods is shown for a specific week of May 2005. In nocturnal boundary layer when it is present a ground level inversion the height of the mixed layer has also been obtained by a manual analysis of the intensity profiles of the echoes measured by the PC-MTSodar. In figure 4 it is also reported the wind speed (at different height) and the values of the σ_w measured by the different instruments.

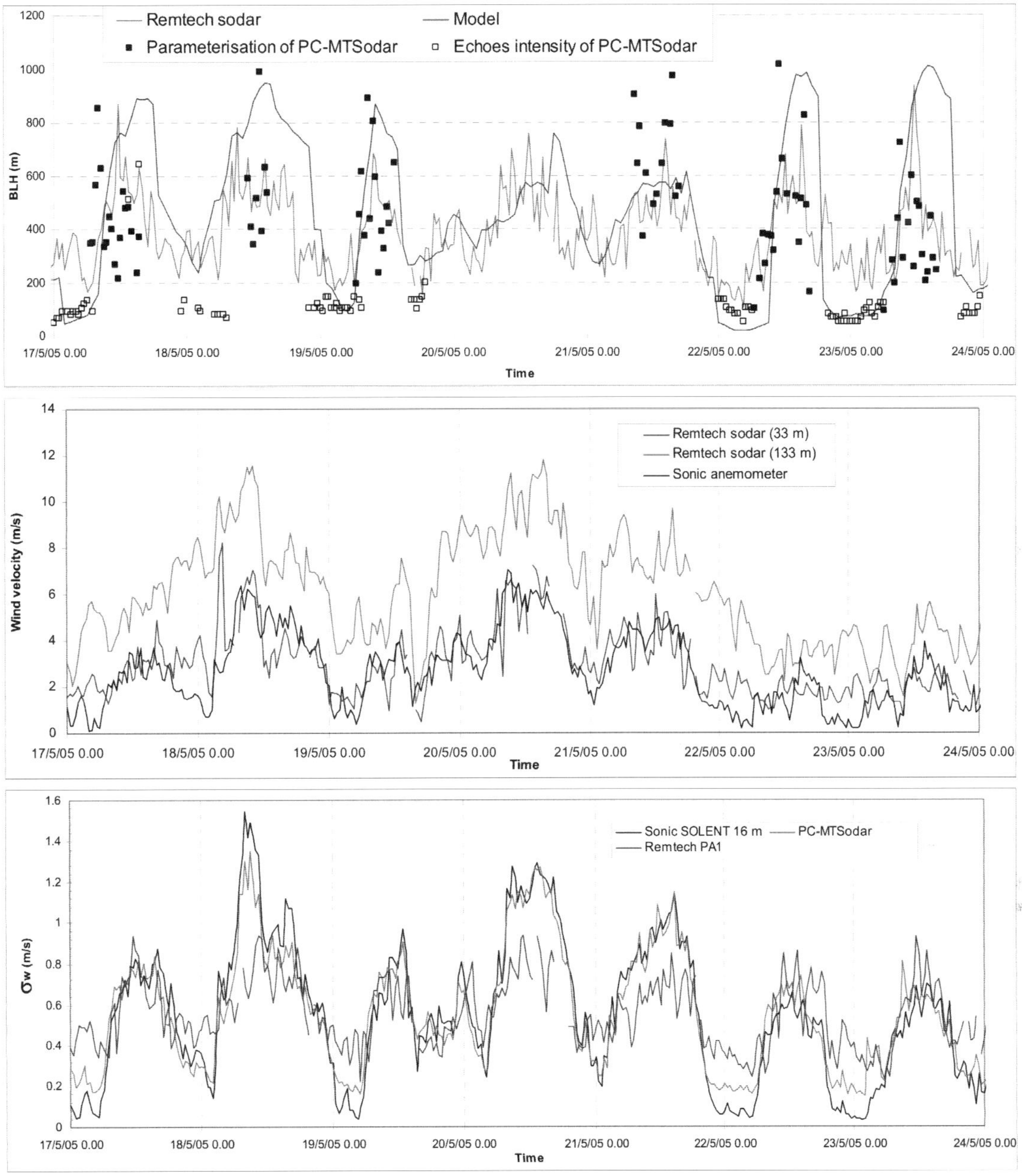

Figure 4. Values of the BLH calculated with the different methods for a specific week (top). Wind velocity (center) and σ_w (bottom).

In unstable diurnal conditions there are differences in the BLH calculated by the different methods and, even if all of them show the same general pattern, only a limited agreement between them is observed. The limited performances of the parameterization through the σ_w are likely due to the scattering in the b-values. Instead the mathematical model calculations seems to be better at high wind speed and the differences in the other cases are likely due to the uncertainty in the lapse rate above the thermal gap that has a relevant effect on BLH calculation especially at low wind speed. In figure 5 a scatter plot between the BLH calculated by the Remtech sodar and the one evaluated through the parameterization of σ_w measured by the PC-MTSodar. Results are reported as a daily average during the diurnal hours of convective conditions. The determination coefficient is relatively low.

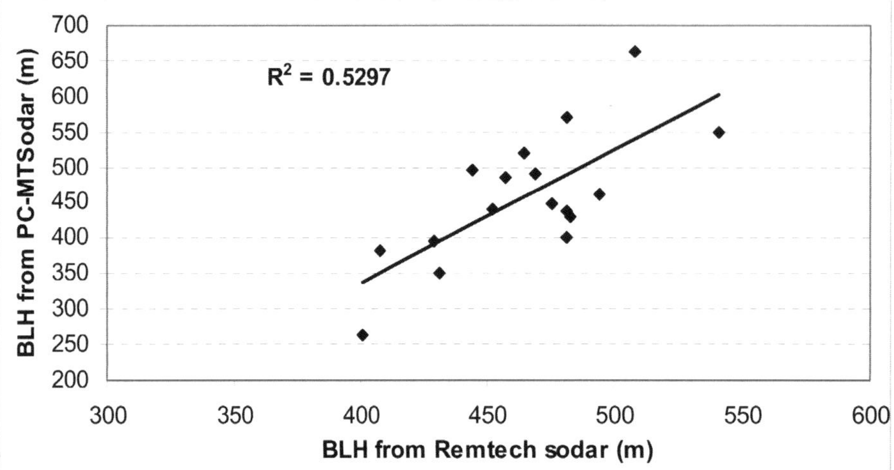

Figure 5. Scatter plot between the average daily values of BLH in convective conditions calculated by the Remtech sodar and by the parameterization of σ_w carried out on the fifth range gate of the PC-MTSodar.

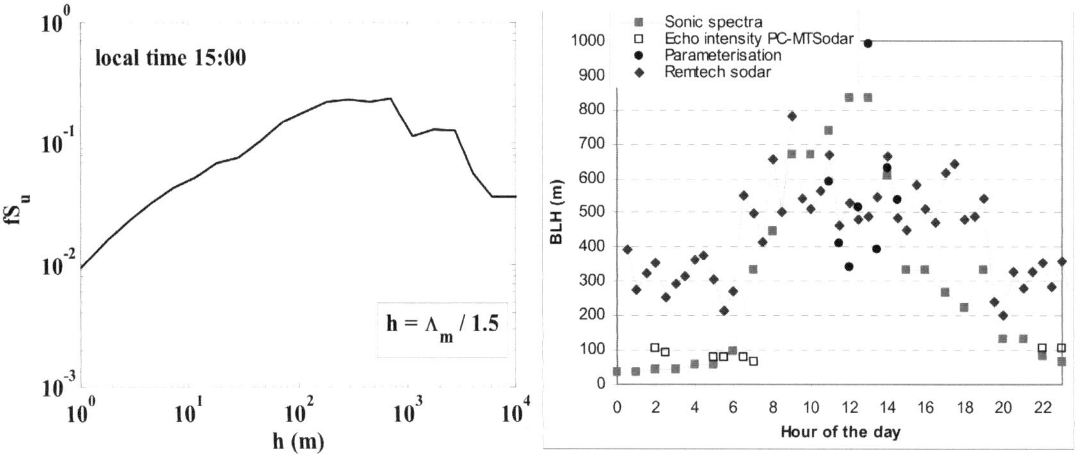

Figure 6. Example of a sonic anemometer spectrum for 18/05/2005 (left) and BLH calculated during the same day.

In nocturnal stable boundary layer the BLH calculated by the Remtech sodar is never lower than about 200 m for all the measurement period. This BLH agrees with that calculated with the mathematical model mainly in certain periods characterised by relatively high levels of vertical turbulence. In other periods the Remtech BLH is higher than the equilibrium height calculated by the mathematical model and also higher than the mixing height evaluated through the analysis of the echoes intensity profiles. This is particularly evident in cases of low wind speed and low vertical turbulence. As shown in figure 4 in these conditions the σ_w measured by the Remtech sodar is highly overestimated probably because of the system errors and of the efficiency of the Doppler-shift retrieval routine according to [7]. It is likely that in these conditions noisy spectra of vertical velocity are obtained with consequent problem in the identification of the large vortexes scale. In figure 6 an example for the 18/05/2005 of BLH calculated from the spectra of the horizontal velocity measured by the sonic anemometer compared with the BLH calculated with the other methods is reported. Results indicate that the BLH calculated by the sonic anemometer is much lower of the one calculated by the Remtech sodar and it is in a better agreement with the calculation carried out from the intensity profiles of the echoes measured by the PC-MTsodar.

Conclusions

The different methods used for the evaluation of BLH show, in unstable conditions, only an agreement of the general patterns indicating a relatively low boundary layer even when strong surface heat fluxes are measured. This is a consequence of the presence of intense wind with advection of air masses from the sea that contribute to keep the BLH low. In particular the parameterization of the σ_w presents a large scattering in the data and the calculations with the mathematical model suffer from the lack of accurate data regarding the temperature lapse rate above the thermal gap especially in low wind speed conditions. In stable conditions the Remtech sodar routine seems to give an overestimation of the BLH especially in low vertical turbulence conditions that could be due to the difficulties of an automatic interpretation of the spectra of vertical wind velocity in presence of noise in measurements of nocturnal σ_w. The evaluation of BLH from the spectra of the horizontal wind measured by the ultrasonic anemometer seems to give better performances in stable boundary layer, although it has been calculated with a parameterization that was originally developed for convective conditions. The conclusion is that further research work is needed on the topic and that the most promising method for automatic detection of the BLH is a spectral method providing that a suitable parameterisation for stable boundary layer is developed.

Acknowledgments

Financial contributions of MIUR, through grant "Bando 1105/2002, progetto n. 245 and of the Interreg IIIA Grecia-Italia through the project ERESIA (Environment Remote Sensing Interregional Agency) are thankfully acknowledged. Authors wish to thank G. Mastrantonio (ISAC-CNR) for useful discussion about interpretation of sodar data.

References

[1] Melas D, Abbate G, Haralampopoulos D and Kelesidis A, 2000. *J. Appl. Meteor.*, 34, 4, 509-515.
[2] Beyrich F, 1997. *Atmospheric Environment*, 31, 3941-3953
[3] Contini D, Donateo A, Belosi F and Prodi F, 2006. Scientific Research and Safeguarding of Venice, Vol. V, edited by CORILA, 183-196.
[4] Seibert P, Beyrich F, Gryning S E, Joffre S, Rasmussen A and Tercier P, 2000. *Atmospheric Environment*, 34, 1001-1027.
[5] Donateo A, Contini D and Belosi F, 2006. *Atmospheric Environment* 40, 1346-1360.
[6] Contini D, Grasso F M, Dargaud G, Mastrantonio G, Viola A and Bonafè U, 2006. Proceedings of the INTERNATIONAL SYMPOSIUM FOR THE ADVANCEMENT OF BOUNDARY-LAYER REMOTE SENSING, 66-68.
[7] Contini D, Grasso F M, Mastrantonio G, Viola A P and Martano P, 2007. *Z. Meteorol.* 16, 357-265.
[8] Mastrantonio G and Fiocco G, 1982. *Journal of Applied Meteorology* 21, 820-830.
[9] Contini D, Mastrantonio G, Viola A and Argentini S, 2004. *Journal of Atmospheric and Oceanic Technology* 21, 1532-1544.
[10] Remtech, 2000. The operation and maintenance manual, REMTECH S.A.
[11] Baumann K and Piringer M, 2001. *Physical Chemical Earth*, 26, n.3, 205-211

[12] Keder J, 1999. *Meteorology and Atmospheric Physics*, 71, pp. 133-138.
[13] Asimanopoulos D N, Helmis C G and Michopoulos J, 2004. *Meteorology and Atmospheric Physics,* 85, 85-92
[14] Kaimal J C, Abshire N L, Chadwich R B, Decker M T, Hooke W H, Kropfli R A, Neff WD and Pasqualucci F, 1982. *J. Appl. Meteor.*, 21, 1123-1129.
[15] Martano P, 2002. *J. Appl. Meteor.* 41, 351-354.
[16] Martano P and Romanelli A., 1997. *Bound-Layer Meteor.*, 82, 105-117.
[17] Tennekes H, 1973. *J. Atmos. Sci.*, 30, 558-567.
[18] Steyn D G and Oke TR, 1992. *Bound-Layer Meteor.*, 24, 161-180
[19] Zilitinkevich S, 1975. *J. Atmos. Sci.* 32, 991-992.
[20] Luhar A K, 1998. *Bound-Layer Meteor.*, 88,103-120.
[21] Zilitinkevich S, Baklanov A, Rost J, Smedman A, Lykosov V and Calanca P, 2002. *Q.J.R. Meteor. Soc.* 128, 25-46.
[22] Nieuwstadt F T M and Tennekes H, 1981. *J. of Atmos. Sci.*, 38, 1418-1428.
[23] Melas D, 1993. *Applied Physics* B, 57, 11-17
[24] Melas D, Kambezidis H D, 1994. *International Journal of Remote Sensing*, 15, 499-505
[25] Angevine W M, Doviak R J and Sorbjan Z, 1994. *J. Appl. Meteor.*, 33, 977-983.
[26] Melas D, 1990. *Atmospheric Environment*, 24A, 2847-2853.
[27] Liu X and Ohtaki E, 1997. *Bound-Layer Meterol.*, 85, 497-504.
[28] Cinque G, Zauri R, Di Carlo P, Iarlori M, and Rizi V, 2000. *Il Nuovo Cimento* C, 23: 331-350.

The structure of the lower ABL over antarctic oasis during the summer

Rostislav D. Kouznetsov

Obukhov Institute for Atmospheric Physics, Pyzhevskii per., 3, 119017, Moscow, Russia

E-mail: roux@ifaran.ru

Abstract. The vertical Doppler sodar has been operated during summer 2006-2007 at the Russian Antarctic station Novolazarevskaya at Schirmacher oasis. The typical structures of the ABL over the oasis are classified. Sodar echogrammes reveal wavy structures in the ABL, caused by local orography.

1. Introduction

The single-antenna version of LATAN-3M sodar has been operated during summer 2006-2007 at the Russian station Novolazarevskaya. The station is located 80 km apart from the coast between continental and shelf glaciers at Schirmacher oasis. The structure of the atmospheric boundary layer over the oasis has been studied with a non-Doppler sodar for several years at the Indian antarctic station Maitri [1, 2, 5] which is located about four kilometers to the wet from the Russian station.

In this paper several results of the ABL structure studies with the Doppler mini-sodar are presented.

2. Measuring site and equipment

The Novolazarevskaya station is located at 70°46′S 11°58′E at Schirmacher oasis in Queen Maud land in Antarctica. The oasis is a rocky terrain between the the edge of the Polar cap ice and the ice shelf. The most part of the oasis is elevated over the shelf ice by 50-100 meters.

The sodar has a single 60cm parabolic dish antenna. It was installed close to the glacier slope (Fig. 1). The sodar have been operated with frequency-coded sounding pulse (8 50ms pulses at frequencies 3.32, 3.46, 3.58, 3.66, 3.76, 3.9, 4.02 and 4.13 kHz) with 10m vertical resolution, acoustic power 1W and pulse repetition period 5 seconds. The details on the signal processing can be found in [4]. The data of the automatic weather station have been used in this study as well.

3. Summertime boundary layer structure over Schirmacher oasis

During the summer time the states of the ABL over the Schirmacher oasis can be divided into four classes. The echogrammes and vertical velocity charts for these classes are shown in Fig. 2. The range-corrected echo intensity in dB is plotted in time-height coordinates. Corresponding values of the vertical velocity component are shown with small strokes with black for updrafts

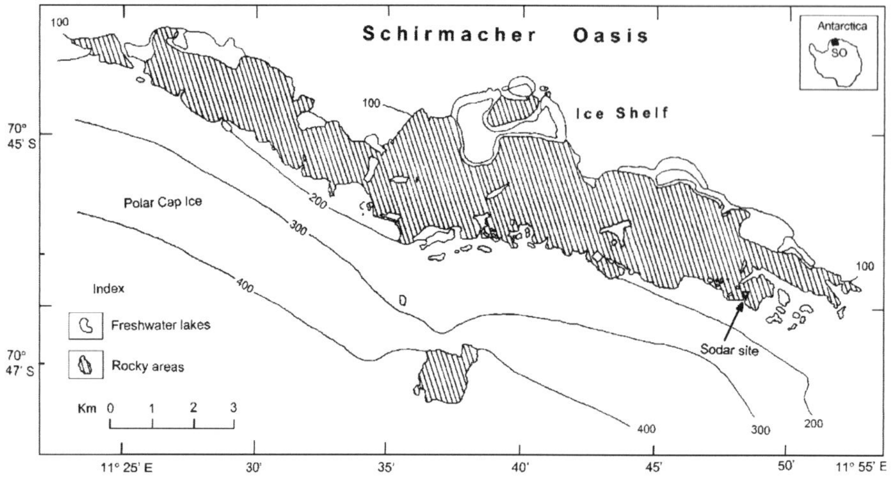

Figure 1. The Schirmacher oasis (adapted from [3])

and grey for downdrafts. The length of a stroke corresponds to the instantaneous value of vertical velocity. These classes can be clearly identified by the data of weather station as well.

The strongest summertime winds (~ 10..20m/s) have south-eastern direction. This is the direction of very strong wintertime winds associated with the polar vortex. These winds are usually accompanied by cloudy weather. They form the least favourable conditions for the sodar operation due to the thermal homogeneity of the ABL and wind noise which clearely appears at the upper part of the echogramme (Fig. 2a). The turbulence in the ABL is purely shear-driven and no visible structures appear in vertical velocity chart. The state of the ABL under such conditions indicate no pronounced diurnal cycle.

At the calm weather daytime a strong convection develops over the oasis(Fig. 2b). The sodar echogramme clearly identifies the thermals associated with updrafts. The dry air causes the strong attenuation of sound at sodar frequencies, thus not much backscattering is seen in between thermals. Note the coming of even more dry air mass at 15:00 that caused the additional decrease of the echo-signal intensity.

At the calm weather night time at weak synoptic winds the strong katabatic winds form at the slope of the continental glacier. These winds can reach up to 10..15m/s. No clear structures appear on the echogrammes under these conditions (Fig. 2c). The vertical wind chart indicates the dominating of the negative vertical velocity. The main difference of the katabatic wind from the synoptic wind is that it exhibits a clear diurnal cycle.

When the humid and warm sea air mass comes from the North the fog forms. This situation is analysed in [3]. The upper boundary of a fog is seen as an elevated layer in the echogramme (Fig. 2d). The waves are formed by the orographic step at the edge of the oasis. The vertical velocity chart clearly indicates the updrafts and downdrafts in the wavy motion. The typical values of the vertical velocities are about 1..1.5m/s, thus the waves observed in the echogramme are propagating rather than advected structures.

4. Conclusion

The sodar has been used to monitor the summertime ABL steructure over the Schirmacher oasis. We identified the four types of weather system: the synoptic wind, the convection, the katabatic wind and the air mass coming from the sea.

The main factor that limited the range of the sodar during our measurements is the attenuation of sound. At 0°C and 35% relative humidity the attenuation at 3.8kHz exceeds 3 dB

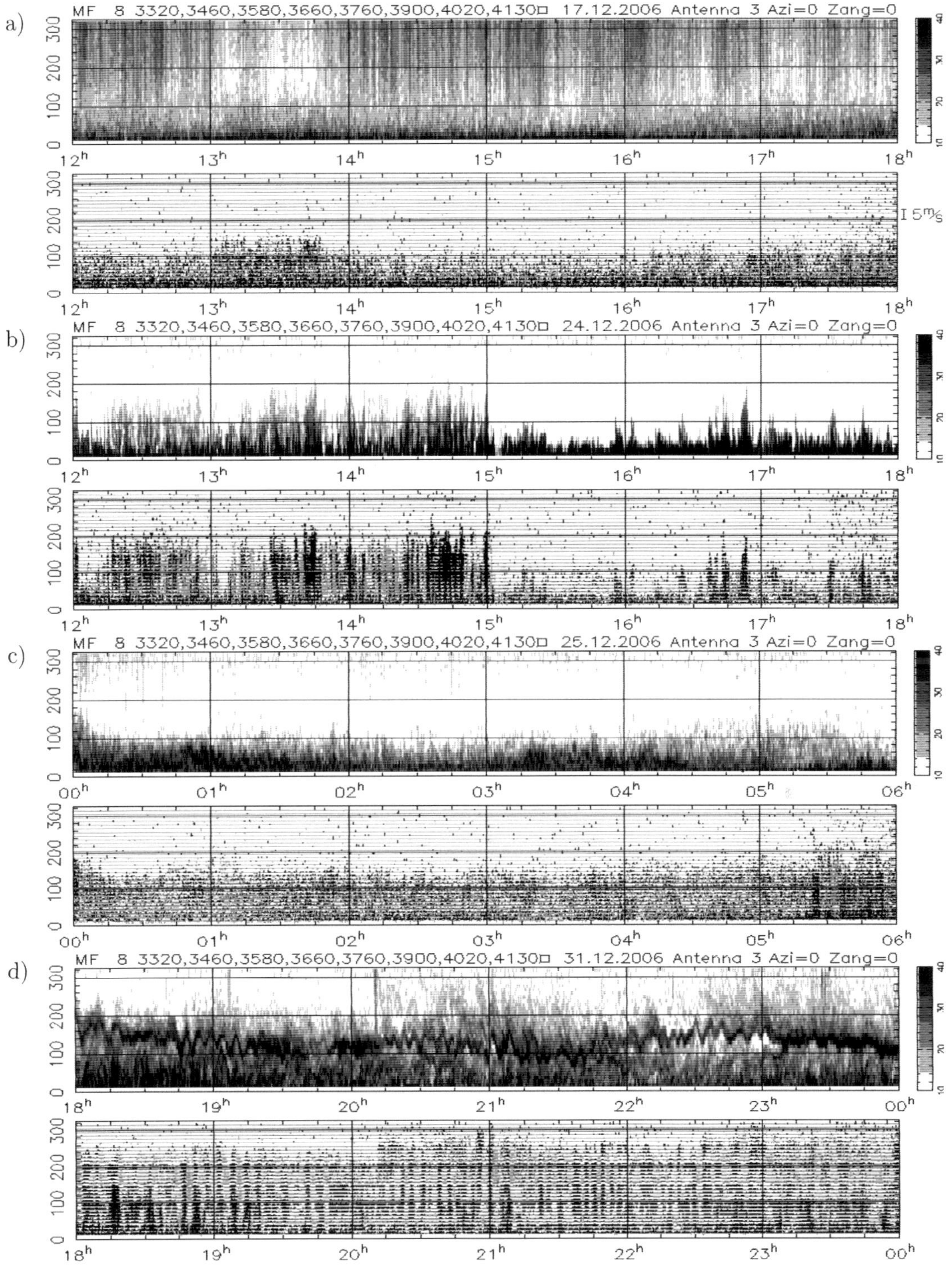

Figure 2. The echogramme and the vertical velocity charts corresponding to four weather types. The heights are in meters a.g.l., the times are GMT. a) strong synoptic wind, b) convection, c) nocturnal katabatic flow, d) humid sea air advection

per 100 meters. In these conditions the use of bigger sodars with lower sounding frequencies is more appropriate and would probably result in a significantly higher data availability.

Acknowledgments
I thank Dr. I. Repina for her help in organization of the expedition, Dr. A. Smirnov and the staff of 52-nd Russian Antarctic Expedition for their help on site. This work is supported by the Russian Foundation for Basic Research through grants 06-05-08086, 06-05-65270, 07-05-00521 and 08-05-00671.

References
[1] H. N. Dutta, P. K. Pasricha, J. Naithani, M. Chatterjee, and B. Reddy. Surface layer characteristics under various atmospheric conditions at the indian pics station maitri. In *Proc. 6-th Int. Symp. Acoust. Rem. Sens*, pages 301–306, Athens, 1992.

[2] K. Gajananda, H. N. Dutta, and V. E. Lagun. An episode of coastal advection fog over East Antarctica . *Current Science*, 93(5):654–659, 2007.

[3] K. Gajananda, A. Kaushik, and H. N. Dutta. Thermal convection over East Antarctica: Potential microorganism dispersal . *Aerobiologia*, 20:21–34, 2004.

[4] R. D. Kouznetsov. Multifrequency acoustic sounding of the abl with parabolic dish antenna. In *Proc. 14-th Int. Symp. Acoust. Rem. Sens.*, pages ***–***, Roskilde, Denmark, 2008.

[5] J. Naithani and H. N. Dutta. Acoustic sounder measurements of the planetary boundary layer at Maitri, Antarctica. *Boundary-Layer Meteorology*, 76:199–207, Oct. 1995.

IOP Conf. Series: Earth and Environmental Science **1** (2008) 012036 doi:10.1088/1755-1307/1/1/012036

Compensation of Vector and Volume Averaging Bias in Lidar Wind Speed Measurements

P J M Clive

Technical Development Officer, SgurrEnergy Ltd, 79 Coplaw Street, Glasgow G42 7JG, UK

Email: peter.clive@sgurrenergy.com

Abstract. A number of vector and volume averaging considerations arise in relation to remote sensing, and in particular, Lidar. 1) Remote sensing devices obtain vector averages. These values are often compared to the scalar averages associated with cup anemometry. The magnitude of a vector average is less than or equal to the scalar average obtained over the same period. The use of Lidars in wind power applications has entailed the estimation of scalar averages by vector averages and *vice versa*. The relationship between the two kinds of average must therefore be understood. It is found that the ratio of the averages depends upon wind direction variability according to a Bessel function of the standard deviation of the wind direction during the averaging interval. 2) The finite probe length of remote sensing devices also incurs a volume averaging bias when wind shear is non-linear. The sensitivity of the devices to signals from a range of heights produces volume averages which will be representative of wind speeds at heights within that range. One can distinguish between the effective or apparent height the measured wind speeds represent as a result of volume averaging bias, and the configuration height at which the device has been set to measure wind speeds. If the wind shear is described by a logarithmic wind profile the apparent height is found to depend mainly on simple geometrical arguments concerning configuration height and probe length and is largely independent of the degree of wind shear. 3) The restriction of the locus of points at which radial velocity measurements are made to the circumference of a horizontally oriented disc at a particular height is seen to introduce ambiguity into results when dealing with wind vector fields which are not irrotational.

1. Introduction

Lidar is a mature remote sensing technology being adopted by the wind power industry for measuring wind velocity, complementing conventional methods. A compact portable ground based device measures wind velocities by the detection of scattered laser emissions. The velocity is determined from the radial velocities measured in different directions relative to the device. In addition the measurements are sensitive to a range of heights determined by the particular design of the device. Both of these considerations mean Lidar wind velocity measurements are vector averages taken over a volume, as distinct from the scalar averages acquired at a point characterizing conventional cup anemometry. This implies care must be taken when treating data obtained using Lidar and conventional methods as equivalent in wind resource calculations. Sources of Lidar bias compared to conventional methods can be identified and are evaluated here.

IOP Conf. Series: Earth and Environmental Science 1 (2008) 012036 doi:10.1088/1755-1307/1/1/012036

Three sources of bias in Lidar measurements are discussed. Vector averaging bias arises as a result of obtaining a vector from the averages of its components and occurs in remote sensing because of the measurement of radial velocities in different directions. Secondly, the sensitivity of measurements at a stated configuration height to wind speeds at a range of heights about that point means that non-linear wind shear will lead to biased results relative to equivalent conventional measurements at the stated configuration height. Finally, the restriction of radial velocity directions to those made available by the scanning of the laser beam about a single axis limits the ability of the device to adequately represent distortions of flow in the averaging volume, leading to bias.

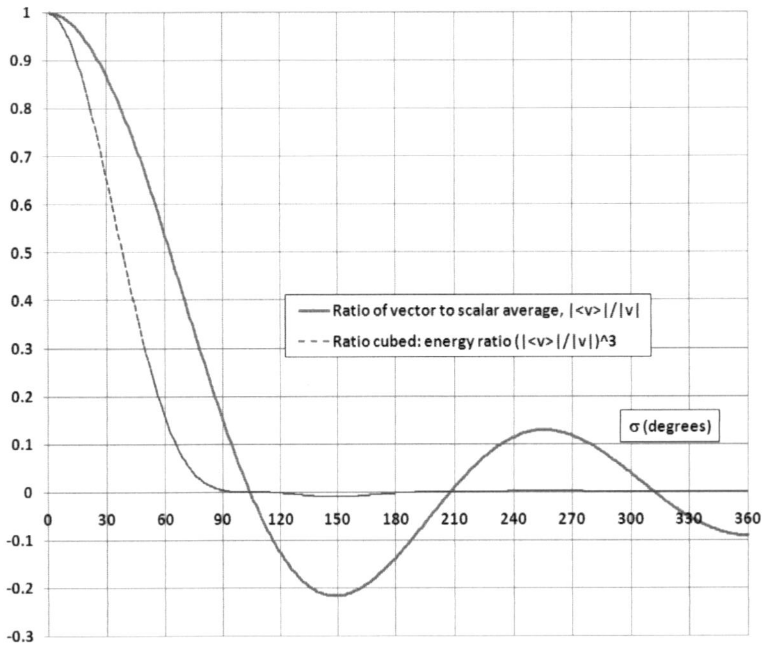

Figure 1. Ratio of the vector to the scalar average as a function of the standard deviation of the wind direction distribution during the averaging interval.

2. Vector averaging and wind veer

A vector which is composed of individual components which are themselves averages acquired over a finite averaging interval is a vector average. The magnitude of the vector average will differ from average magnitude of the vector, i.e. the scalar average. An extreme but instructive example is the situation in which the wind blows at a constant speed from the North for half of an averaging interval and blows at the same constant speed from the South for the other half. The average magnitude, or scalar average, over this interval would be equal to the constant speed, whereas the vector average would have a magnitude of zero. Remote sensing devices accumulate Doppler spectra in a set of discrete directions from which the radial velocity in each of those directions is determined in a process that is equivalent to obtaining average values of vector components from which the vector can be reconstructed. Therefore remote sensing devices acquire vector averages. Cup anemometers, on the other hand, are insensitive to wind direction (other than flow inclination) and measure the scalar average of the wind speed regardless of wind direction. In order to accurately compare the measurements, the difference between vector and scalar averages must be understood.

We can approximate the extent of vector averaging bias using an analytical solution that applies under simple circumstances. Consider wind of constant speed $|\mathbf{v}|$ whose velocity $\mathbf{v}$ is uniformly distributed in direction θ about a mean value of zero such that $-\sqrt{3}\sigma < \theta < \sqrt{3}\sigma$, where σ is the standard deviation of direction in line with the properties of uniform distributions. The magnitude of the vector average $|<\mathbf{v}>|$ is then the magnitude of the average component of the wind velocity in the mean direction $\theta = 0$, which is

$$|<\mathbf{v}>| = <|\mathbf{v}|\cos\theta> = |\mathbf{v}| <\cos\theta>$$

For a uniformly distributed direction θ the probability of direction θ is $P(\theta) = \dfrac{1}{2\sqrt{3}\sigma}$.

Integrating gives an expected value of $\cos\theta$ of

$$<\cos\theta> = \int_{-\sqrt{3}\sigma}^{\sqrt{3}\sigma} P(\theta)\cos(\theta)d\theta = \frac{1}{2\sqrt{3}\sigma}\int_{-\sqrt{3}\sigma}^{\sqrt{3}\sigma}\cos(\theta)d\theta = \frac{\sin\sqrt{3}\sigma}{\sqrt{3}\sigma}$$

So the ratio of vector to scalar average in this instance is given by a Bessel function

$$\boxed{\frac{|<\mathbf{v}>|}{|\mathbf{v}|} = \frac{\sin\sqrt{3}\sigma}{\sqrt{3}\sigma}}$$

This is illustrated in figure1. Empirical data, obtained using a prop and vane anemometer, were used to compare the measured ratio of vector to scalar averages with the theoretical relationship discussed above, to investigate how good an approximation the Bessel function is. These results are shown in Figure 2. A good degree of agreement is observed between the measured and predicted ratios: a linear regression of one upon the other gives $R^2 = 0.86$. The averaging interval was 10 minutes and the data that were averaged were acquired at a rate of 1 Hz.

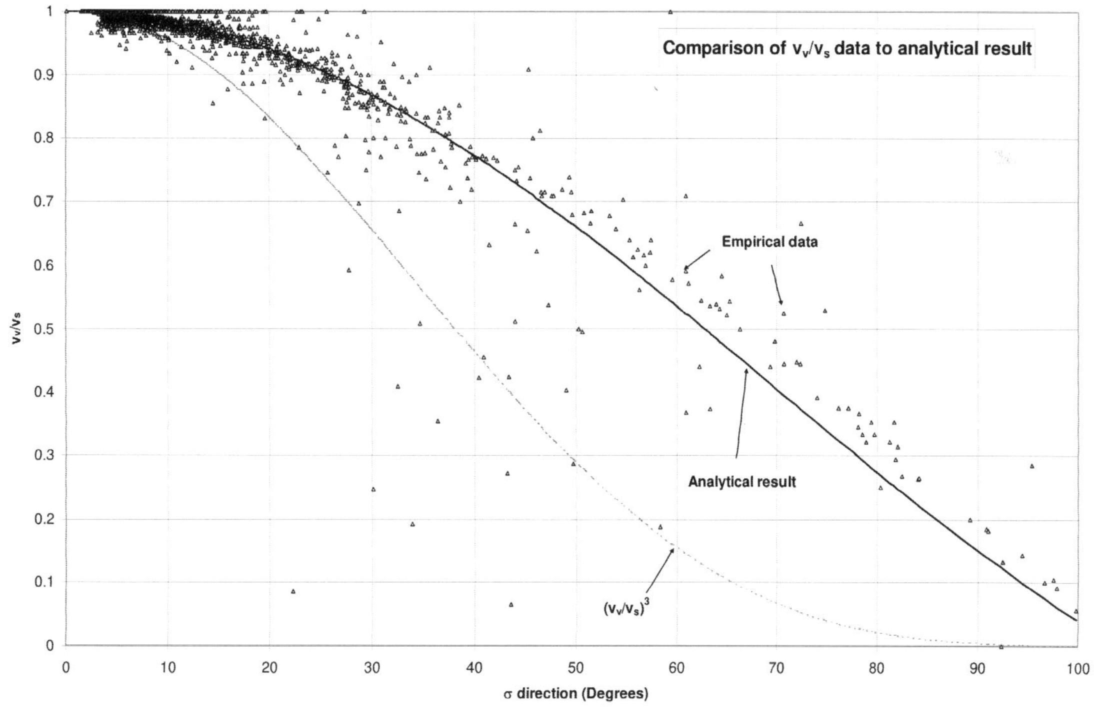

Figure 2. Comparison of theoretical and empirical ratios of vector to scalar averages.

The significance of vector averaging bias is slight for Lidar because the use of light as a probe means the timescales over which the vector averages are acquired are very short compared to the timescales over which the wind direction varies sufficiently to incur a vector averaging bias. However, remote sensing devices that rely on sound as a probe of the wind velocity field are characterized by the accumulation of average vector components over a longer timescale, and further investigation of the impact of vector averaging on SODAR measurements is warranted.

Vector averaging effects are relevant when one considers the response of wind turbines to variations in wind direction with height, or wind veer, across the rotor diameter. A turbine is directed into the wind and so will be sensitive to variations in wind direction across its rotor diameter, producing energy from the component of the wind in the same direction as it is pointing. Turbine output is more directly related to the magnitude of the vector average over the range of directions impinging upon it across the entire rotor diameter than the scalar wind speed at hub height which approximates it. In this instance the vector average is obtained over space rather than over time as was previously the case in our consideration of vector averaging effects in remote sensing. The cube of the ratio of the vector to scalar average, that is, the ratio of the energies associated with the vector and scalar averages discussed earlier were shown in Figure 1 to indicate the significance of vector averaging for energy yields. The ratio of the energies drops off more rapidly with the increasing extent of the variation in wind direction and varies as the cube of the ratio of the vector and scalar average wind speeds. Variations in wind direction might be expected to arise from phenomena such as the Ekman spiral, as well as from the proximity of topographic features which distort wind flow, which may be expected in upland areas.

The impact of vector averaging effects upon turbine output merits further investigation. This has hitherto been neglected because of the limitations imposed on wind direction measurements by the available instrumentation prior to the advent of remote sensing, and in particular, Lidar. Available data were generally inadequate for the purposes of properly assessing the extent of wind veer and predicting its impact on energy yields, limitations that are alleviated by Lidar, which should now enable the inclusion of veer studies in the assessment of wind resource and the prediction of energy yields.

3. Volume averaging, effective probe length, wind shear and apparent height

The use of remote sensing devices such as SODAR and Lidar to obtain wind speed data is gaining widespread acceptance. These devices acquire a volume average of the wind speeds around a given height, in contradistinction to a cup anemometer which operates approximately at a point in space in comparison. Remote sensing devices use sound or light as a probe of the air above them and detect wind speed conditions from the effect these have on the returned signals, such as imposing a Doppler shift on the signals due to the movement of the features in the air from which they are returned. These remote sensing devices have an effective probe length corresponding to a finite height resolution which determines the size of the volume over which wind speeds are averaged. This is determined in the case of pulsed devices such as SODARs and pulsed Lidars by the duration of the output pulse, and in the case of continuous wave (cw) devices such as cw Lidars by the depth of focus of the detector optics used to select the measurement height.

Performing these volumes averages of non-linear wind profiles such a logarithmic wind profiles over a range of heights determined by the height resolution of the device will impose a bias on the measurement due to the non-linearity of the wind profile. The magnitude of this bias can be indicated by finding the ratio of the volume average, which a remote sensing device sensitive to signal returns from a range of heights detects, to the point value in the middle of the height range averaged over, such as would be measured by a cup anemometer, for a non-linear wind profile. This ratio is

$$\beta_{\log,z} = \frac{< u_z >}{u_z}$$

where u_z is the wind speed at height z, $<u_z>$ is the wind speed averaged over the specified height range centered on height z, and $\beta_{log,z}$ is the ratio of these two quantities, assuming a logarithmic wind profile, indicating the extent to which a volume average at this height is biased with reference to a point measurement at the height in the centre of the range

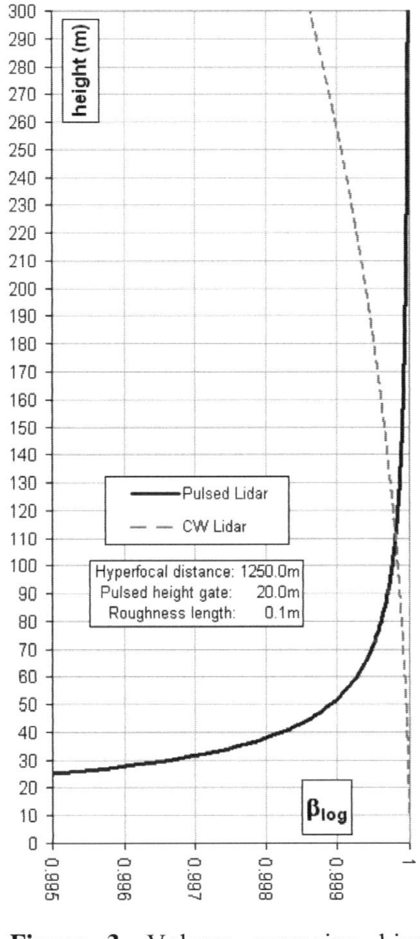

Figure 3. Volume averaging bias arising from finite effective probe length and a logarithmic wind profile.

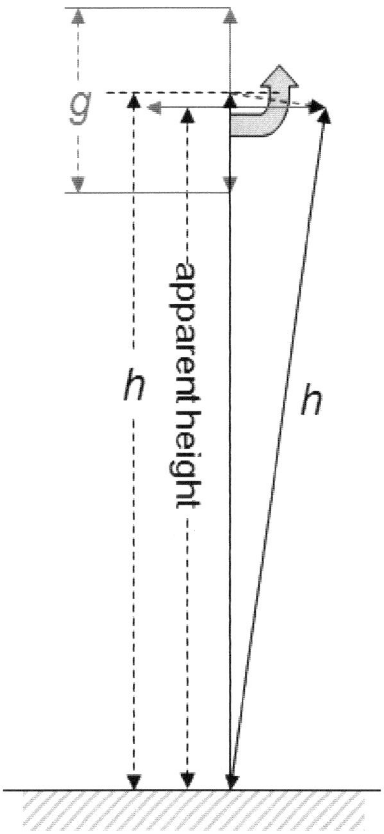

Figure 4. Geometric interpretation of volume averaging bias arising from finite effective probe length and logarithmic wind shear.

Neglecting zero-plane displacement and assuming neutral stability, setting $u_0 = \frac{u_*}{\kappa}$ gives a logarithmic wind profile $u_z = u_0 \ln \frac{z}{z_0}$ where u_z is the wind speed at height z, z_0 is the roughness length, κ is von Karman's constant and u_* is the friction velocity. We can consider the effects of averaging this over a range of heights z by calculating the expected value of u_z for configuration height h thus:

$$< u_h >= \frac{u_0}{g} \int_{h-\frac{1}{2}g}^{h+\frac{1}{2}g} \ln \frac{z}{z_0} dz$$

where g is the range of heights averaged over (the height gate or effective probe length), and h is the height at which the configuration of the remote sensing device has been set to measure. Then

$$< u_h >= \frac{u_0}{g}\left[z \ln\frac{z}{z_0} - z \right]_{h-\frac{1}{2}g}^{h+\frac{1}{2}g} = u_0 \left\{ \frac{h}{g}\ln\frac{h+\frac{1}{2}g}{h-\frac{1}{2}g} + \ln\frac{\sqrt{h^2-\left(\frac{1}{2}g\right)^2}}{z_0} - 1 \right\}$$

Therefore $$\beta_{\log,h} = \frac{< u_h >}{u_h} = \left(\frac{h}{g}\ln\frac{h+\frac{1}{2}g}{h-\frac{1}{2}g} + \ln\frac{\sqrt{h^2-\left(\frac{1}{2}g\right)^2}}{z_0} - 1 \right)\left(\ln\frac{h}{z_0} \right)^{-1}$$

A typical value of g for SODARs and pulsed Lidars is 20m [1]. Figure 3 shows $\beta_{\log,h}$ for $g = 20$m and $z_0 = 0.1$m. Also shown is $\beta_{\log,h}$ for a cw Lidar whose detector optics have a hyperfocal distance H of 1250m. If the value of g, the effective probe length, or size of the height gate, is approximated by the focal depth when the optics are focused at configuration height h, it is not constant as with a pulsed device, but varies according to

$$g = \frac{2Hh^2}{H^2 - h^2}$$

and increases with height. In figure 3 $\beta_{\log,h}$ is closer to unity for a cw Lidar than for a pulsed device at heights where the focal depth is less than the pulse length. At heights where the focal depth exceeds the pulse length the pulsed devices exhibit less bias arising from finite effective probe length.

A geometric interpretation of this result suggests itself from the observation that the first term in the numerator of the expression for $\beta_{\log,h}$ is close to unity except at low heights and so cancels with the third. The behaviour of $\beta_{\log,h}$ at most heights of interest is then dominated by the term

$$\beta_{\log,h} \sim \left(\ln\frac{\sqrt{h^2-\left(\frac{1}{2}g\right)^2}}{z_0} \right)\left(\ln\frac{h}{z_0} \right)^{-1}$$

which is the ratio between the point value at the apparent height equivalent to a side of a right angled triangle whose hypotenuse is the configuration height h and whose other side is half the height gate g, and a point value at the height h. This is shown in figure 4. So if a remote sensing device is configured to measure wind speeds at height h and has an effective probe length of $g = 2\delta$, i.e. is sensitive to wind speeds at $h \pm \delta$, then the effective or apparent height h_{app} at which it is measuring a logarithmic wind profile, that is, the height that the measured wind speeds are representative of a logarithmic wind profile, can be defined to be

$$\boxed{h_{app} = \sqrt{h^2 - \delta^2}}$$

The variation in the concentration of aerosols with height will also affect the returned signal and may be more significant in relation to volume averaging bias that the bias incurred by finite effective probe length. The geometrical arguments above do not consider variations in the concentration of the airborne particulates from which the laser light is being scattered. A generalization of the method above may be able to accommodate this. When dealing with arbitrary wind shear profiles the volume averaging bias can be deconvolved from the measurements using digital Fourier transforms of the measurements and the effective probe lengths. If aerosol concentration data or a suitable proxy is also acquired variations in concentration can be accommodated in the analysis using a similar method. The application of digital Fourier transforms to the deconvolution of volume averages in remote sensing arising from both finite effective probe length and variations in aerosol concentration requires further investigation.

4. Averaging across the scan disc and measurement ambiguity

Remote sensing devices acquire radial velocities in multiple directions from which the wind vector is deduced. The beams emitted by a monostatic device penetrate a volume and assumptions are conventionally made about the uniformity of flow to enable a unique solution for the vector field in that volume to be found on the basis of the measured radial velocities. The consequences of these assumptions are investigated here.

The co-ordinate system which will be used to describe the radial vectors and the points at which radial velocities are measured is shown in figure 5. The device lies at the origin O. The beam is tilted away from the vertical by a zenith distance α, typically fixed at 30°. The beam can be rotated azimuthally about its single degree of freedom by θ to measure radial velocities at point L in the direction OL a distance l from O.

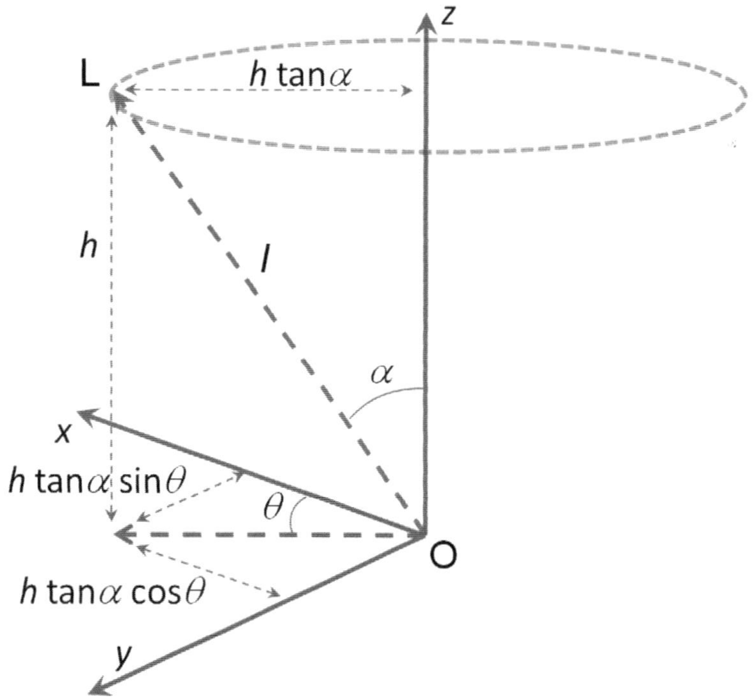

Figure 5. Co-ordinate system for describing Lidar radial vectors.

As the Lidar is swept around its scan the co-ordinates of the point in space L at which it is acquiring radial velocity data are

$x = h \tan\alpha \cos\theta \qquad y = h \tan\alpha \sin\theta \qquad z = h$

where h is the height at which the measurements are being made. The components of the unit vector **e** in the direction of the Lidar beam as it sweeps through azimuthal angles θ are

$e_1 = \sin\alpha \cos\theta \qquad e_2 = \sin\alpha \sin\theta \qquad e_3 = \cos\alpha$

The wind flow can be characterized by a vector $\mathbf{v} = (v_1, v_2, v_3)$. The absolute magnitude of the radial velocity $\mathbf{v}_r$ in the direction in which the Lidar is emitting and detecting light is then given by

$$|\mathbf{v}_r| = |\mathbf{v}.\mathbf{e}| = |\, v_1e_1 + v_2e_2 + v_3e_3 \,|$$

For example, uniform inclined flow in the x,z-plane described by $\mathbf{v}$ = (a, 0, c) gives radial velocities $|\mathbf{v}_r| = |\text{ a} \sin\alpha \cos\theta + \text{c} \cos\alpha\, |$. Plotting $|\mathbf{v}_r|$ against azimuth θ on a polar plot gives the familiar figure-of-eight, or VAD (Velocity Azimuth Display). In practice a function of the form

$$|\mathbf{v}_r| = |\text{ A} \cos(\theta - \text{B}) + \text{C}\, |$$

is fitted to the measured radial velocities from which horizontal velocity a = A/sinα, vertical velocity c = C/cosα and bearing (North of East or West of North) B are deduced [2]. It can be seen from the expressions for the Cartesian co-ordinates of L and the components of the unit vector in the direction OL that

$$\boxed{he_1 = xe_3} \quad \text{and} \quad \boxed{he_2 = ye_3}$$

allowing substitutions to be made. If one relaxes the constraint that the flow is uniform and allows flow in which, in general, is described by

$v_1 = f_1(x, y, z) \qquad v_2 = f_2(x, y, z) \qquad v_3 = f_3(x, y, z)$

and in particular is *not* irrotational, then the ability to make substitutions means that there is no unique solution that describes a correspondence between the observed variation in $|\mathbf{v}_r|$ with azimuthal angle θ and a unique wind flow vector field. So, to give a simple unrealistic but instructive example, flow described by the vector field $\mathbf{v} = (0, 0, \text{m}x)$ gives

$$\mathbf{v}.\mathbf{e} = \text{m}xe_3 = \text{m}he_1$$

That is, measuring the radial velocities at points L around the circumference of the disc at height h does not allow us to distinguish between wind with a constant horizontal component of mh and wind with a variable vertical component of mx. These two cases are shown in figure 6. The differentiation between the two cases is conventionally made on the basis of an assumption of uniform irrotational flow in favour of the former; however this assumption will not always be valid. The assumption is implicit in the form of the function that is conventionally fitted to the measured radial velocities. If the wind vector field is not irrotational, and in particular, if the z-component, that is, the component in the direction of the axis about which the beam is scanned azimuthally, is a function of x and/or y, the substitutions discussed above will mean that measurements are ambiguous.

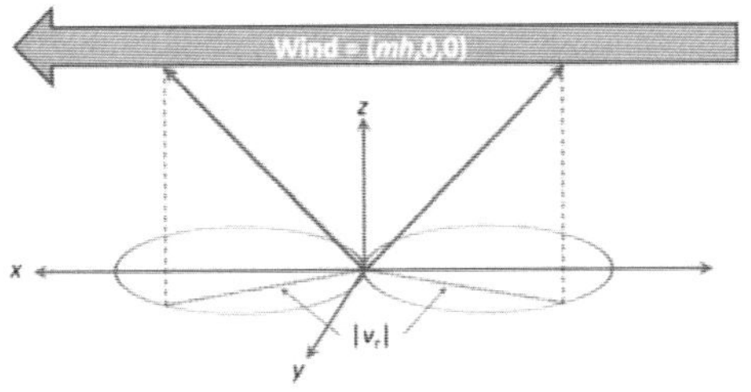

Figure 6a. Uniform horizontal flow.

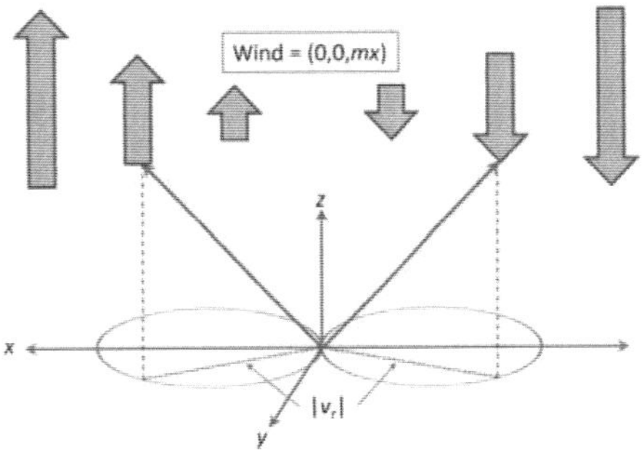

Figure 6b. Variable vertical component.

Distorted wind flow can be expected as a result of complex topography. This will incur volume averaging bias relative to instruments that do not obtain volume averages across the scan disc but rather acquire point measurements at the centre of the scan disc. This limitation can be alleviated if one relaxes the constraints on the scan that allow the substitutions to be made, and one way of doing this is to allow the beam to be scanned about two axes, one which varies the azimuth θ and one which varies the zenith distance α. The radial component of a horizontal flow will have a sine dependence on zenith distance, whereas the radial component of a vertical flow will have a cosine dependence on zenith distance. This allows the two cases previously discussed to be unambiguously resolved. In order to properly represent wind flow in three dimensions it is necessary to acquire measurements of radial velocities at points varying in all in three dimensions, including height. In general measurements should be made at points within a volume, rather than restricted to the circumference of a horizontal disc, by using a pulsed device that varies both azimuth θ and zenith distance α, such that

$$x = l\sin\alpha\cos(\theta - b) \qquad y = l\sin\alpha\sin(\theta - b) \qquad z = l\cos\alpha$$
$$e_1 = \sin\alpha\cos(\theta - b) \qquad e_2 = \sin\alpha\sin(\theta - b) \qquad e_3 = \cos\alpha$$

where b is the bearing, an offset which aligns the prevailing wind direction with the x-axis. Then a more general solution for the wind vector field $\mathbf{v} = (v_1, v_2, v_3)$ such as, for example,

$$v_i = \prod_j K_{ij} x_j \exp\left(\frac{\left(x_j - k_{ij}\right)^2}{2\sigma_{ij}^{\ 2}}\right)$$

where $x_1 = x$, $x_2 = y$ and $x_3 = z$, may be fitted to the observed magnitudes of radial velocity $v_r = \mathbf{v.e}$ as a function of both θ and α to give the parameters K_{ij}, k_{ij}, σ_{ij} and b, such that the wind speed at point P = (x, y, z) can be calculated e.g. for comparison with a cup anemometer measurement at $(0, 0, h)$. Rather than employing a VAD, or Velocity Azimuth Display, the more general VAZDD, or Velocity Azimuth Zenith Distance Display, should be used.

5. Conclusions

Remote sensing devices acquire vector averages over a volume in space. Comparison of these with scalar averages acquired at a point in space requires an understanding of the relationship between vector and scalar averages and the relationship between volume and point measurements. The relationship between vector averages and scalar averages is seen to be well approximated by a Bessel function of the degree to which the wind direction varies during the averaging interval. This may have consequences for the accuracy of SODAR measurements since the timescales associated with the acquisition of vector averages by these devices is comparable to the timescales over which wind direction can vary significantly. However the timescales associated with the acquisition of vector averages by Lidar is sufficiently short for vector averaging to incur no significant bias.

The reliability of Lidar devices in relation to vector averaging effects makes them uniquely suitable for investigating situations where vector averaging effects may have other consequences. In particular, instances where wind direction varies with height across the rotor diameter of a wind turbine can be studied and the hitherto neglected impact of vector averaging on output investigated.

The effective probe length of remote sensing devices means that when they are configured to measure at a certain height the measurements they acquire are indicative of the wind speeds at a range of heights about the configuration height. Non-linear wind shear will cause the average wind speed obtained by integrating over this range of heights to be biased relative to the wind speed at the configuration height. It is possible to define an *apparent* height; this is the height corresponding to the average wind speed measured over the effective probe length if the wind speeds are described by a logarithmic wind profile. The apparent height approximates the height at which the measured wind speed occurs under these circumstances, and is not equal to the configuration height.

In section 4 it is seen that the restriction of the locus of points at which radial velocity measurements are made using Lidar to the circumference of a horizontally oriented disc at a specific height is adequate when characterizing wind speeds that are uniform, and also when characterizing wind vector fields in which the x-, y- and z-components of velocities vary only with z (i.e. can be described by simple flow inclination and wind shear). However this restriction will produce ambiguous results if the z-component is allowed to vary as a function of x and/or y. Such variation can reasonably be expected over terrain whose topographical complexity is typical of upland areas where the installation of wind farms may be proposed. This ambiguity may lead to erroneous results if the measurements of the radial velocities at the edge of the disc are used to predict wind velocities at other points within the scanned volume, such as cup anemometer or wind turbine hub locations. Under these circumstances the relaxation of the constraint on the Lidar scan described above to allow the acquisition of radial velocities as a function of both Lidar beam azimuth and beam zenith distance provides data to which a more general solution can be fitted unambiguously, allowing more accurate prediction of wind speeds at points within the scanned volume.

Acknowledgements

The author would like to thank Kathleen Moore who provided the data that enabled the comparison illustrated in figure 2.

References

[1] Crescendi, H.G., 1997, "A look back of two decades of Doppler sodar comparison studies", Bulletin of the American Meteorological Society, 78, 651-673

[2] Banakh VA, Smalikho IN, Köpp F, Werner C. "Representativeness of wind measurements with a CW Doppler lidar in the atmospheric boundary layer", Applied Optics 1993 **34:** 2055-2067

IOP Conf. Series: Earth and Environmental Science 1 (2008) 012037 doi:10.1088/1755-1307/1/1/012037

Field statistics of antenna arrays of equipment for remote sensing of the atmosphere

Y Shifrin[1], **Y Ulyanov**[2], **N Maksimova**[3]
[1]Chief Scientist, Kharkiv National University of Radio Electronics, Kharkiv, Ukraine
E-mail: shifrin@kture.kharkov.ua
[2]Senior researcher, Kharkiv National University "KhPI", Kharkov, Ukraine
E-mail: ulyanov@kpi.kharkov.ua
[3]Senior researcher, Kharkiv National University of Radio Electronics, Kharkiv, Ukraine
E-mail: shifrin@kture.kharkov.ua

Abstract. Simple formulas for estimation of statistical effects in antenna arrays of the equipment for remote sensing of the atmosphere (RSA) in the presence of random errors of excitation of their radiators are given. It is shown on an example of two concrete arrays that even small phase errors in radiators noticeably influence on the level of side radiation and the directivity of arrays. Therefore, account of random errors is a necessary stage when designing and developing high-quality antenna arrays for RSA systems.

1. Introduction

Recently in systems of remote sensing of the atmosphere (RSA), using the sound radiation (sodar and RASS) phased antenna arrays (PAA) find more and more application as antennas of acoustic and radio channels [1-5]. Application of PAA allows appreciably widening the possibilities of sodar and RASS both in their characteristics control and in the signal processing. In this context they have much common with arrays used in radar.

Along with this, there are significant distinctions between these types of arrays. One of them is in that arrays used in systems of RSA are much stronger imposed by the harmful influence of the randomness (errors) in the distribution of phases and amplitudes of excitation in separate radiators on their characteristics. The first reason of this consists in more (compared with radar arrays) errors of excitation of radiators in PAA of RSA systems. Especially this concerns acoustic PAAs. The phase dispersion in radiators of such arrays reaches tens of degrees. [3]. Another reason of the high "sensibility" of arrays of RSA systems to the randomness in them is the relatively small number of radiators in them. The number N of radiators in moderate PAA used in radar constitutes a value of the order of several hundreds – thousands, whereas that in RSA systems is usually less by the order. However, as is well known, statistical effects, connected with the randomness in the antenna, strengthen with the decrease in N [6]. Unfortunately, in literature known for us, questions of the PAA field statistics in RSA systems practically were not considered.

A goal of the paper is to fill up this gap. The expediency of such an analysis is dictated by two circumstances. On one side, requirements to a side radiation level (SRL) of antennas of RSA systems, defined in much namely by randomness in the antenna, are rather strict. On other side, the solution of the problem being of interest for us, lies almost on the surface, as it is one application of the SAT – the theory of antennas with random sources, developed long ago [6].

2. Statement of problem. Main relationships

There are considered linear and quadratic equidistant arrays. In the absence of errors, arrays are assumed to be synphasal with the uniform amplitude distribution (AD). Let us restrict ourselves by consideration of the most important case – small phase errors (the dispersion of errors is $\alpha = \sigma_0^2 \ll 1$ where σ_0 is the standard deviation of errors). Assume errors in different radiators (designate their total number as N) to be independent and distributed by the normal law.

The field statistics of arrays is developed in many works (see, e.g., works [6-8] and literature given in them). It is shown, in particular, in the works that at the uniform amplitude distribution (AD) in the array, the mean background of side radiation, created by errors in the normalized mean radiation pattern (RP), is $\bar{\Phi}_s = \alpha / N$, whereas the reduction of the mean directivity is

$$\Delta = \frac{D_0 - \bar{D}}{D_0} = 1 - \frac{\bar{D}}{D_0} \approx \alpha$$

where D_0 and $\bar{D}$ are the antenna directivity in the absence of errors and the mean directivity in their presence, respectively, at the wide range of varying the inter-element spacing in the array.

A simple formula takes place also for the deviation of the main maximum direction (MMD) of the array RP, occurring in the presence of errors. For example, for the linear array with the uniform AD, the relative deviation of MMD (the ratio of the standard deviation of MMD to the RP width) is defined (at the sufficiently large N) as [6]

$$\delta_{MMD} = 0.6\sqrt{\frac{\alpha}{N}}\,.$$

The same formula is valid also for the MMD deviation in main planes of the quadratic array [10]. A growth of side radiation, reduction of the directivity and MMD deviation are most unpleasant consequences of occurrence of small phase errors in the antenna.

The relationships given above allow obtaining estimates of the three effects. In practice, these formulas can be used at values of α not exceeding 0.6…0.7. At larger values of errors dispersions of these or those statistical effects, it is possible to use results of numerical calculations available in literature on SAT (see, e.g., [6, 10]). Therewith one should pay attention to other statistical effects – RP widening, displacement of extremes in the mean RP, distortions of RP shape, etc., which at large enough errors can be significant and be of interest when estimating characteristics and possibilities of PAA of the RSA system.

3. Examples.

Relationships given for estimates of the mean SRL, reduction of the directivity and MMD deviation were used for estimation of these effects in two acoustic PAA. The corresponding results are presented in table 1 for PAA described in [3] consisted of 40 piezoelectric loudspeakers. Values of the loudspeaker phase standard deviation, experimentally measured at frequency changes within limits of from 2000 to 4000 Hz, were varying from 14.7 to 47 degrees.

Table 1. Statistical effects for PAA, described in [3], consisted of 40 piezoelectric loudspeakers.

σ_0, degrees	α	Characteristics of PAA [3]		
		$\bar{\Phi}_S$, dB	Δ,%	δ_{MMD},%
15	0.069	-27.7	6.9	2.49
30	0.275	-21.6	24.0*	4.98
50	0.762	-17.2	53.4*	8.3

* Values of the directivity reduction marked by asterix are calculated by the formula $\bar{D}/D_0 = e^{-\alpha}$ valid at any α at independency of errors in radiators [6].

As is seen from table 1, at σ_0 of the order of 50°, reduction of the directivity exceeds 50 %. More serious statistical effect consists in the SRL growth. Even comparatively small errors of excitation of radiators with σ_0 = 15°, at which reduction of the directivity is insignificant (7 %), sometimes origins an absolutely inacceptable background of the mean SRL equal to -28 dB. Requirements to errors $\sigma_{0\ req}$ at the given level of the mean background $\bar{\Phi}_{sgiv}$ are defined as

$$\sigma^0_{0req} = 57\left[N \lg \frac{\bar{\Phi}_{sgiv}}{10}\right]^{1/2}.$$

It is easy to obtain the order of the value of σ_{0req} built by this formula from figure 1. In particular, figure 1 shows that at the given SRL of the order of -50 dB the required value σ_{0req} at the wide range of values of the number of elements (from 8 to 128) constitutes the value of the order of 0.5-2°.

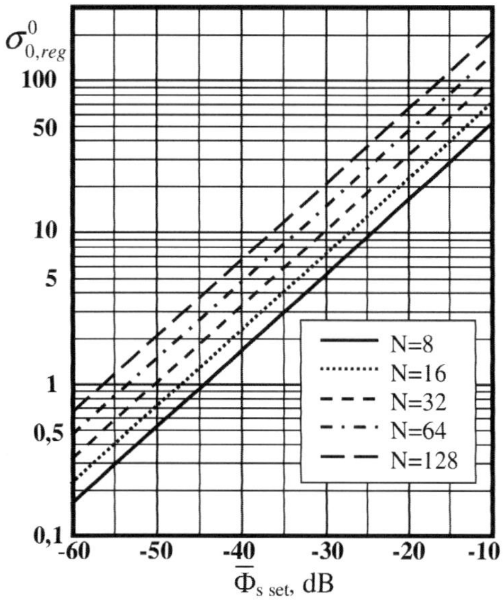

Figure 1. Dependence of σ_{0req} on set SLL for the number of elements in the array of 8 to 128.

The MMD deviation, as is seen from Table 1, is insignificant. Note that at calculations of MMD, errors in radiators were assumed to be independent. The error correlation over the aperture will lead to the increase in the MMD deviation.

By the second example let us consider the acoustic array of RASS developed at the Problem Laboratory of Atmosphere Sounding of the Kharkov National University of Radio Electronics. The quadratic array incorporated 64 horn electromagnetic loudspeakers. The processing of histograms for the phase of loudspeakers has yielded the value $\sigma_0 = 23^0$ [9]. At such a value of σ_0, reduction of the directivity is $\Delta = 16$ %, the level of the mean background is $\bar{\Phi}_S$ =-26 dB, $\delta_{MMD} \approx 3\%$.

The processing of histograms of a spread in loudspeakers powers of these arrays has shown that this spread is equivalent to phase errors with $\sigma_0 = 3°$ [9]. This result has confirmed the statement known from works on SAT, that usually most dangerous are phase errors of excitation of radiators.

4. Conclusion

As is seen from the results of this work, in particular, calculation of characteristics of two concrete acoustic arrays, the random errors in excitation of the array radiators can significantly change real parameters of PAA and of the RSA system as a whole. Therefore, regard of the influence of random errors on characteristics of the RSA system PAA is a necessary stage when designing and developing such PAAs. More in detail these questions are considered in [9, 11]. In particular, it is shown there that due to the influence of randomness in the array, application of the decaying or Chebyshev AD in arrays of RSA systems for lowering SRL can appear evidently inexpedient.

5. References

[1] Wolf D E, Lataitis R J, Weber B L et al 1997 Design and preliminary field test of a combined RF/acoustic phased-array antenna for profiler operation *Ext. Abstracts of COST-76 Profiler Workshop V* **2** Engelberg Switzerland pp 151-154

[2] Khanna M, Bahl M, Batra S et al. 1999 Design of a high efficiency acoustic phased array antenna for an acoustic wind profiler *Meteorol. Atmos. Phys.* **71** pp 35-41.

[3] Krasnenko N P, Kushner A A 2005 Influence of a spread in the parameters of antenna elements on the characteristics of an acoustic array *XVI Session of the Russian Acoustic Society,* November 14-18 pp.376-378 (in Russian)

[4] Vogt, S, and Thomas, P 1994 Test of a phased array sodar by intercomparison with tower data *Journal of Atmos. Oceanic. Technology* **11** pp 94–102

[5] 5. Hashiguchi, H., Fukao, S, Moritani, Y et al. 2004. A lower troposphere radar: 1.3-GHz active phased-array type wind profiler with RASS *Journal of the Meteorological Society of Japan* **82 3** pp 915–931

[6] Shifrin Y S 1970 *Questions of statistical antenna theory* (Sovetskoe radio in Russian) Eng. Trans. 1971 *Statistical antenna theory* (Golem Press Co USA)

[7] Shifrin Y S, Kornienko L G 2000 Statistics of antenna array field. *Antennas* **1** pp 1–26 (in Russian) Eng. Trans. 2002 Statistics of antenna array field *Telecommunications and radio engineering* **58 8** pp 2-33

[8] *Microwave scanning antennas* Ed. by R C Hansen 1966 Volume 1 (New-York and London, Academic Press) Rus. trans. 1969 *Microwave Scanning Antennas* ed by Markov G T and Chaplin A F Volume II M (Sovetskoe radio) 496 p

[9] Shifrin Y S 1997 *Statistical antenna theory* (Handbook on antenna techniques Volume 1 ed by Feld Y N and Zelkin E G Moscow – Chapter 9) 57 p

[10] Shifrin Y S, Ulyanov U N, Maksimova N G On side radiation of antennas of the equipment for acoustic and radioacoustic sounding of the atmosphere 2007 **66(18)** pp 1657 – 1665

[11] Shifrin Y S, Ulyanov U N, Maksimova N G Field statistics of acoustic array of the equipment for remote sensing of the atmosphere 2008 *Telecommunications and Radio Engineering* **67(4)** pp 293 - 308

IOP Conf. Series: Earth and Environmental Science **1** (2008) 012038 doi:10.1088/1755-1307/1/1/012038

Multi-frequency acoustic sounding of the ABL with a dish antenna

Rostislav D. Kouznetsov
Obukhov Institute for Atmospheric Physics, Pyzhevskii per., 3, 119017, Moscow, Russia
E-mail: roux@ifaran.ru

Abstract. We describe the signal processing of the LATAN-3M sodar developed at the Obukhov Institute for Atmospheric Physics. This sodar uses the frequency-coded sounding signal and able to operate the antennae in parallel. The incoherent spectral integration procedure for components of multi-frequency echo-return is described.

1. Introduction

The method of a frequency-coded sounding signal becomes popular in modern sodars. This method is a way to get several echo returns from the same volume within a single sounding cycle. It can be implemented in modern sodars with digital signal processing without significant hardware changes. The multi-frequency sounding gives an opportunity increase the accuracy and noise resistance of sodar measuremens due to increase of return signal statistics.

The multi-frequency mode is implemented in the LATAN-3 sodar developed at the Obukhov Institute of atmospheric physics. The new version is called LATAN-3M. The multi-frequency mode has been used during Dec 2006 – Jan 2007 at Russian Antarctic station Novolazarevskaya for vertical sodar and during summer 2007 in desert region of Kalmykia (Russia). In this paper we describe the basic principles of multi-frequency sounding, the details of signal processing algorithms in LATAN-3M sodar and show several results of experimental tests of the system. The same echo-signal processing routine has been used with two types of antennae 0.6m and 1.2m diameter with corresponding nominal frequencies 4kHz and 2kHz.

2. Basic principles

Consider a sounding cycle of conventional single-frequency sodar (Fig. 1) in time-height coordinates. The antenna emits a sounding signal of frequency f during the interval τ. The signal propagates upward with speed of sound c and small fraction of it scatters back within the atmosphere. After the burst signal the antenna requires some time before receiving of echo-signal due to transitional processes. After that, at the time t_1 the scatter comes from the range gate h_1 (heights $ct_1/2..c(t_1+\tau)/2$), at t_2 from h_2 and so on. The vertical resolution of the sodar is determined by the spatial extent of the sounding signal.

If the sodar emits n (for example $n = 3$) sounding pulses of duration τ at different frequencies consequently before the receiving, the echo can be received at these frequencies simultaneously (Fig. 2). This increases the duration of sounding cycle by $(n-1)\tau$ as compare to single-frequency mode, but gives the opportunity to get n independent echo-signals from higher range gates, where

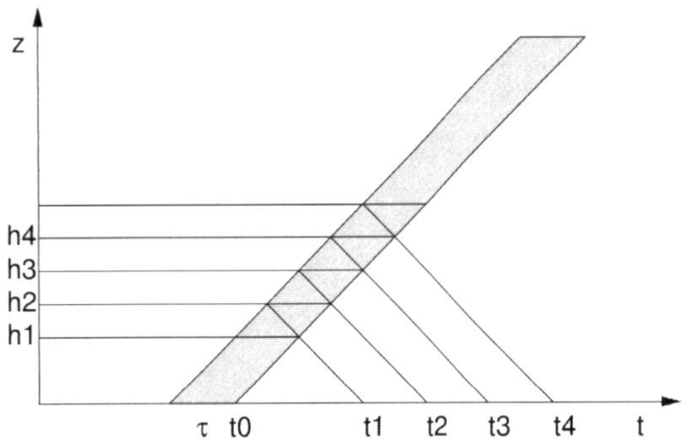

Figure 1. Single-frequency sounding

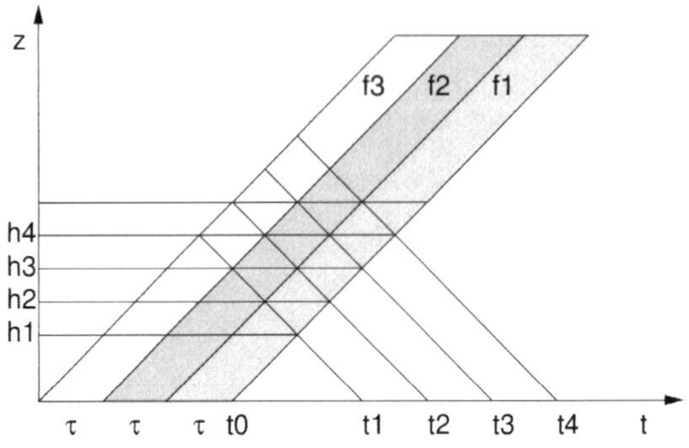

Figure 2. Frequency-coded sounding

the signal is weaker. At any time the received signal comes from all heights reached by the pulse train. However, since sampling starts after the emission of of the last pulse the received scattering is only available for the most recentlyemitted pulses. At the time t_1 the antenna receives the echo from range gates h_1, h_2, and h_3 at frequencies f_1, f_2, and f_3 correspondingly. At time t_2 – from h_2, h_3, and h_4 and so on. Thus it gets one return signal from range gate h_1, two from h_2 and all three from higher range gates.

Most of modern sodars have phased array antennae. It provides a reliable wind profiling with a relatively small setup due to the ability of a phase array antenna to perform a sounding in different directions. The drawback of phased array antenna is its inability to perform a sounding in different directions in parallel. Moreover, most of phased-array sodars have a frequency-dependant tilt angle of beams. For such sodars the returns at different frequencies can not be considered as obtained from the same volume. For instance the centers of sampling volumes for frequencies that differ by factor of 1.5 are separated by angles that differ by the same factor. If the tilt angle for lower frequency beam is 20°, the tilt angle for higher frequency is 30°. The same factor applies to the horizontal wind derived from Doppler shifts at different frequencies.

In case of a dish antenna the tilt angle of the beam is determined by the direction of antenna axis. The beam width, however depends on the frequency for all antenna types. For dish antennae the frequency dependence of sampling volume is a second order effect. Since the time τ is small ($\sim$ 0.1s) as compare to characteristic times at the scales of sampling volume, the returns from the same range gate can be considered as independent samplings of the same volume at the same time. Moreover, the set of frequencies used in sounding signal can be used to identify the echo-signal in ambient noise. The incoherent spectral integration procedure can be used for components of multifrequency echo-return when using the dish antenna.

3. Data processing technique

To achieve maximal temporal resolution the sodar uses no temporal averaging of signal/spectra. The averaged data can be obtained by off-line processing of the raw data. The procedure of data processing of single multi-frequency echo-signal is described below.

The purpose of processing of echo-signal is to get for each range gate the Doppler velocity VR and intensities in signal S and noise N bands. The signal to noise ratio is used to evaluate the validity of data for particular range gate. The basic idea of the processing technique is to synthesize the Doppler spectrum ("instantaneous spectrum") for particular range gate in terms of intensity-Doppler-shift averaged over sounding frequencies and to extract the desired values from it.

The combined spectrum is calculated from the individual spectra estimated for each range gate. The Hanning window in time domain is used to avoid the aliasing. The parts of Fourier

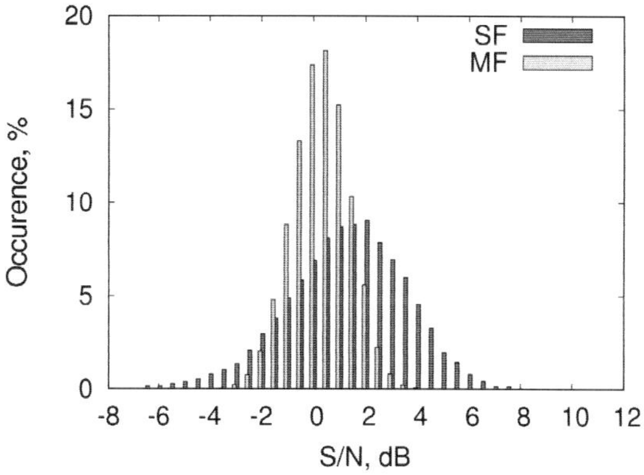

Figure 3. The distribution of the ratio of intensities in signal and noise bands in the absence of signal for conventional (SF) and frequency-coded (MF) mode. $\sim 10^5$ samples.

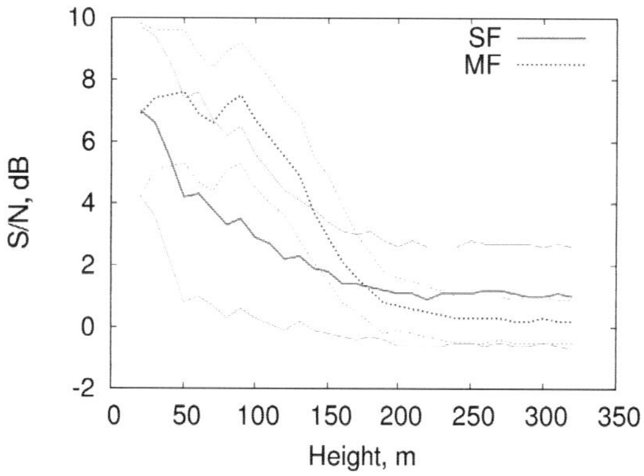

Figure 4. The range distribution of the ratio of intensities in signal and noise bands for real measurements for for conventional and frequency-coded mode.

(FFT) spectra corresponding to particular range gate return from different sounding frequencies after resampling in frequency domain are averaged in logarithmic scale. The resampling procedure is performed to adjust Doppler shifts to the same scale. The combined spectra are processed in a manner similar to that used in single-frequency routines of LATAN-3 [1]. To avoid the fixed echoes the part of the spectrum with zero Doppler shift is masked for inclined antennae.

The range gates are processed sequentially starting from the bottom one. The maxima in the spectra are estimated in two steps. During the first step the Doppler shift is estimated as center of the mass of the spectrum within "search area" ($\pm$10m/s around zero velocity for the first range gate and $\pm$3m/s around the velocity of the previous "good" range gate of for others). The final estimate of Doppler shift is performed in a similar manner within $\pm$1.5m/s around the first proxy. The echo signal intensity is taken as the average power spectral (PSD) density within a spectral width of sounding signal around the frequency found. The intensity of noise is taken as the average PSD in two bands of twice the a spectral width of sounding signal adjacent to the signal band. If the signal to noise ratio is "good" (i.e. $S/N > 2$dB) the found center of a signal is used as a center of a band for preliminary Doppler shift estimate in the next range gate for inclined antennae.

The sodar outputs the values of intensities in "signal" and "noise" bands of received sound (S and N correspondingly). Note that they are not values of signal and noise, but rather estimates of "signal+noise" and "noise". The signal processing routine picks the maximum of spectrum as a signal, thus S is in average slightly higher then N even in the absence of echo-return in the received signal. If the estimates were precise, S whould be equal to N. The systematic overestimate of S depends on statistical stability of instantaneous spectrum estimate. Thus some positive threshold of S/N is required to estimate the validity if samples and either accept or discard them. This threshold is discussed in the next section.

4. Experimental tests

To estimate the improvement of the echo-signal evaluation quality due to the multiple-frequency (MF) mode, consider the statistics of the ratio of intensities in signal and noise bands in the absence of signal for MF and single-frequency (SF) mode. In Fig. 3 the histograms of this ratio for SF (4.1kHz, vertical resolution 10m) and MF (8 frequencies 3.32, 3.46, 3.58, 3.66, 3.76, 3.9, 4.02 and 4.13 kHz) are shown. The distributions illustrate the statistical uncertainty in the estimates of the intensities in the signal and noise bands for two modes of operation. The MFA mode provides much less offset and scattering of the estimates. The S/N threshold of acceptance for MF mode can be taken less than for SF mode. For instance, if the allowed fraction of falsely accepted samples is taken to be 5%, then the threshold should be taken to be 5dB in SF mode and 2dB in MF mode. Thus the MF mode increases the effective sensitivity of the sodar at given noise level by 3dB. Note that this increase comes not from the increase of real signal-to-noise ratio but from the increase in the accuracy of estimate of intensities in the signal and noise bands.

The sample profiles of statistics of S/N measured in the real atmosphere with weak backscattering and strong attenuation in two mentioned above modes are shown in Fig. 4 (1 hour series at 5 s sampling period). In both cases $\tau = 50$ms and acoustic power is about 1W. The bold lines are medians and the thin lines are quartiles of the distribution of S/N. The distance between quartilles corresponds to the width of the distribution at corresponding height. At the bottom level the processing gives the same result since the routines are equivalent. In the SF mode the S/N starts to decrease immediately. This is not the case for MF mode, because the new frequencies are accounted at the heigher ranges. Starting from 90 meters the whole set of frequencies is in processing and the S/N starts to decrease. For upper range gates in both modes the S/N decrease to their values corresponding to the absence of the echo signal.

To estimate the increase of sounding range due to MF mode suppose that we took the mentioned above thresholds. The upper levels for 25% and 50% data availability in Fig. 4 are 50 and 100 meters a.g.l. for SF and 160 and 180 meters for MF mode.

5. Discussion

The right choice of a frequency set is of critical importance for the success of multi-frequency sounding. The attenuation is stronger at higher frequencies and they provide better estimate of a Doppler shift at fixed spatial resolution. Thus the pulses should be ordered with increasing frequency. A sodar antenna has a limited frequency band. All sounding frequencies should be within $\pm 20..25\%$ of the antenna optimal frequency, otherwise the sounding is inefficient. The interval between the frequencies used should be large enough to avoid the wrong identification of the echo-returns. In our experiments the interval of 10m/s was found to be adequate for vertical antenna, so about 8 frequencies can be used. The Doppler shift for inclined antennae might be larger. Under weak wind conditions we used 6 frequencies with increased interval between them for lower range gates (at the end of the train) where a strong wind shear was expected. This choice is hardly suitable under strong winds. In general case the frequency sat has to be either smaller or adaptive depending on wind conditions.

The frequency coding increases the tolerance of a sodar to narrow-frequency disturbances (eg. birds) since the echo-signal has a signature. We experienced a strong interference of the harmonics of low-frequency noises at the equally spaced frequency set. The choice of the set quoted in the previous section has solved the problem.

6. Conclusion

The use of multiple frequency sounding improves the sensitivity of the sodar and the robustness of the sample quality evaluation. The increase of the sensitivity involved the nesesity to introduce the fixed echo cancellation procedure that masks sounding frequencies in the spectrum.

The signal processing procedure quoted above has shown its reliability in several experiments [2, 3]. The separate processing of echo-returns at different frequencies can provide the additional information on the ABL and on the structure of noises. The adaptive frequency set can futher improve the sounding performance at the cost of increased complexity of the signal processing.

Acknowledgments
This work is supported by the Russian Foundation for Basic Research through grants 06-05-08086, 06-05-65270, 07-05-00521 and 08-05-00671.

References
[1] R. D. Kouznetsov. The new PC-based sodar LATAN-3. In *Proc. 13-th Int. Symp. Acoust. Rem. Sens.*, pages 97–98, Garmisch-Partenkirchen, Germany, 2006.
[2] R. D. Kouznetsov. The structure of the lower abl over antarctic oasis during the summer. In *Proc. 14-th Int. Symp. Acoust. Rem. Sens.*, pages ***–***, Roskilde, Denmark, 2008.
[3] V. F. Kramar. Sodar study of the characteristics of big toroid structures above an overheated surface. In *Proc. 14-th Int. Symp. Acoust. Rem. Sens*, pages ***–***, Rosklide, Denmark, 2008.

Quantitative estimate of the role of temperature gradients in sodar echo signal formation

Rostislav D. Kouznetsov

Obukhov Institute for Atmospheric Physics, Pyzhevskii per., 3, 119017, Moscow, Russia

E-mail: roux@ifaran.ru

Abstract. We estimate the intensity of a sodar echo signal caused by thin layer with sharp gradient of a temperature. This intensity under some circumstances may exceed the intensity of backscattering from small-scale turbulent inhomogeneities by orders of magnitude.

1. Introduction

The presence of very thin ($<$ 1m) layers with strong temperature gradients in the stable ABL is reported recently in several studies [6, 4, 1]. These layers are difficult to detect experimentally since they can be located at height of several hundred meters agl with vertical extent of several centimeters. Strong temperature gradients are very effective barriers for the dispersion of atmospheric admixtures. They are also natural media for gravity waves propagation.

It is commonly considered that the sodar return signal is backscattered from small-scale temperature turbulence. Below we show that local temperature gradients can play a dominant role in the echo signal formation. In this paper we estimate the intensity of return signal from a temperature profile measured *in situ* in the real nocturnal boundary layer.

2. Volume scattering and partial reflection

2.1. Volume backscattering

The backscattering cross section density of a small-scale turbulence for acoustic waves is [8]:

$$\sigma = 0.0039k^{\frac{1}{3}}\frac{C_T^2}{T^2}, \tag{1}$$

where k is a wave number, T and C_T^2 are temperature and its structure parameter.

Let the axial-symmetric acoustic source with directional intensity distribution $I(\theta)$ (power per solid angle unit) emits sound into a turbulent medium. The intensity of backscattered signal received at a source from a small volume $r^2dr\sin\theta d\theta d\phi$ at a distance r

$$\delta I_r = \frac{I(\theta)}{r^2}\sigma r^2 dr \sin\theta d\theta d\phi \frac{1}{r^2} \tag{2}$$

Sodar emits a burst pulse during the time interval τ and then receives the backscattered signal. At the moment t after the end of a burst the sodar receives the signal from ranges

IOP Conf. Series: Earth and Environmental Science 1 (2008) 012039 doi:10.1088/1755-1307/1/1/012039

between $R = ct/2$ and $R + c\tau/2$. The received signal intensity is

$$I_r = 2\pi\sigma \int\limits_{R}^{R+c\tau/2} \frac{dr}{r^2} \int\limits_{0}^{\pi} I(\theta)\sin\theta d\theta \tag{3}$$

In a simplest case of sodar antenna pattern

$$I(\theta) = \begin{cases} I, & \theta \leq \theta_0; \\ 0, & \theta > \theta_0, \end{cases} \tag{4}$$

$$I_r = \frac{2\pi\sigma c\tau I}{R^2}\sin^2\frac{\theta_0}{2}. \tag{5}$$

2.2. Partial reflection

The partial reflection of an incident wave normal to the reflecting layer is quantified by the reflection coefficient V, which is a ratio of (complex) amplitudes of incident and reflected waves. The reflection coefficient for acoustic wave in horizontally homogeneous medium with density $\rho(z)$ and speed of sound $c(z)$ can be found from a Riccati differential equation [2]:

$$\frac{dV}{dz} = -2i\beta V + \gamma(1 - V^2), \tag{6}$$

where $\beta = k_0 n(z)$, $n = c_0/c(z)$, $\gamma = \frac{1}{2}\frac{\rho}{\beta}\frac{d}{dz}\left(\frac{\beta}{\rho}\right)$. The boundary condition is $V \to 0$ at $z \to \infty$. This equation is equivalent to [2]:

$$\frac{d}{dz}\left(\exp\left\{2i\int_{z_0}^{z}\beta dz\right\}V(z)\right) = \gamma(1 - V^2)\exp\left\{2i\int_{z_0}^{z}\beta dz\right\} \tag{7}$$

This equation can be written in integral form

$$\exp\left\{2i\int_{z_0}^{z}\beta dz\right\}V(z) = -\int_{z}^{\infty}\gamma(1 - V^2)\exp\left\{2i\int_{z_0}^{z}\beta dz\right\}dz \tag{8}$$

In case of weak reflection the successive approximation method can be used. Zeroth approximation (no reflection) is $V = 0$. The first approximation is [2]:

$$V_{(1)}(z) = -e^{-is(z)}\int_{z}^{\infty}\gamma(z)e^{is(z)}dz, \tag{9}$$

where $s(z) = 2\int_{z_0}^{z}\beta(z)dz$.

In the atmospheric boundary layer the main parameter determining air density ρ and speed of sound c is temperature T.

$$c = c_0\sqrt{\frac{T}{T_0}}, \qquad \rho = \rho_0\frac{T}{T_0}, \tag{10}$$

where T_0 is some reference temperature and $c_0 = c(T_0)$ and $\rho_0 = \rho(T_0)$. The functions $\beta(z)$ and $\gamma(z)$ in equation (6) can be expressed in terms of temperature as well:

$$\beta = k_0\left(1 - \frac{\Delta T}{2T_0}\right); \qquad \gamma = \frac{1}{2}\frac{\rho}{\beta}\frac{d}{dz}\left(\frac{\beta}{\rho}\right) = -\frac{3}{4}\frac{1}{T}\frac{dT}{dz} \tag{11}$$

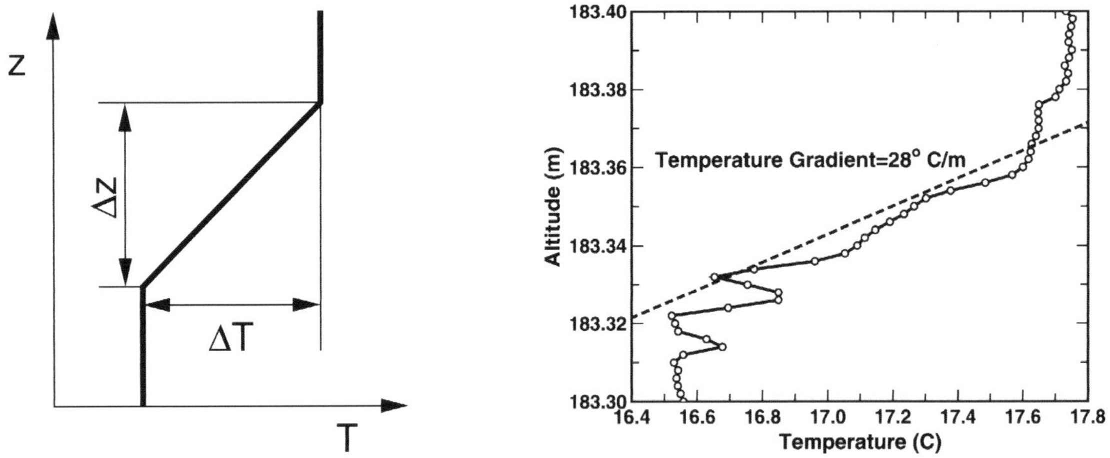

Figure 1. The model temperature profile used for reflection estimate (left) and the measured profile of temperature (right, adapted from[1])

Let the temperature profile be like that shown in Fig. 1. For typical sodars $k = 30..60\text{m}^{-1}$. If $\Delta T \lesssim 1°\text{K}$, then $k_0(n-1)\Delta z \ll 1$ for $\Delta z \lesssim 1\text{m}$. In this case eq. (9) gives:

$$|V| \approx \left| e^{-ikz} \int_0^{\Delta z} \frac{3}{4}\frac{1}{T}\frac{\Delta T}{\Delta z} e^{ikz} dz \right| = \frac{3}{4}\frac{1}{T}\frac{\Delta T}{\Delta z} \left| \frac{e^{ik\Delta z} - 1}{ik} \right| \tag{12}$$

Thus

$$|V|^2 = \frac{9}{16}\left(\frac{\Delta T}{T}\right)^2 \frac{\sin^2 \frac{k\Delta z}{2}}{(\frac{k\Delta z}{2})^2} \tag{13}$$

In a particular case of very sharp step in the temperature profile when $\Delta z \ll \lambda = 2\pi/k$

$$|V|^2 = \frac{9}{16}\left(\frac{\Delta T}{T}\right)^2. \tag{14}$$

The intensity of echo-signal received by vertical sodar from the layer with reflection coefficient V is

$$I_r = \frac{V^2 I}{4R^2} = \frac{9}{64}\left(\frac{\Delta T}{T}\right)^2 \frac{I}{R^2}. \tag{15}$$

3. Application to real measured temperature profile in the ABL

To evaluate the intensity of the echo-signal reflected from a sharp temperature gradient we use the equivalent value of C_T^2 that would give the same echo-signal intensity and compare it to typical values of C_T^2 for the real ABL. As an extreme case we take the temperature profile (Fig. 1) measured by kite-borne set of fast thermometers [1]. This profile has $\Delta z = 4\text{cm}$ and $\Delta T = 1°\text{K}$ in terms of our model.

A sodar with carrier frequency $f = 1.7\text{kHz}$ (wavelength $\lambda = 0.2\text{m}$) has a wave number $k = 30\text{m}^{-1}$. In this case the simple formula (14) gives accuracy of reflection intensity about 10% as compare to (13). The equivalent C_T^2 that would provide the same echo intensity as this layer can be found from (5) and (15):

$$C_T^2 = \frac{\frac{9}{64}(\Delta T)^2}{0.0039 \cdot 2\pi k^{1/3} c\tau \sin^2 \frac{\theta}{2}}. \tag{16}$$

Given $k = 30\text{m}^{-1}$, $c = 340\text{m/s}$, $\tau = 0.1\text{s}$, and $\theta = 6°$, the $C_T^2 = 40\text{K/m}^{2/3}$. It is about two to three orders of magnitude higher than the highest observed values of C_T^2 in the nocturnal ABL [3].

4. Discussion and conclusion

We have shown that the partial reflection from strong temperature gradients that can occur in the nocturnal ABL can dominate the sodar echo signal formation. The horizontal extent of these gradients that is necessary to produce such strong reflection of sodar signal can be as small as several meters. The measurements [1] report the horizontal extent of these layers more then several hundreds meters. It is not clear, how often such reflecting layers occur in the real ABL.

The steady horizontal layers have to produce strong aspect sensitivity of the echo signal intensity. In case of ideal antenna pattern (4) the echo-signal should vanish at zenith angle more then θ_0. The observed ratio of nocturnal echo signal intensities between vertical and 10° off-zenith direction is up to 1.5...2 [7, 5]. This does not contradict the idea that the layers play a role in the echo signal formation. In the case of wavy layers or variation of the layer height, the measured aspect sensitivities can be low.

The layers with thickness comparable to the wavelength of sounding signal can be detected by means of multifrequency sounding since the reflection coefficient (13) in this case has strong wave number dependence. A thin echoing layer that exhibits strong frequency dependence of reflectivity can be interpreted as a sharp temperature gradient.

Acknowledgments

This work is supported by the Russian Foundation for Basic Research through grants 06-05-08086, 06-05-65270, 07-05-00521 and 08-05-00671.

References

[1] B. B. Balsley, R. G. Frehlich, M. L. J. Y. Meillier, and A. Muschinski. Extreme gradients in the nocturnal boundary layer: Structure, evolution, and potential causes. *Journal of Atmospheric Sciences*, 60:2496–2508, 2003.

[2] L. M. Brekhovskikh and O. A. Godin. *Acoustics of Layered Media I: Plane and Quasi-Plane Waves*. Springer, 1990.

[3] S. Danilov, A. Gur'yanov, M. Kallistratova, I. Petenko, S. Singal, D. Pahwa, and B. Gera. Acoustic calibration of sodars. *Measurement Science and Technology*, 3:1001–1007(7), 1992.

[4] E. E. Gossard, J. E. Gaynor, R. J. Zamora, and W. D. Neff. Finestructure of Elevated Stable Layers Observed by Sounder and In Situ Tower Sensors. *Journal of Atmospheric Sciences*, 42:2156–2169, Oct. 1985.

[5] M. A. Kallistratova and I. V. Petenko. Aspect sensitivity of sound backscattering in the atmospheric boundary layer. *Applied Physics B: Lasers and Optics*, 57:41–48, July 1993.

[6] A. Muschinski and C. Wode. First In Situ Evidence for Coexisting Submeter Temperature and Humidity Sheets in the Lower Free Troposphere. *Journal of Atmospheric Sciences*, 55:2893–2906, Sept. 1998.

[7] W. D. Neff. Quantitative evaluations of acoustic echoes from the planetary boundary layer. Technical report, NOAA, Boulder CO, 1975.

[8] V. I. Tatarski. *Effects of the Turbulent Atmosphere on Wave Propagation*. Keter Press, Jerusalem, 1971.

IOP Conf. Series: Earth and Environmental Science 1 (2008) 012040 doi:10.1088/1755-1307/1/1/012040

Investigation of Low-Level-Jets over rural and urban areas using two sodars

M A Kallistratova

Obukhov Institute of Atmospheric Physics RAS, 3 Pyzhevskii, Moscow 119017, Russia

E-mail: margo@ifaran.ru

Abstract. This paper is devoted to the experimental study of jet streams at heights up to 400 m agl (the so called Low-Level Jet, or LLJ) with the help of sodars. These LLJs, which are not connected either to catabatic winds or to synoptic front passing, but are regularly observed in the mid-latitudes over a flat terrain under nocturnal inversion conditions, are investigated. Preliminary results of simultaneous investigations of LLJs over the center of the Moscow megalopolis and over a nearby rural area are presented. Continuous sodar data on wind speed profiles collected during 27 days in July 2005 were analyzed together with *in situ* data from a meteorological mast. Empirical distributions of the maximum jet speed and the height of jet axis were obtained. The relationships between the LLJ parameters and some characteristics of the surface layer were found. Over the urban area, the jets appear less frequently and at greater heights than over the countryside.

1. Introduction

Distinct narrow maxima of wind speed in the nocturnal atmospheric boundary layer (ABL) are frequently observed over a flat terrain beyond any connection with orographic or synoptic effects. The origin of the nocturnal jets (called Low-Level Jets, LLJs) was explained in the middle of the last century [1] by the boundary layer "decoupling" from surface frictional forces at the sunset, when the turbulence dies down. The LLJs can play a noticeable role in the mass/heat transfer in the ABL and influence the performance of wind energy systems as well.

Standard aerological sounding cannot register the LLJs, because they occur at low heights above the ground level (< 400-500 m a.g.l.) and have a small vertical size (50-200 m); so, for this purpose, special means are necessary. An intensive study of the LLJs began in the 1980s, when new tools for *in situ* and remote measurements of wind speed and temperature in the ABL were developed, although the very phenomenon was observed almost a century ago [2]. Currently, in addition to tethered balloons [3] and specialized low-level radiosondes [4], experimental studies of LLJs use sodars [5-7], radars [8], the combinations of minisodar and radar [9], and Doppler lidars [10-13].

In the papers mentioned above, numerous examples of LLJs were demonstrated, and the distributions of jet characteristics were presented. The most part of the investigations was carried out over non-urban areas, except in [13], where 3 short episodes of LLJs over Oklahoma City were described.

Our goal was to take simultaneous, long-term measurements of LLJ characteristics inside and outside a city in order to investigate an urban influence on the jets. Sodar, which is the most simple and economic tool among all means of ground-based sensing, was used to this aim. All the conclusions below refer to the summer season only.

2. Instrumentation

Continuous sodar measurements of wind velocity profiles were carried out in July 2005 [14]. Two sodars operated simultaneously: in the center of the Moscow megalopolis (at the Obukhov Institute of Atmospheric Physics (IAPh)) and at the Zvenigorod Scientific Station (ZSS) of the IAPh, which is located in a rural area 45 km to the west from Moscow. The distance was long enough to exclude the influence of the city on wind field at the ZSS for any wind direction. The identical sodars LATAN-3 developed at the IAPh [15,16] were used in both the sites. The carrying frequency is 1700 Hz, the vertical resolution is 20 m. The sodar measurements at the ZSS were accompanied by *in situ* measurements at a 56-m meteorological mast with the help of two ultrasonic thermometers-anemometers METEK USA-1 ("sonics" for short). Figure 1 shows the location of the LATAN-3 acoustic antennas at the two measuring sites.

Figure 1. The antennas of sodar LATAN-3 at the two measuring sites. Left: on the roof of a three-story IAPh building in the center of Moscow. Right: on the roof of a two-story building at the ZSS; the 56-m meteorological mast equipped with two sonics of German production is seen at the background.

An example of the intensity of sodar return signal ("sodar echogram") and wind speed profiles, which are obtained simultaneously for Moscow and the ZSS, is presented in Figure 2. In order to highlight the steady LLJs and exclude short-term wind blasts, the speed profiles were averaged over 1 hour. The filled circles mark the most reliable portion of profile, where less than 75% of individual values of speed at each height level was rejected due to a poor signal/noise ratio. The continuous lines without circles mark the profile portions with a larger body of rejected data (from 75 to 85%). Data with a low statistical reliability (less than 15% of the averaging time) are eliminated from the further processing.

This example shows a typical difference in the height and strength of the LLJs between urban and rural areas. Besides, a certain restrictions on the use of sodar for LLJ study are seen in Figure 2. These restrictions are owing to the fact that the intensity of temperature fluctuations (which serve as scatterers for a monostatic sodar) rapidly decreases above the inversion level, and, thus, Doppler measurements of speed at the top of the jet have a lower statistical reliability. However, a LATAN-3 sodar has a sufficient sensitivity and interference immunity to detect the minimum of wind speed above the jet axis by using residual turbulence.

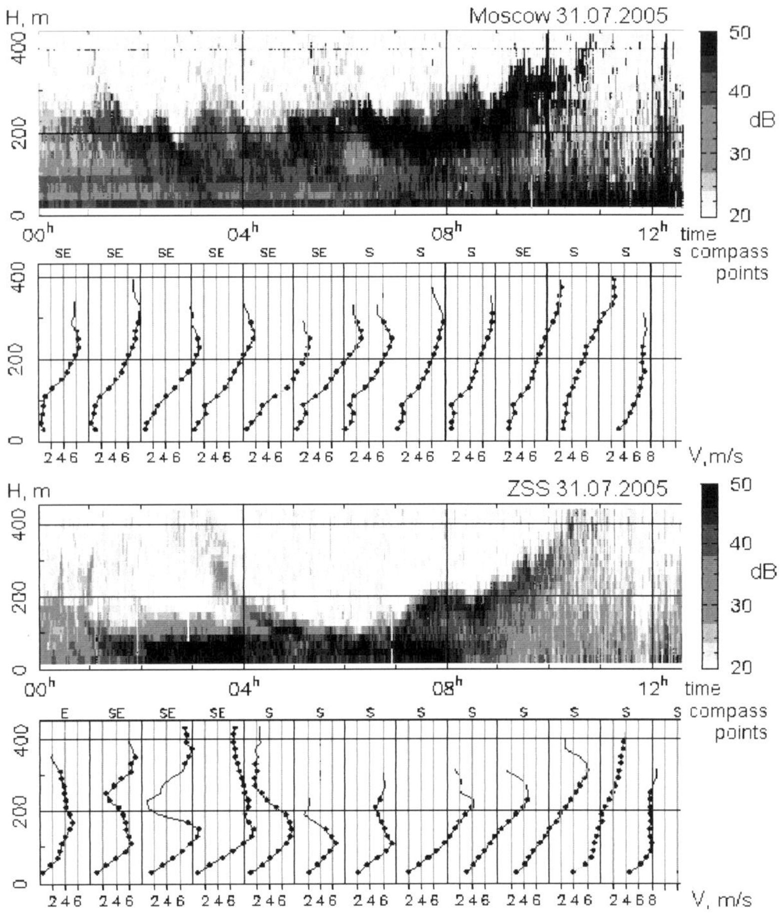

Figure 2. A fragment of sodar echograms and speed profiles, $V(z)$, obtained synchronously on July 31, 2005, in the center of Moscow (top) and at the ZSS (bottom). Each profile is averaged over 1 hour (over 360 pulse shots).

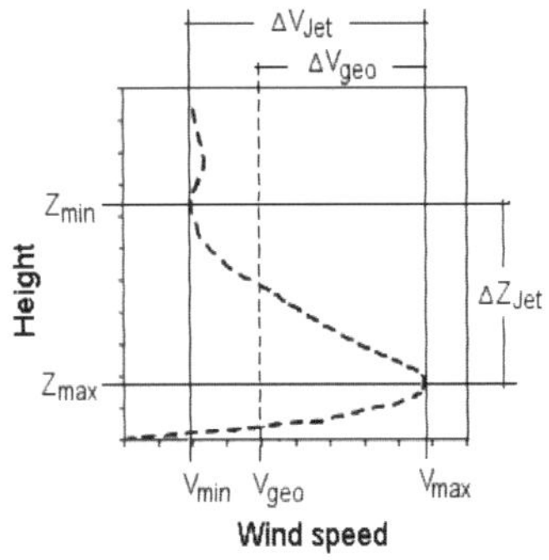

Figure 3. Scheme of LLJ. Z_{max} – height of the jet axis; Z_{min} - height of the first minimum; ΔZ - vertical width of the jet; V_{max} - wind speed at the jet axis; V_{min} – the first minimum of speed above the jet axis; V_{geo} – geostrophic wind; ΔV_{Jet} - strength of the jet; ΔV_{geo} – difference between jet speed and geostrophic wind.

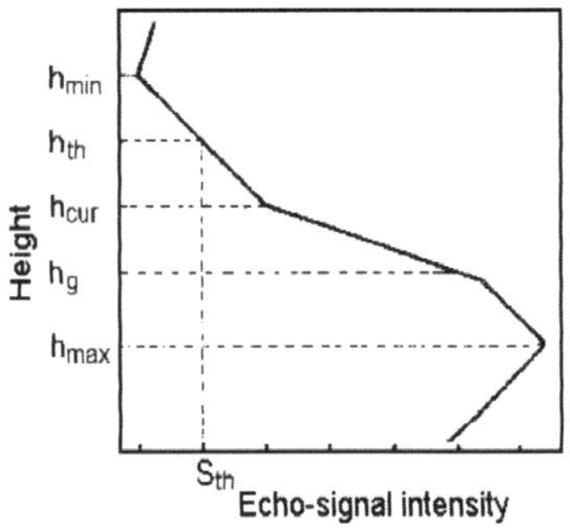

Figure 4. Scheme of the nocturnal profile of the intensity, S, of sodar return (adapted from [18]). h_{max} - height of the S maximum; h_g - height of the maximum gradient of S; h_{cur} - height of the maximum curvature of S-profile; h_{th} - height, at which the intensity is less than a specified threshold, S_{th}; h_{min} - height of the S minimum value.

In data processing, the LLJ characteristics shown in Figure 3 were being investigated. The most important of them are: jet-axis height, Z_{max}; jet speed, V_{max}; and jet strength, ΔV_{Jet}. Our attempts to develop objective criteria for the jet recognition and to apply the automatic processing to wind velocity profiles were not successful. Only the most pronounced jets with the strength $\Delta V_{Jet} > 3$ m/s could be automatically selected. Therefore, we used a visual inspection of each profile, as it was made in [4, 5, 10]. The wind speed profile, which shows a local maximum that is 1 m/s higher than speeds both above and below it, was assumed to be a jet.

The location of the upper boundary of the inversion layer is determined by the profile of the intensity of sodar echo-signal [17]. As shown in [18], different points of the profile can be used to characterize the inversion height. These points are marked in Figure 4. Proper selection of a point for comparison with Z_{max} is justified below in the discussion of Figure 10.

3. Results and discussion

In the rural area, $N = 217$ jets, which lasted at least 1 hour each, were recorded during 27 July days. These jets amount to about 70% of twilight and night observations, and more than 30% of the total observing period. During the same period, $N = 124$ jets were recorded over the city, which amount to 40% of measurements under stable stratification, and about 20% of measurements during 24 hours. The empirical distributions of the jet axis height and maximum wind speed for Moscow and the ZSS are presented in Figures 5 and 6.

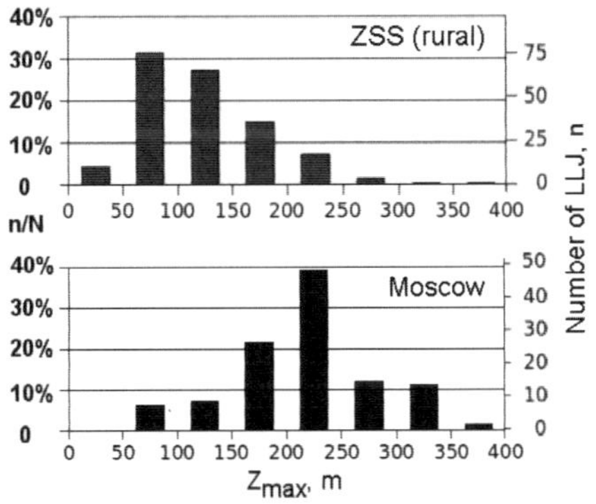

Figure 5. Distributions of jet-axis altitude, Z_{max}, over the rural area (top) and over the city center (bottom).

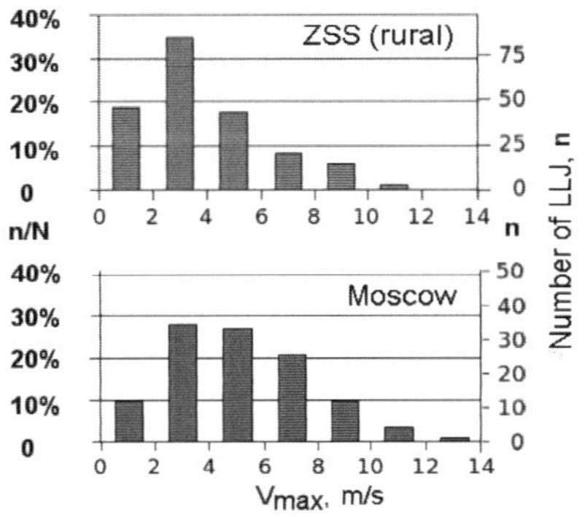

Figure 6. Distribution of the maximum speed, V_{max}, over the rural area (top) and over the city center (bottom).

Note that the shapes of the distributions of Z_{max} and V_{max} for the countryside, which we obtained with the help of sodar, is close to the similar distributions obtained in Kansas [10] with the help of 2-μ Doppler lidar (where a 15-min averaging was used). However, the mean value of V_{max} in the Moscow region is considerably lower. This, apparently, is related to the climatic difference in the general atmospheric circulation between the two regions. Figures 5 and 6 show that the mean height of the LLJ axis over the city is 100-150 m higher than over the rural area, and the mean maximum speed is also higher over the city.

The long period of observations allowed us to estimate the time distribution of the number of LLJs observed at night. Figure 7 shows the time distributions for the city and countryside. Over the rural area, the number of LLJs is almost evenly distributed throughout the dark time of day, while, over the city, the most part of the jets is observed during pre-dawn hours. This is apparently caused by the microclimatic properties of cities, where the stable ABL is formed later than over countrysides due to a stronger heating and slower cooling of buildings and paved surfaces.

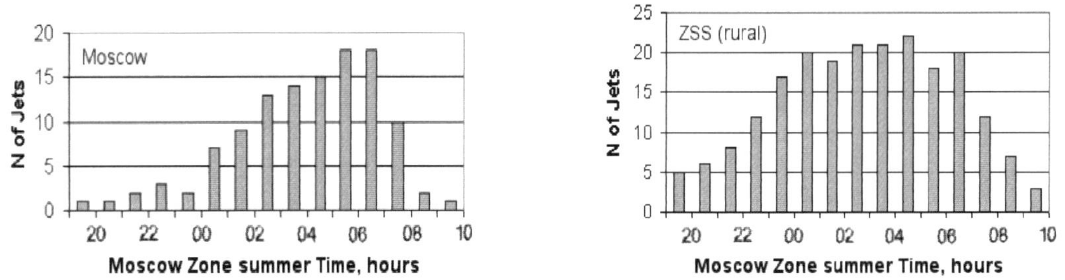

Figure 7. Time distribution of number of the jets over Moscow (left) and ZSS (right).

In addition to the statistical properties of the LLJs observed, the correlation between jet characteristics and meteorological parameters is of great interest. It was investigated only at the ZSS, where *in situ* measurements at the 56-m meteorological mast were carried out synchronously with the sodar operation. The separation of the mast from the sodar was 40 m. The dependence of the jet strength, ΔV_{Jet}, on the diurnal amplitude of air temperature is presented in Figure 8. The amplitude was determined at a height of 56 meters as the difference between the maximum temperature on the day preceding the overnight measurements ($T_{\mathrm{max,\ day}}$) and the minimum temperature during the night measurements ($T_{\mathrm{min,\ night}}$). The noticeable correlation is seen for both the largest $(\Delta V_{\mathrm{Jet}})_{\mathrm{max}}$ and the mean $\langle \Delta V_{\mathrm{Jet}} \rangle$ strengths of jets observed at night. This is consistent with the Blakadar theory of the semi-diurnal inertial oscillations of the ABL wind, which are generated by alternation of the force of the Reynolds turbulence friction, when stratification is abruptly changed.

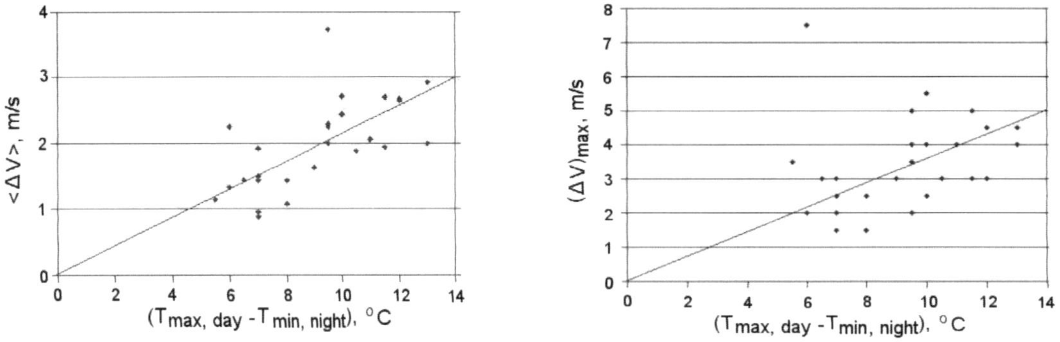

Figure 8. Dependence of the largest strength $(\Delta V_{\mathrm{Jet}})_{\mathrm{max}}$ observed during night (left) and the mean intensity $\langle \Delta V_{\mathrm{Jet}} \rangle$ averaged over the same night (right) on the amplitude of daily changes in temperature for the ZSS. The medians of the data are shown by the sloping lines.

At the same time, it was revealed that both the Z_{max} and ΔV_{Jet} values do not depend on the mean temperature gradient at the bottom of jet. Figure 9 shows that Z_{max} and ΔV_{Jet} are distributed more or less evenly throughout the range of difference (T_{56m} - T_{6m}). Obviously, the characteristics of LLJs should depend rather on the Richardson number (Ri) than on the gradient of temperature. Note, however, that in [4], the dependence of Z_{max} on Ri was not revealed either.

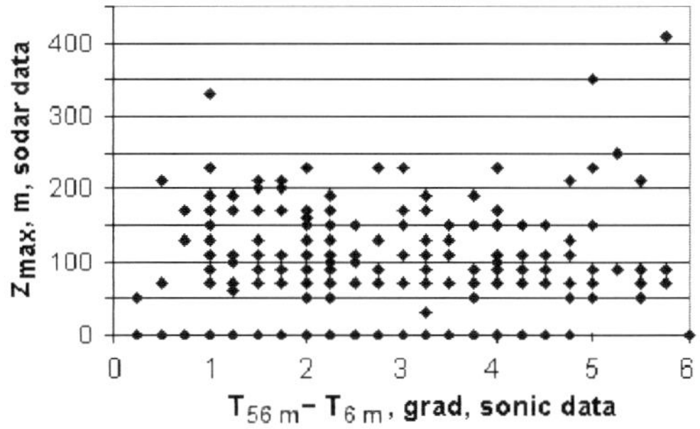

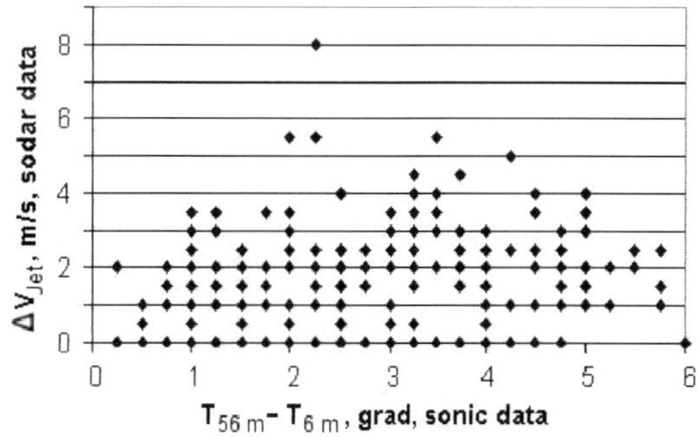

Figure 9. Dependence of the height of the jet axis Z_{max} (left) and the strength ΔV_{Jet} (right) on the difference between temperatures at altitudes of 56 m and 6 m, measured by sonics at the ZSS mast synchronously with the measurements of wind velocity profile. Zero value on the ordinate axis corresponds to a lack of maximum in wind speed profile.

One of the advantages of using sodar in the LLJ research is a possibility to determine the height of the upper border of the inversion, h_i, from echo-signal intensity [17]. Essentially, that the sodar measurements of LLJ characteristics and h_i coincide in space and time. The height h_i can visually be determined from sodar echograms, but the use of digital profiles of the intensity of echo-signal, S, provides a more comprehensive information. However, this raises the question of what point of the profile should be used as h_i. In our comparisons between Z_{max} and h_i, we inspected several options. The best agreement was obtained by using the h_g height, which corresponds to the maximum gradient of echo-signal intensity (see Figure 4). The comparison presented in Figure 10 shows that the agreement between the position of the jet axis and the upper border of the inversion is reasonably good.

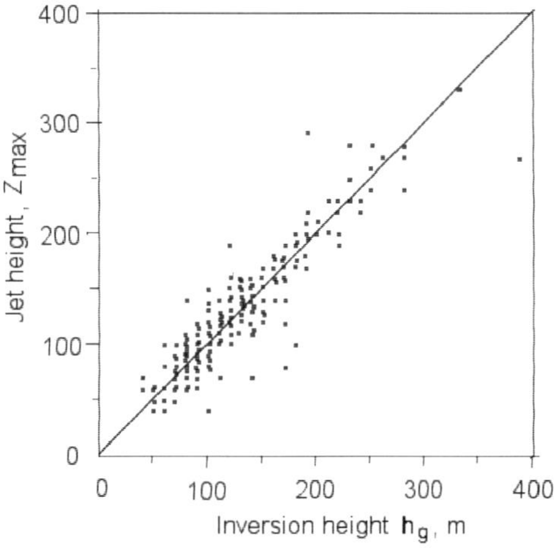

Figure 10. Comparison between the LLJ and ground-based inversion heights over the ZSS. The value of h_g was used to characterize the inversion height (see Figure 4). The line corresponds to the coincidence of Z_{max} and h_g.

4. Conclusion

Despite certain limitations to the application of sodar for the LLJ study, which are caused by a lower statistical reliability of acoustical Doppler data in the vicinity of the top of jet, sodar is the most convenient tool for long-term observations of LLJs and for determination of their main characteristics.

A great variety of the ABL jet stream patterns, which often strongly differ from an idealized form, makes the automated analysis of speed profiles difficult. Only about 20% of the total number of LLJs, which are identified visually, were recognized automatically.

Distributions of maximum speed and height of the jet axis obtained by sodar over the Moscow region are close to those obtained during the CASES-99 in Kansas [10], although the mean value of V_{max} for the Moscow region is considerably lower. Obviously, the distinction is caused by climatic differences in geostrophic wind speeds for two regions. The jet strength increases with an increase in the diurnal amplitude of air temperature, but does not show any dependence on the temperature gradient at the bottom of jet.

A significant difference in the frequency of occurrence of LLJs, their diurnal distribution, and the distribution of jet heights between the two locations (the center of the city and the rural area) was revealed. Over the city, LLJs occur less frequently at high altitudes and are observed mainly during pre-dawn hours, while, outside the city, the number of LLJs is almost evenly distributed over the dark time of day.

A strong correlation is found between the height of jet and surface inversion according to sodar return, which confirms the results [5]. Thus, the distinction in the LLJ behaviors between urban and rural areas is, apparently, due to the general difference in stratification caused by the urban heat island.

Acknowledgments

We are grateful to Dr. R.M. Banta whose brilliant presentation at ISARS2006 on the lidar study of LLJ inspired our interest to this phenomenon, and also to Dr. R.D. Kouznetsov and Mr. D.D. Kuznetsov for wind profiles measurements and data processing.

This research is supported by the Russian Foundation for Basic Research, grant no. 07-05-00521.

References

[1] Blackadar A K 1957 Boundary layer wind maxima and their significance for the growth of nocturnal inversions *Bull. Am. Met. Soc.* **38** 283-90

[2] Richardson L F 1923 Wind above the night-calm at Benson at 7 a.m. *Quart. J. RMS* **49** 34-38

[3] Kraus H, Malcher J and Schaller E 1985 A nocturnal low level jet during PUKK *Bound.-Layer Met.* **31** 187-95

[4] Andreas E L, Claffey K J and Makshtas A P 2000 Low-level atmospheric jets and inversions over the Western Weddell Sea *Bound.-Layer Met.* **97** 459-86

[5] Coulter R L 1981 Nocturnal wind profile characteristics *Proc. 1st Int. Symp. on Acoustic Remote Sensing (Calgary, Canada, 22-25 June 1981)* pp VI-1 – VI-10

[6] Sadani L K and Murthy B S 1995 Detection of the nocturnal jet by Doppler sodar *Proc. 4th National Symp. Advanced Technologies in Meteorology (Hiderabad, India, 8-11 Feb 1995)* C6P05

[7] Reitebuch O, Strassburger A, Emeis S and Kuttler W 2000 Nocturnal secondary ozone concentration maxima analyzed by sodar observations and surface measurements *Atmos. Environ.* **34** 4315-39

[8] Mitchell M J, Arritt R W and Labas K 1995 A climatology of the warm season Great Plains low-level jet using wind profiler observations *Weath. Forecast.* **10** 576-91

[9] Coulter R L and Kallistratova M A 1999 The role of acoustic sounding in a high-technology era *Meteorol. Atmos. Phys.* **71** 3-13

[10] Banta R M, Newsom R K, Lundquist J K, Pichugina Y L, Coulter R L and Mahrt L 2002 Nocturnal low-level jet characteristics over Kansas during CASES-99 *Bound.-Layer Meteor.* **105** 221-52

[11] Banta R M, Mahrt L, Vickers D, Sun J, Balsley B, Pichugina Y L and Williams E J 2007 The very stable boundary layer on nights with weak low-level jets *J. Atmos. Sci.* **64** 3068-90

[12] Banta R M 2008 Stable-boundary-layer regimes from the perspective of the low-level jet *Acta Geoph.* **56** 58-87

[13] Wang Y, Klipp C L, Garvey D M, Ligon D A, Williamson C C, Chang S S, Newsom R K and Calhoun R 2007 Nocturnal low-level-jet-dominated atmospheric boundary layer observed by a Doppler lidar over Oklahoma City during JU2003 *J. Appl. Meteor. Climat.* **46** 2098-109

[14] Kallistratova M, Kouznetsov R, Pekour M, Yushkov V and Svertilov A 2006 Comparison of wind fields in the ABL over big city and nearby rural site by simultaneous sodar measurements. *Extend Abstr. 13th ISARS* (*Garmisch-Partenkirchen, Germany, 18-20 July 2006*) ed S Emeis pp 106-08

[15] Kouznetsov R.D. 2006 New PC-based sodar LATAN-3 *Extend Abstr. 13th ISARS,* (*Garmisch-Partenkirchen, Germany, 18-20 July, 2006*) ed S Emeis pp 97-99

[16] Kouznetsov R, Kramar V and Kallistratova M 2007 The vertical structure of turbulent momentum flux in the lower part of the atmospheric boundary layer *Meteorol. Zeit.* **16** 367-73

[17] Beyrich F 1997 Mixing height estimation from sodar data - A critical discussion *Atmos. Environ.* **31** 3941-53

[18] Beyrich F and Weill A 1993 Some aspects of determining the stable boundary layer depth from sodar data *Bound.-Layer Met.* **63** 97-116

IOP Conf. Series: Earth and Environmental Science 1 (2008) 012041 doi:10.1088/1755-1307/1/1/012041

Quasi-wavelet formulations of turbulence and wave scattering

D. Keith Wilson,[1] Vladimir E. Ostashev,[2,3] and George H. Goedecke[3]

[1] U.S. Army Engineer Research and Development Center, Hanover, NH 03755, U.S.A.
[2] NOAA/Earth System Research Laboratory, Boulder, CO 80305, U.S.A.
[3] Physics Department, New Mexico State University, Las Cruces, NM 88003, U.S.A.

E-mail: `D.Keith.Wilson@usace.army.mil`

Abstract. Quasi-wavelets (QWs) are particle-like entities similar to customary wavelets in that they are based on translations and dilations of a spatially localized parent function. The positions and orientations are, however, normally taken to be random. Random fields such as turbulence may be usefully represented as ensembles of QWs with appropriately selected size distributions, number densities, and amplitudes. This paper provides an overview of previous results concerning QWs. The points of emphasis are the following. (1) Self-similar ensembles of QWs with rotation rates scaling according to Kolmogorov's hypotheses naturally produce classical inertial-subrange and von Karman-like spectra. (2) The spatially localized nature of QWs can be advantageous in wave-scattering calculations and other applications. The scattered wavefield from a single QW can be readily derived and then integrated over scale and volume to obtain expressions for the total scattering cross section. (3) Anistropy, and momentum and heat transfer, in surface-layer turbulence can be described by introducing preferred orientations and correlations among QWs representing temperature and velocity perturbations. (4) Unlike Fourier modes, QWs can be naturally arranged in a spatially intermittent manner. Models for both local (intrinsic) and global intermittency are described.

1. Introduction

Turbulence is often conceived as an assemblage of coherent, eddy-like structures. Townsend [1, Sec. 5.8], for example, proposed modeling a turbulent surface layer with inclined cylindrical vortices, and subsequently linked this description to observed properties of the correlation function of the velocity field. In such a description, the underlying eddy-like structures (the model "building blocks") conceptually have both a spatial scale and position. This contrasts to more common spectral decompositions, for which the basis functions have only an associated scale. We use the term *quasi-wavelets* (QWs) [2–4] for a description of turbulence and other random fields that is conceptually similar to Townsend's in the sense that the field is built from eddy-like structures localized in scale and position. However, the structures are generally simple variations of familiar wavelets. The QW description differs from conventional wavelets in that the positions and orientations are random. As yet, there is no QW transform in the sense of a discrete or continuous wavelet transform of a signal. However, a random ensemble of QWs is similar to a continuous wavelet transform from which a number of wavelets have been systematically removed, with the remaining ones rescaled to preserve the energy of the original field.

This paper provides an overview of recent research with quasi-wavelets, with emphasis placed on application to turbulence. The ensembles of QWs are constructured from self-similarity principles, as will be described in Sec. 2. Previously, it has been shown that a construction in which the QWs have rotation rates scaling with size according to Kolmogorov's original hypotheses [5] produces velocity fields with realistic spectral properties [3]. Models for turbulence spectra, including inhomogeneity and anisotropy, can be developed from QW constructions. Recently, it was shown that the superposition of multiple, correlated QW fields, with suitably selected QW parent functions and probability distribution for the orientations, produces models with anisotropy and correlations among several properties of random fields [6]. Turbulent fluxes can then readily be incorporated into the spectral model. Such extensions would be difficult (if not impossible) to achieve systematically by Fourier methods. When spectral models are kinematically constructed in such a manner, subsequent comparisons with experimental data may provide further insight as to which aspects of the random field can be explained by geometric considerations, and which aspects require more complicated, dynamical explanations.

One of the most promising applications for QWs lies in modeling the spatial unevenness, or intermittency, of activity in random fields. This could be accomplished by clustering the QWs in a manner consistent with the cascade process underlying energy transfer within the random field. Kitagawa and Nomura [7] have explored such a concept using one-dimensional orthogonal wavelets. For synthesizing non-Gaussian random fields, wavelet and QW approaches might prove a useful alternative to methods based on filtering and translating harmonic representations (e.g., [8; 9]). This idea is further explored in Sec. 3.

Previous research has also demonstrated the utility of QWs for describing scattering of sound waves by turbulent eddies in a manner analogous to scattering from particles [10–12]. The QW formulation allows the dependence of the scattering process on both position and particle size to be readily assessed. This research is summarized in Sec. 4.

2. Self-similar ensembles of QWs

2.1. Monopole, dipole, and vector QWs

Like customary wavelets, QWs are derived from translations and dilatations of a dimensionless, spatially localized parent function. We assume that the parent function, $f(\xi)$, is spatially symmetric. Here ξ is the magnitude of the vector $\boldsymbol{\xi} \equiv (\mathbf{r} - \mathbf{b}^{\alpha n})/a_\alpha$, $\mathbf{r}$ is the spatial coordinate, $\mathbf{b}^{\alpha n}$ is the center of the QW, and a_α is its size. The index α indicates the size class of the QW, with $\alpha = 1$ being the largest size and $\alpha = N$ the smallest. The index n indicates a particular QW within a size class. The size a_1 is associated with the outer scale of the turbulence and a_N with the inner scale. (The notational convention followed here is that a double superscript, e.g. αn, indexes a particular QW, whereas a single subscript, e.g. α, indicates a property associated with all QWs of a particular size class.)

The field perturbation associated with a single, scalar (e.g., temperature) QW is written

$$T^{\alpha n}(\mathbf{r}) = \tau^{\alpha n} f\left(\frac{|\mathbf{r} - \mathbf{b}^{\alpha n}|}{a_\alpha}\right), \tag{1}$$

where $\tau^{\alpha n}$ is a random coefficient that may be positive or negative. The $\tau^{\alpha n}$ are assumed to have the following statistical properties: their expected values (means) are zero, the expected values of their squares are the same for all members of a given size class, and the coefficients of the individual QWs are statistically independent. Mathematically, $\langle \tau^{\alpha n} \rangle = 0$ and $\langle \tau^{\alpha n} \tau^{\beta m} \rangle = \delta_{\alpha\beta}\delta_{nm}\tau_\alpha^2$. Here, the angle brackets indicate expected value, δ_{nm} is the Kronecker delta (equals 1 when $m = n$ and 0 otherwise), and τ_α^2 is the expected value of the squared amplitude for size class α [$\tau_\alpha^2 \equiv \langle (\tau^{\alpha n})^2 \rangle$]. Note that we have only prescribed the mean and variance of the $\tau^{\alpha n}$; multiple choices for pdfs satisfying these conditions are possible. For example, the $\tau^{\alpha n}$ could have a fixed absolute value and be positive or negative with equal probability, or they could have a normal distribution.

Typically, we have used a Gaussian parent function, $f(\xi) = \exp\left(-\xi^2/2\right)$. This function is convenient for analysis and rapid calculation on computers. Unlike parent functions for customary wavelets, however, it is not zero mean. This is not necessarily a problem in the QW model, since the zero-mean property of the coefficients $\tau^{\alpha n}$ ensures that the random field will have zero mean. However, the Gaussian parent function has the potential drawback of not being well localized in the spectral sense. This behavior might make a different parent function, such as a "Mexican-hat," more desirable in some applications [6]. We have also considered an exponential parent function, and one involving Bessel functions that leads exactly to a von Kármán spectral model [3].

We refer to Eq. (1) as a *monopole* scalar model, since each QW is spherically symmetric. A *dipole* model can be constructed by differentiating $f(\xi)$ along an axis passing through the QW origin. This creates a QW with a single axis of symmetry. On one side of the plane normal to this axis, and passing through the QW origin, the field is entirely positive. On the other side of the plane it is entirely negative. Designating the unit vector directed along this axis as $\mathbf{d}^{\alpha n}$, we set

$$T^{\alpha n}(\mathbf{r}) = \tau_\alpha a_\alpha \left(\mathbf{d}^{\alpha n} \cdot \nabla\right) f(\xi). \tag{2}$$

Figures 1(a) and (b) compare the monopole and dipole scalar QWs.

Construction of a solenoidal (non-divergent) velocity field can be conveniently based on the curl of a scalar parent function. Specifically, $\mathbf{A}^{\alpha n}(\mathbf{r})$ is a vector potential given by

$$\mathbf{A}^{\alpha n}(\mathbf{r}) = \mathbf{\Omega}^{\alpha n} a_\alpha^2 f(\xi), \tag{3}$$

where $\mathbf{\Omega}^{\alpha n}$ is the angular velocity vector of the QW, and the velocity field is $\mathbf{v}^{\alpha n}(\mathbf{r}) = \boldsymbol{\nabla} \times \mathbf{A}^{\alpha n}(\mathbf{r})$. The presence of a_α^2 in the definition provides dimensional consistency. Writing out the curl of the potential leads to the following result for the rotational velocity field associated with the QW:

$$\mathbf{v}^{\alpha n}(\mathbf{r}) = \mathbf{\Omega}^{\alpha n} \times \left(\mathbf{r} - \mathbf{b}^{\alpha n}\right)\left(-\xi^{-1}\partial f/\partial \xi\right). \tag{4}$$

A velocity QW constructed in this manner is shown in Fig. 1(c).

2.2. *Coupled QWs and fluxes*

Consider now a second scalar field $Q(\mathbf{r})$ that exists concurrently with $T(\mathbf{r})$. Suppose further that there is a one-to-one correspondence between the QWs comprising each of these fields: for each QW contributing to $T(\mathbf{r})$, there is a corresponding QW at the same position and of the same class contributing to $Q(\mathbf{r})$. The field $Q^{\alpha n}(\mathbf{r})$ associated with a particular QW has the same form as Eq. (1), namely

$$Q^{\alpha n}(\mathbf{r}) = \gamma^{\alpha n} g\left(\frac{|\mathbf{r} - \mathbf{b}^{\alpha n}|}{a_\alpha}\right), \tag{5}$$

where $\gamma^{\alpha n}$ is a random coefficient and g is the parent function, which may be different from f. To model correlated random fields, Ref. [6] suggests introducing a correlation between $\gamma^{\alpha n}$ and $\tau^{\alpha n}$, specifically $\left\langle \tau^{\alpha n}\gamma^{\beta m}\right\rangle = \delta_{\alpha\beta}\delta_{nm}\rho_\alpha\tau_\alpha\gamma_\alpha$,where $-1 \le \rho_\alpha \le 1$ is a correlation coefficient for the size class α and $\gamma_\alpha^2 = \left\langle \left(\gamma^{\alpha n}\right)^2\right\rangle$. For self-similar processes, we describe ρ_α with a power law:

$$\rho_\alpha = \rho_m \left(\frac{a_\alpha}{a_m}\right)^\nu, \tag{6}$$

where ρ_m is the correlation coefficient for $\alpha = m$ and ν is an adjustable exponent.

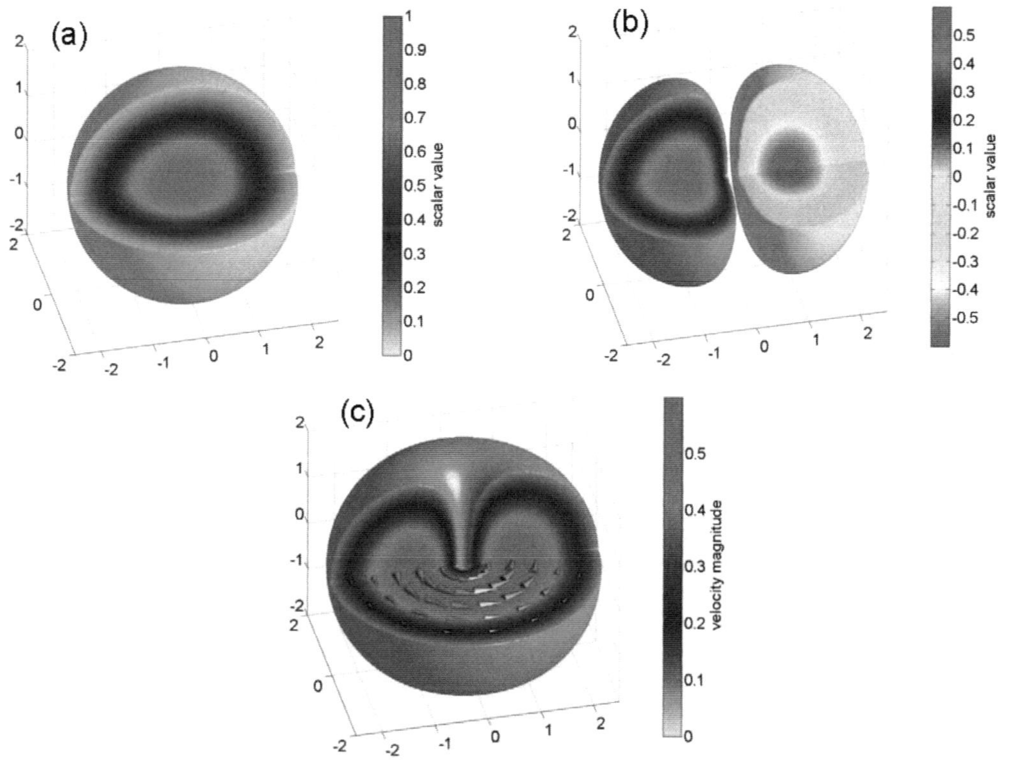

Figure 1. Quasi-wavelets derived from a Gaussian parent function. (a) Scalar, monopole QW. (b) Scalar, dipole QW (orientation horizontal). (c) Velocity QW. Spatial coordinates are in a normalized system ($\boldsymbol{\xi}$ in the text). Part of each QW has been cut away to show the internal structure.

This idea of describing multiple fields with correlated QWs can be extended to dipole and velocity QWs. With the dipole and velocity QWs, however, orientation must also be considered. A model in which monopole QWs are paired off with either dipole or velocity QWs would involve specification of three separate correlation functions. A model in which dipole or velocity QWs are paired off with other dipole or velocity QWs would involve nine separate correlation functions. As shown in Ref. [6], this provides much flexibility in modeling anisotropically correlated fields. In particular, turbulence models with vertical heat and momentum fluxes can be readily constructed, as is appropriate for the atmospheric surface layer.

It can be shown that the coupling of a monopole scalar QW with a velocity QW produces no net flux. One must couple a dipole scalar QW with a velocity QW to produce a flux. This can be understood from Fig. 1. When the dipole in Fig. 1(b) is combined with the velocity field in (c), equal, positive fluxes occur on both the left and right sides. A statistical preference for such orientations produces a net flux.

Velocity correlations (such as a momentum flux) can actually be introduced with preferred orientations of velocity QWs. For example, QWs with rotational axes in the xz-plane, pointing between 0° and 90° and between 180° and 270°, produce a momentum flux toward the ground as occurs in the atmospheric surface layer. This construction is consistent with conceptual pictures of coherent structures in a sturbulent urface layer, e.g., Townsend's inclined vortices [1] or horseshoe vortices [13].

QW models can also be devised with stretching and other distortions to the parent functions. In Ref. [14], rapid-distortion theory was used to derive characteristic stretching in a shear layer and its resulting effect on the turbulence spectrum. This approach is similar to Mann's [15], who also used rapid-distortion theory to calculate shear-layer spectra, although this was done entirely in the spectral domain.

2.3. Ensembles of QWs

The total field is found by summing the contributions of the individual QWs. For a scalar field, for example,

$$T(\mathbf{r}) = \sum_{\alpha=1}^{N} \sum_{n=1}^{N_\alpha} T^{\alpha n}(\mathbf{r}), \tag{7}$$

where N_α is the number of QWs for the size class α. Equations for second-order statistics readily follow and can always be written as a sum over size class. For example, for the spectrum of the monopole model, we have

$$\Phi_T(\boldsymbol{\kappa}) = \sum_{\alpha=1}^{N} \Phi_T^\alpha(\boldsymbol{\kappa}), \tag{8}$$

where

$$\Phi_T^\alpha(\boldsymbol{\kappa}) = (2\pi)^3 n_\alpha a_\alpha^6 \tau_\alpha^2 F^2(\kappa a_\alpha),$$

$\boldsymbol{\kappa}$ is the wavenumber, $n_\alpha = N_\alpha / V$ is the number density of the class α, and

$$F(y) = \frac{1}{(2\pi)^3} \int d^3\xi \, f(\xi) \, e^{-i\mathbf{y}\cdot\boldsymbol{\xi}} \tag{9}$$

is the non-dimensional Fourier transform of the parent function, called the *spectral parent function.* Spectra for the dipole and velocity QWs have similar expressions.

Now let us consider representations where the parameters of the size classes, n_α, τ_α, and Ω_α, scale in a manner consistent with Kolmogorov's original (non-intermittent) hypotheses [5]. First, define the *packing fraction* as

$$\phi_\alpha = n_\alpha a_\alpha^3, \tag{10}$$

For now, we assume ϕ_α is a constant, ϕ, for all QW sizes. The rotation rate Ω_α is proportional to v_α / a_α, where v_α is a characteristic velocity scale for the size class. In the inertial subrange, Ω_α should depend only on the eddy size (a_α in our notation) and the dissipation rate of specific turbulent kinetic energy, ϵ (dimensions length^2 divided by time^3). By dimensional analysis, we must have

$$\Omega_\alpha = c_\Omega a_\alpha^{-2/3} \epsilon^{1/3}, \tag{11}$$

where c_Ω is a constant. Similarly, τ_α should depend only on the eddy size, dissipation rate, and the scalar destruction rate χ_T (dimensions scalar^2 divided by time). Hence

$$\tau_\alpha = c_T a_\alpha^{1/3} \chi_T^{1/2} \epsilon^{-1/6}, \tag{12}$$

where c_T is a constant (not to be confused with the structure-function parameter). As with conventional wavelet analysis, we assume that the ratio of one size class to the next is constant. The following equation has this property:

$$a_\alpha = a_1 e^{-\mu(\alpha-1)}. \tag{13}$$

Here, μ controls the spacing between the size classes. After some manipulations and conversion of the summations to integrals (by taking the limit $\mu \to 0$), one finds

$$\Phi_T(\boldsymbol{\kappa}) = \frac{(2\pi)^3 \varphi}{\mu} c_T^2 \chi_T \epsilon^{-1/3} \kappa^{-11/3} \int_{\kappa a_N}^{\kappa a_1} dy\, y^{8/3} F^2(y), \tag{14}$$

and, for the spectrum of specific turbulent kinetic energy (TKE),

$$E(\kappa) = \frac{32\pi^4 \varphi}{3\mu} c_\Omega^2 \epsilon^{2/3} \kappa^{-5/3} \int_{\kappa a_N}^{\kappa a_1} dy\, y^{14/3} F^2(y). \tag{15}$$

For wavenumbers such that $a_1 \gg \kappa^{-1} \gg a_N$, the integrals are nearly constant. Hence $\Phi_T(\mathbf{k}) \sim \chi_T \epsilon^{-1/3} \kappa^{-11/3}$ and $E(k) \sim \epsilon^{2/3} \kappa^{-5/3}$ in this region, which corresponds to the inertial subrange. The QW spectra also exhibit realistic roll-offs for small ($\kappa a_1 \ll 1$) and large ($\kappa a_N \gg 1$) wavenumbers, similar to the behavior of the von Kármán spectrum in the production and dissipation subranges, respectively.

The ratio ϕ/μ plays an important role in QW models. Second moments are conserved by keeping the ratio ϕ/μ constant; that is, if the packing fraction is halved, the ratio between adjacent sizes should also be halved. In this manner, the total number of QWs is essentially preserved. However, it should be kept in mind that altering the ratio ϕ/μ does alter moments higher than second order. As shown in Ref. [4], a small value of ϕ/μ leads to a large kurtosis; large ϕ/μ results in kurtosis near 3.

3. Intermittency

Intermittency refers to the tendency of turbulence to occur in spatial and temporal bursts of activity. The localized nature of QWs makes them natural for building models with spatial intermittency. Up to this point, we have assumed that the QW positions (the $\mathbf{b}^{\alpha n}$) were smoothly distributed in space. In intermittent turbulence, however, the eddies tend to lump into patches of high activity, particularly at smaller scales. The result is an uneven dissipation rate for TKE [16; 17].

Mahrt [18] has proposed classifying intermittency as *intrinsic* or *global.* The former occurs on scales less than the outer scale of turbulence (the scale of largest eddies) as the turbulent cascade process progressively concentrates turbulent energy dissipation into smaller regions of space [19]. Global intermittency occurs on scales larger than the outer scale and may have several causes (such as topographic flow, mesoscale coherent structures, uneven heating of the ground, and breakdown of stable layers) related to uneven production of turbulence in a particular environment. The terms *small-scale* and *large-scale* are also used to describe intrinsic and global intermittency, respectively [18; 20].

Let us first consider intrinsic intermittency and its modifications to the scaling relationships for the QW model parameters. Regardless of whether intermittency is present, we anticipate that the turbulence is a self-similar cascade process and thus the distribution of size classes continues to be dictated by Eq. (13). The hypothesis that eddies become less space filling with decreasing size amounts to assuming that the packing fraction decreases with increasing α:

$$\phi_\alpha = \phi(a_\alpha / a_1)^\lambda. \tag{16}$$

Here, ϕ is a constant equal to the packing fraction ϕ_1 of the largest QWs and λ controls the intermittency. Increasing λ results in more intermittent turbulence; $\lambda = 0$ corresponds to the non-intermittent case.

The cascade mechanism by which the small eddies become less space-filling controls the spatial distrbition of the QWs. In the turbulence literature, the cascade mechanism is often conceived as a process repeated from one "generation" of eddies to the next: parent eddies produce children eddies that are contained within, yet do not entirely fill, the space of the parents. For example, a parent eddy of size a may break down into several eddies of size $a/2$. This description coincides to the well known beta model of Frisch et al.[19], which defines the parameter β as the number of children eddies divided by 2^3. Since $\beta = \phi_{\alpha+1}/\phi_\alpha = (a_{\alpha+1}/a_\alpha)^\lambda$ when $a_{\alpha+1}/a_\alpha = 1/2$, it follows that $\beta = 2^{-\lambda}$. Frisch et al. [19] show that $\beta = 2^{D-3}$, where $D \approx 2.5$ is the fractal dimension. Hence $\lambda = 3 - D \approx 0.5$.

Conceivably, the beta or a related model could be adapted to organize the QWs spatially according to a cascade process. However, there are many challenging details to devising such a spatial organization, such as how to provide a continuous spatial placement of the QWs and how to derive statistical models when the size classes are not independent. A preliminary method for spatial organization of the QWs was described in Ref. [21].

Intrinsic intermittency of turbulence can also affect the scaling relationships for QW amplitudes, τ_α and v_α. In Ref. [22], it was shown that a cascade process conserving TKE and scalar variance leads to

$$v_\alpha = v_1(a_\alpha/a_1)^{(1-\lambda)/3}, \text{ and } \tau_\alpha = \tau_1(a_\alpha/a_1)^{(1-\lambda)/3}. \tag{17}$$

The eddies (QWs) spin relatively faster, and have a stronger scalar amplitude, to compensate for the decreasing packing fraction. As a result, the temperature spectrum was found to go as $\Phi_T(\mathbf{k}) \sim \kappa^{-11/3-\lambda/3}$ in the inertial subrange. This relationship provides a correction to the classical (non-intermittent) power law.

As discussed earlier, global intermittency involves externally imposed variations in turbulence intensity over regions larger than the outer scale of the turbulence. Many and varied statistical models for global intermittency have been considered in the literature. Mahrt [18], Petenko and Shurygin [23], and other authors use dichotomous regions of strong and weak activity. Antonia et al. [24] considered a linear ramp model. Tatarskii and Zavorotnyi [25] considered a gamma pdf, and Frehlich [26] a log-normal pdf, for the structure-function parameter. This diversity of approaches to modeling global intermittency reflects the different mechanisms producing it as well as the challenges of describing the underlying physics with tractable models.

Conceptually, we model global intermittency by partitioning the overall volume $\mathcal{V}$ occupied by the turbulence into many subvolumes V_i with characteristic dimensions larger than a_1. Within each of these subvolumes, QWs occur as described in the preceding discussion, although the parameters may vary from one subvolume to another. Variations in the parameters τ_1, v_1, and a_1 are of particular interest since they depend on the external mechanism creating the turbulence. We thus require statistical models for the variations in these parameters, or for quantities dependent upon them, such as structure-function parameters. One possible approach, based on a two-state (active/inactive) Markov process model, was described in Ref. [22]. This approach appears to be useful for describing global intermittency along a particular path through space (one-dimensional intermittency), but must be generalized in some manner to describe global intermittency in multiple dimensions.

Some example realizations of random, turbulent temperature fluctuations are shown in Fig. 2. The realizations all use $a_N = 0.2$, $a_1 = 10$ m, $\phi = 0.0023$, $\lambda = 0.5$, and $\tau_1 = 0.5$ K, but vary in the degree of global intermittency. The top panel has no global intermittency. The middle one has strong global intermittency, generated with a single Markov process having a characteristic length scale of 10 m and a 25% activity level. The bottom panel has moderate global intermittency, generated with four, superimposed Markov processes, each having a characteristic length scale of 10 m and a 50% activity level.

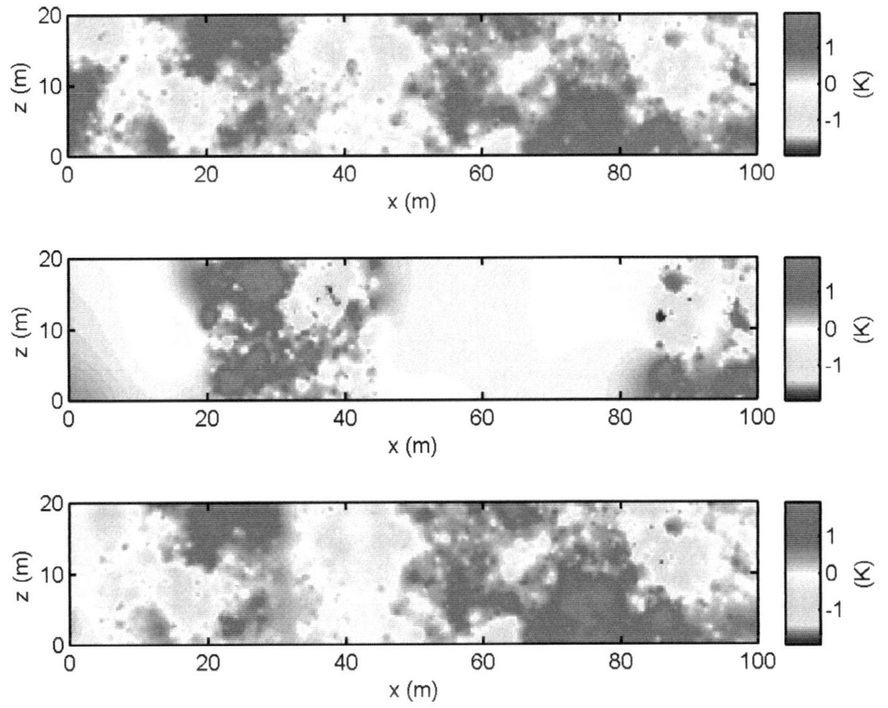

Figure 2. Turbulent temperature fields synthesized from random QWs. The three realizations show varying degrees of global intermittency as described in the text. Top is non-intermittent, middle has strong intermittency, and bottom has moderate intermittency.

4. Application to wave scattering

QWs were originally developed for calcualtions of acoustic wave scattering by turbulence [2; 10]. They are particularly convenient for this purpose, as they allow a particle-like description of turbulent scattering. In Ref. [11], the following expression was derived for single-scattered sound pressure field from an individual velocity QW:

$$\begin{aligned} p_S^{\alpha n}(\mathbf{R}) =& \frac{i4\pi^2 A k^2 a_\alpha^5 F(Ka_\alpha)\cos\theta}{c_0 R R_0} \\ & \times \exp\left[ik\left(R + R_0 - (\mathbf{n} - \mathbf{n}_0)\cdot \mathbf{b}^{\alpha n}\right)\right]\left[\mathbf{n}_0 \cdot \left(\mathbf{K} \times \boldsymbol{\Omega}^{\alpha n}\right)\right]. \end{aligned} \tag{18}$$

Here $\mathbf{K} = \mathbf{k}_0 - \mathbf{k} = k(\mathbf{n}_0 - \mathbf{n})$ is the scattering vector, $\mathbf{R}$ is the vector from the origin to the receiver, $\mathbf{R}_0$ is the vector from the origin to the source, $\mathbf{n} = \mathbf{R}/|\mathbf{R}|$, $\mathbf{n}_0 = -\mathbf{R}_0/|\mathbf{R}_0|$, $\mathbf{k} = k\mathbf{n}$, $\mathbf{k}_0 = k\mathbf{n}_0$, and θ is the angle between $\mathbf{n}$ and $\mathbf{n}_0$. The acoustic wavenumber k is $2\pi f/c_0$, where f is frequency and c_0 is the background sound speed. The pressure field associated with the wave propagating spherically from the source is $p_0(\mathbf{R} - \mathbf{R}_0) = A\exp(ik|\mathbf{R} - \mathbf{R}_0|)/|\mathbf{R} - \mathbf{R}_0|$, where A is the pressure amplitude 1 m from the source. This result assumes validity of the approximations $kR_1' \simeq kR - \mathbf{r}\cdot\mathbf{k}$ and $kR_0' \simeq kR_0 + \mathbf{r}\cdot\mathbf{k}_0$ (valid for $R, R_0 \gg V^{1/3}, kL_0V^{1/3}$, where L_0 is the outer length scale of turbulence). Expressions for the scattered wavefields from other QW models, such as monopole scalars, and for other wave types, such as electromagnetic, can also be derived.

Eq. (18) serves as a starting point for deriving other results for scattering by turbulence. For example, by summing over all QWs of the size class α (index n), and randomizing the orientations, the scattering cross section for the size α (which depends on the acoustic wavelength) follows. The total scattering cross section follows by summing contributions from the size classes. Such a procedure leads to equations in agreement with those derived directly from spectra [11]. Scattering from moving QWs, including Doppler shifts, has also been studied [10].

QW models can also be used to study the effect of intermittency on wave scattering. Ref. [22] considered line-of-sight wave propagation through turbulence with intrinsic and global intermittency. The coherence function between two sensors separated by a distance r (perpendicular to the direction of propagation) was derived as

$$\overline{\Gamma}(x,r) = p_0^2 \exp\left[-\overline{D}_\phi(x,r)/2\right] \left\langle \exp\left[-\frac{\overline{D}_\phi(x,r)}{2x}\int_0^x \zeta(x')\,dx'\right]\right\rangle, \tag{19}$$

where x is the horizontal propagation distance and $\overline{D}_\phi(x,r)$ is the average phase structure function. The quantity $\zeta(x)$ is the normalized fluctuation of the variance or structure-function parameter due to global intermittency. The initial exponential decay factor in Eq. (19) represents the familiar decay of coherence in non-intermittent turbulence. The following part of the expression, within the angle brackets, represents the global intermittency effect. It is bounded as follows:

$$1 \le \left\langle \exp\left[-\frac{\overline{D}_\phi(x,r)}{2x}\int_0^x \zeta(x')\,dx'\right]\right\rangle < \exp\left[\overline{D}_\phi(x,r)/2\right]. \tag{20}$$

The first inequality follows from Jensen's inequality, as previously noted in Refs. [25] and [27]. The second inequality follows from the argument that the path average of $\zeta(x)$ cannot be smaller than -1, since that would correspond to a state of complete inactivity. Thus

$$1 \ge \overline{\Gamma}(x,r)/p_0^2 \ge \exp\left[-\overline{D}_\phi(x,r)/2\right]. \tag{21}$$

Hence intermittency always *increases* the average coherence. When global intermittency is very strong, as might occur in stably stratified, nighttime conditions, signal phase fluctuations become highly non-Gaussian and the coherence is dominated by episodes with little turbulence [22].

5. Conclusions

QW formulations readily accomodate realistic features of turbulence such as anisotropy, correlations, and intermittency. Correlated fields can be created by associating multiple properties of the field with each QW. The anisotropic effects are created through preferred orientations of the QWs; when the orientations between scalar and vector field properties are correlated, anisotropic fluxes result.

Both intrinsic and global intermittency can be built into a QW model. The former corresponds to a cascade process in which the turbulent energy is progressively concentrated into smaller regions of space. Intrinsic intermittency modifies the power law in the inertial subrange. We have characterized global intermittency here as an uneveness in the production mechanism of turbulence, which affects the spatial concentration of the largest eddies (QWs).

The main applications of QWs considered in this review were to random field synthesis and wave scattering. QW approaches could conceivably be further developed for other problems such as identification of patterns in random fields (e.g., coherent structures in turbulence), particle interaction-like simulations of dynamics in random fields, subgrid-scale models in turbulence simulations, turbulent flow initialization [28], synthesis of random terrain elevations and geologic structure, and electromagnetic volume and surface-wave scattering.

References

[1] Townsend A A 1976 *The Structure of Turbulent Shear Flow* (Cambridge: Cambridge University Press)

[2] Goedecke G and Auvermann H J 1997 *J. Acoust. Soc. Am.* **102** 759–771

[3] Goedecke G, Ostashev V E, Wilson D K and Auvermann H J 2004 *Boundary-Layer Meteorol.* **112** 33–56

[4] Goedecke G, Wilson D K and Ostashev V E 2006 *Boundary-Layer Meteorol.* **120** 1–23

[5] Kolmogorov A N 1941 *Dokl. Akad. Nauk SSSR* **30** 299–303 english transl.: Proc. R. Soc. London, Ser. A 434, 9–13 (1991)

[6] Wilson D K, Ostashev V E and Goedecke G H 2008 *J. Prob. Eng. Mech.* in review

[7] Kitagawa T and Nomura T 2003 *Journal of Wind Engineering and Industrial Aerodynamics* **91** 934–964

[8] Gioffré M, Gusella V and Grigoriu M 2000 *Prob. Eng. Mech.*

[9] Masters F and Gurley K R 2003 *J. Eng. Mech.* **129** 1418–1428

[10] Goedecke G H, Wood R C, Auvermann H J, Ostashev V E, Havelock D I and Ting C 2001 *J. Acoust. Soc. Am.* **109** 1923–1934

[11] Wilson D K, Ostashev V E, Goedecke G H and Auvermann H J 2004 *Appl. Acoust.* **65** 605–627

[12] Ostashev V E and Wilson D K 2004 *J. Acoust. Soc. Am.* **115** 497–506

[13] Cantwell B J 1981 *Ann. Rev. Fluid Mech.* **13** 457–515

[14] Goedecke G H, Moore S, Auvermann H J and Ostashev V E 2001 *Proceedings of Battlespace Atmospheric and Cloud Impacts on Military Operations Conference* (Ft. Collins, CO)

[15] Mann J 1994 *J. Fluid Mech.* **273** 141–168

[16] Kolmogorov A N 1962 *J. Fluid Mech.* **13** 82–85

[17] Obukhov A M 1962 *J. Fluid Mech.* **273** 77–81

[18] Mahrt L 1989 *J. Atmos. Sci.* **46** 79–95

[19] Frisch U, Sulem P L and Nelkin M 1976 *J. Fluid Mech.* **87** 719–736

[20] Muschinski A, Frehlich R G and Balsley B B 2004 *J. Fluid Mech.* **515** 319–351

[21] Wilson D K, Goedecke G H and Ostashev V E 2006 *17th Symposium on Boundary Layers and Turbulence* (San Diego: American Meteorological Society)

[22] Wilson D K, Ostashev V E and Goedecke G H 2008 *J. Acoust. Soc. Am.* in review

[23] Petenko I V and Shurygin E A 1999 *Boundary-Layer Meteorol.* **93** 381–394

[24] Antonia R A, Chambers A J and Bradley E F 1982 *Boundary-Layer Meteorol.* **23** 395–403

[25] Tatarskii V I and Zavorotnyi V U 1985 *J. Opt. Soc. Am. A* **2** 2069–2076

[26] Frehlich R 1994 *Applied Optics* **33** 5764–5769

[27] Gozani J 1999 *Optics Letters* **24** 108–110

[28] Jarrin N, Benhamadouche S, Laurence D and Prosser R 2006 *International Journal of Heat and Fluid Flow* **27** 585–593

Application of microwave radiometer and wind profiler data in the estimation of wind gust associated with intense convective weather

P W Chan[1] and K H Wong[2]

[1] Hong Kong Observatory, 134A Nathan Road, Kowloon, Hong Kong, China

[2] City University of Hong Kong, Tat Chee Avenue, Kowloon, Hong Kong, China

E-mail: pwchan@hko.gov.hk

Abstract. Estimates of the wind gusts associated with intense convective weather could be obtained using empirical relationships such as GUSTEX based on radiosonde measurements. However, such data are only available a couple of times a day and may not reflect the rapidly changing atmospheric condition in spring and summer times. The feasibility of combining the thermodynamic profiles from a ground-based microwave radiometer and wind profiles given by radar wind profilers in the continuous estimation of wind gusts is studied in this paper. Based on the results of a 4-month trial of a microwave radiometer in Hong Kong in 2004, the estimated and the actual gusts are reasonably well correlated. It is also found that the wind gusts so estimated provide better indications of the strength of squalls compared with those based on radiosonde measurements and with a lead time of about one hour.

1. Introduction

Situated in a subtropical coastal area, Hong Kong occasionally experiences intense wind gusts associated with convective weather, such as downburst of thunderstorm and gust front, in the spring and summer every year. In the forecasting of wind gusts, reference has been made by the weather forecasters at the Hong Kong Observatory (HKO) to the radiosonde measurements and gust estimates are determined using empirical relationships like GUSTEX [1]. However, radiosonde data are only available a few times a day and may not be updated fast enough to catch the rapidly evolving atmospheric conditions in intense convective weather.

Measurements by remote-sensing meteorological instruments provide basically continuous monitoring of the dynamic and thermodynamic conditions of the troposphere. In Hong Kong, a network of boundary-layer type radar wind profilers (locations in Figure 1) measures the upper-air winds every 10 minutes up to a height of 9 km above ground. Continuous measurement of the temperature and humidity within the troposphere has also been made in field experiments of ground-based, multi-channel microwave radiometers. For instance, in the study of 2004 [2], the radiometer data were demonstrated to vividly reveal the rapid evolution of humidity profiles of the atmosphere (up to 10 km above ground) and the thermodynamic data turned out to be very useful in the nowcasting of the occurrence of thunderstorms and the number of lightning flashes.

The radiometer data in 2004 are examined in this paper from the perspective of nowcasting of the wind gusts associated with intense convective weather. For that purpose, they are combined with the upper-air wind data from the network of radar wind profilers in Hong Kong to provide continuous update of GUSTEX (every 10 minutes). This gust estimate is then compared with the actual gust measurements by the dense network of surface anemometers in Hong Kong (locations in Figure 1).

This paper is organized as follows. Section 2 gives a brief account of the gust estimation based on radiometer and wind profiler data. Section 3 compares the gust estimates with the actual gust measurements in the thunderstorm days during the field experiment of the radiometer in Hong Kong in 2004. Two case studies are given in Section 4. The conclusions of the paper are drawn in Section 5.

2. Wind gust estimate

GUSTEX is based on WINDEX developed in [3]. WINDEX is modified in [1] and given by:

$$WI = 5\ [H_m R_q (L_{max}^2 - 30 + q_l - 2\ q_m)]^{0.5} \quad (1)$$

where WI is the WINDEX in knots, H_m is the height of the melting layer in km, L_{max} is the maximum lapse rate from the ground to the freezing layer in K/km, q_l is the mean mixing ratio of water vapour between the ground and 1 km above in g/kg, q_m is the mixing ratio of water vapour at the melting layer in g/kg, and $R_q = q_l/12$.

GUSTEX also considers the downward transport of horizontal momentum in the upper air. It is modified in [1] and given by:

$$GU = \alpha_1\ WI + \alpha_2\ \rho U_{max} \quad (2)$$

where GU is GUSTEX in knots, U_{max} is the maximum wind speed between 900 and 500 hPa levels in knots, $\rho=\rho_{max}/\rho_{990}$ in which ρ_{max} is the air density at the pressure level at which U_{max} occurs and ρ_{990} is the air density at 990 hPa level. α_1 and α_2 are constants to be determined empirically.

In [1], the empirical constants in equation (2) were determined from the high wind gust cases at the Hong Kong International Airport between 2001 and 2005. The upper-air temperature, humidity and wind data were taken from the twice-daily radiosonde measurements. Besides determining the empirical constants from the least-square linear fit between GUSTEX and the actually measured wind gust from the surface anemometer stations, the upper bound GU_{max} and the lower bound GU_{min} of the wind gust estimate are also calculated. The relevant equations are summarized below:

$$GU = 0.12\ WI + 0.93\ \rho U_{max} \quad (3)$$
$$GU_{max} = 0.30\ WI + 0.93\ \rho U_{max} \quad (4)$$
$$GU_{min} = -0.05\ WI + 0.93\ \rho U_{max} \quad (5)$$

In equation (1), if L_{max}^2 is less than 30, WI is taken as 0 and, from equations (3) to (5), $GU = GU_{max} = GU_{min}$ in this case.

In the present study, the zenith mode measurements by the microwave radiometer are used to retrieve the temperature and humidity profiles up to 10 km above ground. These profiles are then employed to determine the thermodynamic parameters in equations (3) to (5), such as the height of the melting layer, water vapour mixing ratio, lapse rate, etc. The only complication in this calculation is that the radiometer data are only available at fixed heights above ground, instead of fixed pressure levels. The height and the pressure are related using hydrostatic approximation. The maximum wind speed and the corresponding height above ground are obtained by considering the wind profiles from all the four radar wind profilers in Hong Kong. At a particular moment, the maximum wind among the four profilers is selected in the calculation of GUSTEX as well as its upper and lower bounds.

3. Comparison between wind gust estimate and the actual gust measurements

The performance of the wind gust estimate determined from the above method (viz. radiometer and wind profiler data) is compared with the actually measured gust by the dense network of surface anemometers in Hong Kong in 14 thunderstorm episodes between March and June 2004. A thunderstorm episode is defined as a period of time (normally a few hours or more) in which thunderstorm has been reported by the weather observers at HKO Headquarters and/or Hong Kong International Airport. In general, the GUSTEX itself (equation (3)) could underestimate the actual gust measurement at times. For practical applications, the upper bound of wind gust estimate (equation (4)) may be more useful in giving the weather forecaster an idea about how large the gust could attain. As such, GU_{max} is considered in the comparison.

The scatter plot of the maximum wind gust actually measured by the surface anemometers and the largest value of GU_{max} within an hour prior to the maximum gust occurrence for the thunderstorm episodes under study is given in Figure 2(a). The two datasets are found to have good correlation and the latter is only larger than the former by about 20% on average from the least-square linear fit. There are two outlier points in Figure 2(a) (encircled in red in the scatter plot). In both cases, the upper-air winds from the wind profilers are too strong to result in over-estimation of wind gusts.

For comparison of performance, the corresponding scatter plot based on the GU_{max} estimated from radiosonde measurements is shown in Figure 2(b). In this plot, the upper bound of GUSTEX determined from the radiosonde data just prior to the maximum gust occurrence is used, which could be as many as 12 hours earlier (because there are only two radiosonde ascents every day in Hong Kong with both wind and temperature/humidity profiles). Compared to the radiometer/wind profiler plot, the radiosonde plot has weaker correlation and on average the wind gust estimate is 63% larger on average. There are also two outlier points in Figure 2(b) (encircled in red in the scatter plot). In both cases, the upper-air winds measured by the radiosonde are too strong to result in over-estimation of the wind gusts. Overall speaking, the GU_{max} estimated from radiometer/wind profiler data has superior performance and could be useful in the nowcasting of wind gust associated with intense convective weather with a lead time of about one hour.

4. Case studies

Two convective weather cases within the study period in 2004 are discussed in detail in this Section on how the wind gust estimate performs. The first case occurred on 8 May 2004. Synoptically, a strong southerly airflow between a westerly wave to the west of Hong Kong and the subtropical ridge to the east affected the Pearl River Estuary (location in Figure 3(a)) within the boundary layer. At 200 hPa, divergent flow occurred over the south China coast (location in Figure 3(a)) ahead of a deep westerly wave. From the 256-km radarscope, bands of strong echoes with southwest-to-northeast orientation moved from western coast of Guangdong across Hong Kong. Thunderstorms were reported over the territory between 2 and 8 a.m. on that day (Hong Kong time, which is equal to UTC + hours). The first band of heavy rain affected Cheung Chau (location in Figure 1) at about 3:42 a.m. (Figure 3(a)), bringing wind gust of 18.9 m/s, the largest over Hong Kong. The second band affected Waglan Island (location in Figure 1) at about 7:04 a.m. (Figure 3(b)), bringing the maximum wind gust of 26.2 m/s. The time series of maximum wind gust over the surface weather stations in Hong Kong in that morning as well as the GUSTEX, GU_{max} and GU_{min} are given in Figure 3(c). For the two events of the passage of strong radar echoes, the maximum wind gusts lie between the largest values of GU and GU_{max} within an hour prior to each event. As such, the wind gust estimates based on radiometer and wind profilers are considered to be useful in nowcasting the actual strength of the gusts.

It is noted that, after the occurrence of squalls at the surface, the wind gust estimates may remain at rather high values, such as between 4 and 4:30 a.m. on 8 May 2004. This is because of the appearance of strong winds at the upper air after the passage of the squall near the surface.

The second case occurred on 17 April 2004. Synoptically, a trough of low pressure at the surface over the south China coast brought unsettled weather to the region. On the 256-km radarscope, a band of intense echoes with north-northeast to south-southwest orientation moved across Hong Kong in the afternoon on that day. Thunderstorms were reported in Hong Kong at 5 and 6 p.m. At the times shown in Figures 4(a) and (b), high wind gusts were recorded at Cheung Chau (15.1 m/s) and Waglan Island (17 m/s) respectively. The time series of the highest wind gust in Hong Kong together with the various wind gust estimates are given in Figure 4(c). The largest values of GUSTEX and GU_{max} within an hour before the high wind gust events provide reasonable estimates on the gusts that could be attained. As in the previous case, after the occurrence of squalls at the surface, the wind gust estimates may remain at rather high values, such as after 6:30 p.m. on 17 April 2004. This is again because of the appearance of strong winds at the upper air after the passage of the squall near the surface.

5. Conclusions

The performance and application of wind gust estimates based on remote-sensing instruments, viz. radiometer and radar wind profilers are studied in this paper. The GUSTEX originally developed for radiosonde data is adopted here. Compared to the twice-daily balloon ascents, the continuous measurements of temperature, humidity and wind from the remote-sensing instruments better capture the rapid evolution of the troposphere in convective weather conditions. As such, the wind gust estimates so determined has improved correlation with the maximum gusts actually measured by the network of surface anemometer stations in Hong Kong. From two case studies of typical intense convection in the summer, the gust estimates are found to be useful in the nowcasting of high winds associated with thunderstorms.

The present study is based on the field experiment data of a ground-based microwave radiometer in Hong Kong in 2004 only. The amount of data involved is rather limited. A permanent unit of radiometer has recently been set up in Hong Kong and a larger set of data accumulated over the years would then be employed to evaluate the performance of the wind gust estimates determined from radiometer and wind profiler measurements for operational use.

References

[1] Lee OSM 2007: Forecast of strong gusts associated with thunderstorms based on data from radiosonde ascents and automatic weather stations. *21st Guangdong – Hong Kong – Macao Technical Seminar on Meteorological Science and Technology*, Hong Kong, China, 24 – 26 January 2007 (paper in Chinese with English abstract)

[2] Chan PW and CM Tam 2005: Performance and application of a multi-wavelength, ground-based microwave radiometer in rain nowcasting. *Ninth Symposium on Integrated Observing and Assimilation Systems for Atmosphere, Oceans, and Land Surface*, American Meteorological Society, San Diego, California, USA.

[3] McCann DW 1994: WINDEX – A new index for forecasting microburst potential. *Wea. Forecasting*, **9**, 532-541

IOP Conf. Series: Earth and Environmental Science 1 (2008) 012042 doi:10.1088/1755-1307/1/1/012042

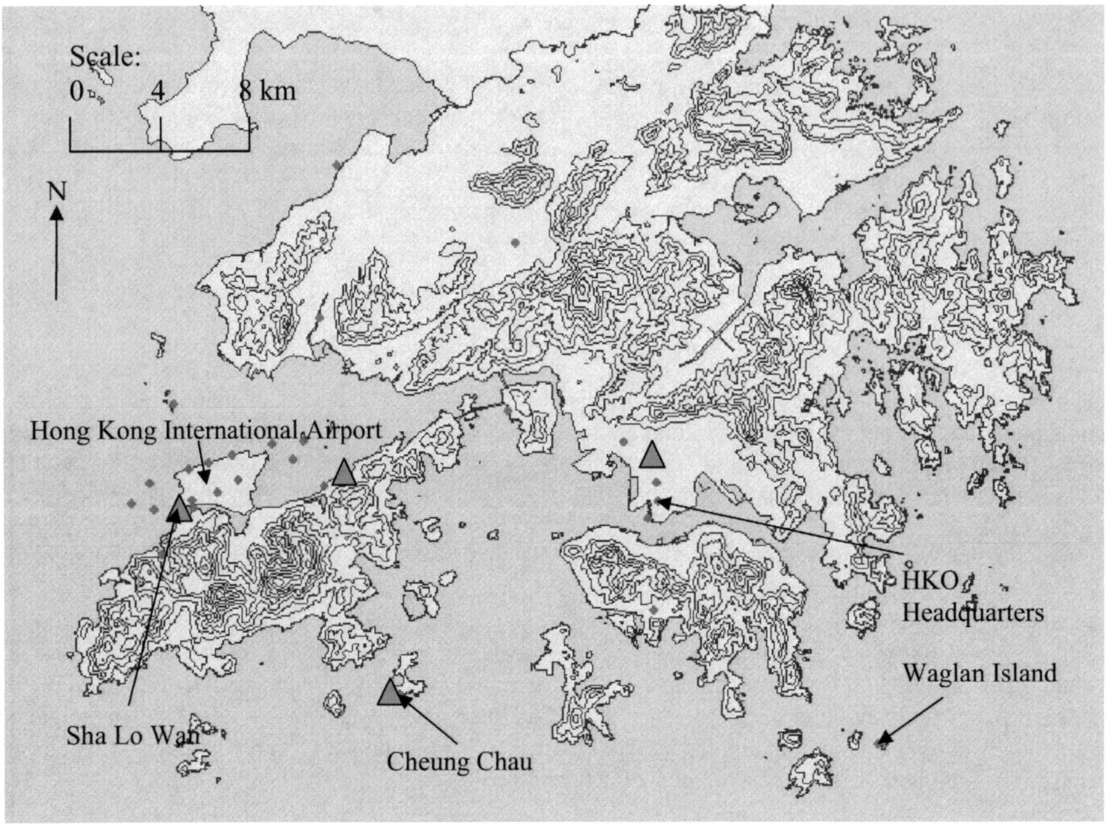

Figure 1. The network of surface anemometers (red dots) and radar wind profilers (green triangles) in Hong Kong. The radiometer has been placed at HKO Headquarters, Cheung Chau and Sha Lo Wan during the experiment. Height contours are in 100 m.

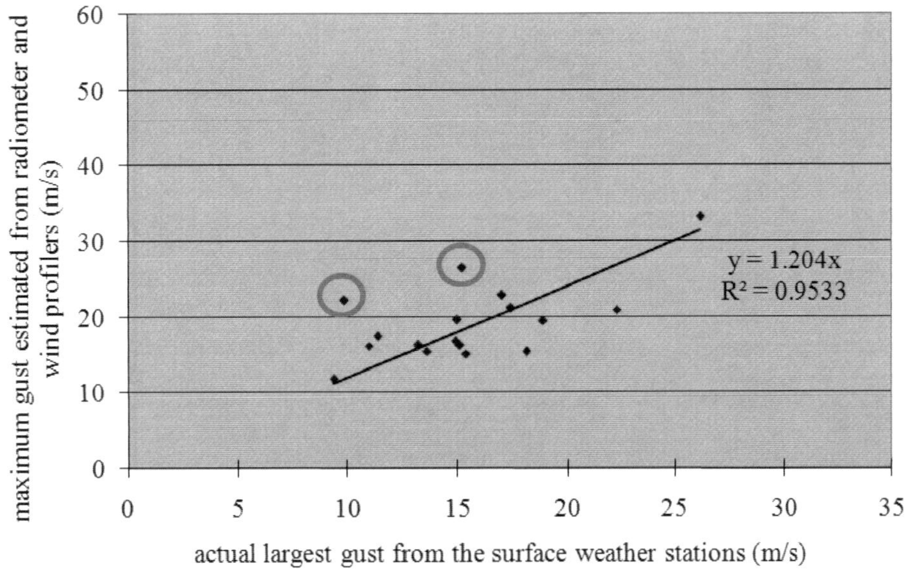

(a)

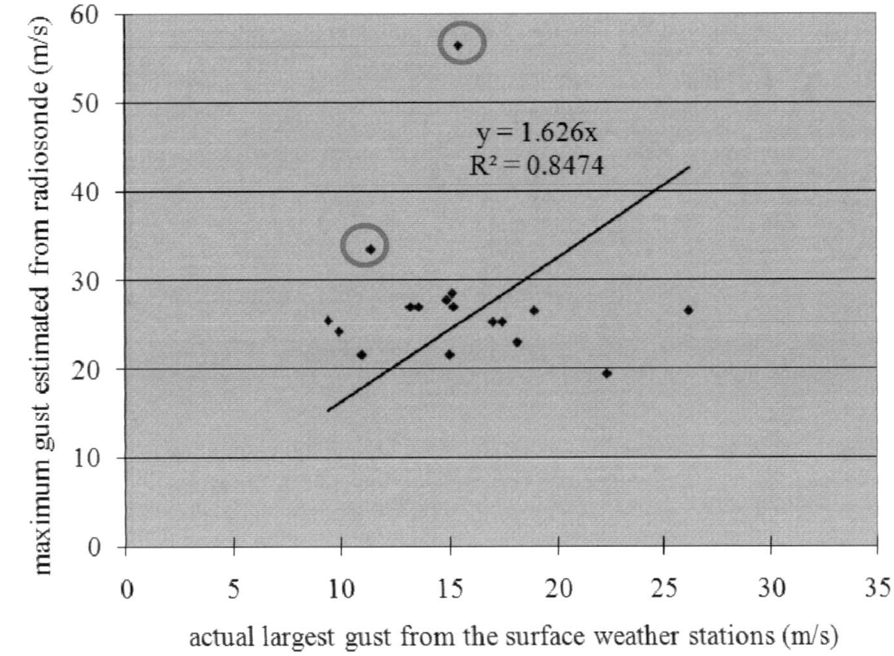

(b)

Figure 2. (a) The scatter plot of the maximum gust estimated from microwave radiometer and radar wind profiler data against the largest value of the actually measured gust from the surface anemometer stations in the 14 thunderstorm episodes in 2004 of the current study. (b) is the corresponding scatter plot based on the maximum gust estimated from the radiosonde data. In (a) and (b), the outliers are encircled in red.

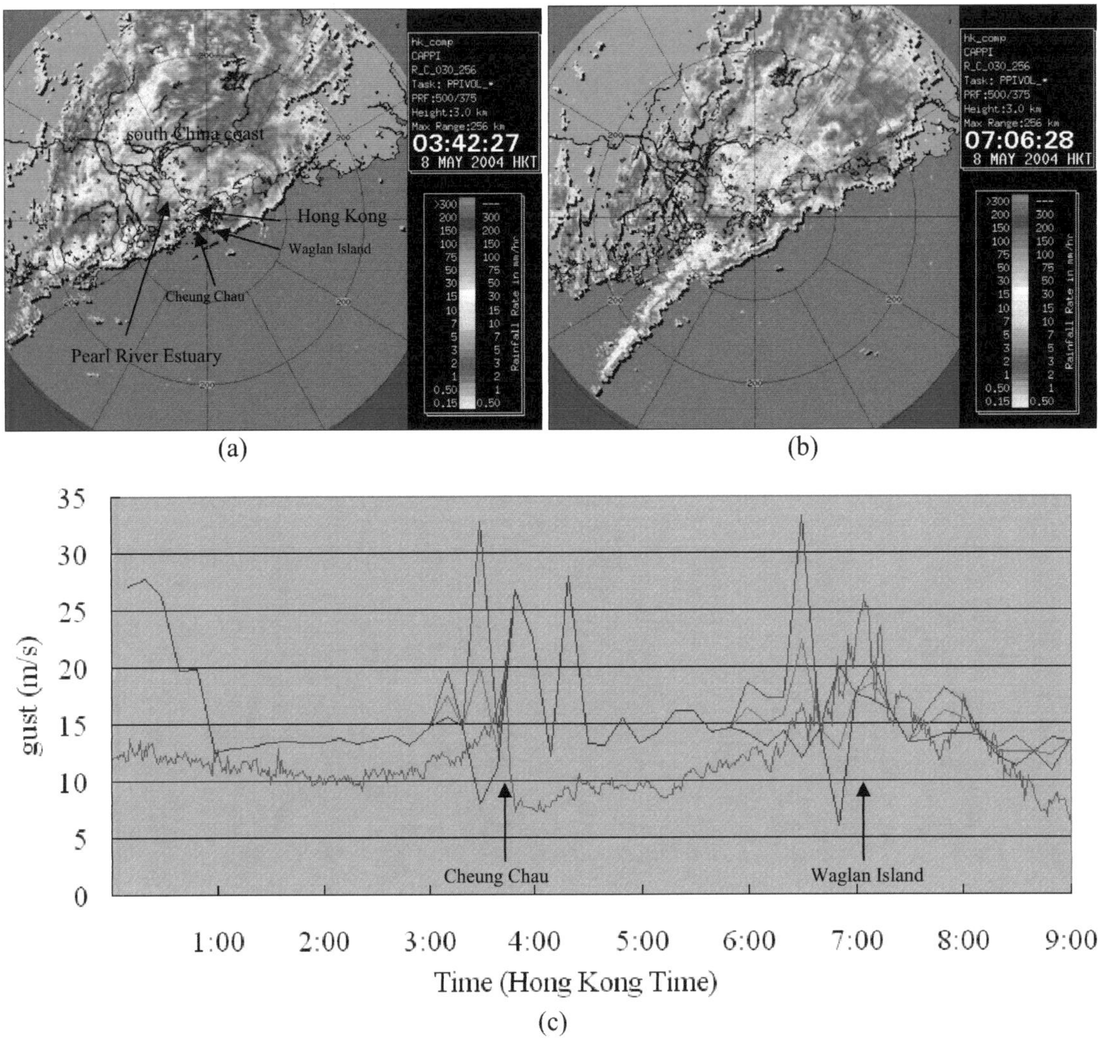

Figure 3. (a) and (b) are the 256-km radarscope images of Hong Kong at 3 km above mean sea level at 3:42 and 7:06 a.m., 8 May 2004 respectively. (c) is the time series of the largest value of the actually measured wind gust from the surface anemometers (green), GU_{min} (blue), GU (red) and GU_{max} (brown).

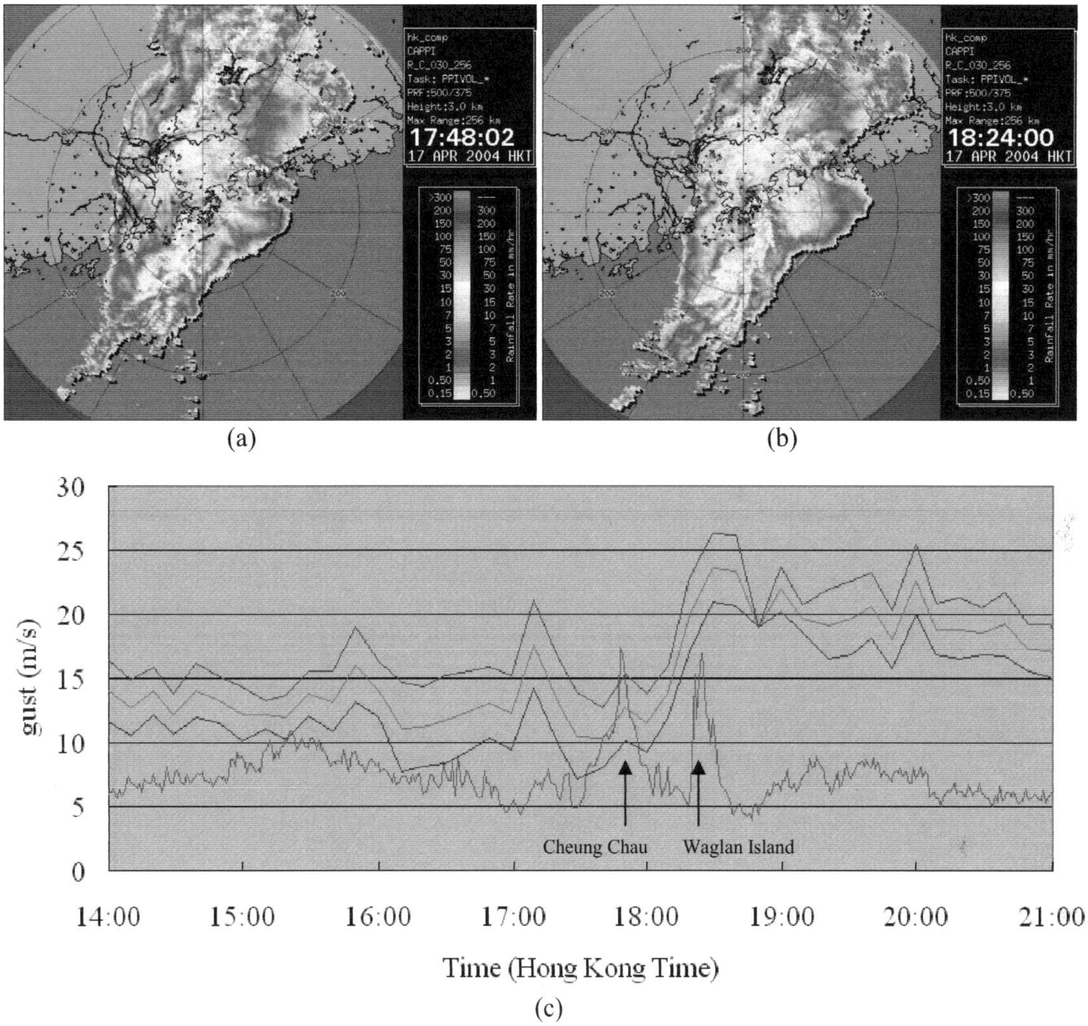

Figure 4. (a) and (b) are the 256-km radarscope images of Hong Kong at 3 km above mean sea level at 5:48 and 6:24 p.m., 17 April 2004 respectively. (c) is the time series of the largest value of the actually measured wind gust from the surface anemometers (green), GU_{min} (blue), GU (red) and GU_{max} (brown).

IOP Conf. Series: Earth and Environmental Science 1 (2008) 012043 doi:10.1088/1755-1307/1/1/012043

Determination of Richardson number profile from remote sensing data and its aviation application

P W Chan

Hong Kong Observatory, 134A Nathan Road, Kowloon, Hong Kong, China

E-mail: pwchan@hko.gov.hk

Abstract. A number of remote-sensing wind-measuring instruments, such as LIDAR and radar wind profilers, have been set up in the vicinity of the Hong Kong International Airport (HKIA) for monitoring of low-level windshear and turbulence. The combined usage of wind data from a radar wind profiler and boundary layer temperature profiles from a ground-based microwave radiometer in turbulence application is studied in this paper. Data collected in a field experiment of the radiometer conducted in Hong Kong in February 2006 were used to obtain the Richardson number profile up to 1.5 km above ground. Two examples of atmospheric turbulence reported by the pilots are considered. The Richardson number profiles are found to capture the turbulence events reasonably well. The evolution of these profiles and airflow disruption by the nearby terrain as a possible cause of the turbulence will also be discussed.

1. Introduction

The Hong Kong International Airport (HKIA) is situated in an area of complex terrain (Figure 1). To its south is the mountainous Lantau Island with peaks rising to about 1000 m above mean sea level with valleys as low as 400 m in between. When winds from the east to southwest climb over Lantau Island, turbulent airflow may occur in the vicinity of the airport. Turbulence could have significant impact on aviation safety. A dense network of anemometers is operated by the Hong Kong Observatory (HKO) to monitor the turbulence condition in the airport area [1]. Studies have also been conducted on the use of measurements from radar wind profilers [1], Doppler LIght Detection And Ranging (LIDAR) systems [2] and Terminal Doppler Weather Radar (TDWR) [3] to monitor turbulence. The locations of these instruments are shown in Figure 1.

The above monitoring methods are based on the fluctuations of the wind field as revealed in the spectrum width or the structure function of the radial velocity provided by the remote-sensing, wind-measuring equipment. Richardson number [4] is another useful parameter for turbulence monitoring and it combines both the thermodynamic and the dynamic profiles in the troposphere. Whilst the wind profiles are measured basically continuously by radar wind profilers in Hong Kong, temperature profiles are only available twice a day from radiosonde.

In 2006, a field experiment was conducted in Hong Kong on the study of aviation applications of a ground-based, multi-channel microwave radiometer [5]. The radiometer gave temperature profiles up to 10 km in the atmosphere nearly continuously. This paper examines the combination of wind profiles from radar wind profilers with the temperature profiles from the radiometer during the field experiment in the calculation of Richardson number and discusses its potential application in the monitoring of turbulence to be encountered by the aircraft.

Section 2 describes the instruments used in this study. Calculation of Richardson number from the instrumental data is discussed in Section 3. Two case studies of the application of Richardson number profiles are then presented, one for a departure runway in Section 4, and another for an arrival runway in Section 5. Conclusions of the paper are drawn in Section 6.

2. Instruments

The vertical wind profiles in this study are provided by the two radar wind profilers near HKIA, namely, Siu Ho Wan and Sha Lo Wan profilers (locations in Figure 1). They belong to boundary-layer type and operate at a frequency of 1299 MHz. The Siu Ho Wan profiler measures in both the low and high modes: the low mode starts at 116 m above ground, with a height increment of 60 m up to about 1500 m; the high mode starts at 260 m above ground, with a height increment of 200 m up to about 6000 m. The Sha Lo Wan profiler operates in the low mode only. Wind profiles are available every 10 minutes. They are obtained by processing the spectrum data from the wind profilers using NCAR Improved Moments Algorithm (NIMA) [6].

The microwave radiometer uses 7 frequencies in the oxygen channel to measure the vertical profile of the temperature. It works in two modes, viz. elevation mode for boundary-layer scanning and zenith mode for measuring in the whole troposphere. The present paper only uses the data of the former mode. It gives temperature data every 50-100 m in the first 1000 m above ground and every 150 – 200 m between 1000 and 2000 m. Temperature profiles were updated every 16 minutes or so during the field experiment in 2006, and more frequent update is possible by changing the scan strategy. The radiometer was located at Siu Ho Wan (Figure 1) during the experiment. Since there was only one radiometer in the study, its data will be used in this study of turbulence both to the east and to the west of HKIA.

3. Calculation of Richardson number

As discussed in [5], the Richardson number Ri is defined as:

$$Ri = \frac{\dfrac{g}{\theta}\dfrac{\partial \theta}{\partial z}}{\left[\left(\dfrac{\partial U}{\partial z}\right)^2 + \left(\dfrac{\partial V}{\partial z}\right)^2\right]} \qquad (1)$$

where g is the acceleration due to gravity, θ the potential temperature, U the east-west component of the wind, V the north-south component of the wind, and z the height. The atmospheric condition is favourable for the occurrence of turbulence when $Ri <$ the critical Richardson number $Rc = 0.25$.

The vertical gradients of wind components are readily calculated from the wind measurements of the low-mode and the high-mode (when available) of the radar wind profiler (in order to get to the height of the pilot turbulence report). To minimize the effect of rapid fluctuations of the wind components on the vertical gradient, the wind profiles have been processed by binomial smoothing for two times. Similar smoothing has also been carried out in the study of Scorer parameter profile for mountain waves [7].

The potential temperature profile and its gradient are computed from the temperature data of the radiometer. Since pressure data above ground are not available, the pressure measurement at the surface and the upper-air temperature data are used to calculate the upper-air pressure values by hydrostatic approximation. Because the temperature profile from the radiometer is generally rather smooth, no further smoothing is applied to the temperature data in the calculation of vertical gradient.

4. Case 1 – 26 February 2006, over a departure runway

Two aircraft departing to the east from the north runway of HKIA, viz. over 07LD runway corridor (Figure 1) reported light to moderate turbulence between 2500 feet (about 760 m) and 5000 feet (about 1500 m) at 20:56 and 21:13 UTC, 26 February 2006. The Richardson number profiles at the closest times of wind and temperature measurements to these two events, namely, 20:39 and 21:09 UTC on that day are evaluated. To examine the evolution of the atmospheric conditions for this case, the profiles at 3-4 hours before and after the events are also considered, namely, at 17:11 UTC, 26 February and 01:03 UTC, 27 February. In this study, *Ri* profiles separated by several hours are considered in order to see the evolution of the chance of occurrence of turbulent airflow due to changes of the mesoscale circulation and the thermodynamic structure.

The potential temperature profiles at the above four times are shown in Figure 2(a). The atmosphere between 500 and 2000 m is generally stable at all times with the potential temperature increasing with height. Among the four times, the profile at 21:09 UTC is slightly more stable between 600 and 1200 m or so with a larger vertical gradient of potential temperature.

The vertical profiles of U and V (without smoothing) are given in Figures 2(b) and (c) respectively. For U component, the profiles at the times of the turbulence events have larger gradient between 600 and 1100 m in comparison to the other two profiles. For V component, the vertical gradients at 20:39 UTC, 26 February between 1000 and 1500 m, and that at 21:09 UTC, 26 February between 900 and 1200 m, are larger than the other two times. Though the wind is generally from the southeast in the lower boundary layer and veers to westerly at about 2000 m for all the four times under consideration, the vertical gradients of the wind components do not stay more or less constant, but show variation with time.

The larger wind component gradients result in the occurrence of *Ri* less than the critical value of 0.25 for some altitudes during the turbulent airflow events over 07LD runway corridor. The Richardson number profiles at the above four times are shown in Figure 2(d). At 20:39 UTC, *Ri* is less than the critical value between 700 and 1300 m. The corresponding altitude range at 21:09 UTC is between 800 and 1100 m over which the wind gradient offsets the slightly more stable boundary layer leading to generally smaller absolute values of *Ri* among the four times. These results are largely consistent with the reported heights of turbulent airflow by the pilots. The *Ri* values at 17:11 UTC, 26 February and 01:03 UTC, 27 February are generally larger than 0.25 over these altitudes.

It is also noted that, at the above four times, the *Ri* could be less than 0.25 below 400 m or so, especially the occurrence of negative *Ri* value. This feature is mainly related to turbulent airflow associated with thermodynamic instability near the ground. However, there were no pilot reports of turbulence at such altitudes closer to the ground.

The appearance of turbulence in the airport area is also examined using the LIDAR data. The conical scan at 4.5-degree elevation angle from the horizon by the first LIDAR (location in Figure 1) is shown in Figure 2(e), depicting the prevalence of southeasterly flow of ~17 m/s within the boundary layer. Unfortunately, the LIDARs and the TDWR did not good signals at the locations of the reported turbulence in this case (at about 4 nautical miles, or 8 km or so, away from the eastern end of the north runway of HKIA). The best available wind data at the event locations come from the Tung Chung Gap vertical-slice scan of he first LIDAR (scan location in Figure 1). The radial velocity data from this scan at 20:41 UTC, 26 February is shown in Figure 2(f). The blockage of the laser beam at about 500 m above mean sea level over Lantau Island in this scan appears to be due to clouds associated with the airflow climbing over the gap from the southeast. Downstream of this airflow, there are a number of waves/vortices between 700 and 1600 m high with the dimensions of several hundred metres in height and a couple of kilometres in the horizontal. Following the argument similar to that in [8], if such features exist downstream of the gap to the west of Sunset Peak – Lin Fa Shan (locations in Figure 1), they may be expected downstream of the gap to the east of these hills as well, which is just near the locations of the turbulence reports. The physical mechanism for turbulence in the present case would require further studies. A possible reason is the shear instability associated with quick veering of the wind with height.

5. Case 2 – 27 February 2006, over an arrival runway

Two aircraft landing at the north runway of HKIA from the west, viz. over 07LA runway corridor (Figure 1), reported moderate turbulence at 11:21 and 11:34 UTC, 27 February 2006 at a height of 800 feet (about 240 m) and 1300 feet (about 390 m) respectively. The Richardson number profile at the wind and temperature measurements closest to these events, namely, 11:20 UTC, 27 February, is considered. Similar to the previous Section, the profiles at 3-4 hours before and after the events, namely, 08:10 and 15:10 UTC, 27 February, are also examined.

The potential temperature profiles at the above three times are shown in Figure 3(a). The gradient of potential temperature between 250 and 400 m at 11:20 UTC is slightly smaller than the other two times over the same range of altitude.

The vertical profiles of U and V (without smoothing) are shown in Figures 3(b) and (c) respectively. With smoothing (not shown), the U profiles at the three times have very small vertical gradients only between 250 and 400 m. The gradients of the smoothed V profiles have larger values for 08:10 and 11:20 UTC over this range of altitude, whereas the corresponding value at 15:10 UTC remains rather small.

Figure 3(d) gives the Richardson number profiles at the three times. The *Ri* at 11:20 UTC below ~600 m is less than 0.25, thus favouring the occurrence of turbulent flow. This is largely consistent with the pilot turbulence reports. Both dynamic and thermodynamic instability could play a role in this case because *Ri* could be positive and negative below ~600 m. On the other hand, the *Ri* values in this range of altitudes at 08:10 and 15:10 UTC are generally larger than 0.25. At the three times, the general veering of the wind with height is the same, namely, the east to southeasterly flow near the surface veers to southerly flow at about 1500 m above mean sea level. However, there are differences in the gradients of the wind components as well as the potential temperature, leading to rather different behaviour of *Ri* values.

TDWR image of radial velocity in the conical scan of 0.6-degree elevation angle from the horizon at 11:21 UTC, 26 February is given in Figure 3(e). For an aircraft landing at 07LA runway, it would first encounter the mountain wake (decelerated flow downstream of the hills on Lantau Island) and then the east to southeasterly jet within the first couple of nautical miles from the runway end. The mountain wake area and the interface between the wake and the jet have rather turbulent flow, as shown by the spectrum width data of TDWR (Figure 3(f)). The spectrum width could get as large as 5.5 – 6.5 m/s. The relationship between TDWR's spectrum width and the cube root of eddy dissipation rate (EDR), the internationally adopted metric of turbulence for aviation application, has not yet been established, but based on correlation equations established in limited case study results [4], the above range of spectrum width value corresponds to $EDR^{1/3}$ much larger than 0.3 $m^{2/3}s^{-1}$, i.e. moderate turbulence or above. The EDR profiles obtained by the glide path scans of the first LIDAR (not shown) following the method in [3] also have maximum values of about 0.28 $m^{2/3}s^{-1}$. Both TDWR and LIDAR measurements support the occurrence of turbulent airflow over 07LA runway corridor.

6. Conclusions

The aviation application of Richardson number profile calculated from remote sensing data, namely, monitoring of turbulence encountered by the aircraft, is discussed in this paper through two examples. The profile is determined from wind data from radar wind profilers as well as temperature data from a ground-based microwave radiometer. The continuous availability of the measurements from these instruments enables the frequent update of the profile. Besides the Richardson number profiles near the times of the turbulence reports, those at 3-4 hours before and after the events are also considered to examine the evolution of the atmosphere in the mesoscale. It turns out that, while the general veering of the wind with height remains about the same over many hours, the vertical gradients of the wind components could be quite different, leading to different behaviour of the *Ri* values. The changes in the vertical gradient of potential temperature also contribute to the evolution of Richardson number profile. In general, the height ranges over which *Ri* less than the critical value of 0.25 are generally consistent with the turbulence reports from the pilots.

The present paper only considers the Richardson number profile over a limited period of time due to the short-term availability of the microwave radiometer in the field experiment in 2006. HKO has recently installed a permanent unit of radiometer at HKIA. With this installation, more data will become available to study the operational application of Richardson number profile in the monitoring of turbulence.

References

[1] Chan PW and Chan ST 2004: Performance of eddy dissipation rate estimates from wind profilers in turbulence detection. *11th Conference on Aviation, Range, and Aerospace Meteorology*, American Meteorological Society, Hyannis, MA, USA, 4-8 October 2004

[2] Kwong KM and Chan PW 2007: LIDAR-based turbulence intensity calculation along glide paths. *14th Coherent Laser Radar Conference*, Snowmass, Colorado, USA, 8-13 July 2007

[3] Chan PW, Li CM and Wong KH 2007: Use of spectrum width data from the Terminal Doppler Weather Radar (TDWR) for monitoring turbulence intensity: case studies. *First Asian Conference on Radar in Meteorology and Hydrology*, Hubei, China, 22-24 October 2007

[4] Stull RB 1988: *An Introduction to Boundary Layer Meteorology*, Kluwer Academic Publishers, 670 pp.

[5] Chan PW, Wu KC and Shun CM 2006: Applications of a ground-based microwave radiometer in aviation weather forecasting. *13th International Symposium for the Advancement of Boundary Layer Remote Sensing*, Garmisch-Partenkirchen, Germany, 18-20 July 2006

[6] Morse CS, Goodrich RK and Cornman LB 2002: The NIMA method for improved moment estimation from Doppler spectra. *J. Atmos. Oceanic Technol.*, **19**, 274–295.

[7] Chan PW 2007: Atmospheric waves associated with a valley of Lantau Island: observation, theory and numerical simulation. *12th Conference on Mesoscale Processes*, American Meteorological Society, Waterville Valley, New Hampshire, USA, 6-9 August 2007

[8] Banta RM and co-authors, 2006: *Detection and diagnosis of wind shear and turbulence using Doppler LIDAR at Hong Kong International Airport.* ESRL/NOAA and HKO, 130 pp.

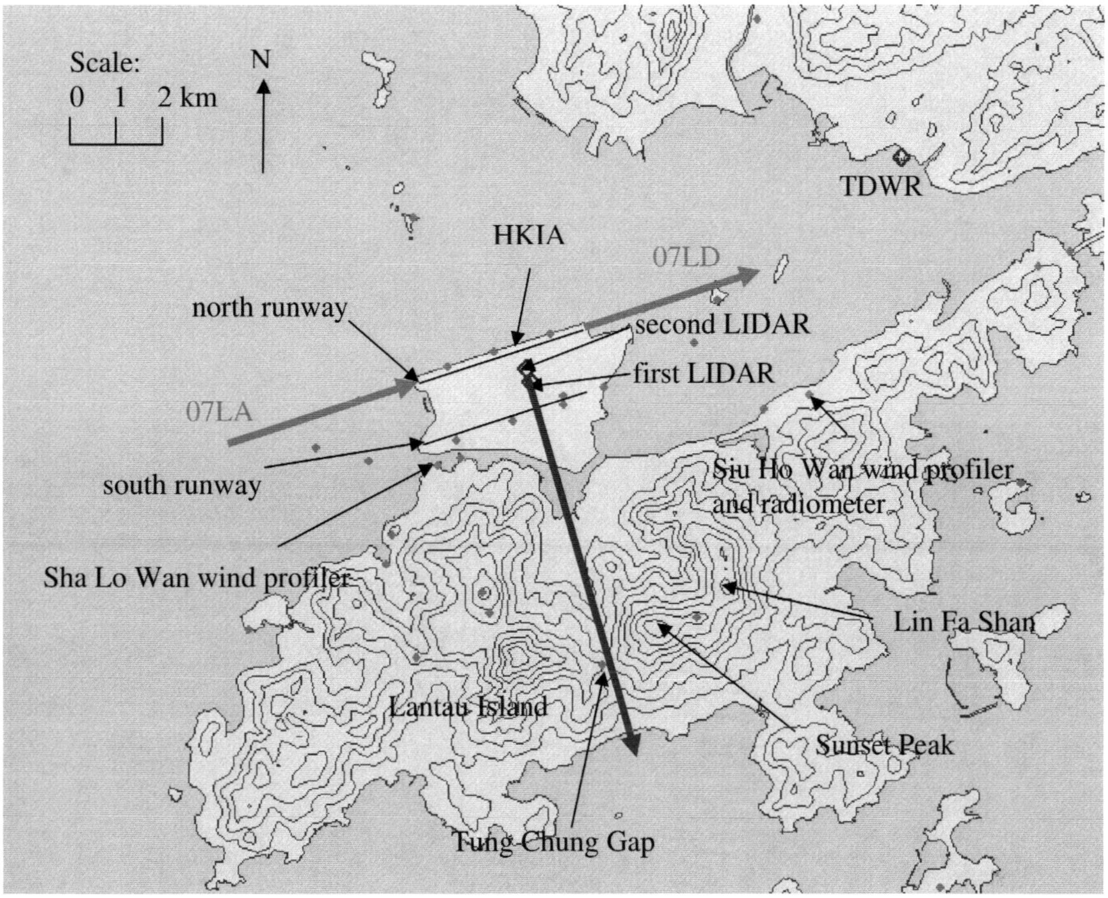

Figure 1. Geographical situation near HKIA. Height contours are in 100 m. Locations of the meteorological instruments considered in this paper are indicated. Red dots are the surface anemometers and weather buoys. The two runway corridors of HKIA discussed in the main text are depicted by red arrows, showing the directions of aircraft movements. Blue arrow is the direction of Tung Chung Gap vertical slice scan of the first LIDAR.

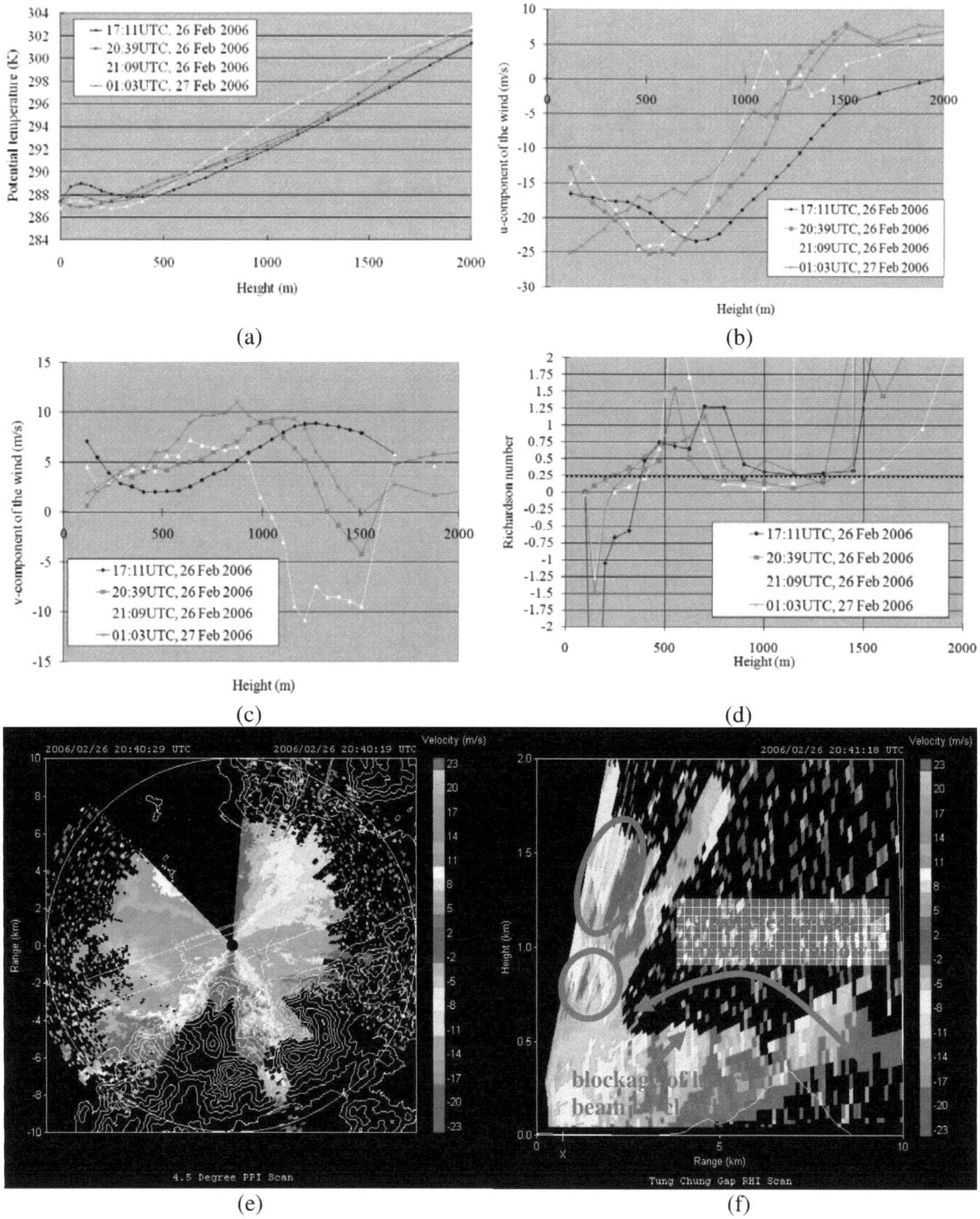

Figure 2. For the case study on 26 February 2006: (a) potential temperature profiles, (b) profiles of east-west (*U*) component of the wind, (c) profiles of north-south (*V*) component of the wind, (d) Richardson number profiles, (e) 4.5-degree conical scan of radial velocity by the first LIDAR, with the velocity scale on the right hand side; the white "skeletons" from the runways are the extended centrelines with tick marks at every nautical mile up to 3 nautical miles away, and (f) Tung Chung Gap vertical slice scan of the first LIDAR, with the velocity scale on the right hand side; the waves/vortices

discussed in the main text are enclosed in red ellipses.

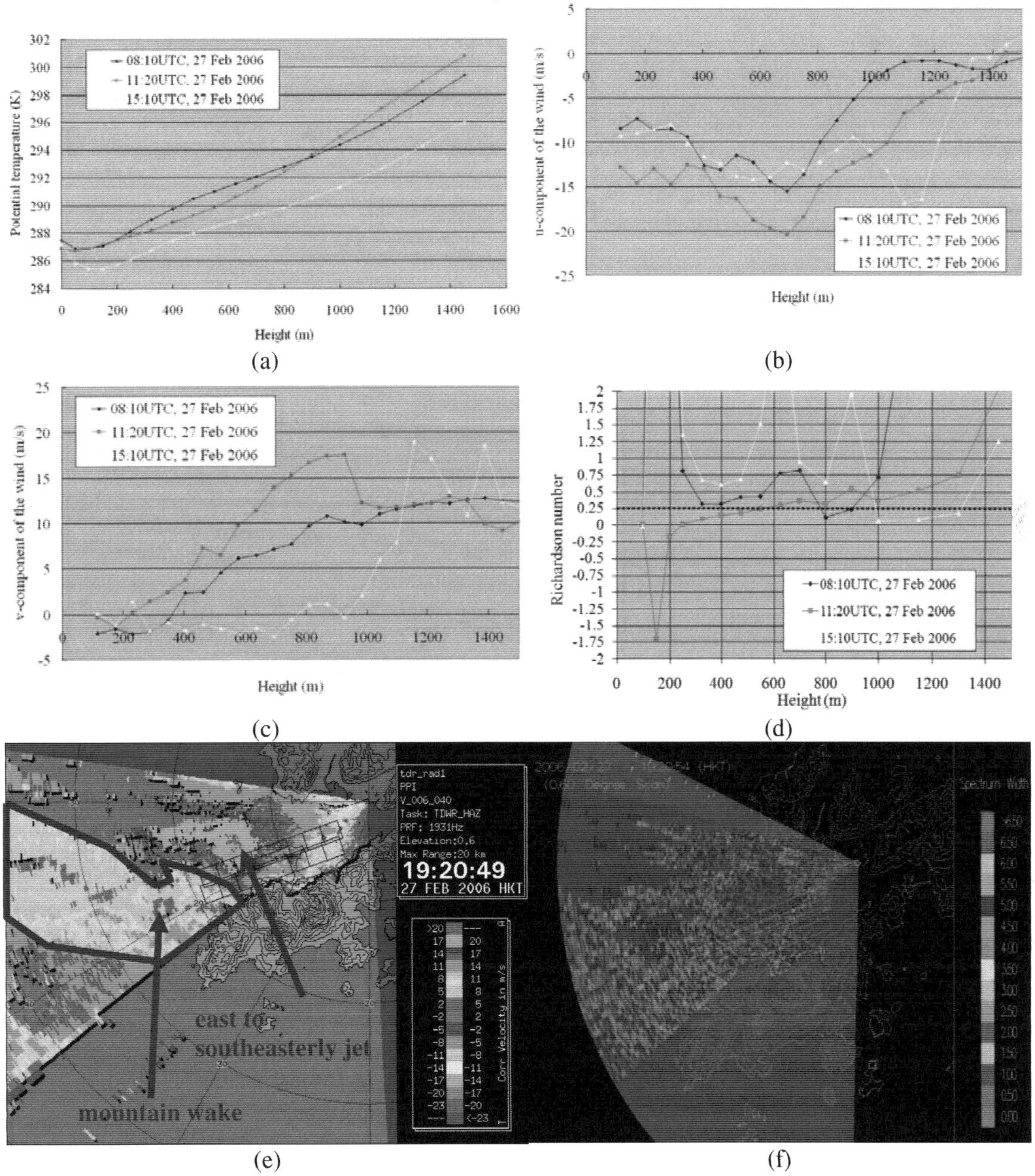

Figure 3. For the case study on 27 February 2006: (a) to (d) same as those in Figure 2, (e) TDWR image of radial velocity at the surveillance scan of 0.6-degree elevation angle from the horizon, (f) the spectrum width image of TDWR at the time corresponding to (e).

IOP Conf. Series: Earth and Environmental Science 1 (2008) 012044 doi:10.1088/1755-1307/1/1/012044

The Development of a Transponder Based Technique for the Acoustic Calibration of SODARs

B. Piper[1] **and S. von Hünerbein**

Acoustic Research Centre, G11 Newton Building, University of Salford, Salford, M5 4WT

E-mail: b.j.piper@pgr.salford.ac.uk

Abstract. The accuracy and comparability of wind measurements by different SODARs is an unresolved issue. A technique is being designed using a laptop based transponder system which can simulate the Doppler shifted return echo signals generated by certain meteorological models. This system uses real-time analysis of the SODAR's output combined with the generation of return signals based upon both the theory of atmospheric absorption and turbulent scattering. This system has been tested with a METEK PCS.2000/24/MF in a controlled environment in various configurations. The results obtained show that this technique is effective in such an environment. The next step in the development of this system is to eliminate influencing factors such as the real atmospheric echo and background noise so that it can be used 'in situ'.

1. Introduction

SODARs are useful tools for examining meteorological trends but their accuracy is not known with different SODARs giving different but comparable results in the same location. SODAR measurements are often compared to those from anemometers to examine their accuracy. This method can only give a partial calibration and the differences between making a point measurement and a volume measurement are likely to cause errors. An independent traceable technique is required. A method of auditing the performance of SODAR instrument based upon an Acoustic Pulse Transponder (APT) system is shown in Baxter [1-4]. This technique is the foundation for the transponder development discussed in this paper where a more complete calibration technique is aimed for.

.

2. Transponder Design

The principle idea behind the development of a transponder based calibration technique is that by simulating the effects of given atmospheric conditions on a pulse of sound it is possible to create a realistic return signal and thus allow a comparison between the SODAR's measured data and the simulated atmosphere. In order to create a realistic return signal an atmospheric model needs to be created which takes into account the effects of wind velocity, beam forming, air absorption, scattering and temperature.

[1] To whom any correspondence should be addressed.

IOP Conf. Series: Earth and Environmental Science 1 (2008) 012044 doi:10.1088/1755-1307/1/1/012044

2.1. Atmospheric Models

Simple atmospheric models have been created which have constant or linearly increasing wind velocity with height. In these models humidity, scattering cross section and temperature are assumed to be constant. As progress is made a more complex model will be created in which these variables alter with height and the effects of ambient noise, fixed echoes and the illumination of the SODAR's receiver are included.

2.2. Prototype set-up and Operation

The current transponder prototype consists of a soundcard attached to a laptop running MatLab. One horn loudspeaker is connected to the soundcard input. This horn is positioned very close to the SODAR's output. Its purpose is to receive the signal emitted by the SODAR instrument. The second horn loudspeaker is connected directly to the soundcard output and is positioned facing the SODAR instrument from a distance which allows maximum illumination of the SODAR's receivers. Using MatLab the signals emitted by the SODAR instrument are recorded and analysed. Return signals are created using information from the input analysis and variables stored in the atmospheric model. These are played back to the SODAR instrument when triggered by an input. As the transponder is developed a pulse will be created individually for each SODAR output and played back in real-time thus accounting for any changes in the input signal. This requires a very quick programming speed and is yet to be achieved.

2.3. Signal Generation

Two techniques of signal generation have been used in the transponder system. These are Sine Wave Synthesis (SWS) and Single Sideband (SSB) modulation. For the SWS technique a return signal is created by generating a sine wave that has a length matching that total height range considered. The frequency is assigned according to FFT analysis of the input signal, where the peak frequency of the pulse is identified, and the required Doppler shift according to the atmospheric model. For the SSB modulation technique the original signal is copied and shifted in time so that one copy corresponds to each range gate. The copies are all summed and then frequency modulation is applied in the time domain by multiplying the summed signal by a cosine wave, which has a frequency of the required Doppler shift, and summing this with the imaginary part of the Hilbert transform of the signal multiplied by a sine wave, which has a frequency of the required Doppler shift. This is shown figure 2.1 from Oppenheim and Schafer [5].

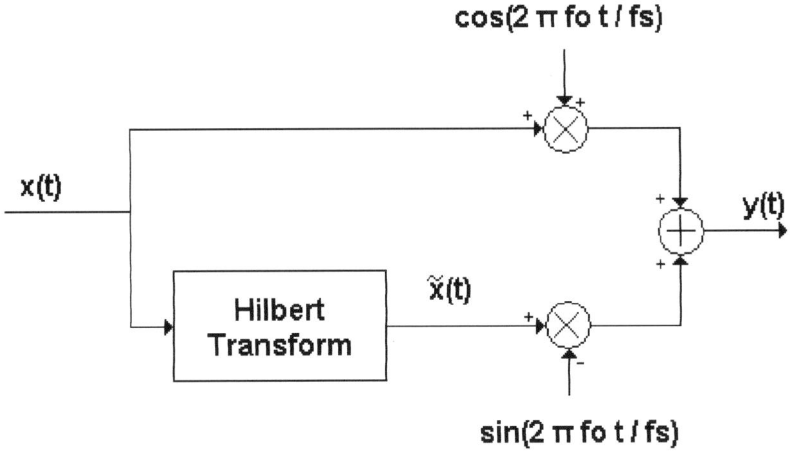

Figure.2.1 – Flow diagram for Single SideBand modulation of a Signal [5]

The reason for using this technique rather than simply multiplying by a cosine carrier wave is that when a signal is multiplied by a cosine carrier two signals are created; one at the desired frequency and one a reflection of this about the original signal. Using SSB modulation eliminates the reflected wave and maintains the energy ratio. This effect is shown in figure 2.2 for a -50 Hz frequency shift.

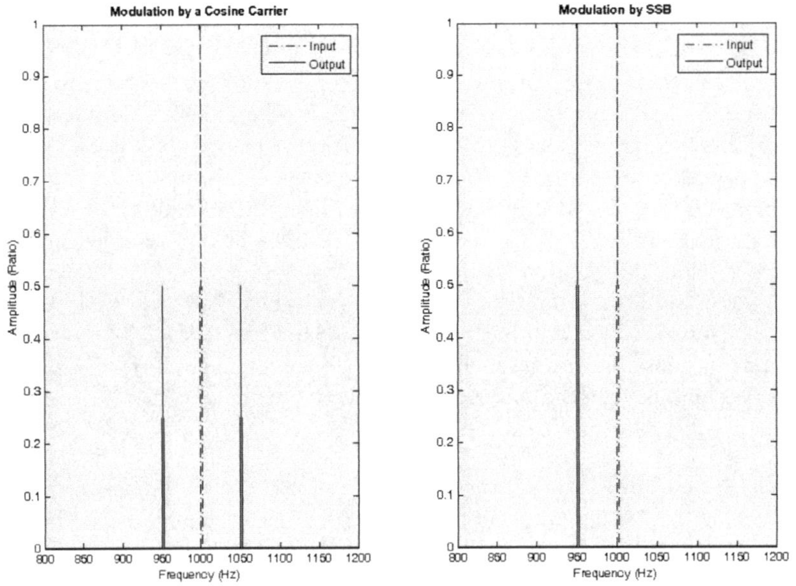

Figure 2.2 – Demonstration of a 50Hz modulation using a Cosine Carrier and SSB

3. Testing

Two periods of testing have been carried out using the calibration transponder in a semi-anechoic chamber; One using the SWS techniques and one using SSB modulation techniques. The physical set-up for both is shown in figure 3.1. In each period of testing the two horizontal components of the wind vector, U & V, were increased in increments of 1 ms^{-1} between -10 and 10 ms^{-1} and the SODAR's responses were recorded. In each case the vertical component, W, was set to 0 ms^{-1} and the SODAR's response for this component was noted as an error if more than +/- 0.01 ms^{-1}.

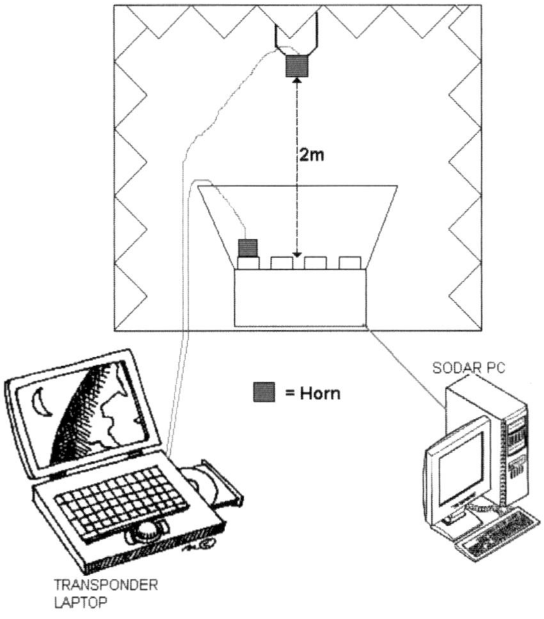

Figure 3.1 – The Physical setup of the Calibration Transponder whilst testing in a semi-anechoic chamber.

3.1. SWS Results

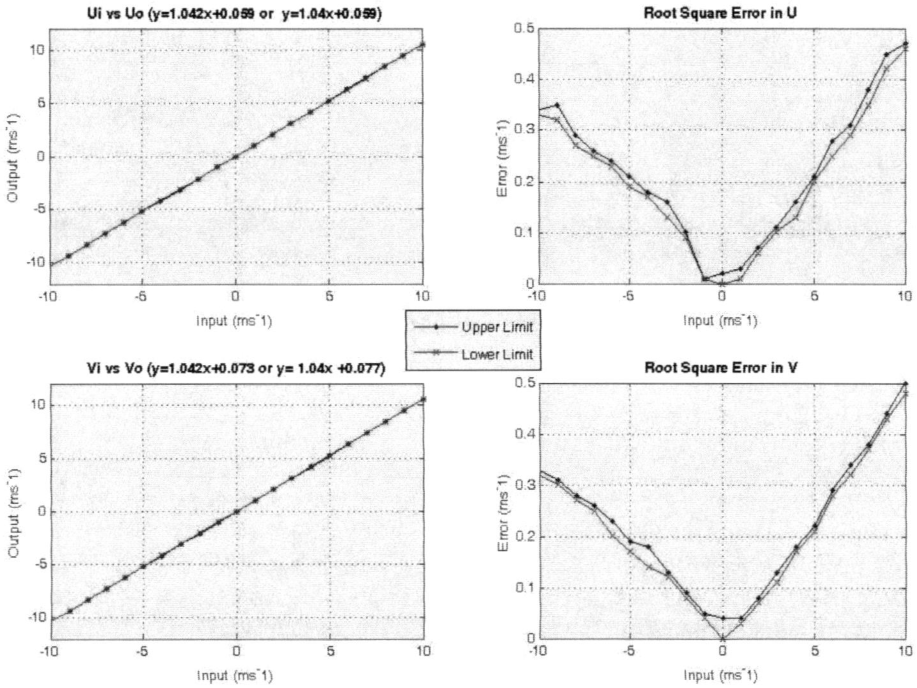

Figure 3.2 – Results from calibration testing using Sine Wave Synthesis Techniques

The results shown in figure 3.2 prove that using the SWS approach it is possible to have errors of less than 5%. When using this technique the SODAR instrument often gives an error code relating to the narrow spectral width of the return echo. This problem arises because the formants of the input signal and spectral spreading due to atmospheric effects are ignored in the synthesis process. This method is limited because it relies on the assumption that the SODAR's output is stable and predictable.

3.2. SSB Results

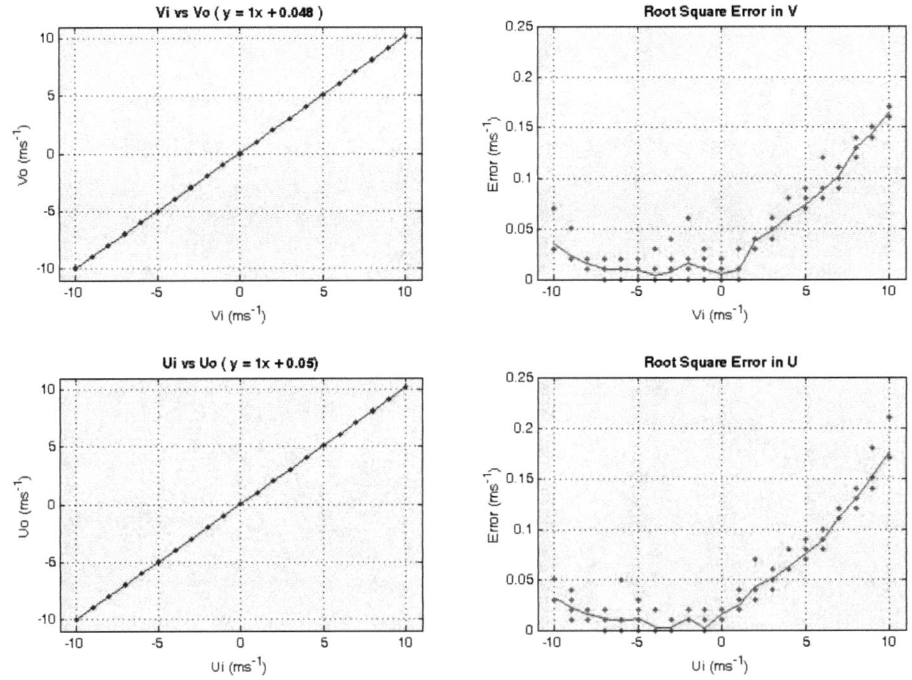

Figure 3.3 – Results from calibration testing using Single Sideband Modulation Techniques

The results shown in fig 3.3 are similar to those found using the SWS technique however there are two notable improvements. Firstly the linear correlation line is y = 1x plus a small offset. Secondly the errors are much smaller. This proves that this signal generation technique is a more accurate approach. It is noticeable in both sets of results that the errors are generally larger for positive input speeds. This issue is unresolved but it's likely cause is incorrect illumination of the SODAR array by the transponder.

4. Further Development Plan

Several improvements need to be made to the transponder in order to accomplish reliable SODAR calibrations. The complexity of the atmospheric model needs to be increased to include temperature, humidity and scattering cross-section changes as well as realistic wind models such as an 'Ekman Spiral' and a several different shear effects. The programming processes need to be quickened to allow for each return pulse to correspond to its trigger rather than being based on a set off pre-recorded pulses. Methods for dealing with in-situ problems such as the real return echo and background noise need to be developed and adaptability must be created so that the transponder can work with all types of SODAR.

IOP Conf. Series: Earth and Environmental Science **1** (2008) 012044 doi:10.1088/1755-1307/1/1/012044

5. Conclusion

A transponder based technique for the acoustic calibration of SODARs has been created to such a level that its potential is clear. Two methods of signal generation have been trialed with the SSB modulation technique showing greater accuracy in results in comparison to the SWS technique whilst using details of the actual output of the SODAR and allowing for more manipulation in terms of the spectral width of the return. In order to move forward with this transponder system a method for processing returns in real-time needs to be developed as well as increasing the complexity models and methods for implementing the transponder 'in-situ' for a range of SODAR types need to be created.

References

[1] Baxter, R. A. (1994). "Development of a Universal Acoustic Pulse Transponding System for Performance Auditing of SODARs." Engineering-Science, Inc.

[2] Baxter, R. A. (1994). "*Performance of SODAR Audits Under the New USEPA Guidance.*" Engineering-Science, Inc.

[3] Baxter, R. A. (1995). "Performance Auditing of SODARs Under the New USEPA Guidance – Applications of a New Instrument to Quantitatively Assess SODAR Performance." Engineering-Science, Inc.

[4] Baxter, R. A. (1995). "Quality Assurance of Data Produced from SODARs -- Application of a New Instrument for Field Performance Auditing." Engineering-Science, Inc.

[5] Oppenheim, A. V. & Schafer, R. W. (1999). "*Discrete-Time Signal Processing (2nd Edition)*". (Upper Saddle River: Prentice Hall International) p798.

IOP Conf. Series: Earth and Environmental Science 1 (2008) 012045 doi:10.1088/1755-1307/1/1/012045

Sodar Study of the Characteristics of Presumable Big Toroid Structures Above an Overheated Surface

Kramar Valery

A.M. Obukhov Institute of Atmospheric Physics, Russian Academy of Sciences, 3 Pyzhevskii lane, Moscow 109017, Russia

vkramar@ifaran.ru

Abstract. A set of sodars allowed to investigate the spatial variability of meso-scale organized structures under conditions of strong convection and weak wind. The distribution of velocity within the structures and the structure typical sizes have been determined as a result of processing of experimental data.

1. Introduction

Large-scale vortex structures arising over overheated surface in the atmospheric boundary layer at relatively low average wind velocities are described in different researches [1,2]. Two forms of convective motion being realized are differentiated in such works. One of the forms is a cellular convection when vortex coherent structures are similar to toroids. The other one is roll-like structures when the motion are realized in the form of long vortex rolls. Showing some different features the structures nevertheless have the following similar behavior. Toroids usually exist at the boundary layer when average wind velocity is less than 5m/s. Roll structures are observed with the wind velocity growth. The rolls are destroyed when wind velocity becomes greater than approximately 10m/s. Both structures have kilometers in their cross section, and the length of roll can be up to hundreds kilometers [1,2].

The results of the research [1] may be used to determine the following, important for the present work, characteristics of vortex convective structures. Calculated vertical velocities in a structure are significantly less than 1m/s. In accordance with [1] typical altitudes for which the velocity in a structure is mainly vertical are from 150-200m to 400-500m. Horizontal directions of velocity of the air flows generated by large-scale vortex motion prevail lower and upper of the indicated levels in the ABL.

The present work is devoted to some results of the field experiment. One of the goals was to research the inner structure of large-scale convective vortex motions that are formed in the atmospheric boundary layer over the overheated surface. Primary data were collected with the system of four spatially distributed synchronized three-components Doppler mono-static sodars. The research was implemented in the frame of a complex field experiment 'Har-Gzyr 2007'. The experiment was devoted to sub-micron aerosol particles transfer into the upper layers of ABL study. The experiment took place in the desert region at Komsomolskii township, Kalmykia Republic, at the southern part of the national park 'Black Sands' in July-August 2007. Description, approaches as well as some results of the experiment concerning transfer of sub-micron aerosol are given in [3].

2. Instrumentation and site

Figure 1 shows the area on which the instruments (points 1-3) where disposed. The schematic image is composed with the help of the Google Earth system.

Figure 1. Mini-sodars location (points 1-3) in the field experiment 'Har-Gzyr 2007' (July-August 2007). Long-wavelength sodar is additionally disposed in the point 1. Mast (15m height) with two ultra-sonic thermometers-anemometers is additionally located in the point 2. Distances on the figure: 3-2 is 3.2km; 1-3 is 1.4km and 1-2 is 3.5km.

Technique of multi-frequency sounding was used in the mini-sodars. Sounding impulses emitted by all three antennas of mini-sodar simultaneously. This methodology made possible both to increase measurement accuracy of a single 'shot' and to get three times larger samples for the following statistical analysis. Long-wavelength three-component sodar emitted multi-frequency sounding impulses sequentially. All sodars made in Obukhov IAPh are deeply improved versions of the Latan-3 unit. Detailed description of the instruments and their characteristics can be found in [4]. Ultrasonic thermometers-anemometers registered the heat fluxes from the ground up to 150-250W/m^2.

3. Data processing

One of the characteristics showing interdependence and mutual influence of motion in different points of a structure is the autocorrelation coefficient. Figure 2 shows autocorrelation coefficient for a subset of data of measurements in point 2 July 28, 2007, 16-17 LMT. The fragment was selected visually according the dependence of vertical wind velocity on time.

Let us compare the figure with the correlation coefficient of the average vertical velocities with the toroid center velocities as a function of distance from the structure center as calculated using large eddy simulations (LES) data in the research [1]. The estimation of the typical dimensions of the structure may be got by measuring the distance from the center of the structure to the first minimum and the second maximum of the correlation coefficient values. The first distance may be taken as ¼ of the width of the structure. For the figure in [1] the dimension is approximately 750m. Accordingly, the whole size of the structure is near 3km. For the figure 2 of the present paper the ¼ of the width is 550s and accordingly the typical size is 2200s. This value is the estimation of the time while the observed phenomenon runs along the point of observation the distance equal to its own size. Taking the dimension of the structure being 3km, (as mentioned above) one can calculate the velocity of the structure moving. It will be about 1.5m/s.

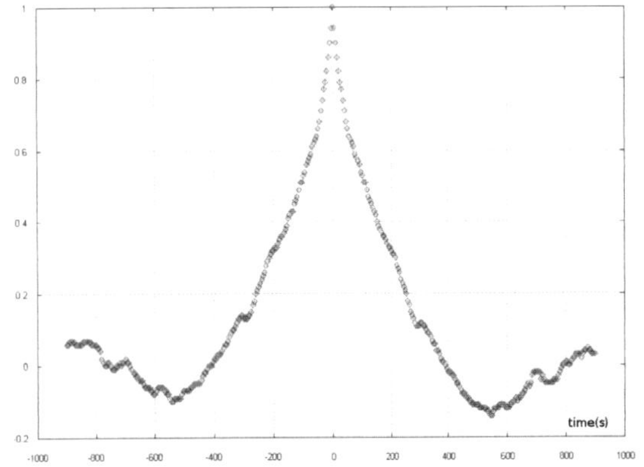

Figure 2. Autocorrelation coefficient of the vertical velocity. Point 2 for 100m level July 28, 16:10 LMT. One hour period is taken as a length of the pattern.

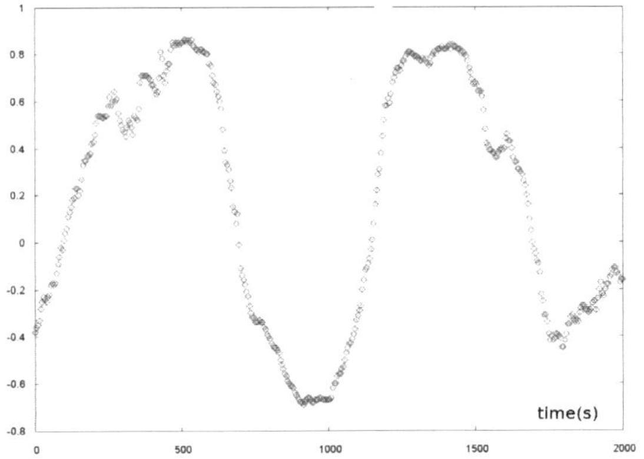

Figure 3. Cross-correlation coefficient of vertical velocity. Points 1-2 for 100m level, July 28, 16:10 LMT. Ten minutes period is taken as a length of the pattern.

Fig. 3 shows the phenomenon from the other point of view. Cross-correlation coefficient of the taken ten minutes period for the point 1 with the data of the instrument disposed in the point 2 is given here. It may be proposed that the behavior of values of vertical velocities in point 1 and 2 is similar with time shift about 1400s. Taking into account the distance 1-2 being 3.5km, one can calculate the possible velocity of the phenomenon horizontal moving. From such estimation it will be about 2.5m/s.

4. Discussion

Though rather disputable assumptions have been made to process the data of the experiment the results show extreme complexity of the phenomenon. All theoretical models including LES strongly simplify the phenomenon. Exact formal definitions of a set of structure parameters should be introduced to use them in data processing of the experiment. A big amount of experimental data needs to be processed comprehensively.

Acknowledgements

The work was supported by Russian Fund of Basic Researches through grants 04-05-64167, 06-05-08086, 07-05-00521, 08-05-00671, 06-05-65216. The author is grateful to Dr. R.Kouznetsov for discussion and critical remarks.

References

[1]Lappen C-L and Randall D 2005 Using Idealized Coherent Structures to Parametrize Momentum Fluxes in a PBL Mass-Flux Model *Journal of the Atmospheric Sciences* **62** 2829-2846

[2]Mikhailova L A and Ordanovich A E 1991 Coherent Structures in the Atmospheric Boundary Layer (review) *Izv. Acad. Sci. USSR. Atmos. Oceanic Phys. 27 593-613*

[3]Kramar V and Granberg I 2008 Research of the Arid Aerosol Carrying Out With a Help of a Set of Sodars *This collection*

[4]Kouznetsov R 2008 Multi-frequency acoustic sounding of the ABL with dish antenna *This collection*

Study of wind variability over the Moscow city by sodar

V P Yushkov

Faculty of Physics, Lomonosov Moscow State University, Lenin Hills, Moscow, 119992, Russia

E-mail: yushkov@phys.msu.ru

Abstract. We used sodar data to obtain spatial, diurnal and seasonal variability of wind speed variances. Comparison of measurements at two sites in Moscow megalopolis and at the rural site (45 km from Moscow) was carried out. A good agreement between sodar and in-situ measurements by ultrasonic anemometer in rural and urban observations was obtained. Variances of radial velocities measured at sodar inclined antennae were compared with vertical wind variance as well as with wind components variances by sonic data. Experience in measuring of meso-scale wind variability by averaged on short time intervals data is demonstrated. Measurement of other statistical characteristics like vertical wind variances is discussed.

1. Introduction

Remote sensing instruments for monitoring of atmospheric boundary layer (ABL) allow collecting detailed statistics of wind variability in ABL for different heights, time intervals, seasons and observing sites [1]. This paper presents such statistics from continuous sodar measurements in Moscow megalopolis (more then 10 millions inhabitants).

To estimate an impact of urban environment on atmospheric pollutants transport, heat and momentum fluxes not only detailed measurements in short field experiments but statistics for long time are also required. The feature of this statistics is meso-scale variability, beyond inertial interval. Till now processes between synoptic scale (tens kilometres and tens minutes) and inertial interval of isotropic and homogeneous turbulence (centimetres and seconds) is researched insufficiently, statistics at this scale is fragmentary in spite of numerous field experiments and routine measurements.

Experimental studies of atmospheric turbulence usually use high-frequency measurements in steady and homogeneous conditions. In field experiments, if it is possible, a flat terrain is used and data are divided by thermal stratification. So, in urban environment and if monitoring instruments have limited ability, adaptation of theoretical models to practice require the experimental estimation of measurement correctness. Measurements which are carried out on buildings, near buildings, above buildings are not allowed by meteorological requirements so urban environment impact has to be investigated in experiments.

2. Instrumentation and measurement sites

Our measurements are carried out by PC-based mono-static three-component sodars Latan-3 developed at the Obukhov Institute of Atmospheric Physics (IAPh) for scientific purposes [2]. Detailed description of measuring procedure and accuracy can be found in [3, 2]. Two sodars

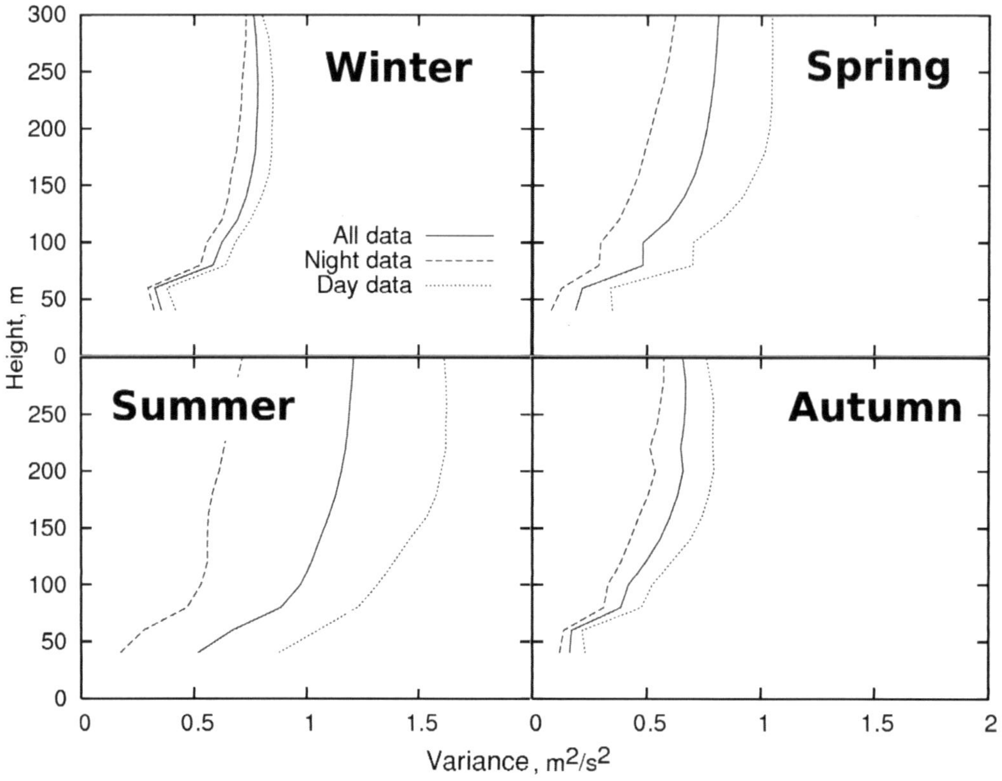

Figure 1. Mean seasonal profiles of vertical velocity variance (σ_w^2) in centre of Moscow for measurements in 2007.

are synchronously operated at two sites: at the IAPh building in Moscow downtown and at Lomonosov Moscow State University (MSU) Physics Faculty building in the south-west district of Moscow. Continuous measurements are carried out since April 2005. Sodar in downtown is placed at the height of 13 m agl (about 195 m asl), in MSU sodar placed at 40 m agl. Distance between two sites is about 7 km. Mean height of Lenin Hills above Moscow River is about 70 m, however our measurements show that it has a matter because of layering of atmospheric flows. Common difference is same as difference in wind profiles and surface inversion thickness determined from echograms in our and earlier experiments [4]. Simultaneous measurements were carried out in rural area at Zvenigorod Scientific Station (ZSS of IAPh) in 45 km from Moscow in June 2005.

Height range was chosen from 40 to 300 meters for data averaging and comparability in two urban and rural locations (at first level buildings influence is considerable). Vertical resolution of LATAN-3 is 20 meters. Carrying frequency is 1700 Hz, sounding frequency is 4x3 pulses per minute.

Variances are calculated over 30 min intervals then means and distributions are computed for a long time (month, season, year). Daytime and night-time averaged data are also calculated. Night interval is from 0 to 6 hour of local standard time (LST) and day interval is from 10 to 16 hour. Data availability in series and at different height intervals is calculated as well as data control of rain and noise impact is carried out. It allows to control noisy environment and to make measuring in heavy traffic and rain conditions [3].

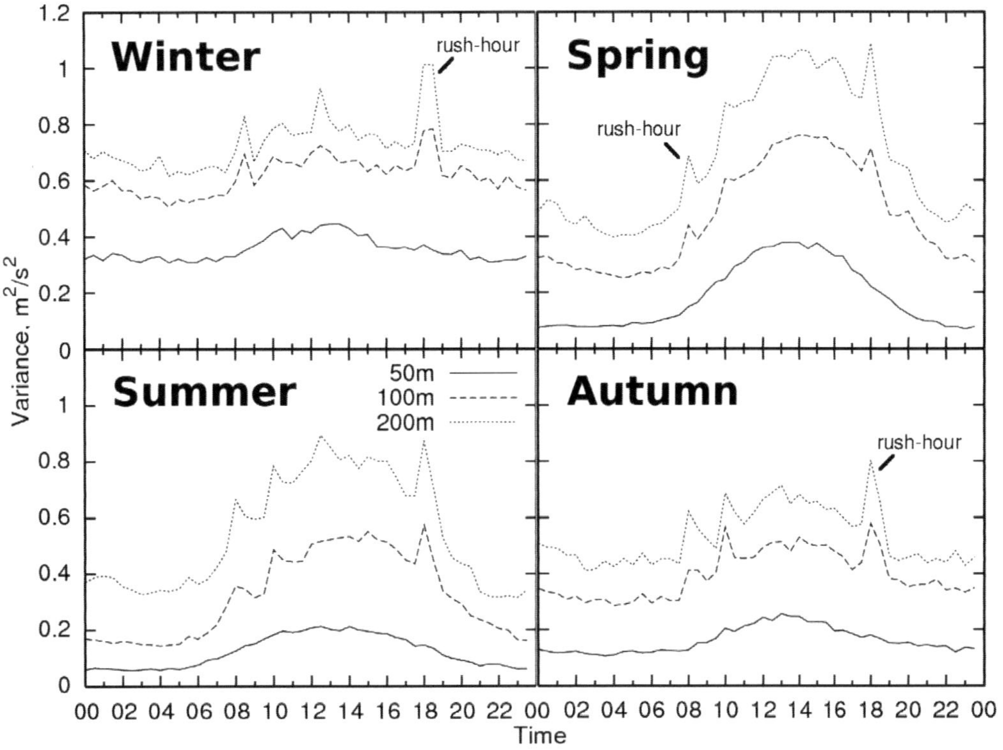

Figure 2. Averaged diurnal course of σ_w^2 at 3 heights: 50, 100 and 200 m for each season. Moscow downtown, 2007.

3. Results

3.1. Vertical velocity variance

Vertical velocity variance (σ_w^2) is most accurately measured by sodar characteristics of turbulent pulsations, however measurements of σ_w^2 demand special equipment (not all commercial sodars

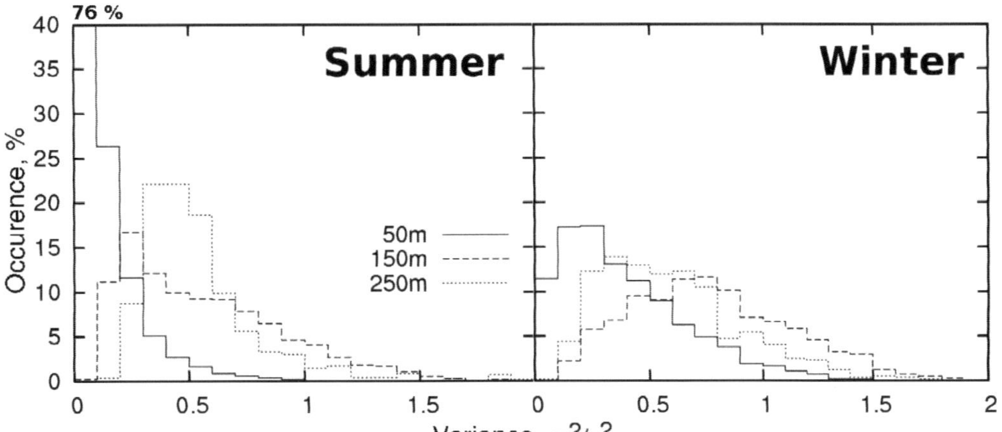

Figure 3. Distribution of σ_w^2 in summer and winter at 50, 150 and 250 m heights. Moscow downtown, 2007.

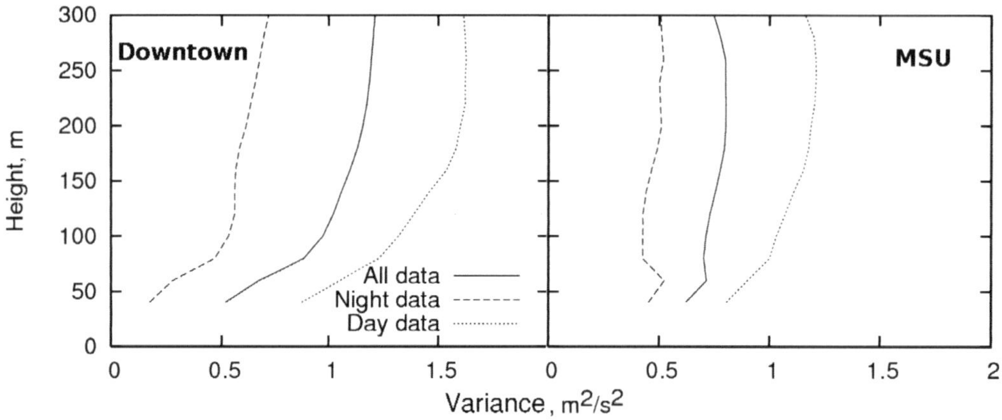

Figure 4. Mean summer profiles of σ_w^2 at IAPh and MSU sites. Summer 2007 (June- August).

can measure this value) and careful control of results [5]. Fig.1 shows seasonal means of σ_w^2 profiles for each season in Moscow downtown. Data for 2007 year (December 2006-November 2007) averaged by 3 months are presented in this figure. Both all-data means and night-time – daytime data are displayed.

Main feature of variance profiles is considerable increasing of diurnal course in summer time. Minimum of diurnal course is observed in winter and autumn due to continuous cloudiness. Another feature of variance profiles is a weak variability with height above 150 m, i.e. main increase of σ_w^2 is observed within sodar range and below. Similar results were demonstrated in Hanover measurements (Fig.5 in [6]).

Averaged diurnal course for each season (about 90 values for seasonal means per one year) and its annual variability is presented in Fig.2. In summer time diurnal amplitude of σ_w^2 in order of magnitude is larger then in winter.

Distribution of wind velocity variance shows a synoptic variability of stratification. Fig. 3 demonstrates probability distributions (histograms) of σ_w^2 at three heights (chosen by profiles varying): 50, 100 and 200 m. Histograms display occurrence of strong and weak stable conditions in night-time as well as strong and weak unstable stratification in daytime. Increase of wind variance with height and stabilization of its growing above 150 m are seen. Thus, statistical generalization presents common pattern of turbulence distribution in urban air.

3.2. Comparison of urban turbulence at two measurement sites

Heterogeneity of turbulence within urban megalopolis is difficult to study, so a possibility of result extrapolation upon whole urban area demands careful investigation because of big city orography and "heat island" deform both mean flows and turbulence characteristics. Comparing of mean profiles of σ_w^2 is shown in Fig. 4. Fig.5 demonstrates mean diurnal course of σ_w^2 by measurements in Moscow downtown and at MSU site. Heights are shown from sodar placement level so height difference of IAPh and Physics Faculty buildings must be taken into account as well as MSU site elevation above Moscow River.

An example of week course in Fig. 6 demonstrates this difference: at the left panel σ_w^2 course at one height is displayed and at right panel - at different heights (with 100 m spread). Taking this difference into account accordance of diurnal course is good.

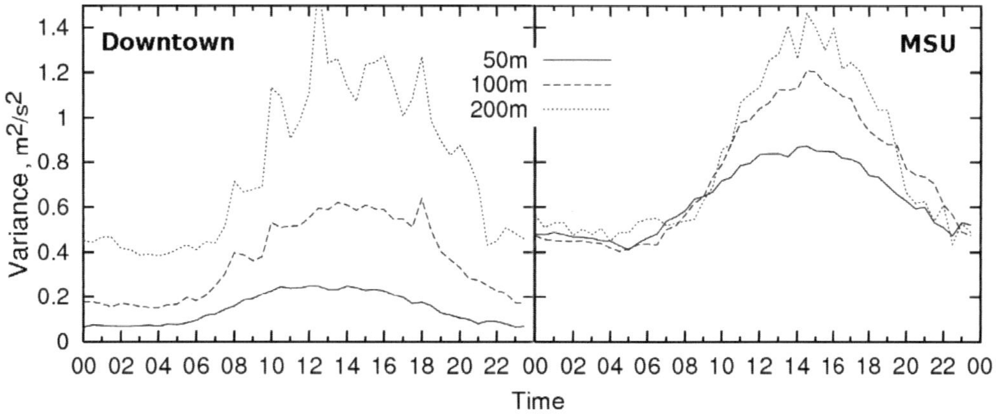

Figure 5. Averaged diurnal course of σ_w^2 at 3 heights: 50, 150, and 250 at two urban sites. Summer 2007.

3.3. Comparison of urban and rural measurements

Comparison of wind velocity variance at urban and rural area plays a major role in estimation of influence of urban heating and urban roughness on turbulence in ABL. Fig.7 shows comparison of monthly averaged σ_w^2 profiles (July 2005). Fig. 8 demonstrates averaged diurnal course at the same heights: 50,100 and 200 m.

Similarity of diurnal course in the centre of Moscow and above rural area is surprised as urban heat flux and urban roughness are larger considerably. Example of week course of σ_w^2 at 100 m at two sites illustrates this phenomenon in Fig. 9.

Probably, wind variances and momentum flux as well as heat flux are more independent then in analytical relations due to synoptic and meso-scale variability.

3.4. Comparison of in-situ and sodar measurements in urban area

Measurements of wind velocity variance by sodar LATAN-3 with in-situ ones were compared at MSU site to estimate wind and turbulence perturbations due to buildings in urban area.

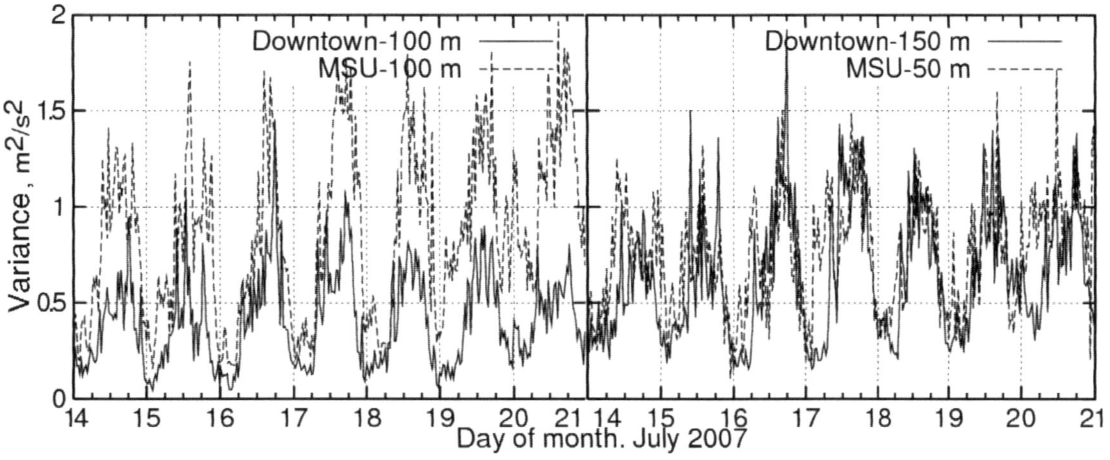

Figure 6. Week example of σ_w^2 at downtown and MSU site at one (left) and at different (right) heights.

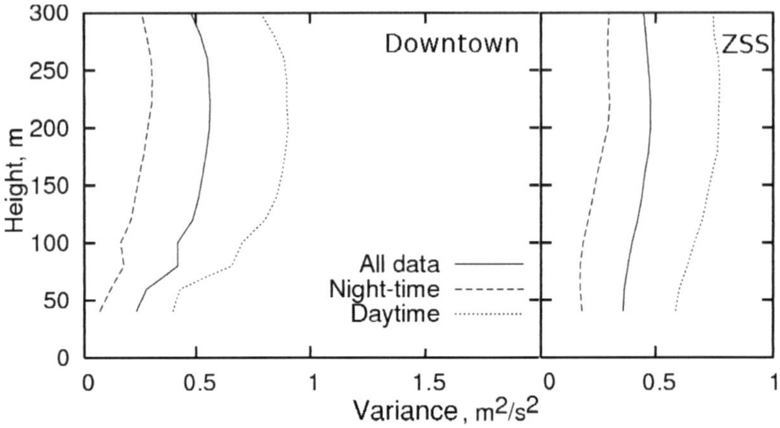

Figure 7. Monthly mean profiles of σ_w^2 in Moscow downtown and at ZSS rural site for July 2005.

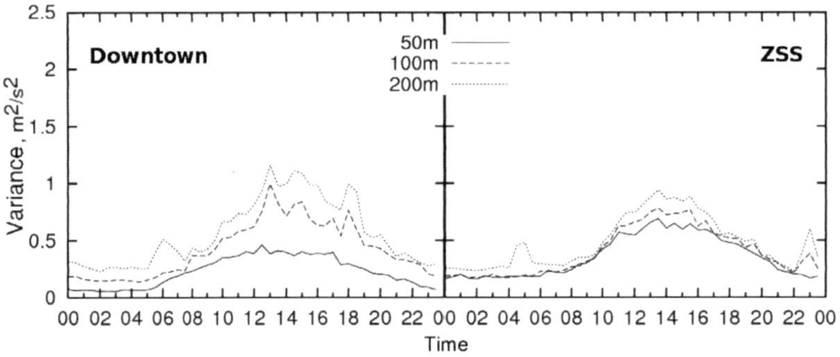

Figure 8. Averaged diurnal course of σ_w^2 at 3 heights in centre of Moscow and at rural site.

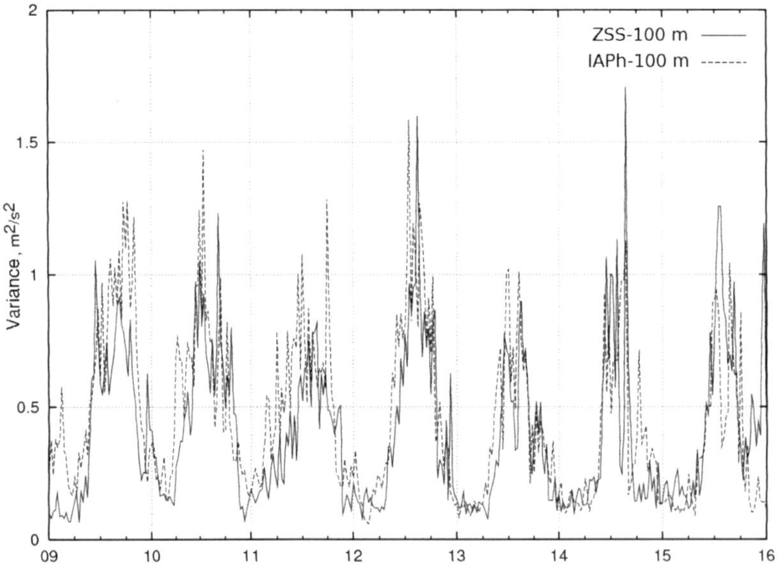

Figure 9. Example of week course of σ_w^2 at 100 m height in Moscow downtown and at ZSS site. 9-16 July 2005.

Ultrasonic anemometer (USA-1, Metek) was mounted at the mast, 10 m above roof level. Data analysis shows increase of mean vertical velocity for east wind direction due to influence of Physics Faculty building shape and also weak alteration of wind direction for north wind due to influence of Main MSU building. Comparison of σ_w^2 measured by sodar at the lowest level (50 m) and in-situ one shows a key role of height above roof (10 m) and above ground level (+40 m) in data analysys.

3.5. Wind variances in inclined antennae.

In-situ measurements in labs and field experiments show stability-dependent relations between longitudinal, transversal and vertical variances of velocity [7]. Measuring of vertical velocity variance by sodar is widely used and using of radial velocity variances by inclined antennae demand realization.

Preliminary experiments were carried out to calculate a ratio of variance components by in-situ measurement in urban environment (at MSU site) and ratio of radial velocity variances by sodar measurement. Qualitative accordance between components of variances by in-situ and sodar measurements was detected. Fig. 10 illustrates these relations in a week series in summer-time. Upper panel shows sodar radial velocity variances calculated over 30 min intervals. Antennae 1 and 2 inclined to vertical at 30° and so (neglecting momentum flux) $\sigma_1^2 = 0.75\sigma_w^2 + 0.25\sigma_{u1}^2$ and $\sigma_2^2 = 0.75\sigma_w^2 + 0.25\sigma_{u2}^2$. Second panel shows variance of wind components (not components of variance). Stability-dependent relation between components is seen. Bottom panel shows wind velocity averaged over 5 minute at the lowest level to compare variances in inertial interval (sonic, 1 sec), at its border (sodar, 15 sec) and at meso-scale (300 sec).

3.6. Meso-scale variability of horizontal wind

Meso-scale variations of all wind components may be estimated by sodar measurements using averaging on short time intervals (3-5 minute). Open architecture of LATAN-3 allows varying of sounding parameters and of collected data processing. Three components of wind velocity vector averaged over 5 minutes (up to 20 measurements) were calculated. Then each variance components (longitudinal, transversal and vertical) for long time interval (60-90 minutes) may be studied as well as their vertical profiles, diurnal course and distributions.

Fig. 11 demonstrates ability of such calculation and qualitative behaviour of horizontal wind variances. Two summer days of measurements by short series are shown in Fig. 11. Meso-scale and synoptic variability is seen. At high levels also data gap and poor data availability are present. Thus difficulties of direct measurement of meso-scale variations are remained but efforts in this direction continue.

4. Conclusion and outlook

Ability to measure of vertical wind variances (σ_w^2) by sodar with high accuracy permits to study a diurnal variability of turbulence and its annual course and to collect a climatological statistics. Statistics of wind variance for a long time in continuous measurements earlier was not analysed. Joint efforts in long-term sodar measurements [6] will be continued.

Main increasing with height of vertical wind velocity variance and total kinetic energy of turbulence (if TKE connected with σ_w^2 by stability- dependent function) take place within the lowest 150-250 m and so sodars will play a major role in ABL turbulence investigating.

Main feature of turbulent exchange in ABL is diurnal course and its annual variations. Vertical velocity variance (and TKE) increases in summer days in 3-4 times to night ones. Diurnal amplitude (difference between day and night values) increases in order of magnitude towards winter value.

IOP Conf. Series: Earth and Environmental Science 1 (2008) 012046 doi:10.1088/1755-1307/1/1/012046

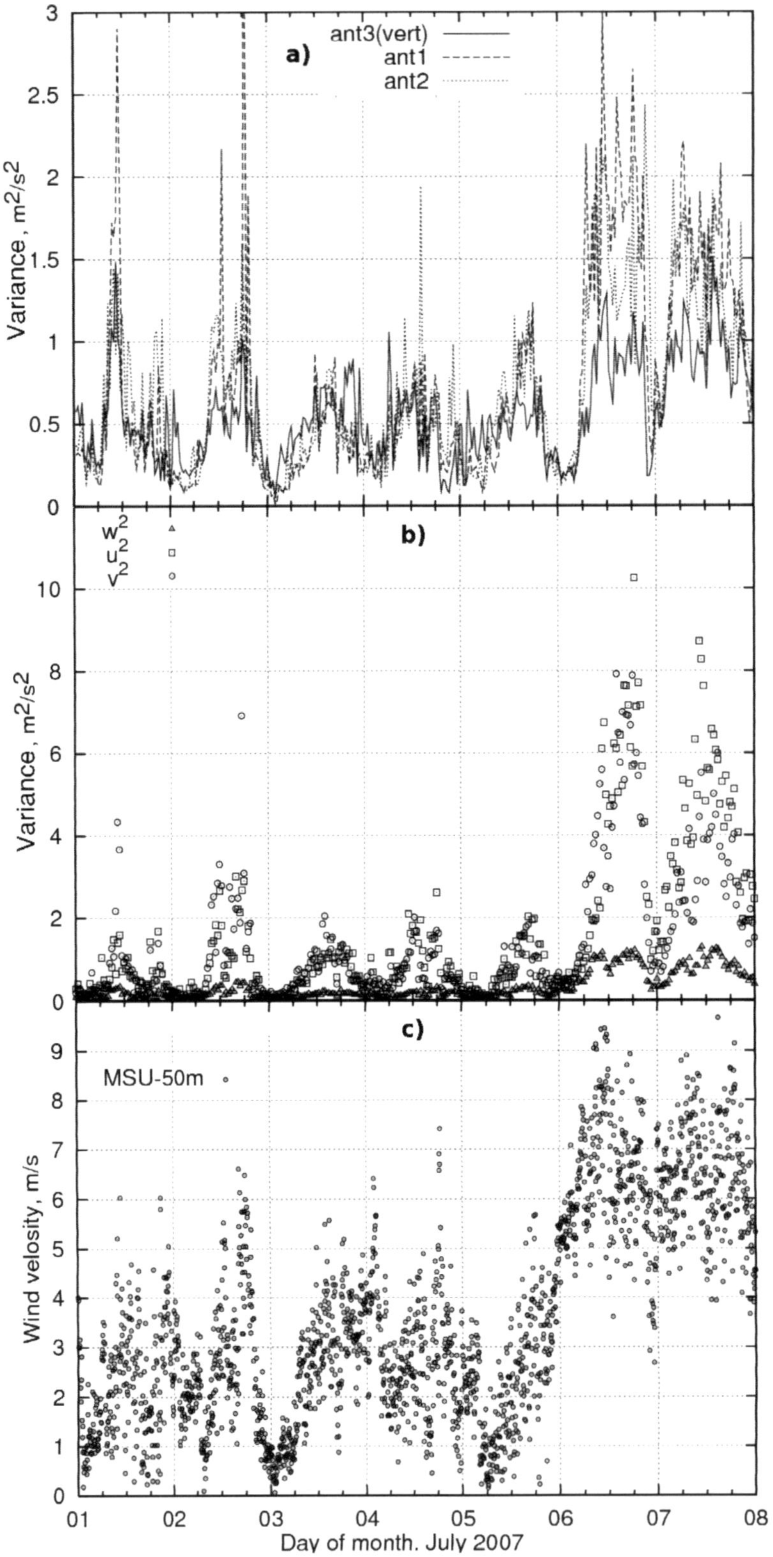

Figure 10. Comparison of sodar radial velocity variances (top panel), in-situ measured variances of wind components (middle) and variability of horizontal wind speed averaged over 5 minutes (bottom). MSU site, 1-8 July 2007.

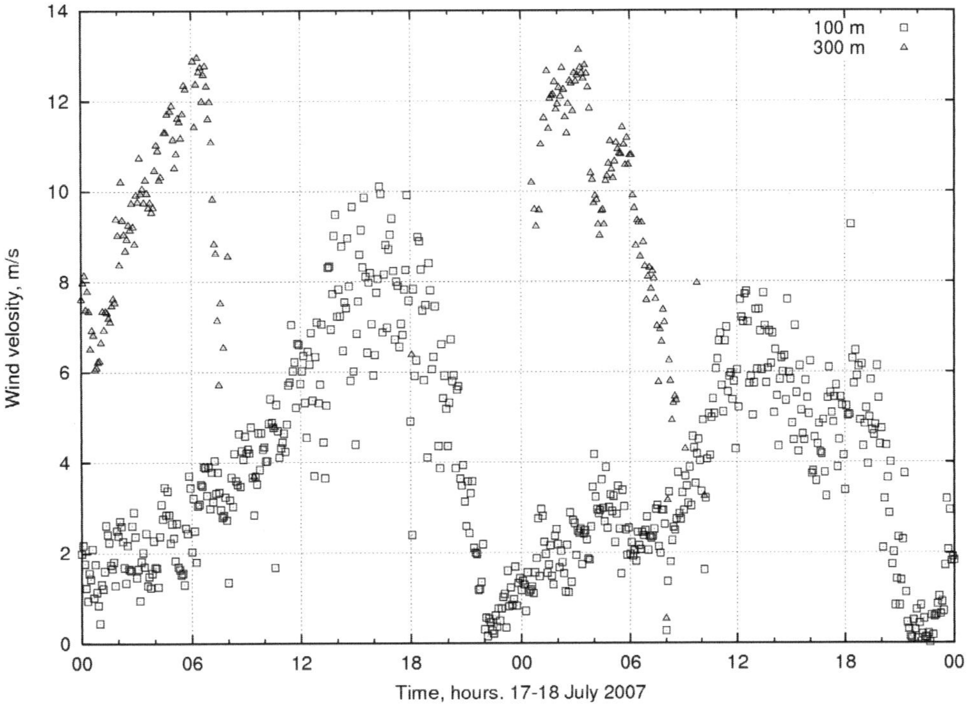

Figure 11. Horizontal wind velocity averaged over 5 minutes at 2 levels: 100 and 300 m. 17-18 July 2007, Moscow downtown. X-axe shows LST.

Probably, Moscow city isn't a fine example of turbulence variability in winter period. More significant diurnal amplitude may be obtained in South Europe cities or under breeze circulations, i.e. regional climatology of turbulence remains important area of research.

Turbulence statistics demonstrates wide synoptic variability and confirms stabilization of turbulence above 150 meters. From 50 to 150 m distribution of variances show clear tendency, that can be analytically described.

Wind velocity variance above 150 m in urban area is strongly varied, perhaps, due to influence of urban noise. Our experiments show that mean values of σ_w^2 averaged for long interval as well as distributions are sensitive to noise filtration procedure. So, comparison of different methods of remote measuring of wind variances must be performed at heights where data availability decreases (above 200 meters).

Important part of turbulence increasing takes place below the lowest measuring level by sodar, so for more detailed study of canopy and wake layers in urban area combination of remote and in-situ (at mast) measurement is actual (or using additionally of minisodar).

Comparison of measurements in two urban points confirms major role of lower part of ABL in turbulence increase. Measurements at Physics Faculty building show that turbulence changing takes place below sodar bottom range. In-situ measurements at mast on the roof of building confirm this.

Diurnal course of variance in MSU demonstrates main influence of unstable stratification on increase of σ_w^2 with height. Comparison of diurnal course in IAPh and in MSU shows that increase of wind variance depends on stratification: in night-time variance in MSU increases faster.

Comparison of urban and rural measurements demonstrates that diurnal variability in centre of Moscow and at rural site in summer-time is very similar at least at 100 m height. Probably, roughness and heating differences manifest itself in canopy and wake layers below 100 m.

Comparison of radial velocities variances on inclined antennae qualitatively confirms the ratio between longitudinal, transversal and vertical components, obtained in laboratories and field experiments over flat terrain. Method suggested in[8] allows calculating of both σ_w^2 and TKE from inclined antennae measuring.

Sodar Latan-3 permit to measure three components of wind velocity over short time intervals (3-5 minutes). Variability at this scale allows calculating of three components of meso-scale variance and their ratio; however, calculation must consider a rapid change of average velocity in synoptic processes, and decrease of data availability with height.

Vertical velocity component is also convenient to measurement of autocorrelation time and spectrum, because of vertical velocity doesn't have a meso-scale (at tens minutes) variability. Wind velocity correlations between nearby "pulses" show variability at scale of tens seconds and spatial correlation between adjacent heights gives estimation of variability at scale of tens meters. Obtained data permit to calculate these characteristics.

Acknowledgments

Measurements in IAPh and at ZSS cite were carried out by Rostislav Kouznetsov, we thank him also for helpful discussion. We are grateful to M.A. Kallistratova for study initialization and numerous remarks. This work is supported by Russian Foundation for Basic Research, projects nos. 05-05-64786, 07-05-13610 and 08-05-00984.

References

[1] Kallistratova M A and Coulter R L 2004 Application of sodars in the study nad monitoring of the environment *Meteorol. Atmos. Physics* **85** 21–37

[2] Kouznetsov R D 2006 The new PC-based sodar LATAN-3. *Extended Abstr. 13th Int. Symp. On Advancement of Boundary Layer Remote Sensing(ISARS-2006), 18-20 July 2006, Garmisch-Partenkirchen, Germany* 97–98

[3] Yushkov V P, Kallistratova M A, Kuznetsov R D, Kurbatov G A and Kramar V F 2007 Experience in measuring the wind-velocity profile in an urban environment with a Doppler sodar *Izvestiya Atmospheric and Oceanic Physics* **43** No. 2 168–80

[4] Pekour M S and Kallistratova M A 1993 Study of the boundary layer over Moscow for air-pollution application *Appl. Phys. B* **57** 49-55

[5] Coulter R L and Kallistratova M A 2004 Two-decade progress in sodar techniques: a review of 11 ISARS proceedings *Meteorol. Atmos. Physics* **85** 3–19

[6] Emeis S, Baumann-Stanzer K, Piringer M, Kallistratova M, Kouznetsov R and Yushkov V 2007 Wind and turbulence in the urban boundary layer - analysis from acoustic remote sensing data and fit to analytical relations *Meteorologische Zeitschrift* **16** No 4 393–06

[7] Kouznetsov R D, Kramar V F, Beyrich F and Engelbart D 2004 SODAR-Based Estimation of TKE and Momentum Flux Profiles in the Atmospheric Boundary Layer: Test of a Parameterization Model *Meteorol.Atmos. Phys* **85** 93–99

[8] Kouznetsov R D 2003 Estimates of Vertical Turbulence Structure by Sodar in the Urban Air *Proceedings of 5th International Conference on Urban Climate, Lodz, Poland* **2** 345–48

IOP Conf. Series: Earth and Environmental Science **1** (2008) 012047 doi:10.1088/1755-1307/1/1/012047

Comparison of nocturnal inversion characteristics obtained by sodar and microwave temperature profiler

V P Yushkov[1], **I N Kouznetsova**[2]

[1] Faculty of Physics, Lomonosov Moscow State University, Lenin Hills, Moscow, 119992, Russia

[2] Hydrometeorological Centre of Russia, 11-13, B. Predtechensky per., Moscow, 123242, Russia

E-mail: yushkov@phys.msu.ru

Abstract. Sodar echogram is a usual means to estimate the height of temperature inversion. Our data of long-term synchronous measurements with sodar and microwave profiler in urban and rural areas showed an intricate relation between intensity of backscattering and potential temperature profiles. Qualitative relationships between echogram pattern, wind speed and temperature profiles were obtained. Distributions of temperature gradients in rural and urban measurements for summer and winter time were calculated.

1. Introduction

For description of stable boundary layer (SBL) meteorologists sometime use a simple characteristics of inversion layer - its height (IH),i.e. height where temperature profile become isothermal (below $\frac{\partial T}{\partial z} > 0$ and upper $\frac{\partial T}{\partial z} < 0$) A difference between temperature of isothermal layer and standard temperature (at 2 m height) is inversion depth. Fig. 1 shows two examples of temperature profiles (discussed below) in rural ABL for strong and weak stability (in night-time) in summer.

But turbulence study and climate or weather prediction models use a mixing height (MH) as characteristics of ABL. In acoustic sounding for that purpose a height of surface layer (in SBL) of strong backscattering of acoustic signal is used [1] (see Fig. 2). In theoretical approach dimensionless Richardson number and its critical value are used for MH determining [2]. Numerous field experiments for comparing of MH and IH were carried out but uncertainty in their closeness continue.

Currently used UHF temperature profilers permit to study this problem more carefully. Continuous measurements [3] unlike occasional ones by radiosondes allow comparing variations of MH and IH. In this paper comparison of temperature profiles and profiles of temperature gradient measured by profiler with MH observed by sodar were carried out.

2. Instrumentation and sites

Simultaneous measurements of temperature profiles by UHF profiler MTP-5 (ATTEX, Russia) and scattering intensity by sodar LATAN-3 (IAPh, Russia) were carried out in rural area at

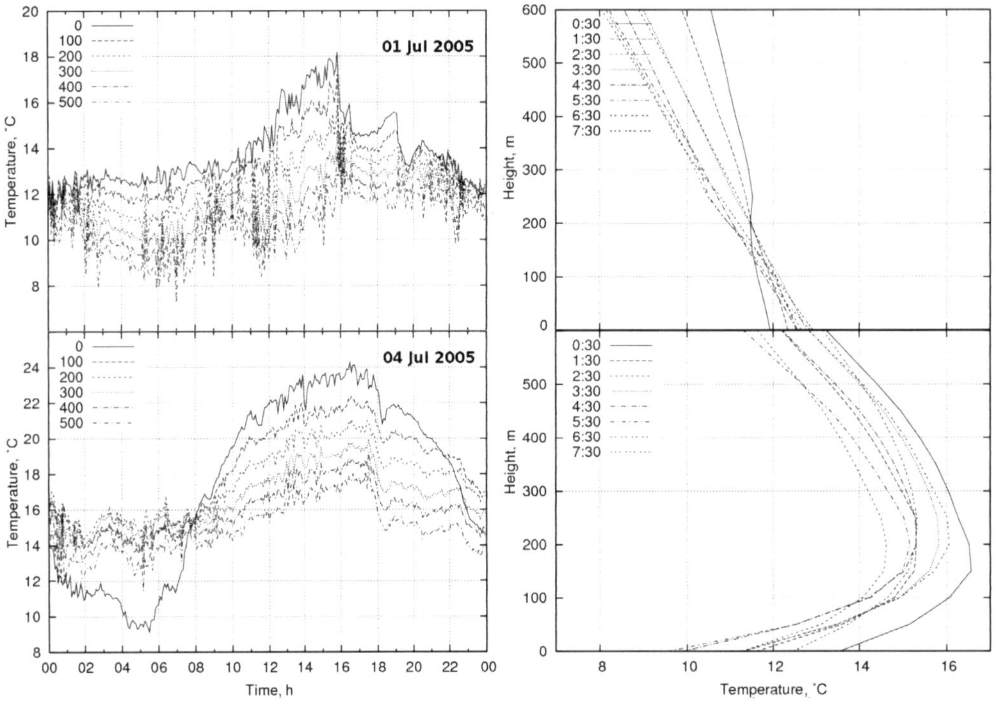

Figure 1. Diurnal course of temperature at different heights (left panel) and profiles of temperature averaged over 30 min in night-time (right panel) measured by MTP-5 profiler at rural area (ZSS of IAPh) for two cases of night ABL stability: weak (01 July 2005, above) and strong (04 July 2005, below).

Zvenigorod scientific station (ZSS) in 45 km from Moscow in July 2005. In Moscow downtown these measurements are continuous.

Profiler MTP-5 developed at the Central Aerological Observatory [4, 3] allows measuring of temperature profiles in ABL up to 600 m height. This remote instrument reconstructs temperature by inclined integral measurements so temperature profiles are smoothed but it allows easy measuring of averaged temperature lapse rate by difference at adjacent levels. Fig.1 shows examples both time series at different heights and mean vertical profiles over 30 min intervals.

LATAN-3 is Doppler mono-static three-component sodar described in [5, 6]. Examples of scattering intensity (echograms) and wind velocity profiles are presented in Fig.2 for those days like in Fig.1.

3. Methods

Observations show that height of isothermal level is not suitable for practical application due to it's not narrow and considerably grows under slightly stable stratification (see Fig.1 above).

Theoretical approach suggests two stability criterions: potential temperature gradient (lapse rate) and Richardson number. Potential temperature gradient (PTG) near surface is connected by bulk relation with IH and inversion depth: $\frac{\partial T}{\partial z} \approx \frac{\Delta T}{H}$, $\frac{\partial \Theta}{\partial z} = \frac{\partial T}{\partial z} + 1°K/100m$, where Θ - potential, T - thermodynamic temperature and ΔT - difference between surface (at standard height) temperature and temperature of isothermal layer, H - the height of inversion layer - IH.

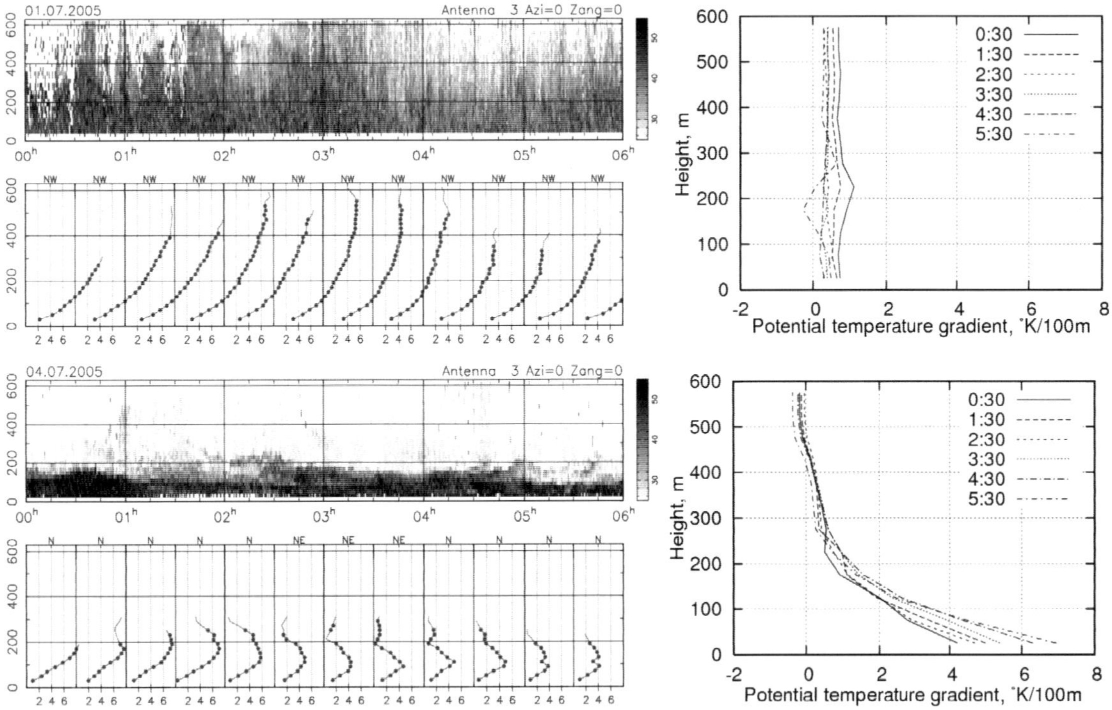

Figure 2. Echograms of acoustic sounding and wind speed profiles (left panel) measured by LATAN-3 sodar and profiles of potential temperature gradient (right panel) in night-time for those dates as in Fig.1. Gray scale on echograms is proportional of scattering intensity (C_T^2) in logarithmic units (shown on the right side). On the left side height is shown, below echograms LST is pointed, below wind profiles scale of wind speed is dislpayed as well as direction is shown above

Richardson number $Ri = \dfrac{\beta(\frac{\partial\Theta}{\partial z})}{(\frac{\partial U}{\partial z})^2 + (\frac{\partial V}{\partial z})^2}$ is dynamic criterion and preferable for MH determination. However, using of model parameters such as Richardson number is not applicable due to synoptic and stochastic variability at those heights where $\frac{\partial\Theta}{\partial z}$ or $(\frac{\partial U}{\partial z})^2 + (\frac{\partial V}{\partial z})^2$ are small. Profiler MTP-5 doesn't allow measuring of temperature gradient with sufficient accuracy above 300 m but profiles of $\frac{\partial\Theta}{\partial z}$ show that at higher levels in night-time state of atmosphere is often slightly-stable, so if we consider that $\frac{\partial\Theta}{\partial z} \approx const$ then inverse Richardson number has a minimum (and critical value) at height where wind speed has a maximum . It confirms well-known fact that MH in SBL is close to height of wind speed maximum. But Richardson number doesn't specify where maximum is reached.

Calculating of PTG by profiler measurements shows correlation of low MH by sodar data with strong stability by profiler data (see Figs. 1 and 2). Increase of stability leads to MH decrease, especially in rural area in summer time.

Calculation of MH by sodar data is carried out visually and accuracy of MH determining fluctuates near 20-100 m because of variations of top boundary of SBL and percentage of error is about 20-25%. Comparison of temperature gradients and MH was carried out when radiation

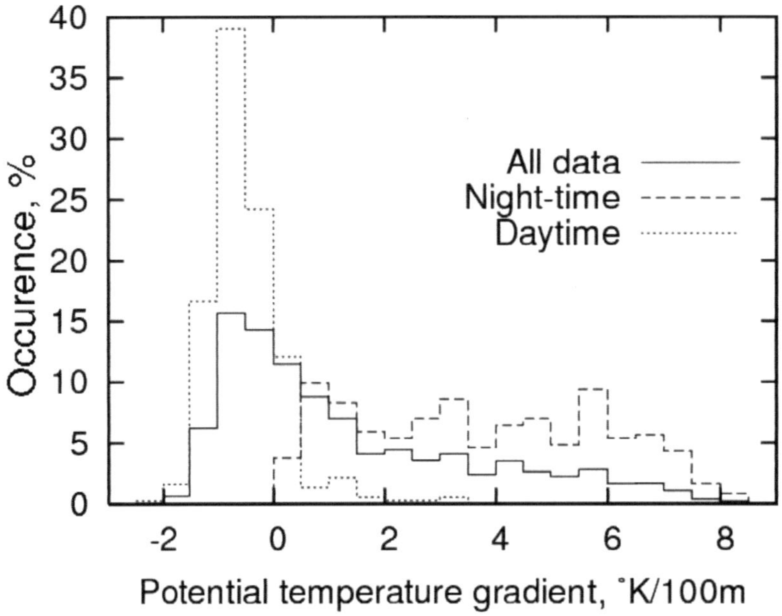

Figure 3. Distribution of PTG in the lowest part of ABL in °K/100m. 0 corresponds to dry-adiabatic temperature gradient(-10°K/km) and 1 is isothermal one.

balance generally is stable and traffic noise is insignificant. For analysis a night-time data from 0 to 6 hour of LST were selected.

For automatic measuring of MH by sodar data well-known property of acoustic backscattering in SBL may be used: above top border of SBL intensity of scattered signal decrease considerably due to neutral stratification and weak wind shear. A few algorithms were tested for determining of MH: by decreasing of intensity, by decreasing of data availability, by vertical integral of scattering intensity. All tested methods work well if height of SBL is intermediate (200 - 300 m) and discrepancy is significant under very stable or neutral stratification because of, in first case, littleness of MH and insufficiency of sodar vertical resolution, in second case, limitation of height range and fuzzy border of SBL.

4. Results

4.1. Distribution of surface temperature gradients

Difference between two lowest levels: 0 and 50 m above profiler displacement height (about 10 m) was used to calculate "surface" potential temperature gradient (sPTG). Surface value is convenient because of larger sensibility to temperature variations. This approach is alike suggested by Holzworth for convective ABL [7]. As mentioned above $\frac{\partial \Theta}{\partial z} \approx (T_{50} - T_0) * 2 + 1$. Distribution of sPTG from one month of summer measurements at ZSS site was calculated. Fig.3 shows general occurrence and depth of surface inversions in night-time for summer conditions at rural area (without urban heating influence). There is wide range of stability in summer connected with large surface radiation emission in clear nights.

Comparison of sPTG with MH by sodar for one month of continual measurements shows clear correlation between sPTG and MH in wide range of stability. This comparison permits to suggest a simple estimation of MH in rural area in middle latitudes: $H \approx 500/\frac{\partial \Theta}{\partial z}$. As Figs. 1 and 2 show if sPTG is about 1°K/100m (isothermal surface layer) then MH is about 500 m,

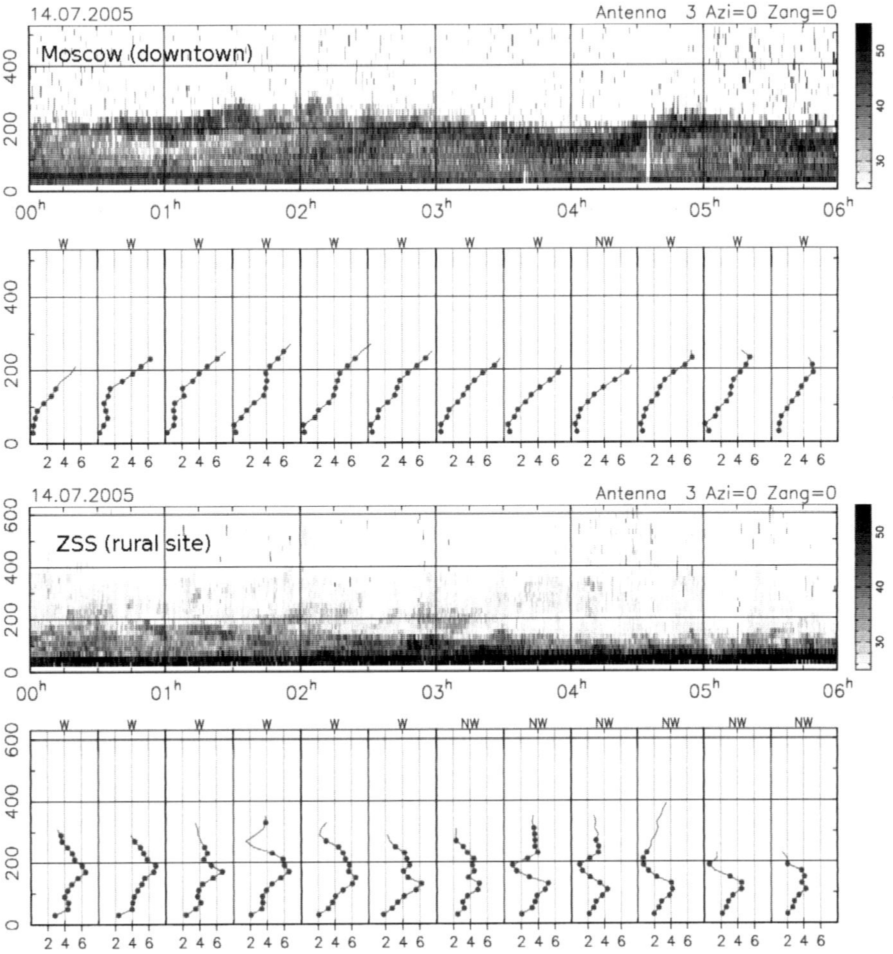

Figure 4. Simultaneous echograms and wind speed profiles in urban and rural environment. See Fig. 2 for explication.

when $\frac{\partial\Theta}{\partial z} > 5^{\circ}K/100m$ MH falls below 100 m. Increasing of MH often results to lesser intensity of scatter signal due to lesser PTG but scattering intensity (C_T^2) also depends on wind velocity gradient.

4.2. Comparison of urban and rural measurements

Comparison of MH by sodar data in rural area and in Moscow downtown [1, 6] shows considerable influence of urban heat island on ABL stability and MH. At night-time thickness of "surface inversion layer" by sodar measurements at urban site is sometimes larger then those at rural site sometime at 50-150 m. It amounts up to 100% of MH in rural area. One example of MH comparison by sodar echograms in urban and rural area in summer time is presented in Fig. 4. Larger MH and wind velocity maximum level is observed over urban area. Comparison of IH by profiler data shows corresponding tendency: PTG is much higher at rural areas in night-time.

In Moscow downtown surface inversions in summer nights by profiler measurements are rare due to urban heating and heat storage. Often PTG profiles have an elevated maximum (Fig. 5). Acoustic measurements show elevated or multilayer inversions with larger scattering intensity.

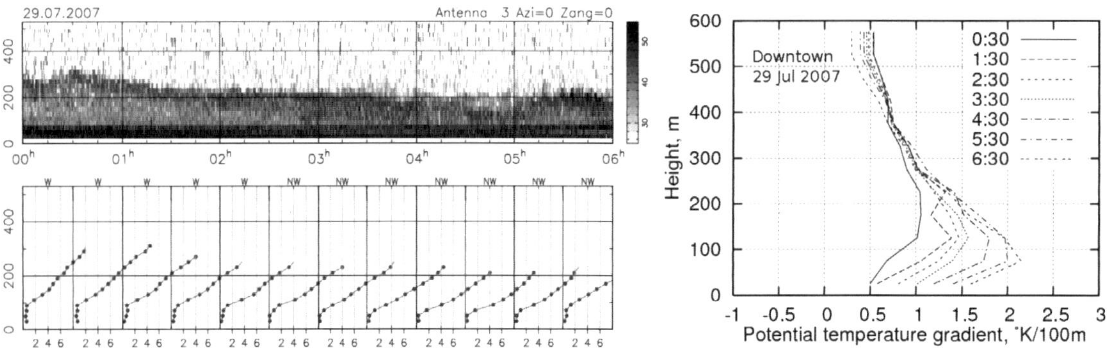

Figure 5. Comparison of MH by sodar echograms and PTG by profiler measurements for vary stable situation in urban area. (Explication is shown in Fig. 2)

The strongest stability in Moscow urban area in summer is about $2-3$°K$/100m$ at 100 m level, as at rural site PTG achieves $7-8$°K$/100m$ at the lowest level. The finest surface layers in Moscow downtown in summer 2007 by sodar data are about 150-200 m.

Thus if stable stratification is present outside the town and wind speed is large at top border of SBL (10-15 m/s) city heat up moved stable-stratified air mass. However, "elevated inversions" on sodar echograms determined by larger scattering intensity at the top of SBL don't correspond to height of PTG maximum. Comparison with wind profiles shows that larger scattering intensity fits to larger wind speed shear, larger turbulence mixing and so larger temperature fluctuations(C_T^2). If wind velocity is small at the top of SBL, scattering of acoustic signal is weaker for equal PTG values.

Because of PTG profiles in urban area in stable state are nonlinear, a simple relation between MH and sPTG is absent. Summer observations in June-August 2007 in Moscow downtown show

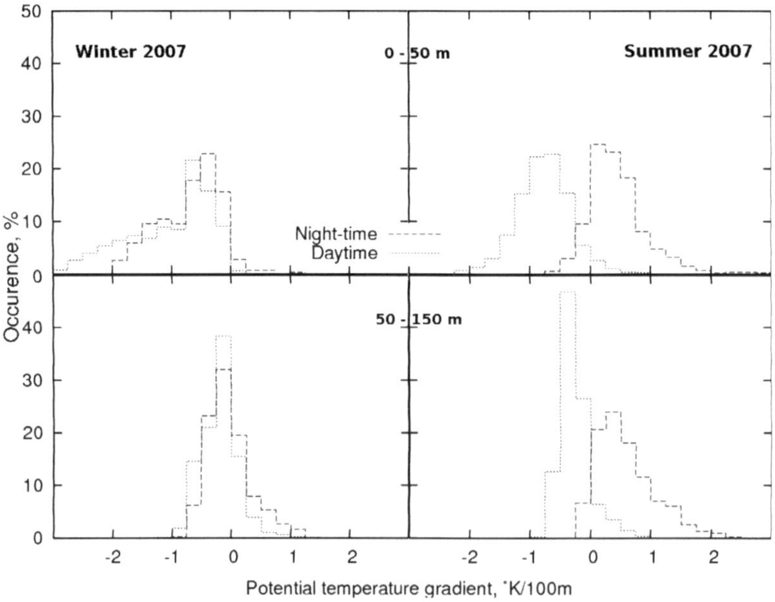

Figure 6. Distributions of PTG for cold and warm seasons at two levels: 0-50 and 50-150 m. Moscow downtown, summer and winter 2007 (December- February).

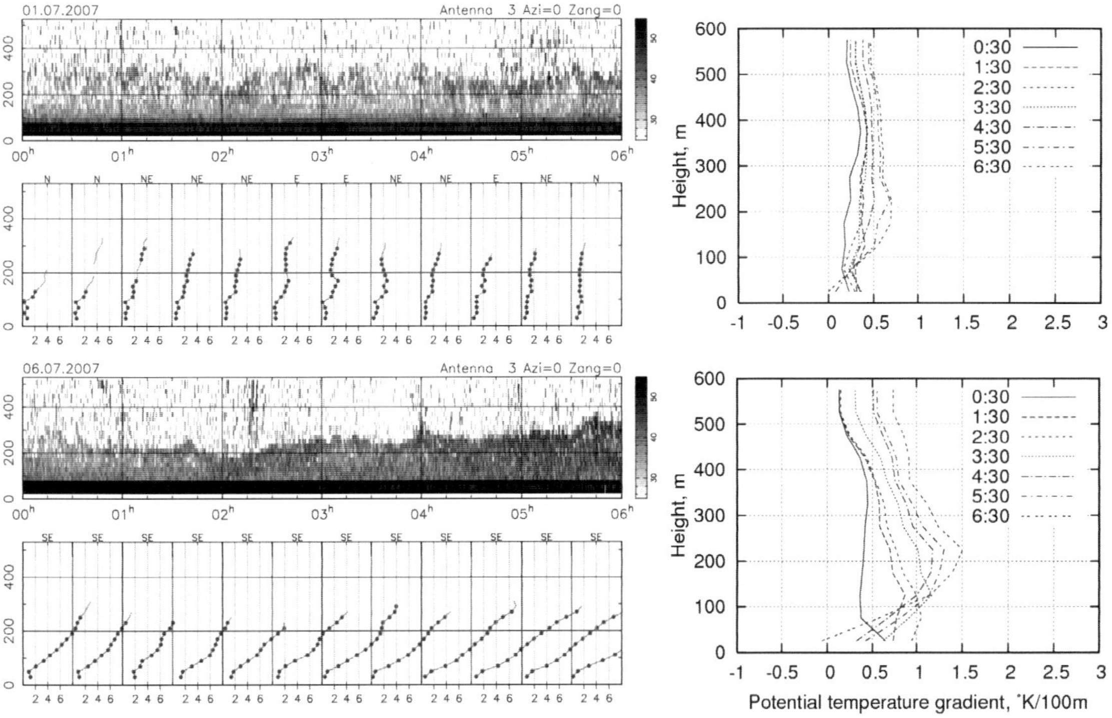

Figure 7. Comparison of echograms and wind profiles in centre of Moscow for weak and strong wind speed. At the right panels profiles of PTG are displayed.

rough empirical estimation of correlation between MH and maximum of PTG: $H \approx 400/\frac{\partial\Theta}{\partial z}_{max}$

4.3. PTG distribution in urban environment

Based on continuous measurements in 2007 distributions of PTG for different heights were calculated. Statistical analysis confirms high sensitivity of sPTG to radiation processes and heat exchange with surface and sensitivity decreases with height. Summer time allows studying different stability of ABL but statistical research demands annual generalization. Calculating and comparison of PTG distribution for summer and winter, for night and day and for different heights were carried out. This analysis demonstrates a convenient of sPTG for statistical stability description.

Fig.6 demonstrates PTG distribution for the lowest level (between 0 and 50 m) and for 100 m level (as difference between 50 and 150 m). Distributions at night-time are evidently contrasted with those at daytime as well as distributions for summer are differ from those for winter. Continuous cloudiness in winter time almost suppresses a diurnal course of PTG and urban heating displace surface inversions from urban area in winter time. Temperature lapse rate in the lowest layer of ABL has the largest diurnal and annual course.

4.4. MH and wind speed

Role of wind speed at the top of SBL in case of weak stability is very important. Small wind speed makes difficult MH determining due to fuzzy border of slightly stable ABL. Fig. 7 shows two examples of weak and strong wind speed at the top of SBL. If wind is strong top border of SBL is sharp due to larger mixing and in most cases the stable surface layer is connected with large wind shear (see Figs. 7 (bottom panel) and 2 (upper panel)).

5. Summary

Continual observations show that temperature profiler MTP-5 allows reliable measuring of averaged temperature gradient in ABL. Comparison with sodar echograms and wind profiles confirms accordance of temperature and wind measurements. The largest sensitivity for this instrument is obtained at the lowest level. Observations in rural area show wide range of stability in night SBL in summer time. Height of SBL - mixing height is surely determined only by acoustic sounding due to high sensitivity of scattered signal to mixing of "air particles". Uncertainty of MH by sodar measurements is 20-100 m. Richardson number connects MH with fast decreasing of wind speed lapse rate at the top border of SBL. Continual measurements show that MH correlates with stability of SBL measured by sPTG because of sPTG is sensible to surface thermal balance. Observations in rural area give a simple empirical estimation of MH and sPTG correspondence. Weak wind speed leads to fuzzy SBL border with lesser MH. In urban area maximum of PTG is often elevated. Empirical estimation connects MH with maximum of PTG. Distributions of PTG at urban and rural areas are differ considerably. This distribution demonstrates urban impact onto air mass.

Acknowledgments

We are grateful to M.A. Kallistratova and R.D. Kouznetsov for helpful discussions. This work is supported by Russian Foundation for Basic Research, projects nos. 07-05-13610 and 08-05-00984.

References

[1] Pekour M S and Kallistratova M A 1993 Study of the boundary layer over Moscow for air-pollution application *Appl. Phys. B* **57** 49-55
[2] Zilitinkevich S and Baklanov A 2002 Calculation of the height of the stable boundary layer in practical applications *Boundary-Layer Meteorology* **105** 389
[3] Khaikine M N, Kuznetsova I N, Kadygrov E N and Miller E A 2006 Investigation of thermal-spatial parameters of an urban heat island on the basis of passive microwave remote sensing. *Theoretical Applied Climatology* **84** No 1-3 161–69
[4] Kadygrov E N, Shur G N and Viazankin A S 2003 Investigation of atmospheric boundary layer temperature, turbulence, and wind parameters on the basis of passive microwave remote sensing*Radio Science* **38** No 3 13-1
[5] Kouznetsov R D 2006 The new PC-based sodar LATAN-3. *Extended Abstr. 13th Int. Symp. On Advancement of Boundary Layer Remote Sensing(ISARS-2006), 18-20 July 2006, Garmisch-Partenkirchen, Germany* 97
[6] Yushkov V P, Kallistratova M A, Kuznetsov R D, Kurbatov G A and Kramar V F 2007 Experience in measuring the wind-velocity profile in an urban environment with a Doppler sodar *Izvestiya Atmospheric and Oceanic Physics* **43** No. 2 168–80
[7] Holzworth G C 1962 A study of air pollution potential for the western United States *J.Appl.Meteor.* **1** 366–82

IOP Conf. Series: Earth and Environmental Science 1 (2008) 012048 doi:10.1088/1755-1307/1/1/012048

Remote sensing of the nocturnal boundary layer for Wind Energy applications.

Pichugina Y. L. [1,2], R. M. Banta[2], N. D. Kelley[3], W. A. Brewer[2], S. P. Sandberg[2], J. L. Machol[1,2], and B. J. Jonkman[3]

[1] Cooperative Institute for Research in Environmental Sciences (CIRES), Boulder, CO, U.S.A.
[2] Earth System Research Laboratory (ESRL), NOAA, Boulder, CO, U.S.A.
[3]National Wind Technology Center/National Renewable Energy Laboratory Golden, CO, U.S.A.

Yelena.Pichugina@noaa.gov

Abstract. The fine temporal and spatial resolution of Doppler lidar observations has been highly effective in the study of wind and turbulence dynamic in the nocturnal boundary layer during Lamar Low-Level Project in 2003. The High-Resolution Doppler Lidar (HRDL), designed and developed at the National Oceanic and Atmospheric Administration (NOAA) Earth System Research Laboratory (ESRL), measures range-resolved profiles of line-of sight (LOS) Doppler velocity and aerosol backscatter with a pulse repetition frequency of 200 Hz, velocity precision about 10 cm s^{-1}, and a very narrow beam width. The majority of the lidar-measured wind speed and variance profiles were derived using a vertical-scan mode and the application of a vertical binning technique. The profile data were used to calculate quantities important for wind energy applications, including turbulence intensity, wind and directional shear through the layer of the turbine rotor. Profiles of all quantities show a strong variation with height. The mean wind fields, the turbulence, and turbulence intensities show a good agreement with sonic anemometer sodar high confidence (high SNR) measurements. The ability of HRDL to provide continuous information about wind and turbulence conditions at the turbine height and above the range of the tower measurements made HRDL as a powerful instrument for studies of the nighttime boundary layer features. Such information is needed as turbine rotors continue to rise higher into the boundary layer.

1. Introduction

The current fast-growing demand for alternative wind energy requires developing new wind farms and increasing the capacity of wind turbines by capturing stronger winds at higher altitudes. Accurate estimates of wind resource potential and turbulence structure of the boundary layer at the heights of turbine rotors is very important as the height reached by commercial turbines increases up to 200-250 m to take advantage of stronger wind speeds at higher altitudes. The estimation of nighttime wind profiles by rawinsonde or radar-wind-profiler data is often limited by biases caused by times of observations, vertical resolution, migrating birds, or occasional unavailability of data due to dry conditions and weak signals or noisy environments such as urban centers. Tall instrumented towers used in wind-energy studies may reach 120-200 m. Such towers are expensive to deploy. Extrapolation of the data to higher altitudes is not very reliable and does not accurately reflect the true vertical structure of the wind-field.

The High-Resolution Doppler Lidar (HRDL), a scanning system designed and developed at the National Oceanic and Atmospheric Administration (NOAA) Earth System Research Laboratory (ESRL), has been highly effective in the study of dynamic processes in the atmospheric boundary layer (ABL) because of its temporal and spatial resolution.

This paper presents HRDL measurements of wind characteristics important for the wind energy industry during the Lamar Low Level Jet Program in September 2003, and describes the advantage of HRDL measurements for wind-resource potential assessment in the atmospheric layer occupied by turbine rotors up to heights of 250 m or more above ground level. The observations of these characteristics at different levels above the heights of routine tower measurements makes it possible to validate the typical approach of extrapolating wind speed profiles from near-surface measurements to higher levels using a logarithmic or "normal" wind speed profile.

2. Lamar Low Level Jet Program (LLLJP)

During 2001-2003 a coordinated effort between the U.S. Department of Energy (DOE) National Renewable Energy Laboratory (NREL) and General Electric (GE) was made to establish a Lamar Low-Level Jet Program to study the wind and turbulence environment at a site located about 20 miles south of the town Lamar, Colorado. The site is characterized by frequent, strong winds during all seasons of the year, and thus has a high wind resource potential to drive wind turbines. This part of Colorado is now the site of the 166- MW Green Wind Plant. The instrumentation in this campaign included a 120-m tall meteorological tower installed by General Electric Wind Energy (GE Wind) and an acoustic wind profiler (sodar) operated by the National Renewable Energy Laboratory (NREL). NREL also collected and processed data from sonic anemometers installed at four tower levels of 54-, 67-, 85-, and 120-m. NOAA/ERSL joined the program for the first half of September 2003 and deployed HRDL to the site to determine mean and turbulent wind structure, and LLJ properties at heights of interest for wind energy. Detailed information on the observational site, instrumentation, and preliminary results can be found in Kelley et al. (2004, 2007), and Pichugina et al. (2004, 2005).

3. HRDL Measurements

HRDL is a scanning, remote sensing lidar that measures range-resolved profiles of Doppler velocity and aerosol backscatter. The lidar operated with a pulse repetition frequency of 200 Hz, typically averaging results from 100 pulses to form range-resolved, line-of-site (LOS) velocity estimates twice per second with a range resolution of 30 m.

Over the last decade, HRDL has been highly effective in the study of dynamic processes in the ABL because of its high temporal and spatial resolution, narrow beam width, and capability to measure the component of the wind velocity parallel to the lidar beam with a precision of ~10 cm s^{-1} (Grund et al. 2001). Detailed descriptions of HRDL operating characteristics can be found in Banta et al. (2002, 2006), and Poulos et al. (2002). During LLLJP, HRDL operated mostly at nighttime, producing vertical-slice, conical, and staring scans to address different objectives of the project. The measurement error was estimated by analyzing fixed-beam scans, and data were quality controlled by removing hard target returns and measurements with low signal-to-noise ratio, as described in Newsom and Banta (2004) and Pichugina et al., (2007).

Various methods have been proposed to determine the components of the wind field from lidar LOS velocity observations. A technique for the calculation of the horizontal velocity component, aligned with the prevalent wind flow (streamwise velocity) and variance using elevation or vertical-slice scans, was proposed in Banta et al. (2002, 2006; Emeis at al., 2007). Results obtained for each individual vertical-slice scan were sorted into bins of certain vertical size. Mean horizontal velocity (U_H) and variance (σ_U) were computed within each horizontal bin to form vertical profiles.

During the period of the HRDL observations at the Lamar site, for most of the nights, the wind speeds ranged from 5 m s^{-1} to 25 m s^{-1} with a strongest mode of 17-18 m s^{-1} in the distribution histogram (Pichugina et al., 2004). Most of the nights were characterized by strong (>15 m s^{-1}) winds from late evening (1-2 UTC) up to early morning (~10 UTC). The rest of the nights show periods of moderate (10-15 m s^{-1}) and lighter (<10 m s^{-1}) winds.

Examples of time-height cross sections of mean wind for the (a) strong-wind and (b) low-moderate-wind nights are given in Figure 1. Horizontal dotted lines in each panel indicate the level of 45-115 m occupied by turbine rotors at the Lamar site.

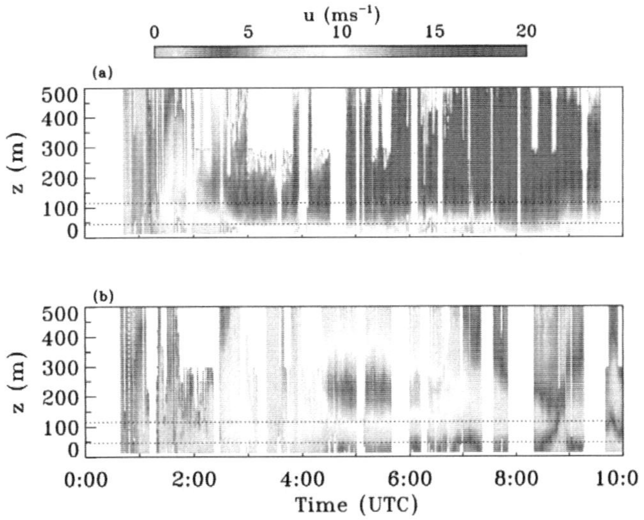

Figure 1. Sample time-height cross sections of mean wind calculated from HRDL vertical-slice scans during (a) strong-wind night of September 15 and (b) low-moderate-wind nights of September 16. Each vertical line represents a vertical profile of the UH derived from individual vertical-slice scans by averaging data over 1 min and within 5-m vertical bins.

This figure illustrates significant differences in the magnitude of U_H between two nights. On (top) September 15 the wind speed was greater than 15 m s^{-1} most of the time and on (bottom panel) September 16 the wind speed was between 5 and 10 m s^{-1} with a short period (~4:40-6:30) of stronger (~12 m s^{-1}) winds at 100-300 m. These contrasting nights, identified as "high-wind" and "low-moderate-wind" nights according to the classification in Banta et al. (2002), were selected to illustrate results obtained during the period of HRDL measurements.

Results of the intercomparison of the mean wind speeds measured by sonic anemometers at four tower heights, sodar and HRDL data obtained at these heights from vertical scans are discussed in Kelley at al. (2007). They found a large and distinct bias of −1.02 m s^{-1} between the mean values of U_H determined from the HRDL vertical-scans and the wind speeds measured by sonic anemometers with the lidar reading lower than the sonic anemometers. In case described in Kelley et al. (2007), a mean U_H was computed for each height bin of 10 m over the length of the record (10 minutes) by averaging data from various range gates (up to 50 gates) and different elevation angles (between 2-20°). Some potential sources for error include a lack of horizontal homogeneity within a bin, variations in radial velocity over the observed range gates, sparse spatial sampling at high elevation angles, and the influence of the vertical wind component not being taken into account when computing the values of U_H. As shown in Pichugina et al. (2007), averaging data within smaller size vertical bins and over appropriate time scales may significantly improve results of the intercomparison of data obtained by different instruments.

To provide additional information on mean wind speed and wind direction during the night, vertical-slice scans were interspersed with conical scans, when HRDL beam sweeps the atmosphere in the full range 0-360° at fixed shallow (2-16°) elevation angles. HRDL data obtained from conical scans were also compared with the appropriate variables measured by the sonic anemometers mounted on the meteorological tower and by the Doppler sodar and quantative data on such comparisons can be find in (Kelly et al., 2007).

An example of visual comparison of wind speed and wind direction observed by the lidar and sodar during the night of September 15, 2003 is shown in Figure2. The top panel in this figure shows time-height cross sections of mean wind speed, obtained from lidar conical scans and processed with a technique described in Banta et al. (2002). The bottom panel shows time-height cross sections of mean wind speed, obtained from sodar data. The arrows in all panels indicate the direction of wind flow, and the color indicates wind speed magnitude from 0 (green) to 20 m s^{-1} (red). Dashed lines on both panels show the height of turbine rotors (45-115 m) that are operated at the Lamar site.

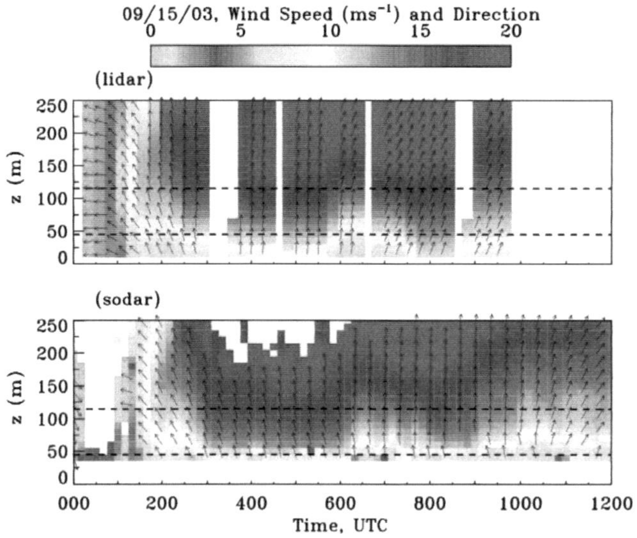

Figure 2. The mean wind speed (color) and direction (arrows), calculated (top) from HRDL conical scans, (bottom) from sodar. In both panels data averaged over 1-min time intervals and over 10 m in vertical (Pichugina et al., 2007).

Both instruments show approximately 6 hours (from 0300 till 0900 UTC) of very strong wind above the level of turbine top, and noticeable veering of wind direction from south to southwest, which occurred on most study nights and also was observed in the long-time (~2 years) of sodar observations on the site (Kelley et al., 2004).

4. Wind characteristics

For further analysis, we used the HRDL data to determine how wind speeds were distributed as a function of height over height intervals that would affect wind turbine operations. Quantities of interest include mean wind speed, wind and directional shear, turbulent velocity variances, and turbulence intensity.

Profiles of U_H, σ_U, and turbulence intensity ($TI= \sigma_U / U_H$) for the night of September 15 are shown in Figure 3. These profiles were composited from 10-min profiles obtained from HRDL vertical-slice data. Vertical axes show height normalized by the height of the LLJ maximum. A strong relationship between the height of the LLJ maximum and the height of the minimum turbulence was shown by Banta et al.(2006) along with a proposition of using LLJ speed maximum U_X as a scaling parameter.

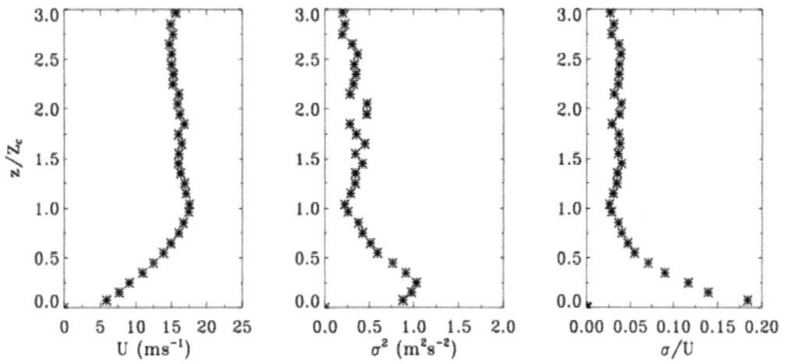

Figure 3. Composite profiles for the night of September 15 showing (from the left to the right) wind speed U_H, streamwise variance σ_U, and turbulence intensity TI (σ_U /U_H). All heights are normalized by the height of the LLJ maxima (Z_C).

Since the wind speed increases in height through the subjet layer and the variance is largest near surface, TI is largest near the surface, when it can exceed 0.1, but it rapidly decreases with height up to the height of the LLJ maximum.

In Figure 4, a turbulence intensity, plotted against the horizontal wind speed for the another strong-wind night of September 5, is shown by dots. The wind speed and standard deviation of the wind speed were averaged over 10 min for the duration of night (100-1000 UTC). Because the wind speed increases monotonically with height, the abscissa of this plot is essentially a scaled height coordinate, and the behavior of the TI is as shown in the right panel of Figure 3.

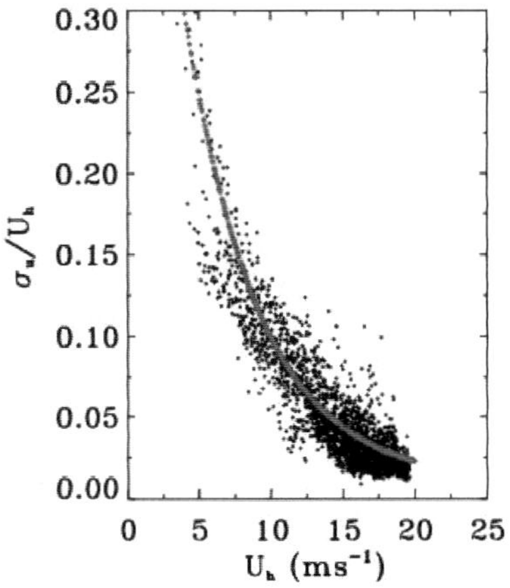

Figure 4. Turbulence intensity, plotted against the horizontal wind speed for the night of September 5. Solid line is a second degree polynomial fit. The wind speed and standard deviation of the wind speed were averaged over 10 min for the duration of night (1:00-10:00 UTC).

Analysis of the turbulence intensities measured by sonic anemometers at the 54-, 67-, 85-, and 116-m levels and HRDL measurements at the same levels show strong decrease of TI with height. Similar results were found for the longer period (~2 years) of sonic anemometer measurements (Kelly et al., 2004).

Histograms of the mean wind speed, standard deviation, and turbulence intensity for the layers of 45-115, 45-200, and 45-250 m, combined for the period of HRDL measurements, are shown in Figure 5 from top to bottom, respectively. All three variables show a near normal distribution in the layer of the current height (45-115) of turbine rotors at the Green Wind Plant near Lamar. In the thicker layers of 45-200 and 45-250 m, wind speeds increase and the distributions show the largest modes of 14-16 m s-1 compare to 8-10 m s-1 for the layer at the turbine rotor height. Distributions of σ_U and TI in both layers of 45-200 and 45-250 m are left-skewed and also decrease with height. The largest mode in speed variance (turbulence intensity) distributions decreased from about 0.8 (0.06-0.065) to 0.3-0.35 $m^2 s^2$ (0.02-0.025) for the thicker layers. These results also support the idea that the flow tends to be smoother at higher levels of the nocturnal boundary layer.

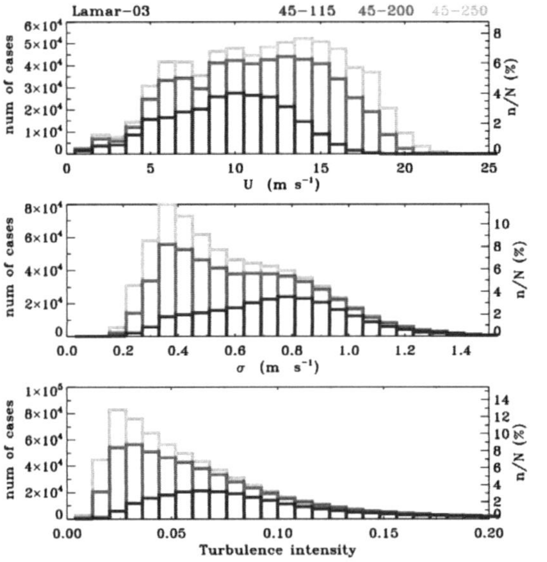

Figure 5. Distribution of the wind speed (top) show a greater wind resource potential for the layers of 45-200 m, and 45-250 m compare to the layer of 45-115 m currently occupied by turbine rotors. Standard deviation (middle) and turbulence intensity (bottom) also increase for the thicker layers and the shapes of distributions changed from (blue) normal to (red and green) skewed (Pichugina et al., 2005)

5. Vertical shear

The information on wind and directional shear across the entire layer occupied by turbine rotors is very important, because modern turbine rotors are so large that wind conditions can be different above and below turbine hub. To estimate wind and directional shear we used profiles of these quantities obtained from HRDL conical scans. Hourly profiles of (left panels) mean wind speed and (right panels) direction were computed using the VAD technique (Banta et al., 2002) for two contrasting nights are shown in Figure6.

Profiles of the wind speed in this figure also show strong (>15 m s^{-1}) winds most of the time during

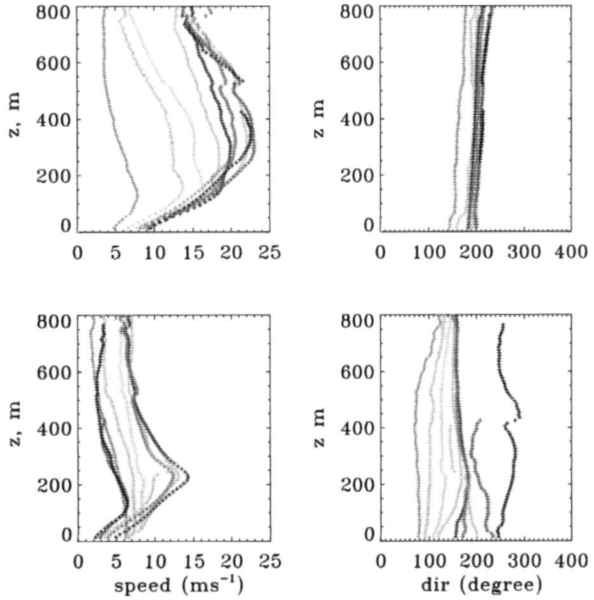

Figure 6. Hourly profiles of the mean wind speed (left) and direction (right), calculated from HRDL conical scans for the nights of September (top) 15 and (bottom) 16.

The night of September 15, and weaker wind speeds (< 15 m s^{-1}, and < 10 m s^{-1}) for most of the night of September 16. Wind directions during the night of September 15 varied from 140° to 200°, while the range of wind directions for the night of September 16 extended from 80° to 250°. Hourly profiles for both nights show small (~0.2-0.5 degrees per meter) directional shear trough the layer of the turbine rotor, with an occasional occurrence of greater directional shear in some of the individual 10-min profiles with wind speeds of ~5 m s^{-1} (not shown).

As shown in Figure 6, the strong acceleration of the flow aloft through the nights was not reflected very strongly at the surface. The wind profile beneath the LLJ was close to linear, rather than showing a "normal" "power law" profile.

The assumption of a normal wind profile or the power law relation: $U_2=U_1(z_2/z_1)^{\alpha}$ is a common approach used in the wind energy industry to estimate the wind speed U_2 at a higher elevation (z_2) using surface (usually at 10 m) or tower measurements of wind speeds U_1 at height z_1. The shear exponent (α) is typically assumed to be equal to 0.2 and rated wind speed for turbine operations near 12 m s^{-1}.

The shear beneath the LLJ is shown in Figure 7, which shows a mean wind profile for each of the 11 nights of LLLJP. Most profiles show strong almost constant shear below the LLJ.

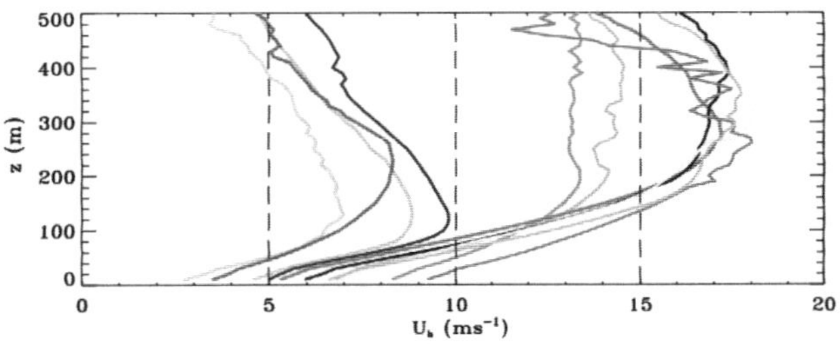

Figure 7. Profiles of mean horizontal velocity composited for each night of HRDL observations during LLLJP, showing almost linear shear up to 100-200 m (Pichugina et al., 2007).

The logarithmic approach exhibits a systematic tendency to underestimation, which increases with height. If an exponential function is forced on the data at two levels, a value for α can be calculated. Distributions of the shear exponent between heights of 45-115, 45-200, and 45-250 m, based on 10-min means of wind speed for the entire period of HRDL observations (Figure 8), indicate that in most cases α>0.2, with a dominant mode of 0.3-0.36, and a secondary mode of 0.145 on each panel.

The results are different from those that were obtained at the Lamar site in September 2002 as described by Kelley et al. (2004). The fact that this exponent varies with height suggests that the power-law profile is inappropriate for stable conditions. The shear exponent calculated between 52 and 113 m using cup anemometer data, showed a maximum peak at 0.2 and a slight secondary maximum at about 0.4. The differences may be explained by the frequent presence of strong low level jets during all eleven study nights of September 2003.

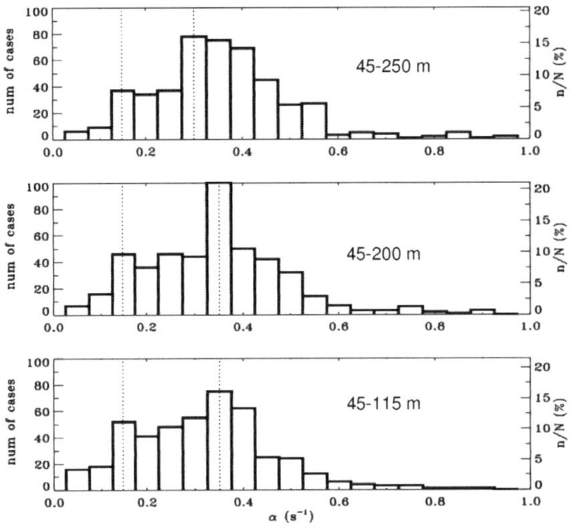

Figure 8. Distribution of the power law exponent (α) between heights of 20-115, 20-200, and 20-250 m, based on 10-min means of wind speed for the entire period of HRDL observations, show that in most cases α>0.2, as specified by International Electrotechnical Commission for Normal wind profile (Pichugina et al., 2005).

6. LLJ created turbulence

Shear developed between the surface and the maximum of the LLJ generates turbulence and turbulent fluxes in this layer (Mahrt and Vickers 2002; Banta et al. 2003, 2006). The magnitude of this turbulence can be estimated by HRDL-measured values of the streamwise velocity variance, which was shown to be numerically equivalent to turbulence kinetic energy (TKE) for stable conditions (Banta et al., 2006). They also showed a very strong relation between height of the LLJ maximum and the height of the minimum value of the streamwise velocity variance profile with proportionality coefficient close to one, suggesting that the LLJ height is generally a measure of the height of the nighttime boundary layer.

Figure 9 illustrates the relation between HRDL streamwise velocity variance and LLJ wind speed maxima. A time-height cross-section of the streamwise velocity variance for the night of September 15 is shown in color, and the heights of the LLJ wind maxima are shown by plus signs. The figure shows that the LLJ maximum serves as an upper bound to the layer of strong turbulence.

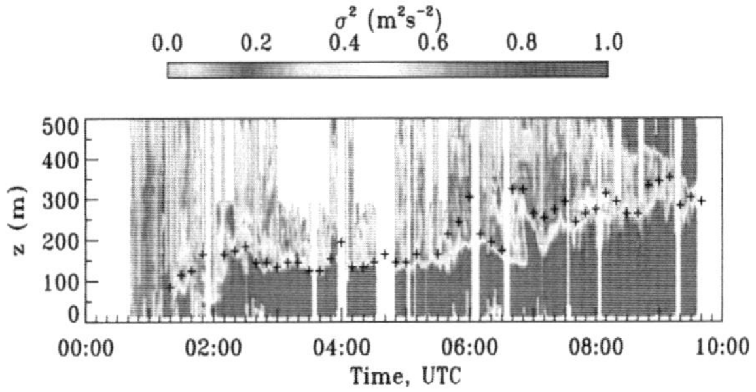

Figure 9. Sample time-height cross sections of the streamwise velocity variance calculated from HRDL vertical-slice scans during night of September 15. Vertical profiles of the streamwise velocity variances were shown to be numerically equivalent to turbulence kinetic energy (TKE) for stable conditions. The height of the LLJ wind maximum is indicated by plus signs (Pichugina et al., 2007)

7. Conclusions

The high temporal and spatial resolution of the HRDL data allows investigation of wind-speed and turbulence conditions in great detail within the nighttime stable boundary layer, including the atmospheric layer occupied by wind- turbine rotors and higher.

Quantities of interest that can be easily monitored using Doppler lidar include low-level jet properties (speed, height, direction) and characteristics of the turbulence below the jet, including estimates of TKE profiles.

The distribution of the shear exponent over the layers of 20-115, 20-200, and 20-250 m computed by lidar measurements of horizontal wind speed during 11 nights show the largest mode between shear exponents of 0.3-0.35 and a slight secondary peak of 0.136.

The LLJ maximum and the height of the minimum value of the streamwise velocity variance profile are strongly related, and the LLJ maximum serves as an upper bound to the layer of strong turbulence. Secondary turbulence maxima may occur above the LLJ if sufficient shear is present.

Profiles of mean horizontal velocity composited for each night of HRDL observations during LLLJP show almost linear shear up to 100-200 m.

Acknowledgments

Field data acquisition and much of the analysis for this research were funded by the National Renewable Energy Research Laboratory (NREL) of the U.S. Department of Energy (DOE) under Interagency Agreement DOE-AI36-03GO13094 (Lamar), and the U.S. Army Research Office (Dr. Walter Bach) of the Army Research Laboratory under Proposal No. 43711-EV (CASES-99) and CIRA Center for Geoscience/ Atmosphere Research. Analysis and manuscript preparation were also supported by the NOAA Health of the Atmosphere Program. We thank our colleagues Raul Alvarez, Lisa Darby, Joanne George, Jennifer Keane, Janet Machol, Brandi McCarty, Andreas Muschinski, Ron Richter, Scott Sandberg, and Ann Weickmann from ESRL, and the following from NREL: J. Adams, Dave Jager, Mari Shirazi and S. Wilde.

References

Banta, R.M., Y.L. Pichugina, and W.A. Brewer, 2006: Turbulent velocity-variance profiles in the stable boundary layer generated by a nocturnal low-level jet. J. Atmos. Sci., 62, 2700-2719.

_____, R.K. Newsom, J.K. Lundquist, Y.L. Pichugina, R.L. Coulter, and L. Mahrt, 2002: Nocturnal low-level jet characteristics over Kansas during CASES-99. Boundary-Layer Meteor., 105, 221-252.

_____ Y.L. Pichugina, and R.K. Newsom, 2003, Relationship between low-level jet properties and turbulence kinetic energy in the nocturnal stable boundary layer. J. Atmos. Sci., 60, 2549-2555.

Emeis, S., M. Harris, and R.M. Banta, 2007: Boundary-layer anemometry by optical remote sensing for wind energy applications. /Meteorologische Zeitschrift/, **16**,337-347.

Grund, C. J., R. M. Banta, J. L. George, J. N. Howell, M. J. Post, R. A. Richter, A. M. Weickmann, 2001: High-resolution Doppler lidar for boundary layer and cloud research. J. Atmos. Oceanic Technol., 18, 376-393.

Kelley, N.D., B.J. Jonkman, G.N. Scott, and Y.L. Pichugina, 2007:. Comparing Pulsed Doppler LIDAR with SODAR and Direct Measurements for Wind Assesement. Awea 2007 Windpower conference, Los Angeles, California. NREL/CP-500-41792

__________N, M. Shirazi, D. Jager, S. Wilde, J. Adams, M. Buhl, P. Sullivan, and E Patton, 2004. Lamar Low-Level Jet Project Interim Report. NREL/TP-500-34593.

Mahrt, L., and D. Vickers, 2002: Contrasting vertical structure of nocturnal boundary layers. Bound.-Layer meteor., 105, 351-363.

Newsom, R.K., and R.M. Banta, 2004: Assimilating Coherent Doppler Lidar Measurements into a Model of the Atmospheric Boundary Layer. Part I. Algorithm, Development and Sensitivity to Measurement error. J. Atmos. Sci., 21, 1328-1345

Pichugina, Y.L., R. M. Banta, N. D. Kelley, S. P. Sandberg, J. L. Machol, and W. A. Brewer, 2004: Nocturnal Low-Level Jet characteristics over southern Colorado. Preprints, 16th Symposium on Boundary Layer and Turbulence, Portland MA.

_____________, R. M. Banta, N. D. Kelley, 2005: Application of high-resolution Doppler lidar data for wind energy assessment. Preprints, 2nd Symposium on Lidar Atmospheric Applications, San Diego CA.

______________, R. M. Banta, and W. A. Brewer, 2006: Vertical profiles of velocity variances and TKE using Doppler-Lidar scan data.17 Symposium on Boundary Layers and Turbulence, San-Diego, CA

______________., R. M. Banta, N. D. Kelley, 2007: Analysis of the southern Colorado Low-Level jet by high resolution Doppler Lidar data. Comparison to the Great Plains LLJ climatologies. 3d Symposium on Lidar Atmospheric Applications, San Antonio, TX.

______________, R. M. Banta, N. D. Kelley, B. J. Jonkman, W. A. Brewer, S. P. Sandberg, and J. L. Machol, 2007: Advantage of the High Resolution Doppler Lidar measurements fornighttime boundary layer study and wind-energy applications. Preprints, 14th Coherent Laser Radar Conference, 8-13 July, 2007, Snowmass, CO, 4 pp.

Poulos, G.S., W. Blumen, D.C. Fritts, J.K. Lundquist , J. Sun, S. Burns, C. Nappo, R.M. Banta, R.K. Newsom, J. Cuxart, E. Terradellas, B. Balsley, M. Jensen, 2002: CASES-99: A comprehensive investigation of the stable nocturnal boundary layer. Bull. Amer. Meteor Soc., 83, 555-581

IOP Conf. Series: Earth and Environmental Science 1 (2008) 012049 doi:10.1088/1755-1307/1/1/012049

Influence of the synoptic circulation on the local wind field in a coastal area of the Tyrrhenian Sea

G Mastrantonio[1], **I Petenko**[1,3], **A Viola**[1], **S Argentini**[1], **L Coniglio**[1], **P Monti**[2] and **G Leuzzi**[2]

[1]Istituto di Scienze dell'Atmosfera e del Clima, CNR, Via Fosso del Cavaliere 100, 00133 - Roma, Italy

[2]Università di Roma "La Sapienza" – DITS, Via Eudossiana 18, 00184, Rome, Italy

[3]IAPh RAS, Pyzhevskiy, 3, 109117, Moscow, Russia

E-mail: g.mastrantonio@isac.cnr.it

Abstract. In this paper the statistical analysis of the wind field measured by a Doppler sodar is used to characterize the local circulation at the Tyrrhenian sea coastal site of the Pratica di Mare airfield. Daily behaviour of the wind direction and intensity, as a function of the season, is highlighted and the existence of two nocturnal alternative components of the local circulation is evidenced. Rawinsonde data were used to determine geostrophic wind values. The comparison between low level and geostrophic wind directions allowed to establish the influence of the synoptic scale forcing in determining both which of the two nocturnal currents can be observed at the site and how their directions evolve during the day. Numerical simulations carried out by using the Colorado State University Mesoscale Model (CSUMM) confirm the statistical analysis results and allow to depict the low level wind field patterns in the area. The use of data from ECMWF analysis is proposed in cases in which rawinsonde data are not available to enquire on the synoptic scale forcing.

1. Introduction

According to climatological studies carried out by some of the authors of this study [1,2], in the coastal regions of the Mediterrenean sea, the local circulation is usually predominant for most period of the year. Palau et al. [3] mention how air quality problems in southern Europe are governed by meteorological processes with marked diurnal cycles. The same authors observe that the properties of the lower troposphere over complex terrain strongly depend on the thermally induced local circulation which develops under clear sky conditions. Savijarvi and Liya [4] note that the local mesoscale air flows, such as sea breezes, mountain, valley and slope winds and the urban heat-island circulations, partly control the local weather and pollutant transport in many parts of the world. Barros et al. [5] remark that factors controlling transport and diffusion in the coastal cities of Lisbon and Barcellona are largely on the local scale. Besides the main role on air pollution transport and dispersion, local circulation has an effect also on agriculture, aviation operations and safety, sports, tourism, services, and energy (windmills) [6,7]. An accurate characterization of local circulation may also be useful for evaluating the performances of numerical models in reproducing the local meteorological systems.

Local wind field is the result of a non linear interaction between large scale and local circulation and it

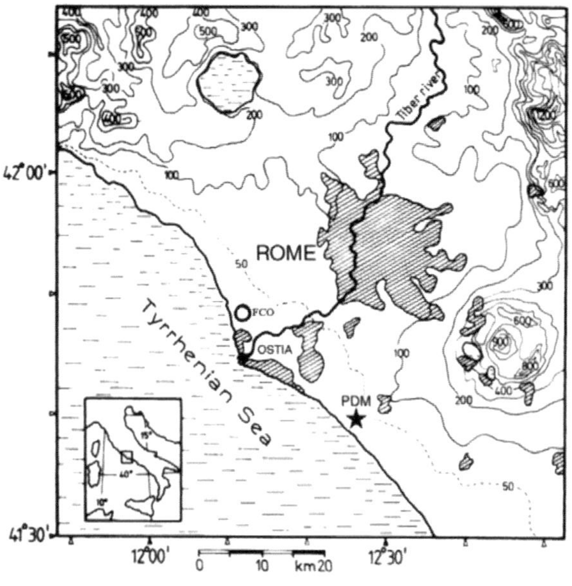

Figure 1. Geographical map of the area, with the locations of Pratica di Mare airport (PDM: full star), and of Fiumicino airport (FCO: open dot). Terrain elevation is also indicated (c.i. = 100 m), with dashed line for the 50 m contour. Densely populated areas are dashed.

is not easy to filter out the component due to synoptic scale forcing. In a previous paper, Mastrantonio et al. [8] showed that an appropriate statistical analysis of the low level wind measurements lets the main characteristics of the local circulation to come on. In what follows, after a short description of the site, of the experimental setup and of the data analysis procedures, the way in which synoptic scale forcing influences the daily pattern circulation at the Pratica di Mare airfield will be highlighted.

2. Experimental and numerical setup

In July-August 1996 a triaxial monostatic Doppler sodar system was operational inside the military airport of Pratica di Mare (PDM), a site 20 km south of Rome and 2 km east from the Tyrrhenian coast (latitude 41°51' N, long. 12°26'E, 20 m a.s.l.), at the southern boundary of "Tenuta di Castel Porziano". In figure 1 the map of the area is presented. The area is flat in the northerly and southerly directions, while it is bounded by the Colli Albani hills 10 km to the E-NE. The sodar, an APM-301 system commercially available from Multimicro, radiates acoustic bursts in three directions simultaneously, each at a different frequency (namely: 1750, 2000 and 2250 Hz). Mastrantonio and Fiocco [9] described the procedure for obtaining the Doppler shifts along the three acoustic beams (one vertical and two inclined 20° off the vertical in perpendicular directions). The signals recorded by the three antennas, after suitable filtering, are sampled and a fast Fourier transform is applied to the digitised return signal, 256 points at a time, corresponding to a height interval of approximately 27 m. 32 intervals are analysed for each burst, the lowest starting at a height of 30 m. Each interval (range gate) is denoted by the height of its center above the surface. The first interval is at 44 m. A signal-to-noise parameter (F) is used as a criterion for accepting or rejecting the datum [9]. Ten minutes averaged wind profiles were inferred from the Doppler shifts measured along the three acoustic beams.
Note that PDM is also a conventional rawinsonde site of the European network. The recorded profiles were used for this study. Moreover, the meteorological information reported in the METAR (the international code to report routine, hourly weather conditions at air terminals), collected at the Fiumicino airport, were used to select the different meteorological conditions and to reject data which were not associated to fair-weather conditions.
The Colorado State University Mesoscale Model (CSUMM, version 2.0, developed originally by Pielke [10]) was used for the evaluation of the wind velocity, temperature and specific humidity in a gridded domain. CSUMM is a hydrostatic, incompressible, prognostic model able to predict the evolution of the atmosphere by the space-time integration of the budget equation of mass, momentum, heat, moisture

and turbulent kinetic energy. At the ground surface the temperature is evaluated from a heat balance, while the moisture flux is proportional to an assigned moisture availability parameter.

The modelling domain is centred about 20 km East from PDM and consists of a 51×51 horizontal grid points with a constant grid size of 4x4 km^2. In order to take the strong surface layer gradients into account, the vertical grid levels are unevenly spaced, i.e. the lower 11 levels are z=10, 20, 44, 100, 300, 500, 700, 900, 1200, 1500 and 2000 m, while the other 7 levels are equally spaced up to 9000 m with a grid size of 1000 m.

CSUMM requires as input data several soil parameters as density, specific heat, thermal diffusivity, moisture availability, roughness and albedo. To take into account their strong space variability, the numerical code has been modified in order to make the setting of these parameters at each grid point possible (only the roughness and albedo have been considered constant all over the domain). The CORINE Land Cover data base was used to assign the soil parameters. CSUMM also requires the knowledge of the large scale geostrophic wind and the initial profiles of temperature and specific humidity. These were taken from PDM rawinsonde profiles. Since in the lower layers winds recorded by the radiosonde are generally representative of local mesoscale circulation, rather than by large scale flow as requested to initialise the numerical model, in the height range $0 < z < 200$ m the wind data were taken from a spatial averaging of ECMWF data set. The temperature of the sea-surface was set to 300.5 K.

3. Data analysis and discussion

In a previous paper [8] the results of the statistical analysis of the wind velocity recorded during fair weather conditions were reported, that show some of the characteristics of the local circulation at the PDM site and how it is observed for most of the year.

In figure 2 the statistics of wind values as a function of time is shown, for two range-gates, as contour maps in which darkness is proportional to the frequency with which each value occurs at the specified

Pratica di Mare: Diurnal behaviour for July and August 1996: night-time directions 0°-360°

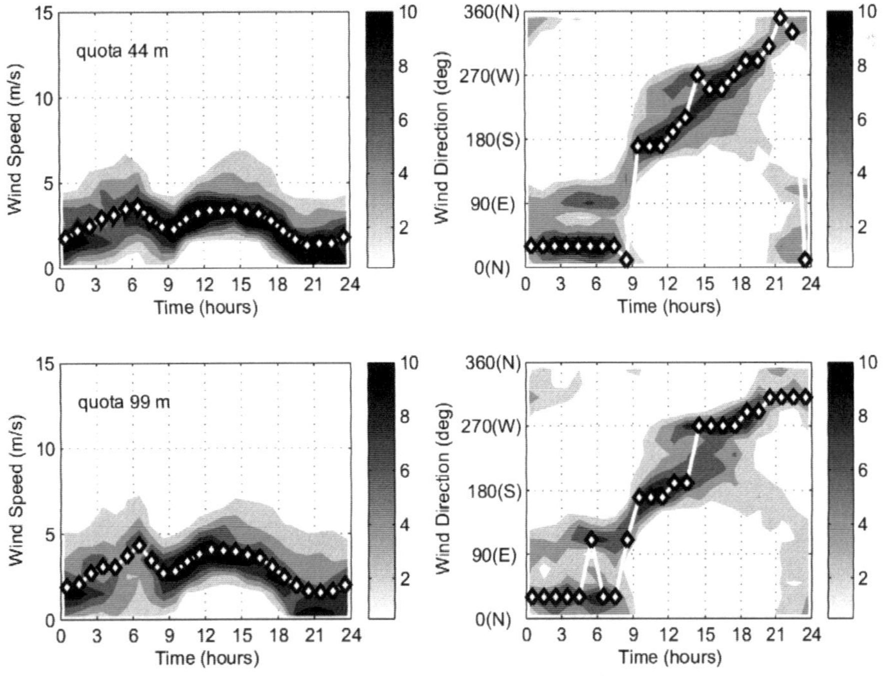

Figure 2. Diurnal behavior of wind, speed and direction, at Pratica di Mare in the months July and August 1996, during clear days.

Pratica di Mare: Diurnal behaviour for July and August 1996: night-time directions 0°-70°

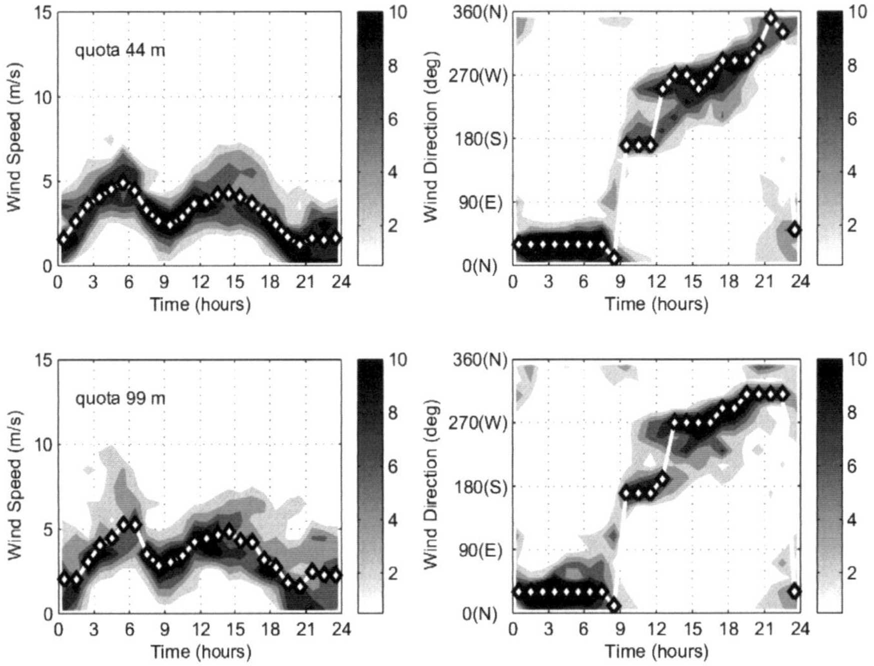

Figure 3. Diurnal behavior of wind, speed and direction, at Pratica di Mare in the months July and August 1996, during clear days with prevailing night-time wind from NNE (sector 0°-70°)

Pratica di Mare: Diurnal behaviour for July and August 1996: night-time directions 80°-150°

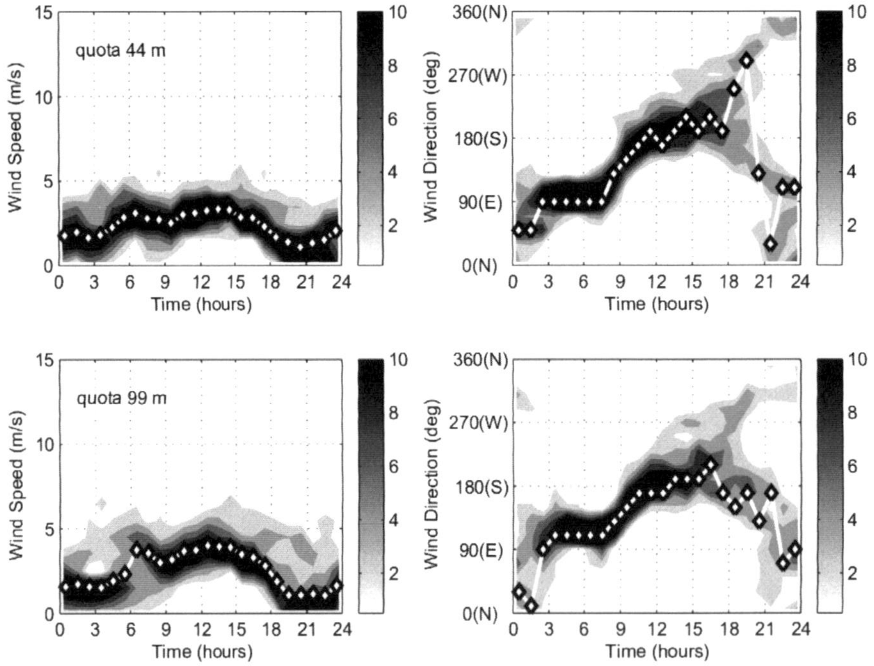

Figure 4. Diurnal behavior of wind, speed and direction, at Pratica di Mare in the months July and August 1996, during clear days with prevailing night-time wind from ESE (sector 80°-150°)

time, in arbitrary units. Hourly averaged values of speed and direction are also shown on these plots as superimposed lines and symbols. The average speed values were calculated as the arithmetic mean. The average values of wind direction were calculated from the median values of wind velocity components. Two alternative patterns already shown [8] clearly emerge. Night-time, two different types of gravity currents may be observed: the first (a drainage flow: direction sector 0°-40°) originating in the low Tiber Valley and flowing over the city of Rome, the second (a katabatic/land breeze flow: direction sector 80°-130°) from the hills East of PDM. Moreover, the complex diurnal pattern that links up with the nocturnal one follows two different paths: in the first, the pattern in the evening links up with the nocturnal regime by a veering of the wind direction, the second by a backing. In order to check the hypothesis that the alternative patterns result from the influence of synoptic scale forcing, and to establish the relationship between them, a procedure that in a first approximation separates the two patterns was adopted. The procedure assumes that in most of the selected cases the synoptic scale forcing does not vary too much during the day.

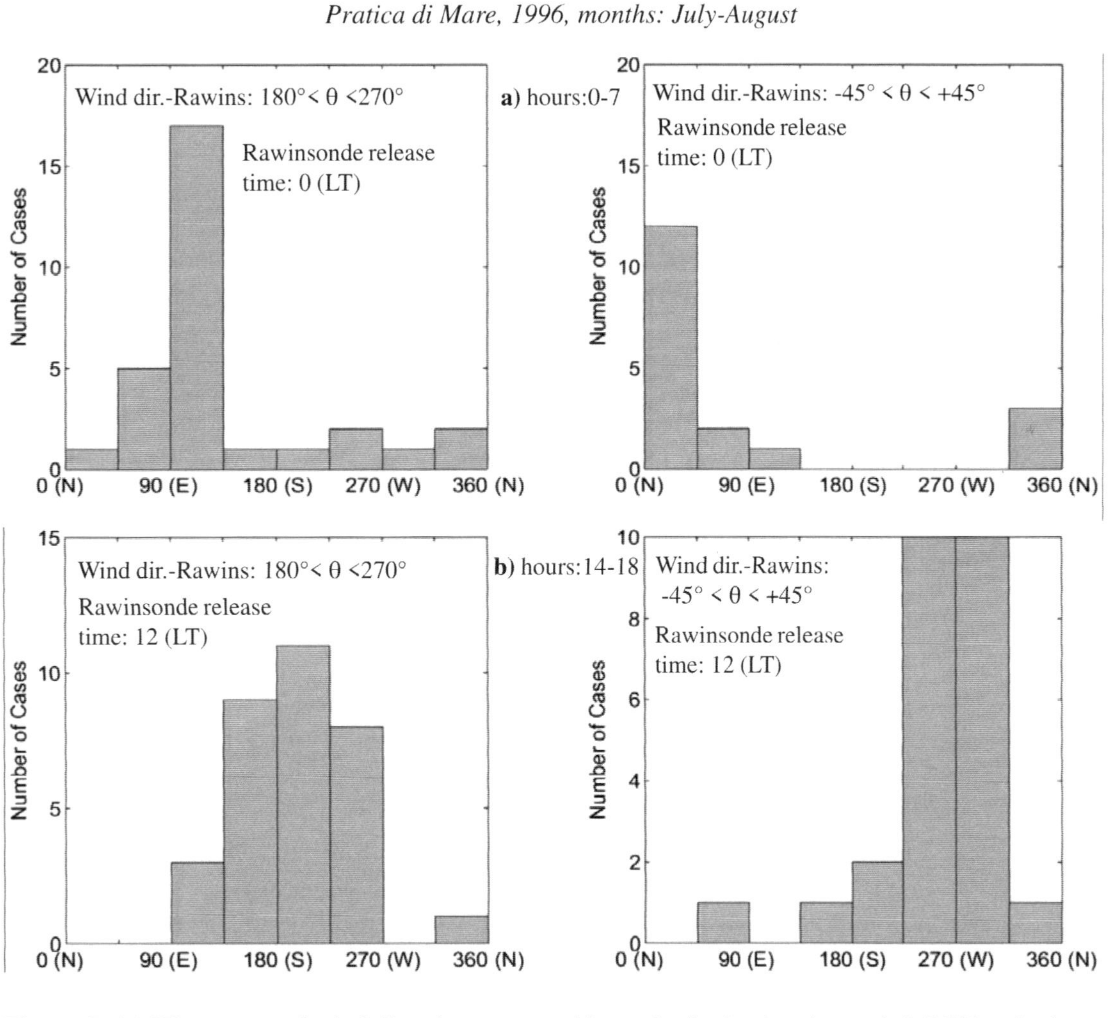

Figure 5. (a) Histograms of wind direction measured by sodar in the time interval 0-7 LT at the layer between 30-150 m, for wind direction measured at 0 LT, at the layer centred at 1500 m in the sector 180°÷270° (left) and in the sector -45°÷45°(right); **(b)** same as in (a) but the wind at 1500 m is measured at 12 LT and the time interval of sodar measurements is in the time interval 14-18.

Two datasets were obtained by using sodar data recorded during fair weather days in the lowest range-gates. In the first dataset, the records in which in the interval 03-06 LT the wind was from the sector 0°-70°, the second in which in the same interval the wind was from the sector 80°-150°. The results of the statistical analyses are shown in figures 3 and 4. The comparison of the two figures demonstrates that the procedure is able to separate the two local circulation patterns. The first, (night-time direction 0°-70°) has a night-time wind speed behaviour that may be clearly associated to a well established flow originated by radiative surface cooling and that has a peak just before the hour 06 (LT) in the morning. Daytime the same background conditions lead to a flow that in the evening joins up the nocturnal pattern by a veering of the wind direction. The second (night-time direction 80° - 150°), unlike the previous case, has not a nocturnal wind speed definite pattern, though a relative maximum is noticeable right after 06 (LT) in the morning. The daytime behaviour shows a pattern that joins up the night-time circulation by a backing of the wind direction.
To investigate the influence of the synoptic scale forcing, the rawinsonde wind profiles collected routinely in the site were used. A scatter diagram (not shown) of wind directions below 150 m (sodar data) versus the ones at 1500 m (rawinsonde data) was inspected and two clusters of low level wind directions were associated to two geostrophic wind prevailing directions (-45 +45°) and (180° ÷ 270°). In figure 5 histograms of wind directions measured by sodar in the range 30-150 m are shown. In figure 5a sodar data were recorded in the time interval 0-7 LT, while in figure 5b data were recorded in the time interval 14-18 LT. The influence of the geostrophic wind on the surface wind pattern clearly emerges. When for the night-time regime geostrophic wind blows from the sector 180° ÷ 270°, the low level wind blows from ESE (figure 5a-left); when geostrophic wind blows from the sector -45° ÷45°, the low level wind blows from NNE (figure 5a-right).
During the day the above conditions as for the wind at 1500 m lead to a more likely backing (figure 5b left) or veering (figure 5b right) of the wind direction in joining up to the nocturnal pattern.
Numerical simulations were carried out to reproduce the low level wind field for four days characterized by fair weather conditions. In two of them (August 6 and 14) the wind at 1500 m was blowing from SW, in the others (August 19 and 23) the wind was from NNW. In figure 6 the wind field pattern at 10 m above surface as results from numerical simulation for August 6 at 06 (LT) and for August 19 at 03 (LT) are shown. Both directions and wind speeds are in good agreement with the observations at Pratica di Mare. However, the simulations of August 6 (and 14 not shown) fail in reproducing the drainage flow from the Tiber valley that was actually measured in the Rome area.

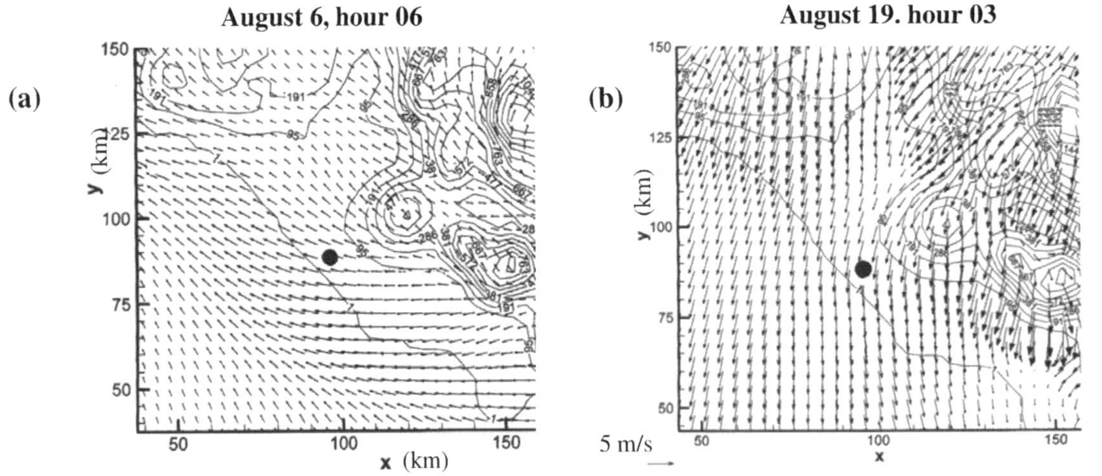

Figure 6. Wind field at 10 m as reproduced by numerical simulations for August 6, when geostrophic wind at 1500 m was from 232° (a) and for August 19 when wind was from 330° (b)

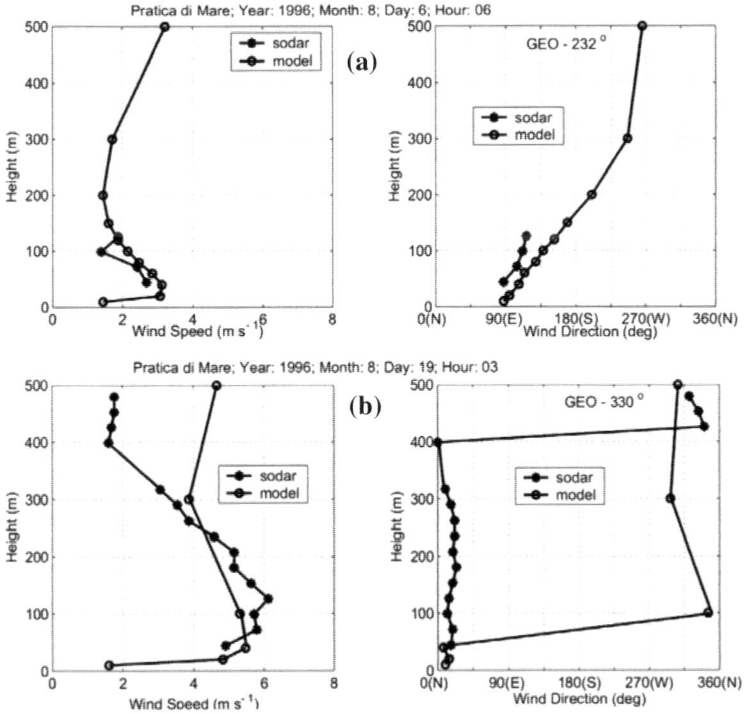

Figure 7. Comparison between profiles measured by sodar and simulated by the CSUMM numerical model at Pratica di Mare for the cases described in figure 6. (a) August 6, 1996 at 06; (b) August 19, 1996 at 03.

In figure 7 the profiles obtained by the numerical model CSUMM for the cases reported in figure 6 are shown. The agreement is satisfactory. In both cases the wind speed profile typical of a gravity current is evident. The relations between the wind direction at 1500 m and in the lowest layers is confirmed by the numerical simulation carried out for the selected days.

4. Conclusions and remarks

A suitable statistical analysis of the low level wind field at the Pratica di Mare airfield allowed to highlight the main characteristics of the local circulation and of its evolution during the day. Even if some of the synoptic scale forcing was filtered by selecting only data recorded in fair weather days, a residual influence is still present and determines which of the nocturnal gravity currents is recorded at the site: the drainage flow from the Tiber valley (NNE) or the catabatic/land breeze flow (ESE). By analysing the correlation between the geostrophic wind obtained by rawinsonde data and the low level wind, it was found that, when the geostrophic wind is from the northern sectors, the drainage flow from the Tiber valley reaches the Pratica di Mare site. In this case the low level wind speed increases during the night up to approximately 5 m/s just before 06 LT. Then it decreases, reaching a minimum at 09 LT, when the diurnal circulazion (sea breeze) begins.

According to the statistical analysis of sodar data two patterns of diurnal wind regime may be observed: the first, blowing from the sector 0° ÷ 70° during the night and veering in the evening to link up to the nocturnal regime; the second blowing from the sector 80° ÷ 180° during the night and backing in the evening to match the nocturnal regime.

The existence of the first or of the second regime correlate with the direction of the geostrophic wind. The first is observed mainly when geostrophic wind is from the sector -45 ÷ +45°; the second when geostrofic wind is from the sector 180° ÷ 270°.

It is then possible to more accurately forecast the diurnal pattern of the local circulation if the relationship between geostrophic wind and the low level circulation is established. A problem arises when it is not

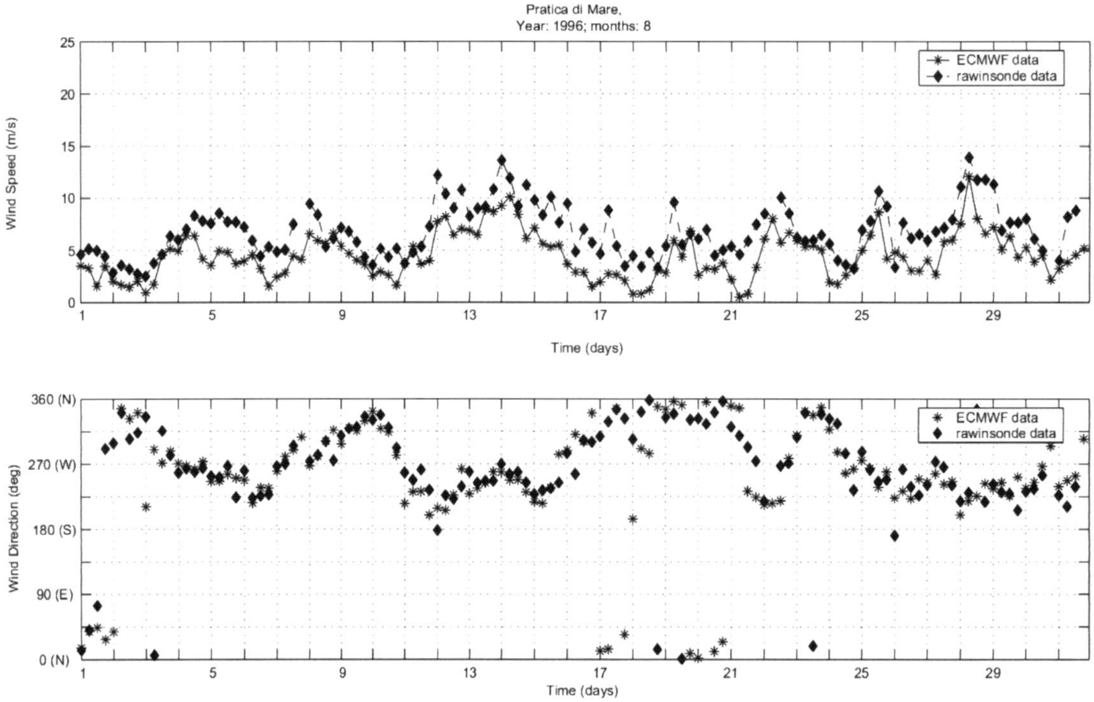

Figure 8. Comparison between wind speed and direction measured by rawinsonde and derived by ECMWF analysis at 1500 m above Pratica di Mare airfield are shown for August 1996.

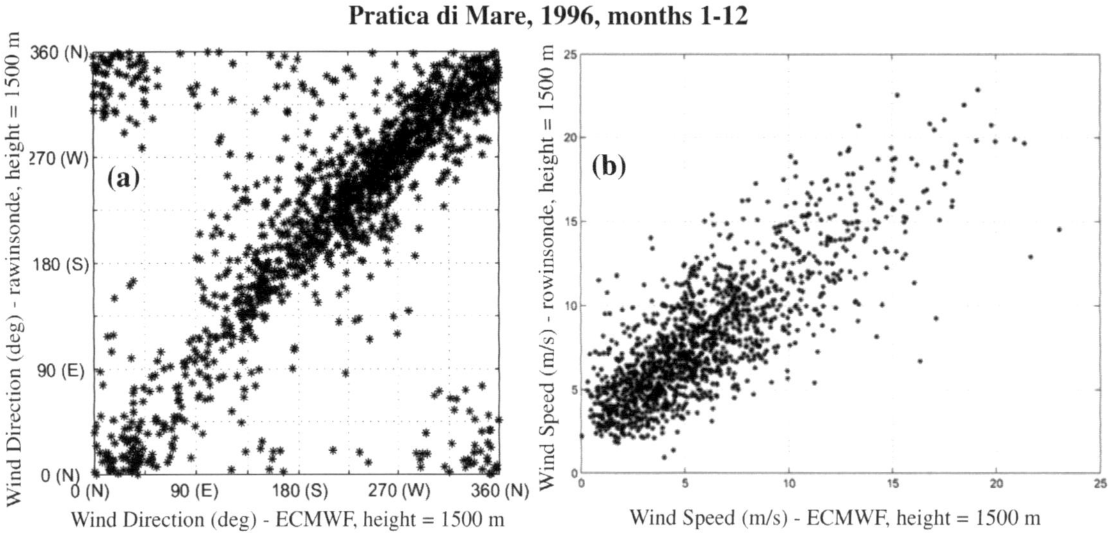

Figure 9. Correlation between wind speed (a) and direction (b) measured by rawinsonde and derived by ECMWF analysis at 1500 m above Pratica di Mare airfield for the year 1996.

possible to rely on nearby rawinsonde data. In this case we verified that similar information may be derived by ECMWF (European Centre for Medium-Range Weather Forecasts) data analyses, at least as for the wind direction. In figure 8 the wind at 1500 m above the observation site, derived by the ECMWF data analyses, is compared with rawinsonde wind data. As for the direction, the agreement is quite satisfactory, while for the wind speed, the ECMWF data underestimate. The same conclusions may be obtained if the analysis is extended to the entire year and a correlation is carried out between the two sets of data. In figure 9a the wind direction correlation is shown. By looking at this figure it should be taken into account that, when wind speed is low, the error in the direction increases and this is very likely at the origin of the spread of most of the points far from the regression line. In figure 9 b the wind speed correlation is shown. The underestimation of ECMWF data is evident.

Acknowledgments
The authors thank the "Segretariato Generale della Presidenza della Repubblica" and the "Commissione Tecnico-Scientifica di Castelporziano" for the support. This work was financially supported by the Accademia Nazionale delle Scienze. The authors gratefully thank Dr. S. Federico for providing with ECMWF wind data.

References

[1] Mastrantonio G, Viola A, Petenko I, Argentini S, Conidi A, Coniglio L 2006, Caratterizzazione della circolazione locale mediante analisi di dati di vento. *Il sistema ambientale della tenuta presidenziale di Castel Porziano.* II Serie. Accademia Nazionale delle Scienze detta dei Quaranta. "Scritti e Documenti" XXXVII, Roma, 13-49.

[2] Martano, P, Mastrantonio G, Miglietta M, Moscatello A, Petenko I, Prodi F, Viola A 2005, Studio di climatologia dei sedimi aeroportuali di Catania, Genova, Napoli, Reggio Calabria, Fiumicino, Malpensa, Olbia. *Relazione dell'attività prevista nella lettera d'ordine, prot. GC/BDR/5007600/ 604 del 31 dicembre 2004,* ENAV O.d.A. 200401526, pp 493.

[3] Palau J L, Péerez-Landa G, Diéguez J J , Monter C, and Millan M M, 2005, The importance of meteorological scales to forecast air pollution scenarios on coastal complex terrain. *Atmos. Chem. Phys.*, **5**, 2771-85.

[4] Savijarvi H, and Liya J, 2001: Local winds in a valley city. *Bound. Lay. Meteorol.,* **100**, 3001-319.

[5] Barros N, Borrego C, Toll I, Soriano C, Jiménez P, Baldasano J M, 2003, Urban Photochemical Pollution in the Iberian Peninsula: Lisbon and Barcelona Airsheds, *Air&Waste Manage. Assoc.*, **53**, 347-59.

[6] Azorin-Molina C, Baldi M, Dalu G, 2008: An analytical study of the inland penetration of sea breezes in presence of complex topography. *Geoph. Res. Abstr.*, **10**, EGU2008-A-05678

[7] Sutton O. G., 1977: *Micrometeorology*, Robert Krieger Pub. Co., Huntington, NY, pp. 333.

[8] Mastrantonio G, Viola A., Petenko I, Argentini S, 2006: *Characteristics of the Local Circulation in the Low Tiber Valley Inferred by the Statistical Analyses of Wind Measurements.* Proc.13th Int. Symp. Adv. Bound. Layer Rem. Sens., Garmish-Partenkirchen, Germany, 18-20 July 2006, 112-14.

[9] Mastrantonio G, Fiocco G, 1982: Accuracy of wind velocity determinations with Doppler Sodar, *J.Appl.Met.*, **21**, 820-30.

[10] Pielke R A,1974: A three-dimensional numerical model of the sea breeze over south Florida, *Monthly Weather Rev.*, **102**, pp. 115-39.

IOP Conf. Series: Earth and Environmental Science **1** (2008) 012050 doi:10.1088/1755-1307/1/1/012050

Laser measurements of flow over a forest

Jakob Mann, Ebba Dellwik, Ferhat Bingöl and Ole Rathman

Risø National Laboratory/DTU, Wind Energy Department,
Roskilde Denmark

jakob.mann@risoe.dk

Abstract. It is estimated that 20-30% of the total European wind energy growth takes place in areas where the wind flow is affected by forests. The description of the wind conditions near and above forests poses a challenge, since assumptions of classical boundary-layer theory are violated. Turbines are designed for maximal turbulence intensity and wind profile gradient. In forested areas, these limits are often violated possibly leading to reduced turbine life-time. In this paper we investigate the mean wind profile and turbulence statistics above an 85 years old dense beech forest by use of a laser Doppler anemometer and compare the profiles with a CFD model specifically made for the modelling of flow over vegetation canopies.

1. Introduction

Wind turbine sites in or near forest are problematic because of the high turbulence intensity, implying larger structural loads, and lower average winds. However, the advantage is less visual impact and lower prices of the land, and despite the drawbacks an increasing number of turbines are erected in or close to forests. In northern Europe more than half of all turbines are erected in forested areas.

Over ordinary land surfaces the mixing length is proportional to the distance to the surface, but over a forest the situation is more complicated. Also the correlation coefficient of the horizontal and vertical velocity fluctuations is larger right over the forest compared to an ordinary surface [1]. The purpose of this work is to quantify properties of the atmospheric flow over a forest relevant for loads on wind turbines by an atmospheric field experiment using sophisticated laser anemometry and to compare them to a κ-ε models of [2].

2. Experiments

Two experiments are a part of our effort to understand aspects of flow over forest with relevance for wind turbines.

The Sorø site is located in an 85 year old beech forest called Lille Bøgeskov (55°29'9.00" N, 11°38' 40.73 E) on the island Zealand, Denmark. The forest extends 1 km in the east-west direction and 2 km in the north-south direction (see figure 1, left).

The area surrounding the forest is dominated by agricultural fields and the area is flat. The western sector of the Sorø site is homogeneous with relatively constant tree top height and is the sector chosen for modeling. The 57 m tall mast is located in the middle of the forest with a fetch of approximately 500 m both to the west and the east. The beech trees are on average 25 m tall, but the forest also contains scattered stands of conifers. Mean leaf area index for the main footprint of the forest is 5 m^2/m^2. The leaf area index is approximately constant between June and September and drops slowly during the autumn. Displacement height and roughness length are d= 20.6 ± 4 m and z_0 = 1.8 ± 0.7 m [3] based on measurements taken during early autumn.

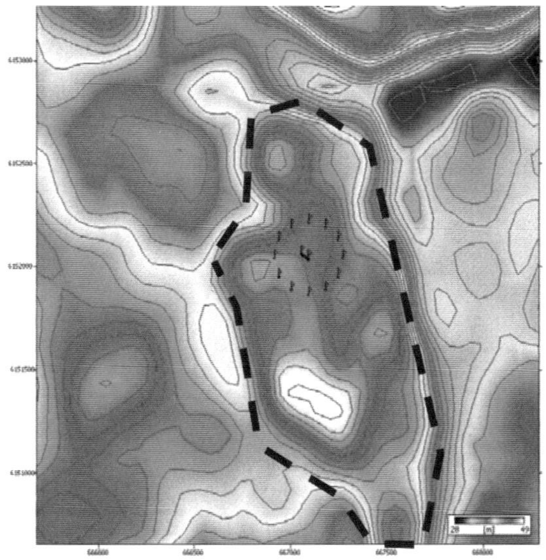

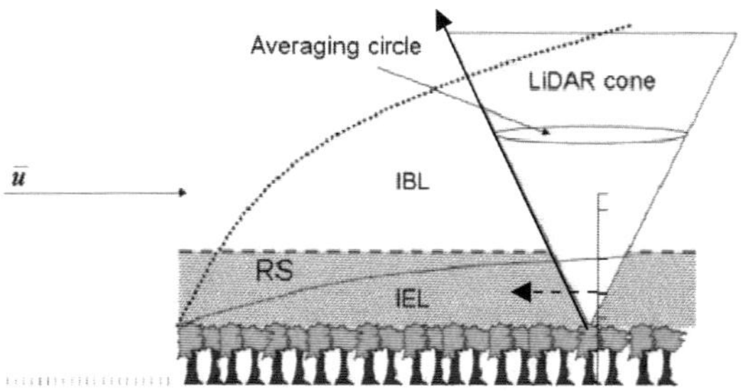

Figure 1: (Top) Sketch of the field experiment with a met mast and laser anemometer situated 0.5 km from the nearest forest edge. IBL: internal boundary layer, RS: roughness sub-layer, IEL: internal equilibrium layer. (Left) Elevation map of the Lille Bøgeskov with the experimental site indicated by the flags. Black line is the forest boundary. Centre flag is the position of mast and the LiDAR. Circle shape is the maximum radius of the conical scan. Truncate site to the NW from the mast is the horizontal scan setup.

In conical scan experiment, next to the mast there is a 24m tall scaffolding tower, where the QinetiQ ZephIR Doppler LiDAR was placed at the level of the tree tops, see figure 1, top. This LiDAR is described in more detail in [4]. Wind measurements in the mast consist of cup anemometers at 37, 41, 48 and 57 m above ground level (Cup P2244, Risø National Laboratory, Denmark). Two sonic anemometers are mounted at 43 and 48 m (Solent 1012R2, Gill Instruments Ltd., Lymington, UK). Wind vanes (Risø National Laboratory, Denmark) are placed at 43 m and 57 m height. Information on other measurements taken at the Sorø site can be found in [5].

In second experiment the heterogeneity over the forest has been tried to measure. The LiDAR has taken from the tower and mounted on to the met. mast at 38.5 m height looking at 300°. The instrument was used without the wedge so it was measuring in beam direction, horizontally, between 5 and 45 m in 5 m steps.

3. CFD model description

SCADIS, which is an acronym for Scalar Distribution, is a κ-ε type flow model for the time averaged Navier—Stokes equations. It has mainly been used for estimation of the source area, the so called footprint, for scalar measurements of carbon dioxide. For this study we disregard the footprint calculator and focus on the flow model output. The SCADIS flow model is especially designed for modelling the flow in and above vegetation canopies. It is well documented in the scientific literature [2,6,7,8] and has been verified against numerous field measurements. The model exists in both 2D and a 3D version. The user-friendly 2D version can be run at a normal PC, where a model run takes approximately an hour. The vertical domain and resolution, which both are fixed, are 0-3025m and 75 points respectively. The default horizontal resolution is 50~m and the PC version allow for 50 points.

Model input is one sided total leaf area index (LAI), LAI vertical distribution, vegetation height and ground/soil roughness length. This information is given for each vegetation type in the modeling domain. Subsequently the different vegetation types are placed on a matrix, representing a map. Model input for the western sector of the Sorø site is given in table 1. We model both a typical winter case and a summer case with a higher LAI in order to improve our understanding of how the LAI parameter affects the flow.

Table 1: Input parameters for SCADIS runs. Run 1 and 2 correspond to typical winter and summer conditions respectively. The LAI is given in m2/m2, leaf area distribution (LAD) in m2/m3 and the number corresponds to a fix setting in SCADIS, h is vegetation height in m and z0 is the ground/soil roughness length also in m.

Run	Agricultural fields				Forest			
	LAI	LAD	*h*	z_o	LAI	LAD	*h*	z_o
1	0	0	0	0.05	1.5	5	25	0.1
2	4	3	0.5	0.1	5	5	25	0.1

4. Results

From December 2006 and the first half of January 2007 we select a number of half hour periods from conical scans in order to compare the measured flow with the model. We select neutral atmospheric stratification $|(z-d)/L| < 0.05$, where the displacement length is 20.6 m We choose a narrow westerly sector defined by the 57 m vane readings being between 270° and 290°. Moderately strong winds are selected: The cup 57 m measures a half hour mean wind speed between 6 and 8 m/s. The LiDAR measuring heights are 48, 57, 65, 76, 92, 113, 140, 175 m. We had to exclude data from one day in December due to lacking LiDAR data at high altitude. The reason of this absence of data is currently not known. This selection procedure leaves us with 78 half hour periods.

The average spectra of all half hour runs as functions of direction shows that the wind speed increases with height (figure 3). Also, the wind speeds of individual half hour runs are chosen together with the corresponding directions. The heights are chosen such that if the wind profile is logarithmic the wind speeds will be spaced equidistantly. This seems often to be the case. The wind directions are seen to turn slightly clockwise as height increases for most of the half hour runs. This is qualitatively what should be expected from the Ekman turning. It would be difficult, if not impossible, to align ordinary vanes in a mast to measure the same, and it is something that can be compared with model calculations, if these include a Coriolis force.

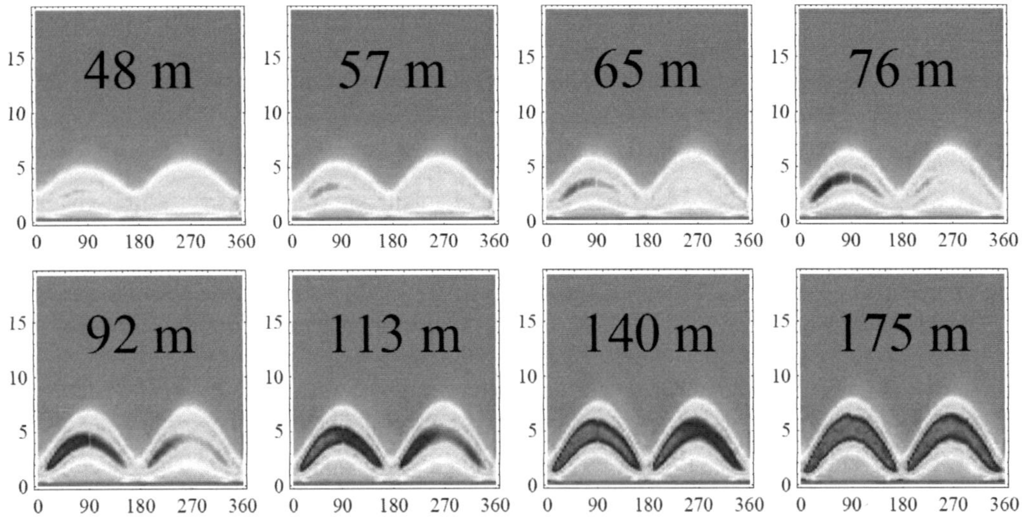

Figure 2: Average spectra at eight heights above ground level shown as functions of direction in ° and radial wind speed in m/s. The upwind direction is around 270°.

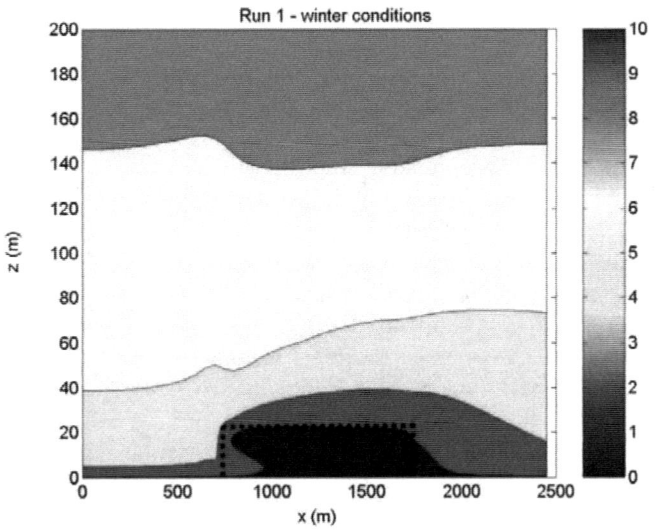

Figure 3: SCADIS 2D calculations of wind speed in m/s for the winter conditions. The extend canopy is shown with a dashed line.

Choosing the appropriate LAI for the winter period is not a straight forward matter. On one and the measured and actual leaf area is low, but on the other hand the supporting structure of the leaves (trunks, branches, and twigs) which absorb most of the momentum in the flow, is the same all year round. In this sense, one could argue that it is the leaf area's supporting structure which is the most relevant parameter. The effect of the leaves on the flow is poorly investigated and more research is needed.

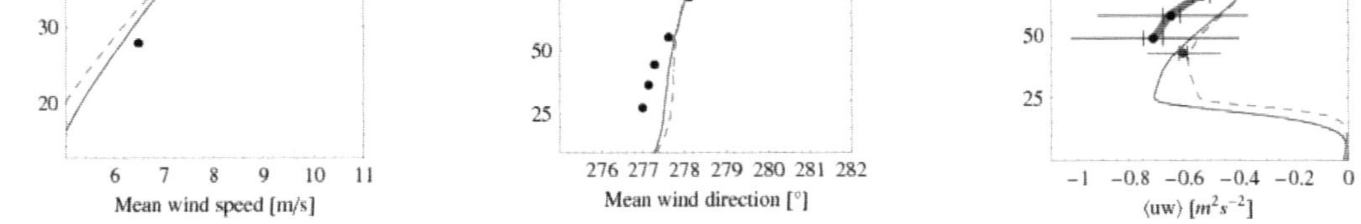

Figure 4: (Left) Measured half hour mean wind speeds (dots) compared to SCADIS calculations for winter conditions (solid, blue) and summer conditions (dashed, red). (Middle) Mean wind directions compared to simulations. (Right) Momentum flux measurements. The blue lower point is from a sonic, while the rest of the points are from the LiDAR. The broad solid curve is another way of calculating the momentum flux from the LiDAR data. SCADIS simulations also shown.

The average of all comparison runs are shown in figure 4. We are quite confident about the mean wind speed and direction profiles, but the momentum flux needs some discussion. First of all, it seems that the LiDAR overestimates slightly the momentum flux as compared to the sonic. We have made a similar comparison of momentum fluxes at another site (Høvsøre) with a newer version of the LiDAR, which indicates the same. More detailed analysis is needed to explain the difference. Another obvious difference between LiDAR and sonic momentum fluxes is the scatter. The standard deviation of all half hour runs is shown as a horizontal line for both instruments. The shorter interval is the standard deviation of the mean, i.e. the standard deviation divided by the square root of the number of half hour runs. The standard deviation of the LiDAR is twice as large as the sonic. One cause for this could be that the LiDAR measures only for less than three minutes at each height for each half hour run, which must imply are larger uncertainty.

The SCADIS simulations compare very well with the mean wind speed and direction profiles. Two curves are shown in figure 4; one for winter conditions, which is most relevant to compare to the measurements, and one for summer conditions. The momentum flux profiles compare less well.

From January 2007 to March 2007 we selected a number of half hour periods from the horizontal wind profile measurements. The data is split into day and night time in order to see the homogeneity in both conditions. Unfortunately, there were not much amount of data from expected direction and only couple of day and night time profiles was available. Using measurement from other wind directions would not be very realistic due to the fact that LiDAR only measures the wind speed in beam direction. In all horizontal profiles, all the measurements from the focus distances are quite close to each other and correlate with the cup anemometer at 38.5 m. This shows that the measurement distances are under the boundary layer that as a theory the wind is homogeneous over the forest at related height therefore we did not try SCADIS model on this setup.

5. Conclusion

We can measure vertical profiles with a LiDAR both of the mean wind speed, the wind direction, and, with more uncertainty, the momentum flux. The calculations with the SCADIS model for the vertical profile are very encouraging. Especially the mean wind and direction profiles are predicted well. For the horizontal wind profile measurements, according to the results, only conclusion can be made as; the boundary layer is higher than our measurement height from the edge of the forest.

Acknowledgments

We acknowledge support from the Committee for Energy and Environment under the Danish Strategic Research Council through grant no. 2104-05-0076.

References

[1] Raupach M R, Finnigan J J & Brunet Y 1996 Boundary-Layer Meteorol. 56, 163–195.
[2] Sogachev A & Panferov O 2006 Boundary-Layer Meteorol. 121(2), 229–266.
[3] Dellwik E & Jensen N O 2005 Boundary-Layer Meteorol. 115(2), 179–204.
[4] Mann J, Dellwik E, Bingol F, Rathmann O 2007 *J. Phys.: Conf. Ser.* 75 012057 (7pp)
[5] Pilegaard K, Mikkelsen T N, Jensen C B N O, Ambus P & Ro-Poulsen H 2003 Boreal Environment Research 8(4), 315–333.
[6] Sogachev A, Leclerc M Y, Karipot A, Zhang G & Vesala T 2005 Agricultural and Forest Met. 133(1–4), 182–196.
[7] Sogachev A, Menzhulin G V, Heimann M & Lloyd J 2002 Tellus B 54(5), 784–819.
[8] Sogachev A, Panferov O, Gravenhorst G & Vesala T 2005 Theor. and Appl. Climat. 80(2–4), 169–185.

IOP Conf. Series: Earth and Environmental Science 1 (2008) 012051 doi:10.1088/1755-1307/1/1/012051

Time series analysis of continuous-wave coherent Doppler Lidar wind measurements

M. Sjöholm, T. Mikkelsen, J. Mann, K. Enevoldsen and M. Courtney

Wind Energy Department, Risø DTU National Laboratory for Sustainable Energy, P. O. Box 49, DK-4000 Roskilde, Denmark

Email: mikael.sjoeholm@Risoe.dk

Abstract. The influence of spatial volume averaging of a focused 1.55 μm continuous-wave coherent Doppler Lidar on observed wind turbulence measured in the atmospheric surface layer over homogeneous terrain is described and analysed. Comparison of Lidar-measured turbulent spectra with spectra simultaneously obtained from a mast-mounted sonic anemometer at 78 meters height at the test station for large wind turbines at Høvsøre in Western Jutland, Denmark is presented for the first time.

1. Introduction

Remote sensing of the atmosphere is an appealing approach compared to in-situ sensing since it circumvents the need to place an instrument at the actual probing height and thus does not require high supporting masts, aircraft platforms or similar expensive structures. This advantage of remote sensing is particularly beneficial in the context of wind speed and turbulence measurements for wind energy applications where the constantly increasing sizes of wind turbines demand higher, more expensive and practically cumbersome meteorology masts for supporting the standard cup anemometers and sonic anemometers that are commonly used for in-situ wind measurements. However, it is important to be aware of the differences between the wind speed observed by remote sensing instruments and the wind speed measured by common standard devices if one wants to be able to compare results obtained by the different methods.

Here, we will deal with the volume averaging effect of a continuous-wave (cw) coherent coaxial Doppler LIght Detection And Ranging (Lidar) device. This instrument senses remotely the wind velocity component along the emitted beam of 1.55 μm laser radiation by extracting the wind speed induced Doppler shift of the radiation field. The ability to measure the wind at various locations along the laser beam is in the present case of a cw Lidar obtained by setting the focus of the beam to the location of interest. Thus, this method does not provide a true point-measurement of the radial wind speed, but rather the wind speed measured is a weighted average of velocities along the beam where velocities at locations near the focus location contributes more to the observed velocity. In the following sections we will describe a method for doing an inter-comparison between a remote sensing wind measurement instrument prone with a non-zero volume averaging effect and a device such as a sonic anemometer that is close to an ideal point measurement instrument in comparison. In particular, results from a recently conducted Lidar vs. mast measurements inter-comparison field campaign (The December 2007 Musketeer-experiment) in Western Jutland (Høvsøre), Denmark are presented.

2. The field measurement campaign

During a measurement campaign at the Risø DTU operated test station for large wind turbines, Høvsøre, in Western Jutland, Denmark, conducted in December 2007, a ground based cw coherent wind Lidar (ZephIR prototype manufactured by QinetiQ, Malvern, U.K.) was set to stare (not scan) towards a Metek sonic anemometer positioned at 78 meters height above ground. The Lidar beam was focused at a distance of 138 m and 56° down from zenith in the vicinity of the sonic anemometer, which was situated in a direction of 26° East of North.

Time series of the 10-minutes averages of the remotely sensed radial wind component along the line-of-sight of the Lidar is presented in figure 1 together with the absolute value of the projection in the same direction of the velocity vector measured by the sonic anemometer. The curves follow each other fairly well although some deviations do occur. For instance during episodes of low clouds the cw Doppler Lidar can pick up erroneous contributions to the wind speeds at the cloud base, even if it is far from the focus location, due to the significant higher backscatter from the cloud base compared to the backscatter from aerosols at the aimed measuring height. This cloud problem can in principle be dealt with by a proper cloud-correction algorithm [1], although that was not available for implementation in the Lidar used during the particular measurement campaign presented here. Instead, the cloud base was monitored by a ceilometer and only periods of relatively high cloud base altitude was included in this study for more detailed inter-comparison.

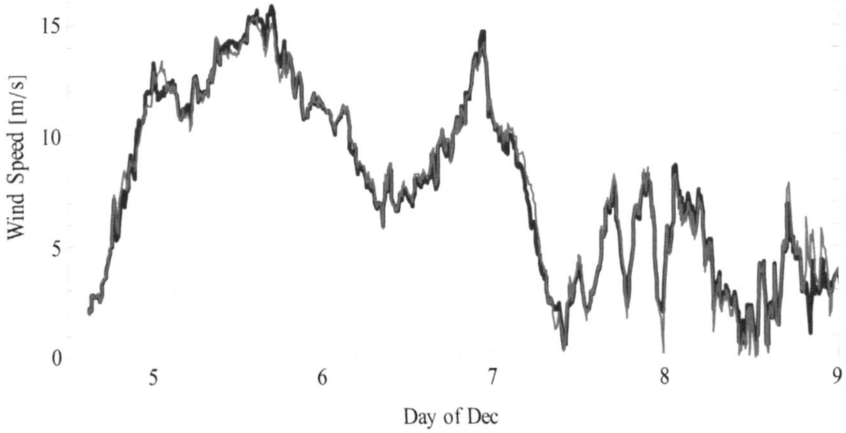

Figure 1. Time series of 10 minute averaged radial wind speed measured by a sonic anemometer (Red curve) and a focused cw coherent Doppler Lidar (Blue curve) at the Risø DTU operated test station for large wind turbines at Høvsøre in Western Denmark.

3. The spatial volume averaging effect of a focused continuous-wave (cw) Doppler Lidar

The wind speed $v(s)$ measured by a cw Lidar focused at a location x can generally be expressed by an integration along the laser beam of the wind field $\vec{u}$ projected to the direction of the beam,

$$v(x) = \int_{-\infty}^{\infty} \varphi(s)\vec{n} \cdot \vec{u}(s\vec{n} + x)ds \, , \tag{1}$$

where $\vec{n}$ is a unity vector along the beam and $\varphi(s)$ is the spatial volume averaging function.

This is a general formulation and for a particular instrument the spatial volume averaging function, $\varphi(s)$, needs to be specified. For a focused cw coherent Doppler Lidar $\varphi(s)$ is shown, for instance by [2], to be well described by a Lorentzian-shaped spatial weighting function

$$\varphi(s) = \frac{1}{\pi} \frac{\frac{1}{z_R}}{1+(s/z_R)^2}, \tag{2}$$

where s is the distance from the focus point along the beam and z_R is the so-called Rayleigh length, related to the depth of focus, which depends quadratically on the focus distance, which means that the spatial confinement becomes wider to the second power, the farther away the measurement volume is centered.

Assuming the beam to be in the direction of the mean wind speed, the counterpart in wave-number domain to the cw Lidars line-of-sight spatial averaging function, equation (2), is an exponential low-pass filter-function for the wind speed power spectrum,

$$L_{Lorentzian}(k_1) = \exp(-2z_R|k_1|), \tag{3}$$

where k_1 is the wave number along the mean wind direction. Here, we can notice a similarity with an exponential radiative decay, which is well-known to result in radiation with a Lorentzian spectral line profile that becomes broader, the shorter the lifetime of the decaying radiative state is.

The transformation between the measured frequency and the in equation (3) applied wave number is based on the Taylor's frozen-wave hypothesis [3], which means that the average wind speed in the mean wind direction U is the velocity used for the coupling between time and space, i.e., between wave number and frequency ($\omega/k \approx U$). Thus, if this hypothesis is applicable, the ratio between a wind spectrum measured by a Lidar directed along the mean wind direction and a wind spectrum measured simultaneously by an almost point measuring sonic anemometer is basically the spectral filter function, e.g. equation (3).

If however the line-of-sight of the Lidar does not coincidence with the mean wind direction, the ratio between the volume-averaged Lidar-observed wind speed spectrum and the spectrum of the corresponding wind speed component measured by a sonic anemometer, will in general then be dependant on the entire three-dimensional structure of the turbulence. Thus, the full 3-D spectral tensor has to be considered and it can be calculated by a model which estimates it from time series of the three wind components provided by a sonic anemometer [4, 5].

4. Experimental results in comparison to the theoretical spatial filter model

Wind data from a period with no cloud interference was selected for further analysis of the spatial volume averaging effect of the Lidar. Here some preliminary results of the analysis of wind data from such a period of about 5 hours in the morning of December 5th, 2007 are presented. Figure 2 shows an example of 20 Hz sampled time series of the cw Lidar-measured wind speed together with the corresponding sonic-measured wind velocity projected along the line-of-sight of the Lidar. It can be seen that the time series of the wind measured by the sonic anemometer displays more high-frequency wind fluctuations than the corresponding time series measured by the cw Lidar. This is in intuitive agreement with the expected filtering effect of the Lidar as discussed in section 3.

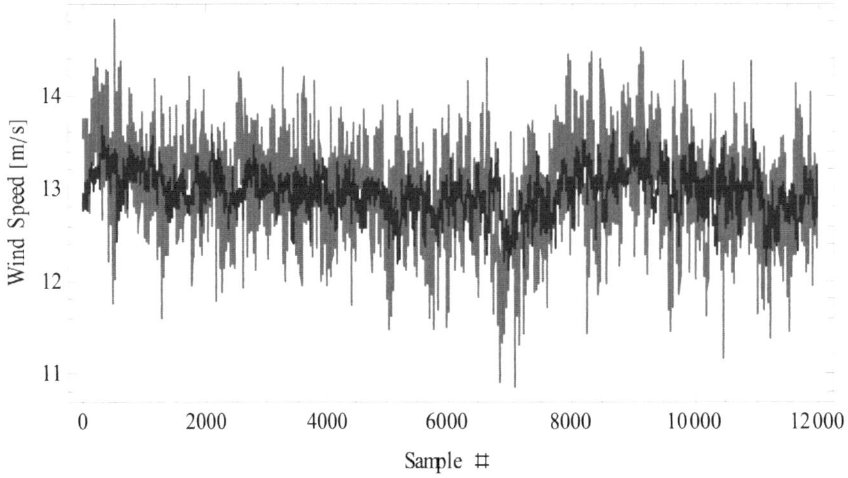

Figure 2. An example of wind time series measured by a sonic anemometer (Red curve) and a focused cw coherent Doppler Lidar (Blue curve). The data presented were sampled at 20 Hz for a period of 10 minutes.

Spectra of the time series of the wind component along the line-of-sight of the Lidar obtained by the Lidar itself and as a projection of the wind velocity measured by the sonic anemometer were averaged during the five hours period selected and are displayed in figure 3. It can be seen that the Lidar-measured spectrum (Blue curve) at high frequencies (wave numbers) falls off more rapidly than the sonic anemometer based spectrum (Red curve). However, at the very highest frequencies (above 2-3 Hz) the Lidar spectrum is seen to increase again, which is likely due to noise, since the particular prototype ZephIR's sensitivity was relatively low during the measurement campaign due to elevated power loss in the optical fiber connectors. It should also be noted that the atmosphere was rather strong stably stratified during the 5 hour period analyzed, implying that a rather small longitudinal turbulence length scale was present, and therefore the effect of the spatial volume averaging on the total measured variances was relatively large.

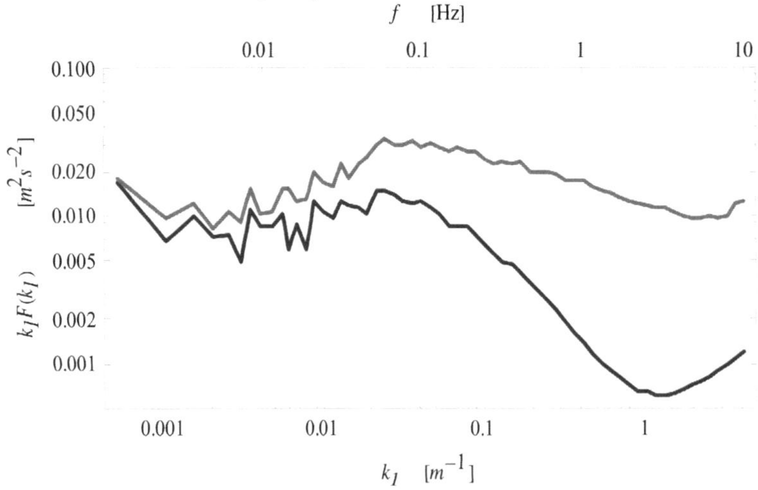

Figure 3. The Lidar-measured wind spectrum (Blue curve) and the corresponding projected sonic-measured wind spectrum (Red curve) from a measurement period of ~5 hours obtained in the morning of December 5th, 2007 at the Risø DTU operated test station for large wind turbines at Høvsøre.

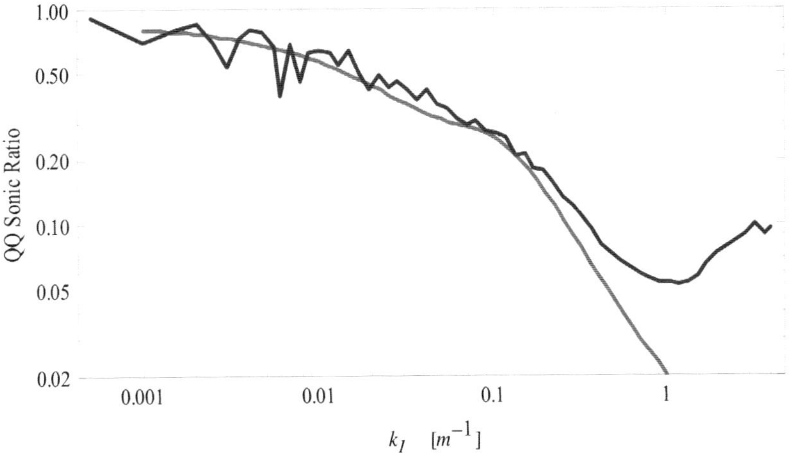

Figure 4. The predicted spectral transfer function, i.e., the ratio between the modelled Lidar spectrum and the sonic spectrum as a function of wave-number (Red curve) and the measured spectral transfer function, i.e., the ratio between the Lidar-measured and the sonic-measured power spectra (Blue curve) from the same measurement set as in figure 3.

In order to compare the experimental results to the theoretical spatial volume averaging function, equation (3), the entire three-dimensional structure of the turbulence, i.e., the 3-D spectral tensor, has to be known. This was estimated by Mann's spectral tensor estimator [4] from the wind velocities measured by the sonic anemometer during the time period of interest. The predicted ratio between the spectrum of the modeled Lidar-observed velocity, corresponding to integration according to equation (1) of the modeled velocity field, and the spectrum of the corresponding sonic-measured wind velocity projected along the line-of-sight of the Lidar is represented by the red curve in figure 4. The blue curve in figure 4 is the ratio between the spectrum of the Lidar-measured velocity and the spectrum of the corresponding wind velocity, projected along the line-of-sight of the Lidar, simultaneously measured by the sonic anemometer. Although the curves deviate slightly for higher frequencies due to the influence of noise, the general agreement between the modeled spectral ratio and the experimental spectral ratio is, as figure 4 shows, surprisingly good.

5. Conclusions

The influence of the spatial volume averaging effect of a focused cw Doppler Lidar on the time series of the wind speed observed has been investigated experimentally as was also recently done for a pulsed Doppler Lidar [5]. The preliminary results indicate that the procedures presented, with a more extensive analysis of similar data, has the potential to reveal the details of the spatial volume averaging function of a remote sensing coherent Doppler Lidar.

6. Acknowledgement

The authors gratefully acknowledge Bjarne Sønderskov and Anders Ramsing Vestergaard for mounting the instruments and all the other participants of the Musketeer experiment at Høvsøre in December 2007 for their support during the inter-comparison campaign. This work was financially supported by the Committee for Energy and Environment under the Danish Strategic Research Council through grant no. 2104-05-0076.

IOP Conf. Series: Earth and Environmental Science **1** (2008) 012051 doi:10.1088/1755-1307/1/1/012051

References

[1] Courtney M, Wagner R 2008 Testing and comparison of lidars for profile and turbulent measurements in wind energy *Proceedings of ISARS 2008*
[2] Sonnenschein C M, Horrigan F A 1971 *Applied Optics* **10** 1600-1604
[3] Panofsky H A and Dutton J A 1984 *Atmospheric Turbulence* (New York: John Wiley & Sons)
[4] Mann J 1994 *J. Fluid Mech.* **273** 141–168
[5] Mann J, Cariou J-P, Courtney M, Parmentier R, Mikkelsen T, Wagner R, Lindelöw P, Sjöholm M, Enevoldsen K 2008 Comparison of 3D turbulence measurements using three staring wind lidars and a sonic anemometer *Proceedings of ISARS 2008*

IOP Conf. Series: Earth and Environmental Science **1** (2008) 012052 doi:10.1088/1755-1307/1/1/012052

Analysis of Current Status of Theory and Practice of Radio-acoustic Sounding Systems

Kartashov V.M., Babkin S.I., Volokh A.V.
Kharkov National University of Radio Electronics
14, Lenin Avenue, 61166 Kharkov, Ukraine; tel. +38 (057) 70-21-587

E-mail: res@kture.kharkov.ua

Abstract. The analysis of a current status of theory and practice of radio-acoustic sounding systems (RASS), which have been under development since 1961 for remote sensing the profiles of vertical air temperature and humidity and wind velocity and direction, is made. A traditional construction scheme of the RASS, as implemented in practically all known systems, is singled out. Methods used for radio-acoustic atmosphere sounding for the purpose of collecting relevant meteorological information are examined. It's pointed out that all problems of system, technical nature, arising in the course of development of the RASS, were resolved before 2000 by engineering, experimental way. Later on creation of the RASS theory started, and problems of synthesis and optimization of systems, devices, signals were formulated. In this connection an adequate constructive model of a radio-acoustic location channel and synthesized signal processing algorithms are under examination. For determining the form of base oscillations of a multi-channel processing unit the scattering function should be used.

1. Introduction

Presently the radio-acoustic sounding systems enable to remotely determine basic meteorological values (air temperature and humidity, wind velocity and direction) and deem to be promising tools for meteorological observation purposes. Imperfection of the existing RASS results from insufficient level of development of the RASS theory and methodical support to development engineers. Features of such systems as compared to other various location systems are stipulated, first of all, by uniqueness of the scattering volume – acoustic wave packet, on which radio-wave scattering is observed. In particular, owing to peculiarities of the scattering volume the existing methods of synthesis and analysis of radio systems and devices are in demand to significant extent while designing and optimizing the RASS.

2. Analysis of RASS structure

The first RASS used impulse radio and acoustic signals and were based on a monostatic scheme, in which the phase centres of all antennas coincide in one point. The radio-wave length of temperature-wind sounding system EMAC [1], developed in 1961, was 3 cm, acoustic wave length –1.5 cm, and maximum range of action – only 30 m. The obtained range disappointed the developers and the next RASS was designed as late as in 1972. This installation of temperature sounding was called RASS and had the following parameters [1]: radio wave length – 8.15 m; acoustic wave length – 4.075 m. This installation made it possible to get a reflected signal from the altitude of 1.5 km, but the minimum altitude of sounding was 600 m, and the space resolving power - 200 m. Most of the later RASS used pulse acoustic emission and continuous monochromatic radio emission and separate (placed apart at some distance) transmitting and receiving radio antennas. The acoustic antenna, as a rule, is placed in the middle of the distance between the radio antennas. Such layout plan of the antennas as a whole with the abovementioned combination of sounding radio and acoustic signals (system construction scheme) was called <<basic>>.

IOP Conf. Series: Earth and Environmental Science **1** (2008) 012052 doi:10.1088/1755-1307/1/1/012052

By the mid-1980s, when, the principle of construction and structure of the Doppler RAS system mainly formed, approximately 10 operating installations were known in the world, most of which were based on the basic scheme [1]. The differences of such known installations were the following: usable wavelength ranges, emission powers and numbers of wavelengths in the acoustic impulse, the time length of which defines the space resolving power of measurements. All existing at that time systems (exclusive of RAS-10) were fixed. The transmitting and receiving radio antennas most often were dish antennas, and the sound-emitting devices were loudspeaker grills. Some structural features are characteristic for RAS-10 designed by Kharkov Institute of Radio Electronics (currently – Kharkov National University of Radio Electronics (KNURE)) [2]. In this system the radio and acoustic antennas are placed in a manner as prescribed by the basic chart, but on a rotating platform, and the acoustic antenna, moreover, may move in the direction that is at right angle to the line, which connects the centres of the radio antennas. This is made for the purpose of compensation within small limits of wind carryover of the sound package. The operability of the installation in case of a displaced acoustic antenna is provided by way of rotating the platform until the reflected radio signal arrives in the receiver. Then the acoustic antenna is placed at a longer distance in the line and it's necessary to ensure receipt of the signal from the next profile point. This enables to increase the operating range of the system and to perform measurements of velocity and direction of the horizontal wind. Subsequent RASS [3,4,5,] has no principal differences. However, due to improved capacity of computers, such systems offer additional capabilities connected with sounding results processing and displaying. If at earlier time (in 1970-80s) most RASS were developed by groups of experimentalists and manufactured as single units, later (alongside with unique, single designs) small-batch production of these stations started. The designers of the RASS, intended for commercial manufacture, are, as a rule, the same research teams, which make experimental installations for <<themselves>>.

The radio channel of the RASS is realized according to the <<true coherency>> scheme, when the output signal of the radio transmitter and all heterodyning voltages of the receiver form from a single high-stability master generator by way of multiplying its frequency by a required number of times. Such construction scheme enables, if a master generator with sufficiently short-term frequency stability is used, to get the required accuracy of estimations of sound velocity in the atmosphere. The Doppler frequency meter may be implemented both in analogue or digital form. The computer performs the functions of synchronization, control and adaptation, as well as metering results processing. The peculiarity of the RASS is the necessity to fulfil the Bragg condition as dictated by both energy and information considerations. In case of using simple acoustic and electromagnetic sounding signals, change of temperature and radial wind velocity at varying heights leads to breach of the Bragg ratio. A sufficiently strong reflected signal exists only within a certain range of altitudes, and, in this case, the longer the acoustic pulse is, the less this range is [1]. Adjustment to the Bragg condition at higher temperature gradients for each profile point (<<site>>) manually by way of changing the frequency of the emitted sound signal takes quite much time, which, depending on a number of profile points, may be 0.1- 3 hours [1,2]. The used measurement algorithm significantly limits promptness of obtaining the profiles of meteo values – a feature, being one of the basic merits of the RAS technique. In addition, such time measures alike the time of quasi-stationarity of processes in the atmosphere, within which it's possible to do averaging. In this connection, from the very beginning of development of the RAS technique the attempts to get profiles <<by single pulsing >> were made [1], with setting the acoustic generator frequency for the beginning, middle and end of the entire pathway [2]. However, a level of the signal, which is sufficient for confident recording, even in favourable meteorological conditions may be obtained only for a number of nearby profile points. Any breach of the Bragg condition in the end points, in which it's possible to record the signal, results in errors in determining the temperature in the order of 0.5-1.0 ^{0}C [1]. Such error is within the scope of random errors of radio sounders, but turns to be correlated to the sound velocity gradient, in which case the gradient is always reduced in respect to the absolute value. And although the total temperature profile measurement time by such technique reduced significantly, a non-random error, which cannot be reduced by way of averaging and correction of the measurement results, occurs. The cardinal solution of the problem of improving the operational efficiency of the RASS and accuracy of measuring the meteo values may be adjustment of the electromagnetic radiation frequency to the Bragg condition so far as a simple acoustic impulse spreads in the atmosphere. Such techniques are described in publications [1,5] and some others. Since it turns to be impossible to provide fulfilment of the Bragg condition within the entire acoustic wavelength change range by way of adjusting the radio signal frequency, it's expedient to compensate slow (seasonal and daily) changes of meteo values by adjusting the frequency of the sound generator. Such periodical adjustment of the sound frequency may be carried out basing on results of direct measurements of wind temperature and velocity at the ground level (automatically or using an operator). Any compensation of rapid changes of the acoustic wave length (within the term of spread of the packet by the sounding pathway) should be carried out by way of changing the radio emission frequency. However, no evidence of successful implementation of similar procedures is available in practice due to technical complexity of the problem, on the one hand, and imperfectness of the sounding signal frequency control algorithms, on the other hand. The discrete receiving radio antennas may be used in the RASS, including, without limitation, big antennas for sounding up to high altitudes. For efficient use of the grill aperture the receiving electromagnetic and acoustic antennas may in this case move in conditions of changing meteo conditions, holding a position windward, or a number of transmitting antennas, which move, may be used [6]. In addition, relevant signal processing algorithms for such scheme are developed [6]. They allow, in particular, to estimate coordinates of the centre of the spot of scattered signals, which moves in the process of measurements within the grill aperture.

3. Engineering approach to designing the RASS

It should be noted that all problems of system, technical nature, arising while designing the RASS, were solved before 2000 by engineering, experimental way. Upon accepting the <<basic>> scheme of construction of such installations the efforts of all designers were aimed at perfecting different elements of the structure. The possibility of use of different types of antennas, acoustic emitters, radio receivers, Doppler frequency signal emission devices, spectral analysis schemes and algorithms were under consideration. Rather successful combination of the sounding signals, which is widely spread in practice: the pulse acoustic signal with sinusoidal infilling and continuous monochromatic radio signal was also proposed by the engineers. A number of publications of research nature [7,8,9], which are devoted to researching the properties of some types of the sounding signals, appeared later. All results obtained evidenced rightness of the engineering intuition, suggesting use of this combination of signals, and to some extent widened apprehension of properties and capacities of other types of sounding oscillations. The mentioned publications used a sufficiently complex mathematical apparatus that rendered difficult to apprehend the given material by engineers dealing with the equipment, and made problematic its development and consideration of signals of other types. Just these circumstances are the reasons of any errors and delusions, encountered in the mentioned literature, despite high qualification of the authors. Some actual delusions are also set forth in the work [1]. The apparatus problems are not given much attention in the publications devoted to the atmosphere sounding problems using the sound waves, and this attention reduces continuously [1,5]. If before 1983 more than 50% reports given at the ISARS symposia researched different technique- and equipment-related problems, at the next symposia there were less than 10-15% of such reports (the main attention in them was paid to wind profilers and combined systems). The rest of the reports were devoted to different geophysical messages. The reports of leading scientists repeatedly gave opinions that development of the aspects of construction of the RASS practically finished and was on the verge of practical use. Thus, at a certain stage of development in the field of design of the RASS there was a situation, when the main design methods were engineering intuition, based on experience of analogous and similar solutions in other fields, and on the experiment. No works aimed at creating the theory of radio acoustic systems actually were done. And no problems of synthesis and optimization of systems, devices, signals were set. The current situation is stipulated, evidently, by the fact that leading positions and the scientific authority in the area of atmosphere sounding are held by atmosphere specialists – physicists, and, especially, radio physicists, who research the problems of propagation and scattering of waves in the atmosphere. A number of the RAS installations, especially sodars, were designed under the conceptual direction of specialists in this field. All problems, connected with design of equipment, were solved mainly by way of borrowing some known solutions from radio location taking the meteorological conditions and peculiarities of radio-acoustic sounding. The approach used to significant extent predetermined all success achieved at a certain stage of development of the technique and RASS. However, it also slowed further development of this field, because far from all problems of system technical nature in this field may be solved efficiently in such way due to specifics of the RASS. Substantial success was achieved in a field that is sufficiently close to the RAS systems – development of radiolocation of wind profilers (RWP). The wind profiler, in essence, is an electromagnetic channel of the radio acoustic system, designed and built in such manner, so that it could receive weak signals from turbulent irregularities distributed in space. Naturally, as a complex fixed (expensive) facility, the RWP is capable to receive a radio signal scattered at a coherent acoustic wave pulsing (AWP). However, for the purpose of bringing into effect, first of all, all underlying power capacities while using the AWP, it's necessary, in the process of design (or additional development) to take into consideration the occurring system-formation links. Since the RWP is a RLS with high energy potential, which is intended for operation with volume-distributed targets with a very little radar echoing area (REA), while using in them, as a scattering object of the coherent space structure, the AWP, some <<excess>> of potential forms, which enables to use a complex acoustic signal. The radio emission frequency in the RWP may be set as constant (but not varying for meeting the Bragg condition) [11]. The most frequent approach is to use in the RWP the most evident engineering solution – a sound signal of linear frequency modulation (LFM). The acoustic signals with a wide spectrum are presently used also in a number of the RAS installations with <<ordinary>> lower power potential of the radio channel.

4. Basic provisions of the RASS theory

For the purpose of creation of an efficient RASS it's necessary to perform deep theoretical working-out of all relevant equipment-related problems with use of adequate approaches, as many problems of this field are not represented in the theory of radiolocation systems. The basic theoretical results, obtained recently, first of all, by the scientists of Kharkov National University of Radio Electronics, are given briefly below. There is a regularity of forming a radio signal, scattered on the sound, as a mutual autocorrelation function, with regard to the range of emitted electromagnetic and acoustic oscillations. This regularity originates from known ratios obtained in a theoretical way, and is evidenced by numerous experimental results. Adequate constructive models of information channels (radio acoustic and acoustic) are created. A two-dimension cross-correlation function of acoustic and radio signals – scattering function – is introduced. Its main properties, expressing the most common characteristics of the scattered signals, are determined in a theoretical manner. The problem of scattering the electromagnetic wave on the sound is presented using the signal theory and functional analysis apparatus, in which case the functionality of development improves substantially. The function of scattering, <<coopting>> the properties of the interaction environment and characteristics of the sounding acoustic and electromagnetic oscillations, makes it possible to quite simply find a type of the scattered signal, corresponding to various conditions. This opens wide opportunities to research various types of sounding oscillations and development of the theory of the systems under consideration in whole. The sounding acoustic and electromagnetic research techniques are developed. Representation of the scattering function in different mathematical forms and graphically in the form of surfaces – scattering bodies – enables, using different types of body sections, to perform efficient analysis of sounding vector radio acoustic signals. Types of signals that are most frequently used in practice and promising ones are studied. It's shown that the received signal processing algorithms, used currently in the radio acoustic systems, which are borrowed from radiolocation, are not adequate to the processes occurring in the radio acoustic location channel, as they take in no account a change of the signal structure in conditions of scattering on the sound pulsing. This leads to substantial deterioration of quality of the stations, in particular – accuracy of temperature measurements (the error may reach degree units). For obtaining the estimations of maximum credibility of the measured parameter, the reference signals in different points of the range of possible values of this parameter should differ as for type, form, but should not be a copy of the emitted radio signal, as is customary. All forms of reference oscillations of a multi-channel processing device should be determined using the scattering function, reflecting the peculiarities of transformation in the channel. The synthesized signal processing algorithms may be considered as generalization of the radio system theory results in the case, when the signal form at scattering transforms according to the deterministic law. If the scattering volume degenerates into the point target, the received and reference signals become a copy of the sounding, emitted radio signal. The proposed signal processing procedures reduce in this case to the previously known algorithms of estimation of any oscillation parameters.

References

[1] Kallistratova M A, Kon A I Radio-acoustic Atmosphere Sounding. 1985 (M.: Nauka) 196
[2] Babkin S I, Kutsenko V I, Pakhomov Yu A et al System of Radio-acoustic Sounding of Atmosphere of Centimetre Wave Band// 5 All-Union Symposia on Laser and Acoustic Atmosphere Sounding: Abstracts of Report, Part 3. 1978 (Tomsk: TFSO AA USSR) 143-6
[3] P Trivero, A Masorati, P Marcacci, G Bonino Improvements of a Demetric RASS// 1998 Proc.of ISARS'98.(Vienna, Austria) 196-9
[4] Y Akai, T Kanzaki The Application of a Mobile RASS to Observation of an Urban Heat Island// (as above) 200-3
[5] Vogt S Advantes in RASS since 1990 and practical application of RASS to air pollution and the ABL studies// 1996 Proc.of ISARS'96 (Moscow) 37- 50
[6] Kartashov V M Peculiarities of Processing the Radio Signal Scattered by Acoustic Wave Packet //1998 *Radiotekhnika* (Kharkov) № 105 75-9
[7] Kon A I, Nalbandyan O G Radio-acoustic Atmosphere Sounding by Continuous Sound Emission // 1981 *Izv. AS USSR. Atmosphere and Ocean Physics* **17**, № 8 824-36.
[8] Gurvich A S, Kon A I, Nalbandyan O G et al Radio-acoustic Atmosphere Sounding Techniques. 1976 Prepr. IFA AS USSR (M.: Nauka) 43
[9] Azizyan G V Scattered Signal Frequency Spectrum in Radio-acoustic Atmosphere Sounding Systems// 1981 *Izv. AS USSR. Atmosphere and Ocean Physics* **17**, № 8 883-6
[10] Petrov A V, Yakovlev A A Radio Facility Analysis and Synthesis 1984 (M.: Radio and Communication) 246
[11] Wind profiler assessment report 1994 NOOA, Silver Spring, Maryland 141

IOP Conf. Series: Earth and Environmental Science **1** (2008) 012053 doi:10.1088/1755-1307/1/1/012053

Development of Radio Acoustic Sounding Method in Kharkov National University of Radio Electronics

Ye.G.Proshkin, V.M. Kartashov, S.I. Babkin
Kharkov National University of Radio Electronics
14, Lenin Ave., 61166 Kharkov, Ukraine; tel.: +38 057 70-21-333

E-mail: res@kture.kharkov.ua

Summary. The analysis of the role of Kharkov National University of Radio Electronics in developing the atmosphere radio acoustic sounding (RAS) method within the period from 1968 to 2008 was carried out. As a part of the investigation program six experimental models of the sounding radio equipment were developed and manufactured. The atmosphere sounding methods were developed for measuring the base meteorological values. For the first time in the world practice, relevant comparative measurements of air temperature, wind velocity and direction were performed on a short base (about 150 m) using the centimetre RAS equipment and standard sensors of a high (300 m) meteorological mast. The RAS equipment was used for the purpose of meteorological support to investigations in the field of the atmosphere physics and applied problems. All instrumental, atmosphere and social factors, affecting operation the RAS systems, were generalized. It is shown that compact and mobile systems for remote monitoring of the atmospheric boundary layer with possible prompt obtaining of relevant information about base meteorological values in large volumes can be based on the RAS equipment.

1. Introduction

The history of development of the atmosphere radio acoustic sounding (RAS) method began at the end of 1950s. A number of review works devoted to historical, scientific and applied aspects of development and application of the atmosphere radio acoustic sounding method are known. However, these papers insufficiently cover researches in the field of the RAS method development that have been performed in the Soviet Union and, in particular, at Kharkov National University of Radio Electronics, where this method has been developing in the most intensive manner since 1965.

2. Development of Radio Acoustic Sounding Modes and Equipment

During those years when the RAS method was recognized as unpromising, it made good headway in KNURE. In 1965-1966 an experimental monostatic RAS system, which conceptually was identical with EMAC [1,2], was created at the Department of Radio Equipment Design for the purpose of investigating the electromagnetic radiation scattering on the acoustic waves and for testing the Bragg diffraction critical values while conducting the RAS of the atmosphere. The system consisted of a Doppler radar with continuous radiation of radio waves on the frequency of 10 GHz and a sound parcel former (loaded on Hartman whistle), radiating sound waves on the frequency of 22 kHz. The latter one was placed in the focus of a radar parabolic antenna, which formed directional diagrams of radio and sound radiation. Experimental researches of this system repeated results obtained using the EMAC system: the maximum sounding range was 40 m. Inefficiency of jet-edge sound converters was ascertained. Thus, in the course of further development of the RAS systems, electrodynamic sound radia-

tors were used, as they ensured high stability of frequency, operation in pulse mode and practically inertialess frequency adjustment for meeting the Bragg condition while sounding. The theoretical investigations of the RAS method, carried out within the period of 1966-1970, showed the following results:
1. It is established that the cause of short range of detection of radio signals, scattered by a sound wave, by EMAC and RAS-1 systems lies in substantial decay of the used frequency of sound waves in the atmosphere.
2. An oblique incidence sounding, which makes it possible to determine high-altitude distribution of air temperature, wind velocity and direction using one RAS system, is developed.
3. A radio acoustic sounding range equation is obtained. It takes into account all technical characteristics of the system, decay of sound waves in the atmosphere and their wind refraction.

Based on these results, mobile bistatic system RAS-2 was designed by 1969 with the Doppler radar operating in a continuous mode on the frequency of 3 GHz. An acoustic array, consisting of horn-type loudspeakers, was used as a sound radiator. The system was intended for remote measurement of high-altitude distribution of air temperature, wind velocity and direction in the lower 100 m layer of atmosphere with high spatial resolution. The equipment was placed in the cargo body of a truck, and an antenna was mounted on its roof. The results of the field tests of the system evidenced prospectivity of the RAS method for monitoring an atmospheric boundary layer.

Based on positive results of theoretical and experimental researches, in 1971 an atmosphere sounding research & development laboratory, which kept on development of the RAS method and systems, was organized. By September 1974 researchers of this laboratory finished design of a model of the mobile 10 cm RAS-3 system with internal coherence and the following technical parameters: radio transmitter power – 0.5 W; radio receiver susceptibility – 10^{-11} W; maximum electric capacity of acoustic pulse generator – 1000 W. The antenna arrangement – transmitting and receiving reflector-type antennas of the radio channel and an electro-acoustic transducer as an array consisting of 56 electrodynamic radiators - was placed on a rigid spatial metallic truss mounted on a turntable platform. This RAS system had two operation modes: vertical temperature sounding and oblique incidence temperature and wind sounding. From November 1974 to March 1975, for the first time in the world practice the correct comparative recording of vertical profiles of air temperature, wind velocity and direction was carried out by the 10 cm RAS system and by standard sensors of a high-altitude 300 m meteorological mast (HMM) of the Institute of Experimental Meteorology (IEM, Obninsk) on a short base (the compared measuring devices were placed at a distance of around 150 m). The maximum altitude of recording the vertical temperature profile was 180 m, and the type of a device meant for accumulation of sounding data was the small-sized printer MPU16-2. Root-mean-square deviations of the comparative measurements: air temperature – (0.1 – 0.5) K; wind velocity _ (0.2 – 0.6) m/s; direction – $(1.5\text{-} 5)^0$. Such errors of the RAS method and equipment meet the requirements for the meteorological observations practice. Considerable variability of single vertical profiles of meteorological values at the time of recording by the RAS equipment as compared to profiles obtained by meteorological sensors was noticed. From this point on this 10 cm RAS system underwent major modernization that resulted in achieving the following technical parameters: maximum radio transmitter power – 5 W; radio receiver susceptibility – 10^{-14} W; type of sounding data accumulation – perforated tape. From November 1977 to March 1978 this system was used for the second run of comparative measurements of meteorological values on the territory of the HMM. These experiments resulted in achieving the maximum altitude of air temperature recording – 300 m, and wind velocity and direction – 200 m. Based on results of comparative measurements, sounding methods were worked out under various meteorological conditions, including cases of increased wind velocities (up to 8–10 m/s).The achieved results of development of this RAS method enabled to use a modernized RAS-3 system for research purposes in the field of atmospheric physics and some applied problems. Thus, in August 1979 the laboratory with such system took part in the expedition investigation of pollution of air environment in the vicinity of aluminium smelter. In these experiments all data of temperature sounding performed using the RAS-3 system were compared with helicopter sounding data. From July to October 1980 the modernized

IOP Conf. Series: Earth and Environmental Science 1 (2008) 012053 doi:10.1088/1755-1307/1/1/012053

RAS-3 system was used for works relating to active effects on hydrometeorological processes (continental fog dispersion) under the aegis of IEM. In November 1980 development of sounding methods and creation of a model of the dual-frequency RAS-4 system started for experiments connected with remote measurement of one more important meteorological value – air humidity. By the end of 1982 such model of the RAS-4 system was created with the following characteristics: operating radio channel wavelength – 10 cm and 20 cm (with possible separate operation of channels); maximum power of 20 cm radio transmitter channel - 30 W; sounding result management and processing device – microcomputer D3-28 "Electronics". The system provides for an opportunity of partial compensation of the sounding acoustic package drift by wind. From this point on, the sounding methods were worked out for the purpose of synchronous recording the air temperature and humidity under different meteorological conditions. From then on a new RAS system, operating in the radio wave band of 1.76 GHz (acoustic radiation frequency 3.5 – 4.3 kHz) was developed at the Department. These both systems, for the purpose of improving the isolation between the radar transmitting and receiving antennas, used "edging" signal compensators, composed of a coaxial T-piece, attenuator and phase shifter. Compensation of a phantom signal is realized by way of sending an additional signal, which is equal to the phantom one as regards amplitude and opposite as regards phase, to the radio receiver input. A control and computation system based on the microcomputer "Electronics-NC-OZT" was used for the installation control and measurement data processing in the course of sounding. Proper interfaces were developed for interfacing the microcomputer trunk line with interfaces of measuring and recording devices. This system allowed to record vertical profiles of air temperature and wind velocity within the altitude ranges of 30-300 m. From May 1984 to May 1992 the model of the dual-frequency RAS-4 system was used in warm sea fog dispersion studies at the land-sea boundary. According to the radio acoustic sounding data, the experiments showed explicit transformation of the vertical temperature profile at different fog evolution stages: onset, existence and break-up. While performing these studies relevant comparative measurements of air temperature and humidity, wind velocity and direction were carried out by way of vertical sounding using RAS equipment and automated sensors of the 50 m meteorological mast belonging to Odessa Hydrometeorological Institute. At the same time, integration of the model of the dual-frequency RAS system with the acoustic radar of vertical sounding, belonging to the Institute of Radio Electronics of the Academy of Sciences of the USSR, was carried out (with the aim to increase volume and quality of all meteorological data obtained). From April 1988 to October 1990 a 20 cm channel of the dual-frequency RAS system was used in the investigation of effect of breeze circulation on vertical distribution of the refraction index for radio waves in the littoral zone (temperature-wind sounding). The RAS-4 system sounding altitude turned out to be insufficient for confident recording of the breeze cell emergence (especially its upper boundary). From the beginning of 1993 to the end of 1995 the laboratory made a model of a stationary RAS-6 system integrated with acoustic radar and intended for studying the thermodynamic processes in the atmospheric boundary layer. Main technical parameters: operating radar frequency – 440 MHz, operating acoustic radar frequency – 1 kHz, sounding data management and processing device – personal computer AT XT 286. From this point on, until September 1999 the experimental optimization of overall vertical sounding of the atmosphere was performed using this equipment. Since 2000 transition from the model-hardware modelling of the RAS system to the mathematical modelling has been completed using personal computers on the basis of the developed RAS systems theory [3].

3. Basic Statements of the Theory

For the purpose of creation of efficient RAS systems, it's necessary to perform deep theoretical elaboration of all relevant equipment-related problems using adequate approaches, as many problems of this field are not stated in the theory of radiolocation systems. The basic theoretical results, obtained recently, first of all, by the scientists of Kharkov National University of Radio Electronics, are given briefly below. There is a regularity of forming a radio signal, scattered on the sound, as a mutual autocorrelation function, with regard to the range of emitted electromagnetic and acoustic oscillations. This regularity originates from known ratios obtained in a theoretical way, and is evidenced by numerous experimental results. Adequate constructive models of information channels (radio acoustic and acoustic) are created. A two-dimension cross-correlation function of acoustic and radio signals – scattering function – is introduced. Its main properties, expressing the most common characteristics of the scattered signals, are determined in a theoretical manner. The problem of scattering the electromagnetic wave on the sound is presented using the signal theory and functional analysis apparatus, in which case the functionality of development improves substantially. The function of scattering, <<co-opting>> the properties of the interaction environment and characteristics of the sounding acoustic and electromagnetic oscillations, makes it possible to quite simply find a type of the scattered signal, corresponding to various conditions. This opens wide opportunities to research various types of sounding oscillations and development of the theory of the systems under consideration in whole. The sounding acoustic and electromagnetic research techniques are developed. Representation of the scattering function in different mathematical forms and graphically in the form of surfaces – scattering bodies – enables, using different types of body sections, to perform efficient analysis of sounding vector radio acoustic signals. Types of signals that are most frequently used in practice and promising ones are studied. It's shown that the received signal processing algorithms, used currently in the radio acoustic systems, which are borrowed from radiolocation, are not adequate to the processes occurring in the radio acoustic location channel, as they take in no account a change of the signal structure in conditions of scattering on the sound parcel. This leads to substantial deterioration of quality of the stations, in particular – accuracy of temperature measurements (the error may reach degree units). For obtaining the estimations of maximum credibility of the measured parameter, the reference signals in different points of the range of possible values of this parameter should differ as for type, form, but should not be a copy of the emitted radio signal, as is customary. All forms of reference oscillations of a multi-channel processing device should be determined using the scattering function, reflecting the peculiarities of transformation in the channel. The synthesized signal processing algorithms may be considered as generalization of the radio system theory results in the case, when the signal form at scattering transforms according to the deterministic law. If the scattering volume degenerates into the point target, the received and reference signals become a copy of the sounding, emitted radio signal. The proposed signal processing procedures reduce in this case to the previously known algorithms of estimation of any oscillation parameters.

4. Discussion of the Accumulated Experience and Conclusions

In the paper [4] the main methods of radio acoustic sounding, developed in KNURE for remote recording of the main meteorological values, are set forth: air temperature and humidity, wind velocity and direction in the atmospheric boundary layer. In the course of realization of these methods using the models of the RAS equipment a great experience was accumulated while setting up all relevant equipment, in the course of experimental sounding and when performing overall works in the field of atmospheric physics and pollution. The thorough analysis of efficiency of the RAS equipment under different meteorological conditions and circumstances made it possible to single out a number of factors (methodical, technical, atmospheric, social, etc.) that restrain capacities of this method. The joint use of acoustic and electromagnetic waves, focusing of electromagnetic waves by acoustic ones and the Bragg scattering can be referred to the methodical factors. Technical factors – the necessity to maintain the Bragg conditions along the whole sounding path and ensuring a high degree of isolation between the transmitting and receiving antennas of the radar. The main atmospheric factors, effecting efficiency of the RAS systems, include existence of daily and seasonal progress of meteorological values and occurrence of turbulence. The social factor manifests itself as noise pollution of an area of location of the RAS system.

Generalization of results of more than 40 years of investigations, carried out in KNURE and aimed at promoting the RAS method, enables to formulate a number of principles giving an idea of the contribution of the experts from KNURE to the world achievements in this research field.

1. The theoretical researches, clarifying the cause of small range of the RAS systems at the initial stage of researches, were carried out. The methods of measuring the wind parameters using one system and the range equation for the RAS of the atmosphere were proposed.
2. The results of correct comparative measurements of air temperature, wind velocity and direction using standard sensors of a high-altitude meteorological mast and the RAS equipment located on a short base were obtained for the first time.
3. The methods of dual-frequency sounding of the atmosphere for air humidity measurement purposes were developed and used in real conditions.
4. The conclusion was made about optimality of realization of the RAS systems in the centimetre wave band that were intended for monitoring the atmospheric boundary layer.
5. The RAS system theory, regarded as a variation of the radar system theory, was created.

References

[1] Smith P L Remote measurements of wind velocity by the electromagnetic-acoustic probe. I. System analysis. 1961. In: Conf. proc. 5th Annu. convention on military electronics. Wash (D.C.), rep № 419, 43-53
[2] Fetter R V II. Experimental system. (Ibid), 54-9
[3] Kartashov V.M. Development of theory and improvement of radio acoustic and acoustic atmosphere sounding systems. 2003. Author's abstract to the Doctor's Degree Dissertation. (Kharkov: KNURE), 35
[4] Remote methods and Earth atmosphere process investigation tools /Edited by Kashcheyev B.L., Proshkin Ye.G., Lagutin M.F. Section 2. Radio acoustic sounding of atmospheric boundary layer. 2002 (Kharkov: Collegium), 44-98

IOP Conf. Series: Earth and Environmental Science 1 (2008) 012054 doi:10.1088/1755-1307/1/1/012054

Study of the development of the sea breeze and its micro-scale structure at a coastal site using a Multi-Tone Sodar system

L. De Leo[1,2], S. Federico[1,2], A.M. Sempreviva[1], L. Pasqualoni[2], E. Avolio[1,2], C. Bellecci[2,3]

E-mail: s.federico@isac.cnr.it

1) Italian National Research Council - Institute of Atmospheric Sciences and Climate (CNR-ISAC), Lamezia Terme (CZ)
2) CRATI s.c.r.l., c/o University of Calabria, Rende (CS)
3) Department of Engineering, University of Rome "Tor Vergata", Rome

Abstract The purpose of this paper is to study the development of the coastal flow i.e. its micro-scale characteristics within the surface layer at the sea-land transition and to investigate the development of inland formation of cloudiness during merging of sea/land and valley-mountain breeze regimes. The area is the mountainous southern tip of Italy, about 50 km wide and 300 km long, in the Central Mediterranean Basin. Here, thermal meso-scale circulations and cloud formations on the mountains are pronounced because of their peculiar elongated shape. Upslope flow is very strong, due to the steepness of Apennines, and sea-breeze circulation is supported by the thermally forced meso-scale flows. Here, we present preliminary results from a case study during six consecutive days where sea/land breeze circulation was well developed and discuss the effect on the formation of inland clouds. Sea breeze days are selected by a sodar and surface meteorological station measurements. Cloud development is analyzed by subjective inspection of METEOSAT MSG images in the visible channel. The WRF (Weather Research and Forecasting) model is used to analyze the meteorological situation. The Institute of Atmospheric Sciences and Climate CNR-ISAC in house modular Multi-Tone Sodar system (PC-MTSodar) and the surface meteorological station are installed and operated at a coastal site located at about 300 m from the coastline. Preliminary results show that the methodology is viable to study the development of thermally forced meso-scale circulations over Calabria at micro and large scale and that these circulations can be intense over the peninsula thanks to its peculiar shape. It is shown that the formation of afternoon cumulus clouds, which may result in rainfall, is modulated by synoptic scale conditions. Further development will be a climatic assessment of the role of breezes on local climate.

1. Introduction

The sea/land breeze circulation regime is a characteristic feature of coastal areas, typically during fair weather conditions. Sea/land breeze circulation is generated by the thermal difference between an air column over the sea and an air column over the land. For a peninsula, two breeze cells develop on both sides and can interact, provided that the peninsula width is less than two times a Rossby radius (100 km for each kilometer of CBL depth ([1])). The sea breeze intensity, that can reach 10 ms^{-1}, varies depending on several factors such as large scale flow, atmospheric static stability, cloud coverage etc., conversely, land breeze is weaker than sea breeze.

Furthermore, whenever mountain valley orographic contrast is present, a slope flow i.e. anabatic or katabatic winds develops in clear sky during day and night respectively. Also those winds are determined by the thermal gradient existing between an air column above mountain and an air column above valley. Whenever a mountainous peninsula is considered both mountains valley and sea land breeze winds develop. Several studies can be found in the literature on sea breeze and local winds, some of them being reported in [2]. The Calabria region (Figure 1) is a mountainous peninsula located at the southern tip of Italy, about 50 km wide and 300 km long, elongated in the north-south direction in the Central Mediterranean Basin.

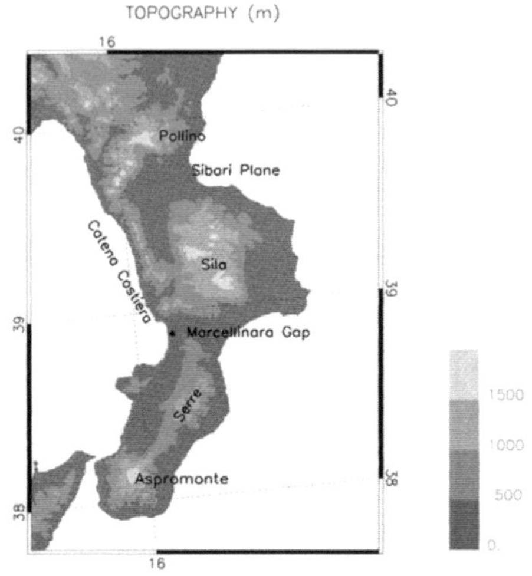

Figure 1 Main topographical features in Calabria. The dot indicates the experimental site.

The region is particularly interesting for studying the development of such thermally forced meso-scale circulations (TFMCs). Owing to the specificity of the Mediterranean climate and to the topographic characteristics of the peninsula, i.e. the presence of high mountains parallel to the coast, intense local wind circulations can develop, mainly in fair weather conditions, which are common during summertime in the Mediterranean Basin ([3]). The high water vapor atmospheric content in summer, due to evaporation over the sea, is reorganized by the TFMCs; because of the peculiar shape of the peninsula, the sea-land and mountain-valley circulations, associated with thermal gradients, interact generating wide and intense local circulations. Humid marine air masses, which come from lower levels, are advected and lifted upward by breeze circulation and can determine the onset of cumulus clouds over the main peaks, with chances to trigger afternoon thunderstorms. A comprehensive study can only be conducted using spatial and temporal high resolution information by integrating experimental data and modeling. In this paper, we investigate the effects of breeze circulations during a week in summer 2007, using measurements, model simulation and satellite images. Experimental data sets consists of surface measurements by a meteorological station and vertical profiles of vertical velocity w and its standard deviation σ_w by a Sodar located at about 300 m from the shoreline; cloud development has been analyzed by subjective inspection of METEOSAT MSG images in the high resolution visible channel. WRF (Weather Research and Forecasting) model is used to analyze the regional circulation. The interaction between breezes and large scale circulation and their impact on cumulus cloud formation is further investigated for 3 and 4 August.

2. Experimental setup

The Sodar and the meteorological station are both installed and operated routinely within the area of the ISAC-CRATI research centre at a coastal site located at about 300 m from the coastline at the west end of the only sea-to-sea valley about 30 km long at the narrowest part of the region along West-East direction, i.e. Marcellinara gap (Figure 1).

2.1. Sodar setup

The in house modular Multi-Tone Sodar system (PC-MTSodar) developed and assembled by the Institute of Atmospheric Sciences and Climate CNR-ISAC at the Unity of Lecce ([4] and [5]), is installed on the roof of the ISAC-CRATI building at a height of 7 m. The PC-MTSodar system is routinely used in mini-sodar configuration with a minisodar antenna, vertically oriented (77 cm in diameter and 140 cm tall), equipped with an Electro-Voice (DH7) transducer and two other antennas for the horizontal wind speed. The system is based on and entirely controlled by a Personal Computer (PC) and uses two synchronized commercial I/O cards (NI-PCI4451) that operate synchronously and a specifically developed software to handle all system functions. Each card controls two sodar channels and has two analog inputs, two analog output channels and several digital I/Os used to control the antenna preamplifiers. A 16 bit I/O converter is used to record the antenna signal associated to each channel, so that a maximum dynamic range of about 96 dB may be obtained. Two interconnection boxes provide electrical test points to monitor the signals exchanged between the NI-PCI4451 cards and the other parts of the system. Each interconnection box forwards the DC electrical power from the main board of the PC to two directly connected preamplifiers. The system emits pulses with a 3 s repetition rate and at 3500, 4000, and 4500 Hz for the z, y and x antennas respectively. The sodar is programmed to acquire 28000 points at 20000 Hz and a decimation factor of 6 is used after filtering with a Butterworth digital filter (30th order) with a pass-band between 3200 Hz and 3800 Hz, between 3700 Hz and 4300 Hz, and between 4200 Hz and 4800 Hz, for each antenna. The evaluation of the Doppler shift was performed on 9 range gates with a vertical resolution of about 13.2 m (first available range gate is 23.9 m above the roof). The procedure, first localize the spectrum of the echo in the band in which it can be moved by the Doppler shift (300 Hz in our case). The actual value of its first moment is then determined using only the narrower frequency band (130 Hz) in which the echo is limited. The rate of the integrals over the narrower and the entire frequency band is used as an indication of the Signal-to-Noise (S/N) ratio; the Doppler shift and S/N were calculated using Fast Fourier Transform on 256 points following the two-step procedure described in [6]. The software allows the presence of periods of passive acquisitions in which the transfer function (TF) of each transducer is evaluated. The TF are successively used to take into account the variation of the transducer response as a function of the frequency during calculation of radial wind velocities. This procedure has several advantages and in particular it can reduce random errors on velocity measurements, especially when mean vertical wind velocity is needed ([7]). Only data with S/N greater than 0.75 is retained in the elaborations and the 30-minute averages have been considered valid if at least 70 % of the data were available in a specific range gate.

2.2. The meteorological system

The meteorological surface station includes a cup anemometer and a wind vane at 10 m height for measuring wind speed and direction, and some other sensors for measuring air temperature, atmospheric pressure, relative humidity, rainfall and solar radiation. These instruments have been manufactured by Adcon Telemetry. Data are acquired every 15 minutes. Moreover a ground station to receive METEOSAT-MSG images in real time is available to examine the formation of cumulus clouds over the country in the High Resolution Visible channel, which has about 2 km horizontal resolution over Calabria.

3. Model setup

The meteorological situation for 3 and 4 August is analyzed by the WRF (version 2.2) model, a next-generation mesocale numerical weather prediction system designed to serve both operational forecasting and atmospheric research.

The multiple-nest capability of the model with two-way interaction between successive nesting layers is used in order to get realistic terrain features and explicit representation of meso-sale processes. For the present simulation we use three nested domains under a stereographic projection. The grid points number for the three domains are, respectively, 140x110x24, 112x100x24 and 81x129x24. The horizontal resolutions for the three domains are 30 km, 10 km and 2.5 km while the terrain following σ-coordinate is used in the vertical. The pressure at the model top is 50 hPa. The shortwave radiation scheme is based on [8]. It is a simple downward integration of solar flux, accounting for clear-air scattering, water vapor absorption ([9]), and cloud albedo and absorption. Longwave radiation scheme is based on [10], and it is a spectral-band scheme using the correlated-k method. It uses pre-set tables to accurately represent longwave processes due to water vapor, ozone, CO_2 as well as accounting for cloud optical depth. The surface layer paramaterization uses stability functions from [11] and [12] to compute surface exchange coefficients for heat, moisture, and momentum. Explicit microphysics is represented by the Kessler scheme ([13]), which includes the following microphysical processes: the production, fall and evaporation of rain, the accretion and autoconversion of cloud water, and the production of cloud water from condensation. Convective precipitation is parameterized by the modified version of the Kain-Fritsch scheme ([14] and [15]) available in WRF. Convective precipitation is parameterized for the first and second domain and we assume that it is explicitely resolved for the third domain. Initial and dynamic boundary condition are derived from ECMWF analysis, available every six hours. Simulation starts at 12 UTC of 2 August and lasts sixty hours. First twelve hours are discarded from the analysis (spin-up time).

4. Results and discussion

Figure 2 shows the measured wind speed and direction for the period 2-7 August 2007. The abscissa is the Local Time (LT).

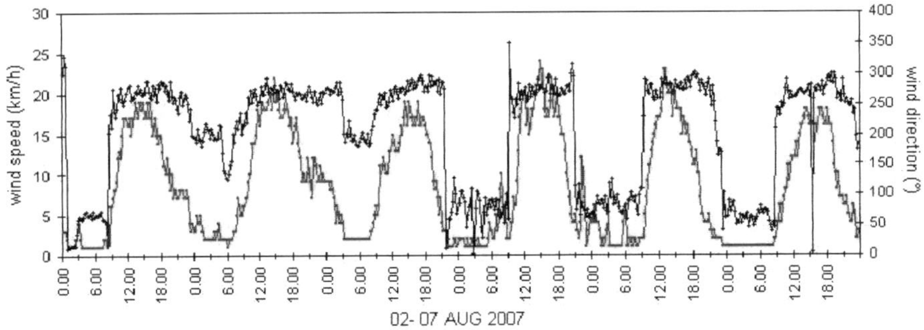

Figure 2: Wind speed and direction measured by the surface station from 02 August 2007 (00:00 LT) to 7 August 2007 (23:45 LT).

For August, the conversion between LT and UTC is LT = UTC + 2 h. Wind speed reach values of 7 ms^{-1} during the central hours of the day and early afternoon while at night the winds are light (< 2 ms^{-1}). Wind direction is from west during the day and from east at the night, if we exclude the nights between 2-3 August and 3-4 August, when the light wind is from S-SW. Considering the location of the experimental site, the measurements in figure 2 are typical of sea breeze development for the whole period. Figure 3 shows the vertical velocity measured by the sodar from 2 to 7 August. It is positive during the day and negative at night, noticeably for 2, 5 and 6 August.

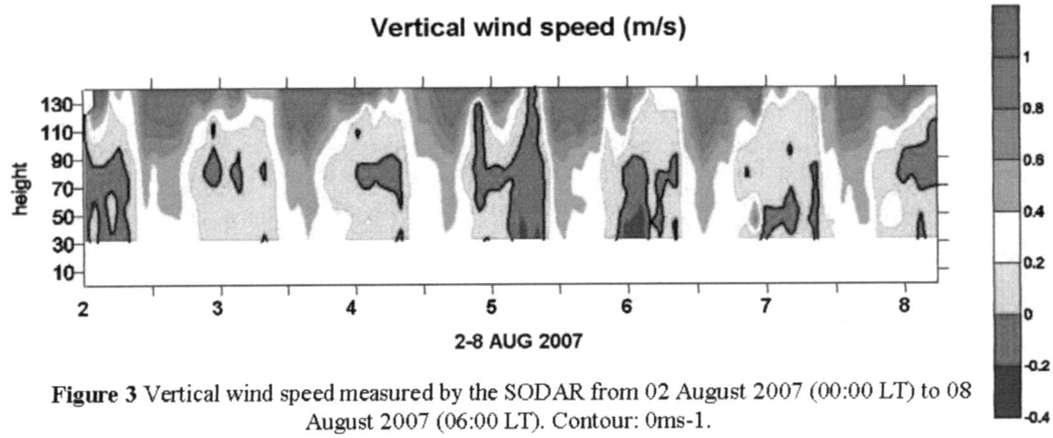

Figure 3 Vertical wind speed measured by the SODAR from 02 August 2007 (00:00 LT) to 08 August 2007 (06:00 LT). Contour: 0ms-1.

Diurnal values are larger than 1 ms^{-1} above 100 m and confirm the full convection development above the surface layer (about 100 m). The sea breeze front overpass the coastal site between 06:00 and 09:00 AM LT. As expected for TFMCs the absolute value of vertical velocity is higher during the day. This is evident from figure 4 that shows the vertical velocity variance. Values between 0.5-1.0 m^2s^{-2} are measured for the central hours of the day while they are nearly zero at night. Figures 2, 3 and 4 show the occurrence of rather intense breeze circulations over Calabria, as expected considering the peculiar shape of the peninsula and the characteristic of the Mediterranean climate in summer.

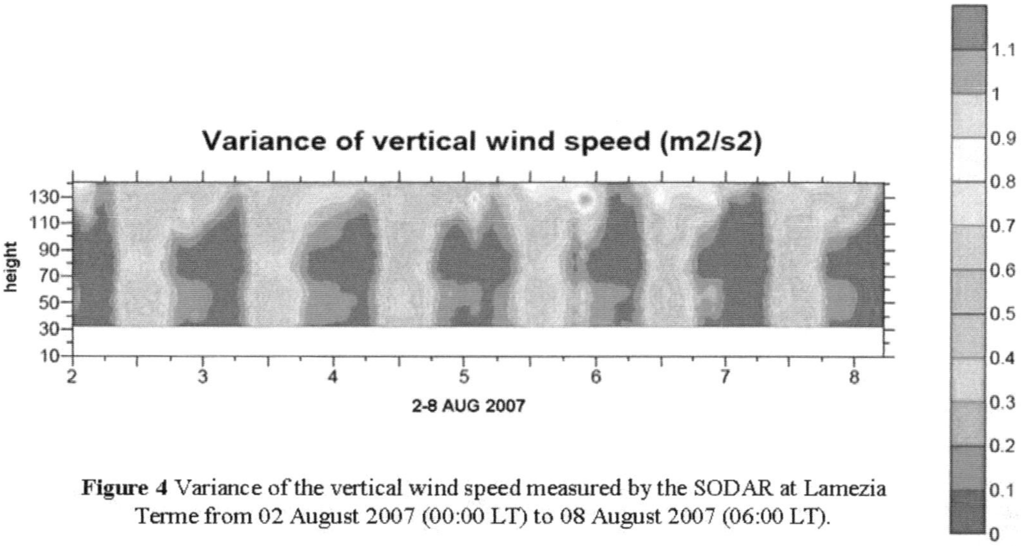

Figure 4 Variance of the vertical wind speed measured by the SODAR at Lamezia Terme from 02 August 2007 (00:00 LT) to 08 August 2007 (06:00 LT).

An interesting issue about breeze circulations is their ability to form cumulus cloud over the land during the day and to produce afternoon thunderstorms ([2]). Indeed, the importance of the sea breeze on the precipitation pattern has been known for many years ([16] and [17]) and the modification of their pattern and intensity can result in major changes of local climate and hydrologic cycle ([18]). Figures 5 and 6 show the METEOSAT-MSG images at 11:57 UTC (13:57 LT) for 3 and 4 August.

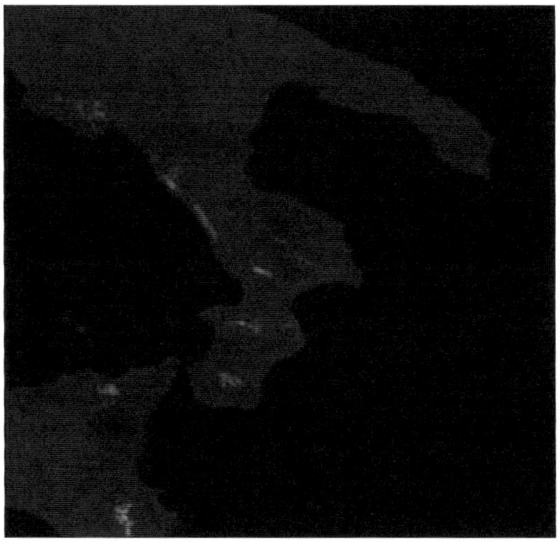

Figure 5: METEOSAT-MSG image for the High Resolution Visible channel at 11:57 UTC (13:57 LT) 03 August 2007.

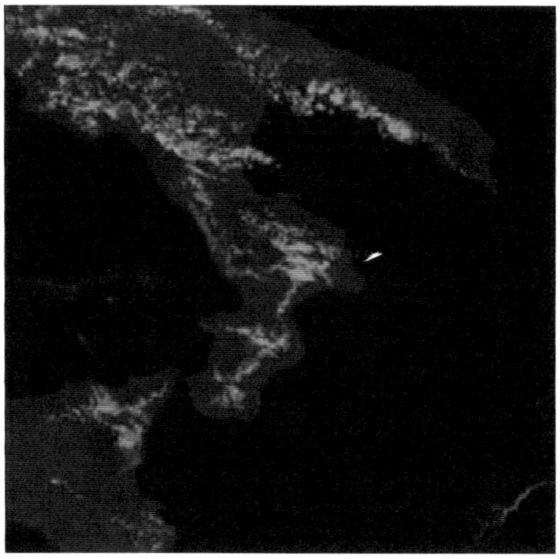

Figure 6: METEOSAT-MSG image for the High Resolution Visible channel at 11:57 UTC (13:57 LT) 04 August 2007.

Few clouds are present over main peaks for 3 August while clouds development is evident for 4 August. It is worth noting that cumulus clouds are associated with both sea/land breeze and mountain/valley flows. Cumulus clouds in the Marcellinara gap are the result of the sea breeze fronts convergence from Tyrrhenian and Ionian Sea, while mountain valley flows are responsible for the development of cumulus clouds over the main peaks and mountain chains. According to a previous energetics study ([1]) the intensity of mountain valley and sea breeze circulations in Calabria are comparable each other, hence the local meso-scale flow is driven by both factors. Precipitation associated with cumulus cloud is recorded for only two stations over Sila on 4 August (not shown). The cumulus cloud formation for 5 August (not shown) is similar to figure 6 while precipitation is reported in few stations in southern Calabria (Figure 7).

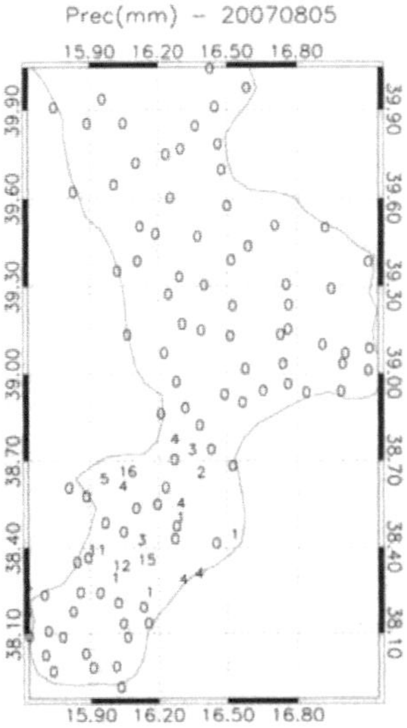

Figure 7 Precipitation recorded on 05 August 2007 by the Regional Meteorological Network of Civil Protection.

Accordingly, measurements and METEOSAT MSG images suggest that TFMCs could be important not only because they determine local circulations, but also because they can trigger cumulus cloud development and rainfall. We can only speculate about this last issue because our analysis refers to five days only, while a climatological study ([19]), based on 30 years wind records in seven surface stations over Calabria, shows that more than 70% of summer day circulations are characterized by TFMCs. To explain the difference between the cumulus clouds development for 3 and 4 August we analyze the meteorological situation by WRF model. Figure 8a shows the sea level pressure and 925 hPa temperature at 14:00 LT on 3 August. A high pressure extends over the Western Mediterranean and Western Europe with a maximum over the Biscay Bay. The 1017 hPa isobar is aligned with the Calabrian west coast and the pressure field over the central Mediterranean is rather flat. These surface conditions lead to calm synoptic scale winds, which favor breezes development. The corresponding situation at 500 hPa is shown in figure 8b.

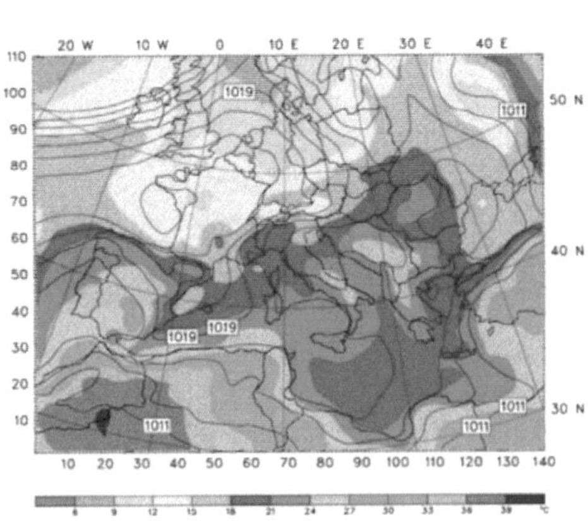

Figure 8a Sea level pressure (hPa, black solid contours) and 925hPa temperature (°C, filled contours) at 14:00 LT 03 August 2007.

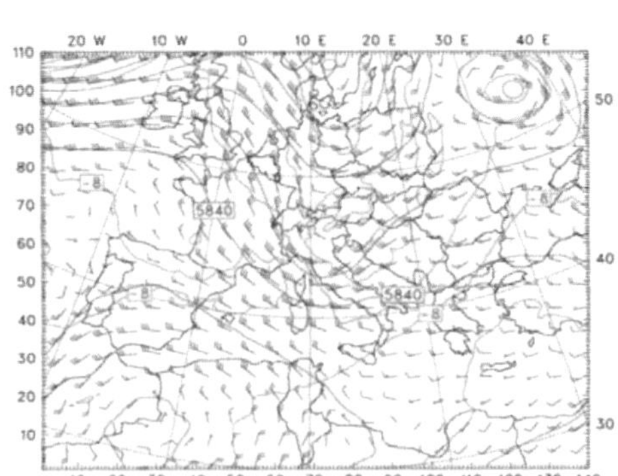

Figure 8b Geopotential height (blue contours) and temperature (°C, red contours) at 500hPa at 14:00 LT 03 August 2007.

A pronounced mid-tropospheric trough extends from Scandinavia to the Mediterranean while a ridge is well evident over western Africa and Europe. The trough is modified by the interaction of synoptic scale winds with Alps and deepens over northern Italy. This baric configuration determines winds from W over Calabria at 500 hPa. An inspection of model output for 850 hPa and 750 hPa shows that the wind direction is from W-NW over Calabria. The following day, the maximum surface pressure moves from the Biscay Bay to central Europe (Figure 9a). The pressure over the Central Mediterranean weakens and the 1015 hPa contour is aligned with the Calabria west coast. A sea level pressure decrease of 2 hPa between 3 and 4 August is also confirmed by measurements at the experimental site surface station (not shown). The pressure gradient, however, is weak over the central Mediterranean and favors the development of TFMCs. At 500 hPa (Figure 9b), the trough evolved in a cold cut-off low over Eastern Europe while the ridge over Africa expands toward Northeast.

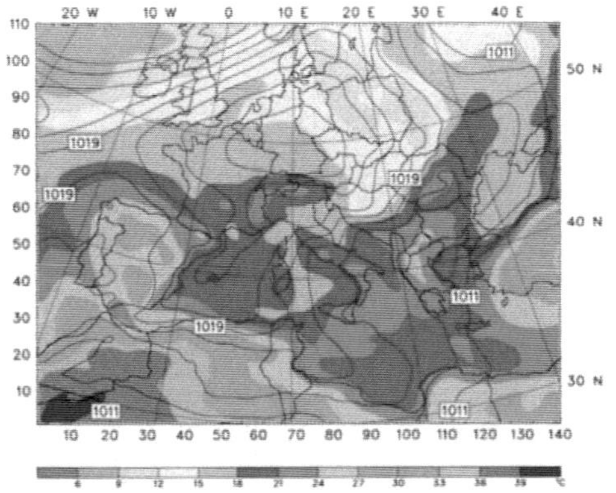

Figure 9a Sea level pressure (hPa, black solid contours) and 925hPa temperature (°C, filled contours) at 14:00 LT 04 August 2007.

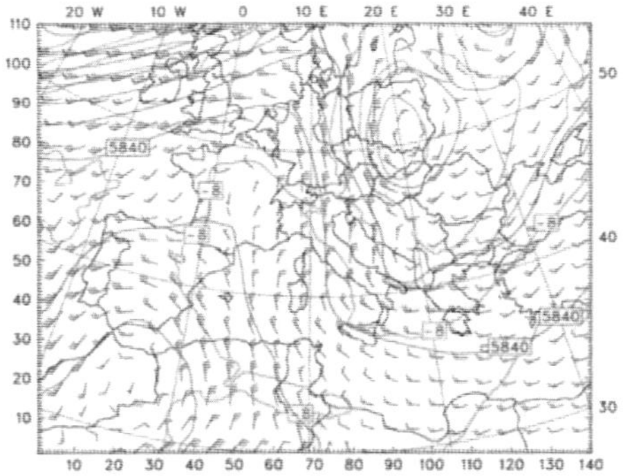

Figure 9b Geopotential height (blue contours) and temperature (°C, red contours) at 500hPa at 14:00LT 04 August 2007.

Compared to the previous day, the winds over Central Mediterranean turn clockwise and advect colder and drier air masses over Calabria. This change in wind direction is very important for cumulus cloud formation. Indeed, an inspection of model output over Calabria shows that temperature decreases over the region. This is partially visible at 925 hPa (Figures 9a and 9b) and is confirmed by the model output for the third grid and by surface station measurements. So, relative humidity increases between the two days and cumulus clouds develop on 4 August because of the temperature decrease. It could be argued that air masses transported over Calabria on 4 August are also drier compared to the previous day, which should decrease the relative humidity. Nevertheless, an inspection of model output shows that the decrease of specific humidity is negligible at least under 700 hPa. This is not surprising because humidity is mainly produced by sea evaporation and transported at higher levels by TFMCs and its amount at lower tropospheric levels mainly depends on sea surface temperature, surface winds and humidity of air masses above the sea. These physical factors did not change sizably between 3 and 4 August. Moreover, the humidity was transported upward at a faster rate on 4 August, because of the stronger development of TFMCs. Therefore, specific humidity was almost constant for the whole period, at least below 700 hPa, and relative humidity was higher for 4 August because of the temperature decrease. This process lead to the formation of several cumulus clouds on 4 August (Figure 6).

The humidity reorganization by TFMCs is shown in figure 10 for the early afternoon of 4 August. Several specific humidity local maxima at 750 hPa are evident over the main peaks, especially over Sila, while the advection determines asymmetries in the field, which is higher downwind. The maxima over the main peaks are determined by TFMCs: humid air masses from both side of the peninsula are transported into the interior where, owing to the convergence, are uplifted. This peculiar local circulation generates convective clouds, eventually transported downwind. Several convective clouds are simulated over Sila and other main peaks (Figure 10). Figure 11 shows a longitude-height cross section over the Sila. The convergence of air masses from both sides of the peninsula is evident over the main peak of the cross-section. Cumulus clouds forms in the convergence area and are transported downwind. These clouds can determine rainfall, as reported in figure 7 for 5 August.

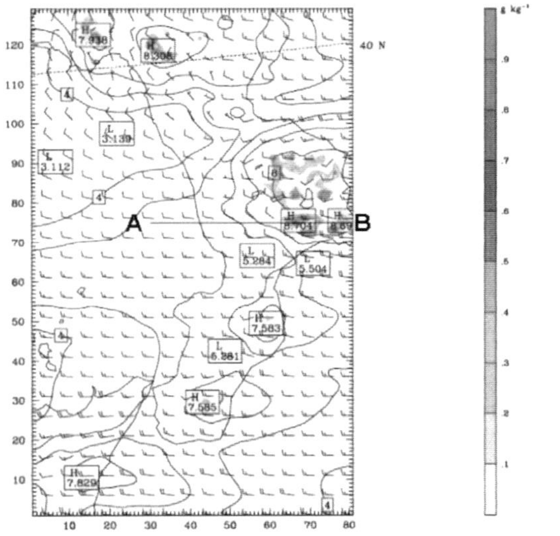

Figure 10 Horizontal section of water vapour (contours, gr/kg) and cloud water (filled contours gr/kg) at 750hPa on 04 August 2007 14:00 LT. Cross section of figure 11 is taken along the AB line.

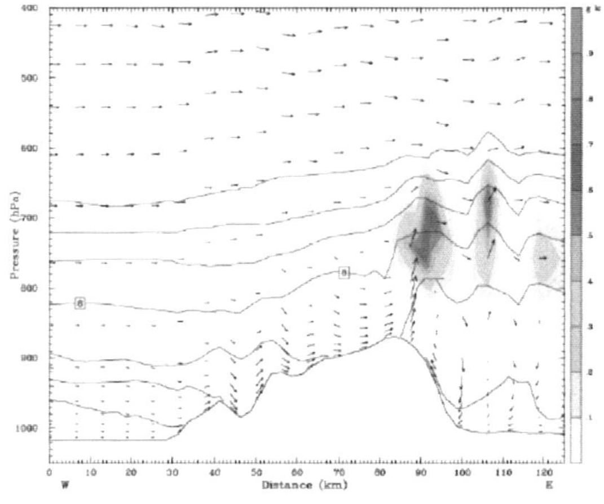

Figure 11 Vertical cross section of water vapour (blue contours, gr/kg), cloud water (filled contours gr/kg) and wind vectors on 04 August 2007 14:00 LT. The cross section is taken along the AB line of figure 10.

5. Conclusions

In this paper, we investigated the breeze circulations over Calabria for a week in summer 2007, when conditions for the onset of thermally forced meso-scale circulation are more favorable, using experimental measurements, satellite images and modeling. Days characterized by breeze circulations have been identified by sodar and meteorological ground station measurements at the experimental site of Lamezia Terme.

The pronounced thermally forced meso-sale circulations in Calabria are related to two basic features: first Calabria is located in the Central Mediterranean Basin where synoptic scale conditions are favorable to breezes development, second the peculiar shape of the peninsula induces strong sea/land breezes and mountain/valley flows which acts synergically.

The water vapor is organized by local circulations and transported in the center of the peninsula where, owing to convergence is uplifted. This process can produce cumulus clouds and trigger rainfall in the afternoon if suitable conditions are met.

Results are promising and show the capability of the methodology for studying thermally forced meso-scale circulations over Calabria. As next step, we are considering a whole year period where sodar data are available to assess the impact of breezes on Calabria climate.

6. Acknowledgments

This work was partially funded by "Ministero dell'Università e della Ricerca Scientifica" in the framework of the projects "SAI - Messa a punto di un Sistema per l'Allerta precoce di Incendi e per la minimizzazione dei falsi allarmi" and "TAI – Tecniche Agronomiche Innovative per la valorizzazione delle produzioni ed il miglioramento della qualità ambientale". We are grateful to "Protezione Civile della Calabria" for rain gauge data.

7. References

[1] S. Federico, G.A. Dalu, C. Bellecci, M. Colacino, 2000: Meso-sale energetics and flows induced by the sea-land and mountain-valley contrast, Ann. Geophys. 18, 235-246.
[2] J.E. Simpson, 1996: Sea breezes and local winds, Cambridge University Press, 234pp.
[3] M. Colacino, 1992: Mediterranean meteorology. In winds and currents of the Mediterranean basin, Proc. NATO-ASI, Ed H. Charnock, Reports in Meteorology and Oceanography 40, 1-38.

Harvard University Press, USA, 1992.

[4] D. Contini, F.M. Grasso, G. Dargaud, G. Mastrantonio, A.P. Viola, U. Bonafè: 2006, A modular PC-based multi-tone sodar system. Proceedings of International Symp. for the Advancement of Boundary-Layer Remote Sensing, ISAR 2006, Garmisch-Partenkirchen, Germany, 66–68.

[5] D. Contini , F.M. Grasso, G. Mastrantonio, A.P. Viola and P. Martano: 2007, Performances of a modular PC-based Multi-Tone Sodar system in measuring vertical wind velocity, Meteorologische Zeitschrift, 16-4, pp. 357-365(9).

[6] G. Mastrantonio, G. Fiocco, 1982: Accuracy of wind velocity determinations with Doppler Sodar, J. Appl. Meteor. 21, 820–830.

[7] D. Contini, G. Mastrantonio, A.P. Viola, S. Argentini: 2004: Mean Vertical Motions in the PBL Measured by Doppler Sodar: Accuracy, Ambiguities, Possible Improvements, J. Atmos. Ocean. Technol. 21, 1532–1544.

[8] J. Dudhia, 1989: Numerical study of convection observed during the winter monsoon experiment using a meso-sale two-dimensional model, J. Atmos. Sci., 46, 3077-3107.

[9] A.A. Lacis and J. E. Hansen, 1974: A parameterization for the absorption of solar radiation in the earth's atmosphere, J. Atmos. Sci., 31, 118-133.

[10] E.J. Mlawer, S.J. Taubman, P.D. Brown, M.J. Iacono, and S.A. Clough, 1997: Radiative transfer for inhomogeneous atmosphere: RRTM, a validated correlated-k model for the long-wave, J. Geophys. Res., 102 (D14), 16663-16682.

[11] C.A. Paulson, 1970: The mathematical representation of wind speed and temperature profiles in the unstable atmospheric surface layer, J. Appl. Meteor., 9, 857-861.

[12] A.J. Dyer and B.B. Hicks, 1970: Flux-gradient relationships in the constant flux layer, Quart. J. Roy. Meteor. Soc., 96, 715-721.

[13] E. Kessler, 1969: On the distribution and continuity of water substance in atmospheric circulation, Meteor. Monogr., No. 32, Amer. Meteor. Soc., 84pp.

[14] J.S. Kain and J. M. Fritsch, 1990: A one-dimensional entraining/ detraining plume model and its application in convective parameterization, J. Atmos. Sci., 47, 2784-2802.

[15] J.S. Kain and J. M. Fritsch, 1993: Convective parameterization for meso-sale models: The Kain-Fritcsh scheme. The representation of cumulus convection in numerical models, K.A. Emanuel and D.J. Raymond, Eds., Amer. Meteor. Soc., 246 pp.

[16] R.A. Anthes, 1984: Enhancement of convective precipitation by mesoscale variations in vegetative covering in semiarid regions, J. Climate Appl. Meteor., 23, 541–554.

[17] R.W. Arritt, M. Segal, M. Leuthold, C.J. Anderson, and R.W. Turner, 1996: An observational study of the Lake Okeechobee lake breeze and its effect on deep convection, Preprints, Conf. on Coastal and Oceanic and Atmospheric Prediction, Atlanta.

[18] R.A. Pielke, 2002:Mesoscale Meteorological Modeling, Academic Press., 702pp.

[19] M. Colacino, M. Conte, and E. Piervitali, 1997: Elementi di climatologia della Calabria, CNR-IFA Roma, 200pp.

IOP Conf. Series: Earth and Environmental Science **1** (2008) 012055 doi:10.1088/1755-1307/1/1/012055

Emission rates with the boundary layer budget method supported by acoustic remote sensing

S Emeis

Institut für Meteorologie und Klimaforschung – Atmosphärische Umweltforschung, Forschungszentrum Karlsruhe, Garmisch-Partenkirchen, Germany

stefan.emeis@imk.fzk.de

Abstract. Remote sensing can help to derive turbulent fluxes in the atmosphere even if the remote sensing technique used itself is not able to resolve turbulent fluctuations. Turbulent fluxes are important components of atmospheric energy and substance budgets and they play a dominant role in the surface-atmosphere exchange. Apart from high-resolution in situ measurements it is difficult to determine them. This paper presents an inverse method to derive a regionally-averaged vertical turbulent flux of methane from the surface into the atmospheric boundary layer. The determination of aggregated surface emission fluxes of methane from an agricultural area in Southern Germany on an intermediate scale between the local and the regional scale is based on using a boundary layer budget method combined with surface-based acoustic remote sensing of the mixing-layer height with a sodar. The order of magnitude of the resulting methane emission fluxes coincide well with the national methane emission inventory.

1. Introduction

The supervision of the emission rates of greenhouse gases has become an important task for all nations complying with the Kyoto Protocol. This task is difficult if the sources are not well-known or if the sources are diffuse without fixed boundaries. One method to derive such diffuse emission fluxes is to measure the concentration of the emitted substance in question upwind and downwind of the supposed emission source and then – afterwards – to calculate the emission source strength via an inverse (i.e. backward) dispersion modelling ([1], [2]).

If the diffuse emission takes place into a well-defined volume of air then the emission could be easily computed from the concentration increase within this air volume by a residual technique ([3], [4]). Such a well-defined air volume could be the air underneath a nocturnal near-surface inversion. The problem is then shifted to the task to monitor the height of this inversion. Here, remote sensing can be a good tool to do so ([5], [6], [7], [8]).

In principle, different methods are available for the remote sensing of the mixing-layer height and nocturnal inversions. In [8] and [9] the abilities of a sodar and a ceilometer have been demonstrated. Here, a sodar was available as remote sensing device to monitor the height of the nocturnal inversion. The methane concentrations have been measured by a near-by surface-based in-situ instrument. This paper will describe in more detail the residual technique and will show the feasibility of this remote sensing-based method with an example already sketched in [10]. Section 2 shortly introduces the data, section 3 presents the residual technique, and section 4 displays some results.

2. Data

The data have been taken in the framework of the ICAROS NET project in May 2003 in Southern Germany. The ICAROS NET was designed to deliver mixing-layer height and air quality data for the intercomparison with satellite data (see [11] for an evaluation of the correlation between air pollution and mixing-layer height from this project). The methane concentrations have been measured with 30 min resolution by a HORIBA APHA-350E HC analysator in Maisach west of Munich in a few metres height above ground. The sodar was a three-antenna METEK DSDR3x7 Doppler sodar with a maximum range of 1300 m which was operated with 30 m vertical resolution and 10 minutes time resolution. The spatial distance between the two instruments was about 2 km. The measurement site was an essentially flat rural area with a few isolated pieces of forest. Munich is situated about 25 km east of the site. Between the two instruments was the short-cut meadow of a no longer used military airfield.

3. Principle of the residual method

Let *e* be an atmospheric property – here the concentration of the greenhouse gas methane - and $\vec{v}$ the vector of the three-dimensional motion, and let us denote the mean quantity by an overbar ($\bar{\ }$) and the deviation from this mean by a prime (‘). Then $\bar{e}\bar{\vec{v}}$ is the mean flux of the property *e* and $\overline{e'\vec{v}'}$ the turbulent (sub-synoptic) flux of this property. The actual flux of the property *e* therefore is the sum out of two parts:

$$\overline{e\vec{v}} = \bar{e}\bar{\vec{v}} + \overline{e'\vec{v}'} \qquad (1)$$

In areas – like those close to the lower boundary of the atmosphere - where mean vertical motions are negligible vertical transports of any property are completely governed by turbulent fluxes of this property. Basis of all residual techniques is the continuity equation for an atmospheric property e that has sources Q and/or sinks R in the volume of interest V.

$$\int_V \frac{\partial e}{\partial t} dV = -\int_V \nabla e\vec{v} dV + \int_V Q dV - \int_V R dV \qquad (2)$$

By the virtue of the Gauss theorem the first volume integral on the right-hand side of (2) can be replaced by an integral over the surface S surrounding this volume (v_n is the velocity component perpendicular to this surface).

$$\int_V \nabla e\vec{v} dV = \int_S e v_n \cdot dS \qquad (3)$$

Following (1) we can rewrite the right-hand side of (3). Hereby we imply that the spatial dimensions over which the integrals are computed are much larger than the spatial scale designed by the overbar.

$$\int_S e v_n dS = \int_S \bar{e}\bar{v}_n dS + \int_S \overline{e'v'_n} dS \qquad (4)$$

Putting (2) to (4) together we can solve for the integral over the turbulent fluxes $\overline{e'v'_n}$ through the surface enclosing the volume of interest:

$$\int_S \overline{e'v'_n} dS = -\int_V \frac{\partial e}{\partial t} dV - \int_S \overline{e}\overline{v}_n dS + \int_V Q dV - \int_V R dV \tag{5}$$

Neglecting horizontal turbulent fluxes, which are usually much smaller than their respective mean fluxes, denoting the vertical wind component by w and introducing a surface flux $\overline{e'w'_{surf}}$ we get for the vertical turbulent flux through the horizontal surface S_1 at the top of the volume V:

$$\int_{S_1} \overline{e'w'} dS = -\int_V \frac{\partial e}{\partial t} dV - \int_S \overline{e}\overline{v}_n dS - \int_{S_{surf}} \overline{e'w'} dS + \int_V Q dV - \int_V R dV \tag{6}$$

4. Determination of spatially aggregated surface fluxes

Imagine now an atmospheric boundary layer over an area with inhomogeneous land use patterns, which is covered by a thermal inversion that suppresses all turbulent motions at the height of the inversion. In this case, the left-hand side of (6) would be zero. In this case equation (6) can be evaluated in a residual way ([3], [4]). If we further assume that horizontal advection can be neglected, then the second term on the right-hand side of (6) vanishes, too. The same applies to the fourth and the fifth term on the right-hand side if we stipulate the absence of any sources and sinks within the boundary-layer air. We then get:

$$\int_{S_{surf}} \overline{e'w'} dS = -\int_V \frac{\partial e}{\partial t} dV \tag{7}$$

This means that a spatially aggregated surface flux of a property e into the atmospheric boundary layer can be determined from the increase of the density of e within the boundary-layer air in a volume $V = S_{surf} \cdot z$ if the boundary layer is horizontally homogeneous in the respect that horizontal advection is negligible and if it is covered by an inversion in a known height z. Above the height z we assume to have background concentrations of methane. If the inversion moves upward this would mean entrainment of air with this background concentration from above. We will evaluate (7) for the time from 10 p.m. to 6 a.m. During this time the inversion will be in its lowest position (see Fig. 2, especially in the night from 23 May to 24 May and the following one). Therefore we will neglect entrainment when evaluating Eq. (7) in this paper. A possible imbalance in Eq. (7) may be given by the measurement errors for the density of the property e and the height of the volume z, the correctness of the assumption of negligible horizontal advection, and the correctness of the assumption of perfect vertical mixing in the layer below the inversion. Probably the last point has the largest influence on the result. The absolute value of the methane concentration is not really important, only the nocturnal increase has been used. This increase is believed to be correct with a few percent. The precision of the height z from the sodar data is limited by the vertical resolution used during the sodar operation. Thus the lowest height may be in error by up to 30%. With increasing height z this error decreases linearly. Therefore, finally, the methane surface flux values may be in error of about 50%.

The algorithm used here for the determination of the mixing layer height z from sodar data ([7], [8]) searches for:

- the height of a turbulent layer (H1) characterised by high acoustic backscatter intensities R(z) due to thermal fluctuations (therefore having a high variance of the vertical velocity component σ_w),
- several lifted inversions (H2_n) characterized by secondary maxima of acoustic backscatter due to a sharp increase of temperature with height and simultaneously low σ_w, and

- the height of a surface-based stable layer (H3) characterised by high backscatter intensities due to a large mean vertical temperature gradient starting directly at the ground and having a low variance of the vertical velocity component.

Details for the determination of H1 and just one lifted inversion H2 have been described in [7]. In extension to the method described in [5] the variance of the vertical velocity component is used here as a second variable in addition to the acoustic backscatter intensity. Acoustic remote sensing by a SODAR also yields wind fields that in principle could be used for determination of the mixing layer height, too. Because the wind information above the inversion is not regularly available from SODAR measurements, wind data have not been included into the scheme for the determination of the mixing layer height.

The height H1 corresponds to a sharp decrease of the acoustic backscatter intensity R(z) with height z usually indicating the top of a turbulent layer:

$$H1 = z, \text{ if } (R(z) < 88 \text{ dB and } R(z+1) < 86 \text{ dB and } R(z+2) < 84 \text{ dB}). \qquad (8)$$

The dB values are derived from an arbitrary scale because the received acoustic backscatter intensities from a SODAR cannot be calibrated. Therefore, the R(z) values are specific for the type of SODAR used. Therefore, the numbers in (8) can be taken for illustration purposes only.

Elevated inversions are diagnosed from secondary maxima of the backscatter intensity that are not related to high turbulence intensities. For elevated inversions increase in backscatter intensity below a certain height z = H2 and a decrease above is stipulated while the turbulence intensity is low:

$$H2\_n = z, \text{ if } (\partial R/\partial z|_{z+1} < 0 \text{ and } \partial R/\partial z|_{z-1} > 0 \text{ and } \sigma_w < 0.70 \text{ ms}^{-1}) \qquad (9)$$

for n = 1, …, N. In [8] N was chosen to be five. Here, for the boundary-layer budget method, N=1 is sufficient. The determination of the height of the stable surface layer H3 is started if the backscatter intensity in the lowest range gates is above 105 dB while σ_w is smaller than 0.3 ms^{-1}. The top of the stable layer H3 is at the height where either the backscatter intensity sinks below 105 dB or σ_w increases above 0.3 ms^{-1}.

$$H3 = z, \text{ if } (R(z) > 105 \text{ dB and } R(z+1) < 105 \text{ dB and } \sigma_w(z) < 0.3 \text{ ms}^{-1})$$
$$\text{or if } (\sigma_w(z) < 0.3 \text{ ms}^{-1} \text{ and } \sigma_w(z+1) > 0.3 \text{ ms}^{-1} \text{ and } R(z) > 105 \text{ dB}) \qquad (10)$$

The dB values in (10) are given for the same illustrative purpose as in (8). The search for H1 to H3 can be integrated into one search algorithm. If a surface-based stable layer with height H3 has been detected, the search for the first height H2_1 is started at the first range gate above H3, otherwise it starts at the lowest range gate. The search for elevated inversion is stopped when N elevated inversions have been found or the highest range gate of the available data has been reached.

The σ_w values used in (9) and (10) have been determined by optimizing the automatic application of the detection algorithm. In doing so it turned out that no lifted inversions occurred with a variance σ_w lower than 0.7 ms^{-1} and that the variance σ_w in nocturnal stable surface layers was below 0.3 ms^{-1}. The first σ_w threshold made it possible to distinguish between inversions and elevated layers of enhanced turbulence. The latter σ_w threshold made it possible to differentiate between nocturnal stable surface layers and daytime super-adiabatic surface layers although both types of surface layers yield more or less the same level of backscatter intensity. Finally the mixing-layer or inversion height z from the acoustic remote sensing is determined as the minimum of H1, H2_1, and H3:

$$z = \min(H1, H2\_1, H3) \qquad (11)$$

An example for the application of (7) together with (11) is given here from data from a measurement campaign from 13 to 30 May 2003 which took place west of Munich, Germany. In contrast to [3] and [4], in the dataset used here the most important parameter in (7), the mixing-layer height z, is known continuously from surface-based acoustic remote sensing with a sodar.

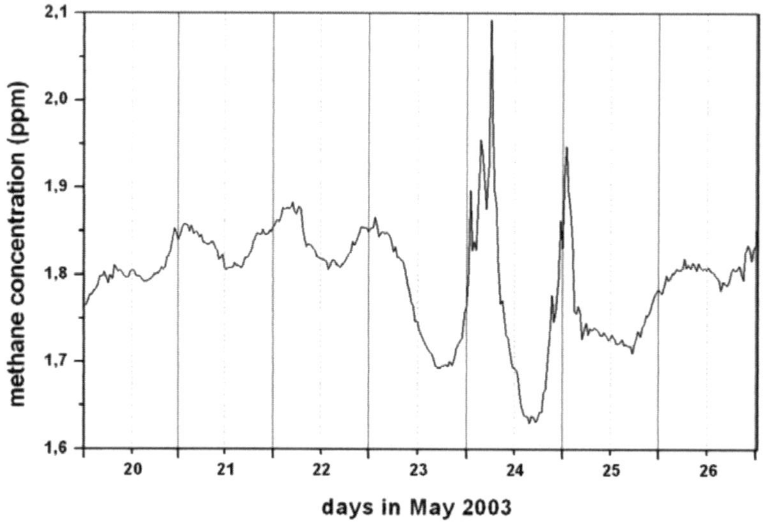

Figure 1. Variation of the near-surface concentration of methane (in ppm) in a rural area west of Munich during the week from May 20 to May 26, 2003.

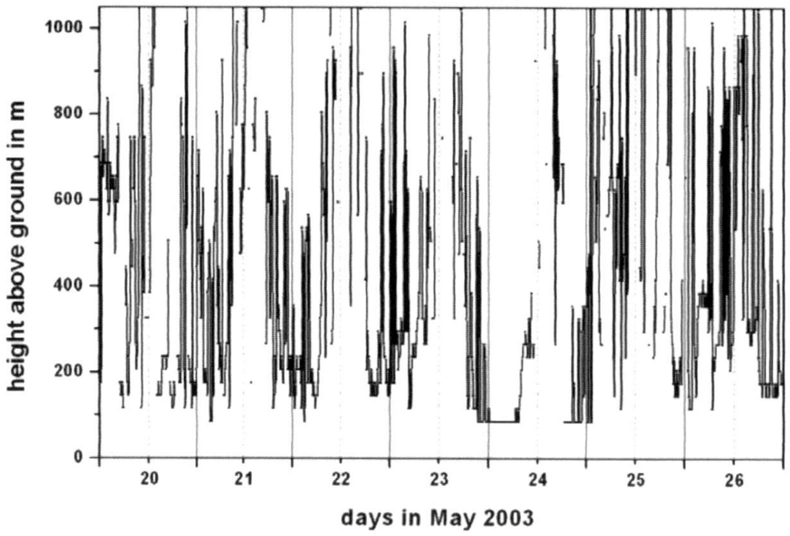

Figure 2. Fig. 1, but for the mixing-layer height derived from surface-based acoustic remote sensing (in m).

Fig. 1 shows a part of the time series of the near surface concentration of methane. It is remarkable in this time series that the concentration always increased in the evening and in the night and that it then decreased again during the next day. Apart from slightly changing wind speeds the depth of the nocturnal stable surface layer seems to be the most important factor controlling these concentration increases in the night. Fig. 2 displays the variation of the mixing layer height derived from sodar data (hourly values) for the same time period. In the nights to May 24 and 25 the mixing-layer heights took their lowest values and remained at these low values for six (May 25) to twelve (May 24) hours. Consequently the nocturnal increase was largest in these nights. Fig. 3 illustrates the overall correlation between the data shown in Figs. 1 and 2. There is a clear relation showing increasing methane concentrations with decreasing inversion height z although the squared regression coefficient is low due to the large scatter of the data. Therefore, for the further analysis we will concentrate on the overall nocturnal increase of the methane concentration over eight hours (roughly from 10 p.m. to 6 a.m.) and compare this increase to the mean MLH for the same period.

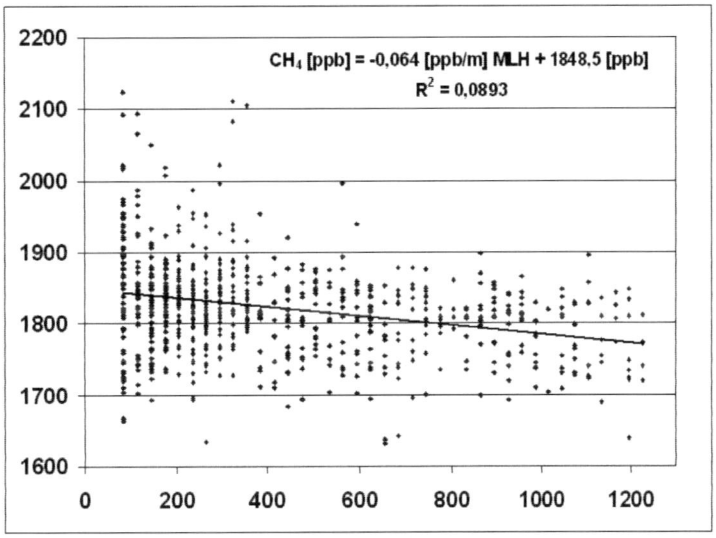

Figure 3. Overall correlation between the methane concentration and MLH based on half-hourly values for the period 10 May to 30 May 2003. A linear regression line is plotted and the equation for the regression line is given.

We now assume that the methane concentrations are equally dispersed within the shallow nocturnal boundary layer and that the measured mixing-layer height is representative for the region. We use (7) to convert the measured methane concentrations into spatially aggregated surface methane emission rates. The result is displayed in Fig. 4 for the whole measurement period of two weeks. A mean mixing-layer height for each night has been derived from the data shown in Fig. 2 and has been related to the mean nocturnal rate of increase in the methane concentration. The array of hyperbolae in Fig. 4 indicates emission rates which would explain the assumed correlation between the concentration increase and the mixing-layer height if the adopted conditions would last for about eight hours. It turns out that most data points lie between 0.1 and 1 μg/(m^2 s). If one assumes that vertical mixing takes place within one hour (from the definition of the term mixing-layer adopted in [6]) then moderate horizontal winds of 3 to 5 m/s would lead to horizontal transports of 11 to 18 km during this hour. From this estimation it can be inferred that emission rates calculated with this boundary-layer budget scheme are representative for an area of about 15 by 15 km^2.

The emission rate of 0.1 to 1 µg/(m^2 s) can be compared to the total methane emissions in the Federal Republic of Germany which is 2.2 10^{18} µg per year ([12]). Normalized to the area of 3.57 10^{11} m^2 this results in an average German methane emission rate of 0.2 µg/(m^2 s). Two thirds of these emissions ([12]) are supposed to be from the agricultural sector (i.e. 0.13 µg/(m^2 s). Since the measurements above have been made in a predominantly agricultural area, values a bit higher than the national average agricultural emission rate, as displayed in Fig. 4, can be regarded as realistic. On the other hand, incomplete vertical mixing in the nocturnal stable surface layer below the inversion would cause – under the assumptions used here - higher values emission values than the true value. Thus, some part of the higher emission rate found here could be due to incomplete vertical mixing. Seasonal variations in the methane emissions have been disregarded in this comparison.

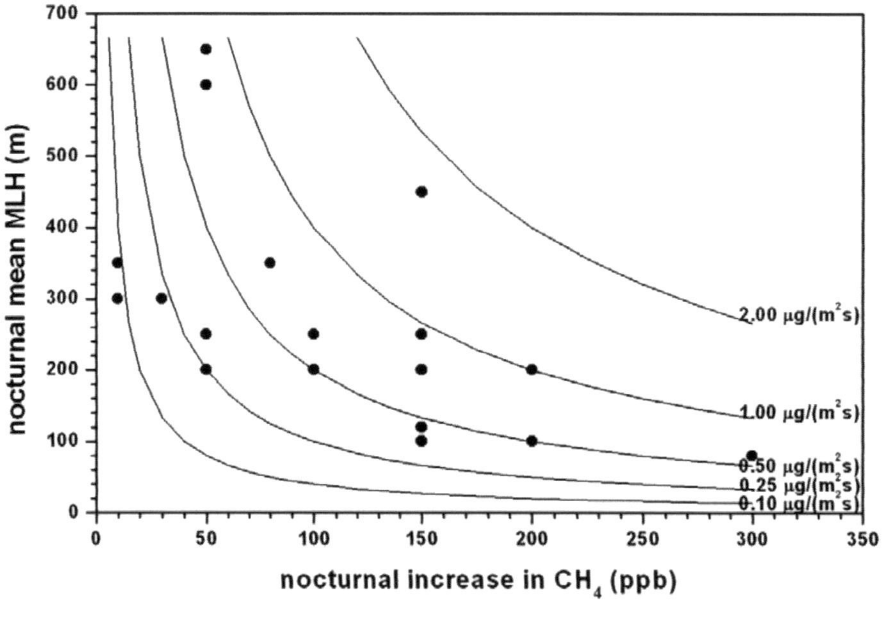

Figure 4. Correlation between the nocturnal increase of the near-surface methane concentration (x axis, in ppb) and the nocturnal mean mixing-layer height (y-axis, in m) in a rural area west of Munich for 18 nights between May 13 and May 30, 2003 (dots). Curves give mean emission rates for constant conditions over about eight hours.

5. Conclusions

The feasibility of remote sensing to help to derive aggregated diffuse emission fluxes from larger areas has been demonstrated. In-situ measurements of such turbulent fluxes in the free atmosphere are difficult. Therefore other methods had to be developed to monitor turbulent fluxes. Naturally, these methods are indirect ones based on some sort of inversion algorithms. Here, a residual technique (or budget method) has been presented in chapters 2 and 3. Remote sensing of the mixing-layer height can deliver the most important parameter which is needed to apply successfully the nocturnal boundary-layer budget method addressed in chapter 3.

The acoustic remote sensing technique alone is probably sufficient as long as the heights of nocturnal inversions have to be determined. If inversion or the mixing-layer heights during daytime are to be monitored the vertical range of a sodar can be insufficient, especially in spring and summer. In this case it is advisable to run a ceilometer in parallel to the sodar. The combined evaluation of both remote sensing devices can yield a quite complete picture of the vertical structure of the atmospheric boundary layer ([11]).

A good knowledge and a full understanding of the vertical turbulent fluxes in the atmosphere is a prerequisite for the assessment of air-surface exchange and regional and global energy budgets. In future, both the development of ground-based remote sensing and of budget methods has to be continued in order to further advance reliable weather and climate prediction techniques.

6. Acknowledgements

The measurements west of Munich have been made in the project ICAROS NET (IST-2000-29264) within the 5th Framework Programme of the European Union.

References

[1] Schürmann G, Schäfer K,Jahn C, Hoffmann H, Groma V, Török S and Emeis S 2006 Airport air quality and emission studies by remote sensing and inverse dispersion modelling *Proceedings of SPIE, Vol. 6362: Remote Sensing of Clouds and the Atmosphere XI, Slusser J R, Schäfer K, Comerón A, Editors,* 63621E. (ISBN 0-8194-6457-0. ISSN 0277-786X. DOI: 10.1117/12.695821)

[2] Schürmann G, Schäfer K, Jahn C, Hoffmann H, Bauerfeind M, Fleuti E and Rappenglück, B 2007 The impact of NOx, CO and VOC emissions on the air quality of Zurich airport *Atmos. Environ.* **41** 103-18

[3] Denmead O T, Raupach M R, Dunin F X, Cleugh H A and Leuning R 1996 Boundary-layer budgets for regional estimates of scalar fluxes *Global Change Biol.* **2** 255-64

[4] Denmead O T, Leuning R, Griffith D W T, Jamie I M, Esler M B, Harper L A and Freney J R 2000 Verifying inventory predictions of animal methane emissions with meteorological measurements *Bound.-Lay. Meteorol.* **96** 187-209

[5] Beyrich F 1997 Mixing height estimation from sodar data - a critical discussion *Atmos. Environ.* **31** 3941-53

[6] Seibert P, Beyrich F, Gryning S-E, Joffre S, Rasmussen A and Tercier P 2000 Review and intercomparison of operational methods for the determination of the mixing height *Atmos. Environ.* **34** 1001–27

[7] Emeis S and Türk M 2004 Frequency distributions of the mixing height over an urban area from SODAR data *Meteorol. Z.* **13** 361-67

[8] Emeis S, Jahn C, Münkel C, Münsterer C and Schäfer K 2007 Multiple atmospheric layering and mixing-layer height in the Inn valley observed by remote sensing *Meteorol. Z.* **16** 415-24

[9] Emeis S and Schäfer K 2006 Remote sensing methods to investigate boundary-layer structures relevant to air pollution in cities *Bound-Lay. Meteorol.* **121** 377-85

[10] Emeis S 2008 Examples for the determination of turbulent (sub-synoptic) fluxes with inverse methods *Meteorol. Z.* **17** 3-11

[11] Schäfer, K., Emeis, S., Hoffmann, H., Jahn, C.: Influence of mixing layer height upon air pollution in urban and sub-urban area. Meteorol. Z. 15 (2006), 647-658.

[12] UBA 2007 Nationaler Inventarbericht zum Deutschen Treibhausgasinventar 1990-2005 *Umweltbundesamt, Dessau. Schriftenreihe Climate Change* **04/07** (ISSN 1862-4359), 538 pp. (available from http://www.umweltdaten.de/publikationen/fpdf-l/3239.pdf)

A 100 m × 10 m Sonic to observe area averaged wind and temperature data in comparison to FTIR line integrated measurements

A Schleichardt[1], M Barth[1], K Schäfer[2], A Raabe[1]

[1] University of Leipzig, Institute for Meteorology, Stephanstr. 3, 04103 Leipzig, Germany
[2] Forschungszentrum Karlsruhe GmbH, Institute for Meteorology and Climate Research, Kreuzeckbahnstr. 19, 82467 Garmisch-Partenkirchen, Germany

E-mail: schleichardt@uni-leipzig.de

Abstract. An acoustic tomographic system has been used to estimate area averaged wind and temperature data within an area of 97 m × 12 m considering the dependence of sound speed on meteorological conditions. To obtain information about vertical structure of meteorological data, eight sound sources and receivers were placed in two different heights above the ground (0.5 m and 2.7 m). Spatially, the acoustic measurements correspond to line integrated N_2O concentration measurements (98 m) using FTIR-spectrometers. Taking stability of atmospheric layering into account, acoustic tomographic measurements serve as basis for estimating vertical fluxes of momentum and sensible heat.

1. Introduction

Greenhouse gases effect an increase of global temperature by partly absorbing infrared radiation emitted by the earth's surface. Important greenhouse gases are water vapor, carbon dioxide (CO_2), methane (CH_4), and nitrous oxide (N_2O). The longest lifetime of the previously mentioned gases has N_2O (114 years) [1].

Even though man-made emissions of N_2O, mainly derived from agriculture (e.g. animal breeding, use of fertilizers), are known, sources and sinks quantitatively are not understood very well [2]. Thus, transport and conversion processes of N_2O have to be investigated in more detail. Up to now, a large amount of trace gas emission measurements has been done using different sized chambers [3][4] or in situ ground surveys.

To get more detailed information to characterize the emission of the ground it is also necessary to consider the influence of atmospheric processes. In the following article, two systems, the acoustic tomography and two ultrasonic anemometers (USAs) are described, which were used to achieve more information about the atmospheric conditions. In order to describe vertical changes of meteorological parameters, the meteorological data have to be measured spatially resolved. A method, which is capable of providing such area-averaged (line-integrated) data of temperature and wind within a certain area, is the acoustic travel-time tomography (A-Tom). This method uses the dependency of sound speed on temperature and wind along the propagation path of acoustic signals. The ultrasonic anemometers are used to achieve wind and temperature data at a certain point (small volume). The estimate vertical gradients of the meteorological parameters, which are the bases for calculating vertical fluxes of momentum and sensible heat, were compared with each other.

The results of the N_2O flux determination are under evaluation up to now and will be presented after further experimental work.

2. Measuring systems and calculation methods

Meteorological quantities are estimated based on sound propagation properties in the atmosphere by A-Tom and the two USAs. The difference between the two systems, is the method, which can be used to calculate turbulent parameters. But first of all, the two systems are described in more details subsequently.

2.1. Acoustic tomography

Acoustic tomography uses the dependency of sound speed in air on meteorological quantities, particularly on temperature and wind, along the propagation path. There, the effective sound speed c_{eff} is given by

$$c_{eff} = c_L + v_R, \tag{1}$$

where v_R is the wind speed along the sound ray and c_L is the Laplace's sound speed in m/s

$$c_L = \sqrt{401.87\,[J/(kg\ K)] \cdot T_{av}}, \tag{2}$$

which depends on the acoustic virtual temperature T_{av} in K.

The latter one is linked with the air temperature T dependend on the specific humidity q by

$$T_{av} = T \cdot (1 + 0.51 \cdot q). \tag{3}$$

To estimate meteorological quantities, travel-time measurements of acoustic signals are carried out along different propagation paths within an area under investigation. Combining these data with information about the positions of the sound sources (loudspeakers) and receivers (microphones), which have to be known exactly, effective sound speed values can be calculated. In order to separate the scalar and vectorial influence of the temperature (acoustic virtual temperature) and wind respectively, bi-directional sound propagation is considered. Thus, for a pair of sources and receivers, which are positioned at a distance d, c_L and v_R can be calculated from travel-time measurements tt in forward (for) and backward (back) direction

$$c_L = \frac{d}{2}\left(\frac{1}{tt_{for}} + \frac{1}{tt_{back}}\right) \quad \text{and} \quad v_R = \frac{d}{2}\left(\frac{1}{tt_{for}} - \frac{1}{tt_{back}}\right). \tag{4}$$

To obtain information on the distribution of meteorological quantities from line-integrated measurements along different propagation paths, tomographic techniques are used. Advantages of this method compared to conventional measurements are, that the area under investigation can be investigated remotely without inserting devices into it. Furthermore, the method is capable of providing information about the spatial distribution of meteorological parameters within a certain volume [5][6][7][8].

2.2. Ultrasonic anemometer

An ultrasonic anemometer records temperature and wind data with high temporal resolution. This system works with ultrasonic pulses along three non-coplanar ray paths. The ultrasonic anemometer has three sound source-receiver-pairs. The propagation path between the pairs is about 175 mm. So the measurements are at the same volume.

By using the dependence from the meteorological parameters to the effective sound speed c_{eff} as mentioned before the USA's measured the acoustic virtual temperature and the wind speed. The point measurements take place by in-situ-method. To estimate fluxes like momentum flux and sensible heat flux the eddy covariance methode were used [9].

2.3. Profile similarity and eddy covariance method

The profile similarity is a procedure to determine the flux (momentum flux, sensible heat flux) in the atmosphere by measuring the quantity (wind speed) in different heights assuming a logarithmic profile of the quantity (wind speed, temperature) [10]. The fluxes calculated from the data measured by A-Tom were estimate using the profile similarity.

The eddy covariance method is based on the estimation of perturbation values of the measured data. Therefore, the measured quantity (e.g. the temperature T) is divided into a mean value $\overline{T}$ and perturbations from the mean T'

$$T = \overline{T} + T' . \tag{5}$$

The values u' and w' are fluctuations of the present horizontal wind u and the vertical wind w respectivly, compared to the mean wind value

$$u = \overline{u} + u' \tag{6}$$

$$w = \overline{w} + w' . \tag{7}$$

In both cases, an overbar indicates a temporal average of the quantity.

The time series of temperature perturbations together with the vertical wind speed perturbations are used to estimate the kinematic turbulent heat flux in vertical direction. With the perturbation of the horizontal wind speed u' and the vertical wind speed w', one can calculate the friction velocity. This is a simple method for direct measurements to calculate fluxes independent from the height or location. To measure these data a fast response sensor is required. By using slower response sensors the perturbation values are filtered, so that the fluxes would not be correct [11].

2.4. Arrangement of the measuring devices

The investigations were carried out near an agriculturally used land south of Fuhrberg near Hanover (Germany). The spatial distribution of meteorological quantities was measured using 8 sources and 8 receivers, which were placed around the measuring site with an extension of 97.2 m x 13.2 m x 2.2 m (Fig.1). The configuration was carried out in such a way that in each corner of the field one source-receiver-pair was fixed to ensure bi-directional sound propagation. From this arrangement of sound sources and receivers, spatially averaged values of acoustic virtual temperature and wind (speed and direction) could be achieved in different height ranges.

Fig. 1:Arrangement of the measurement systems south of Fuhrberg near Hanover, Germany (52.545°N, 9.855°E). Sensors (points: USA: ultrasonic-anemometers, A-Tom: acoustic tomography) are located in two heights (0.5 m, 2.7 m) each. Lines in the same colour indicate accordant measurement paths (satellite image by Google Earth [12]).

Ultrasonic anemometers, which provide temporally highly resolved acoustic virtual temperature and wind data were positioned at the northwestern boundary of the measuring site.

3. Results:

The combined measurements of A-Tom and USAs took place during late afternoon and night. At first, meteorological conditions are described. Afterwards, stability of atmospheric layering, vertical fluxes of momentum and sensible heat are presented.

3.1. Horizontal wind speed

To compare data with a comparable temporal resolution, for both systems data have been averaged. A single A-Tom measurement was taken every 20 seconds. Thus, 30 values are averaged to have a 10 minutes mean. A single USA measurement was taken every one tenth of a second, thus, a 10 minutes mean was calculated from 6000 values.

The horizontal wind speed is shown in figure 2. From 4:30 pm till 7:30 pm all four wind speed values (top and bottom) are comparably small. Furthermore, the differences between the wind speed values in the two heights at both systems are relatively low. The maximum wind speed during this time amounts to 1.3 m/s. After sunset (clear sky conditions) the wind speed is low and the atmosphere is characterized by absence of (turbulent) vertical and horizontal mixing. Thus, without external forces, mixing only takes place by diffusion processes, which are slow and less effective compared to turbulent transport.

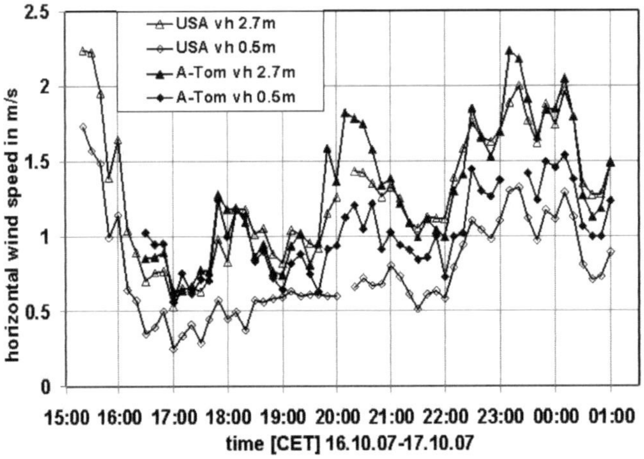

Fig. 2: Time series of the horizontal wind speed measured by the acoustic tomography (A-Tom) in two heights compared to the horizontal wind speed measured by the USAs in two heights.

3.2. Vertical fluxes

In order to characterize vertical exchange of different properties, vertical fluxes like momentum flux and vertical heat flux are estimated from A-Tom and USA measurements. The calculations for the fluxes from A-Tom data base on the profile similarity, while the fluxes calculated from USA data base on the eddy covariance method. To make allowance for the stability of the atmospheric layering, turbulence parameters obtained from the USAs are taken into account.

3.2.1. Momentum flux

The friction velocity u_* is related to the momentum flux τ by

$$\tau = \rho \cdot u_*^2 = -\rho \cdot \overline{u'w'}\,, \tag{8}$$

where ρ is air density and $\overline{u'w'}$ describes the kinematic flux of the horizontal wind in vertical direction. As mentioned before, u' and w' are fluctuations of the horizontal and vertical wind, respectively.
Thus, the friction velocity u_* can be calculated

$$u_* = (-\overline{u'w'})^{\frac{1}{2}} = \frac{k \cdot (u_2 - u_1)}{\ln z_2 - \ln z_1}\,. \tag{9}$$

The first term corresponds to the friction velocity using the eddy covariance method. The kinematic flux of the horizontal wind in vertical direction was obtained by the ultrasonic anemometers by in-situ measurements. The second term describes the friction velocity using the profile similarity for a neutral stratified atmosphere. Here, u_i is the horizontal wind speed in height z_i (0.5 m, 2.7 m) and k is the von Karman constant (0.4). The atmospheric stratification was stable during the measurements. Due to this, the friction velocity, calculated after the profile similarity, has to be stability corrected. Thus, the equation for the friction velocity, including a correction term, can be written

$$u_* = \frac{k \cdot (u_2 - u_1)}{\ln \frac{z_2}{z_1} + \frac{\alpha}{L} \cdot (z_2 - z_1)} \tag{10}$$

where $\alpha = 5$ and L is the averaged Monin-Obukhov-length, which was estimated from the USA data. The parameter L characterizes the ratio of the dynamic buoyancy to the thermal lift and is given by

$$L = -\frac{\overline{T}}{g \cdot k} \frac{(-\overline{u'w'})^{\frac{3}{2}}}{(\overline{T'w'})} . \tag{11}$$

It is a function of vertical transport of momentum $\overline{u'w'}$, the vertical kinematic temperature flux $\overline{T'w'}$, mean temperature $\overline{T}$ and the gravitational acceleration g. As before, k is the von Karman constant [9][13].

Using profile similarity, the friction velocity u_* was calculated from horizontal wind speed data (A-Tom). The correction concerning stability was done using temporal highly resolved data from the ultrasonic anemometers. Time series of u_* of A-Tom in comparison to the u_* values calculated from ultrasonic anemometers at 0.5 m and 2.7 m, using the eddy covariance method, are shown in figure 3.

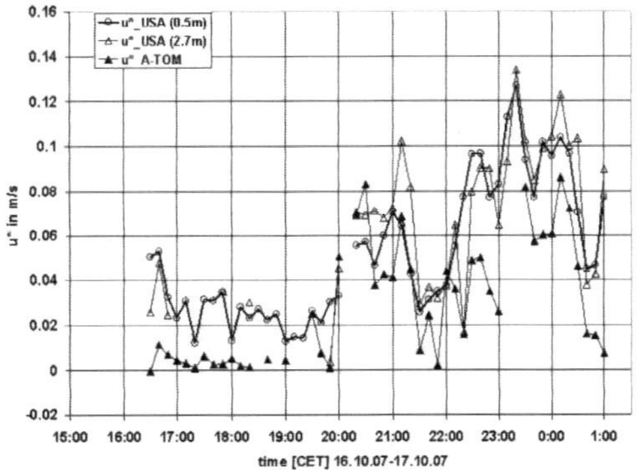

Fig. 3: Comparison between friction velocity from A-Tom data using the profile method and from USA-data using eddy-covariance method (correction due to stable stratification using USA-data for both methods).

By occurring the developing of values one can see at both measurement systems that lower values describe a stable atmosphere and high values explain an atmospheric layer, when turbulence started and the result is vertical mixing of the air. The results between the both methods were at the beginning different in the magnitude.

3.2.2. The eddy diffusivity for momentum

The eddy diffusivity for momentum K_z is calculated from the corrected friction velocity and characterizes the interaction between the horizontal wind speed in two different heights. The relation to the momentum flux and the friction velocity respectively is given by

$$K_z = k \cdot z \cdot u_* = k \cdot z \cdot \sqrt{\frac{\tau}{\rho}} \quad \text{with} \quad z = z_2 - z_1 . \tag{12}$$

An increase of vertical wind speed differences corresponds to an increase of K_z. The parameter K_z is also an identifying feature to characterize the behavior of the atmosphere. In figure 4 one perceive the fluctuations during the process. As mentioned by the friction velocity the time series of the eddy diffusivity for momentum increased. For A-Tom data the eddy diffusivity for momentum increased, when the difference between the horizontal wind speed in the upper and lower height increased (profil similarity). Especially for the acoustic anemometers the parameter K_z depends on the friction velocity. Thus, an increasing K_z is the result of an increasing friction velocity. The atmosphere had turbulent character. At low values for K_z the atmosphere was stable stratified.

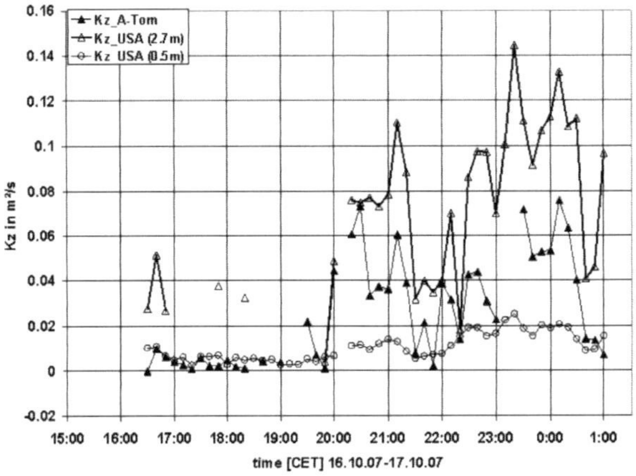

Fig. 4: Time series of the eddy diffusivity for momentum of the acoustic tomography and the USAs.

3.2.3. *The sensible heat flux*

The sensible heat flux H characterizes the transport of sensible heat. In general, during daytime the sensible heat flux is directed from the ground towards the air. This is caused by short wave solar radiation, which reaches the earth's surface and is converted into infrared radiation (heat). The hot ground heats the lowest air, which becomes less dense and rises upwards. A result of this process is a well-mixed atmosphere due to thermal convection. During nighttime, the transport of heat is reversed, because after sunset, the ground is no longer heated and starts to cool down. Thus, during night, sensible heat flux is directed from the atmosphere to the ground. Furthermore, there is hardly any turbulence due to thermal processes during night.

The sensible heat flux H is estimated by

$$H = \rho \cdot c_{\mathrm{p}} \cdot u_* \cdot T_* . \tag{13}$$

In this equation, ρ is the air density, c_{p} is the specific heat at constant pressure, u_* is the friction velocity and T_* is the scale parameter of the heat flux. This coefficient describes the change of temperature depending on height. As well as for u_*, a correction concerning the stable stratification of the atmosphere has to be taken into account when calculating T_*. Thus, the equation for this scale parameter is given by

$$T_* = \frac{k \cdot (T_{\mathrm{av\_2}} - T_{\mathrm{av\_1}})}{\ln \frac{z_2}{z_1} + \frac{\alpha}{L} \cdot (z_2 - z_1)} \tag{14}$$

The time series of *H* calculated from acoustic tomographic measurements and USA measurements are shown in figure 5. From approximately 5:00 pm until 8:00 pm there was only little sensible heat flux, again a characteristic of stable atmosphere. Later when the sensible heat flux increased the atmosphere got more turbulent.

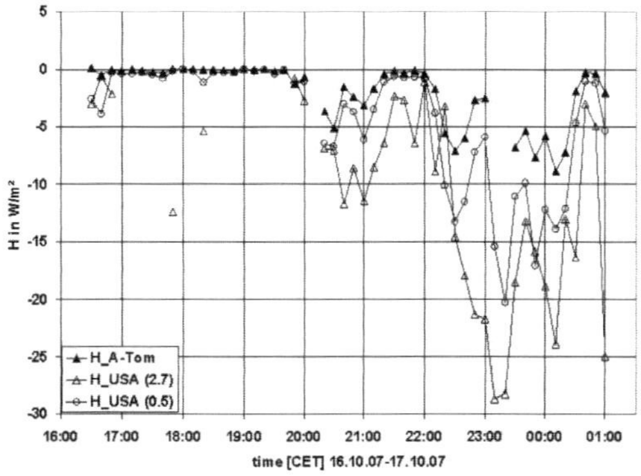

Fig. 5: Time series of the sensible heat flux over night from A-Tom and USAs.

4. Conclusions

Two measurement systems with similar measurement modality and two different method of calculation (for vertical fluxes), namely eddy covariance method and profile similarity, were used to compare the results of the meteorological parameters and turbulent fluxes. Data were obtained in two different heights to enable investigations of vertical gradients of the different properties. Using profile similarity, measured values from different heights (and their gradients) serve as basis for calculating vertical fluxes (sensible heat flux) and the eddy diffusivity for momentum. The calculation (by measuring the perturbation of horizontal wind *u'* and the perturbation *w'* in vertical direction), using the eddy covariance method, is based on the two acoustic anemometers. The figures showed a similar behavior of the different series per measurand. However, differences between area-averaged measurments and averaging data by point measurments were in evidence. Low wind speed values and low gradients enable a stable layering of the atmosphere. When the wind freshens and vertical gradient rise, a vertical mixing of air takes place.

References

[1] Forster, P., V. Ramaswamy, P. Artaxo, T. Berntsen, R. Betts, D.W. Fahey, J. Haywood, J. Lean, D.C. Lowe, G. Myhre, J. Nganga, R. Prinn, G. Raga, M. Schulz, R. van Dorland, 2007: Changes in Atmospheric Constitutents and in Radiative Forcing. In: Climate Change 2007: The Physical Science Basis. Contributions of Working Group I to the Fourth Assessment Report of the Intergovernmental Panel on Climate Change [Solomon, S., D. Qin, M. Manning, Z. Chen, M. Marquis, K.B. Averyt, M. Tignor, H.L. Miller (eds)]. Cambridge University Press, Cambridge, United Kingdom and New York, NY, USA.

[2] Bange, H.W., 2006: New directions: The importance of oceanic nitrous oxide emissions. Atmospheric Environment 40, 198-199.

[3] Amon, B., V. Kryvoruchko, T. Amon, S. Zechmeister-Boltenstern, 2006: Methane, nitrous oxide and ammonia emissions during storage and after application of dairy cattle slurry and influence of slurry treatment. Agriculture, Ecosystems and Environment, 112, 2-3, 153-162.

[4] Denmead, O.T., R. Leuning, I. Jamie, D.W.T. Griffith, 2000: Nitrous oxide emissions from

grazed pasture: measurements at different scales. Chemisphere – Global Change Science, 2, 301-312.

[5] Raabe, A., K. Arnold, A. Ziemann, 2002:Horizontal Turbulent Fluxes Heat and Horizontal Homogeneity in Micrometeorological Experiments.J. Atmos. Oceanic Technol., 19,1225-1230

[6] Arnold, K., A. Ziemann, A. Raabe, 1999: Acoustic Tomography inside the Atmospheric Boundary Layer.Phys. Chem. Earth, 24B, 133-137

[7] Ziemann, A., K. Arnold, A. Raabe, 2002: Acoustic Tomography as a Remote Sensing Method to Investigate the Near-Surface Atmospheric Boundary Layer in Comparison with In Situ Measurements, J. Atmos. Oceanic Technol., 19, 1208-1215

[8] Barth, M., A. Raabe, K. Arnold, C. Resagk, R. du Puits, 2007: Flow field detection using acoustic travel time tomography. Meteorol. Z., 16 (4), 443-450

[9] Foken,T.,2003: Angewandte Meteorologie. Springer Verlag, Berlin Heidelberg New York

[10] Etling,D.,2002:Theoretische Meteorologie. Springer Verlag, Berling Heidelberg New York

[11] Stull, R.B:,1997: An Indtroduction to Boundary Layer Meteorology.Kluwer Academic Publishers, The Netherlands

[12] Google Earth, URL: http://earth.google.com

[13] Monin, A.S., and A.M. Obukhov, 1954:Basic laws of turbulent mixing in the atmosphere near the ground, translate by John Miller 1959, http://stinet.dtic.mil/cgi-bin/GetTRDoc?AD=AD672723&Location=U2&doc=GetTRDoc.pdf

IOP Conf. Series: Earth and Environmental Science 1 (2008) 012057 doi:10.1088/1755-1307/1/1/012057

Dynamics of the thermal structure of the lower atmosphere above Tomsk from the data of acoustic sounding

G V Bukhlova[1], **N P Krasnenko**[1,2], **P G Stafeev**[1]

[1]Institute of Monitoring of Climatic and Ecological Systems of the SB RAS, 10/3, Akademicheskii Pr., 634055 Tomsk, Russia
[2]Tomsk State University of Control Systems and Radioelectronics, Tomsk, Russia

E-mail:galvik@mail.ru, krasnenko@imces.ru, spg@iao.ru

Abstract. In the present work, results of investigations into the thermal structure of the atmospheric boundary layer from the data of acoustic sounding are given. Acoustic sounding was performed by the Zvuk-3 sodar in the suburbs of Tomsk in different seasons of 2005-2007. The stratification type (class of atmospheric stability), mixing layer height, and temperature inversion layers together with their thickness and heights of their upper and lower boundaries were determined. Their statistical characteristics were also calculated. The characteristics of temperature stratification were in good agreement with the previously obtained data. Thus, in winter the number of surface inversions was maximal. In our case, in December of 2005-2007, the surface inversions were observed in more than 50% of cases. The height of the lower boundary of elevated inversions was in the altitude range 50-100 m, the upper boundary of the elevated inversion was in the altitude range 140-200 m. The average thickness of elevated inversions in December was about 90 m. In summer, inversions were observed mainly at night, because unstable stratification prevailed in the afternoon. The average height of the lower boundary of the elevated inversion in June was 70 m, and that of the upper boundary was 230 m. The ABL structure characteristics obtained with the acoustic radar can be used to predict climatic prerequisites for atmospheric pollution events.

1. Introduction

Acoustic remote sounding [1-3] is the most efficient method of investigating the atmospheric boundary layer (ABL) that allows its structure and dynamics to be monitored in real time, the type of the temperature stratification to be determined, and the wind velocity profile and characteristics of atmospheric turbulence such, for example, as structure constants of temperature and wind velocity fluctuations and acoustic refractive index to be measured. The method of acoustic remote sensing is well complemented by the ultrasonic method of *in situ* measurements of the meteorological parameters and turbulence characteristics in the surface layer of the atmosphere.

Long-term investigations of the spatial and temporal variability of the meteorological parameters occupy the central place in the study of climate-forming factors. Moreover, the lower (surface and boundary) layer of the atmosphere is characterized by the greatest variability of its state. The present study is aimed at accumulation of experimental data of monitoring of the lower layer of the atmosphere in Tomsk and interpretation of the data obtained to retrieve the characteristics of the temperature stratification and their interrelation with other characteristics of the atmosphere in different seasons of the year and different times of day.

Experiment was carried out in 2005-2007. The data of the acoustic radar (sodar) were supplemented with the data of the ultrasonic meteorological complex that allowed us to describe in more detail the state of the atmosphere during the period of observations.

2. Instruments and measurement procedure

The Zvuk-3 monostatic acoustic radar and the ultrasonic meteorological complex were used as measuring devices [4, 5]. They were placed in the territory of the Climatic-Ecological Observatory of the Institute in the outskirts of Tomsk Akademgorodok. Their external view is shown in figure 1.

Figure 1. External view of the sodar antenna and ultrasonic meteorological complex mounted at a 5-m mast.

The Zvuk-3 monostatic Doppler acoustic radar is intended for monitoring of the temperature stratification in the atmospheric boundary layer. Operating frequency – 2000 Hz. It determines the stratification type (class of atmospheric stability), mixing layer height, heights of the temperature inversions and convective plumes, structural characteristic of the temperature field, and radial (in the direction of the antenna axis) component of the wind velocity vector in real time. The thermal structure of the atmosphere is estimated from a facsimile record (echogram) of the sodar. The artificial color of the facsimile record is related qualitatively to the intensity of the scattered sound (the temperature structure constant). Here a darker color corresponds to a more intensive signal. The facsimile record of the reflected signal is used to extract information on the thermal structure of the atmosphere, more exactly, it maps the altitude-time distribution of temperature inhomogeneities along the sounding path. The main advantages of facsimile records is the possibility of obtaining information and estimating the dynamics of atmospheric processes in real time.

The ultrasonic meteorological complex comprises and ultrasound measuring head (sensor) and pressure and humidity sensors. The ultrasound sensors are intended for inertialess measurements of the three components of the wind velocity and air temperature.

The main advantage of the ultrasonic measuring devices is that the statistical characteristics and different parameters of temperature and wind turbulence in the surface atmospheric layer, including the total energy of turbulent motion, momentum and heat fluxes, wind and temperature scales, Monin–Obukhov length, etc. are directly calculated from their data. The calculated parameters are sufficient for estimation of the dynamic regime of turbulence in the surface layer of the atmosphere, including the class of stability of atmospheric stratification. On the basis of the Monin–Obukhov similarity theory, prognostic estimates were obtained for the following group of the parameters: local wind velocity and temperature gradients, turbulent momentum and heat exchange coefficients, outer scale of turbulence, energy dissipation rates of wind and temperature fluctuations. The profiles of atmospheric parameters were also reconstructed within the limits of the surface layer.

In combination with lidars, they allow efficient monitoring of the meteorological atmospheric state and prediction of the concentration and propagation regime of gaseous air pollutants.

3. Results of investigations

On the basis of continuous monitoring, results of sodar and ultrasonic investigations of the lower layer of the atmosphere were obtained. Diurnal and seasonal variations of the atmospheric stratification and interrelation of the atmospheric parameters were analyzed.

The characteristics of the ABL temperature stratification were in agreement with their general pattern for mid-latitudes. Thus, in summer inversions were observed mostly at night, because unstable stratification in the afternoon prevailed, especially at noon hours. Therefore, in summer no surface inversions were observed in the afternoon because of contrasts in the thermal and radiation regimes during the day. Synoptic processes also contribute to laws of inversion evolution, though in summer they seldom changed the stratification type.

In winter, elevated inversions prevailed at any time, which was in agreement with observations reported in [6].

In our case, inversions were observed at any time of day from December to January, and surface inversions prevailed mainly in the daytime. Probably, this was connected with the advection processes, since the diurnal temperature variations were less pronounced in winter than in summer because of the snow cover, cloudiness, and other reasons. Advection of heat facilitated the formation of surface inversions, whereas advection of cold interfered with it. The unstable stratification was seldom observed in winter. The elevated inversions were observed in November and December, but much less often than surface ones, and mostly in the afternoon. The surface inversions were well traced at night and in the evening; in the daytime they raised (for example, Fig. 2, Fig. 3).

The recurrence of surface inversions increased under conditions of frosty weather, especially at night when they were recorded almost constantly. As a whole, the elevated inversions were observed in the cold season much more often than in other seasons.

With increase in the wind velocity, the low-level inversions were observed less often. When the wind velocity (at the sensor altitude) exceeded 6 m/s, no inversion was observed. However, in a summer day, the wind velocity at the sensor altitude had essentially no effect on the stratification type; it was important only at night, when it smeared surface inversions.

In our case, the unstable stratification was observed more and more seldom in fall from October to November, and the ABL stratification more typical of winter was observed. Weak elevated inversions prevailed that sometimes intensified in the morning hours.

From May to July, the unstable ABL stratification was most often encountered. As already indicated, this is connected with strong heating of the underlying surface in the daytime. Elevated inversions were observed only at night. The elevated temperature inversions were also typical of the morning hours in spring, after transition of inversions from the surface night-time to the elevated morning-time ones. Isothermal or weakly stable stratifications were most seldom encountered in this season.

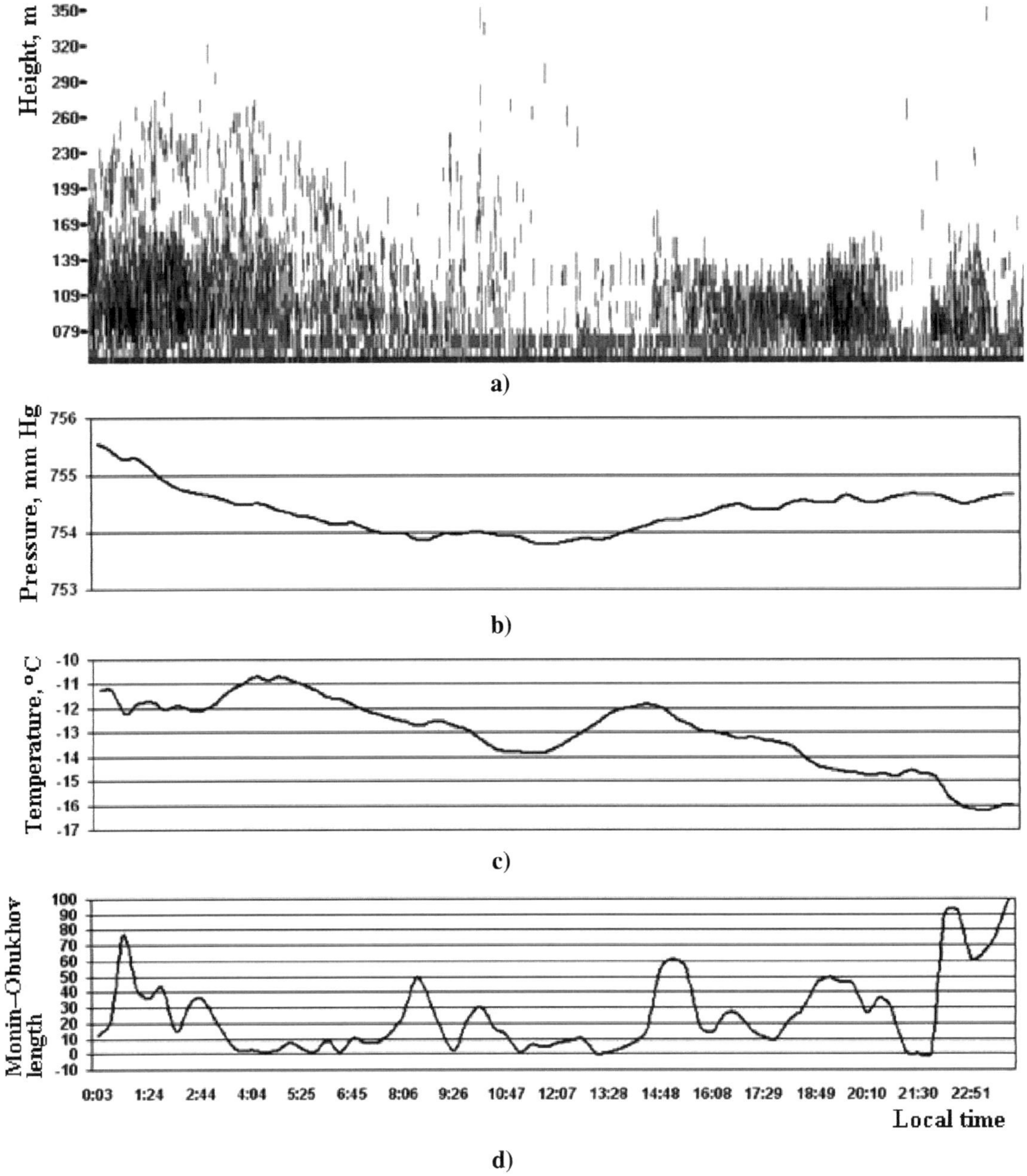

Figure 2. Facsimile record of the ABL structure (a) (sounding height, in meters, is put on the ordinate) and diurnal variations of the atmospheric pressure (b), in mm Hg, air temperatures (c), in °C, and the Monin–Obukhov length (d), in m, in December 23, 2005.

In general, the temperature inversions were mostly observed at night in spring and fall. The elevated inversions were observed, as a rule, in the morning hours in spring, and the stratification was transformed into unstable one during light hours.

As for summer months, inversions were observed mostly at night, since strong convective flows were formed in the atmosphere in the daytime. As the morning hours approached, the inversions rose and became elevated. Then they were smeared when the stratification became unstable.

IOP Conf. Series: Earth and Environmental Science **1** (2008) 012057 doi:10.1088/1755-1307/1/1/012057

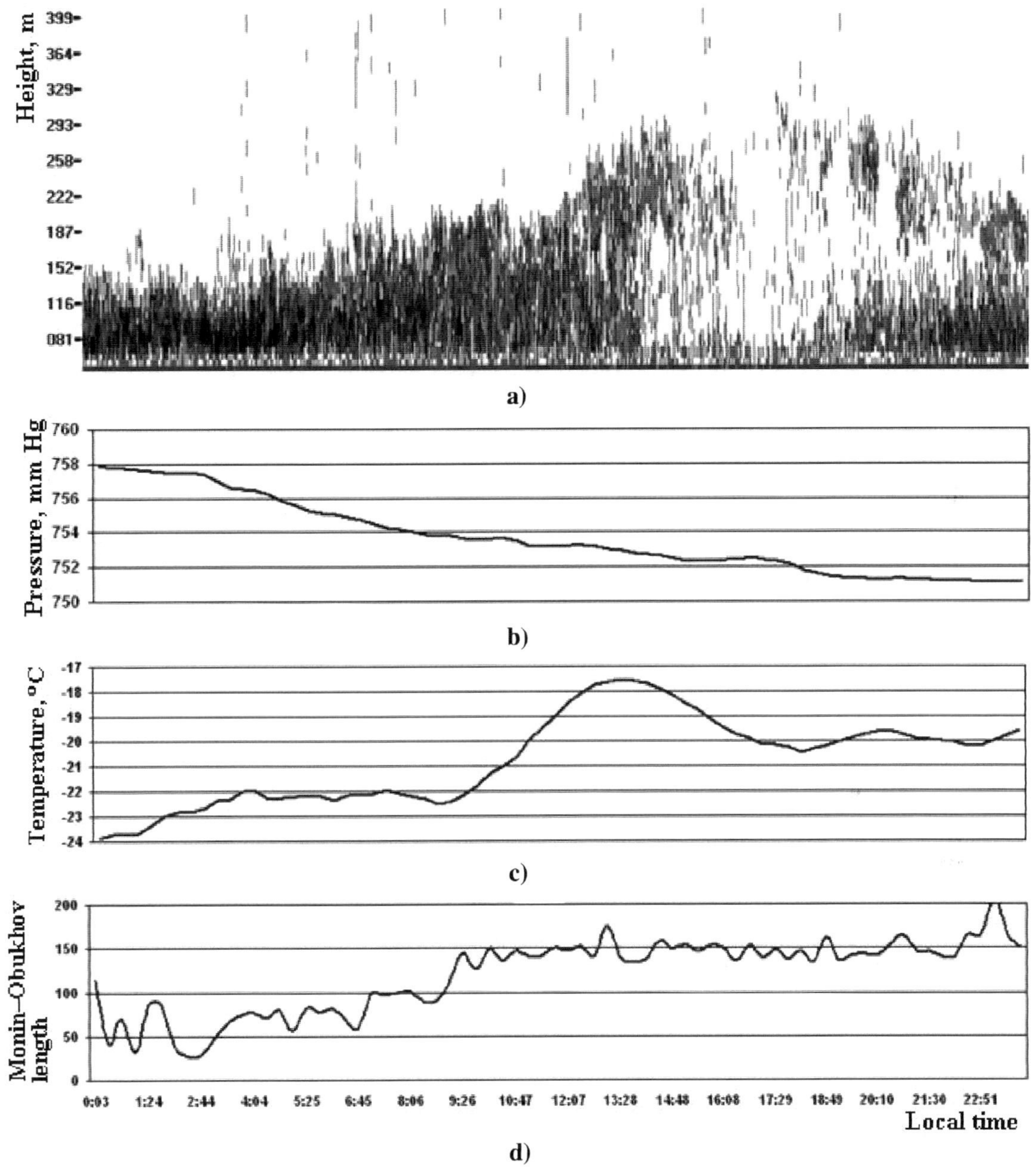

Figure 3. Facsimile record of the ABL structure (a) (sounding height, in meters, is put on the ordinate) and diurnal variations of the atmospheric pressure (b), in mm Hg, air temperatures (c), in °C, and the Monin–Obukhov length (d), in m, in December 27, 2005.

4. General stratification types of the surface atmospheric layer above Tomsk

As for the temperature inversions, repeatability of ground-based inversions in Tomsk was not too different from the data obtained previously for temperate latitudes. In winter, ground-based inversions were more often observed than in summer. If we consider the seasonal distribution of temperature stratifications, in winter it changes only insignificantly in the daytime, because the winter conditions smooth out the diurnal temperature variations. The formation of ground inversions is promoted by heat advection, whereas in summer, the advection processes do not play such a role.

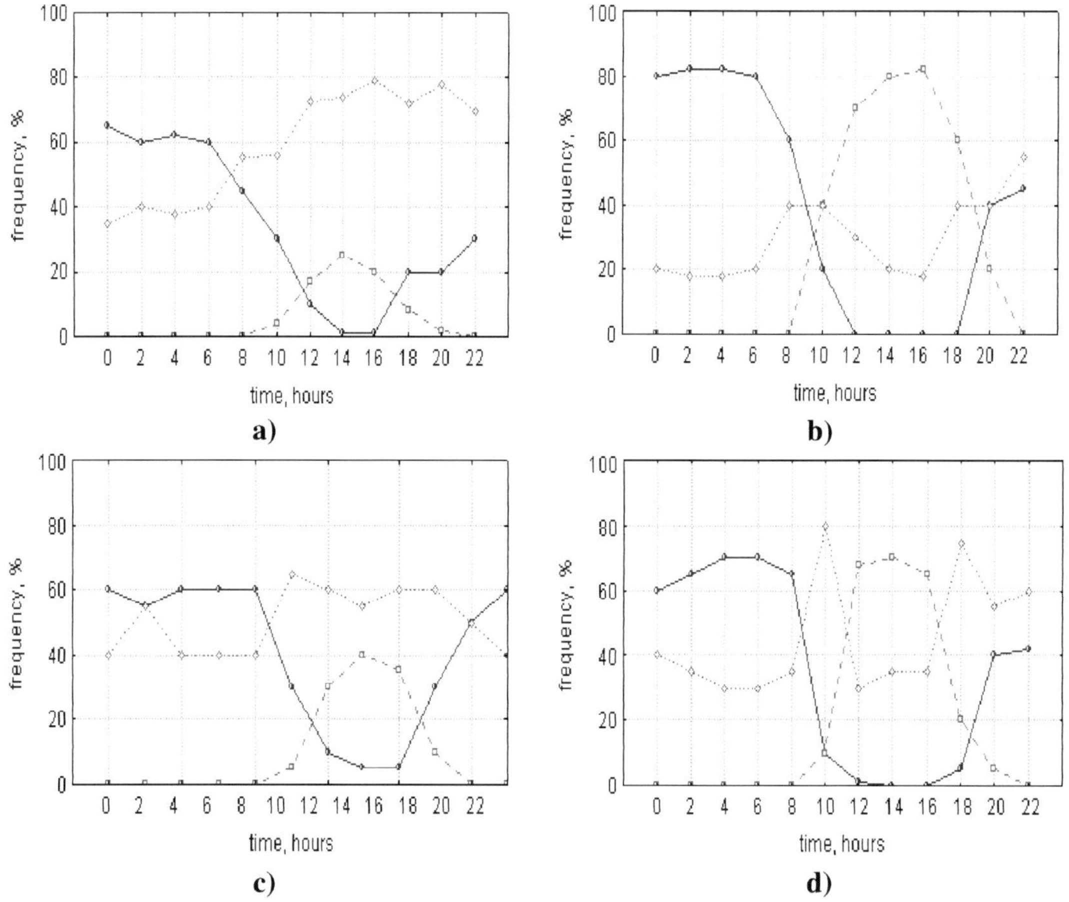

Figure 4. Diurnal variations of the repeatability of temperature stratifications in Tomsk in winter (**a**), summer (**b**), fall (**c**), and spring of 2005–2007 (**d**) retrieved from sodar data:
—○— - stable, -□- - unstable, ◇ - neutral stratification.

The number of cases of stable atmospheric stratification was about 40% for all seasons of year, and in winter, the ground-based inversions were observed in 60% of cases in frosty weather irrespective of the time of day, but the greatest number of ground-based inversions was registered from 0 to 6 h, LT. In the afternoon, weakly stable or neutral stratifications were mostly observed. During the observation period, the upper boundary of the ground-based inversions was always above 80 m; in most cases, it lay between 100–220 m, but could be as great as 350 m. In summer, the ground-based inversions were more often observed at night. Because of the presence of diurnal temperature contrasts, the ground-based inversions were virtually not observed in the daytime. The unstable atmospheric stratification was most often observed in June; in the daytime, it was observed in about 80% of cases and was followed by the neutral or stable stratification in the evening. In winter, the unstable stratification was observed in ~20% of cases. As a rule, it was observed during passage of cyclones. In fall, the unstable stratification was observed in 30% of cases, the ground-based inversions could reach 60% of cases, and their maximum repeatability occurred in the morning hours. In transient seasons, simultaneous ground-based and elevated inversions were frequently observed. High repeatability of unstable stratification in spring months was mainly due to its May values. The greatest number of days and hours with unstable stratification were observed in May.

As for the diurnal variations of the atmospheric temperature stratification in different seasons, from Fig. 4 it can be seen that in winter and fall the temperature inversions are observed at any time. In the daytime, their number decreases several times, but does not vanish, while in spring and summer, no inversions were observed from 12 to 16 h, LT. The greatest repeatability of ground-based inversions was recorded at night and in the morning. This was typical of all seasons, especially for summer, when inversions were not observed at all in the daytime.

The unstable atmospheric stratification, on the contrary, was most often observed in the daytime in all seasons of the observation period, though in winter its repeatability was minimal even in these hours and reached about 20%. From 22 h to 8 h, LT no unstable stratifications were registered. The neutral stratification dominated from 10 to 20 h, LT, in all seasons except for summer when the unstable stratification was dominant.

5. Statistical characteristics of the temperature inversions

Sodar data were displayed on the computer monitor as facsimile records. Inversions types, times of their formation, heights, and thicknesses were measured every 5 min. Our analysis demonstrated that inversions were observed in the atmosphere above Tomsk in more than 40% of cases during the observation period mostly at night and in the morning. The average thickness of the ground-based temperature inversion was 183.5 m at night and 160.5 m in the daytime. The statistical characteristics of the temperature inversion boundaries are tabulated in table 1.

Table 1. Statistical characteristics of temperature inversion boundaries in Tomsk in winter of 2005-2007.

Parameter	Average height, $\overline{H}$, m	Standard deviation σ_H, m	Variation coefficient, $\sigma_H/\overline{H}$
Upper boundary of elevated inversions:			
At night (20.00–6.00)	231.6	38.04	0.16
In the daytime (9.00–16.00)	228.6	67.6	0.3
Lower boundary of elevated inversions:			
At night (20.00–6.00)	41.6	11.7	0.28
In the daytime (9.00–16.00)	41.3	19.6	0.47

The results presented in table 1 demonstrate that the average thickness of elevated inversions in winter was ~190 m at night and in the daytime during the period of observations. This thickness is mostly due to the contribution of January inversions that develop against the background of stable anticyclones. The variability of heights of inversion layer boundaries is somewhat characterized by the variation coefficient. The boundaries of elevated inversions are most stable at night. The ground-based temperature inversions for which the variation coefficient exceeds 0.4 have the least stability. The lower boundaries of the elevated temperature inversions are less stable than their upper boundaries.

6. Conclusions

The ABL characteristics (temperature stratification and especially temperature inversions) obtained with the use of the acoustic radar are important for estimation of climatic prerequisites for atmospheric pollution events. It is demonstrated that the statistical characteristics of the temperature stratification in different seasons differ from each other, which results, for example, in different conditions for pollutant accumulation. The application of the acoustic radar and ultrasonic meteorological station allows a finer structure of the meteorological fields in the atmospheric boundary layer and its variability to be recorded, thereby supplementing significantly the standard meteorological measurements. Our investigations confirm the reliability of sodar data on the thermal structure of the lower atmospheric layer. Our data confirm previous results on the structure of the surface layer at temperate latitudes and can widely be used for practical and scientific purposes.

References

[1] Krasnenko N P 1986 *Acoustic Sounding of the Atmosphere* (Novosibirsk: Nauka, Siberian Branch)
[2] Krasnenko N P 2001 *Acoustic Sounding of the Atmospheric Boundary Layer* (Tomsk: Vodolei)
[3] Lokoshchenko M A 1994 *Meteorology and Hydrology.* 6 54-65
[4] Krasnenko N P and Stafeev P G 2006 *Pribory i Tekhnika Eksperimenta.* 1 160-162
[5] Krasnenko N P, Kudryavtsev A N, Mananko E E and Stafeev P G 2006 *Pribory i Tekhnika Eksperimenta.* 6 144-145
[6] Byzova N L, Ivanov V N and Garger E K 1989 *Turbulence in the Boundary Atmospheric Layer* (Leningrad: Gidrometeoizdat)

IOP Conf. Series: Earth and Environmental Science **1** (2008) 012058 doi:10.1088/1755-1307/1/1/012058

LiDAR error estimation with WAsP Engineering

Ferhat Bingöl[1], Jakob Mann[1] and Dimitri Foussekis[2]

[1]Risø National Laboratory · Technical University of Denmark – DTU
Frederiksborgvej 399 · P.O. 49 · DK-4000 Roskilde, DENMARK
[2]Centre for Renewable Energy Sources (CRES), Wind Energy Department
19th km Marathonos Av., GR-190 09, Pikermi, Attiki, GREECE

ferhat.bingoel@risoe.dk

Abstract. The LiDAR measurements, vertical wind profile in any height between 10 to 150m, are based on assumption that the measured wind is a product of a homogenous wind. In reality there are many factors affecting the wind on each measurement point which the terrain plays the main role. To model LiDAR measurements and predict possible error in different wind directions for a certain terrain we have analyzed two experiment data sets from Greece. In both sites LiDAR and met. mast data have been collected and the same conditions are simulated with Risø/DTU software, WAsP Engineering 2.0. Finally measurement data is compared with the model results. The model results are acceptable and very close for one site while the more complex one is returning higher errors at higher positions and in some wind directions.

1. Introduction

The LiDAR (**Li**ght **D**etection **a**nd **R**anging) instruments are becoming new option in vertical profile measurements for wind assessment. The British company, QinetiQ, has developed a LiDAR called ZephIR in recent years [1]. In profile measurements the ZephIR is used as a vertical scanning tool facilitating focusing of the laser beam at different focus distances (heights) between 10 and 150m. At the given altitude it scans a circular pattern with a cone angle of 30.4° by means of an optical wedge from which the horizontal wind speed can be derived. More information about the working mode of ZephIR, one can read the paper reference [2]. In this kind of working mode the main assumption of the instrument is that the conical scan area is homogenous; that all the measured points on a conical scan circle are the same wind. In reality there are many factors affecting the wind on each point which the terrain plays the main role on error.

To model LiDAR measurements and predict possible error in different wind directions for a certain terrain we have analyzed two experiment data sets from two Greece sites; Lavrio (Λαύριον) and Panachaiko (Παναχαϊκό). At Lavrio, CRES has a test facility and at Panachaiko wind park[3,4]. In both sites LiDAR is far away from any wind turbine and close to terrain reference mast. While the first site has simpler terrain the second site has a very complex structure and weather conditions. In both sites LiDAR and met.mast data have been collected and the same conditions are simulated with Risø, WAsP Engineering with the method described in this paper. Finally, real data is compared with the model results.

2. Experiment Sites

The Lavrio site is located 38km SE of Athens central close to the east coast of Greek Aegean Sea (Figure 1, Left). Experiment has taken place between 2008-December-01 and 2008-January-15. The LiDAR is located nearly 12m N of the mast. The LiDAR measurement heights are at 32 and 76m. The cup and vane measurements on the mast are also at 32 and 76m. The Panahaiko site is located 165km NW of Athens central. The site, from West to East, is 14km away from the Coast of Patras Sea (Figure 1, Right). Experiment has taken place between 2007-09-19 and 2007-10-11. The LiDAR is located nearly 20m WSW of the mast. LiDAR measurement heights are 30 and 53m. The cup and vane measurements on the mast are 30 and 53m. Panahaiko has a very complex terrain and highest point is nearly 2000m. Instrument locations are given in Table 1.

Table 1: Positions of the instruments in both experiments

		Geographic Coordinates		Helenic Coordinate System GRS80	
Experiment	**Site**	**Lat[°]**	**Lon[°]**	**Lat[$_m$]**	**Lon[$_m$]**
Lavrio	**Mast**	37°46'4.02"N	24° 3'43.94"E	505347.100	4179758.338
	LiDAR	37°46'4.44"N	24° 3'43.92"E	505335.236	4179797.113
Panahaiko	**Mast**	38°12'32.57"N	21°52'20.35"E	313567	4230862
	LiDAR	38°12'32.30"N	21°52'19.62"E	313549	4230854

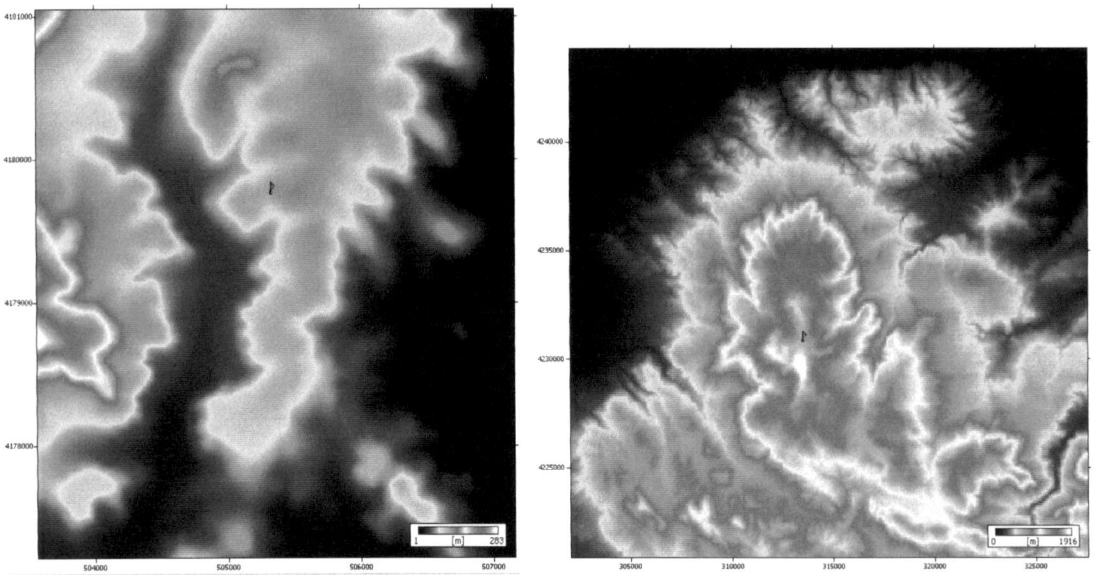

Figure 1: Elevation plots of site Lavrio (Left) and Panahaiko (Right).
Flags are demonstrating the measurement points. The coordinate system is HGRS80.

IOP Conf. Series: Earth and Environmental Science **1** (2008) 012058 doi:10.1088/1755-1307/1/1/012058

3. Theory & Method

Our main goal is to generate v_r (the radial velocity in the direction of the laser beam) for each height with WEng and use these values in standard computation method of QinetiQ ZephIR to calculate horizontal wind speed. To do this, first the computation method is derived to get the v_r values and after that WEng model is put on some tests to see if it can return reasonable outcome to compute v_r.

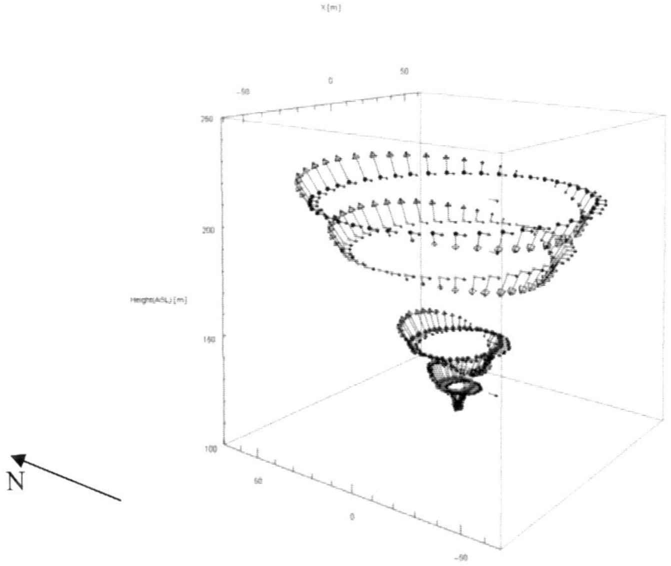

Figure 2: Calculated horizontal speed and LiDAR beam direction projections for Lavrio Test Centre measurements for wind direction is 0° north. Projection vectors have been scaled double size.

3.1. Simulating the LiDAR

The QinetiQ ZephIR makes conical scan in different heights with 50 data points on each circle. This data is put into a fit function and returns the u, w and the wind direction without sign. To simulate this condition in WEng a C# command line tool is coded which uses WEng libraries from outside over .NET environment. The software does below steps for the simulation:

- Real measurement coordinates and measurements heights are set for the mast and the LiDAR.
- The normal vectors for each beam direction on the scanning circle are computed with Equation 1. Φ is the cone angle for the wedge and φ is the scan azimuth angle which is constant as 30.4°.

$$n_{\{x,y,z\}} = \{Sin(\phi), Sin(\phi), Cos(\phi)\}.\begin{Bmatrix} Sin(\varphi_i) \\ Cos(\varphi_i) \\ 1 \end{Bmatrix}$$

- Positions of the measurement conical circle are computed with Equation 2 and horizontal wind speed is calculated via WEng model at these points. p is the position $_{\{x,y,z\}}$, h is the measurement height.

$$p_i = hCos(\phi).n_{\{x,y,z\}} + p_{LiDAR}$$

- The calculations are projected to laser beam direction with Equation 3, where U is the calculated horizontal wind speed at each point, θ is the calculated wind direction and Ψ is the calculated tilt.

$$v_r = n_{\{x,y,z\}} \cdot U \begin{Bmatrix} -Sin(\theta) \\ -Cos(\theta) \\ Tan(\psi) \end{Bmatrix}$$

- The v_r versus φ takes the form of a rectified sine wave, with the values corresponding to upwind and downwind measurements and fits to Equation 4 [5]. In QinetiQ ZephIR, since the sign of the wind vector cannot be seen, the function is absolute and a possible ambiguity of 180° in bearing is resolved with reference to a wind-vane reading. In our model since we know the sign the function gets simplified.

$$v_r = |aCos(\varphi - b) + c|$$

$$u = a / Sin(\phi) \qquad w = c / Cos(\phi) \qquad \theta = b$$

- For the mast simulation, horizontal wind speed, wind direction and tilt values are computed for each cup and vane height.
- The compression method we use between LiDAR and mast is simply dividing LiDAR measurement with mast value. If value is bigger than 1 LiDAR is over estimating if it is below 1 it is lower estimating. This value is symbolized with Greek letter, λ.

$$\lambda = \frac{u_{LiDAR}}{u_{Cup}}$$

3.2. WEng Model

WEng is a linear flow model developed at Risø/DTU since 1999 and uses LINCOM (Linearized Computational Model) [6]. We have tested the model for both sites to see if it can generate reliable information about the terrain effect around the scanning area because the method described above depends on good wind vector tilt predictions.

We used 5m resolution for both sites and generate a transect points from LiDAR upwind position to downwind position passing through the mast location in any case. This transect points are the ones where LiDAR conical scan and cup measurements red in reality. The results are in Figure 3. In all cases wind vector tilt predictions follows the elevation values which shows that they are acceptable in simulation. But it must be noted that this result is only for one direction. The effect of other wind directions will be discussed in conclusion.

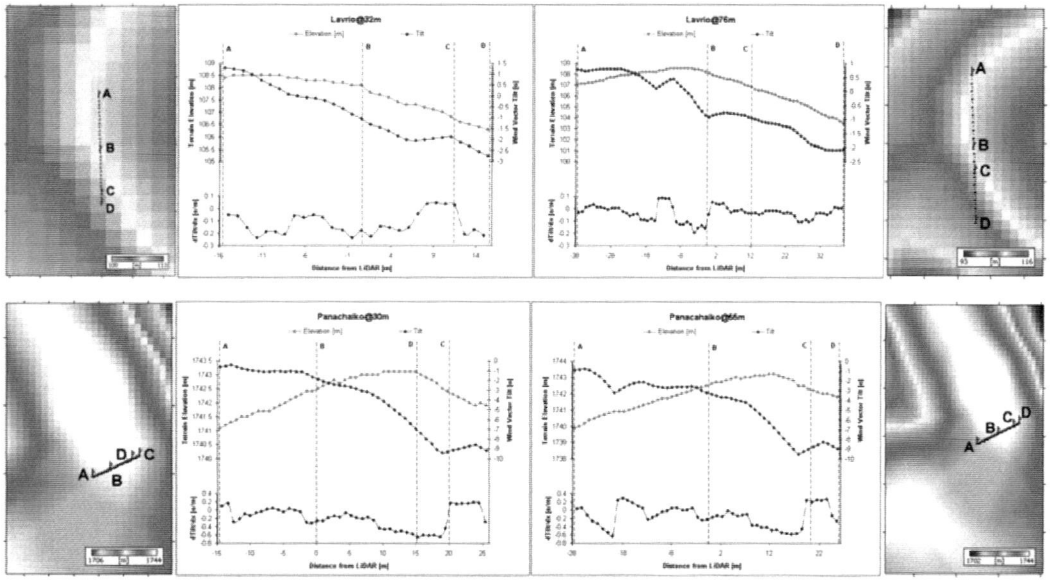

Figure 3: Test run for WEng to see the predictions for wind vector tilt.
A: Upwind position, ***B***: Position of the LiDAR ***C***: Position of the mast ***D***: Downwind position.
At each site wind is coming from the LiDAR (***A***) to the mast (***C***).
WEng model predicts the wind vector tilt correlated to the elevations.

4. Results

4.1. Lavrio

Lavrio is relatively complex terrain. We selected all the data except from 65° to 159° and from 245° to 338° due to the mast shadow before comparing with WEng results (Figure 4). The selected data show good correlation between LiDAR and the mast. Most of the data is from northerly winds which mean it is coming from LiDAR to the mast. Scatter plots are already showing ~4.5% to ~%6 errors in wind speed measurements in overall.

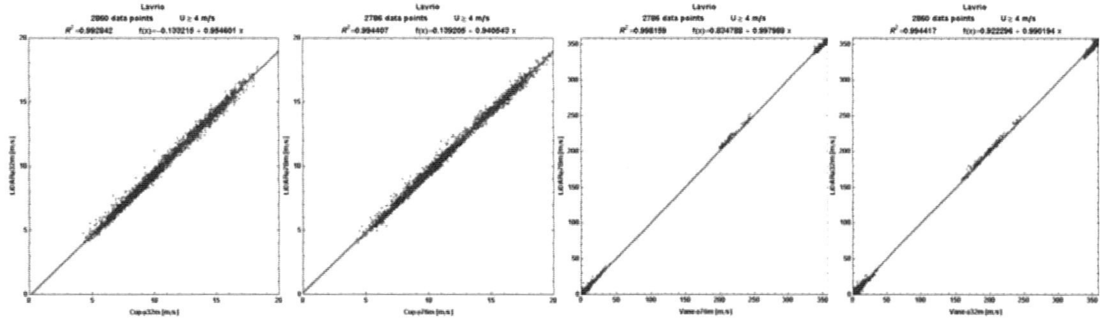

Figure 4: Selected data from Lavrio. Each data points describe 10 minutes statistics. The site has a very good correlation in wind direction and speed. Sectors, from 67° to 159° and from 245° to 338° are ignored due to the mast shadow.

For the WEng model we have selected ~3 km to 3 km map with 4~10 m resolution simulating the wind direction from 0° to 360° with 6° bins. The results are in Figure 5. The plots include all the selected data but correlation between model and the measurements is only good between -20° and +40°. We believe this error comes from the lack of measurement data and the fact that the LiDAR is very close to the mast where focus volume get affected the most when wind blows from South because of the moving objects on the mast, like vanes and cups.

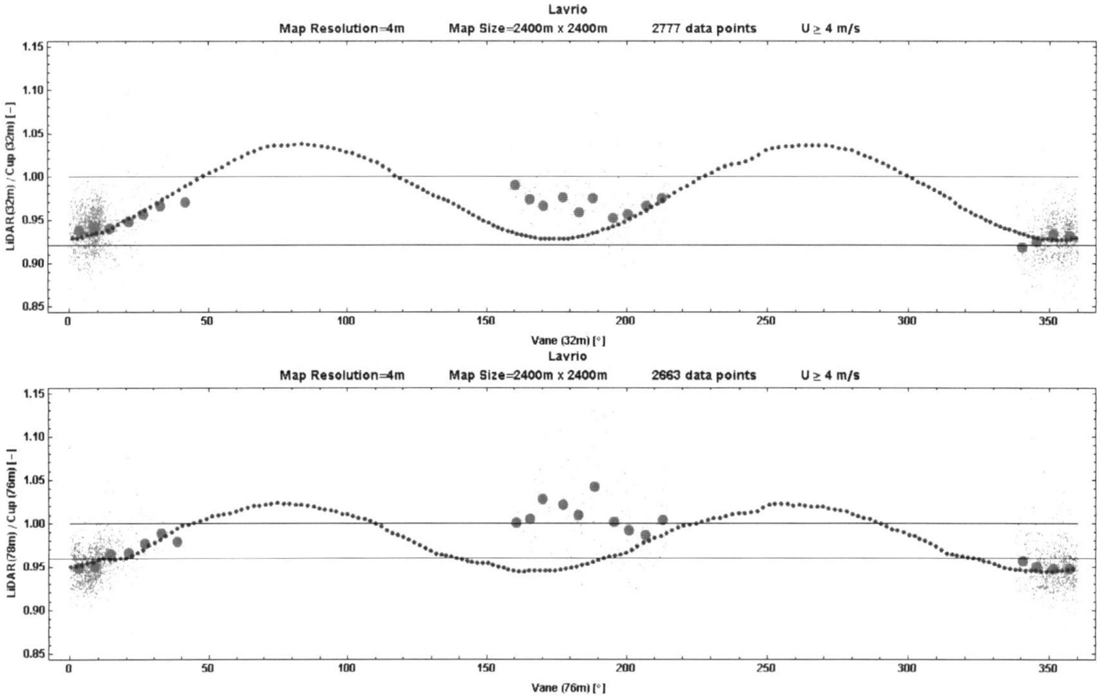

Figure 5: Measurement results and WEng predictions for two different heights. Small size of red dots are the 10 minutes averages. Big red dots are the averages in 6° wind directions. Blue line is the WEng error predictions. WEng results fits best between -20° and +40° in lower heights.

4.2. Panahaiko

Panahaiko is a very complex terrain. We selected all the data except from 132° to 167° and from 192° to 232° due to the mast shadow before comparing with WEng results (Figure 6). The selected data does not show good correlation between LiDAR and the mast this time. Mostly, the SW winds show high errors. Scatter plots are already showing ~10% to ~11% errors in wind speed measurements in overall.

For the WEng model we have selected ~5 km to 5 km map with 10 m resolution simulating the wind direction from 0° to 360° with 6° bins. The size of the map is bigger because the site is more complex. The results are in Figure 7. The plots include all the selected data but correlation between model and the measurements is only at lower height. We believe this error is combination of lack of measurements and WEng model predictions.

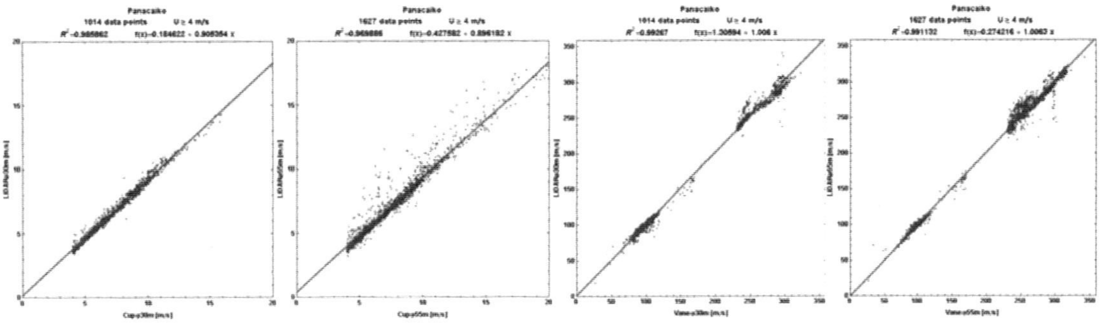

Figure 6: Selected data from both sites. Each data points describe 10 minutes statistics. (Top) Lavrio site has a very good correlation in wind direction and speed. Sectors, from 67° to 159° and from 245° to 338° are ignored due to the mast shadow. (Bottom) Panahaiko is better in SE winds while it is worse in NW winds. Sector from 170° to 230° is ignored due to the mast shadow. Unfortunately there is no data from N and S.

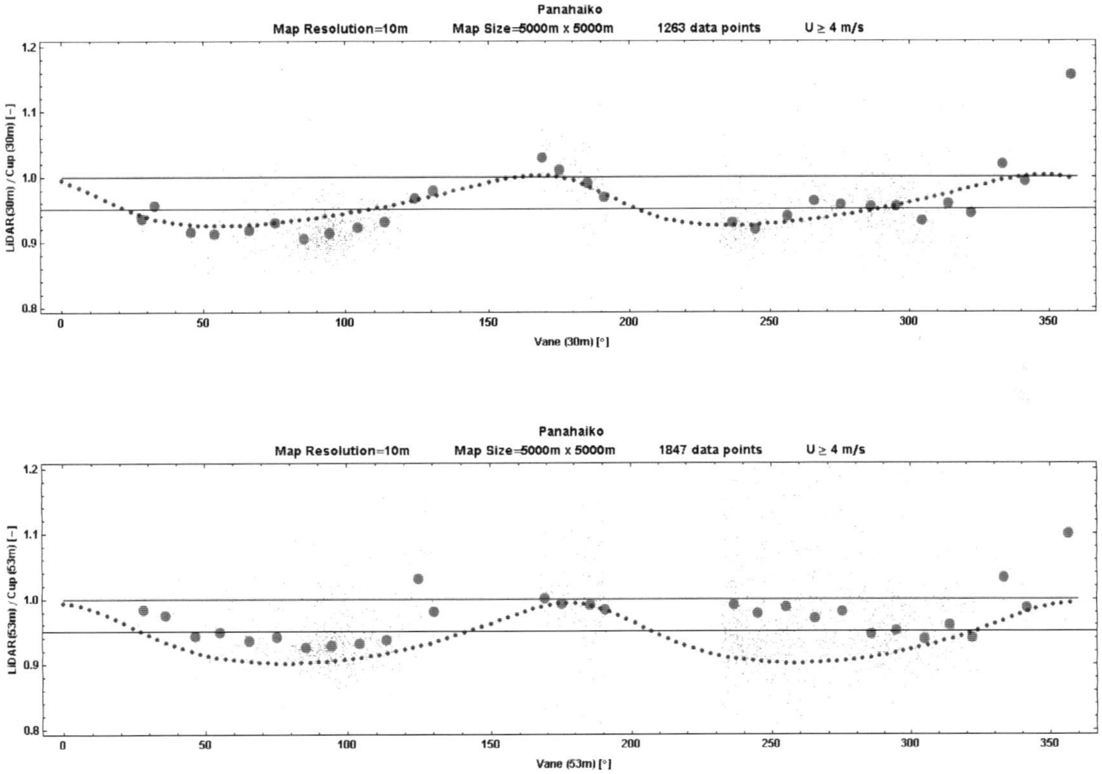

Figure 7: Measurement results and WEng predictions for two different heights. Small sizes of red dots are the 10 minutes averages. Big red dots are the averages in 10° wind directions. Blue line is the WEng error predictions. WEng results are not as precise as the Lavrio results.

4.3. Ruggedness index (RIX)

The ruggedness index concept has been used extensively over the last 10 years in wind resource assessment and sitting studies in complex terrain, especially in terrain which is outside the operational envelope of linearised flow models such as WAsP & WEng. Earlier studies on WAsP [7] shows that by the magnitude of the WAsP prediction error in terrain characterized by RIX values greater than 0 is confirmed to depend strongly on the orographic performance indicator. We believe this result is also the main factor for WEng model. In Figure 8, the RIX values of the two sites can be seen. Center points are the LiDAR positions. The radius size is the map size we have used. The linear relation between the RIX values and the error predictions are not studied in this paper due to fact that available data points and number of masts are not enough.

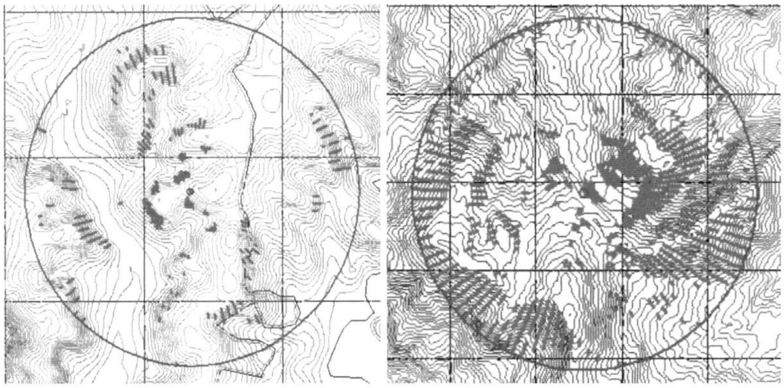

Figure 8: LiDAR position RIX values for both sites.
(Left) Lavrio, (Right) Panahaiko. Total RIX values are ~10% and ~25% respectively.

5. Conclusion

In general, it is not recommendable to apply the WEng model (or any other linear flow model) outside its operational envelope. We can model the LiDAR error in complex terrain with WEng where close area of scanning volume has low RIX values while it cannot predict very precisely with higher RIX values.

For more reliable comparison we need more data which was a problem for the both of the sites in this study. Instead of 1-2 months measurements, more than 4 months measurements are advised where data can be collected from all wind direction sectors, if it is possible. According to the complexity of the terrain 4 to 5 km maps are advised with low resolution values between 5 and 10 m.

References

[1] David A. Smith, Michael Harris, Adrian S. Coffey, Torben Mikkelsen, Hans E. Jørgensen, Jakob Mann, and Régis Danielian. Wind Lidar evaluation at the Danish wind test site Høvsøre. Wind Energy, 9(1–2):87–93, 2006.

[2] Jørgensen, H.; Mikkelsen, T.; Mann, J.; Bryce, D.; Coffy, A.; Harris, M.; and Smith, D. "Site Wind Field Determination Using a CW Doppler Lidar - Comparison with Cup Anemometers at Risø." Proc. EWEA Special Topic Conference: The Science of Making Torque from Wind; April 19-21, 2004, Delft, The Netherlands. pp.261-266.

[3] D. Foussekis, F. Mouzakis, P. Papadopoulos, P. Vionis "Wind Profile Measurements using a LIDAR and a 100m Mast". Conference Proceedings EWEC 2007

[4] Jørgensen, H.E.; Courtney, M.S.; Mikkelsen, T.; Antoniou, I.; Bradley, S.; Hunerbein, S. Von; Perez, I.M.; Arranz, P.G.; Foussekis, D.; Harris, M., UpWind progress in remote sensing of the wind using lidars and sodars. 2008 European Wind Energy Conference and Exhibition,

Brussels (BE), 31 Mar - 3 Apr 2008.
[5] M. Harris, M. Hand, A. Wright "Lidar for Turbine Control" Technical Report NREL/TP-500-39154 January 2006
[6] Jakob Mann, Søren Ott, Bo Hoffmann Jørgensen & Helmuth P. Frank, "WAsP Engineering 2000" Technical Report Risø–R–1356(EN) August 2002
[7] Mortensen, N.G., A.J. Bowen and I. Antoniou (2006). "Improving WAsP predictions in (too) complex terrain" Proceedings of the 2006 European Wind Energy Conference and Exhibition, Athens, Greece, February 27 to March 2.

Remote sensing used for power curves

Rozenn Wagner [1], Hans E Jørgensen [1], Uwe S Paulsen [1], Torben J Larsen[1], Ioannis Antoniou [2] , Leo Thesbjerg [3]
[1] Risø/DTU, Frederiksborgvej 399, Postboks 49, 4000 Roskilde, Denmark
[2] Siemens Wind Power A/S, Measurement Department, PG R319, Dybendalsvænget 3, 2630 Tåstrup, Denmark
[3] Vestas Wind Systems A/S, Alvej 21, 8900 Randers, Denmark

e-mail: rozenn.wagner@risoe.dk

Abstract: Power curve measurement for large wind turbines requires taking into account more parameters than only the wind speed at hub height. Based on results from aerodynamic simulations, an equivalent wind speed taking the wind shear into account was defined and found to reduce the power standard deviation in the power curve significantly. Two LiDARs and a SoDAR are used to measure the wind profile in front of a wind turbine. These profiles are used to calculate the equivalent wind speed. The comparison of the power curves obtained with the three instruments to the traditional power curve, obtained using a cup anemometer measurement, confirms the results obtained from the simulations. Using LiDAR profiles reduces the error in power curve measurement, when these are used as relative instrument together with a cup anemometer. Results from the SoDAR do not show such promising results, probably because of noisy measurements resulting in distorted profiles.

1. Introduction

A measured power curve normally consists of the 10 minutes mean turbine electrical power plotted against the wind speed at hub height. Such a plot usually shows a significant spread of values and not a uniquely defined function, see example in Figure 1. The spread origin of the scatter can mainly be grouped to three categories: the wind turbine components characteristics (e.g. roughness sensitivity, yaw error), the sensors error and the wind characteristics. Within the last group, the standards require to measure the wind speed at hub and the air density at a distance of 2 to 4 rotor diameters [1]. However, other wind characteristics can influence the power production; for example: the wind shear and the turbulence.

The wind shear influence for the intermediate load range was shown after simulations by Wagner et al. in [2]. In order to take it into account in the power curve production, an equivalent wind speed based on wind profiles was defined. The use of such a wind speed reduced the scatter.

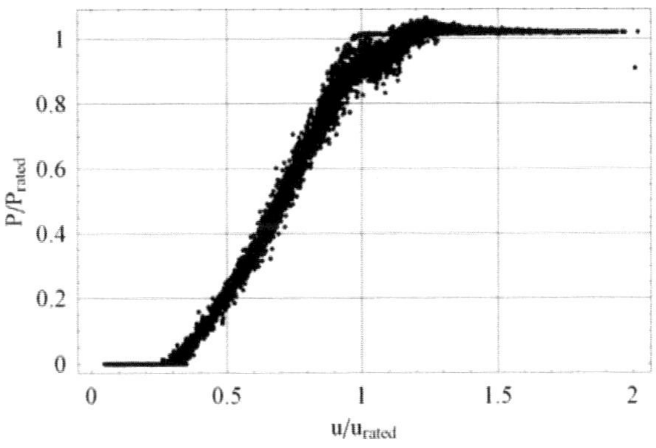

Figure 1. Example of scatter plot of power curve measurement for a multi-megawatt wind turbine

In this paper, this exercise has been repeated but this time using the complete range of hub height wind speeds required to produce a power curve. Based on both sets of results, it is expected that wind profile measurements in front of the wind turbine would improve the power curve measurements. A mast mounted with cups at different heights seems then to be required. However, for large wind turbines the hub height exceeds 80m, and the installation of such a high mast takes time and its instrumentation – requiring several cup anemometers - is expensive. Therefore ground based remote sensing systems like LiDARs and SoDARs appear to be a good alternative. Indeed such devices enable us to measure wind profiles up to 150m and are easy to install. Thus, two LiDARs and a SoDAR were installed in front of a wind turbine, measuring at different heights within the swept rotor area.

After explaining the simulations results, this paper describes the measurement campaign. Then, results using one of the LiDARs are presented and commented in detail. Finally, results from all three instruments are compared. For reasons of confidentiality all the numbers presented here are normalized. However it is fair to say that the data are representative of a modern, multi-megawatt wind turbine.

2. Preliminary work: Aerodynamic simulations

First of all, simulations of a wind turbine subjected to different forms of wind shear were carried out with an aerodynamic model. The following section describes the model.

2.1 Aerodynamic simulations

A simplified version of the aeroelastic model HAWC2, the HAWC2Aero code [3], is used in the analysis. It contains the same aerodynamic and wind modules as the full HAWC2 code and can be considered as a Blade Element Momentum model operating in the time domain, thus capable of including turbulence and non linear effects from aerodynamics as well as influence of the turbine controller. The turbine is a variable speed pitch-regulated turbine.

The main wind related inputs to the model are the mean wind profile and the turbulence intensity. The Mann model of turbulence [4] is used to simulate the wind fluctuations which are added to the mean wind speeds field, and the wind is assumed horizontally homogeneous. For each profile, 10 simulations (implying 10 different turbulent fields) were carried out in order to map the statistical variation in normal turbulence fields expected over a turbine rotor.

The main output is 10 minutes time series of the power produced by the turbine, but also time series of the free wind at 5 different points regularly distributed along a vertical line within the rotor plane: one at hub height and 4 others at symmetrical positions above and below, see Figure 2. The data point height is located at the average height of the segment.

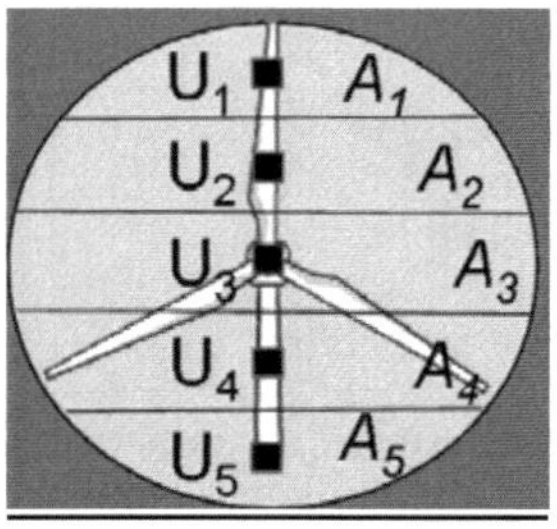

Figure 2. Swept rotor area divided in 5 segments of area A_i;

the wind speed in the middle of the ith area is U_i.

2.2 Preliminary study results

Profiles measurements from the combination of two met masts (one 116.5 m high and the other 160 m high) over a one year period were binned according to the wind speed at hub height (2m/s bins). For each bin, 10 different profiles were chosen in order to cover the whole wind shear range observed for this range of wind speed at hub height. Each set of profiles are normalized so they have the same wind speed at hub height. Examples of normalized profiles are shown in Figure 3. The simulations were carried out using these normalized profiles as input.

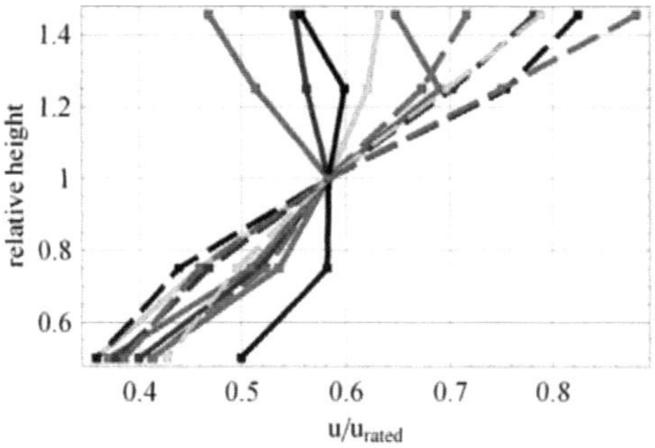

Figure 3. Example of 10 measured wind speed profiles (10 minutes average values) normalized so they all have the same wind speed at hub height (relative height = 1).

In Figure 4, 10 minutes averages of the power output are plotted against the 10 minutes average of the wind speed to give the power curve. For the blue dots and curve, the wind speed corresponds to the wind speed at hub height. Since for each set of input profiles, the hub height wind speed was the same for all the profiles, the points are aligned vertically. This means that even though the wind speed at hub height is the same, the wind turbine power production varies; and these variations are due to the wind shear.

Therefore, in order to take the wind shear into account when measuring the power curve, an equivalent wind speed was defined as the average of the wind speed at different heights weighted by the area of the corresponding segment of the swept rotor area [2].

$$U_{eq} = \frac{1}{A}\sum_i \overline{U_i} \cdot A_i \qquad (1)$$

where $\overline{U_i}$ is the 10 minutes mean wind speed at the ith height, A_i is the area corresponding at the specific data point height and A the swept rotor area. Figure 2 shows an example for the swept rotor area divided in 5 segments.

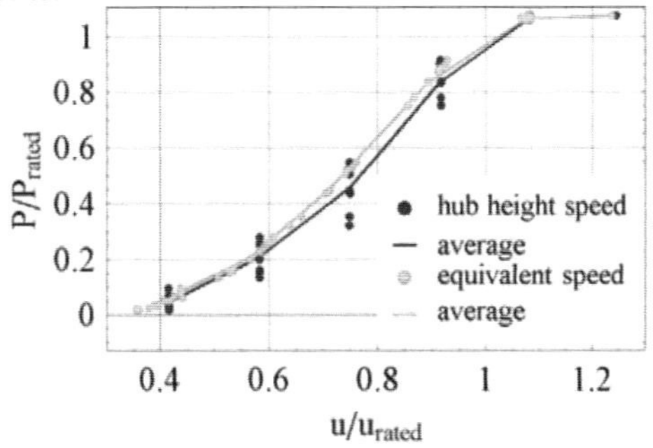

Figure 4. Simulated power curves. Blue: power as a function of the wind speed at hub height. Green: power as a function of the equivalent wind speed calculated from profiles.

In Figure 4, for the green dots and line, the wind speed is the equivalent wind speed calculated according to (1). As the power values are the same (between the two cases only the wind speeds change), plotting them as a function of the equivalent wind speed instead of hub height wind speed shifts the points to the right when $U_{eq} > U_{hub}$ (usually corresponding to high power value) and to the left when $U_{eq} < U_{hub}$ (usually corresponding to low power value). As a consequence, the points are almost aligned with a curve, corresponding to a power curve independent of the wind shear; and the power fluctuations (due to wind shear) around the average curve are reduced compared to the first method – using hub height wind speed.

A way to quantify these fluctuations is to bin the points according to the wind speed and calculate the standard deviation of the power in each bin. Thus Figure 5 left displays the power standard deviation for each bin of 1m/s, and Figure 5 right shows the mean of these values. In these two figures it is easily seen that the use of the equivalent wind speed reduces the power curve uncertainty.

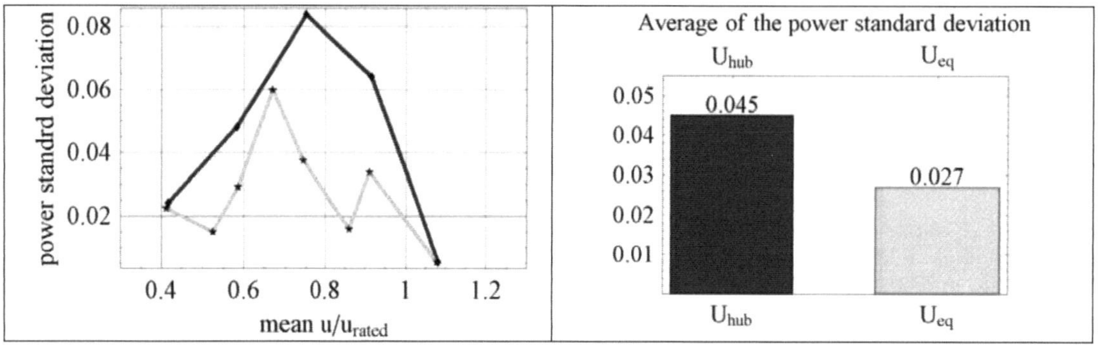

Figure 5. left: power standard deviation as a function of hub height wind speed (blue line), of equivalent wind speed (green line); right: average power standard in both cases.

From these results, profile measurements in front of the turbine are expected to improve power curve measurement. As high met mast installation is long and expensive, ground based remote sensing systems like LiDARs and SoDARs appear to be a good alternative. Thus a measurement campaign including 3 remotes sensing devices was set up. This is described in the next section.

3. Description of the measurement campaign

A ZephiR LiDAR, a Windcube LiDAR and an AQ500 SoDAR were placed close to each other at about 300m from a large wind turbine as shown in Figure 6.

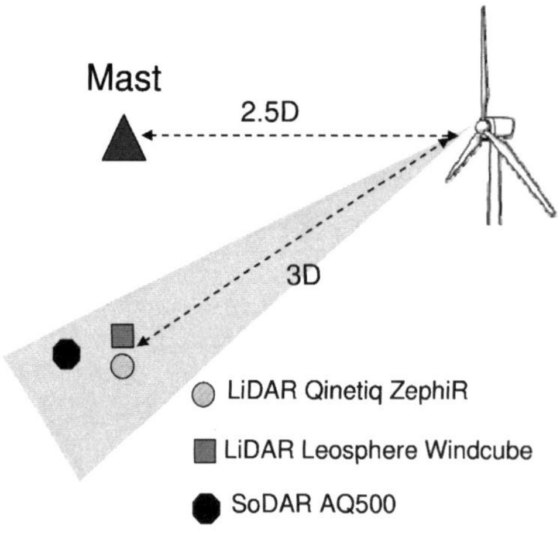

Figure 6. Plan view of the measurement campaign with two lidars and a sodar in front of a large wind turbine (D is the rotor diameter).

The QinetiQ ZephIR is a continuous wave LiDAR scanning conically with an angle of 30 degrees from the vertical and with a sample frequency of 50Hz [5]. Measurements can be done at different heights by varying the focus; however this implies a measuring volume increasing with altitude. Each height is scanned for 3 seconds. Thus a radial wind speed is available every 7.2

degrees and a wind vector is computed from 150 radial speeds. The ZephiR LiDAR measured in this location for 5 weeks.

The Leosphere WindCube is a pulsed LiDAR. The height discrimination is then inherently defined by the time taken by the backscatter to come back to the receptor. This implies that the measuring volume height is constant and about 25m. The radial wind speed is measured at all heights simultaneously. 4 line-of-sights at 30 degrees from the vertical (each separated by 90 degrees of azimuth) are sequentially scanned and a new wind vector is computed at the end of each scan, about every 1.5s. The WindCube LiDAR measured in this location for 2 weeks.

Both the ZephIR and the Windcube are extensively described and compared in [4].

The AQ500 SoDAR is a 3-components monostatic SoDAR, meaning that a pulsed sound signal is sent in 3 directions with a beam angle of 15 degrees. As for the pulsed LiDAR, the measuring volume height is constant and about 5m. However the sound intensity decreases with height. The AQ500 SoDAR has been measuring at this location for some months.

All of the 3 instruments measure at the same 5 heights equally distributed within the swept rotor area: one at hub height and 4 others at symmetrical positions above and below, as shown in Figure 2. The cup anemometer mounted on the top of the mast at hub height (for standard power curve measurement purpose) is used as a reference.

The test site is flat, surrounded by grassland and 1.7 km from a coast. The measurements were selected for a narrow wind sector around the instrument-turbine direction (30 degrees), so the wind speed measured by the instruments corresponds to the wind speed going directly to the wind turbine and the best correlated to the wind seen by the turbine. This wind sector presents a very large range of wind shears, ranging from "nose" profiles (profiles presenting a local maximum) to very large shear.

Moreover, the data were selected for rain-free periods since rain can disturb the measurements of all 3 instruments. In addition, as the ZephiR is disturbed by clouds, therefore only 10 minutes data with no low clouds were taken into account.

4. LiDAR profile measurements

In this section, ZephIR measurements are used as an example, but what follows can actually be done for each one o the three remote sensing instruments involved in this campaign.

4.1 Profile correction

LiDAR measurements at a given height are generally not exactly equal to cup measurements at the same height. This is due to several factors, the most important being that a LiDAR measures over a volume whereas a cup anemometer gives a point measurement. Then LiDAR measurement error seems to depend on the wind speed and the wind gradient. That is why the LiDAR here is used as a relative instrument together with a cup anemometer at hub height, considered as an absolute instrument. Thus for each ten minutes average measurement, the wind profile measured by the LiDAR is shifted so the hub height wind speed from the LiDAR is equal to the cup measurement, see Figure 7. I n average the shift corresponds to 7% of the wind speed.

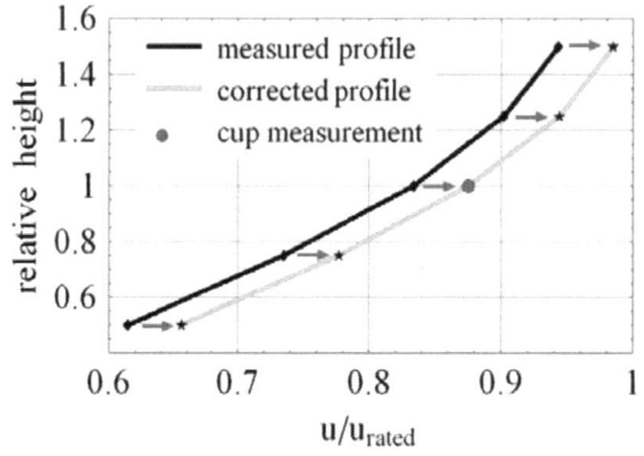

Figure 7. The profile measured by the lidar is shifted so that the wind speed at hub height (relative height = 1) is the same as the one measured by the cup anemeometer.

4.2 Power curve

The corrected profiles are used to compute the equivalent wind speed according to equation (1). As for the simulations results, we plot the power curve using the wind speed at hub height (in blue) – this is the procedure described in the standards [1]- on one hand; and on the other hand the power curve as a function of the equivalent wind speed (in green), see Figure 8. We observe that the results qualitatively agree with the simulations. First, the mean curve is slightly shifted to the left. Indeed, most of the profiles are not linear and present a smaller wind speed and a larger shear below hub than above. Consequently the equivalent wind speed is lower than the wind speed at hub height. Besides, for nose profiles, the equivalent wind speed is obviously smaller than the wind speed at hub height. These measurements confirm that the power standard deviation is reduced when using the equivalent wind speed based on profiles, see Figure 9.

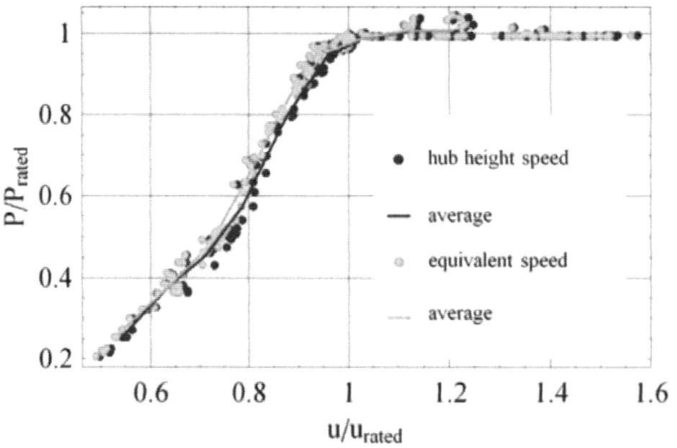

Figure 8. Power curve measured using the cup anemometer at hub height (blue) and the equivalent wind speed computed from the lidar profile measurements(green).

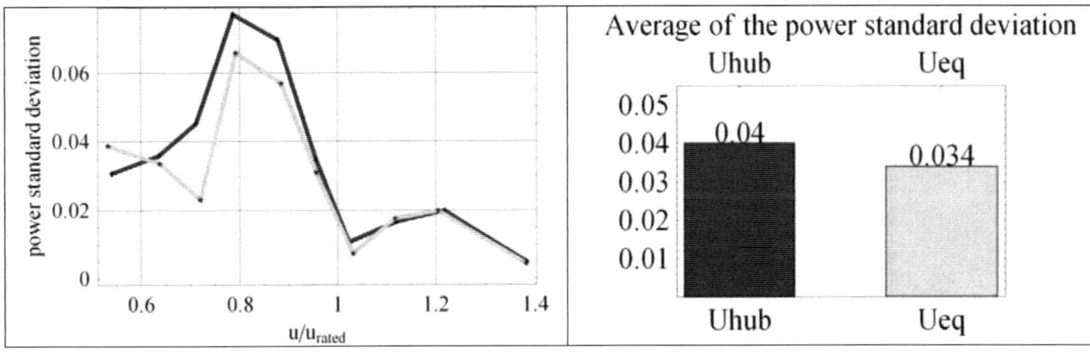

Figure 9. left: power standard deviation as a function of hub height wind speed (blue line), of equivalent wind speed (green line); right: average power standard (for $u \leq 1.1u_{rated}$) in both cases.

4.3 Necessity to measure above hub height

Since it is now a fact that the use of an equivalent wind speed taking the wind shear into account reduces the error in the power curve; and since we noted that the equivalent wind speed seems to be influenced more by the profiles below hub than the part above hub, one could think that two wind speed measurements below or at hub height would be sufficient. Indeed, it is then easy to reproduce a whole profile by fitting the 2 values to a model. The power law profile for example enables to represent different wind shear with an exponent:

$$U(z) = U_r \left(\frac{z}{z_r} \right)^{\alpha} \quad (2)$$

Where U is the wind speed, z the altitude, z_r a reference altitude and $U_r=U(z_r)$, and α the shear exponent.

This procedure was applied to the corrected lidar measurements using hub height as z_r and the lowest height to determine α. For some cases, this model fits the measured profile very well, but for 70% of the profiles selected here, the model gives an error larger than 10 % in the wind speed at the highest height. Figure 10 shows 2 examples of measured profiles and their corresponding modeled profiles (called theoretical profiles).

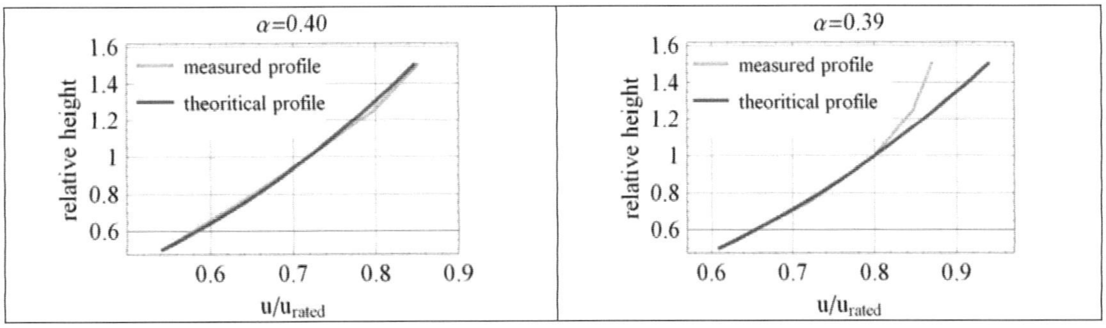

Figure 10. Two examples of (corrected) lidar profiles and their fit to the power law model (using the wind speed values at hub height and at the lowest height). The model fits very well the measured profile on the left, but it does not work for the profile on the right.

The equivalent wind speed was then calculated using the power law profiles and these wind speed values were used to plot the power curve, Figure 11 compares the power standard deviation obtained with modeled profiles to the ones obtained with measured profiles. Firstly, it shows that the use of fitted profiles reduces the power standard deviation as well, and so it confirms again that the use of equivalent wind speed taking the shear into account improves the power curve measurement. Secondly, by using measured profiles (measurement at every height) the standard deviation is reduced compared to the use of modeled profiles. This is simply due to the fact that the model is not right in every case (for example it cannot represent nose profiles) and in this case the equivalent wind speed value does not match with the turbine power. The main conclusion of this exercise is that measurements below hub height are not sufficient and measurements are necessary over the whole rotor height.

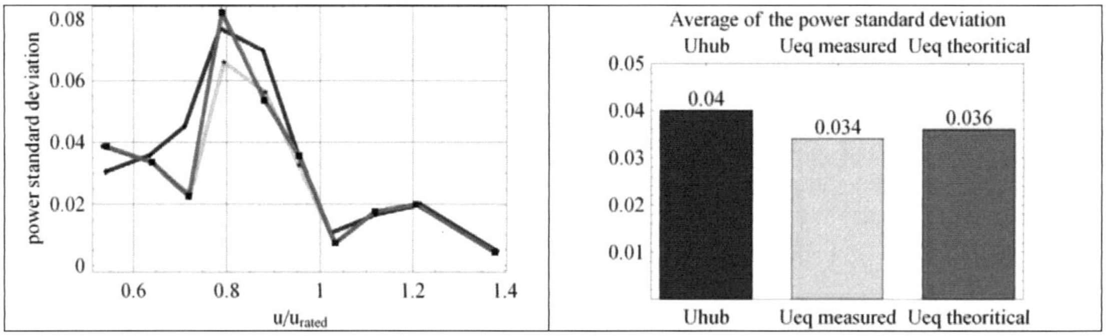

Figure 10. left: power standard deviation as a function of hub height wind speed (blue line), of equivalent wind speed computed with lidar measured profiles(green line), of equivalent wind speed computed with modelled profiles(purple line); right: average power standard in the 3 cases (for $u \leq 1.1u_{rated}$).

5. Comparison of results from measurement from the 3 instruments: ZephiR LiDAR, Windcube LiDAR, AQ500 SoDAR

In this section we are going to compare the power curve measurement from the three instruments. A common data set has been selected (profiles measured at the same time). As the instruments were installed at the measurement site at different dates, and because of the data filtration required for rain and cloud, the common data set counts only 89 profiles. Therefore, this data set enables us to observe trends but might not be suitable for quantification.

For each instrument, each profile is corrected as explained in 4.1 and an equivalent wind speed is computed according to equation (1). Four power curves are then obtained: one using the cup at hub height and one for each remote sensing instrument (using the equivalent wind speed values). Figure 12 compares the power standard deviation for these 4 cases.

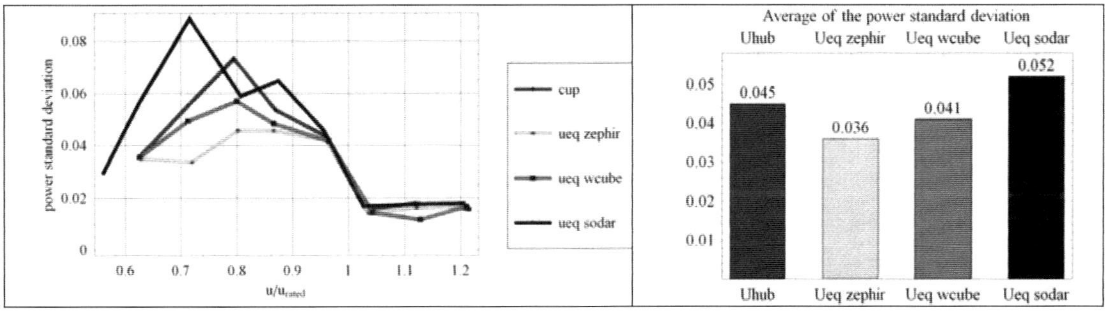

Figure 11. left: power standard deviation as a function of hub height wind speed (blue line), of equivalent wind speed computed with ZephIR LiDAR measured profiles(green line), of equivalent wind speed computed with WindCube LiDAR measured profiles (red line) , of equivalent wind speed computed with AQ500 SoDAR measured profiles (black line); right: average power standard deviation in the 4 cases.

As observed previously, the profiles from the ZephIR LiDAR reduce the power curve error. The same observation can be made for the WindCube LiDAR; although it is not as good as the ZephiR. This can be due to different parameters, among others: up to hub height the ZephIR's measurement volume height is smaller than the WindCube's; furthermore an error in the wedge angle was detected after the measurement campaign began. Regarding the AQ500 SoDAR, the power standard deviation is bigger than the one obtained with a cup anemometer. This is probably due to the fact that SoDAR measurements are very noisy and the profiles are often distorted.

6. Discussion

This investigation shows that profile measurements can improve the power curve measurement; but isalso a first step for the use of remote sensing instrument in power curve measurement. In this investigation, each LiDAR and SoDAR has been used as a relative instrument together with a calibrated cup anemometer, because the objective was to show the necessity of measuring profile instead of wind speed at hub height, and, for now, only the cup anemometer is considered as an absolute instrument. However LiDARs and SoDARs are expected to be used on their own at some point. In order to make this realizable, more research must be done to fully understand what makes the difference between a wind speed measurement from a cup anemometer and a measurement from a LiDAR/SoDAR. Indeed, this difference is not necessarily an error from the remote sensing instrument. We must keep in mind that they measure over a volume whereas a cup anemometer measures at one point. The focus volume or the pulse length, according to the system, implies a weighted average in the vertical direction, which is therefore influenced by the wind shear and the turbulence as well. This could actually be better for power curve measurement purposes since this volume averaged wind speed might be better representative of the wind speed over a rotor segment area A_i than the wind speed measured at a point in the center of the segment.

It would not make sense to have very large measurement volume, especially in the vertical direction, however the fact that the SoDAR measurements were performed with 5 m high volume measurement might be an explanation for the distorted profiles. Indeed, averaging over a higher volume would give more robust wind speed (better signal to noise ratio) and therefore less noisy profiles.

Furthermore, in this investigation, 10 minutes mean values were considered because it is the averaging time required in the standards. However modern megawatt turbines respond quite fast to the wind fluctuations and two wind speed time series with the same statistics over 10 minutes can in fact be very different. Therefore, a better reduction of the uncertainty of the power curve measurement could be expected from using shorter averaging periods.

7. Conclusions

Aerodynamic simulations have shown that wind turbine power production is influenced by the wind shear. The use of an equivalent wind speed, taking the wind shear into account, could reduce the fluctuation in power when plotting the power as function of this speed. This is now confirmed by profile measurements.

LiDARs and SoDARs, being able to measure wind profile quite accurately, are a good alternative to met mast anemometry; especially knowing that the several measurements not only below but also above hub height can reduce the error even more.

References

[1] IEC 61400-12-1. Wind Turbines – Part 12-1: *Power performance measurements of electricity producing wind turbines*; 2005.
[2] Wagner R, Antoniou I, Pedersen S M, Jørgensen H E, *Wind profile influence on turbine performance*, submitted to the Wind Energy in July 2007.
[3] Larsen T J, Wawc2aero, the user's manual, Risø-R-1631, Jan 2008.
[4] Mann J, *Wind field simulation*, Probabilistic Engineering Mechanics, Oct 1998; **13**: (4) 269-282.
[5] Smith D A, Wind LiDAR evaluation at the Danish test site of Høvsøre, Wind Energy, Jan 2006, **9**: (1-2) 87-93.
[6] Courtney M, Wagner R, Lindelöw P, *Testing and comparison of LiDARs for profile and turbulence measurements in wind energy*, EWEC 2008.

IOP Conf. Series: Earth and Environmental Science **1** (2008) 012060 doi:10.1088/1755-1307/1/1/012060

Arrangements for enhanced measurements of a large turbine near-wake using LiDAR from the nacelle

J J Trujillo, A Rettenmeier and D Schlipf

Endowed Chair of Wind Energy, Institute of Aircraft Design, University of Stuttgart, Allmandring 5B, 70569 Stuttgart, Germany

E-mail: juan-jose.trujillo@ifb.uni-stuttgart.de

Abstract. New LiDAR techniques are being tested and developed to support the development of large offshore wind turbines. Our interest in this paper is concentrated in wake measurements; therefore, a pulsed standard LiDAR is adapted for fullscale wind field measurements from the nacelle of a large wind turbine. We show the conceptual framework for planned adaptations to a Windcube® LiDAR for operation at the nacelle of a 5 MW wind turbine. The standard scanning mode is to be modified to properly obtain downstream and also upstream wind speeds. The wind field measurements are intended for verification of models for near-wake wind speed, wake meandering and new predictive control estrategies.

1. Introduction

The wind energy community has noticed a need of full scale wake experimental data. The available wake models have to be further developed and verified with improved measurements [1]. The models which have been developed in the past were based at best on measurements of medium-sized wind turbines in non-complex terrain [2]. Moreover, today attention has been drawn to dynamic wake effects. The so called wake meandering is detrimental for some components of downwind turbines and a thorough understanding and quantitative estimation of its effects is needed [3].

The best approach to understand these phenomena is the analysis of fullscale wind field measurements. Unfortunately, conventional wind measurements based on meteorological masts (metmast) are unpractical. Light detection and ranging (LiDAR) technology has been applied for some decades in the field of atmospheric remote sensing. However, its application for wind energy purposes can be said to be in its early stages[4]. Nevertheless, the developments have been rapidly evolving mainly oriented towards replacement of standard cup anemometry [5]. The great advantage of such measurements is the capability of realizing enhanced wind field mesaurements. Inflow measurements from the nacelle for improvement of wind turbine control are envisaged and have been already tested [6, 7].

Another particular implementation is the measurement of wake wind field. This is specially needed in order to observe the scaling of wake properties with wind turbine scale. The short range LiDAR systems commercially available at the moment, ZephIR® from QinetiQ® and Windcube® from Leosphere®, are capable of scanning at a downstream distance around two diameters ($\sim 2\mathcal{D}$) behind the turbine. This is is normally taken as the distance at which the near-wake vanishes [2]. This means that for modelling purposes the pressure gradients are assumed to be negligible. Moreover, it is assumed that the wind speed presents a particular profile which is used as initial conditions for calculation of the downstream flow field. Therefore, it is an important parameter dependent on the wind turbine operational characteristics. This has to be verified and the most promising way at the moment is using LiDAR measurements.

Experience has been gathered with far–wake measurements from the wind turbine nacelle[8]. The results have given confidence on the ability to not only observe statistics of wake flow but also dynamic behaviour of the wake. The so called meandering of the wake has been measured. Based on this we will try to measure the behaviour of the wake at the limit of the near-wake. Here we explain the procedure to be followed in order to prepare our measurement campaign planned for the first quarter of 2009. The measurements are planned to take place from the nacelle of a Multibrid M5000 wind turbine. This 5 MW turbine has a rotor diameter of 116 m and a hubheight of 102 m.

The measurement campaigns with the short range system will be accompanied by measurements with a long range LiDAR from the German Aerospace Center (DLR) . This LiDAR system has been succesfully used for measurements of wake vortices of large transport aircrafts from ground and on flight [9]. This long range system will serve for comparison purposes during non-wake and wake situations.

2. LiDAR wind measurements

2.1. Standard vertical profiles

The LiDAR system to be used is a modified Windcube®. This commercial system is manufactured by the french company Leosphere®. It is a pulsed Doppler LiDAR designed to measure vertical wind profiles up to around 200 m. Its working principle is the so called doppler effect. Airborne aerosols, which are assumed to flow with the wind, reflect back an emitted laser beam or pulse. The frequency of the received light changes depending on the speed of the reflecting objects (aerosols). In this way the wind speed can be estimated by quantifying the change in frequency from the measured light spectra.

In the commercial systems (Windcube® and ZephIR®) the focus point movement is normally driven by a rotating wedge. The wedge is the result of cutting transversally a cylindrical glass with a slight angle with respect to its axis. The laser beam entering parallel to the axis through the base will be deviated when going out through the inclined face. In this way one complete rotation of the wedge would generate a conus-like domain. If the focus distance is fixed, then the focus point follows a circular path. This is the standard usage for vertical wind profiling where the system is setup to focus at a different height after completing a defined amount of turns at a particular level.

2.2. Line-of-sight wind speed

It must be noted that the measured wind speed at a focus point in a *staring mode* is the so called line-of-sight wind speed (u_{los}). This speed is aligned with the laser beam and in order to obtain the actual, three dimensional, wind vector at a point we would need three non-aligned measurements pointing at the same spatial point. Since we plan to use just one LiDAR, we intend to treat u_{los} in a non-standard way for our nacelle measurements; only the u_{los} will be available and misalignment of the actual and line-of-sight vectors can lead to overestimation or underestimation of the norm of u_{los} (Fig. 1). Nevertheless, we will assume that this value is representative of the main wind speed vector along the wind turbine axis. A proper correction procedure would force not only a geometrical correction but also inclusion of statistics of wind direction change. However, this very complicated procedure is expected to bring a low gain in accuracy. Therefore, we intend to use initially u_{los} without any correction for our analysis but taking into account that in average this value tends to underestimate the norm of the actual instantaneous wind speed vector.

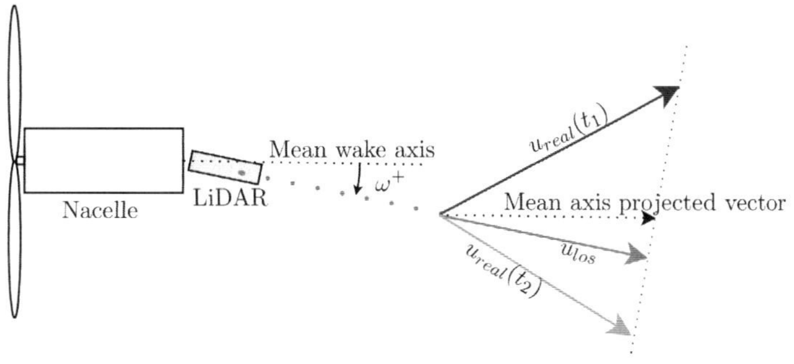

Figure 1. Comparison of measured line-of-sight wind speed(u_{los}) with instantaneous wind vector.

3. Nacelle LiDAR setup

Wind field measurements on wind turbines have been succesfully performed from the nacelle. LiDAR inflow measurement have shown a potential in wind turbine control, performance evaluation and power curve calibration [6, 7].

Moreover, studies on wind turbine wakes have shown the advantages and potential of the flexible LiDAR technologies. A continuum Doppler LiDAR from QinetiQ® has been succesfully used for measurement of far–wake dynamics [8]. The prototype used in those experiments at Risø– DTU test field was modified in order to implement non-standard scanning modes.

For our purposes the standard system from Leosphere® will experience some adaptations. The changes are still in a conceptual stage. The conditions for such adaptations, which apply for inflow and wake measurements, are the following:

- *Multiple point measurements.* The system should be capable of scanning different points on an area transversal to the rotor axis in a quasi-instantaneous sense. In the worst case, the time scale should not exceed the time needed by the wind to travel the equivalent of one rotor diameter ($1\mathcal{D}$).
- *Reasonable measurement point density.* The measured points should give a fair representation of the wind field. Especially, for individual pitch control (IPC) purpsoses the density should be enough to simulate loading at different blade azimuthal positions. As a first approximation, we estimate that the spatial scale(grid resolution) should not exceed a quarter of a rotor diameter ($\frac{1}{4}\mathcal{D}$).
- *Fast estimation of wind field.* This is rather critical for predictive control strategies. The measured inflow wind field should be *available* enough time ahead for a proper prediction horizon. Additionally, the wind field scanning rate should be also high for proper control action. This is achieved by optimisation of two aspects. First, measuring at a further distance in front of the turbine. Second, by estimating the wind field as fast as possible. In the wind turbine scales this is estimated to be in the order of two diameters in front of the turbine and wind fields delivered every 1s, respectively.
- *Flexible scanning modes.* Operation of the system should allow to easily change measurement mode. Two main scanning modes are envisaged, namely profiling and slicing. The first is used to produce horizontal of vertical profiles at a certain distance from the nacelle. The second makes a surface (two-dimensional) scanning of the wind flow mainly perpendicular to the main flow direction.

Different approaches are being studied in order to fullfil these requirements as can be seen below.

Hardware adaptations
For the two-dimensional wake measurements at Risø– DTU test field [8] the standard wedge movement has been changed. A full rotational movement was constrained to a back and forth movement by means of a special mechanism. With the focus distance fixed, they obtain a focus point path that follows an arc. At the end a combination of this movement and horizontal panning of the whole LiDAR, results in a coverage of a surface. This clever solution worked succesfully, however, some inconvenience was found in the measurement point density. The focus points were not evenly distributed over the scanned area. Unfortunately, the scanned points were concentrating on areas of low interest.

The usage of such an approach is being evaluated. However, we are looking at other setups leading to improved scanning. In our case there is a difficulty in the fast movement of the whole system with a pan and tilt system. The casing of the Leosphere® system is larger and heavier than the optical head of the QinetiQ® prototype. Therefore we are looking for purely optical approaches. In the meantime two setups are being evaluated, one with a double counter-rotating wedge and the second with mirrors. Both versions can give a two-dimensional pattern, however, their applicability has to be tested.

Software adaptations
On the other hand there is a need of reevaluating the signal processing strategy to speed up the wind speed estimation process. At the present the Leosphere® is set up to measure four points in a circle. The wedge stops and the system shoots and reads and processes the line-of-sight wind speed simultaneously at different distances, then it rotates a quarter of a circle and repeats the procedure. In fact, the measurements is not performed over a conus but over a pyramide. There is some ability to change this configuration and raise the amount of points, however, this scanning procedure is still very slow for our application and has to be accelerated. Possible solutions we are working on are the reduction of averaged spectra per point, which at the moment is set to 10000. Furthermore, the amount of scanned points has to be raised, while the stations in the line-of-sight are lowered. Improvements to the basic FFT routines have to be also evaluated.

4. Instrument verification
The standard Leosphere® system is being compared with a metmast of 102 m height located at Bremerhaven in the north of Germany. These campaign is testing how the vertical wind speed profiles from Windcube® and conventional metmast compare. For nacelle measurements other verification measurements are planned.

First, comparisons with another pulsed LiDAR are planned. This measurement campaign will be held at the DLR test site at Oberpfaffenhofen in Germany. The 2–μm Doppler LiDAR from the DLR has been succesfully used for mesaurement of wake vortex from aircrafts. This long range LiDAR will be put in parallel with Windcube®in a staring modus and measurements of horizontal wind speed will be compared.

Moreover, we are developing a LiDAR simulator based on synthetic wind fields. The purpose is to test numerically the characteristics of the wind fields obtained with a *perfect* LiDAR system through different scanning modes. The simulator is based on two modules, namely a LiDAR and a wind field module. The first is a generator of focuspoint paths depending on the selected scanning strategy. The second, takes care of the generation of a suitable synthetic wind field with a stochastic wind generator. Moreover it makes a proper interpolation of the line-of-sight wind speed (u_{los}) at the position focus point and direction given by the LiDAR module. It assumes a perfect estimation of the wind speed by the LiDAR. The system will be capable of reproducing the performance of the LiDAR system. The important outcome is the possibility of testing the accuracy of the LiDAR wind field and give insight in the qualitative and quantitative errors by usage of different scanning approaches in inflow and wake.

5. Conclusions

In this paper we show the motivation and work being done at the Endowed Chair of Wind Energy (SWE) in the area of wind field measurements with LiDAR from the nacelle of wind turbines. The research aims at further development of the potentials of LiDAR technologies in the area of improvement of wind turbine performance. The intended measurements in inflow and wake, oblige us to propose adaptations to the commercial system Windcube®from Leosphere®. The suggested hardware and software changes, fullfilling a set of special conditions, are to be performed eventually together with the manufacturer.

Acknowledgements

This research is made in the framework of the German joint research project "Development of LiDAR technologies for the German offshore test field". This project with number 41V4936 is funded by The Federal Environment Ministry of Germany (BMU).

References

[1] Barthelmie R, Rathmann O, Frandsen S T, Hansen K S, Politis E, Prospathopoulos J, Rados K, Cabezón D, Schlez W, Phillips J, Neubert A, Schepers J G and van der Pijl S P 2007 *Journal of Physics: Conference Series* **75** 012049 (9pp)
[2] Ainslie J F 1988 *Journal of Wind Engineering and Industrial Aerodynamics* **27** 213–224
[3] Thomsen K and Madsen H A 2005 *Wind Energy* **8** 35–47
[4] Smith D A, Harris M, Coffey A S, Mikkelsen T, Jørgensen H E, Mann J and Danielian R 2006 *Wind Energy* **9** 87–93
[5] Jørgensen H, Mikkelsen T, Mann J, Bryce D, Coffey A, Harris M and Smith D 2004 *Special topic conference: The science of making torque from wind* (Delft University of Technology) pp 261–266
[6] Harris M, Hand M and Wright A 2006 Lidar for turbine control Tech. Rep. NREL/TP-500-39154 National Renewable Energy qinetiQ Limited Malvern, Worcestershire, United Kingdom National Renewable Energy Laboratory Golden, Colorado
[7] Harris M, Bryce D J, Coffey A S, Smith D A, Birkemeyer J and Knopf U 2007 *Journal of Wind Engineering and Industrial Aerodynamics* **95** 1637–1647
[8] Bingöl F, Larsen G C and Mann J 2007 *Journal of Physics: Conference Series* **75** 012059 (8pp)
[9] Rahm S, Smalikho I and Köpp F 2007 *Journal of Aircraft* **44** 799 – 805

Wind speed errors for LIDARs and SODARs in complex terrain

Stuart Bradley

Physics Department, The University of Auckland, Private Bag 92019, Auckland, New Zealand,

and

School of Computing, Science and Engineering, University of Salford, M5 4WT, UK

s.bradley@auckland.ac.nz

Abstract. All commercial LIDARs and SODARs are monostatic and hence sample distributed volumes to construct wind vector components. We use an analytic potential flow model to estimate errors arising for a range of LIDAR and SODAR configurations on hills and escarpments. Wind speed errors peak at a height relevant to wind turbines and can be typically 20%.

1. The nature of the problem

SODARs and LIDARs generally rely on beams tilted at an angle to the vertical. Scattering back to the source location allows three vector wind components to be estimated from at least three distinct beam directions.

The fundamental assumption is that the wind at each of the locations sampled by the distinct beams is identical. In complex terrain situations this is not true. The effect of wind shear is *particularly bad for LIDARs* since they typically have beams at 30° to the vertical whereas SODARs typically have beams at 15° to the vertical. To put this into context, a LIDAR having a 30° beam angle is sampling at 75 m horizontally away from the vertical at a height of 150 m: this makes total nonsense of the claims that LIDARs can measure closer to a mast or turbine than a typical SODAR.

Here we consider a simple 2D hill model with only 2D flow. The reason for the study is that analytic results are obtained which allow wind errors to be easily characterized through their functional dependencies on factors such as on hill height, hill width, beam tilt angle, beam orientation with respect to the ridge, and instrument distance upstream or downstream of the ridge.

We embed typical 3-beam SODAR and conical scan LIDAR configurations into this flow regime. Such instruments are retrieving three wind components (u, v, and w) rather than the minimum requirement of nine parameters u, v, w, $\partial u/\partial x$, $\partial u/\partial y$, $\partial v/\partial x$, $\partial v/\partial y$, $\partial w/\partial x$, and $\partial w/\partial y$. We also discuss the 5-beam METEK SODAR, which has the potential to partially sense wind shears.

2. Complex Terrain Model for Monostatic SODARs and LIDARs

Consider uniform horizontal flow of speed U in the $+\zeta$ direction incident on a cylinder centred on the origin in a rectangular Cartesian coordinate system (ζ, y, η). The potential flow solution gives a stream function

$$\psi = U\eta\left(1 - \frac{a^2}{\zeta^2 + \eta^2}\right) \tag{1}$$

where a is the radius of the cylinder. Streamlines are defined by

$$\eta_0 = \eta\left(1 - \frac{a^2}{\zeta^2 + \eta^2}\right). \tag{2}$$

These streamlines have a bell-shape with maximum elevation h above η_0 occurring at $\zeta = 0$, with

$$h = \frac{1}{2}\left(\sqrt{\eta_0^2 + 4a^2} - \eta_0\right). \tag{3}$$

Any such streamline can be considered to be a solid surface, thereby defining a useful set of bell-shaped hill profiles having larger width-to-height ratios as η_0 increases.

The hill half-width L at half height is

$$L^2 = \left(\eta_0 + \frac{h}{2}\right)\left(\eta_0 + \frac{3h}{2}\right). \tag{4}$$

Combining (3) and (4) gives

$$\eta_0 = \left(L^2 + \frac{h^2}{4}\right)^{1/2} - h, \qquad a^2 = h\left(L^2 + \frac{h^2}{4}\right)^{1/2}. \tag{5}$$

Consider now a SODAR (or LIDAR) placed at (ζ_s, η_s) on the surface of the hill, and define a SODAR-centred coordinate system

$$x = \zeta - \zeta_s \qquad z = \eta - \eta_s. \tag{6}$$

The x and z velocity components, u and w, are given by

$$\frac{u}{U} = \frac{1}{U}\frac{\partial\psi}{\partial\eta} = 1 - h\left(L^2 + \frac{h^2}{4}\right)^{1/2} \frac{(x+\zeta_s)^2 - (z+\eta_s)^2}{\left[(x+\zeta_s)^2 + (z+\eta_s)^2\right]^2}, \tag{7a}$$

$$\frac{w}{U} = -\frac{1}{U}\frac{\partial\psi}{\partial\zeta} = -2h\left(L^2 + \frac{h^2}{4}\right)^{1/2} \frac{(x+\zeta_s)(z+\eta_s)}{\left[(x+\zeta_s)^2 + (z+\eta_s)^2\right]^2}. \tag{7b}$$

The geometry in shown in Figure 1.

3. Beam orientation model

The remote sensing instrument senses the Doppler shift at a range r using a beam tilted at an angle θ to the vertical and at an azimuth angle ϕ with respect to the flow. The overall geometry for the instrument location and beam orientation is shown in Figure 2.

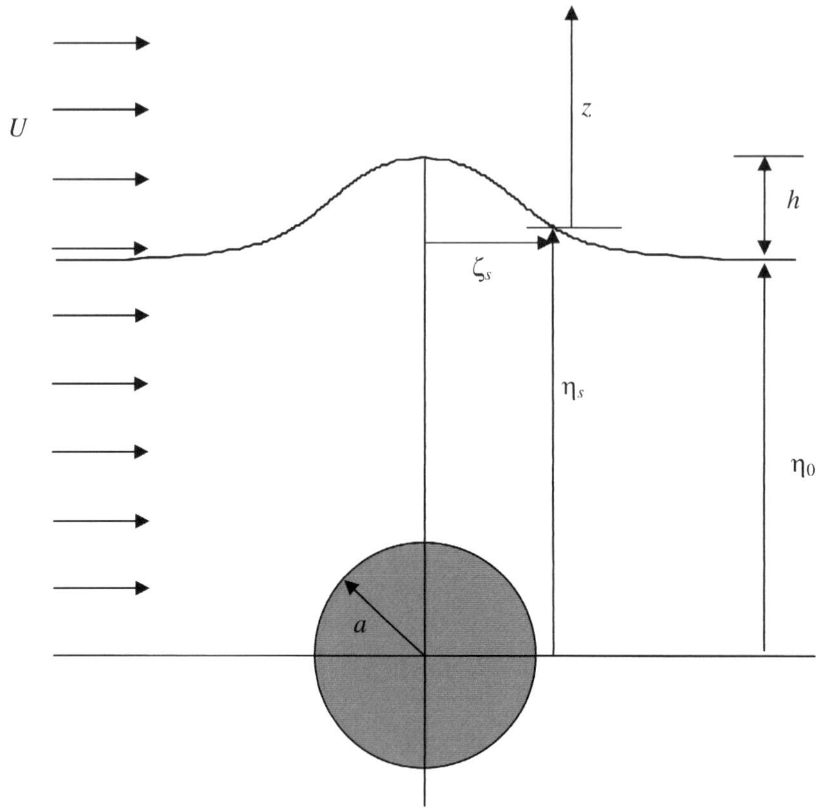

Figure 1. The geometry of the hill flow model.

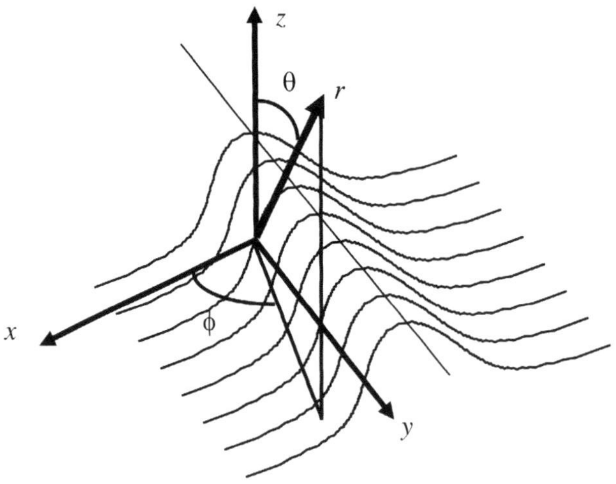

Figure 2. The coordinate system for a SODAR or LIDAR downstream of the ridge

From this geometry,

$$x = r\sin\theta\cos\phi, \quad z = r\cos\theta \tag{8}$$

The radial wind component, as sensed by the instrument, is

$$v_r = u\sin\theta\cos\phi + w\cos\theta. \tag{9}$$

3.1. SODAR having a vertical beam and two tilted beams

A common SODAR design (for example, an ACS SODAR) has a tilted beam (θ, ϕ) producing a radial wind estimate v_1, a second tilted beam (θ, ϕ+π/2) producing estimate v_2, and a vertically-pointing beam providing a radial wind estimate v_3. Three equations give the three radial winds as linear combinations of the three Cartesian wind components, which can be inverted to obtain estimates ($\hat{u}$, $\hat{v}$, $\hat{w}$) as described in [1].

$$\begin{aligned} \hat{u}\sin\theta &= v_1\cos\phi - v_2\sin\phi + v_3\cos\theta(\sin\phi - \cos\phi) \\ \hat{v}\sin\theta &= v_1\sin\phi + v_2\cos\phi + v_3\cos\theta(\sin\phi + \cos\phi). \\ \hat{w} &= v_3 \end{aligned} \tag{10}$$

Each of $\hat{u}$, $\hat{v}$, and $\hat{w}$ are evaluated using v_1 at ($r\sin\theta\cos\phi$, $r\cos\theta$), v_2 at ($-r\sin\theta\sin\phi$, $r\cos\theta$), and v_3 at (0, r) and then compared with the wind components directly above the SODAR. The result is a relative error in estimated wind speed and direction.

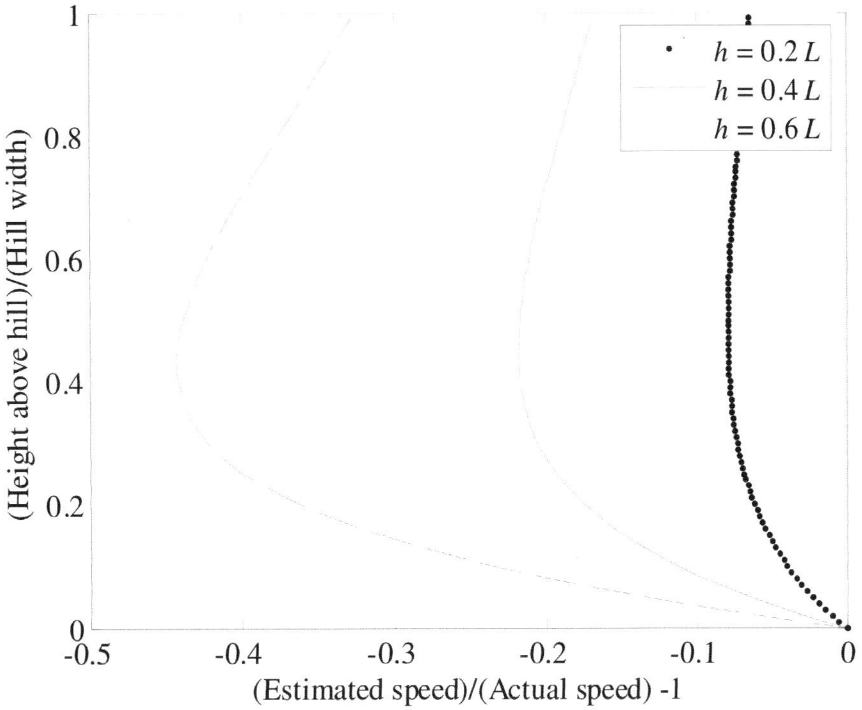

Figure 3. Wind estimation errors for three hill shapes. The ASC beam configuration is modelled, with a beam tilt = 20° and azimuth = 0°

The effect of hill shape is shown in Figure 3. The errors are rather large, with a maximum occurring at a height above the hill of 0.2 to 0.4 hill half-width distances. The maximum occurs because near the surface the three beams are sampling volumes very close together, whereas at large heights the flow is quite uniform. So the greatest error occurs at intermediate heights. Mostly the error maximum is occurring well above the region of interest for wind turbines, so an approximately linear dependence of error on height will prevail.

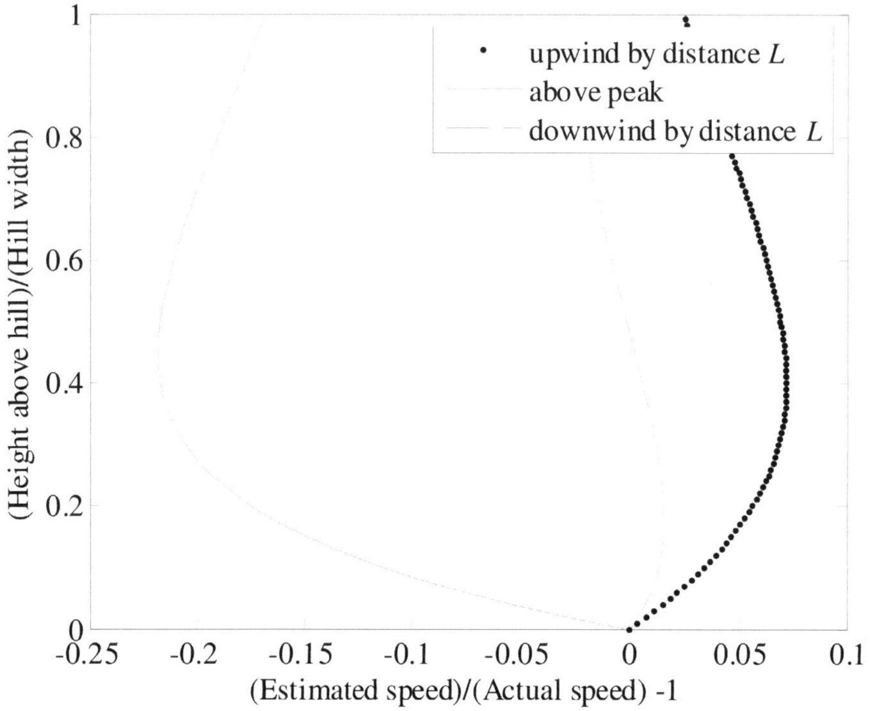

Figure 4. Wind estimation errors for three site locations for the SODAR or LIDAR on a $h = 0.4L$ hill. The ASC beam configuration is modelled, with a beam tilt = 20° and azimuth = 0°

The effect of placing a SODAR upwind, on the ridge, or downwind of the ridge is shown in Figure 4. The result is not symmetric. In particular, the largest errors occur when the SODAR is placed at the ridge, where the speed-up is highest. Both upwind and downwind placements are in regions of lower shear so the errors are less. The positive errors for *both* off-ridge placements occurs because the gradient for vertical velocities is in the same direction, whereas the speed-up in horizontal wind is not as great as for the ridge position.

As expected, the upwind and downwind errors have identical behaviour for and azimuth orientation of 45°, as seen in Figure 5 (the two curves have been separated a little for clarity).

3.2. SODAR having three tilted beams

The AQS SODAR has three beams tilted at the same zenith angle but separated in azimuth by 120° steps, at (θ, ϕ), $(\theta, \phi+2\pi/3)$, and $(\theta, \phi+4\pi/3)$. Solution of the three radial wind equations gives

$$\hat{u}\sin\theta = \frac{1}{3}(2v_1 - v_2 - v_3)\cos\phi - \frac{1}{\sqrt{3}}(v_2 - v_3)\sin\phi$$
$$\hat{v}\sin\theta = \frac{1}{3}(2v_1 - v_2 - v_3)\sin\phi + \frac{1}{\sqrt{3}}(v_2 - v_3)\cos\phi \quad (11)$$
$$\hat{w}\cos\theta = \frac{v_1 + v_2 + v_3}{3}$$

As before, the solutions can be found both using the radial velocities directly above the SODAR and at the positions actually sensed by the tilted beams. Errors are shown in Figures (5a) and (5b) for two orientations of the instrument. Comparison between Figures (6a) and (4) show that the AQS beam configuration gives very similar errors to the ASC configuration.

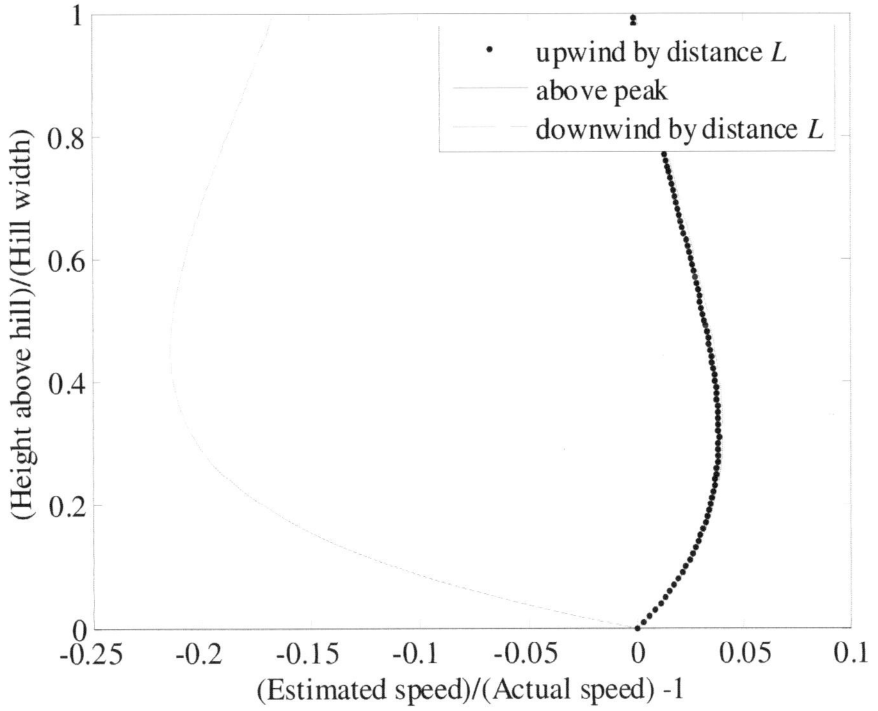

Figure 5. As for Figure 4, but with azimuth orientation of the SODAR at 45°.

3.3. Larger 30° cone angle, as for ZephIR and Wind-cube LIDARS

Simply increasing the cone angle from 20° to 30° did not make much difference to the errors described above, and at this stage we have not attempted to model the LIDAR scanning in any detail.

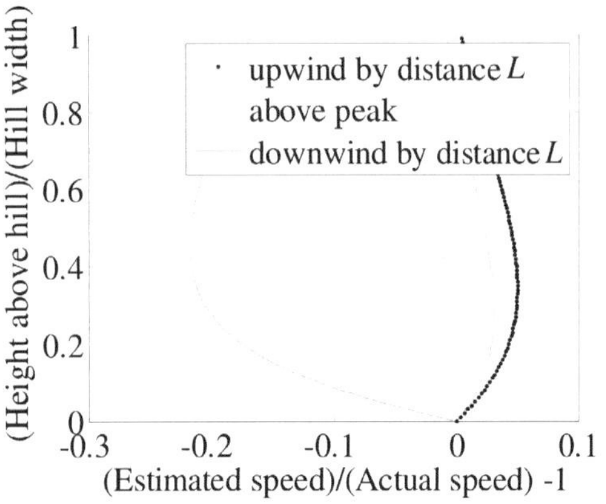

Figure 6a. AQS beam configuration with θ=20° and ϕ = 0° on a $h = 0.4L$ hill.

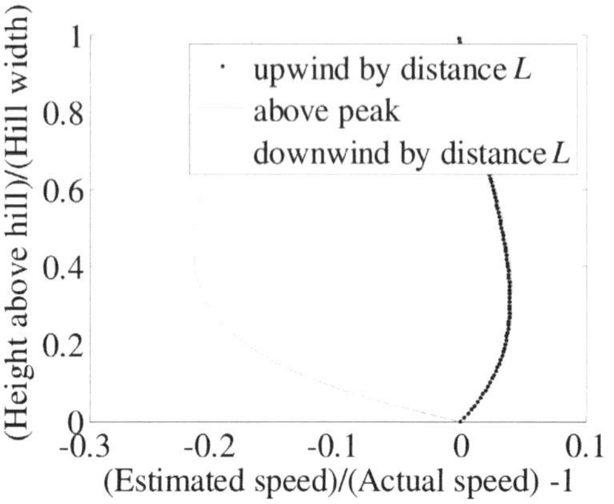

Figure 6b. AQS with azimuth ϕ = 30°.

4. Conclusions

A simple potential flow model has been used to model the flow over hills of approximately Gaussian shape and having varying height/width ratios. Into this model is embedded a mono-static remote sensing instrument. We have used the usual equations by which SODARs estimate 3-component vector winds from three radial velocities obtained from beams pointing in different directions. Comparison is made between winds derived using the actual atmospheric volume locations which would occur in practice, and the wind vectors at the same range but directly above the SODAR. it is, of course, the latter wind speed which is sought by use of this instrument.

The errors found are typically in the range of 5-20%. Such errors are *huge* in comparison with errors with which investigators are usually concerned. The main point of this paper, therefore, is to raise a red flag to indicate that there are going to be very problems obtaining quality wind measurements in complex terrain with such instruments as current mono-static SODARs and LIDARs (although the plots show simulations for SODARs, the problems are likely to be even worse for LIDARs, which typically have large beam cone angles).

There is already field comparison evidence from a number of sources which lends credibility to the analysis in this paper. For example, Jørgensen [2] found V_{LIDAR}=0.9995 V_{MAST} + 0.0194 in flat terrain, whereas with the same model LIDAR, V_{LIDAR} = 0.8753 V_{MAST} + 0.4519 in complex terrain. Both regressions had R^2>0.99. Other results of similar concern are likely to become more numerous as more effort is put into placing remote sensing instruments at complex terrain wind farm sites.

The simple analytic tool described here can readily be used for simulations on beam configurations, and it is also possible to obtain analytic expressions for, as examples, the height of maximum error, the orientation giving minimum error, or the slope of error vs height at low heights.

References

[1] Bradley S. G. *Atmospheric Acoustic Remote Sensing*. Florida, CRC Press/Taylor and Francis Group, 271pp., 2007. ISBN 978-0-8493-3588-4.

[2] Jorgensen, H. UpWind progress in remote sensing of the wind using LIDARs and SODARs. EWEC 2008, Brussels.

IOP Conf. Series: Earth and Environmental Science **1** (2008) 012062 doi:10.1088/1755-1307/1/1/012062

The impact of high resolution surface properties retrieved from Satellite in the urban canopy model

Y M Kim, J Y Byon and Y J Choi

National Institute of Meteorological Research

Korea Meteorological Administration

Seoul, South Korea

yjchoi@kma.go.kr

Abstract. This study aims to improve the atmospheric boundary condition in mesoscale model by satellite derived-land surface temperature. We have performed the WRF-UCM (Urban Canopy Model) using land surface temperature (LST) retrieved from MODIS satellite data. We evaluate the performance of the coupled WRF-UCM modeling system using LST against surface observation in Seoul, South Korea.

1. Introduction

Numerical models with a fine grid can be a useful tool in urban weather forecast which provides external forcing into air dispersion and pollution models. Urban weather simulation and/or prediction may be conducted by using either Computational Fluid Dynamics (CFD) model or mesoscale model. However, a small domain of the CFD model has limitations for studying large scale forcing of the urban environment because of the computational efficiency. In this context, improvements of the computational environment and physics parameterization in mesoscale model can be an alternative way for urban scale prediction with a larger domain. As an example, high-resolution Sea Surface Temperature (SST) retrieved from satellite leads to improvement in temperature forecast in National Institute of Meteorological Research/Korea Meteorological Administration (METRI/KMA) short range forecast (Lee and Ahn, 2003).

In this study, we focus on Land Surface Temperature (LST) rather than SST, as an important factor in the urban scale modelling, modulate the comings and goings of energy and water in land-atmosphere interaction (Earth observatory; NASA, cited 2008). We investigate the impacts of the improved bottom boundary condition, i.e., satellite retrieved LST from the MODIS satellite, on urban weather prediction using a mesoscale model.

2. Land surface temperature data

LST is significantly influenced by land surface characteristics such as land use, soil texture, topography, vegetation, and many others. Even on the same date, LST and its diurnal temperature range (DTR) show wide widths among different stations due to the different land surface characteristics. But, it is difficult to make high resolution data set with the ground station data for urban scale model because weather stations are distributed irregularly and sparsely.

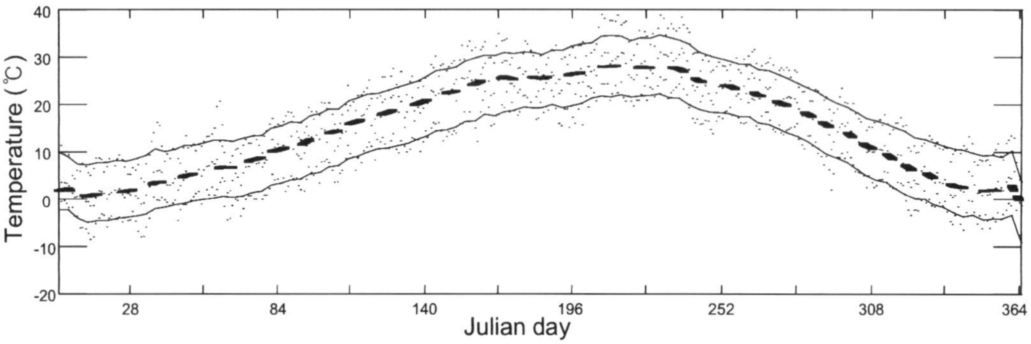

Figure 1. Time series of the station-averaged maximum (upper line), minimum (lower line), and daily mean (thick dashed) LST for 2007. Single dots indicate both the maximum and minimum temperatures at 76 stations in South Korea

On the other hand, satellite derived data provides a high resolution-LST information in a large area in a short time. Figure 2 presents spatial variations of LST in station data and Moderate Resolution Imaging Spectroradiometer (MODIS) LST. Both of them show a high LST in south-west and a low LST in north-east. Compared with ground observed LST, MODIS LST shows detail feature of the surface.

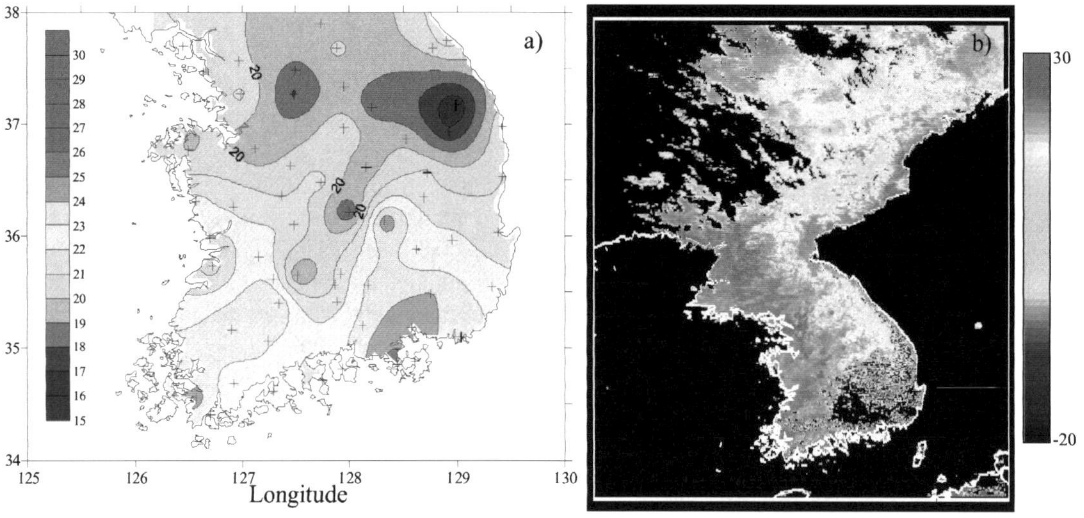

Figure 2. Spatial variation of land surface temperature (°C) a) in 76 station (cross mark) data, and b) MODIS data at 5 October 2007 over South Korea.

The MODIS is an instrument onboard the EOS Terra satellite, launched in December 1999 (Justice et al, 2002). There are seven products in LST products of MODIS and one of them is MOD11_L2, which is a daily LST data at 1 km spatial resolution (Wang, 2006). This data is the result of the generalized split-window method (Wang, 2006; Wan and Dozier, 1996). Urban scale model needs a high resolution data which reveals detail structure of the surface appropriately. This study uses the LST derived from MODIS/terra for the bottom boundary condition.

IOP Conf. Series: Earth and Environmental Science **1** (2008) 012062 doi:10.1088/1755-1307/1/1/012062

3. Methodology

Mesoscale model used in this study is WRF model developed in NCAR. NCAR coupled an urban canopy model (UCM) with land surface model (Noah LSM) in the WRF model to represent the urban by high resolution of land-use and building information realistically. The coupled WRF/Noah LSM/UCM was tested over the urban regions of Houston (Chen et al, 2004), Salt Lake City (Tewari et al., 2005) and Tokyo (Kusaka et al., 2005).

In case of real weather data, it needs pre-processing for a performance of WRF. WPS which mean the WRF preprocessing system is comprised of three programs (Figure 3). First, geogrid part defines simulation domains. Ungrib part transfers a format of meteorological data from grib format to intermediate format. Metgrid works a job that meteorological data are interpolated to simulation grids defined by geogrid (ARW, 2007). We have performed the WRF-UCM using LST retrieved from MODIS/Terra. As MODIS LST having HDF data format is neither grid data nor grib data, we are not able to use ungrib program for exchanging format directly, and therefore it needs another job. After MODIS LST exchange from non-grid data to grid data, it transfers intermediate format.

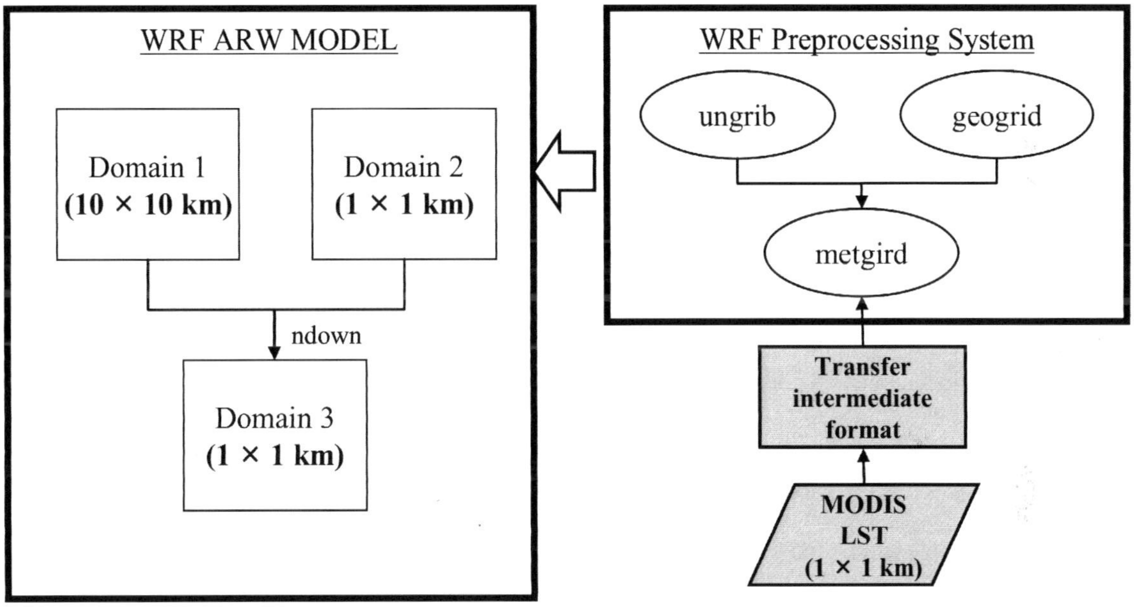

Figure 3. Flow chart of WRF-UCM model process.

WRF model for urban weather forecast is configured by one-way-interactive domain of 10 and 1 km resolutions forced on the Seoul area (Figure 3). Here we used numerical weather prediction data at Korea meteorological administration for initial conditions. Progressing to domain 3 from domain 1 and 2, it used the high resolution-MODIS LST for boundary condition.

The WRF result over Seoul would be presented in the symposium. We evaluated the performance of the WRF-UCM modeling system using against surface observation for Seoul, South Korea. This results will be available for urban weather forecast.

References

[1] Wang W, Barke D, Bray J, Bruyére, Duda M, Dudhia J, Gill D and Michalakes J 2007, *ARW-modeling system user's guide,* version 2 (Mesoscale & Microscale Meteorology Division. National center for atmospheric research)

[2] Chen F, Kusaka H, Tewari M, Bao J-W and Hirakuchi H 2004 Utilizing the coupled WRF/LSM/Urban modeling system with detailed urban classification to simulate the urban heat island phenomena over the Greater Houston Area *5th conference on urban environment* (Vancouver, BC Canada, 23-27 August 2004)

[3] Justice C O, Townshend J R G, Vermote E F, Masuoka E, Wolfe R E, Saleous N, Roy D P and Morisette J T 2002 An overview of MODIS land data processing and product status. *Remote Sens. Environ.* **83** 3-15

[4] Kusaka H , Chen F and Tewari M 2005 Numerical simulation of urban heat island of Tokyo using the next generation Weather Research and Forecasting (WRF) model with an urban canopy model: a preliminary study. *16th regional conference of Clean Air and Environment in Asia Pacific Area* (Tokyo, Japan, 2-4 August 2005)

[5] Lee Y H and Ahn M-H 2001 The impact of high resolution SST on the performance of temperature prediction in a short-range prediction., *J. of the Korean Meteor. Soc.* **37(6)** 607-619

[6] Tewari M., Chen F, Warner T T, Coirier W J and Kim S 2005 Numerical modeling study of wind flow over the Salt Lake City region using integrated WRF-Noah-UCM model at meso-gamma scale. *Joint WRF/MM5 user's workshop* (Boulder, USA, June 2005)

[7] Wan Z and Dozier J 1996 A generalized split-window algorithm for retrieving land-surface temperature from space *IEEE Trans. Geosci. Remote Sens.* **35(4)** 892-905

[8] _____ , 2006: MODIS land surface temperature products user;s guid.

[9] Earth observatory, NASA, cited 2008, [Available online http://earthobservatory.nasa.gov/Observatory/Datasets/lstday.modis.html]

[10] WRF, NCEP, cited 2008 [Available online http://www.mmm.ucar.edu/wrf/users/docs/user_guide/users_guide_chap1.html]

IOP Conf. Series: Earth and Environmental Science 1 (2008) 012063 doi:10.1088/1755-1307/1/1/012063

Development of LiDAR measurements for the German Offshore Test Site

A. Rettenmeier, M. Kühn, M. Wächter , S. Rahm , H. Mellinghoff , B. Siegmeier, L. Reeder

Endowed Chair of Wind Energy, Universität Stuttgart, 70569 Stuttgart, Germany
ForWind, Carl von Ossietzky Universität, 26111 Oldenburg, Germany
DLR, German Aerospace Center, 82234 Wessling, Germany
German Wind Energy Institute (DEWI), 26382 Wilhelmshaven, Germany
Multibrid GmbH, 27572 Bremerhaven, Germany
Federation of German Windpower (FGW), 24103 Kiel, Germany

rettenmeier@ifb.uni-stuttgart.de
kuehn@ifb.uni-stuttgart.de
matthias.waechter@uni-oldenburg.de
Stephan.Rahm@dlr.de
h.mellinghoff@dewi.de
boern.siegmeier@multibrid.com
lr@wind-fgw.de

Abstract. The paper introduces the content of the recently started joint research project "Development of LiDAR measurements for the German Offshore Test Site" which has the objective to support other research projects at the German offshore test site "alpha ventus". The project has started before the erection of the offshore wind farm and one aim is to give recommendations concerning LiDAR technology useable for offshore measurement campaigns and data analysis. The work is organized in four work packages. The work package *LiDAR technology* deals with the specification, acquisition and calibration of a commercial LiDAR system for the measurement campaigns. *Power curve measurements* are dedicated to power curve assessment with ground-based LiDAR using standard statistical methods. Additionally, it deals with the development of new methods for the measurement of non-steady short-term power curves. *Wind field research* aims at the development of wake loading simulation methods of wind turbines and the exploration of loading control strategies and nacelle-based wind field measurement techniques. Finally, dissemination of results to the industry takes place in work package *Technology transfer*.

1. Introduction

LiDAR, as a remote sensing technique with high spatial and temporal resolution, presents a great advantage for offshore wind energy developments [1] [2]. The measurement techniques to be developed will have direct applications for the measurement of power curve and nacelle-based inflow and wake wind fields. Moreover, research-oriented work is done in regards to wake loading simulation and loading control strategies based on inflow measurements [3]. The previously mentioned nacelle-based wind field measurements are used to verify the findings in these two applications.

The LiDAR project lasting from August 2007 to March 2010 is coordinated by the Endowed Chair of Wind Energy (SWE) at the Universität Stuttgart, Germany.

Further experience with power performance measurements is provided by two project partners, the German Wind Energy Institute (DEWI) and the Universität Oldenburg (ForWind). The German Aerospace Center (DLR) contributes with its experience with LiDAR systems. Measurement campaigns onshore will be performed on a 5MW wind turbine owned by Multibrid GmbH, while offshore met mast and LiDAR comparisons are performed at the FINO 1 offshore platform. Dissemination to the German industry and support in the definition of guidelines for power curve measurements with LiDAR are performed by an expert committee of the Federation of German Windpower (FGW) (figure 1).

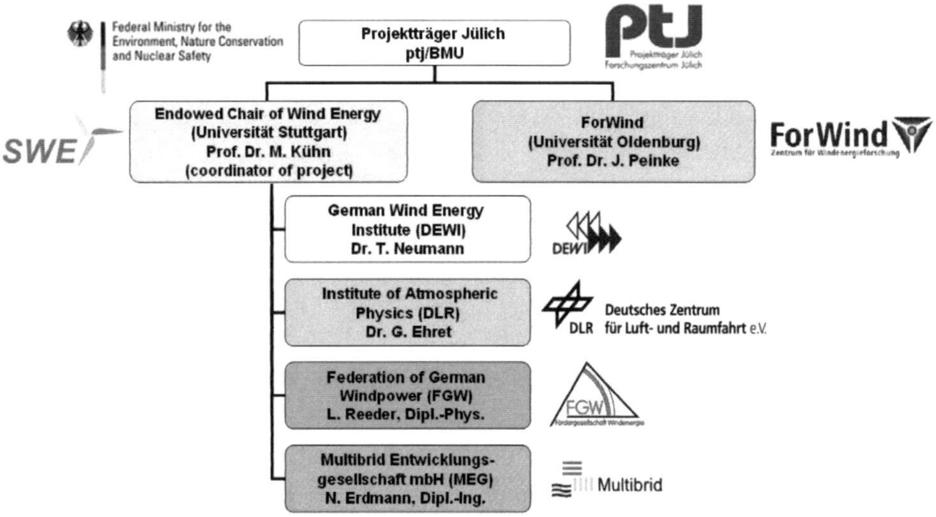

Figure 1. Organisation and participants of the project

2. LiDAR technology

2.1. System procurement

The first step of this work package was to describe the specification of the LiDAR system which has to fulfill the needs of the measurements in the joint project: The ground-based vertical measurements are going to be compared with standardized met mast measurements. Next different horizontal scanning modes of the inflow and wake from the nacelle will be carried out in further work packages. Therefore new developments and add-ons to the standard LiDAR systems that are available on the market had to be specified.

Another decision had to be made concerning the different laser technologies. Which advantages and disadvantages does a continuous laser have versus a pulsed laser beam? Which available system has the longer range with proper accuracy? Together with the DLR, which has experience in long range LiDAR systems for several years, the requirements and comparisons were specified and the decision was made to use the pulsed LiDAR system Windcube® of the French company Leosphere™ (figure 2).

An additional CAN-bus interface, not available in the standard Windcube®, was introduced by the manufacturer in agreement with one of the requirements for the project. This interface is needed for parallel data acquisition from the already existing measurement infrastructure at the met mast at Bremerhaven. The LiDAR system was investigated in an initial test (figure 3), of the system on the roof of SWE during several weeks of winter weather conditions and the occasional winter storm. During this test the project partners, ForWind, DEWI and SWE, could gain their first experience with this technology [4].

2.2. System calibration

From April to July 2008 a comparison of the LiDAR system with the SWE owned 102m met mast will take place at Bremerhaven, close to the Multibrid M5000 prototype. The measurement data of sonic and cup anemometers at three different heights (44, 73, 103m) and wind vanes will be compared with the measurement data from the LiDAR system.

Another measurement campaign will take place in Munich, where the Windcube system and a powerful long–range pulsed LiDAR of the DLR will be compared directly. The aims are to investigate the qualitative and quantitative differences in the measurements and determine where improvements need to be made, e.g. in measurement analysis for the following work steps.

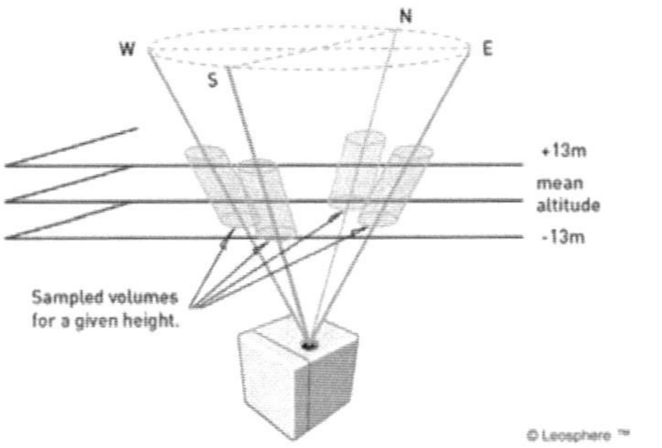

Figure 2. Windcube® System (Leosphere ™)

Figure 3. Initial test of the LiDAR system

2.3. System-Evaluation

All results and experiences will be gathered in this work package. The aim is to work out recommendations concerning the use and the required developments of LiDAR hardware and software, useable onshore and offshore.

3. Power curve measurements

3.1. Stationary measurements of the power curve and offshore capability

In this work package, DEWI will check and contribute to the development of reproducible power curve measurement methods without using a met mast. At the site of Multibrids' M5000 prototype turbine (5MW rated power, 116m rotor diameter) at Bremerhaven the SWE operates a met mast with various meteorological sensors as well as an extensive data acquisition system for data logging (figure 4). The LiDAR system measures wind characteristics up to the rotor blade tips (160m) and this data are logged along with met mast data, nacelle anemometry and turbine controller data. The evaluation of this data to a LiDAR wind speed measurement power curve and the comparison with the standard IEC- power curve results are the first steps [5]. Within the classical approach, the LiDAR data set will also be tested regarding the contributions of the wind profile and turbulence to the power curve. Both are the most interesting parameters for an extended definition of the power curve. Wind profile data of Bremerhaven have so far been only available up to hub height, the upper part of the rotor area has usually not been assessed. The present studies on measured turbulence show that the reported standard deviation in wind speed is smaller than those measured with a cup anemometer [6]. This underlines the contribution of the measurement technology to the measurement result. However, the ideal power curve assessment should be universal. The deviation in the annual energy production (AEP) will be compared as a criterion to calculate the quality of the LiDAR -based power curve to the IEC power curve.

Within a measurement campaign on the research platform FINO I (figure 5), the LiDAR system will be checked concerning the capability to operate in the adjacent offshore test site "alpha ventus" [7] [8]. The German offshore test site, with its twelve 5MW turbines from Multibrid GmbH and REpower AG, will be erected in years 2008 and 2009 in a water depth of approximately 30m. The measurement of the LiDAR system of up to 220 m above sea level (platform height plus LiDAR measurement height) will reveal new aspects concerning wind shear and turbulence in the atmospheric boundary layer offshore.

Figure 4. Base-ground LiDAR comparison measurements with a met mast of 102m height at the Multibrid M5000 prototype

Figure 5. Research platform FINO I, close to the offshore test site "alpha ventus"

3.2. Unsteady measurements of the power curve

The aim is to investigate the capability of LiDAR systems for a new method of power curve determination utilizing the time series information of the measured data set [9] [10]. The typical time increment is on the order of one second or less. In this case, sonic anemometers are installed in addition to the LiDAR system and the met mast with its common standard first class anemometry.

At ForWind, new methods are developed to analyze the dynamics of the power conversion process and derive a power curve from high frequency measurement data. From the LiDAR technology, with its advantageous high spatial and temporal resolution, we expect new insights on how yawed flow, turbulence and wind shear have to be taken into account in the determination of power curves of Multi-Megawatt wind turbines.

4. Wind field research

4.1. Inflow, load and control

This work package intends to develop more information about the loading of the wind turbines supported by LiDAR inflow measurements. Also pilot surveys concerning the control of wind turbines will be carried out.

Work is being done in upgrading the scanning modes of the standard Windcube® system in order to cope with this non-standard way of measuring the inflow wind. Eventually, first evaluation of predictive control strategies based on online LiDAR measurements is planned.

4.2. Measurements of wind fields and simulation of dynamic wake loading of Multi-MW- wind turbines

The development of a physical model for the simulation of dynamic wake loading aims at the development of wind field simulation tools which allow for the complex inflow conditions of huge onshore and offshore wind farms [11] [12]. The LiDAR system enables the measuring of this complex structure of the wind field and the analysis of the measurements is needed to validate the new model.

The German Aerospace Center does research in connection with turbulence of aircraft wake vortices supported by airborne coherent Doppler LiDAR and will assist the measurements with a long-range LiDAR system in Bremerhaven [13]. Their instrument is going to be used for measurements of the far-wake of the 5 MW Multibrid wind turbine which are used for validation of wake models describing the stationary and dynamic characteristics of the flow behind wind turbines.

5. Technology transfer

The establishment of LiDAR wind measurements for onshore and offshore purposes critically depends on upon the development of new guidelines. In the field of power curve measurements with common anemometry the Federation of German Windpower [14] possesses a huge know-how due to the number of members from accredited measurement institutes. This knowledge is needed to evaluate the results of the project and leading, potentially, to a revision of the "Technical Guideline for Wind Turbines Part 2: Determining the Power Performance and Standardised Energy Yields" of FGW with an addition concerning new remote sensing techniques e.g. LiDAR.

The scientific exchange with national and international partners of the wind energy community is a crucial goal, mainly the activities in the range of IEA Wind Annex 23 ("Offshore Wind Energy Technology Deployment"), IEA Topical Expert Meetings, e.g. "State of the art of Remote Wind Speed Sensing Techniques using SoDAR, LiDAR and Satellites", and the IEC committee TC88.

6. Conclusion

The aim of this joint project is to work out recommendations concerning the use and the required developments of LiDAR hardware and software, useable onshore and offshore.

The results of the first work packages aim at the application of LiDAR technology in the offshore test site "alpha ventus", where inflow and wake measurements from the nacelle will be carried out in further RAVE projects, e.g. OWEA.

In future work wind turbine control algorithms and predictive control strategies as well as power curve and load measurements based on online LiDAR measurements are planned to be developed.

Acknowledgements

This research is done in the framework of the German joint research project "Development of LiDAR technologies for the German offshore test field". This project with number 0327642 is funded by the German Federal Ministry for the Environment, Nature Conservation and Nuclear Safety (BMU).

References

[1] C.B Hasager, A. Peña, T. Mikkelsen, M. Courtney, I. Antoniou, S.-E. Gryning, P. Hansen, P.B. Sørensen, "12MW Horns Rev experiment". Risø-R-1506(EN) (2007) 83 p.

[2] D.A. Smith, M. Harris, A.S. Coffey, T. Mikkelsen, H.E. Jørgensen, J. Mann, G. Danielian, "Wind lidar evaluation at the Danish wind test site in Høvsøre". Wind Energy (2006) 9, p. 87-93

[3] M.Harris, M. Hand, A. Wright, "Lidar for Turbine Control", Technical Report NREL/TP-500-39154, January 2006

[4] M. Wächter, A. Rettenmeier, J. Peinke, "Wind velocity measurements using a pulsed LIDAR, system: first results", ISARS, Risoe, 2008

[5] International standard, IEC 61400-12 "Wind turbines: Part 12-1, Power performance measurements of electricty producing wind turbines, First edition 2005"

[6] R. Wagner, H.E. Jørgensen, J. Mann, M.S. Courtney, I. Antoniou, "Investigation of the measurement of the wind speed standard deviation using a LiDAR", EWEC, Brusells, 2008

[7] Research at "alpha ventus" (RAVE): http://www.rave-offshore.de/

[8] Offshore Test Site "alpha ventus": http://www.alpha-ventus.de

[9] E. Anahua, M. Lange, F. Boettcher, S. Barth, J. Peinke, "Stochastic analysis of the power output for a wind turbine", EWEC, London, 2004.

[10] E. Anahua, M. Lange, F. Boettcher, S. Barth, J. Peinke, "Characterization of the Wind Turbine Power Performance Curve by Stochastic Modeling" , EWEC, Athens, 2006.

[11] J.J. Trujillo, M. Kühn, "Aeroelastic Simulation of a Multi-MW Wind Turbine Operating in Wake", DEWEK, Bremen, 2006

[12] J.J. Trujillo, A. Rettenmeier, D. Schlipf, "Enhanced Measurements of a Large Turbine Near-Wake Using LIDAR from the Nacelle", ISARS, Risoe, 2008

[13] S. Rahm, I. Smalikho, F. Köpp. "Characterization of aircraft wake vortices by airborne coherent doppler lidar", Journal of Aircraft, 44:799 – 805, 2007. doi:10.2514/1.24401.

[14] http://www.wind-fgw.de/tr_engl.htm

IOP Conf. Series: Earth and Environmental Science **1** (2008) 012064 doi:10.1088/1755-1307/1/1/012064

Wind Profiles in Moscow city by the Sodar Data

Lokoshchenko M.A., Yavlyaeva E.A.

Department of Meteorology and Climatology, Geography Faculty, Lomonosov Moscow State University. 119991, Lengory, Moscow, Russia
Phone: +7-495-9394284; Fax: +7-495-9392479; E-mail: loko@geogr.msu.su

Abstract. The results of wind measurements in the low 500 m air layer above Moscow by Doppler sodar 'MODOS' have been presented for period from 2004 to 2008. General features of both annual course and the diurnal one of wind profiles have been discussed. The most wind speed values are observed in average in autumn and in winter, the least ones – in spring and in summer. The most frequent wind direction is south-western; the rarest one is northern. The crossover height of wind profiles has been analyzed. As it was found it consists of 60-80 m in average in warm period.
For the first time, mean values of wind turning during passing of atmospheric fronts have been received and studied by sodar data. It was found that right-hand (clockwise) turning of wind at cold fronts is more and occurs more quickly in time in the comparison with warm fronts. The occlusion fronts demonstrate intermediate mean values of wind turning.

1. Introduction

The acoustic remote sensing allows study a fine structure and detailed dynamics in time of wind profiles. Continuous observations by Doppler sodars during a long time give possibility to investigate climatologic features of wind structure of the Atmospheric Boundary Layer (ABL) with an excellent spatial resolution. As one knows first Doppler sodars for measurements of wind profiles have been constructed at the beginning of 1970-s. Several years later, at the end of 1970-s, the first Doppler sodar in Russia and in former Soviet Union was made and first wind profiles were measured in the Institute of Atmospheric Optics (IAO) in Tomsk by Professor Krasnenko and his group [1]. Now in Russia several sodars operate in Tomsk, in Moscow and in Obninsk. They are used both for scientific investigations, and for air quality monitoring. Now there are five main scientific groups in Russia which work about sodar data: IAO in Tomsk, Institute of Optical Monitoring in Tomsk, Institute of Atmospheric Physics in Moscow, Lomonosov Moscow State University in Moscow (presented by authors) and Institute of Experimental Meteorology (RPA 'Typhoon') in Obninsk.
The acoustic remote sensing at the Geography Faculty of Lomonosov Moscow State University (MSU) is carried out since 1988. Now at the Meteorological observatory of MSU two sodars operate synchronously: vertical sodar "ECHO-1" (of GDR production), and "MODOS" sodar (of METEK firm production, Germany). The results of sounding at MSU by sodar 'ECHO-1' have been published many times in Proceedings of all ISARS Symposia since 1996 and in a lot of journals. Among others, long-term data both about the ABL thermal stratification and about profiles of the σ_W (standard deviation of the wind vertical component) received by use of 'ECHO-1' sodar have been summarized and published in [2-4, etc.].
Sodar data about wind structure in the low 500 m air layer in Moscow for last three and half years have been received by use of Doppler sodar "MODOS". It operates at MSU continuously since 2004, and the results of wind measurements have been presented there at first. The operation

Figure 1. The 'MODOS' sodar (made by METEK firm, Germany) at Moscow University.

frequency of 'MODOS' sodar which we use is 2000 Hz, the vertical range is in limits from 40 to 500 m, spatial resolution is 20 m, reliable wind profiles are measured in average for every 10 minutes.
This serial brand of sodars is well-known everywhere. Classic Doppler scheme of three beams with separate antennas for each beam (one vertical and two inclined) is used there. Each antenna consists of seven equally-phased loudspeakers (in all – 21) and the electric transmit power is 1 kW. Reliability of received wind profiles is tested automatically with the account of signal-to-noise ratio value at any separate level of measurements. Wind measurements of 'MODOS' sodars have been multiply compared with *in situ* data on a lot of field experiments (e.g. [5], etc.). As a result, this brand of sodars gets a sertificate as standard tool of wind profiling. An accuracy of wind speed V measurements by 'MODOS' sodar is ±0.24 m/s if $V \leq 6$ m/s and ±4 % of V value if $V > 6$ m/s; an accuracy of wind direction is 3-4°. All conclusions made by authors below are reliable with the account of these accuracy estimations. 'MODOS' sodar is installed at the Meteorological observatory of MSU at south-western periphery of Moscow. This site represents comparatively flat and open area. There are no great vertical obstacles in vicinity of sodar; the only tall tower – the main building of MSU of 250 m height – is situated at 750 m apart from Meteorological observatory (see Fig.1).
Now a lot of sodars produced by METEK firm operate successfully all round the world. Among others, they are used in Germany for studying of the urban ABL wind structure ([6] and others).
The 'MODOS' sodar began to operate at MSU in November of 2004. Since that time till February of 2008 sodar data about wind profiles have been received during 28 full and non-full months. During this period the longest break of sounding by this sodar took place from February to October of 2005. The main purpose of sodar sounding at MSU is long-term studying of climatic features of the ABL wind structure above Moscow. General advantage of sodar data is their excellent spatial (20 m) and temporal (10 min) resolution that allows study the ABL wind structure in details.

2. Wind speed profiles.

The annual course of wind speed V in Moscow at different heights since 2004 till nowadays has been presented by sodar data in Figure 2. As one can see the most V everywhere in the low 200 m air layer has been measured in cold period of year whereas the least one – in warm period. Just above 'dead zone' at the 40 m level V in average varies from nearly of 3 to 4 m/s during a year. The more is height inside sounding range the more is amplitude of the annual course of V. At the levels of 180 and 200 m wind speed varies from nearly of 6 to 8 m/s and even a bit more. So, at the level of 40 m amplitude of the wind speed annual course is 1.2 m/s whereas at the level of 200 m it increases up to 2.0 m/s.

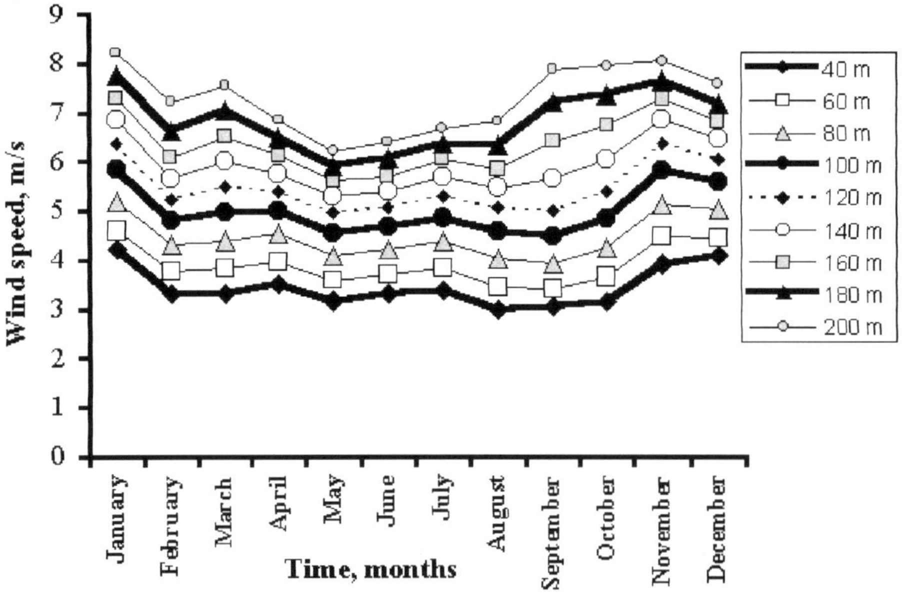

Figure 2. Annual course of average wind speed at different heights by sodar data. Moscow, MSU, period from 2004 to 2008.

As for as additional features of the annual course like some secondary maxima and minima from one month to another they seem to be caused by specific of the atmospheric circulation during last years.
All the measured wind profiles (curves) will probably become the more smoothed the longer averaging periods are applied in the future.
For more detailed illustration of seasonal dynamics the averaged wind profiles for each season have been presented in Figure 3. As it is seen the most values of V are observed both in winter, and in autumn. On the contrary, the least values of V have been measured in summer. Nevertheless, wind speed in spring at any height is only a bit more than in summer. Thus, clear asymmetry between conditions of transitional seasons exists so that an autumn is much more windy time in Moscow than a spring. These seasonal particularities of the wind dynamics seem to be not surprised. In fact, Iceland low which exists during a cold period appears in late autumn already. Strong winds in European Russia are connected as a rule with zone of gradient flows between this depression and ridges of Siberian high. It is well-known that zonal westerlies in mid-latitudes are more intensive in cold period (e.g., [7]).
On the contrary, in summer Western transfer becomes weaken and conditions close to calm often take place in Central region of Russia. So, less average values of V in summer confirm geographic features of atmospheric circulation in mid-latitudes.

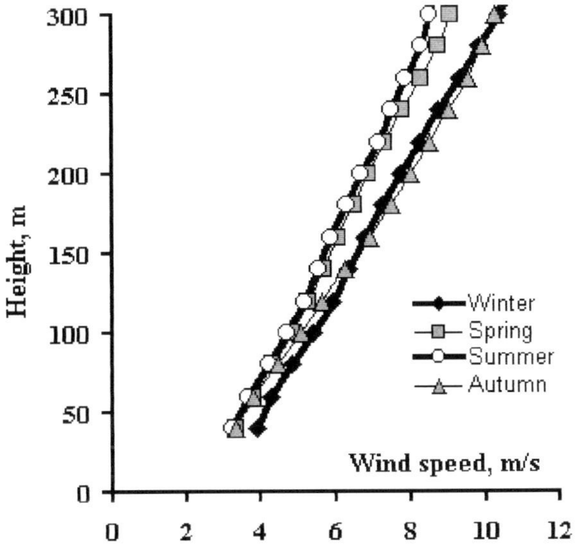

Figure 3. Average wind profiles at various seasons. Moscow, MSU, period from 2004 to 2008.

3.Wind direction.

Wind direction in Moscow has been analyzed as well by 'MODOS' sodar data. It should be noted that measurements of wind direction by routine vanes on the level of 10-15 m height at ground meteorological stations in cities are often erroneous because of multiple vertical obstacles in vicinity of stations. Unlike them sodar data about wind direction in the ABL are much more reliable – of course, in case if any tall towers are absent close to a sodar. In Figure 4 a total wind rose has been demonstrated for a whole period of sounding of 'MODOS' sodar at Moscow University from November 9th of 2004 to March 3rd of 2008. At this figure all separate measurements have been collected – for every interval of 10 min and at all levels covered by reliable data. Total amount consisted of slightly more than 830.000 wind measurements. Color scale represents a range from almost absence of data (white) to 100.000 separate values (black). At the Table a percentage of cases for various compass points have been presented. Margins in azimuth between different sectors are 22.5°, 67.5°, etc. As it is seen the most probable

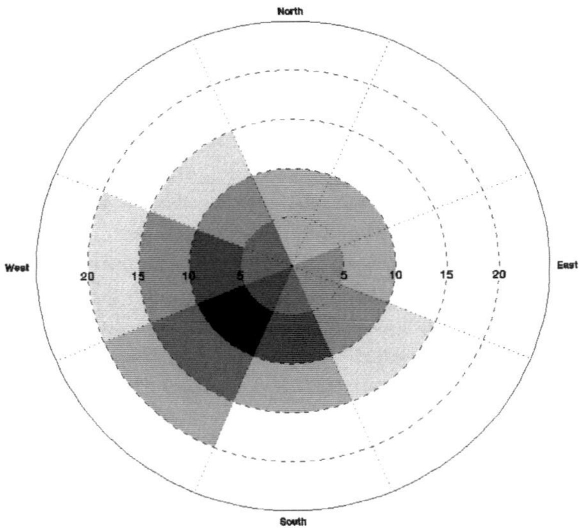

Figure 4. Total wind rose in Moscow by 'MODOS' sodar data for period 2004-2008.

wind direction in Moscow is south-western: 28 %. Except south-western, the most probabilities among the rest sectors are associated with western and southern winds: 18 and 17 % correspondingly. Circles in Figure 4 demonstrate intervals of a wind speed detected at any azimuth sector. One can see that the most often winds are south-western with a speed from 5 to 10 m/s (black color). Total occurrence probability of wind directions from western to southern, i.e. at azimuth from 157.5° to 292.5°, consists of nearly 64 % although this sector is only 37.5 % from all circle.

Thus, sodar data in common confirms well-known climatologic features. In fact, the most often wind direction on geographic region of Central Russia is south-western and the rarest directions are northern and eastern. Accordingly to sodar data, an occurrence probability of northern winds in Moscow is even a bit less than of eastern ones. Possibly, specific of atmospheric circulation during last years influences on this statistics at the Table because period of sounding is not yet sufficiently long. However, in general these data seem to be close already to climatic estimations.

Total occurrence probability for any direction is the most for moderate wind speed from 5 to 10 m/s.

Table. Total statistics about wind direction and wind speed by sodar data, probabilities in %. Moscow, MSU, period from 2004 to 2008.

Compass sectors:	**North**	**North-East**	**East**	**South-East**	**South**	**South-West**	**West**	**North-West**	**Total:**
0-5 m/s	2.9	2.9	3.7	4.1	6.0	7.9	6.2	5.0	38.7
5-10 m/s	2.0	2.3	2.6	3.5	8.3	12.3	7.6	4.7	43.3
10-15 m/s	0.2	0.2	0.4	0.9	2.7	6.1	3.5	0.7	14.7
15-20 m/s	0.0	0.0	0.0	0.2	0.4	1.6	0.7	0.0	2.9
20-25 m/s	0.0	0.0	0.0	0.0	0.0	0.3	0.1	0.0	0.4
25-30 m/s	0.0	0.0	0.0	0.0	0.0	0.0	0.0	0.0	0.0
Total:	5.1	5.4	6.7	8.7	17.4	28.2	18.1	10.4	100.0

4. Crossover height.

As it is known wind speed V in the ground air layer in the afternoon is more as a rule than at night. Since some level – so-called 'crossover height' – daily values of V become, vice versa, less than the nocturnal ones in above air layers. In other words, the crossover height is a point of intersection between daily and nocturnal wind profiles in the low troposphere. The cause of opposite daily courses below and above this point is different intensity of vertical turbulent exchange and of downward transfer of momentum during a day ([8] and others). In the middle of a day unstable stratification leads to more strong vertical mixing. As a result, a daily wind profile is smoothed and the vertical gradient of V is significantly less than at night. Thus, the crossover height is an important parameter of the ABL structure because at this level diurnal course of wind speed in average is absent.

Sodar data due to their excellent spatial resolution allowed detect at the first the crossover height for Moscow conditions with a high accuracy. In fig.5 one can see three-dimensional surfaces of combined values of V calculated both for daily time, and for a date from one month to another at different heights. As an example a period from March to August 2006 has been analysed there. In Figure 5 (a) one can see that that topography of the surface at 40 m has form like a dome so that diurnal values are more than the nocturnal ones. Besides, in 2006 wind speed was slightly more in the middle of warm season than in spring and at the end of summer. As a consequence, the most values of V (more than 3.3 m/s) create closed zone whereas the least ones (less than 2 m/s) lens out. At the next two levels – 60 and 80 m – three-dimensional surfaces are quite plane and not be curved (Fig.5 (b) and (c)). It means that wind speed is nearly the same all round a day. Thus, vertical range from 60 to 80 m represents an average crossover height in Moscow city in spring and in summer. At more height since 100 m wind speed, on the contrary, becomes in the afternoon less than at night. As it is seen a convexity is downward at the surfaces in Fig.5 (d), (e) and (f).

IOP Conf. Series: Earth and Environmental Science 1 (2008) 012064 doi:10.1088/1755-1307/1/1/012064

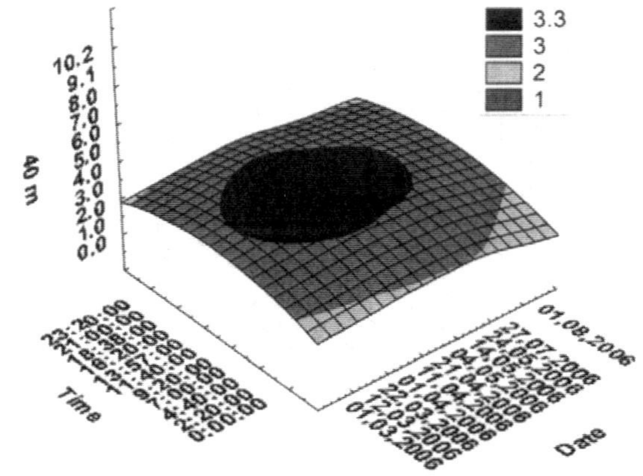

(a) 40 m level;

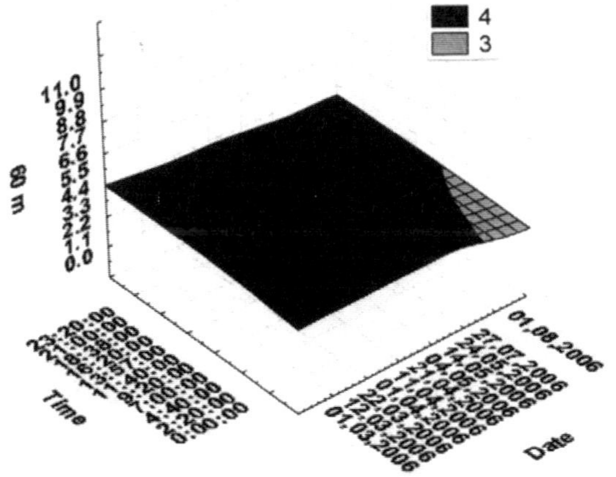

(b) 60 m level;

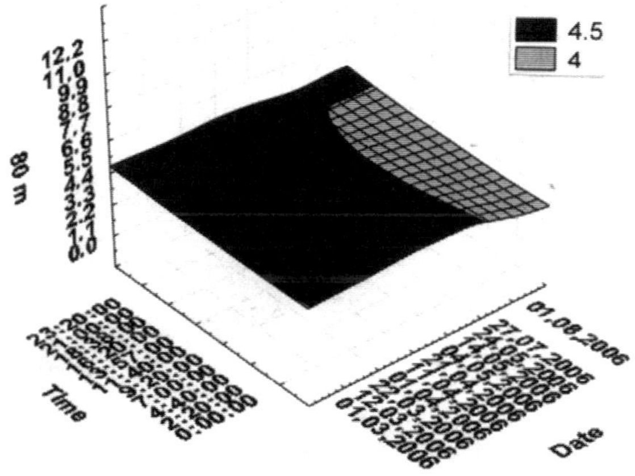

c) 80 m level;

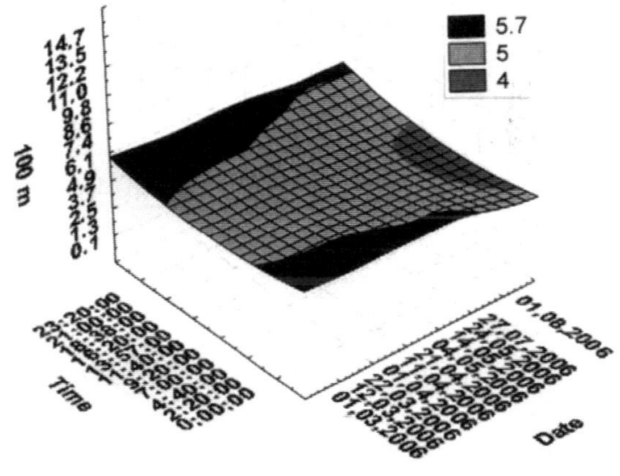

(d) 100 m level;

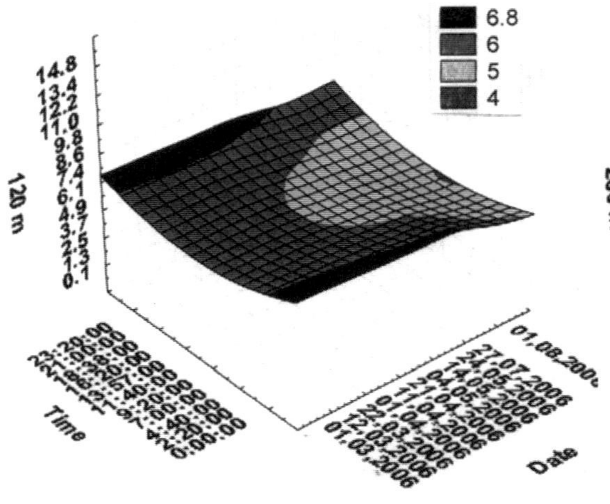

(e) 120 m level;

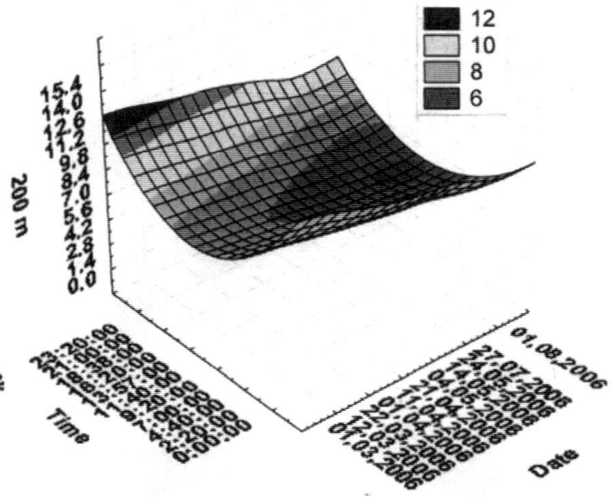

(f) 200 m level;

Figure 5. Distributions of wind speed for hours round the clock (left axis) and for dates (right axis) at different heights by the data of 'MODOS' sodar at MSU in Moscow, period from March to August, 2006.

5. Wind turning at zones of atmospheric fronts.

As it is known the right-hand (clockwise) wind turning occurs as a rule at zones of atmospheric fronts [7, etc.]. For the first time, statistical estimations of this synoptic phenomenon for Moscow conditions have been received with the use of sodar data. Conditions of cold season have been studied. Two separate periods have been taken into account for analysis: from November 2004 to January 2005 and from November 2005 to January 2006. Each case of possible front passing during that time has been tested precisely with the use both of synoptic charts, and of data about dynamics of meteorological parameters near the ground at MSU. Firstly baric tendency, advective air temperature change, type of condensation systems, typical atmospheric phenomena and the amount of precipitation have been taken into account. Then probable events of front passing have been verified by usage of synoptic charts both of surface analysis and of AT_{850}. As a result of this analysis 25 cases of atmospheric fronts have been accepted as undoubtedly reliable. A moment of passing has been detected everywhere by use of barogram (that is weekly recording of standard barograph) with an accuracy of one hour.

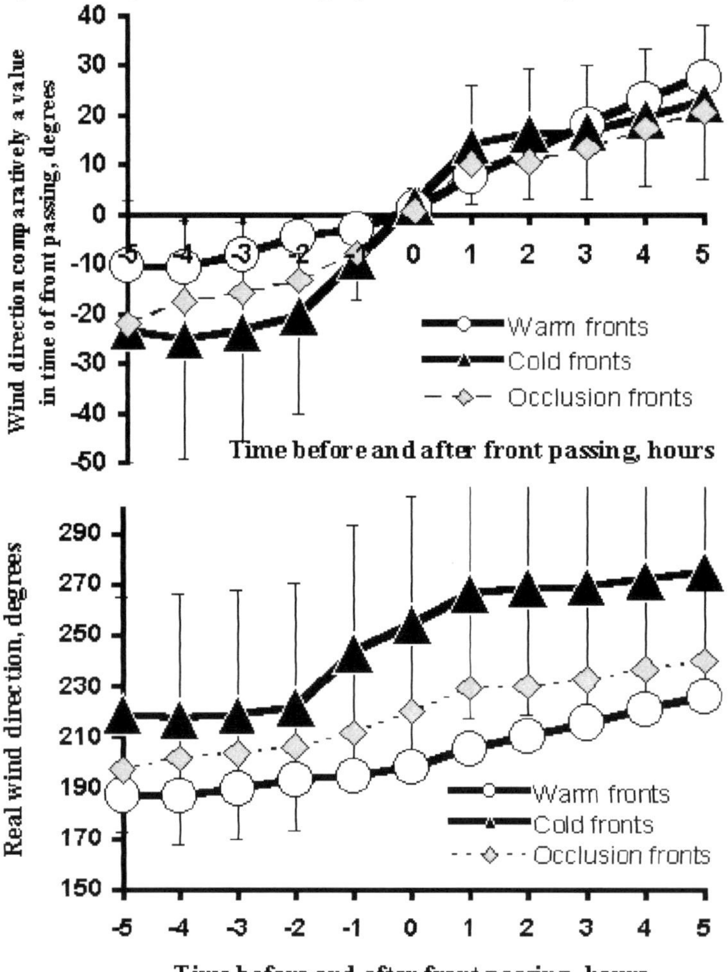

Figure 6. Change of wind direction during passing of different types of atmospheric fronts in the low 500 m air layer in Moscow by sodar data in cold period from 2004 to 2006.
Average values for cold fronts are supplied by confidence intervals of 5 % significance level.

The results of statistical averaging have been presented in Fig.6. On upper part of Fig.6 both horizontal axis and the vertical one have been normalized to a moment of front passing In other words, zero value indicates there a moment of front passing and a wind direction at that time. On bottom part of Fig.6 the same results are shown for real wind direction so that only horizontal axis has been normalized. On upper part of Fig.6 confidence intervals confirm that change of wind direction (wind turning) at frontal zone is statistically significant phenomenon – at least, for cold fronts. It should be noted that a length of confidence intervals is reliable only at edges of upper part and is understated close to zero point because of normalization procedure. As for the bottom part of Fig.6, confidence intervals confirm there statistical significance of a difference between wind directions after passing of cold fronts and warm ones. As one can see the sharpest right-hand turning of V is connected with cold atmospheric fronts – 55° in average. Warm atmospheric fronts demonstrate the least value of the wind turning – nearly of 40° only in average. It is notable that occlusion atmospheric fronts are connected with intermediate average value - 45° of approximately. It is seen as well that wind turning at a frontal zone in any case is right-hand. It means that wind direction at Northern hemisphere changes clockwise during front passing that confirms fundamentals of classic Norwegian synoptic school [e.g., 7]. This qualitative effect is well-known for specialists a long time but statistical estimations of the wind right-hand turning hardly have been investigated before.

Besides, a wind turning at zones of cold fronts occurs much more quickly than in zones of warm ones – in average during 3-4 hours only, as it is seen in Fig.6. It should be noted that more quick wind turning there corresponds to well-known synoptic fact that zones of cold fronts are narrower than of warm ones at the middle latitudes.

Average wind direction at zones of warm fronts consists of 210°, at zones of occlusion fronts - 220°, at zones of cold fronts – 240°. A total increase of wind speed at zones of any fronts has been detected, but it doesn't be clearly localized in time.

6. Conclusions.

1. Wind speed profiles above Moscow in the low 500 m air layer demonstrate in average more values in autumn and in winter and less ones in spring and in summer. Average wind speed at the level of 200 m height varies from 6 m/s in warm period to 8 m/s in cold period of a year.
2. The most often wind direction in Moscow at heights from 40 to 500 m is south-western and the rarest one is northern.
3. The cross-over height above Moscow is equal to 60-80 m in warm period of a year.
4. Right-hand (clockwise) wind turning at cold fronts is more intense and quick in the comparison with warm fronts. The occlusion fronts demonstrate intermediate values of wind turning.

Acknowledgements.

Authors are much grateful to Dr. H.-J. Kirtzel and his colleagues from METEK Company for their careful attention and help and, besides, to Dr. V.G.Perepelkin for his help to providing of continuous sodar observations at MSU.

This work has been supported by Russian Basic Research Foundation (project 07-05-00874).

References.

1. Krasnenko N.P. 1986 *The Acoustic Remote Sensing of Atmosphere,* in Russian. (Novosibirsk: Nauka, Siberian division), 168 p.
2. Lokoshchenko M.A. 2002 *Journal of Atmospheric and Oceanic Technology*, **19**, August, pp.1151-1162.
3. Lokoshchenko M.A., Perepelkin V.G. and Semenova N.V. 2007 *Meteorologische Zeitschrift*, August, **16**, No.4, pp.407-414.
4. Lokoshchenko M.A. 2007 *Russian Meteorology and Hydrology*, No.1.
5. Kirtzel H.-J. et al. 2000 *Proc. 10th ISARS (Auckland, New Zealand, Nov 26th – Dec 1st 2000)* ed St.Bradley pp.35-38
6. Emeis S. 2004 *Meteorologische Zeitschrift*, **13**, 353-359.
7. Chromov S.P. 1948 *Foundations of Synoptic Meteorology*, in Russian. (Leningrad: Gidrometeoizdat), 696 p.
8. Matveev L.T. 1965 *Fundamentals of General Meteorology. Physics of the Atmosphere,* in Russian. (Leningrad: Gidrometeoizdat), 876 p.

Estimates of the height of the boundary layer using SODAR and rawinsoundings in Amazonia

G Fisch[1] and L A R dos Santos[2]

1 Instituto de Aeronaútica e Espaço (IAE/CTA), São José dos Campos, 12228-904, Brazil

2 Instituto Nacional de Meteorologia (INMET), Brasília, 70680-900, Brazil

Emails: gfisch@iae.cta.br, landre@inmet.gov.br

Abstract. During the LBA campaign in Amazonia 2002, simultaneous measurements were made of the boundary layer using different instruments (rawinsoundings and SODAR). The profiles of potential temperature and humidity were used to estimates the height of the boundary layer using 3 different techniques. The SODAR´s measurements did not capture the shallow morning boundary layer observed at the profiles.

1. Introduction

The deforestation of the Amazon tropical forest is being made in order to extract elective wood or for agricultural and cattle activities. This vegetation change can modify the local, regional and global climate. The General Circulation Models can be used to investigate these effects in all scale but they need to represent the structure and the physical processes of the boundary layer. The height of the Convective Boundary Layer (CBL) is one of the key parameters to be given for the GCMs to correct distribute of the energy and water vapor necessary for the weather forecast and climatic studies [1].

The goal of this work is to compare 3 different methods to estimate the height of the Convective Boundary Layer amongst objectives methods (profile and parcel) as well as direct measurements (SODAR).

2. Data set, site and methodology

2.1. Data Set and Site

In order to estimate the height of the CBL sometimes also know as mixed layer depth, measurements were made by commercial DigiCora MW 11 rawinsounding system (sondes RS80-15G from Vaisala Oy, Finland) and the useful information for this analysis were height (m), temperature (°C) and specific humidity (g kg^{-1}). The soundings were made at 08:00 Local Time (GMT – 4 hours), 11:00 LT, 14:00 LT and 17:00 LT in order to represent the time evolution of the CBL. The SODAR data were obtained continuously from an equipment made by Remtech PA2 (Phased Array), programmed to emit pulses each 5 s and the average data were stored at 30 min interval. The SODAR collected data up to an average height of 1500 m. All the measurements were made at a typical pasture site in Amazonia (10° 46'S, 62° 20'W, 292 MSL) at Ji-Paraná city in the southwest of Amazônia [2]. This site is a large pasture area and it is suitable for acoustic measurements because of the absence of obstacle and noise in the surroundings. The field experiment was held from Sept 12 up to 30 (named Phase I) during the end of the dry season and from Oct 15 up to 29 (named Phase II) during the onset of the wet season from 2002. The main objective of this field campaign was to collect meteorological data for understanding the differences in convection over pasture and forest in Amazonia, which is part of a larger international program (name Large Scale Biosphere Atmosphere Experiment in Amazonia – LBA). The latter is being occurring since 1999 in order to investigate how the Amazonia works as a whole system and all its interrelations amongst physical climate, carbon balance, atmospheric chemistry and so on.

2.2 Methodology
For this inter-comparison, three different methods were used: the profile method, the parcel method and the SODAR measurements. The profile method is based on the characteristics of the well mixed layer. As the vortices/eddies mixes the properties (temperature and humidity) inside the CBL, the vertical gradients must be almost null and the properties are constant with height. At the top of the CBL, there is an increase of the temperature and a decrease of the humidity from above (e.g the free atmosphere), which is hotter and dryer [1]. The lowest level of the vertical gradient for temperature (positive gradient) and humidity (negative) identifies the top of the CBL. The parcel method is based on the assumption of lifting of an air parcel following the surface dry adiabatic and its intersection with the real sounding (temperature profile). This represent a height were the growing of an air parcel by the buoyancy should reach. This very simple method is being used in the operational centers due to its practice and simplicity [3]. Both methods are schematic depicted at Figure 1 and they are on the same scale.

The measurements from SODAR are possible due to the reflection (backscatter) of the sound pulses emitted from surface by the intensity of the atmospheric turbulence and thermal structure at the top of the CBL. This type of measurements is being used for operational estimate of the CBL [4-5].

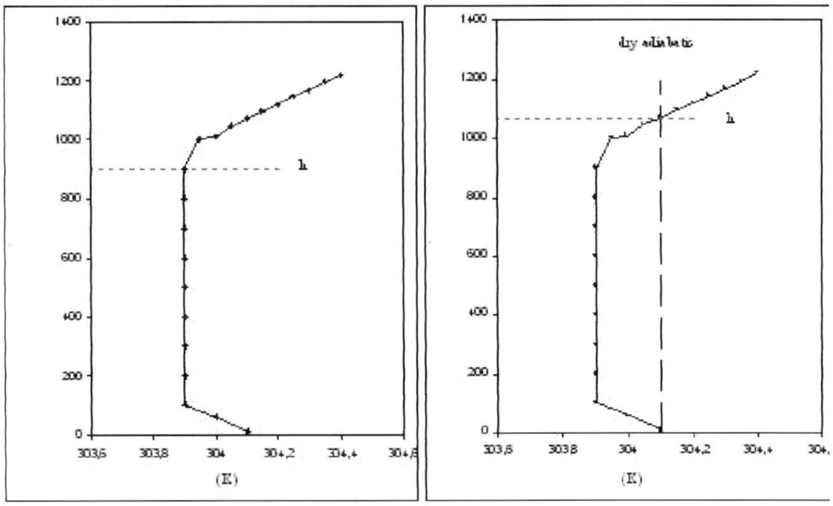

Figure 1. Example how to determine the height of the CBL by the profile (left) and parcel (right) methods.

3. Results
The main results are showed at Figure 2. In the early morning, the SODAR measurements presented data in the range of 400-1000 m. This is not the height of the shallow CBL typically observed in the pasture in Amazonia [2], but the residual layer of the previous day. The parcel method (left panels)

systematic presented higher values of CBL for both Phases (I and II). The profile method showed better agreement, especially during the Phase II, as the surface is wetter due to the rainfall events. Consequently the sensible heat flux from the surface has decreased. This flux is very important in mixing all the CBL. Previous measurements [2] have shown its importance in determining the structure of the CBL. The number of estimated height pairs (N) is shown in the right bottom corner for each panel as well as the line 1:1.

4. Final Remarks

The results showed that the parcel method overestimate the height of the CBL. This overestimates can be associated to the fact that the surface temperature heats stronger and faster than the layer at the top of the CBL. This has also found from [6] comparing the parcel method with LIDAR data set. The measurements from SODAR were systematic lower than the others, especially at the early morning (at 0800 LT). For this situation, the SODAR measured the residual boundary layer instead of the shallow convective boundary layer.

Acknowledgments: The authors would like to thanks Conselho Nacional de Desenvolvimento Científico e Tecnológico (CNPq) for the financial support.

References

[1] Stull R B 1988 *Introduction to Boundary Layer Meteorology* (Dordrecht: Kluwer)
[2] Fisch G and 7 co-authors 2004 *Theor. Appl. Climatol.* 78 47
[3] Seibert P, Beirich F, Gryning S, Joffre S, Rasmussen A and Tercier P 2000 *Atmos. Environ.* 34 1001
[4] Keder J 1999 *Meteorol. Atmos. Phys.* 71 133
[5] Beyrich F 1997 *Atmos. Environ.* 31 3941
[6] Hennemuth B and Lammert, A 2006 *Bound-Layer Meteorol.* 118 242

IOP Conf. Series: Earth and Environmental Science **1** (2008) 012065 doi:10.1088/1755-1307/1/1/012065

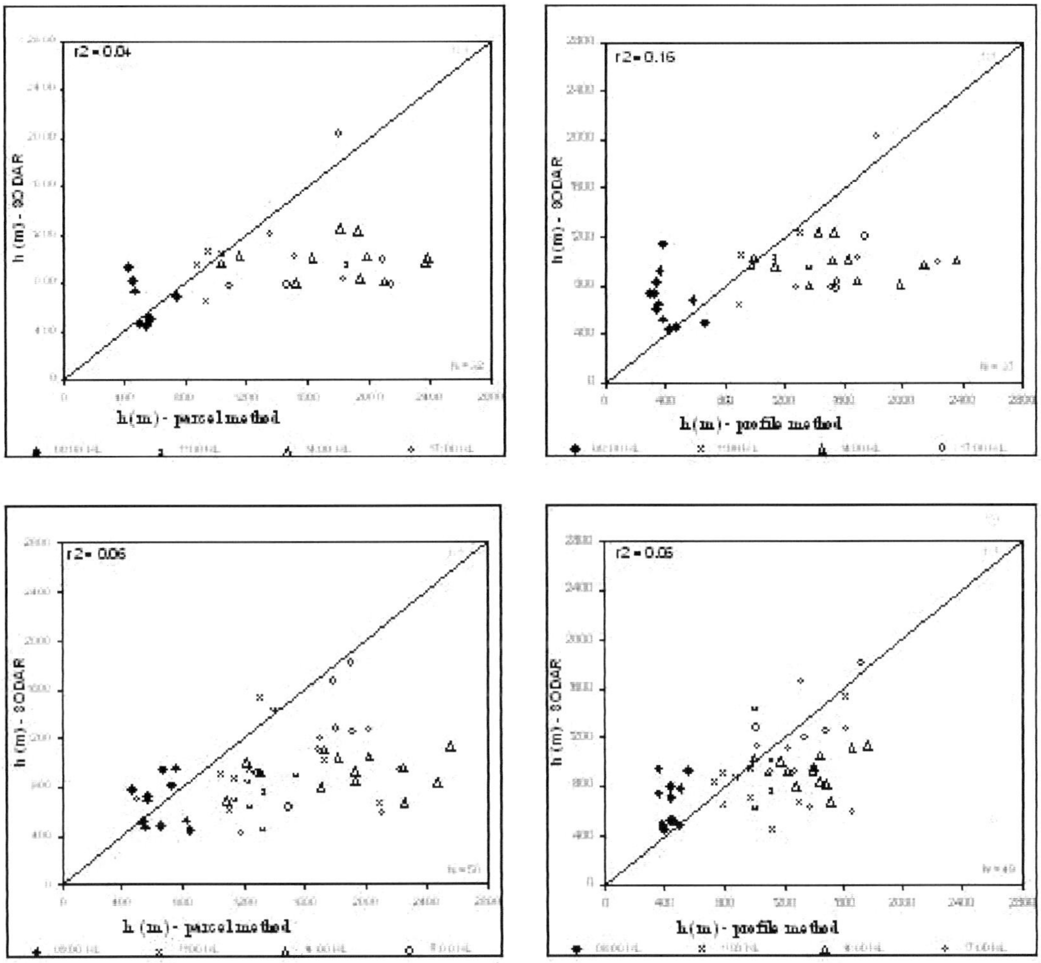

Figure 2. Plots of the intercomparison between heights of the CLB derived by the parcel method (left) for Phase I (top) and Phase II (bottom) and derived by the profile method (right) for Phase I (top) and II (bottom).

IOP Conf. Series: Earth and Environmental Science **1** (2008) 012066 doi:10.1088/1755-1307/1/1/012066

Wind velocity measurements using a pulsed LIDAR system: first results

M Wächter[1], A Rettenmeier[2], M Kühn[3] and J Peinke[4]

[1,4] ForWind Center for Wind Energy Research, University of Oldenburg, Germany
[2,3] Endowed Chair of Wind Energy, University of Stuttgart, Germany

E-mail: [1]matthias.waechter@uni-oldenburg.de, [2]rettenmeier@ifb.uni-stuttgart.de, [3]kuehn@ifb.uni-stuttgart.de, [4]peinke@uni-oldenburg.de

Abstract. Wind velocity measurements were taken using a Leosphere Windcube LIDAR system, which operates as a pulsed laser Doppler anemometer. Here we report on first results, which show typical characteristics of atmospheric wind velocities.

1. Introduction

Measurements of the atmospheric wind field are of essential relevance for wind energy utilization. Different technologies are in use in this field, among them LIDAR (LIght Detection And Ranging) systems. This measurement principle is currently not as well established for wind velocity measurements as other techniques, such as cup or ultrasound anemometers. Nevertheless, the LIDAR technology offers some advantages over more traditional methods, which shall – as well as possible drawbacks – be investigated in the ongoing research project "Further development of LIDAR wind measuring techniques for the offshore test field" funded by the German Federal Environment Ministry.

For these purposes we use a Leosphere Windcube LIDAR system, which operates as a pulsed laser Doppler anemometer. A laser beam of $1.54\,\mu m$ wavelength takes measurements of the wind speed in beamwise direction. To obtain the three-dimensinal wind vector, the beam is inclined by 30° from vertical direction and measurements are taken under four different azimuth angles (see figure 1). As the laser can take measurements, in our case, every $1.5\,s$, a new wind vector can be derived from the four most recent successive measurements at the same frequency. The length of the laser pulses of about $26\,m$ defines the vertical extension of the measurement volume.

As most significant advantages, a LIDAR system can take measurements in large heights (up to about $200\,m$ in our case) from ground level, without the need of a metmast. Additionally, the principle of a pulsed laser Doppler anemometer allows for simultaneous measurements in several height levels (up to ten for the Leosphere Windcube). The system is mobile and can be moved to different locations with only little effort, compared to a metmast.

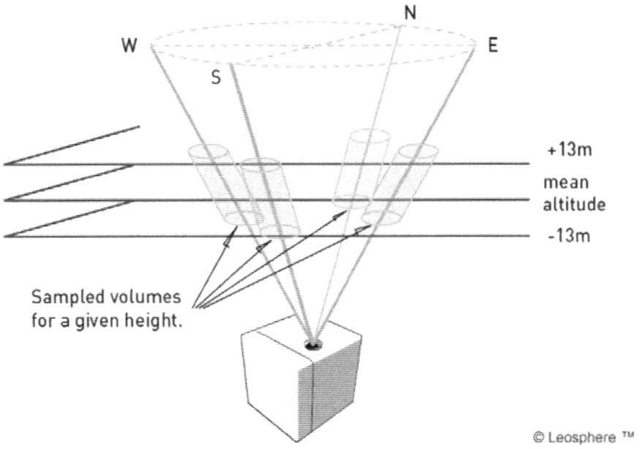

Figure 1. Sketch of the measurement principle of the Leosphere Windcube. An inclined laser beam takes Doppler measurements of the velocity in beamwise direction under four different azimuth angles. The three-dimensional wind velocity vector is then derived from four successive measurements. Up to ten different height levels can be measured simultaneously.

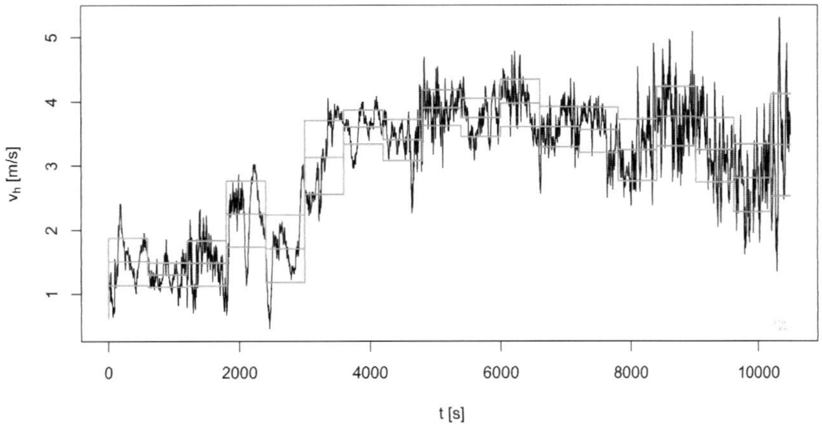

Figure 2. Segment of measured time series of the horizontal wind speed magnitude v_h in $120\,m$ height. Additionally the ten minute mean values μ and the intervals given by $\mu \pm \sigma$, where σ denotes the standard deviation of a ten minute interval, are indicated by grey lines.

2. Wind field characterisation

Atmospheric wind fields are complex structures, fluctuating in time as well as in space. In this paper we will concentrate on the horizontal wind speed magnitude v_h as a frequently used quantity for wind field description. As a simple approach to wind field characterisation, the *ten minute mean value* of the horizontal wind speed magnitude, $\mu = \langle v_h \rangle_{10\,min}$, is widely used together with the *standard deviation* σ with respect to the same time interval, compare figure 2. These quantities are one-point statistics and thus do not reflect the succession of velocity values. Fluctuations on time scales below ten minutes are important characteristics of the wind field, e.g., they determine fast load changes on wind turbines.

To characterize *fluctuations* of v_h, two-point statistics have to be used. Most prominent are the *auto-correlation function* $\langle v_h(t+\tau)v_h(t)\rangle$ or the *power spectral density*, which can under some conditions be understood as the Fourier transform of the auto-correlation function. As these quantities are constructed from mean square values, they reflect second-order statistics,

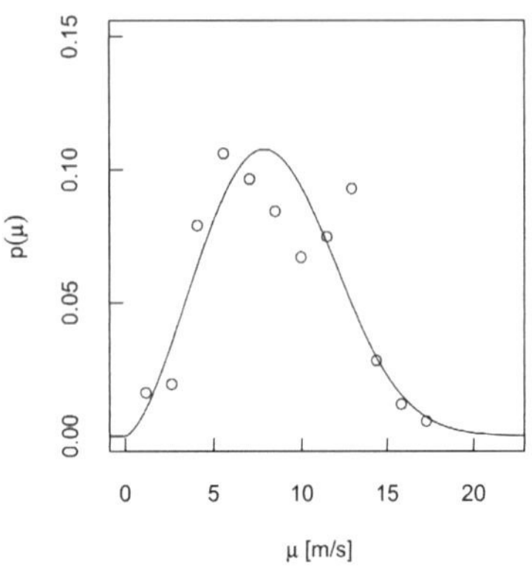

Figure 3. PDF of the ten minute mean values μ of the horizontal wind speed measured at $120\,m$ height. The solid line denotes the fit of a Weibull distribution to the data.

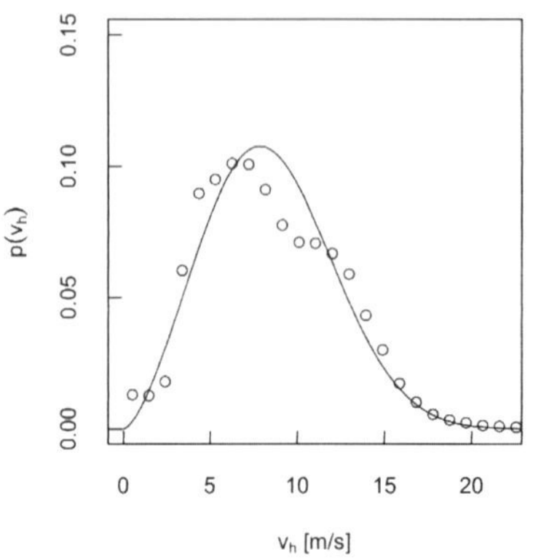

Figure 4. PDF of the horizontal wind speed magnitude v_h measured at $120\,m$ height. The solid line denotes the fit of a Weibull distribution to the data.

and provide information about the second moments of the fluctuations.

A more general approach to two-point (or fluctuation) statistics is the investigation of wind speed *increments*

$$\delta v_h(t,\tau) = v_h(t+\tau) - v_h(t)\;. \tag{1}$$

The moments $\langle \delta v_h(\tau)^k \rangle$ of these increments are called *structure functions* in turbulence research and reflect the fluctuation statistics of order k. It is easy to see that the second-order structure function is directly related to the auto-correlation function and thus also to the power spectral density. Besides the moments, also the probability density functions (PDFs) $p(\delta v_h)$ of δv_h can be investigated, which reflect the fluctuation statistics of all orders.

In [1] it has been shown that the PDFs of atmospheric wind speed increments differ significantly both from Gaussian distributions on the one hand and from PDFs measured in laboratory turbulence experiments on the other hand. Furthermore, in [2] a simple stochastic model was presented, reproducing characteristic properties of different locations, such as offshore sites or onshore sites in different terrain. In this contribution we want analyse the first LIDAR measurements in the light of these publications.

3. Measurement data

First measurements were taken with the Leosphere Windcube device located on the roof of the Institute of Aircraft Design at the university of Stuttgart. The wind velocity was measured at ten height levels between 40 and $220\,m$. Under these conditions the device achieved a sampling frequency of $0.66\,Hz$, and nine files of continuous twelve-hour measurements could be recorded, resulting in about 250 000 samples per height level.

Invalid measurements occured from time to time, especially in large heights. According to the manufacturer these cases are caused by to too low aerosol concentration in the air, which leads to a low SNR. The Windcube is adjusted to evaluate 10 000 laser pulses per measurement and accept values at an SNR of $-22\,dB$ or better [3]. For the analysis therefore only complete, uninterrupted ten-minute intervals were selected. A section of a recorded time series of the horizontal wind speed v_h in $120\,m$ height is presented in figure 2, together with the mean values μ and standard deviations σ of each ten minute interval.

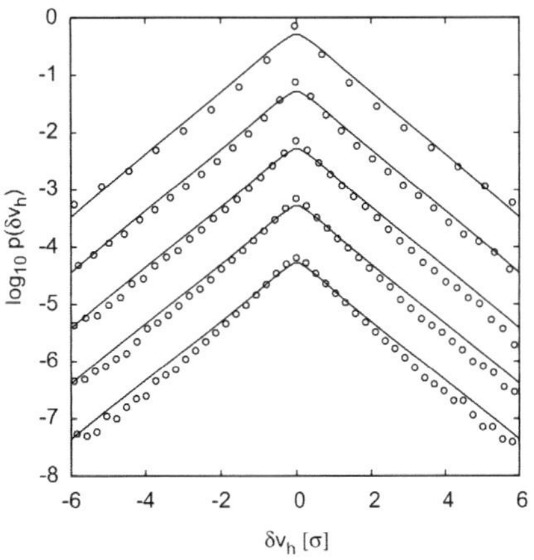

Figure 5. PDFs of horizontal wind speed increments $\delta v_h(\tau)$ (symbols). Scales τ are 1.5, 4.5, 15, 48, and 162 s from top to bottom, and fits according to [2] are shown as solid lines. PDFs are vertically shifted for clarity of presentation.

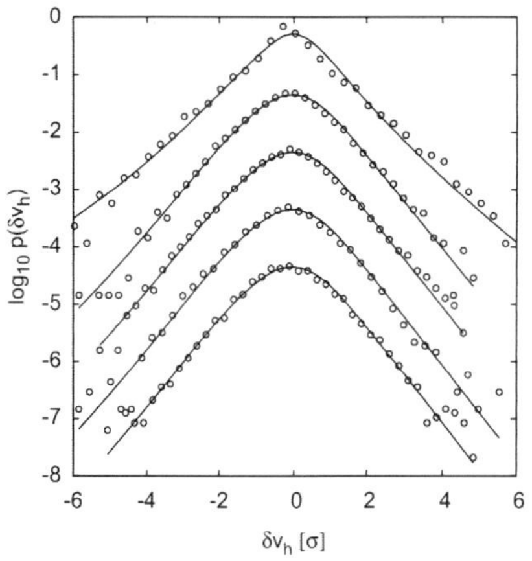

Figure 6. PDFs of horizontal wind speed increments $\delta v_h(\tau)$, conditioned on $\mu = (5.5 \pm 0.5) m/s$ (symbols). Scales τ are chosen as in figure 5, and fits according to [1, 2] are shown as solid lines.

4. PDFs of horizontal wind speed

The PDF of the ten minute mean values μ of the horizontal wind speed measured at 120 m height is shown in figure 3. In meteorology the assumption is well accepted that μ obeys a Weibull distribution for sufficiently long measurement periods [4]. Therefore a fit of a Weibull distribution to the data is added to the plot. It can be seen that in our case the Weibull distribution does not model the empirical distribution satisfyingly. As possible reasons, the measurement time of nine twelve-hour periods may have been too short, and the meteorological situations may not have been well-balanced. In figure 4 the empirical PDF of v_h measured in the same height level is presented, again together with a fit of the Weibull distribution. Here the accordance is a little better, but still not good. While there are much more samples of v_h than of μ, the same reasons for deviations have to be considered as before.

5. Increment PDFs of horizontal wind speed

In figure 5 normalized PDFs of the horizontal wind speed increments $\delta v_h(\tau)$, see equation (1), measured at 120 m height, are shown for different scales τ between 1.5 and 162 s. The shapes of these PDFs clearly show a peak around the mean value and so-called *heavy tails*. These heavy tails denote increased probability of extreme fluctuations δv_h compared to Gaussian statistics. PDFs of this shape are called *intermittent* in turbulence research. It can also be seen from figure 5 that the shape of the PDFs changes only slightly with the scale τ, and does not converge to a Gaussian shape over a wide range of scales. This is a remarkable and characteristic feature of atmospheric wind data [1, 5, 6]. As wind speed increments on rather small scales $\tau \leq 1\,min$ can be interpreted as gusts, these statistics also mean an increased probablity of strong gusts, compared to a Gaussian distribution (cf. [1]).

In reference [1] PDFs of δv_h were also investigated after conditioning on a certain mean value. Here, we follow the same procedure and present in figure 6 the PDFs of δv_h only for those ten minute intervals with $\mu = (5.5 \pm 0.5) m/s$. The same effect as in [1] can be seen also here:

the PDFs now obtain a pronounced intermittent shape only for the smallest scales τ, changing to a more Gaussian-like shape for large scales. The conditioning apparently leads to a similar situation as in laboratory turbulence experiments, where a well-prepared mean velocity assures a stationary flow.

For increment PDFs of laboratory turbulence a description by Castaing's formula [7] is well established. It models the turbulent PDFs as a weighted superposition of Gaussian PDFs, depending mainly on the so-called form parameter λ. In figure 6, we have added fits according to Castaing's formula, which give a good representation of the empirical PDFs. As already stated, for atmospheric turbulence the situation is much more complex than for laboratory experiments. In [2] an explicit formula was developed which describes increment PDFs of atmospheric flows as a superposition of Castaing functions, according to the distribution of the ten minute mean values of the wind speed. We have added fits according to this formula to figure 5, which – at least qualitatively – agree well with the PDFs of the atmospheric wind speed increments.

Both fitting procedures were performed following reference [2], including the simplified parameterizations mentioned there. For details of the procedure, the reader is kindly referred to the original article. The potential of both formulas for a rather simple description of atmospheric flows is confirmed by our data. Possible sources of deviations in figure 5 are the mentioned simplified parameter relations, as in [2]. In particular, it can be seen from figure 3 that the ten minute mean values of our measurements are not optimally described by a Weibull distribution, as assumed by our model.

6. Summary and outlook

In this article we have evaluated first measurements of atmospheric wind velocities, taken by a Leosphere Windcube pulsed LIDAR system. For comparison and consistency with other publications, we concentrated here on the horizontal wind speed magnitudes v_h. Despite the measurement period was rather short, our data show characteristic features of atmospheric flows. In particular these are the pronounced intermittent increment PDFs of v_h on all observed time scales. These PDFs approach a behaviour as it is found for laboratory turbulence when only periods with similar mean wind speed μ are selected, and thus support the (simplified) picture of atmospheric wind fields as a superposition of turbulent flows with different mean velocities, cf. [1, 2].

Further investigations will focus on the spatial and temporal resolution and averaging effects of the Leosphere Windcube. As described in section 1 and figure 1, its principle of measurement uses a rather large spatial and temporal region of the wind field. Quantitative comparison to classical measurement systems, namely cup and ultrasound anemometers, will be of special practical relevance. An additional task in this research project will be to investigate the derivation of characteristic power curves of wind turbines using LIDAR wind speed measurements. Besides the standard procedure, we expect some advantages of a new method which considers the dynamics of the wind power conversion process [8, 9], combined with the flexibility and mobility of a LIDAR system.

Acknowledgments

This work is part of the research project "Further development of LIDAR wind measuring techniques for the offshore test field – »LIDAR«" funded by the German Environment Ministry under the code number 0327642A.

References

[1] Böttcher F, Renner C, Waldl H P and Peinke J 2003 *Bound-Lay Meteorol* **108** 163–173
[2] Böttcher F, Barth S and Peinke J 2007 *Stoch Environ Res Ris Assess* **21** 299–308

[3] Aussibal C 2007 *WINDCUBE*^{TM} *User's Manual* LEOSPHERE LIDAR Environmental Observations Orsay, France
[4] Burton T, Sharpe D, Jenkins N and Bossanyi E 2001 *Wind energy handbook* (New York: Wiley)
[5] Lange B, Barthelmie R J and Højstrup J 2001 Description of the rødsand field measurement Report Ris-R-1268 Risø National Laboratory 4000 Roskilde, DK
[6] Ragwitz M and Kantz H 2001 *Physical Review Letters* **87** 254501
[7] Castaing B, Gagne Y and Hopfinger E J 1990 *Phys D* **46** 177–200
[8] Anahua E, Barth S and Peinke J 2007 *Wind Energy*
[9] Gottschall J and Peinke J 2007 *J Phys: Conf Ser* **75** 012045

IOP Conf. Series: Earth and Environmental Science **1** (2008) 012067 doi:10.1088/1755-1307/1/1/012067

Characterization of PM10 accumulation periods in the Po valley by means of boundary layer profilers

M E Ferrario[1], A Rossa[1], and D Pernigotti[1]

[1]Servizio Centro Meteorologico di Teolo, Dipartimento Regionale per la Sicurezza del Territorio, Agenzia Regionale per la Prevenzione e la Protezione Ambientale del Veneto, via Marconi 55, 35037 – Teolo (PD) Italy.

E-mail: mferrario@arpa.veneto.it

Abstract. Low wind and stable atmospheric conditions cause frequent PM10 episodes in the Po Valley. Such weather conditions are well characterized by temperature and wind profiles. Thermal inversion strength, height and duration, as well as wind profiles are set in relation to daily PM10 increments, and guidelines for forecasting air quality (AQ) in said conditions are derived.

1. Introduction

As laid out in the WHO air quality guidelines it is established that airborne particulate matter (PM) has an adverse impact on public health at exposures that are currently experienced by urban populations in both developed and developing countries. The range of health effects is broad, but are predominantly to the respiratory and cardiovascular systems, where all population is affected, but susceptibility to the pollution may vary with health or age [1].

Air pollution concentration, and that of PM10 (i.e. PM with diameters less than 10um) is strongly subject to meteorological action so that a region's burden not to exceed PM10 concentration limits depends on its climatology. The region of the Po Valley has a particularly unfortunate topographical conformation which favours low wind and strong inversion situations, conditions which can lead to extremely high PM concentrations, especially during the cold season [2]. This causes very frequent exceedance of the thresholds of PM10 concentrations set by the Council Directive 1999/30/EC (50 mg/m3 for single daily averages, 40 mg/m3 for yearly average).

In response to this particular situation of the Po Valley, the Centro Meteorologico di Teolo (CMT-DRST) of the Agenzia Regionale per la Prevenzione e la Protezione Ambientale del Veneto (ARPA Veneto) has recently installed a network of four passive radiometers (R or R*) and four SODAR (S) for air quality monitoring purposes [3] The instruments are all located in Veneto, a region in the northeastern Italy, as shown in figure 1, and were usual in a previous study related to a PM10 episode (Pernigotti [4]). While the network, in the framework of the project DOCUP (DOCumento Unico di Programmazione) Program 2 co-funded by the European Union, national Italian government, and the Veneto regional government, is the first of its kind in Italy, other Italian regional protection Agencies (e.g. Piemonte or Marche) have installed such instruments.

Temporal and spatial resolution of these profilers is significantly higher compared to radio soundings, whereas the vertical range is limited, allowing detailed observations and analyses of Planetary Boundary Layer (PBL). In this contribution the potential of having a network of PBL profilers is demonstrated.

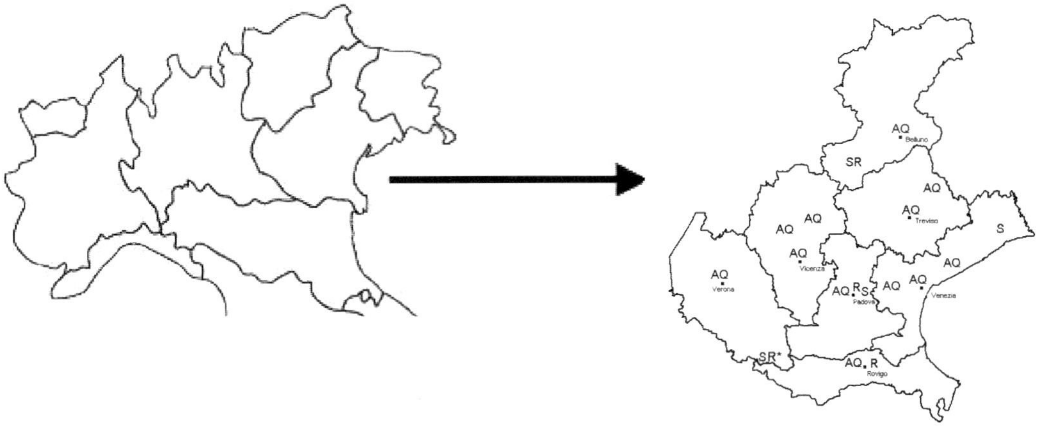

Figure 1. Temperature profilers (R or R*) and SODAR wind profilers (S) and Air Quality (AQ) surface station network of the north-eastern Italian region Veneto (right panel) operated by ARPA Veneto – CMT-DRST. Status is as of the end of 2007; left panel shows where Veneto is located in northern Italy.

Winter 2007-08	Verona 1	Verona 2	Vicenza 1	Vicenza 2	Padova 1	Padova 2	Venezia 1	Venezia 2	Venezia 3	Venezia 4	Treviso 1	Rovigo 1	Belluno 1	Belluno 2
Verona 1	1,00	0,97	0,88	0,91	0,84	0,82	0,78	0,84	0,86	0,87	0,86	0,74	0,73	0,82
Verona 2		1,00	0,86	0,89	0,80	0,76	0,76	0,82	0,81	0,86	0,84	0,65	0,76	0,83
Vicenza 1			1,00	0,98	0,84	0,83	0,86	0,87	0,88	0,96	0,88	0,85	0,74	0,75
Vicenza 2				1,00	0,86	0,85	0,82	0,85	0,86	0,91	0,86	0,86	0,77	0,73
Padova 1					1,00	0,96	0,90	0,94	0,94	0,93	0,91	0,88	0,61	0,69
Padova 2						1,00	0,86	0,89	0,90	0,93	0,89	0,87	0,50	0,59
Venezia 1							1,00	0,95	0,93	0,96	0,88	0,80	0,64	0,67
Venezia 2								1,00	0,98	0,97	0,92	0,81	0,65	0,71
Venezia 3									1,00	0,96	0,91	0,82	0,59	0,66
Venezia 4										1,00	0,84	0,92	0,75	0,82
Treviso 1											1,00	0,76	0,68	0,73
Rovigo 1												1,00	0,48	0,57
Belluno 1													1,00	0,96
Belluno 2														1,00

Figure 2. Correlation in last winter 2007-08 between the AQ stations in Veneto Region. Note the low correlation of Belluno city with the other AQ site, excellent correlation between all the cities that lay in Po valley (figure 1), and a good correlation between Padova and Rovigo (0.87-0.88).

Table 1. Some parameters on instruments settings and data availability, Vertical range for SODAR is the altitude for which data availability drops to 30% of the one of the first level (40m). Data availability doesn't include periods in which the instrument was stopped for software/hardware upgrades.

Profilers	SODAR PCS2000-24		SODAR PCS2000-64		MTP5-HE			HATPRO
Area operation	Belluna Valley	Padova	Legnago	Loncon	Padova	Belluna Valley	Rovigo	Legnago
ID number	17	18	19	20	25	26	27	1
Hight resol.	20m	20m	20m	20m	50m	50m	50m	50-75m
Time resol	15’	15’	15’	15’	5’	5’	5’	20’
Vertical range	~ 500m	~ 500m	~ 700m	~ 700m	1000m	1000m	1000m	2000m
Data 2005	79 %	72 %	78 %	84 %	95 %	76 %	84 %	68 %
Data 2006	83 %	85 %	87 %	87 %	100 %	97 %	83 %	94 %
Data 2007	68 %	80 %	82 %	86 %	94 %	100 %	96 %	93 %

2. Data set

The 3 MTP5-HE Radiometers (R, in figure 1) are manufactured by Attex, Moscow (RUS), and distributed by Kipp & Zonen, Delft (NL). As shown in Kadygrov [5] this instrument reports a good agreement within 0.5-0.8K with a co-located radio sounding found in Payerne, Switzerland, on 63 profiles. Ferrario [6] confirmed this data for the ARPAV MTP5-HE radiometers. All instruments are set to have the first level at 50m.

The Radiometer (R*, in figure 1), is manufactured by Radiometer Physics GmbH, Meckheim (D). It receives radiation emitted by the atmosphere in 14 channels (molecular oxygen and water vapour lines) and converts this data to profiles for temperature and humidity (Rose [7])via a neural algorithm optimized for the measuring site. In addition, the instrument installed in Legnago performs a vertical scan every 20 minutes for temperature with variable vertical resolution (50m up to 2000m) in the PBL, where the declared accuracy is 1K. The measuring sites are flat and rural (Legnago and Loncon), flat and urban (Padova and Rovigo) and in a closed valley with very light winds and strong thermal inversions (Belluna Valley).

The SODAR (S, in figure 1), are two PCS2000-24 and two PCS2000-64 manufactured by Metek, Elmshorn (D). This is well known technology and the manufacturer declared accuracy of the data is of 0.3m/s for wind intensity and 5-8° for wind direction. The vertical range of SODARs is strongly dependent upon the state of atmospheric turbulence which in stable conditions frequently features a strong decrease with increasing altitude with a corresponding decrease in data availability. All the CMT SODARs are set with first level at 40m and temporal resolution of 15 minutes.

In table 1 type and mode of operation of profilers are shown. Moreover real percentage of work are listed (we don't consider the data holes when the instrument was stopped for regular maintenance). Radiometer MTP5-HE are very reliable thermal profilers and when their efficiency drops it is often due to problems with the power supply. The Metek SODARs operated by ARPAV run quite reliably, but quite often can exhibit problems with the plausibility of data in low wind conditions. The HATPRO radiometer, when working, is a powerful thermal profiler but has had many up-grades in software and hardware in the measuring period that has reduced the availability.

3. PM10 Episodes in northern Italy

Episodes of pollutions by particulate matter with diameter less than 10µm (PM10) are quite frequent in the Po Valley, figure 4 (Rossa [8]). From 1980-ies the Regional Agencies for Environment in northern Italy began to check and try to quantify the problem. First type of measurement was TSP (Total Suspended Particulates) variables. TSP decreased significantly during the 1990-ies (figure 3) probably due to improved technology deployed in power plants, engine for transports, and discontinuation of some heavy industry. PM10 is monitored in Veneto only since 2001 and the corresponding air quality monitoring network became stable only in 2007 thanks to the DOCUP project.

We made an attempt to characterize in manner the PM10 episodes by evaluating the whole historical series isolating only periods that exhibited PM10 concentration above the 95th percentile (Figure 5). We catalogued the episodes with a number (the year) and a Letter (A,B,C for early months, X,Y,Z for last months in the year). Results are shown in table 2. Episodes are present from late autumn to spring, but take place especially in the winter season. The evolution of the PM10 concentration in acute winter episodes shows a regular build up as well as spin down phase, whereas episodes in Autumn or Spring tend to end abruptly, say with a major precipitation event linked to a frontal passage. In figure 6 the average temperature profile in different months is shown, i.e. that from the April to September atmosphere with frequent unstable to neutral conditions that establish the end of Acute Episode. Typically 3 to 4 episodes lasting in the order of 10 days take place per cold season, table 2. In 2007 windy conditions above normal gave rise to only 1 episode. Our network of PBL profilers was installed and operates since 2005 Spring.

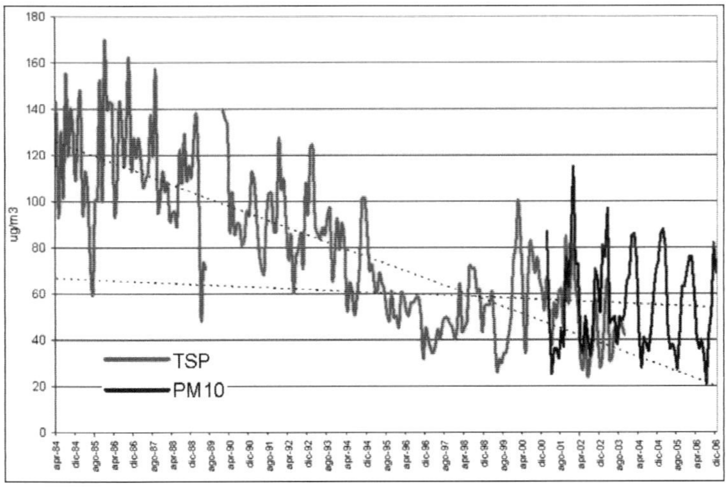

Figure 3. City of Padova, monthly average of TSP (light grey, μg/m^3) and PM10 (dark grey, μg/m^3), dashed line are linear trends, a similar behavior was detected in the other cities of Veneto. Data source DAP of Padova ARPA Veneto (on the x-axis every 8th month is shown starting April 1984, ending December 2006).

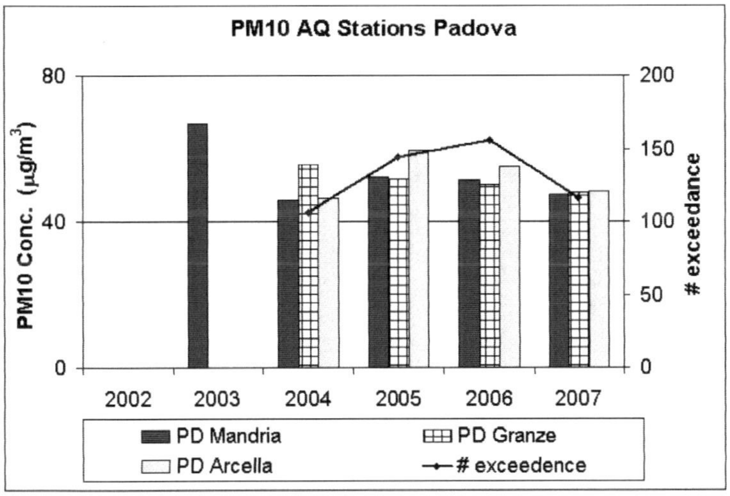

Figure 4. Annual average of PM10 in 3 air quality stations in Padova (always over 40 μg/m^3!) and the number of days for which the 50 μg/m^3 threshold was exceeded (always significantly (over 100 times) more than the 35 allowed) [9].

Table 2. Episodes of PM10 pollution in Veneto region since 2002.

Year	# Ep.	Periods			
2002	1	A 07 Mar.-22 Mar.			
2003	3	A 12 Jan.-24 Jan	B 24 Feb.-13 Mar.	X 13 Nov.-25 Nov.	
2004	4	A 04 Jan.-14 Jan.	B 24 Jan.-12 Feb.	C 13 Mar.-22 Mar.	X 08 Dec. -19 Dec.
2005	5	A 01 Jan.-18 Jan.	B 06 Feb.-14 Feb.	X 22 Oct.-5 Nov.	Y 18 Dec.-27 Dec.
2006	4	A 09 Jan.-22 Jan.	B 06 Feb.-16 Feb.	C 15 Mar-18 Mar.	Y 04 Nov.-22 Nov.
2007	1	A 01 Jan.-22 Jan.			
2008	1	A 10 Feb.-23 Feb.			

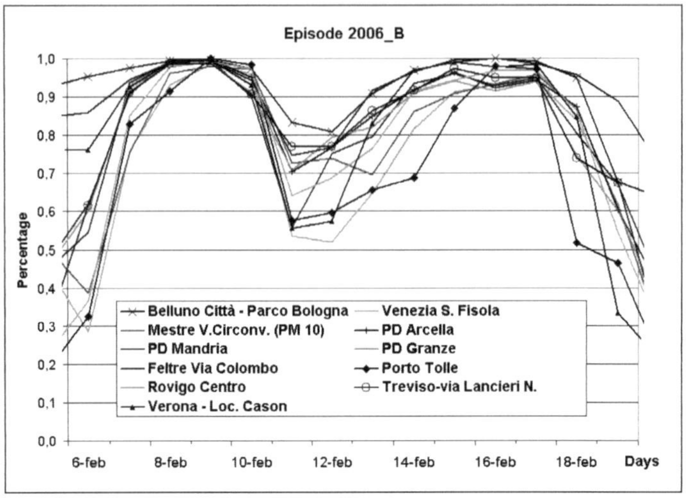

Figure 5. Episode 2006_B note the very high correlation in all the AQ station.

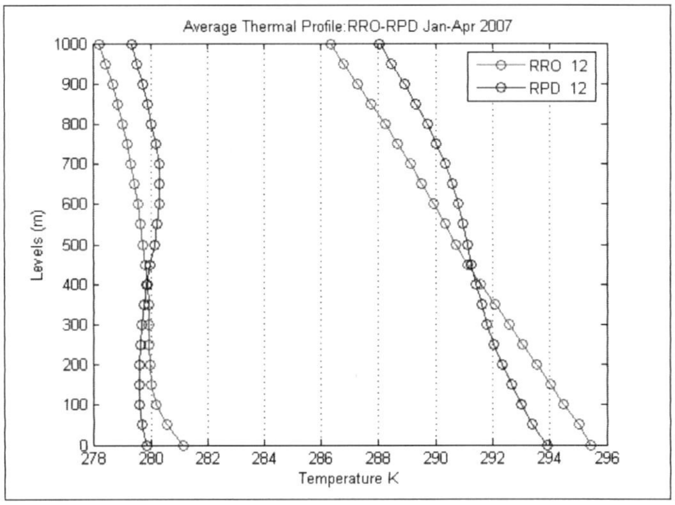

Figure 6. From Winter (January left graph) to Spring (April right graph) the thermal profile became more unstable that establish the end of Acute Episode. (RRO Rovigo MTP5-HE Radiometer, MTP5-HE RPD Padova Radiometer).

most AQ stations were placed in the main cities of Veneto and the profilers were confined by the DOCUP project to stay in a specific list of municipalities, it was not straight forward to have AQ station and profiler collocated. Moreover, because of their very nature SODARs need to stay away from town centre, where a lot of AQ stations are placed. In order to justify the analysis of PM10 and the PBL profiler it is important to note that in all the major cities (except Belluno which is located behind the pre-alpine chain in the Belluna Valley) are located in the Po Valley. In addition, the DOCUP project prescribed the siting of instrumentation in less developed areas of the political region Veneto, thus basically excluding city centers. Placing SODARs on flat, high roof-tops has not yet been explored. In anticylonic conditions in winter this translated into relatively uniform PBL structure (e.g. during a long stable period with high pressure, with no precipitation and only weak winds); PM10 concentrations even from distant AQ station exhibit a high correlation of about 0.9 if the cities are closer (like 30km apart). We therefore think that in episodic conditions the PBL profilers give representative observations for the cities which are 'not too far away', figure 2.

4. Profiler and Acute Episodes

As is well known, for northern Italy most episodes happen during the cold season when there are stable anticyclonic synoptic conditions characterized by absence of precipitation and significant winds like Bora or Sirocco. The important atmospheric phenomena which then govern the accumulation mechanisms for PM10 are mechanically and thermally induced turbulence, both of which can be estimated in a qualitative way by SODARs and radiometers respectively. Figure 7 shows the periodic build up of temperature inversions after sunset and the break up some time after sun rise. During some days the winter sun is not able to disrupt the thermal inversion. Height and strength of the inversion are two important parameters which both can be retrieved from the MTP5-HE.

A typical example is shown in figure 8 for the Padova AQ station from 10-23 February 2008 (2008_A). In the presence of thermal inversions during nighttime there is a successive build up of PM10 which reaches a first peak concentration around the 14th. In the weekend of the 16-17th the stability of the boundary layer completely changed from very stable to almost neutral. The moderate winds (not shown) were efficient in dispersing the particulate matter. Subsequently stability of the boundary layer was re-established leading to some of the highest concentrations since the monitoring started in 2001. Starting from the 21st humidity increased and led to fog and low clouds, a factor which further influences the deposition of PM10.

It is interesting to note that the inversion height seems to play a role in the accumulation process. When the inversion height is lower, i.e. around 200-400m the PM10 concentration often is lower than for situations in which the inversion height is above 600m. However, a systematic study needs to be done in order to understand to what extent PM10 dispersion is taking place for very confined thermal inversion situations. Also, the role of the direction of the typically very weak winds during pollution episodes needs more investigation, as for the Veneto AQ stations westerly flow can imply import of pollution from the industrialized neighboring region Lombardia, whereas easterly flow brings clean air from the northern Adriatic Sea.

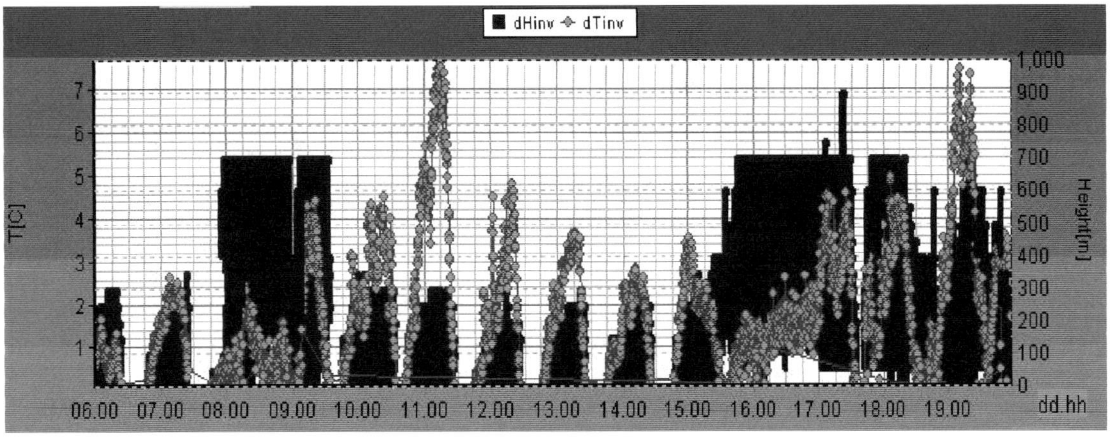

Figure 7. Time series of the temperature inversion as retrieved from the MTP5-HE located in Padova for the period 6-19 February 2006 (2006_B). Light grey dots denotes the inversion strength (C, left axis), while dark grey line denote the portion of the atmosphere in which there is a temperature inversion. The upper border of the blue lines denote the height of the thermal inversion (m, right axis).

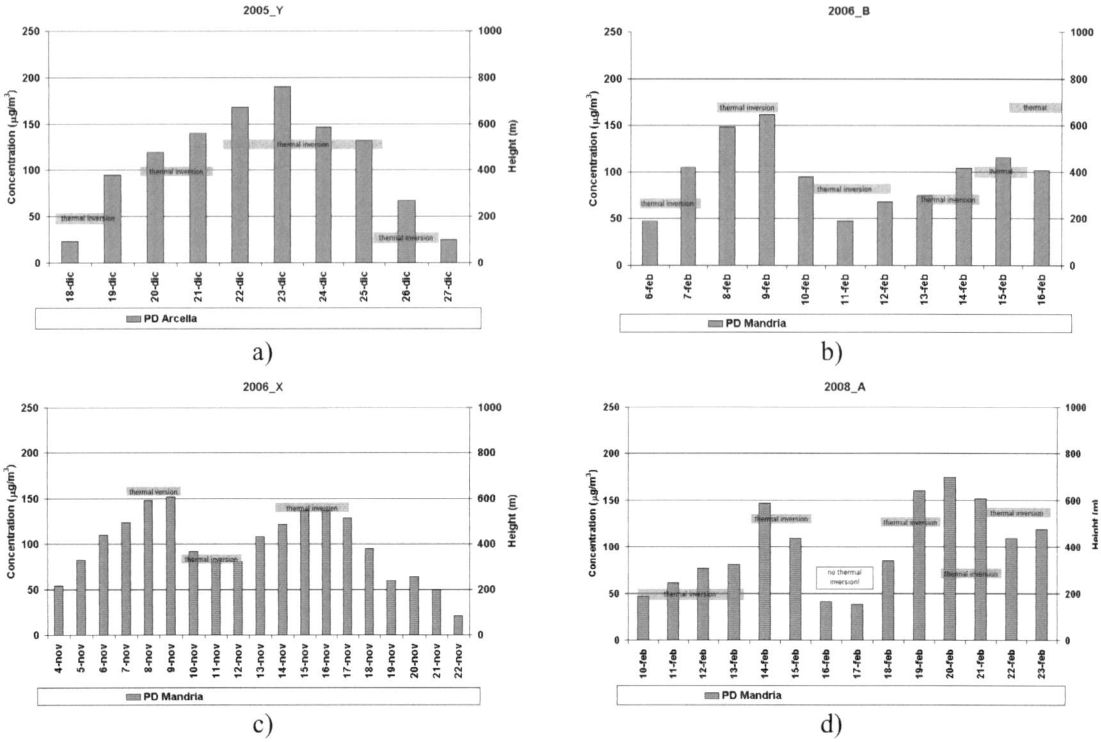

Figure 8. Time series of PM10 concentration (µg/m³) as measured at the Padova AQ station, The horizontal shaded bars denote a representative height of the top of the thermal inversion at the given days; where there is no bar a thermal inversion was absent. Panel a) For episode 18-27 December 2005 (2005_Y). Panel b) episode 6-16 February 2006 (2006_B). Panel c) For episode 4-22 November 2006 (2006_X). Panel d) for episode 10-23 February 2008 (2008_A).

5. Conclusions

An overview of PM10 episodes in the Po Valley was given and related to the observations obtained from the recently installed PBL profiler network of the Centro Meteorologico di Teolo. For meteorological conditions without precipitation, significant winds, and a very moist boundary layer, there seems to be no appreciable accumulation in the absence of a thermal inversion. Moreover, higher reaching inversions appear to be more efficient for the PM10 accumulation than shallow inversions.

From this preliminary analysis it appears that the details of the temperature profile could be important to understand the potential of accumulation of PM10. Therefore, having a network of temperature profilers could unveil regional differences. Either way, the temperature profiles are a valuable piece of information for the AQ forecaster in estimating the evolution of the PM10 concentration.

References

[1] WHO air quality guidelines, Scientific Evidence, Global Update 2005: http://www.euro.who.int/air/activities/20050222_2

[2] Kukkonen J, Pohjola M, Sokhi R S, Luhana L, Kitwiroon N, Fragkou L, Rantamäki M, Berge E, Ødegaard V, Slørdal LH, Denby B and Finardi S 2005: Analysis and evaluation of selected local-scale PM10 air pollution episodes in four European cities: Helsinki, London, Milan and Oslo *Atm. Env.* **39**, 2759-2773.

[3] Data from SODARs and Radiometers profilers on line from network ARPA Veneto: http://www.arpa.veneto.it/bollettini/htm/profilatori.asp

[4] Pernigotti D, Rossa A, Ferrario M E, Sansone M and Benassi A 2007 Influence of ABL stability on the diurnal cycle of PM10 concentration: illustration of the potential of the new Veneto network of MW-radiometers and Sodar. ISARS 2006 Special Issue of the *Meteorologische Zeitschrift* **16** No. 5 505-511

[5] Kadygrov E, Khaikine M, Miller E, Shaposhnikov A and Troitsky A 2005 Advanced atmospheric boundary layer temperature profiling with MTP5-HE microwave system. Teco 2005 Posters

[6] Ferrario M E, Rossa A M, Pernigotti, D, Sansone M and Benassi A 2006 *Proc. of the Int. Conf. on Urban Climate (Goteborg)*. Presentation and first assessment of a radiometer network in the Italian region Veneto p 288-291, www. http://www.altostratus.it/download.htm

[7] Rose T, Crewell S, Löhnert U and Simmer C 2005 A Network suitable microwave radiometer for operational monitoring of the cloudy atmosphere *Atmospheric Research* **75** 183-200

[8] Rossa A M, Pernigotti D, Ferrario M E, Sansone M and A. Benassi 2007 *Proceedings of 6th UAQ International Conference on Urban Air Quality (Cyprus)* Documentation of diurnal cycle of PM10 concentration for the urban site of Venice-Mestre p 272

[9] Data PM10 in Veneto: http://www.arpa.veneto.it/bollettini/htm/riepilogo_sup_pm10.asp

Acknowledgments
This work was supported by the European Community, Italian Government and Padova Municipality trough the DOCUP Project and the 'Air pollution in the city of Padova' Project. Thanks to Mr. Bruno Renon, Dr. Massimo Bressan, Dr. Paolo Frontero, Mr. Luciano Botton, Mr. Silvio Fazion, for their continued first level support to the profilers. A particular thanks to Dr. Alessandro Benassi for having initiated the DOCUP project, and Dr. Marco Monai for revising the manuscript.

AUTHOR INDEX

AUTHOR INDEX